No.	Loading	Equations for Fix
4.	l_1, l_2, w_1, w_2, FM_b, FM_e, b, e, L, FS_b, FS_e	$FS_b = \frac{w_1(L-l_1}{20L^3}$ [illegible] $\{(7L+8l_1) -$ [illegible] $\frac{\quad}{(L-l_1)}$ $\times\left[1+\frac{l_2}{L-l_1}+\frac{l_2^2}{(L-l_1)^2}\right]+\frac{2l_2^4}{(L-l_1)^3}\}$ $+\frac{w_2(L-l_1)^3}{20L^3}\left\{(3L+2l_1)\left[1+\frac{l_2}{L-l_1}\right.\right.$ $\left.\left.+\frac{l_2^2}{(L-l_1)^2}\right]-\frac{l_2^3}{(L-l_1)^2}\left[2+\frac{15L-8l_2}{L-l_1}\right]\right\}$ $FM_b = \frac{w_1(L-l_1)^3}{60L^2}\left\{3(L+4l_1)-\frac{l_2(2L+3l_1)}{L-l_1}\right.$ $\left.\times\left[1+\frac{l_2}{L-l_1}+\frac{l_2^2}{(L-l_1)^2}\right]+\frac{3l_2^4}{(L-l_1)^3}\right\}$ $+\frac{w_2(L-l_1)^3}{60L^2}\left\{(2L+3l_1)\left[1+\frac{l_2}{L-l_1}\right.\right.$ $\left.\left.+\frac{l_2^2}{(L-l_1)^2}\right]-\frac{3l_2^3}{(L-l_1)^2}\left[1+\frac{5L-4l_2}{L-l_1}\right]\right\}$ $FS_e = \left(\frac{w_1+w_2}{2}\right)(L-l_1-l_2)-FS_b$ $FM_e = \frac{L-l_1-l_2}{6}[w_1(-2L+2l_1-l_2)$ $-w_2(L-l_1+2l_2)]+FS_b(L)-FM_b$
5.	l_1, l_2, W, FA_b, FA_e, b, e, L	$FA_b = \frac{Wl_2}{L}$ $FA_e = \frac{Wl_1}{L}$
6.	l_1, l_2, w, FA_b, FA_e, b, e, L	$FA_b = \frac{w}{2L}(L-l_1-l_2)(L-l_1+l_2)$ $FA_e = \frac{w}{2L}(L-l_1-l_2)(L+l_1-l_2)$
7.	l_1, l_2, M_T, FT_b, FT_e, b, e, L	$FT_b = \frac{M_T l_2}{L}$ $FT_e = \frac{M_T l_1}{L}$

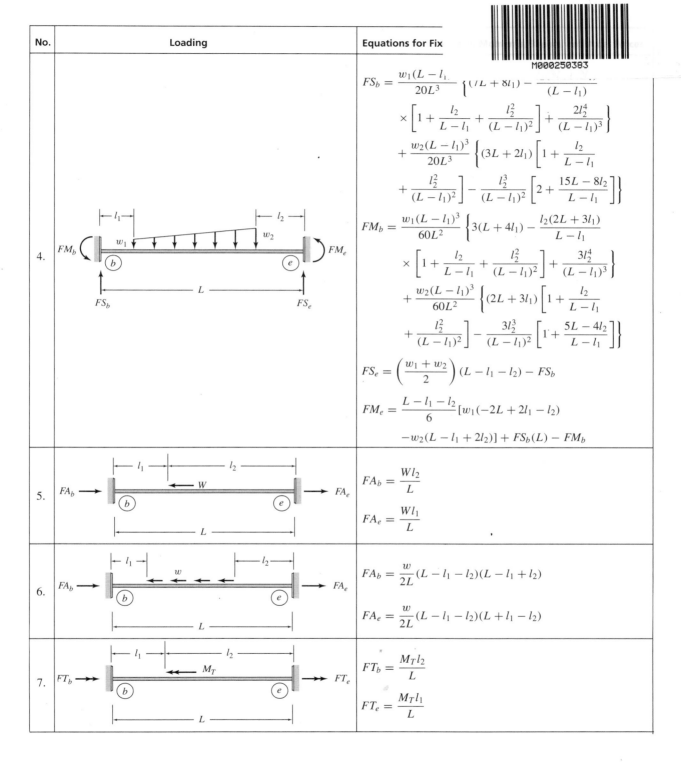

MATRIX ANALYSIS OF STRUCTURES

Second Edition

MATRIX ANALYSIS OF STRUCTURES

Second Edition

ASLAM KASSIMALI

Southern Illinois University—Carbondale

Australia • Brazil • Japan • Korea • Mexico • Singapore • Spain • United Kingdom • United States

Matrix Analysis of Structures,
Second Edition
Aslam Kassimali

Publisher, Global Engineering: Christopher M. Shortt

Acquisitions Editor: Randall Adams

Senior Developmental Editor: Hilda Gowans

Editorial Assistant: Tanya Altieri

Team Assistant: Carly Rizzo

Marketing Manager: Lauren Betsos

Media Editor: Chris Valentine

Content Project Manager: Jennifer Ziegler

Production Service: RPK Editorial Services

Copyeditor: Erin Wagner

Proofreader: Martha McMaster

Indexer: Shelly Gerger-Knechtl

Compositor: MPS Limited, a Macmillan Company

Senior Art Director: Michelle Kunkler

Internal Designer: Carmela Periera

Cover Designer: Andrew Adams

Cover Image: © gary718/Shutterstock; AMA/Shutterstock; ILYA GENKIN/Shutterstock

Text and Image Permissions Researcher: Kristiina Paul

First Print Buyer: Arethea L. Thomas

Library of Congress Control Number: on file

ISBN-13: 978-1-111-42620-0

ISBN-10: 1-111-42620-1

Cengage Learning
200 First Stamford Place, Suite 400
Stamford, CT 06902
USA

Printed in the United States of America
1 2 3 4 5 6 7 13 12 11 10

IN MEMORY OF MY FATHER,
KASSIMALI B. ALLANA

CONTENTS

PREFACE

The objective of this book is to develop an understanding of the basic principles of the matrix methods of structural analysis, so that they can be efficiently implemented on modern computers. Focusing on the stiffness approach, *Matrix Analysis of Structures* covers the linear analysis of two- and three-dimensional framed structures in static equilibrium. It also presents an introduction to nonlinear structural analysis and contains the fundamentals of the flexibility approach.

The book is divided into ten chapters. Chapter 1 presents a general introduction to the subject, and Chapter 2 reviews the basic concepts of matrix algebra relevant to matrix structural analysis. The next five chapters (Chapters 3 through 7) cover the analysis of plane trusses, beams, and plane rigid frames. The computer implementation of the stiffness method is initiated early in the text (beginning with Chapter 4), to allow students sufficient time to complete development of computer programs within the duration of a single course. Chapter 8 presents the analysis of space trusses, grids, and space rigid frames, Chapter 9 covers some special topics and modeling techniques, and Chapter 10 provides an introduction to nonlinear structural analysis. All the relationships necessary for matrix stiffness analysis are formulated using the basic principles of the mechanics of deformable bodies. Thus, a prior knowledge of the classical methods of structural analysis, while helpful, is not essential for understanding the material presented in the book. The format of the book is flexible enough to enable instructors to emphasize topics that are consistent with the goals of the course.

Each chapter begins with a brief introduction that defines its objectives, and ends with a summary outlining its salient features. An important general feature of the book is the inclusion of step-by-step procedures for analysis, and detailed flowcharts, to enable students to make an easier transition from theory to problem solving and program development. Numerous solved examples are provided to clarify the fundamental concepts, and to illustrate the application of the procedures for analysis.

A computer program for the analysis of two- and three-dimensional framed structures is available on the publisher's website www.cengage.com/engineering. This interactive software cab be used by students to check their answers to text exercises, and to verify the correctness of their own computer programs. The MATLAB® code for various flowcharts given in the book is available to instructors for distribution to students (if they so desire). A solutions manual, containing complete solutions to text exercises, is also available for instructors.

A NOTE ON THE REVISED EDITION

In this second edition, while the major features of the first edition have been retained, an introductory chapter on nonlinear analysis has been added because of its increasing use in structural design. In addition, the sections on temperature changes and fabrication errors (Section 7.5), and nonprismatic members (Section 9.8), have been expanded via inclusion of additional examples. The total number of examples has been increased by about 10 percent, and the number of problems has been increased by about 15 percent to 255, of which about 40 percent are new problems. These new problems include some computer exercises intended to familiarize students with the use of the general-purpose structural analysis software. There are many other minor revisions, including some in the computer software, which has been upgraded to make it compatible with the latest versions of Microsoft Windows®. Finally, most of the photographs have been replaced with new ones, some figures have been redrawn and rearranged, and the page layout of the book has been redesigned to enhance clarity.

ACKNOWLEDGMENTS

I wish to express my thanks to Hilda Gowans, Christopher Shortt and Randall Adams of Cengage Learning for their constant support and encouragement throughout this project, and to Rose Kernan for all her help during the production phase. The comments and suggestions for improvement from colleagues and students who have used the first edition are gratefully acknowledged. All of their suggestions were carefully considered, and implemented whenever possible. Thanks are also due to the following reviewers for their careful reviews of the manuscripts of the first and/or second editions, and for their constructive suggestions:

Riyad S. Aboutaha
Georgia Institute of Technology

Osama Abudayyeh
Western Michigan University

George E. Blandford
University of Kentucky

Kenneth E. Buttry
University of Wisconsin-Platteville

Joel P. Conte
University of California, San Diego

C. Armando Duarte
University of Illinois, Urbana-Champaign

Fouad Fanous
Iowa State University

Larry J. Feeser
Rensselaer Polytechnic Institute

Barry J. Goodno
Georgia Institute of Technology

George J. Kostyrko
California State University

Marc Levitan
Louisiana State University

Daniel G. Linzell
The Pennsylvania State University

Vernon C. Matzen
North Carolina State University

Everett E. McEwen
University of Rhode Island

Joel Moore
California State University

Ahmad Namini
University of Miami

Finally, I would like to express my loving gratitude to my wife, Maureen, for her unfailing support and expertise in helping me prepare this manuscript, and to my sons, Jamil and Nadim, who are a never-ending source of love, pride, and inspiration for me.

Aslam Kassimali

1 INTRODUCTION

Beijing National Olympic Stadium—Bird's Nest
(Eastimages / Shutterstock)

Structural analysis, which is an integral part of any structural engineering project, *is the process of predicting the performance of a given structure under a prescribed loading condition.* The performance characteristics usually of interest in structural design are: (a) stresses or stress resultants (i.e., axial forces, shears, and bending moments); (b) deflections; and (c) support reactions. Thus, the analysis of a structure typically involves the determination of these quantities as caused by the given loads and/or other external effects (such as support displacements and temperature changes). This text is devoted to the analysis of *framed structures*—that is, structures composed of long straight members. Many commonly used structures such as beams, and plane and space trusses and rigid frames, are classified as framed structures (also referred to as *skeletal structures*).

In most design offices today, the analysis of framed structures is routinely performed on computers, using software based on the matrix methods of structural analysis. It is therefore essential that structural engineers understand the basic principles of matrix analysis, so that they can develop their own computer programs and/or properly use commercially available software—and appreciate the physical significance of the analytical results. The objective of this text is to present the theory and computer implementation of matrix methods for the analysis of framed structures in static equilibrium.

This chapter provides a general introduction to the subject of matrix computer analysis of structures. We start with a brief historical background in Section 1.1, followed by a discussion of how matrix methods differ from classical and finite-element methods of structural analysis (Section 1.2). Flexibility and stiffness methods of matrix analysis are described in Section 1.3; the six types of framed structures considered in this text (namely, plane trusses, beams, plane frames, space trusses, grids, and space frames) are discussed in Section 1.4; and the development of simplified models of structures for the purpose of analysis is considered in Section 1.5. The basic concepts of structural analysis necessary for formulating the matrix methods, as presented in this text, are reviewed in Section 1.6; and the roles and limitations of linear and nonlinear types of structural analysis are discussed in Section 1.7. Finally, we conclude the chapter with a brief note on the computer software that is provided on the publisher's website for this book (Section 1.8). (www.cengage.com/engineering)

1.1 HISTORICAL BACKGROUND

The theoretical foundation for matrix methods of structural analysis was laid by James C. Maxwell, who introduced the method of consistent deformations in 1864; and George A. Maney, who developed the slope-deflection method in 1915. These classical methods are considered to be the precursors of the matrix flexibility and stiffness methods, respectively. In the precomputer era, the main disadvantage of these earlier methods was that they required direct solution of simultaneous algebraic equations—a formidable task by hand calculations in cases of more than a few unknowns.

The invention of computers in the late 1940s revolutionized structural analysis. As computers could solve large systems of simultaneous equations, the analysis methods yielding solutions in that form were no longer at a disadvantage, but in fact were preferred, because simultaneous equations could be expressed in matrix form and conveniently programmed for solution on computers.

S. Levy is generally considered to have been the first to introduce the flexibility method in 1947, by generalizing the classical method of consistent deformations. Among the subsequent researchers who extended the flexibility method and expressed it in matrix form in the early 1950s were H. Falkenheimer, B. Langefors, and P. H. Denke. The matrix stiffness method was developed by R. K. Livesley in 1954. In the same year, J. H. Argyris and S. Kelsey presented a formulation of matrix methods based on energy principles. In 1956, M. T. Turner, R. W. Clough, H. C. Martin, and L. J. Topp derived stiffness matrices for the members of trusses and frames using the finite-element approach, and introduced the now popular *direct stiffness method* for generating the structure stiffness matrix. In the same year, Livesley presented a nonlinear formulation of the stiffness method for stability analysis of frames.

Since the mid-1950s, the development of matrix methods has continued at a tremendous pace, with research efforts in recent years directed mainly toward formulating procedures for the dynamic and nonlinear analysis of structures, and developing efficient computational techniques for analyzing large structures. Recent advances in these areas can be attributed to S. S. Archer, C. Birnstiel, R. H. Gallagher, J. Padlog, J. S. Przemieniecki, C. K. Wang, and E. L. Wilson, among others.

1.2 CLASSICAL, MATRIX, AND FINITE-ELEMENT METHODS OF STRUCTURAL ANALYSIS

Classical versus Matrix Methods

As we develop matrix methods in subsequent chapters of this book, readers who are familiar with classical methods of structural analysis will realize that both matrix and classical methods are based on the same fundamental principles—but that the fundamental relationships of equilibrium, compatibility, and member stiffness are now expressed in the form of matrix equations, so that the numerical computations can be efficiently performed on a computer.

Most classical methods were developed to analyze particular types of structures, and since they were intended for hand calculations, they often involve certain assumptions (that are unnecessary in matrix methods) to reduce the amount of computational effort required for analysis. The application of these methods usually requires an understanding on the part of the analyst of the structural behavior. Consider, for example, the moment-distribution method. This classical method can be used to analyze only beams and plane frames undergoing bending deformations. Deformations due to axial forces in the frames are ignored to reduce the number of independent joint translations. While this assumption significantly reduces the computational effort, it complicates the analysis by requiring the analyst to draw a deflected shape of the frame corresponding to each degree of freedom of sidesway (independent joint translation), to estimate the relative magnitudes of member fixed-end moments: a difficult task even in the case of a few degrees of freedom of sidesway if the frame has inclined members. Because of their specialized and intricate nature, classical methods are generally not considered suitable for computer programming.

In contrast to classical methods, matrix methods were specifically developed for computer implementation; they are *systematic* (so that they can be conveniently programmed), and *general* (in the sense that the same overall format of the analytical procedure can be applied to the various types of framed structures). It will become clear as we study matrix methods that, because of the latter characteristic, a computer program developed to analyze one type of structure (e.g., plane trusses) can be modified with relative ease to analyze another type of structure (e.g., space trusses or frames).

As the analysis of large and highly redundant structures by classical methods can be quite time consuming, matrix methods are commonly used. However, classical methods are still preferred by many engineers for analyzing smaller structures, because they provide a better insight into the behavior of structures. Classical methods may also be used for preliminary designs, for checking the results of computerized analyses, and for deriving the member force–displacement relations needed in the matrix analysis. Furthermore, a study of classical methods is considered to be essential for developing an understanding of structural behavior.

Matrix versus Finite Element Methods

Matrix methods can be used to analyze framed structures only. Finite-element analysis, which originated as an extension of matrix analysis to surface structures (e.g., plates and shells), has now developed to the extent that it can be applied to structures and solids of practically any shape or form. From a theoretical viewpoint, the basic difference between the two is that, in matrix methods, the member force–displacement relationships are based on the exact solutions of the underlying differential equations, whereas in finite-element methods, such relations are generally derived by work-energy principles from assumed displacement or stress functions.

Because of the approximate nature of its force–displacement relations, finite-element analysis generally yields approximate results. However, as will be shown in Chapters 3 and 5, in the case of linear analysis of framed structures composed of prismatic (uniform) members, both matrix and finite-element approaches yield identical results.

1.3 FLEXIBILITY AND STIFFNESS METHODS

Two different methods can be used for the matrix analysis of structures: the *flexibility* method, and the *stiffness* method. The flexibility method, which is also referred to as the *force* or *compatibility* method, is essentially a generalization in matrix form of the classical method of consistent deformations. In this approach, the primary unknowns are the redundant forces, which are calculated first by solving the structure's compatibility equations. Once the redundant forces are known, the displacements can be evaluated by applying the equations of equilibrium and the appropriate member force–displacement relations.

The stiffness method, which originated from the classical slope-deflection method, is also called the *displacement* or *equilibrium* method. In this approach, the primary unknowns are the joint displacements, which are determined first by solving the structure's equations of equilibrium. With the joint displacements known, the unknown forces are obtained through compatibility considerations and the member force–displacement relations.

Although either method can be used to analyze framed structures, the flexibility method is generally convenient for analyzing small structures with a few redundants. This method may also be used to establish member force-displacement relations needed to develop the stiffness method. The stiffness method is more systematic and can be implemented more easily on computers; therefore, it is preferred for the analysis of large and highly redundant structures. Most of the commercially available software for structural analysis is based on the stiffness method. In this text, we focus our attention mainly on the stiffness method, with emphasis on a particular version known as the *direct stiffness method,* which is currently used in professional practice. The fundamental concepts of the flexibility method are presented in Appendix B.

1.4 CLASSIFICATION OF FRAMED STRUCTURES

Framed structures are composed of straight members whose lengths are significantly larger than their cross-sectional dimensions. Common framed structures can be classified into six basic categories based on the arrangement of their members, and the types of primary stresses that may develop in their members under major design loads.

Plane Trusses

A *truss* is defined as an assemblage of straight members connected at their ends by flexible connections, and subjected to loads and reactions only at the joints (connections). The members of such an ideal truss develop only axial forces when the truss is loaded. In real trusses, such as those commonly used for supporting roofs and bridges, the members are connected by bolted or welded connections that are not perfectly flexible, and the dead weights of the members are distributed along their lengths. Because of these and other deviations from idealized conditions, truss members are subjected to some bending and shear. However, in most trusses, these secondary bending moments and shears are small in comparison to the primary axial forces, and are usually not considered in their designs. If large bending moments and shears are anticipated, then the truss should be treated as a rigid frame (discussed subsequently) for analysis and design.

If all the members of a truss as well as the applied loads lie in a single plane, the truss is classified as a *plane truss* (Fig. 1.1). The members of plane trusses are assumed to be connected by frictionless hinges. The analysis of plane trusses is considerably simpler than the analysis of space (or three-dimensional) trusses. Fortunately, many commonly used trusses, such as bridge and roof trusses, can be treated as plane trusses for analysis (Fig. 1.2).

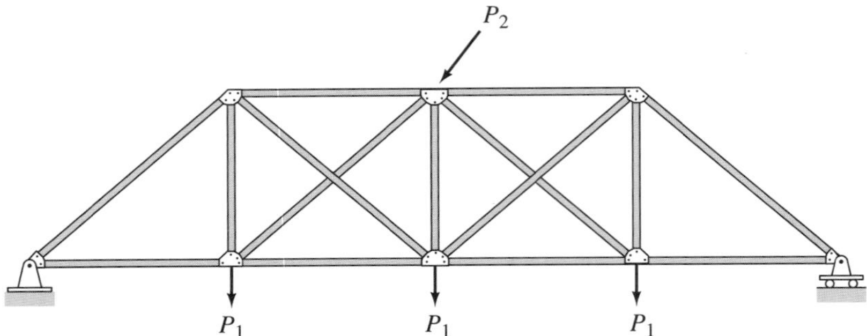

Fig. 1.1 *Plane Truss*

Fig. 1.2 *Roof Truss*
(Photo courtesy of Bethlehem Steel Corporation)

Beams

A *beam* is defined as a long straight structure that is loaded perpendicular to its longitudinal axis (Fig. 1.3). Loads are usually applied in a plane of symmetry of the beam's cross-section, causing its members to be subjected only to bending moments and shear forces.

Plane Frames

Frames, also referred to as *rigid frames,* are composed of straight members connected by rigid (moment resisting) and/or flexible connections (Fig. 1.4). Unlike trusses, which are subjected to external loads only at the joints, loads on frames may be applied on the joints as well as on the members.

If all the members of a frame and the applied loads lie in a single plane, the frame is called a *plane frame* (Fig. 1.5). The members of a plane frame are, in

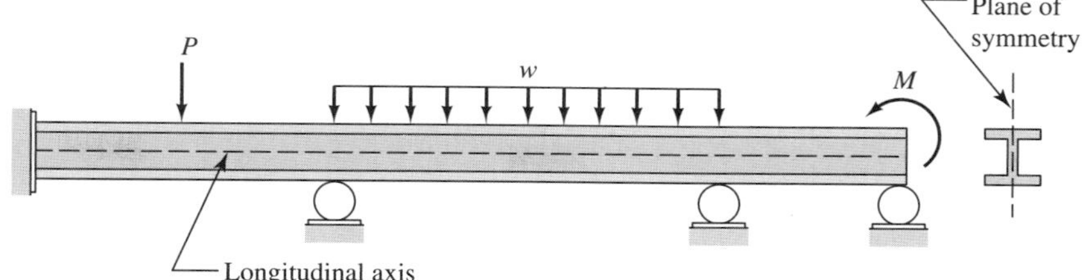

Fig. 1.3 *Beam*

Fig. 1.4 *Skeleton of a Structural Steel Frame Building*
(Joe Gough / Shutterstock)

general, subjected to bending moments, shears, and axial forces under the action of external loads. Many actual three-dimensional building frames can be subdivided into plane frames for analysis.

Space Trusses

Some trusses (such as lattice domes, transmission towers, and certain aerospace structures (Fig. 1.6)) cannot be treated as plane trusses because of the arrangement of their members or applied loading. Such trusses, referred to as *space trusses,* are analyzed as three-dimensional structures subjected to three-dimensional force systems. The members of space trusses are assumed to be connected by frictionless ball-and-socket joints, and the trusses are subjected to loads and reactions only at the joints. Like plane trusses, the members of space trusses develop only axial forces.

Grids

A *grid,* like a plane frame, is composed of straight members connected together by rigid and/or flexible connections to form a plane framework. The

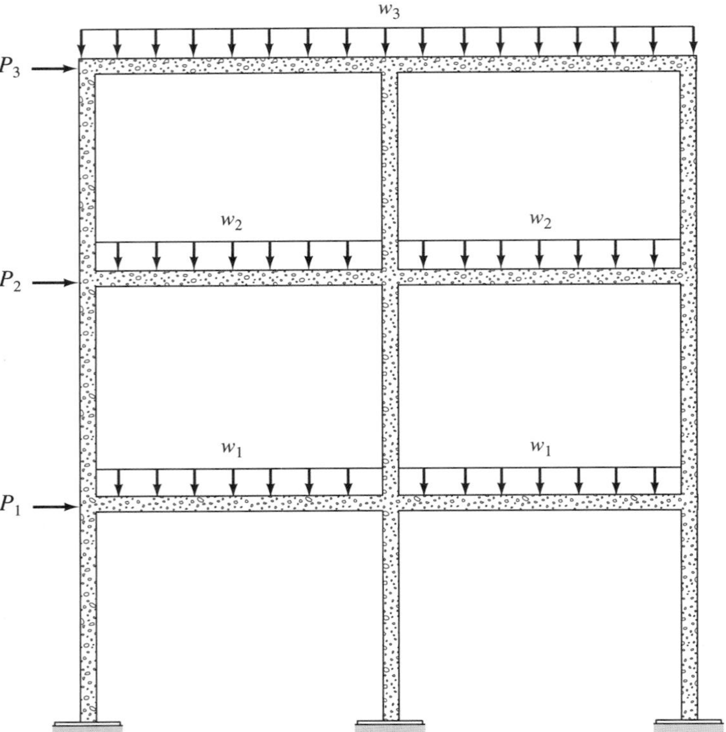

Fig. 1.5 *Plane Frame*

Fig. 1.6 *A Segment of the Integrated Truss Structure which Forms the Backbone of the International Space Station*

(Photo Courtesy of National Aeronautics and Space Administration 98-05165)

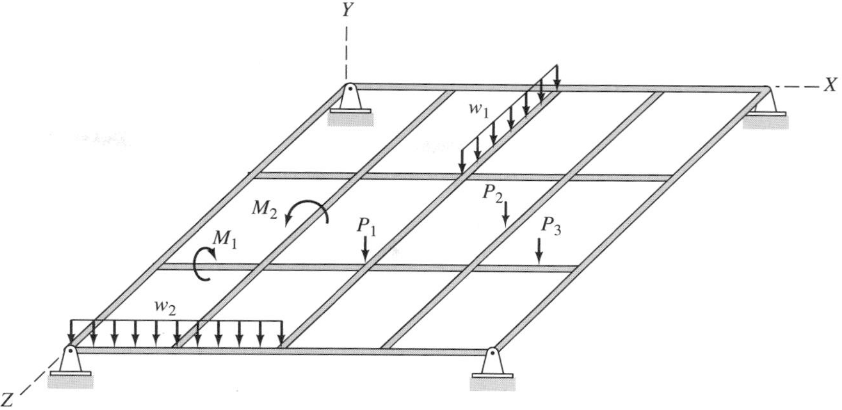

Fig. 1.7 *Grid*

Fig. 1.8 *National Air and Space Museum, Washington, DC (under construction)*
(Photo courtesy of Bethlehem Steel Corporation)

main difference between the two types of structures is that plane frames are loaded in the plane of the structure, whereas the loads on grids are applied in the direction perpendicular to the structure's plane (Fig. 1.7). Members of grids may, therefore, be subjected to torsional moments, in addition to the bending moments and corresponding shears that cause the members to bend out of the plane of the structure. Grids are commonly used for supporting roofs covering large column-free areas in such structures as sports arenas, auditoriums, and aircraft hangars (Fig. 1.8).

Fig. 1.9 *Space Frame*
(© MNTravel / Alamy)

Space Frames

Space frames constitute the most general category of framed structures. Members of space frames may be arranged in any arbitrary directions, and connected by rigid and/or flexible connections. Loads in any directions may be applied on members as well as on joints. The members of a space frame may, in general, be subjected to bending moments about both principal axes, shears in both principal directions, torsional moments, and axial forces (Fig. 1.9).

1.5 ANALYTICAL MODELS

The first (and perhaps most important) step in the analysis of a structure is to develop its analytical model. An analytical model is an idealized representation of a real structure for the purpose of analysis. Its objective is to simplify the analysis of a complicated structure by discarding much of the detail (about connections, members, etc.) that is likely to have little effect on the structure's behavioral characteristics of interest, while representing, as accurately as practically possible, the desired characteristics. It is important to note that the structural response predicted from an analysis is valid only to the extent that the analytical model represents the actual structure. For framed structures, the establishment of analytical models generally involves consideration of issues such as whether the actual three-dimensional structure can be subdivided into plane structures for analysis, and whether to idealize the actual bolted or welded connections as hinged, rigid, or semirigid joints. Thus, the development of accurate analytical models requires not only a thorough understanding of structural behavior and methods of analysis, but also experience and knowledge of design and construction practices.

In matrix methods of analysis, a structure is modeled as an assemblage of straight members connected at their ends to joints. A *member* is defined as *a part of the structure for which the member force-displacement relationships to be used in the analysis are valid.* The member force-displacement relationships for the various types of framed structures will be derived in subsequent chapters. A *joint* is defined as *a structural part of infinitesimal size to which the ends of the members are connected.* In finite-element terminology, the members and joints of structures are generally referred to as *elements* and *nodes,* respectively.

Supports for framed structures are commonly idealized as fixed supports, which do not allow any displacement; hinged supports, which allow rotation but prevent translation; or, roller or link supports, which prevent translation in only one direction. Other types of restraints, such as those which prevent rotation but permit translation in one or more directions, can also be considered in an analysis, as discussed in subsequent chapters.

Line Diagrams

The analytical model of a structure is represented by a *line diagram,* on which each member is depicted by a line coinciding with its centroidal axis. The member dimensions and the size of connections are not shown. Rigid joints are usually represented by points, and hinged joints by small circles, at the intersections of members. Each joint and member of the structure is identified by a number. For example, the analytical model of the plane truss of Fig. 1.10(a) is shown in Fig. 1.10(b), in which the joint numbers are enclosed within circles to distinguish them from the member numbers enclosed within rectangles.

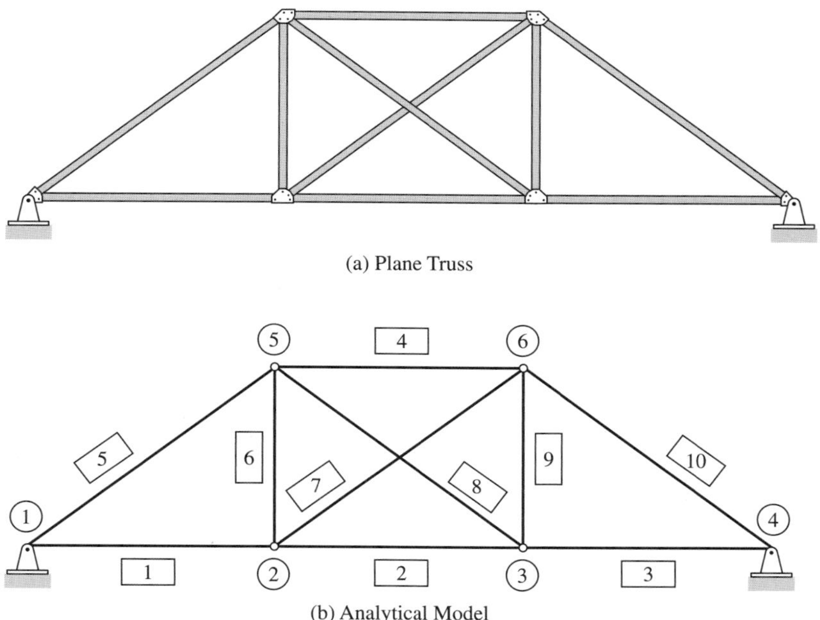

Fig. 1.10

1.6 FUNDAMENTAL RELATIONSHIPS FOR STRUCTURAL ANALYSIS

Structural analysis, in general, involves the use of three types of relationships:

- Equilibrium equations,
- compatibility conditions, and
- constitutive relations.

Equilibrium Equations

A structure is considered to be in equilibrium if, initially at rest, it remains at rest when subjected to a system of forces and couples. If a structure is in equilibrium, then all of its members and joints must also be in equilibrium.

Recall from *statics* that for a plane (two-dimensional) structure lying in the *XY* plane and subjected to a coplanar system of forces and couples (Fig. 1.11), the necessary and sufficient conditions for equilibrium can be expressed in Cartesian (*XY*) coordinates as

$$\boxed{\sum F_X = 0 \qquad \sum F_Y = 0 \qquad \sum M = 0} \tag{1.1}$$

These equations are referred to as the *equations of equilibrium* for plane structures.

For a space (three-dimensional) structure subjected to a general three-dimensional system of forces and couples (Fig. 1.12), the equations of equilibrium are expressed as

$$\boxed{\begin{array}{lll} \sum F_X = 0 & \sum F_Y = 0 & \sum F_Z = 0 \\ \sum M_X = 0 & \sum M_Y = 0 & \sum M_Z = 0 \end{array}} \tag{1.2}$$

For a structure subjected to static loading, the equilibrium equations must be satisfied for the entire structure as well as for each of its members and joints. In structural analysis, equations of equilibrium are used to relate the forces (including couples) acting on the structure or one of its members or joints.

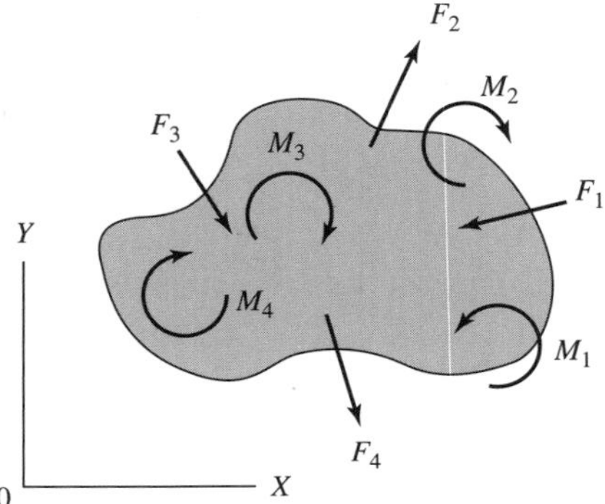

Fig. 1.11

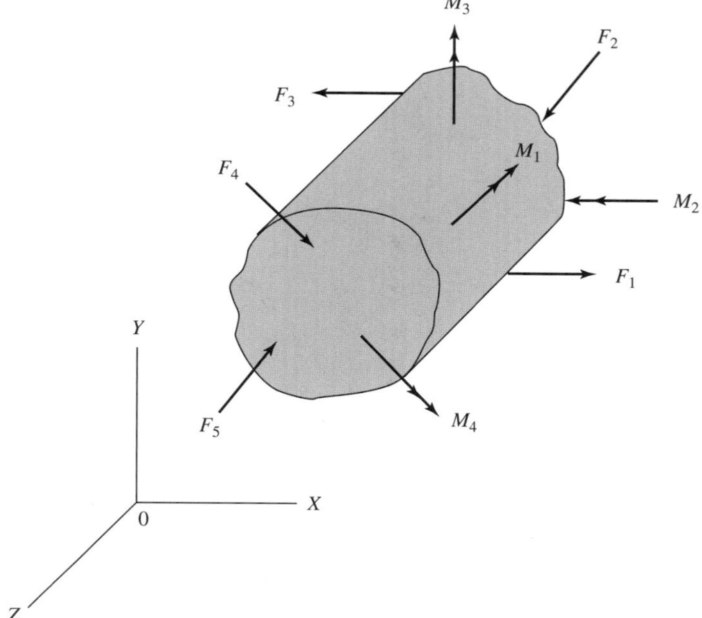

Fig. 1.12

Compatibility Conditions

The *compatibility conditions* relate the deformations of a structure so that its various parts (members, joints, and supports) fit together without any gaps or overlaps. These conditions (also referred to as the *continuity conditions*) ensure that the deformed shape of the structure is continuous (except at the locations of any internal hinges or rollers), and is consistent with the support conditions.

Consider, for example, the two-member plane frame shown in Fig. 1.13. The deformed shape of the frame due to an arbitrary loading is also depicted, using an exaggerated scale. When analyzing a structure, the compatibility conditions are used to relate member end displacements to joint displacements which, in turn, are related to the support conditions. For example, because joint 1 of the frame in Fig. 1.13 is attached to a roller support that cannot translate in the vertical direction, the vertical displacement of this joint must be zero. Similarly, because joint 3 is attached to a fixed support that can neither rotate nor translate in any direction, the rotation and the horizontal and vertical displacements of joint 3 must be zero.

The displacements of the ends of members are related to the joint displacements by the compatibility requirement that the displacements of a member's end must be the same as the displacements of the joint to which the member end is connected. Thus, as shown in Fig. 1.13, because joint 1 of the example frame displaces to the right by a distance d_1 and rotates clockwise by an angle θ_1, the left end of the horizontal member (member 1) that is attached to joint 1

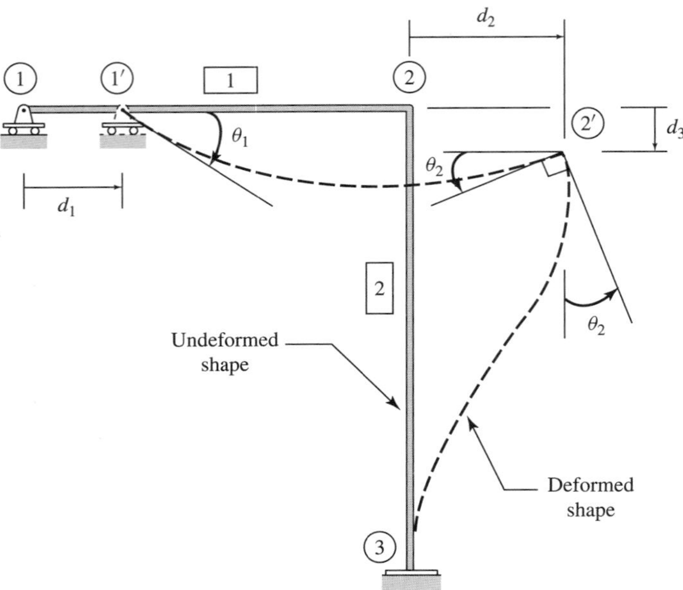

Fig. 1.13

must also translate to the right by distance d_1 and rotate clockwise by angle θ_1. Similarly, because the displacements of joint 2 consist of the translations d_2 to the right and d_3 downward and the counterclockwise rotation θ_2, the right end of the horizontal member and the top end of the vertical member that are connected to joint 2 must also undergo the same displacements (i.e., d_2, d_3, and θ_2). The bottom end of the vertical member, however, is not subjected to any displacements, because joint 3, to which this particular member end is attached, can neither rotate nor translate in any direction.

Finally, compatibility requires that the deflected shapes of the members of a structure be continuous (except at any internal hinges or rollers) and be consistent with the displacements at the corresponding ends of the members.

Constitutive Relations

The *constitutive relations* (also referred to as the *stress-strain relations*) describe the relationships between the stresses and strains of a structure in accordance with the stress-strain properties of the structural material. As discussed previously, the equilibrium equations provide relationships between the forces, whereas the compatibility conditions involve only deformations. The constitutive relations provide the link between the equilibrium equations and compatibility conditions that is necessary to establish the load-deformation relationships for a structure or a member.

In the analysis of framed structures, the basic stress-strain relations are first used, along with the member equilibrium and compatibility equations, to establish relationships between the forces and displacements at the ends of a member. The member force-displacement relations thus obtained are then treated as the constitutive relations for the entire structure, and are used to link the structure's equilibrium and compatibility equations, thereby yielding the load-deformation relationships for the entire structure. These load-deformation relations can then be solved to determine the deformations of the structure due to a given loading.

In the case of statically determinate structures, the equilibrium equations can be solved independently of the compatibility and constitutive relations to obtain the reactions and member forces. The deformations of the structure, if desired, can then be determined by employing the compatibility and constitutive relations. In the analysis of statically indeterminate structures, however, the equilibrium equations alone are not sufficient for determining the reactions and member forces. Therefore, it becomes necessary to satisfy simultaneously the three types of fundamental relationships (i.e., equilibrium, compatibility, and constitutive relations) to determine the structural response.

Matrix methods of structural analysis are usually formulated by direct application of the three fundamental relationships as described in general terms in the preceding paragraphs. (Details of the formulations are presented in subsequent chapters.) However, matrix methods can also be formulated by using work-energy principles that satisfy the three fundamental relationships indirectly. Work-energy principles are generally preferred in the formulation of finite-element methods, because they can be more conveniently applied to derive the approximate force-displacement relations for the elements of surface structures and solids.

The matrix methods presented in this text are formulated by the direct application of the equilibrium, compatibility, and constitutive relationships. However, to introduce readers to the finite-element method, and to familiarize them with the application of the work-energy principles, we also derive the member force-displacement relations for plane structures by a finite-element approach that involves a work-energy principle known as the *principle of virtual work*. In the following paragraphs, we review two statements of this principle pertaining to rigid bodies and deformable bodies, for future reference.

Principle of Virtual Work for Rigid Bodies

The *principle of virtual work for rigid bodies* (also known as the *principle of virtual displacements for rigid bodies*) can be stated as follows.

> *If a rigid body, which is in equilibrium under a system of forces (and couples), is subjected to any small virtual rigid-body displacement, the virtual work done by the external forces (and couples) is zero.*

In the foregoing statement, the term virtual simply means imaginary, not real. Consider, for example, the cantilever beam shown in Fig. 1.14(a). The free-body diagram of the beam is shown in Fig. 1.14(b), in which P_X, and P_Y are the components of the external load P in the X and Y directions, respectively, and R_1, R_2, and R_3 represent the reactions at the fixed support 1. Note that the beam is in equilibrium under the action of the forces P_X, P_Y, R_1, and R_2, and the couple R_3. Now, imagine that the beam is given an arbitrary, small virtual rigid-body displacement from its initial equilibrium position 1–2 to another position 1′–2′, as shown in Fig. 1.14(c). As this figure indicates, the total virtual

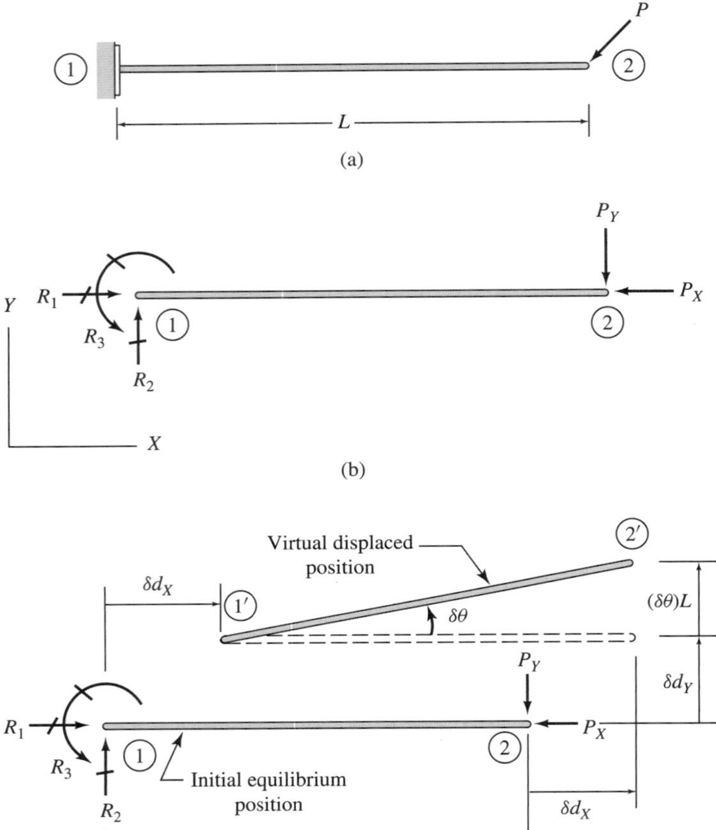

Fig. 1.14

displacement of the beam can be decomposed into rigid-body translations δd_X and δd_Y in the X and Y directions, respectively, and a rigid-body rotation $\delta\theta$ about point 1. Note that the symbol δ is used here to identify the virtual quantities. As the beam undergoes the virtual displacement from position 1–2 to position 1′–2′, the forces and the couple acting on it perform work, which is referred to as the *virtual work*. The total virtual work, δW_e, can be expressed as the algebraic sum of the virtual work δW_X and δW_Y, performed during translations in the X and Y directions, respectively, and the virtual work δW_R, done during the rotation; that is,

$$\delta W_e = \delta W_X + \delta W_Y + \delta W_R \tag{1.3}$$

During the virtual translation δd_X of the beam, the virtual work performed by the forces can be expressed as follows (Fig 1.14c).

$$\delta W_X = R_1 \delta d_X - P_X \delta d_X = (R_1 - P_X)\, \delta d_X = \left(\sum F_X\right) \delta d_X \tag{1.4}$$

Similarly, the virtual work done during the virtual translation δd_Y is given by

$$\delta W_Y = R_2 \delta d_Y - P_Y \delta d_Y = (R_2 - P_Y)\,\delta d_Y = \left(\sum F_Y\right) \delta d_Y \tag{1.5}$$

and the virtual work done by the forces and the couple during the small virtual rotation $\delta\theta$ can be expressed as follows (Fig. 1.14c).

$$\delta W_R = R_3 \delta\theta - P_Y\,(L\delta\theta) = (R_3 - P_Y L)\,\delta\theta = \left(\sum M_{①}\right) \delta\theta \tag{1.6}$$

The expression for the total virtual work can now be obtained by substituting Eqs. (1.4–1.6) into Eq. (1.3). Thus,

$$\delta W_e = \left(\sum F_X\right) \delta d_X + \left(\sum F_Y\right) \delta d_Y + \left(\sum M_{①}\right) \delta\theta \tag{1.7}$$

However, because the beam is in equilibrium, $\sum F_X = 0$, $\sum F_Y = 0$, and $\sum M_{①} = 0$; therefore, Eq. (1.7) becomes

$$\boxed{\delta W_e = 0} \tag{1.8}$$

which is the mathematical statement of the principle of virtual work for rigid bodies.

Principle of Virtual Work for Deformable Bodies

The *principle of virtual work for deformable bodies* (also called the *principle of virtual displacements for deformable bodies*) can be stated as follows.

> *If a deformable structure, which is in equilibrium under a system of forces (and couples), is subjected to any small virtual displacement consistent with the support and continuity conditions of the structure, then the virtual external work done by the real external forces (and couples) acting through the virtual external displacements (and rotations) is equal to the virtual strain energy stored in the structure.*

To demonstrate the validity of this principle, consider the two-member truss of Fig. 1.15(a), which is in equilibrium under the action of an external load P. The free-body diagram of joint 3 of the truss is shown in Fig. 1.15(b). Since joint 3 is in equilibrium, the external and internal forces acting on it must satisfy the following two equations of equilibrium:

$$\begin{aligned} &+\rightarrow \sum F_X = 0 \qquad && -F_1 \sin\theta_1 + F_2 \sin\theta_2 = 0 \\ &+\uparrow \sum F_Y = 0 \qquad && F_1 \cos\theta_1 + F_2 \cos\theta_2 - P = 0 \end{aligned} \tag{1.9}$$

in which F_1 and F_2 denote the internal (axial) forces in members 1 and 2, respectively; and θ_1 and θ_2 are, respectively, the angles of inclination of these members with respect to the vertical as shown in the figure.

Now, imagine that joint 3 is given a small virtual compatible displacement, δd, in the downward direction, as shown in Fig. 1.15(a). It should be noted that this virtual displacement is consistent with the support conditions of the truss in the sense that joints 1 and 2, which are attached to supports, are not displaced. Because the reaction forces at joints 1 and 2 do not perform any work,

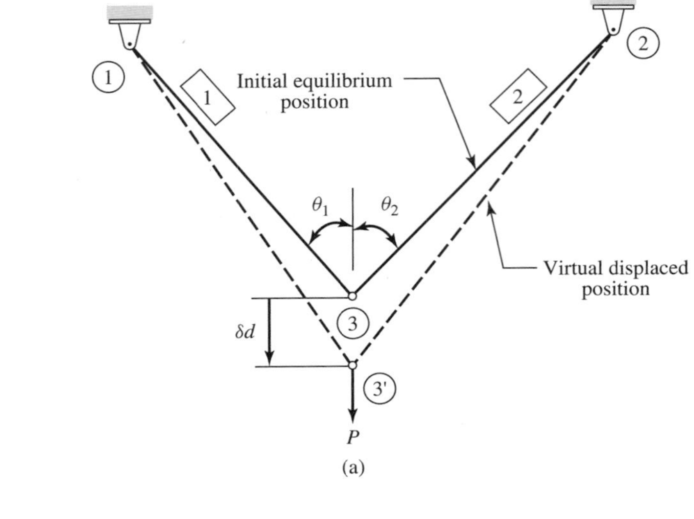

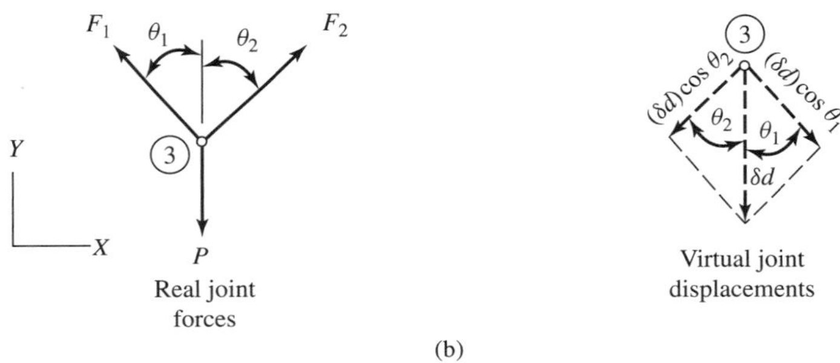

Fig. 1.15

the total virtual work for the truss, δW, is equal to the algebraic sum of the virtual work of the forces acting at joint 3. Thus, from Fig. 1.15(b),

$$\delta W = P\delta d - F_1(\delta d \cos\theta_1) - F_2(\delta d \cos\theta_2)$$

which can be rewritten as

$$\delta W = (P - F_1 \cos\theta_1 - F_2 \cos\theta_2)\ \delta d \tag{1.10}$$

As indicated by Eq. (1.9), the term in parentheses on the right-hand side of Eq. (1.10) is zero. Therefore, the total virtual work, δW, is zero. By substituting $\delta W = 0$ into Eq. (1.10) and rearranging terms, we write

$$P(\delta d) = F_1(\delta d \cos\theta_1) + F_2(\delta d \cos\theta_2) \tag{1.11}$$

in which the quantity on the left-hand side represents the virtual external work, δW_e, performed by the real external force P acting through the virtual external displacement δd. Furthermore, because the terms $(\delta d)\cos\theta_1$ and $(\delta d)\cos\theta_2$ are equal to the virtual internal displacements (elongations) of members 1 and 2, respectively, we can conclude that the right-hand side of Eq. (1.11) represents the virtual internal work, δW_{i}, done by the real internal forces acting through the corresponding virtual internal displacements; that is,

$$\delta W_{\text{e}} = \delta W_{\text{i}} \tag{1.12}$$

Realizing that the internal work is also referred to as the strain energy, U, we can express Eq. (1.12) as

$$\boxed{\delta W_{\text{e}} = \delta U} \tag{1.13}$$

in which δU denotes the virtual strain energy. Note that Eq. (1.13) is the mathematical statement of the principle of virtual work for deformable bodies.

For computational purposes, it is usually convenient to express Eq. (1.13) in terms of the stresses and strains in the members of the structure. For that purpose, let us consider a differential element of a member of an arbitrary structure subjected to a general loading (Fig. 1.16). The element is in equilibrium under a general three-dimensional stress condition, due to the real forces acting on the structure. Now, as the structure is subjected to a virtual displacement, virtual strains develop in the element and the internal forces due to the real stresses perform virtual internal work as they move through the internal displacements caused by the virtual strains. For example, the virtual internal work done by the real force due to the stress σ_x as it moves through the virtual displacement caused by the virtual strain $\delta\varepsilon_x$ can be determined as follows.

$$\begin{aligned} \text{real force} &= \text{stress} \times \text{area} &&= \sigma_x \, (dy \, dz) \\ \text{virtual displacement} &= \text{strain} \times \text{length} &&= (\delta\varepsilon_x) \, dx \end{aligned}$$

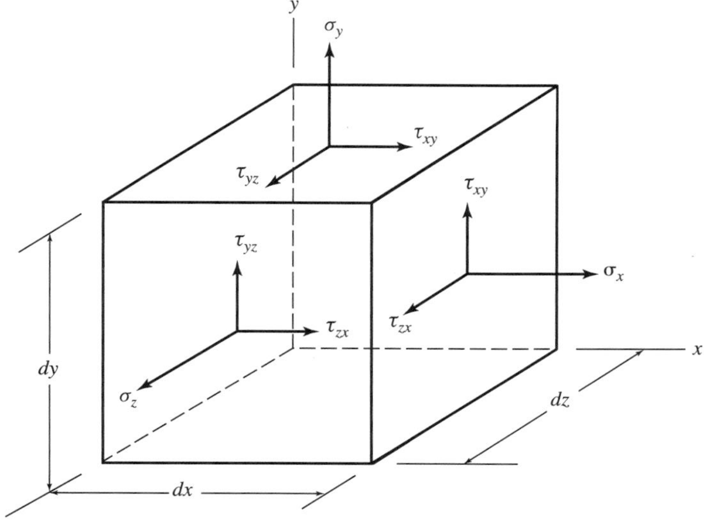

Fig. 1.16

Therefore,

$$\begin{aligned}\text{virtual internal work} &= \text{real force} \times \text{virtual displacement}\\ &= (\sigma_x \, dy \, dz)\,(\delta\varepsilon_x \, dx)\\ &= (\delta\varepsilon_x \, \sigma_x)\, dV\end{aligned}$$

in which $dV = dx\,dy\,dz$ is the volume of the differential element. Thus, the virtual internal work due to all six stress components is given by

$$\begin{aligned}&\text{virtual internal work in element } dV\\ &\quad = (\delta\varepsilon_x\sigma_x + \delta\varepsilon_y\sigma_y + \delta\varepsilon_z\sigma_z + \delta\gamma_{xy}\tau_{xy} + \delta\gamma_{yz}\tau_{yz} + \delta\gamma_{zx}\tau_{zx})\, dV \qquad \textbf{(1.14)}\end{aligned}$$

In Eq. (1.14), $\delta\varepsilon_x, \delta\varepsilon_y, \delta\varepsilon_z, \delta\gamma_{xy}, \delta\gamma_{yz},$ and $\delta\gamma_{zx}$ denote, respectively, the virtual strains corresponding to the real stresses $\sigma_x, \sigma_y, \sigma_z, \tau_{xy}, \tau_{yz},$ and τ_{zx}, shown in Fig. 1.16.

The total virtual internal work, or the virtual strain energy stored in the entire structure, can be obtained by integrating Eq. (1.14) over the volume V of the structure. Thus,

$$\delta U = \int_V \left(\delta\varepsilon_x\sigma_x + \delta\varepsilon_y\sigma_y + \delta\varepsilon_z\sigma_z + \delta\gamma_{xy}\tau_{xy} + \delta\gamma_{yz}\tau_{yz} + \delta\gamma_{zx}\tau_{zx}\right) dV \qquad \textbf{(1.15)}$$

Finally, by substituting Eq. (1.15) into Eq. (1.13), we obtain the statement of the principle of virtual work for deformable bodies in terms of the stresses and strains of the structure.

$$\boxed{\delta W_e = \int_V \left(\delta\varepsilon_x\sigma_x + \delta\varepsilon_y\sigma_y + \delta\varepsilon_z\sigma_z + \delta\gamma_{xy}\tau_{xy} + \delta\gamma_{yz}\tau_{yz} + \delta\gamma_{zx}\tau_{zx}\right) dV} \qquad \textbf{(1.16)}$$

1.7 LINEAR VERSUS NONLINEAR ANALYSIS

In this text, we focus our attention mainly on *linear analysis* of structures. Linear analysis of structures is based on the following two fundamental assumptions:

1. The structures are composed of linearly elastic material; that is, the stress-strain relationship for the structural material follows Hooke's law.
2. The deformations of the structures are so small that the squares and higher powers of member slopes, (chord) rotations, and axial strains are negligible in comparison with unity, and the equations of equilibrium can be based on the undeformed geometry of the structure.

The reason for making these assumptions is to obtain linear relationships between applied loads and the resulting structural deformations. An important advantage of linear force-deformation relations is that the *principle of superposition* can be used in the analysis. This principle states essentially that *the combined effect of several loads acting simultaneously on a structure equals the algebraic sum of the effects of each load acting individually on the structure*.

Engineering structures are usually designed so that under service loads they undergo small deformations, with stresses within the initial linear portions of the stress-strain curves of their materials. Thus, linear analysis generally proves adequate for predicting the performance of most common types of structures under service loading conditions. However, at higher load levels, the accuracy of linear analysis generally deteriorates as the deformations of the structure increase and/or its material is strained beyond the yield point. Because of its inherent limitations, linear analysis cannot be used to predict the ultimate load capacities and instability characteristics (e.g., buckling loads) of structures.

With the recent introduction of design specifications based on the ultimate strengths of structures, the use of *nonlinear analysis* in structural design is increasing. In a nonlinear analysis, the restrictions of linear analysis are removed by formulating the equations of equilibrium on the deformed geometry of the structure that is not known in advance, and/or taking into account the effects of inelasticity of the structural material. The load-deformation relationships thus obtained for the structure are nonlinear, and are usually solved using iterative techniques. An introduction to this still-evolving field of nonlinear structural analysis is presented in Chapter 10.

1.8 SOFTWARE

Software for the analysis of framed structures using the matrix stiffness method is provided on the publisher's website for this book, www.cengage.com/engineering. The software can be used by readers to verify the correctness of various subroutines and programs that they will develop during the course of study of this text, as well as to check the answers to the problems given at the end of each chapter. A description of the software, and information on how to install and use it, is presented in Appendix A.

SUMMARY

In this chapter, we discussed the topics summarized in the following list.

1. *Structural analysis* is the prediction of the performance of a given structure under prescribed loads and/or other external effects.

2. Both matrix and classical methods of structural analysis are based on the same fundamental principles. However, classical methods were developed to analyze particular types of structures, whereas matrix methods are more general and systematic so that they can be conveniently programmed on computers.

3. Two different methods can be used for matrix analysis of structures; namely, the *flexibility* and *stiffness* methods. The stiffness method is more systematic and can be implemented more easily on computers, and is therefore currently preferred in professional practice.

4. *Framed structures* are composed of straight members whose lengths are significantly larger than their cross-sectional dimensions. Framed structures can be classified into six basic categories: plane trusses, beams, plane frames, space trusses, grids, and space frames.

5. An *analytical model* is a simplified (idealized) representation of a real structure for the purpose of analysis. Framed structures are modeled as assemblages of straight members connected at their ends to joints, and these analytical models are represented by *line diagrams*.

6. The analysis of structures involves three fundamental relationships: equilibrium equations, compatibility conditions, and constitutive relations.

7. The principle of virtual work for deformable bodies states that if a deformable structure, which is in equilibrium, is subjected to a small compatible virtual displacement, then the virtual external work is equal to the virtual strain energy stored in the structure.

8. Linear structural analysis is based on two fundamental assumptions: the stress-strain relationship for the structural material is linearly elastic, and the structure's deformations are so small that the equilibrium equations can be based on the undeformed geometry of the structure.

2 MATRIX ALGEBRA

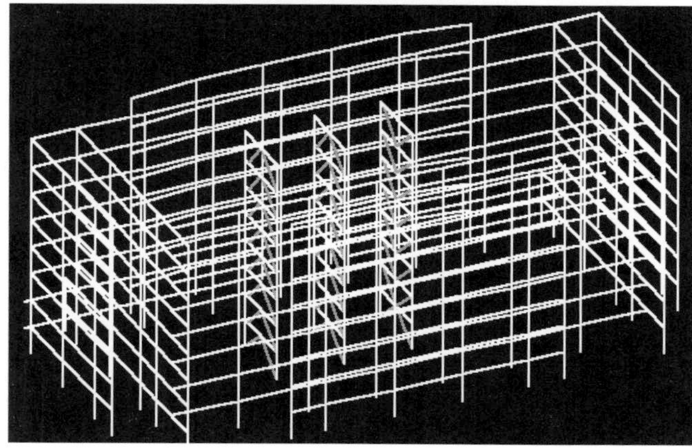

Somerset Corporate Center Office Building, New Jersey, and its Analytical Model
(Photo courtesy of Ram International. Structural Engineer: The Cantor Seinuk Group, P.C.)

In matrix methods of structural analysis, the fundamental relationships of equilibrium, compatibility, and member force–displacement relations are expressed in the form of matrix equations, and the analytical procedures are formulated by applying various matrix operations. Therefore, familiarity with the basic concepts of matrix algebra is a prerequisite to understanding matrix structural analysis. The objective of this chapter is to concisely present the basic concepts of matrix algebra necessary for formulating the methods of structural analysis covered in the text. A general procedure for solving simultaneous linear equations, the *Gauss–Jordan method,* is also discussed.

We begin with the basic definition of a matrix in Section 2.1, followed by brief descriptions of the various types of matrices in Section 2.2. The matrix operations of equality, addition and subtraction, multiplication, transposition, differentiation and integration, inversion, and partitioning are defined in Section 2.3; we conclude the chapter with a discussion of the Gauss–Jordan elimination method for solving simultaneous equations (Section 2.4).

2.1 DEFINITION OF A MATRIX

A *matrix* is defined as *a rectangular array of quantities arranged in rows and columns.* A matrix with m rows and n columns can be expressed as follows.

$$\mathbf{A} = [A] = \begin{bmatrix} A_{11} & A_{12} & A_{13} & \cdots & & \cdots & A_{1n} \\ A_{21} & A_{22} & A_{23} & \cdots & & \cdots & A_{2n} \\ A_{31} & A_{32} & A_{33} & \cdots & & \cdots & A_{3n} \\ \cdots & \cdots & \cdots & \cdots & A_{ij} & \cdots & \\ A_{m1} & A_{m2} & A_{m3} & \cdots & | & \cdots & A_{mn} \end{bmatrix} \begin{matrix} \\ \\ \\ i\text{th row} \\ \\ \end{matrix} \qquad \begin{matrix} j\text{th column} \qquad m \times n \end{matrix} \tag{2.1}$$

As shown in Eq. (2.1), matrices are denoted either by boldface letters (**A**) or by italic letters enclosed within brackets ([A]). The quantities forming a matrix are referred to as its *elements.* The elements of a matrix are usually numbers, but they can be symbols, equations, or even other matrices (called submatrices). Each element of a matrix is represented by a double-subscripted letter, with the first subscript identifying the row and the second subscript identifying the column in which the element is located. Thus, in Eq. (2.1), A_{23} represents the element located in the second row and third column of matrix **A**. In general, A_{ij} refers to an element located in the ith row and jth column of matrix **A**.

The size of a matrix is measured by the number of its rows and columns and is referred to as the *order* of the matrix. Thus, matrix **A** in Eq. (2.1), which has m rows and n columns, is considered to be of order $m \times n$ (m by n). As an example, consider a matrix **D** given by

$$\mathbf{D} = \begin{bmatrix} 3 & 5 & 37 \\ 8 & -6 & 0 \\ 12 & 23 & 2 \\ 7 & -9 & -1 \end{bmatrix}$$

The order of this matrix is 4×3, and its elements are symbolically denoted by D_{ij} with $i = 1$ to 4 and $j = 1$ to 3; for example, $D_{13} = 37$, $D_{31} = 12$, $D_{42} = -9$, etc.

2.2 TYPES OF MATRICES

We describe some of the common types of matrices in the following paragraphs.

Column Matrix (Vector)

If all the elements of a matrix are arranged in a single column (i.e., $n = 1$), it is called a *column matrix*. Column matrices are usually referred to as *vectors,* and are sometimes denoted by italic letters enclosed within braces. An example of a column matrix or vector is given by

$$\mathbf{B} = \{B\} = \begin{bmatrix} 35 \\ 9 \\ 12 \\ 3 \\ 26 \end{bmatrix}$$

Row Matrix

A matrix with all of its elements arranged in a single row (i.e., $m = 1$) is referred to as a *row matrix.* For example,

$$\mathbf{C} = [9 \quad 35 \quad -12 \quad 7 \quad 22]$$

Square Matrix

If a matrix has the same number of rows and columns (i.e., $m = n$), it is called a *square matrix.* An example of a 4×4 square matrix is given by

$$\mathbf{A} = \begin{bmatrix} 6 & 12 & 0 & 20 \\ 15 & -9 & -37 & 3 \\ -24 & 13 & 8 & 1 \\ 40 & 0 & 11 & -5 \end{bmatrix} \tag{2.2}$$

main diagonal

As shown in Eq. (2.2), the *main diagonal* of a square matrix extends from the upper left corner to the lower right corner, and it contains elements with matching subscripts—that is, $A_{11}, A_{22}, A_{33}, \ldots, A_{nn}$. The elements forming the main diagonal are referred to as the *diagonal elements;* the remaining elements of a square matrix are called the *off-diagonal elements*.

Symmetric Matrix

When the elements of a square matrix are symmetric about its main diagonal (i.e., $A_{ij} = A_{ji}$), it is termed a *symmetric matrix*. For example,

$$\mathbf{A} = \begin{bmatrix} 6 & 15 & -24 & 40 \\ 15 & -9 & 13 & 0 \\ -24 & 13 & 8 & 11 \\ 40 & 0 & 11 & -5 \end{bmatrix}$$

Lower Triangular Matrix

If all the elements of a square matrix above its main diagonal are zero, (i.e., $A_{ij} = 0$ for $j > i$), it is referred to as a *lower triangular matrix*. An example of a 4×4 lower triangular matrix is given by

$$\mathbf{A} = \begin{bmatrix} 8 & 0 & 0 & 0 \\ 12 & -9 & 0 & 0 \\ 33 & 17 & 6 & 0 \\ -2 & 5 & 15 & 3 \end{bmatrix}$$

Upper Triangular Matrix

When all the elements of a square matrix below its main diagonal are zero (i.e., $A_{ij} = 0$ for $j < i$), it is called an *upper triangular matrix*. An example of a 3×3 upper triangular matrix is given by

$$\mathbf{A} = \begin{bmatrix} -7 & 6 & 17 \\ 0 & 12 & 11 \\ 0 & 0 & 20 \end{bmatrix}$$

Diagonal Matrix

A square matrix with all of its off-diagonal elements equal to zero (i.e., $A_{ij} = 0$ for $i \neq j$), is called a *diagonal matrix*. For example,

$$\mathbf{A} = \begin{bmatrix} 6 & 0 & 0 & 0 \\ 0 & -3 & 0 & 0 \\ 0 & 0 & 11 & 0 \\ 0 & 0 & 0 & 27 \end{bmatrix}$$

Unit or Identity Matrix

If all the diagonal elements of a diagonal matrix are equal to 1 (i.e., $I_{ij} = 1$ and $I_{ij} = 0$ for $i \neq j$), it is referred to as a *unit* (or *identity*) *matrix*. Unit matrices are commonly denoted by **I** or $[I]$. An example of a 3×3 unit matrix is given by

$$\mathbf{I} = \begin{bmatrix} 1 & 0 & 0 \\ 0 & 1 & 0 \\ 0 & 0 & 1 \end{bmatrix}$$

Null Matrix

If all the elements of a matrix are zero (i.e., $O_{ij} = 0$), it is termed a *null matrix*. Null matrices are usually denoted by **O** or $[O]$. An example of a 3×4 null matrix is given by

$$\mathbf{O} = \begin{bmatrix} 0 & 0 & 0 & 0 \\ 0 & 0 & 0 & 0 \\ 0 & 0 & 0 & 0 \end{bmatrix}$$

2.3 MATRIX OPERATIONS

Equality

Matrices **A** and **B** are considered to be equal if they are of the same order and if their corresponding elements are identical (i.e., $A_{ij} = B_{ij}$). Consider, for example, matrices

$$\mathbf{A} = \begin{bmatrix} 6 & 2 \\ -7 & 8 \\ 3 & -9 \end{bmatrix} \quad \text{and} \quad \mathbf{B} = \begin{bmatrix} 6 & 2 \\ -7 & 8 \\ 3 & -9 \end{bmatrix}$$

Since both **A** and **B** are of order 3×2, and since each element of **A** is equal to the corresponding element of **B**, the matrices **A** and **B** are equal to each other; that is, $\mathbf{A} = \mathbf{B}$.

Addition and Subtraction

Matrices can be added (or subtracted) only if they are of the same order. The addition (or subtraction) of two matrices **A** and **B** is carried out by adding (or subtracting) the corresponding elements of the two matrices. Thus, if $\mathbf{A} + \mathbf{B} = \mathbf{C}$, then $C_{ij} = A_{ij} + B_{ij}$; and if $\mathbf{A} - \mathbf{B} = \mathbf{D}$, then $D_{ij} = A_{ij} - B_{ij}$. The matrices **C** and **D** have the same order as matrices **A** and **B**.

EXAMPLE 2.1 Calculate the matrices $\mathbf{C} = \mathbf{A} + \mathbf{B}$ and $\mathbf{D} = \mathbf{A} - \mathbf{B}$ if

$$\mathbf{A} = \begin{bmatrix} 6 & 0 \\ -2 & 9 \\ 5 & 1 \end{bmatrix} \quad \text{and} \quad \mathbf{B} = \begin{bmatrix} 2 & 3 \\ 7 & 5 \\ -12 & -1 \end{bmatrix}$$

SOLUTION

$$\mathbf{C} = \mathbf{A} + \mathbf{B} = \begin{bmatrix} (6+2) & (0+3) \\ (-2+7) & (9+5) \\ (5-12) & (1-1) \end{bmatrix} = \begin{bmatrix} 8 & 3 \\ 5 & 14 \\ -7 & 0 \end{bmatrix} \qquad \textbf{Ans}$$

$$\mathbf{D} = \mathbf{A} - \mathbf{B} = \begin{bmatrix} (6-2) & (0-3) \\ (-2-7) & (9-5) \\ (5+12) & (1+1) \end{bmatrix} = \begin{bmatrix} 4 & -3 \\ -9 & 4 \\ 17 & 2 \end{bmatrix} \qquad \textbf{Ans}$$

Multiplication by a Scalar

The product of a scalar c and a matrix $\mathbf{A}$ is obtained by multiplying each element of the matrix $\mathbf{A}$ by the scalar c. Thus, if $c\mathbf{A} = \mathbf{B}$, then $B_{ij} = cA_{ij}$.

EXAMPLE 2.2 Calculate the matrix $\mathbf{B} = c\mathbf{A}$ if $c = -6$ and

$$\mathbf{A} = \begin{bmatrix} 3 & 7 & -2 \\ 0 & 8 & 1 \\ 12 & -4 & 10 \end{bmatrix}$$

SOLUTION

$$\mathbf{B} = c\mathbf{A} = \begin{bmatrix} -6(3) & -6(7) & -6(-2) \\ -6(0) & -6(8) & -6(1) \\ -6(12) & -6(-4) & -6(10) \end{bmatrix} = \begin{bmatrix} -18 & -42 & 12 \\ 0 & -48 & -6 \\ -72 & 24 & -60 \end{bmatrix} \qquad \textbf{Ans}$$

Multiplication of Matrices

Two matrices can be multiplied only if the number of columns of the first matrix equals the number of rows of the second matrix. Such matrices are said to be *conformable* for multiplication. Consider, for example, the matrices

$$\mathbf{A} = \underset{3 \times 2}{\begin{bmatrix} 1 & 8 \\ 4 & -2 \\ -5 & 3 \end{bmatrix}} \quad \text{and} \quad \mathbf{B} = \underset{2 \times 2}{\begin{bmatrix} 6 & -7 \\ -1 & 2 \end{bmatrix}} \qquad \textbf{(2.3)}$$

The product **AB** of these matrices is defined because the first matrix, **A**, of the sequence **AB** has two columns and the second matrix, **B**, has two rows. However, if the sequence of the matrices is reversed, then the product **BA** does not exist, because now the first matrix, **B**, has two columns and the second matrix, **A**, has three rows. The product **AB** is referred to either as **A** *postmultiplied* by **B**, or as **B** *premultiplied* by **A**. Conversely, the product **BA** is referred to either as **B** postmultiplied by **A**, or as **A** premultiplied by **B**.

When two conformable matrices are multiplied, the product matrix thus obtained has the number of rows of the first matrix and the number of columns of the second matrix. Thus, if a matrix **A** of order $l \times m$ is postmultiplied by a matrix **B** of order $m \times n$, then the product matrix $\mathbf{C} = \mathbf{AB}$ has the order $l \times n$; that is,

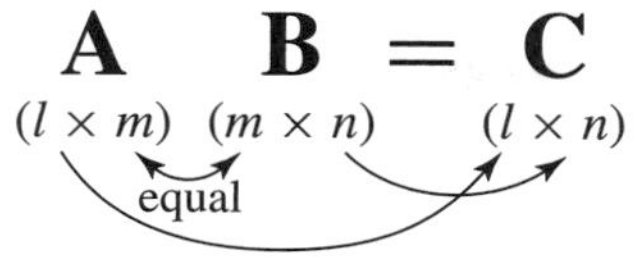

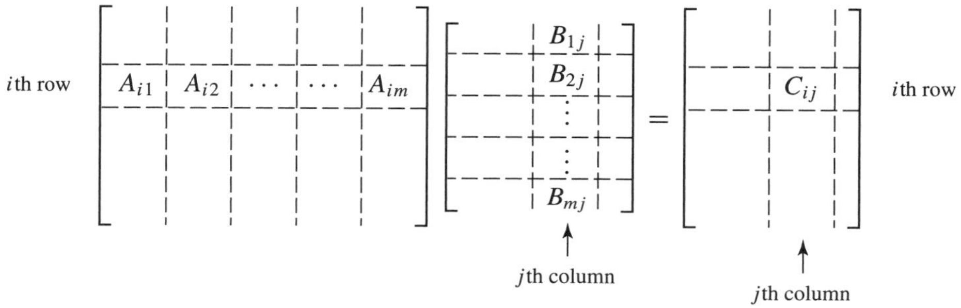

(2.4)

Any element C_{ij} of the product matrix **C** can be determined by multiplying each element of the ith row of **A** by the corresponding element of the jth column of **B** (see Eq. 2.4), and by algebraically summing the products; that is,

$$C_{ij} = A_{i1}B_{1j} + A_{i2}B_{2j} + \cdots + A_{im}B_{mj} \qquad \textbf{(2.5)}$$

Eq. (2.5) can be expressed as

$$\boxed{C_{ij} = \sum_{k=1}^{m} A_{ik}B_{kj}} \qquad \textbf{(2.6)}$$

in which m represents the number of columns of **A**, or the number of rows of **B**. Equation (2.6) can be used to determine all elements of the product matrix $\mathbf{C} = \mathbf{AB}$.

EXAMPLE 2.3 Calculate the product $\mathbf{C} = \mathbf{AB}$ of the matrices **A** and **B** given in Eq. (2.3).

SOLUTION

$$\mathbf{C} = \mathbf{AB} = \underset{(3 \times 2)}{\begin{bmatrix} 1 & 8 \\ 4 & -2 \\ -5 & 3 \end{bmatrix}} \underset{(2 \times 2)}{\begin{bmatrix} 6 & -7 \\ -1 & 2 \end{bmatrix}} = \underset{(3 \times 2)}{\begin{bmatrix} -2 & 9 \\ 26 & -32 \\ -33 & 41 \end{bmatrix}} \qquad \textbf{Ans}$$

The element C_{11} of the product matrix **C** is determined by multiplying each element of the first row of **A** by the corresponding element of the first column of **B** and summing the resulting products; that is,

$$C_{11} = 1(6) + 8(-1) = -2$$

Similarly, the element C_{12} is obtained by multiplying the elements of the first row of **A** by the corresponding elements of the second column of **B** and adding the resulting products; that is,

$$C_{12} = 1(-7) + 8(2) = 9$$

The remaining elements of **C** are computed in a similar manner:

$$C_{21} = 4(6) + (-2)(-1) = 26$$
$$C_{22} = 4(-7) - 2(2) = -32$$
$$C_{31} = -5(6) + 3(-1) = -33$$
$$C_{32} = -5(-7) + 3(2) = 41$$

A flowchart for programming the matrix multiplication procedure on a computer is given in Fig. 2.1. Any programming language (such as FORTRAN, BASIC, or C, among others) can be used for this purpose. The reader is encouraged to write this program in a general form (e.g., as a subroutine), so that it can be included in the structural analysis computer programs to be developed in later chapters.

An important application of matrix multiplication is to express simultaneous equations in compact matrix form. Consider the following system of linear simultaneous equations.

$$\begin{aligned} A_{11}x_1 + A_{12}x_2 + A_{13}x_3 + A_{14}x_4 &= P_1 \\ A_{21}x_1 + A_{22}x_2 + A_{23}x_3 + A_{24}x_4 &= P_2 \\ A_{31}x_1 + A_{32}x_2 + A_{33}x_3 + A_{34}x_4 &= P_3 \\ A_{41}x_1 + A_{42}x_2 + A_{43}x_3 + A_{44}x_4 &= P_4 \end{aligned} \tag{2.7}$$

in which xs are the unknowns and As and Ps represent the coefficients and constants, respectively. By using the definition of multiplication of matrices, this system of equations can be expressed in matrix form as

$$\begin{bmatrix} A_{11} & A_{12} & A_{13} & A_{14} \\ A_{21} & A_{22} & A_{23} & A_{24} \\ A_{31} & A_{32} & A_{33} & A_{34} \\ A_{41} & A_{42} & A_{43} & A_{44} \end{bmatrix} \begin{bmatrix} x_1 \\ x_2 \\ x_3 \\ x_4 \end{bmatrix} = \begin{bmatrix} P_1 \\ P_2 \\ P_3 \\ P_4 \end{bmatrix} \tag{2.8}$$

or, symbolically, as

$$\mathbf{Ax} = \mathbf{P} \tag{2.9}$$

Matrix multiplication is generally not commutative; that is,

$$\boxed{\mathbf{AB} \neq \mathbf{BA}} \tag{2.10}$$

Even when the orders of two matrices **A** and **B** are such that both products **AB** and **BA** are defined and are of the same order, the two products, in general, will

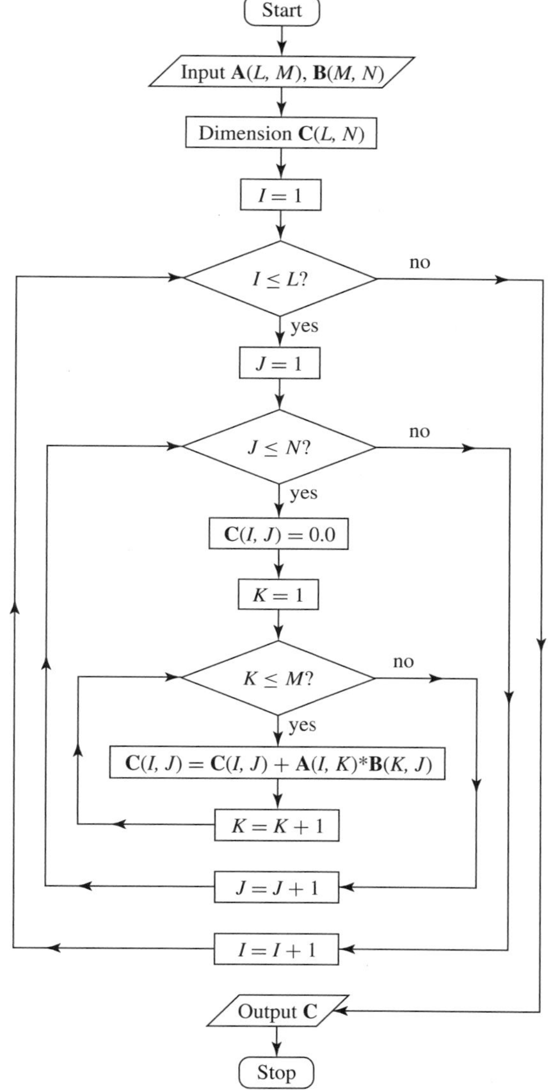

Fig. 2.1 *Flowchart for Matrix Multiplication*

not be equal. It is essential, therefore, to maintain the proper sequential order of matrices when evaluating matrix products.

EXAMPLE 2.4 Calculate the products **AB** and **BA** if

$$\mathbf{A} = \begin{bmatrix} 1 & -8 \\ -7 & 2 \end{bmatrix} \quad \text{and} \quad \mathbf{B} = \begin{bmatrix} 6 & -3 \\ 4 & -5 \end{bmatrix}$$

Are the products **AB** and **BA** equal?

SOLUTION

$$\mathbf{AB} = \begin{bmatrix} 1 & -8 \\ -7 & 2 \end{bmatrix} \begin{bmatrix} 6 & -3 \\ 4 & -5 \end{bmatrix} = \begin{bmatrix} -26 & 37 \\ -34 & 11 \end{bmatrix}$$ **Ans**

$$\mathbf{BA} = \begin{bmatrix} 6 & -3 \\ 4 & -5 \end{bmatrix} \begin{bmatrix} 1 & -8 \\ -7 & 2 \end{bmatrix} = \begin{bmatrix} 27 & -54 \\ 39 & -42 \end{bmatrix}$$ **Ans**

Comparing products **AB** and **BA**, we can see that $\mathbf{AB} \neq \mathbf{BA}$. **Ans**

Matrix multiplication is associative and distributive, provided that the sequential order in which the matrices are to be multiplied is maintained. Thus,

$$\mathbf{ABC} = (\mathbf{AB})\mathbf{C} = \mathbf{A}(\mathbf{BC}) \tag{2.11}$$

and

$$\mathbf{A}(\mathbf{B} + \mathbf{C}) = \mathbf{AB} + \mathbf{AC} \tag{2.12}$$

The product of any matrix **A** and a conformable null matrix **O** equals a null matrix; that is,

$$\mathbf{AO} = \mathbf{O} \qquad \text{and} \qquad \mathbf{OA} = \mathbf{O} \tag{2.13}$$

For example,

$$\begin{bmatrix} 2 & -4 \\ -6 & 8 \end{bmatrix} \begin{bmatrix} 0 & 0 \\ 0 & 0 \end{bmatrix} = \begin{bmatrix} 0 & 0 \\ 0 & 0 \end{bmatrix}$$

The product of any matrix **A** and a conformable unit matrix **I** equals the original matrix **A**; thus,

$$\mathbf{AI} = \mathbf{A} \quad \text{and} \quad \mathbf{IA} = \mathbf{A} \tag{2.14}$$

For example,

$$\begin{bmatrix} 2 & -4 \\ -6 & 8 \end{bmatrix} \begin{bmatrix} 1 & 0 \\ 0 & 1 \end{bmatrix} = \begin{bmatrix} 2 & -4 \\ -6 & 8 \end{bmatrix}$$

and

$$\begin{bmatrix} 1 & 0 \\ 0 & 1 \end{bmatrix} \begin{bmatrix} 2 & -4 \\ -6 & 8 \end{bmatrix} = \begin{bmatrix} 2 & -4 \\ -6 & 8 \end{bmatrix}$$

We can see from Eqs. (2.13) and (2.14) that the null and unit matrices serve purposes in matrix algebra that are similar to those of the numbers 0 and 1, respectively, in scalar algebra.

Transpose of a Matrix

The *transpose* of a matrix is obtained by interchanging its corresponding rows and columns. The transposed matrix is commonly identified by placing a superscript T on the symbol of the original matrix. Consider, for example, a 3×2 matrix

$$\mathbf{B} = \underset{3 \times 2}{\begin{bmatrix} 2 & -4 \\ -5 & 8 \\ 1 & 3 \end{bmatrix}}$$

The transpose of **B** is given by

$$\mathbf{B}^T = \underset{2 \times 3}{\begin{bmatrix} 2 & -5 & 1 \\ -4 & 8 & 3 \end{bmatrix}}$$

Note that the first row of **B** becomes the first column of $\mathbf{B}^T$. Similarly, the second and third rows of **B** become, respectively, the second and third columns of $\mathbf{B}^T$. The order of $\mathbf{B}^T$ thus obtained is 2×3.

As another example, consider the matrix

$$\mathbf{C} = \begin{bmatrix} 2 & -1 & 6 \\ -1 & 7 & -9 \\ 6 & -9 & 5 \end{bmatrix}$$

Because the elements of **C** are symmetric about its main diagonal (i.e., $C_{ij} = C_{ji}$ for $i \neq j$), interchanging the rows and columns of this matrix produces a matrix $\mathbf{C}^T$ that is identical to **C** itself; that is, $\mathbf{C}^T = \mathbf{C}$. Thus, *the transpose of a symmetric matrix equals the original matrix.*

Another useful property of matrix transposition is that *the transpose of a product of matrices equals the product of the transposed matrices in reverse order*. Thus,

$$\boxed{(\mathbf{AB})^T = \mathbf{B}^T\mathbf{A}^T} \qquad \textbf{(2.15)}$$

Similarly,

$$(\mathbf{ABC})^T = \mathbf{C}^T\mathbf{B}^T\mathbf{A}^T \qquad \textbf{(2.16)}$$

EXAMPLE 2.5 Show that $(\mathbf{AB})^T = \mathbf{B}^T\mathbf{A}^T$ if

$$\mathbf{A} = \begin{bmatrix} 9 & -5 \\ 2 & 1 \\ -3 & 4 \end{bmatrix} \quad \text{and} \quad \mathbf{B} = \begin{bmatrix} 6 & -1 & 10 \\ -2 & 7 & 5 \end{bmatrix}$$

SOLUTION

$$\mathbf{AB} = \begin{bmatrix} 9 & -5 \\ 2 & 1 \\ -3 & 4 \end{bmatrix} \begin{bmatrix} 6 & -1 & 10 \\ -2 & 7 & 5 \end{bmatrix} = \begin{bmatrix} 64 & -44 & 65 \\ 10 & 5 & 25 \\ -26 & 31 & -10 \end{bmatrix}$$

$$(\mathbf{AB})^T = \begin{bmatrix} 64 & 10 & -26 \\ -44 & 5 & 31 \\ 65 & 25 & -10 \end{bmatrix} \qquad \textbf{(1)}$$

$$\mathbf{B}^T\mathbf{A}^T = \begin{bmatrix} 6 & -2 \\ -1 & 7 \\ 10 & 5 \end{bmatrix} \begin{bmatrix} 9 & 2 & -3 \\ -5 & 1 & 4 \end{bmatrix} = \begin{bmatrix} 64 & 10 & -26 \\ -44 & 5 & 31 \\ 65 & 25 & -10 \end{bmatrix} \qquad \textbf{(2)}$$

By comparing Eqs. (1) and (2), we can see that $(\mathbf{AB})^T = \mathbf{B}^T\mathbf{A}^T$. **Ans**

Differentiation and Integration

A matrix can be differentiated (or integrated) by differentiating (or integrating) each of its elements.

EXAMPLE 2.6 Determine the derivative $d\mathbf{A}/dx$ if

$$\mathbf{A} = \begin{bmatrix} x^2 & 3\sin x & -x^4 \\ 3\sin x & -x & \cos^2 x \\ -x^4 & \cos^2 x & 7x^3 \end{bmatrix}$$

SOLUTION By differentiating the elements of **A**, we obtain

$$A_{11} = x^2 \qquad \frac{dA_{11}}{dx} = 2x$$

$$A_{21} = A_{12} = 3\sin x \qquad \frac{dA_{21}}{dx} = \frac{dA_{12}}{dx} = 3\cos x$$

$$A_{31} = A_{13} = -x^4 \qquad \frac{dA_{31}}{dx} = \frac{dA_{13}}{dx} = -4x^3$$

$$A_{22} = -x \qquad \frac{dA_{22}}{dx} = -1$$

$$A_{32} = A_{23} = \cos^2 x \qquad \frac{dA_{32}}{dx} = \frac{dA_{23}}{dx} = -2\cos x \sin x$$

$$A_{33} = 7x^3 \qquad \frac{dA_{33}}{dx} = 21x^2$$

Thus, the derivative $d\mathbf{A}/dx$ is given by

$$\frac{d\mathbf{A}}{dx} = \begin{bmatrix} 2x & 3\cos x & -4x^3 \\ 3\cos x & -1 & -2\cos x \sin x \\ -4x^3 & -2\cos x \sin x & 21x^2 \end{bmatrix}$$ **Ans**

EXAMPLE 2.7 Determine the partial derivative $\partial\mathbf{B}/\partial y$ if

$$\mathbf{B} = \begin{bmatrix} 2y^3 & -yz & -2xz \\ 3xy^2 & yz & -z^2 \\ 2x^2 & -2xz & 3xy^2 \end{bmatrix}$$

SOLUTION We determine the partial derivative, $\partial B_{ij}/\partial y$, of each element of **B** to obtain

$$\frac{\partial\mathbf{B}}{\partial y} = \begin{bmatrix} 6y^2 & -z & 0 \\ 6xy & z & 0 \\ 0 & 0 & 6xy \end{bmatrix}$$ **Ans**

EXAMPLE 2.8 Calculate the integral $\int_0^L \mathbf{A}\mathbf{A}^T dx$ if

$$\mathbf{A} = \begin{bmatrix} 1 - \dfrac{x}{L} \\ \dfrac{x}{L} \end{bmatrix}$$

SOLUTION First, we calculate the matrix product $\mathbf{AA}^T$ as

$$\mathbf{B} = \mathbf{AA}^T = \begin{bmatrix} \left(1 - \frac{x}{L}\right) \\ \frac{x}{L} \end{bmatrix} \left[\left(1 - \frac{x}{L}\right) \ \frac{x}{L}\right] = \begin{bmatrix} \left(1 - \frac{x}{L}\right)^2 & \frac{x}{L}\left(1 - \frac{x}{L}\right) \\ \frac{x}{L}\left(1 - \frac{x}{L}\right) & \frac{x^2}{L^2} \end{bmatrix}$$

Next, we integrate the elements of $\mathbf{B}$ to obtain

$$\int_0^L B_{11} dx = \int_0^L \left(1 - \frac{x}{L}\right)^2 dx = \int_0^L \left(1 - \frac{2x}{L} + \frac{x^2}{L^2}\right) dx$$

$$= \left(x - \frac{x^2}{L} + \frac{x^3}{3L^2}\right)_0^L = \frac{L}{3}$$

$$\int_0^L B_{21} dx = \int_0^L B_{12} dx = \int_0^L \frac{x}{L}\left(1 - \frac{x}{L}\right) dx = \int_0^L \left(\frac{x}{L} - \frac{x^2}{L^2}\right) dx$$

$$= \left(\frac{x^2}{2L} - \frac{x^3}{3L^2}\right)_0^L = \frac{L}{2} - \frac{L}{3} = \frac{L}{6}$$

$$\int_0^L B_{22} dx = \int_0^L \left(\frac{x^2}{L^2}\right) dx = \left(\frac{x^3}{3L^2}\right)_0^L = \frac{L}{3}$$

Thus,

$$\int_0^L \mathbf{AA}^T dx = \begin{bmatrix} \frac{L}{3} & \frac{L}{6} \\ \frac{L}{6} & \frac{L}{3} \end{bmatrix} = \frac{L}{6}\begin{bmatrix} 2 & 1 \\ 1 & 2 \end{bmatrix}$$ **Ans**

Inverse of a Square Matrix

The inverse of a square matrix $\mathbf{A}$ is defined as a matrix $\mathbf{A}^{-1}$ with elements of such magnitudes that the product of the original matrix $\mathbf{A}$ and its inverse $\mathbf{A}^{-1}$ equals a unit matrix $\mathbf{I}$; that is,

$$\boxed{\mathbf{AA}^{-1} = \mathbf{A}^{-1}\mathbf{A} = \mathbf{I}} \tag{2.17}$$

The operation of inversion is defined only for square matrices, with the inverse of such a matrix also being a square matrix of the same order as the original matrix. A procedure for determining inverses of matrices will be presented in the next section.

EXAMPLE 2.9 Check whether or not matrix $\mathbf{B}$ is the inverse of matrix $\mathbf{A}$, if

$$\mathbf{A} = \begin{bmatrix} -4 & 2 \\ -3 & 1 \end{bmatrix} \quad \text{and} \quad \mathbf{B} = \begin{bmatrix} 0.5 & -1 \\ 1.5 & -2 \end{bmatrix}$$

SOLUTION

$$\mathbf{AB} = \begin{bmatrix} -4 & 2 \\ -3 & 1 \end{bmatrix} \begin{bmatrix} 0.5 & -1 \\ 1.5 & -2 \end{bmatrix} = \begin{bmatrix} (-2+3) & (4-4) \\ (-1.5+1.5) & (3-2) \end{bmatrix} = \begin{bmatrix} 1 & 0 \\ 0 & 1 \end{bmatrix}$$

Also,

$$\mathbf{BA} = \begin{bmatrix} 0.5 & -1 \\ 1.5 & -2 \end{bmatrix} \begin{bmatrix} -4 & 2 \\ -3 & 1 \end{bmatrix} = \begin{bmatrix} (-2+3) & (1-1) \\ (-6+6) & (3-2) \end{bmatrix} = \begin{bmatrix} 1 & 0 \\ 0 & 1 \end{bmatrix}$$

Since $\mathbf{AB} = \mathbf{BA} = \mathbf{I}$, $\mathbf{B}$ is the inverse of $\mathbf{A}$; that is,

$$\mathbf{B} = \mathbf{A}^{-1}$$ **Ans**

The operation of matrix inversion serves a purpose analogous to the operation of division in scalar algebra. Consider a system of simultaneous linear equations expressed in matrix form as

$$\mathbf{Ax} = \mathbf{P}$$

in which $\mathbf{A}$ is the square matrix of known coefficients; $\mathbf{x}$ is the vector of the unknowns; and $\mathbf{P}$ is the vector of the constants. As the operation of division is not defined in matrix algebra, the equation cannot be solved for $\mathbf{x}$ by dividing $\mathbf{P}$ by $\mathbf{A}$ (i.e., $\mathbf{x} = \mathbf{P}/\mathbf{A}$). However, we can determine $\mathbf{x}$ by premultiplying both sides of the equation by $\mathbf{A}^{-1}$, to obtain

$$\mathbf{A}^{-1}\mathbf{Ax} = \mathbf{A}^{-1}\mathbf{P}$$

As $\mathbf{A}^{-1}\mathbf{A} = \mathbf{I}$ and $\mathbf{Ix} = \mathbf{x}$, we can write

$$\mathbf{x} = \mathbf{A}^{-1}\mathbf{P}$$

which shows that a system of simultaneous linear equations can be solved by premultiplying the vector of constants by the inverse of the coefficient matrix.

An important property of matrix inversion is that *the inverse of a symmetric matrix is also a symmetric matrix.*

Orthogonal Matrix

If the inverse of a matrix is equal to its transpose, the matrix is referred to as an *orthogonal matrix.* In other words, a matrix $\mathbf{A}$ is orthogonal if

$$\mathbf{A}^{-1} = \mathbf{A}^T$$

EXAMPLE 2.10 Determine whether matrix $\mathbf{A}$ given below is an orthogonal matrix.

$$\mathbf{A} = \begin{bmatrix} 0.8 & 0.6 & 0 & 0 \\ -0.6 & 0.8 & 0 & 0 \\ 0 & 0 & 0.8 & 0.6 \\ 0 & 0 & -0.6 & 0.8 \end{bmatrix}$$

SOLUTION

$$\mathbf{AA}^T = \begin{bmatrix} 0.8 & 0.6 & 0 & 0 \\ -0.6 & 0.8 & 0 & 0 \\ 0 & 0 & 0.8 & 0.6 \\ 0 & 0 & -0.6 & 0.8 \end{bmatrix} \begin{bmatrix} 0.8 & -0.6 & 0 & 0 \\ 0.6 & 0.8 & 0 & 0 \\ 0 & 0 & 0.8 & -0.6 \\ 0 & 0 & 0.6 & 0.8 \end{bmatrix}$$

$$= \begin{bmatrix} (0.64+0.36) & (-0.48+0.48) & 0 & 0 \\ (-0.48+0.48) & (0.36+0.64) & 0 & 0 \\ 0 & 0 & (0.64+0.36) & (-0.48+0.48) \\ 0 & 0 & (-0.48+0.48) & (0.36+0.64) \end{bmatrix}$$

$$= \begin{bmatrix} 1 & 0 & 0 & 0 \\ 0 & 1 & 0 & 0 \\ 0 & 0 & 1 & 0 \\ 0 & 0 & 0 & 1 \end{bmatrix}$$

which shows that $\mathbf{AA}^T = \mathbf{I}$. Thus,

$$\mathbf{A}^{-1} = \mathbf{A}^T$$

Therefore, matrix **A** is orthogonal. **Ans**

Partitioning of Matrices

In many applications, it becomes necessary to subdivide a matrix into a number of smaller matrices called *submatrices*. The process of subdividing a matrix into submatrices is referred to as *partitioning*. For example, a 4 × 3 matrix **B** is partitioned into four submatrices by drawing horizontal and vertical dashed partition lines:

$$\mathbf{B} = \left[\begin{array}{cc:c} 2 & -4 & -1 \\ -5 & 7 & 3 \\ 8 & -9 & 6 \\ \hdashline 1 & 3 & 8 \end{array}\right] = \begin{bmatrix} \mathbf{B}_{11} & \mathbf{B}_{12} \\ \mathbf{B}_{21} & \mathbf{B}_{22} \end{bmatrix} \tag{2.18}$$

in which the submatrices are

$$\mathbf{B}_{11} = \begin{bmatrix} 2 & -4 \\ -5 & 7 \\ 8 & -9 \end{bmatrix} \qquad \mathbf{B}_{12} = \begin{bmatrix} -1 \\ 3 \\ 6 \end{bmatrix}$$

$$\mathbf{B}_{21} = [1 \quad 3] \qquad \mathbf{B}_{22} = [8]$$

Matrix operations (such as addition, subtraction, and multiplication) can be performed on partitioned matrices in the same manner as discussed previously by treating the submatrices as elements—provided that the matrices are partitioned in such a way that their submatrices are conformable for the particular operation. For example, suppose that the 4 × 3 matrix **B** of Eq. (2.18) is to be postmultiplied by a 3 × 2 matrix **C**, which is partitioned into two submatrices:

$$\mathbf{C} = \left[\begin{array}{cc} 9 & -6 \\ 4 & 2 \\ \hdashline -3 & 1 \end{array}\right] = \begin{bmatrix} \mathbf{C}_{11} \\ \mathbf{C}_{21} \end{bmatrix} \tag{2.19}$$

The product **BC** is expressed in terms of submatrices as

$$\mathbf{BC} = \begin{bmatrix} \mathbf{B}_{11} & \mathbf{B}_{12} \\ \mathbf{B}_{21} & \mathbf{B}_{22} \end{bmatrix} \begin{bmatrix} \mathbf{C}_{11} \\ \mathbf{C}_{21} \end{bmatrix} = \begin{bmatrix} \mathbf{B}_{11}\mathbf{C}_{11} + \mathbf{B}_{12}\mathbf{C}_{21} \\ \mathbf{B}_{21}\mathbf{C}_{11} + \mathbf{B}_{22}\mathbf{C}_{21} \end{bmatrix} \tag{2.20}$$

It is important to realize that matrices **B** and **C** have been partitioned in such a way that their corresponding submatrices are conformable for multiplication; that is, the orders of the submatrices are such that the products $\mathbf{B}_{11}\mathbf{C}_{11}$, $\mathbf{B}_{12}\mathbf{C}_{21}$, $\mathbf{B}_{21}\mathbf{C}_{11}$, and $\mathbf{B}_{22}\mathbf{C}_{21}$ are defined. It can be seen from Eqs. (2.18) and (2.19) that this is achieved by partitioning the rows of the second matrix **C** of the product **BC** in the same way that the columns of the first matrix **B** are partitioned. The products of the submatrices are:

$$\mathbf{B}_{11}\mathbf{C}_{11} = \begin{bmatrix} 2 & -4 \\ -5 & 7 \\ 8 & -9 \end{bmatrix} \begin{bmatrix} 9 & -6 \\ 4 & 2 \end{bmatrix} = \begin{bmatrix} 2 & -20 \\ -17 & 44 \\ 36 & -66 \end{bmatrix}$$

$$\mathbf{B}_{12}\mathbf{C}_{21} = \begin{bmatrix} -1 \\ 3 \\ 6 \end{bmatrix} [-3 \quad 1] = \begin{bmatrix} 3 & -1 \\ -9 & 3 \\ -18 & 6 \end{bmatrix}$$

$$\mathbf{B}_{21}\mathbf{C}_{11} = [1 \quad 3] \begin{bmatrix} 9 & -6 \\ 4 & 2 \end{bmatrix} = [21 \quad 0]$$

$$\mathbf{B}_{22}\mathbf{C}_{21} = [8][-3 \quad 1] = [-24 \quad 8]$$

By substituting the numerical values of the products of submatrices into Eq. (2.20), we obtain

$$\mathbf{BC} = \begin{bmatrix} \begin{bmatrix} 2 & -20 \\ -17 & 44 \\ 36 & -66 \end{bmatrix} + \begin{bmatrix} 3 & -1 \\ -9 & 3 \\ -18 & 6 \end{bmatrix} \\ [21 \quad 0] + [-24 \quad 8] \end{bmatrix} = \begin{bmatrix} 5 & -21 \\ -26 & 47 \\ 18 & -60 \\ -3 & 8 \end{bmatrix}$$

2.4 GAUSS–JORDAN ELIMINATION METHOD

The *Gauss–Jordan elimination method* is one of the most commonly used procedures for solving simultaneous linear equations, and for determining inverses of matrices.

Solution of Simultaneous Equations

To illustrate the Gauss–Jordan method for solving simultaneous equations, consider the following system of three linear algebraic equations:

$$\begin{aligned} 5x_1 + 6x_2 - 3x_3 &= 66 \\ 9x_1 - x_2 + 2x_3 &= 8 \\ 8x_1 - 7x_2 + 4x_3 &= -39 \end{aligned} \tag{2.21a}$$

To determine the unknowns x_1, x_2, and x_3, we begin by dividing the first equation by the coefficient of its x_1 term to obtain

$$\begin{aligned} x_1 + 1.2x_2 - 0.6x_3 &= 13.2 \\ 9x_1 - x_2 + 2x_3 &= 8 \\ 8x_1 - 7x_2 + 4x_3 &= -39 \end{aligned} \quad \textbf{(2.21b)}$$

Next, we eliminate the unknown x_1 from the second and third equations by successively subtracting from each equation the product of the coefficient of its x_1 term and the first equation. Thus, to eliminate x_1 from the second equation, we multiply the first equation by 9 and subtract it from the second equation. Similarly, we eliminate x_1 from the third equation by multiplying the first equation by 8 and subtracting it from the third equation. This yields the system of equations

$$\begin{aligned} x_1 + 1.2x_2 - 0.6x_3 &= 13.2 \\ -11.8x_2 + 7.4x_3 &= -110.8 \\ -16.6x_2 + 8.8x_3 &= -144.6 \end{aligned} \quad \textbf{(2.21c)}$$

With x_1 eliminated from all but the first equation, we now divide the second equation by the coefficient of its x_2 term to obtain

$$\begin{aligned} x_1 + 1.2x_2 - 0.6\ x_3 &= 13.2 \\ x_2 - 0.6271x_3 &= 9.39 \\ -16.6x_2 + 8.8\ x_3 &= -144.6 \end{aligned} \quad \textbf{(2.21d)}$$

Next, the unknown x_2 is eliminated from the first and the third equations, successively, by multiplying the second equation by 1.2 and subtracting it from the first equation, and then by multiplying the second equation by -16.6 and subtracting it from the third equation. The system of equations thus obtained is

$$\begin{aligned} x_1 + 0.1525x_3 &= 1.932 \\ x_2 - 0.6271x_3 &= 9.39 \\ -1.61\ x_3 &= 11.27 \end{aligned} \quad \textbf{(2.21e)}$$

Focusing our attention now on the unknown x_3, we divide the third equation by the coefficient of its x_3 term (which is -1.61) to obtain

$$\begin{aligned} x_1 + 0.1525x_3 &= 1.932 \\ x_2 - 0.6271x_3 &= 9.39 \\ x_3 &= -7 \end{aligned} \quad \textbf{(2.21f)}$$

Finally, we eliminate x_3 from the first and the second equations, successively, by multiplying the third equation by 0.1525 and subtracting it from the first equation, and then by multiplying the third equation by -0.6271 and subtracting it from the second equation. This yields the solution of the given system of equations:

$$\begin{aligned} x_1 &= 3 \\ x_2 &= 5 \\ x_3 &= -7 \end{aligned} \quad \textbf{(2.21g)}$$

or, equivalently,

$$x_1 = 3; \quad x_2 = 5; \quad x_3 = -7 \tag{2.21h}$$

To check that this solution is correct, we substitute the numerical values of x_1, x_2, and x_3 back into the original equations (Eq. 2.21(a)):

$$5(3) + 6(5) - 3(-7) = 66 \qquad \textbf{Checks}$$
$$9(3) - 5 \quad + 2(-7) = 8 \qquad \textbf{Checks}$$
$$8(3) - 7(5) + 4(-7) = -39 \qquad \textbf{Checks}$$

As the foregoing example illustrates, the Gauss–Jordan method basically involves eliminating, in order, each unknown from all but one of the equations of the system by applying the following operations: dividing an equation by a scalar; and multiplying an equation by a scalar and subtracting the resulting equation from another equation. These operations (called the *elementary operations*) when applied to a system of equations yield another system of equations that has the same solution as the original system. In the Gauss–-Jordan method, the elementary operations are performed repeatedly until a system with each equation containing only one unknown is obtained.

The Gauss–Jordan elimination method can be performed more conveniently by using the matrix form of the simultaneous equations ($\mathbf{Ax} = \mathbf{P}$). In this approach, the coefficient matrix $\mathbf{A}$ and the vector of constants $\mathbf{P}$ are treated as submatrices of a partitioned *augmented matrix,*

$$\underset{n \times (n+1)}{\mathbf{G}} = [\underset{n \times n}{\mathbf{A}} \; \vdots \; \underset{n \times 1}{\mathbf{P}}] \tag{2.22}$$

where n represents the number of equations. The elementary operations are then applied to the rows of the augmented matrix, until the coefficient matrix is reduced to a unit matrix. The elements of the vector, which initially contained the constant terms of the original equations, now represent the solution of the original system of equations; that is,

$$\mathbf{G} = \begin{cases} [\mathbf{A} \; \vdots \; \mathbf{P}] \\ \quad \downarrow \text{ elementary operations} \\ [\mathbf{I} \; \vdots \; \mathbf{x}] \end{cases} \tag{2.23}$$

This procedure is illustrated by the following example.

EXAMPLE 2.11 Solve the system of simultaneous equations given in Eq. 2.21(a) by the Gauss–Jordan method.

SOLUTION The given system of equations can be written in matrix form as

$$\mathbf{Ax} = \mathbf{P}$$

$$\begin{bmatrix} 5 & 6 & -3 \\ 9 & -1 & 2 \\ 8 & -7 & 4 \end{bmatrix} \begin{bmatrix} x_1 \\ x_2 \\ x_3 \end{bmatrix} = \begin{bmatrix} 66 \\ 8 \\ -39 \end{bmatrix} \tag{1}$$

from which we form the augmented matrix

$$\mathbf{G} = [\mathbf{A} \,\vdots\, \mathbf{P}] = \left[\begin{array}{ccc:c} 5 & 6 & -3 & 66 \\ 9 & -1 & 2 & 8 \\ 8 & -7 & 4 & -39 \end{array}\right] \tag{2}$$

We begin Gauss–Jordan elimination by dividing row 1 of the augmented matrix by $G_{11} = 5$ to obtain

$$\mathbf{G} = \left[\begin{array}{ccc:c} 1 & 1.2 & -0.6 & 13.2 \\ 9 & -1 & 2 & 8 \\ 8 & -7 & 4 & -39 \end{array}\right] \tag{3}$$

Next, we multiply row 1 by $G_{21} = 9$ and subtract it from row 2; then multiply row 1 by $G_{31} = 8$ and subtract it from row 3. This yields

$$\mathbf{G} = \left[\begin{array}{ccc:c} 1 & 1.2 & -0.6 & 13.2 \\ 0 & -11.8 & 7.4 & -110.8 \\ 0 & -16.6 & 8.8 & -144.6 \end{array}\right] \tag{4}$$

We now divide row 2 by $G_{22} = -11.8$ to obtain

$$\mathbf{G} = \left[\begin{array}{ccc:c} 1 & 1.2 & -0.6 & 13.2 \\ 0 & 1 & -0.6271 & 9.39 \\ 0 & -16.6 & 8.8 & -144.6 \end{array}\right] \tag{5}$$

Next, we multiply row 2 by $G_{12} = 1.2$ and subtract it from row 1, and then multiply row 2 by $G_{32} = -16.6$ and subtract it from row 3. Thus,

$$\mathbf{G} = \left[\begin{array}{ccc:c} 1 & 0 & 0.1525 & 1.932 \\ 0 & 1 & -0.6271 & 9.39 \\ 0 & 0 & -1.61 & 11.27 \end{array}\right] \tag{6}$$

By dividing row 3 by $G_{33} = -1.61$, we obtain

$$\mathbf{G} = \left[\begin{array}{ccc:c} 1 & 0 & 0.1525 & 1.932 \\ 0 & 1 & -0.6271 & 9.39 \\ 0 & 0 & 1 & -7 \end{array}\right] \tag{7}$$

Finally, we multiply row 3 by $G_{13} = 0.1525$ and subtract it from row 1; then multiply row 3 by $G_{23} = -0.6271$ and subtract it from row 2 to obtain

$$\mathbf{G} = \left[\begin{array}{ccc:c} 1 & 0 & 0 & 3 \\ 0 & 1 & 0 & 5 \\ 0 & 0 & 1 & -7 \end{array}\right] \tag{8}$$

Thus, the solution of the given system of equations is

$$\mathbf{x} = \begin{bmatrix} 3 \\ 5 \\ -7 \end{bmatrix} \qquad \textbf{Ans}$$

To check our solution, we substitute the numerical value of **x** back into Eq. (1). This yields

$$\begin{bmatrix} 5 & 6 & -3 \\ 9 & -1 & 2 \\ 8 & -7 & 4 \end{bmatrix} \begin{bmatrix} 3 \\ 5 \\ -7 \end{bmatrix} = \begin{bmatrix} 15 + 30 + 21 = 66 \\ 27 - 5 - 14 = 8 \\ 24 - 35 - 28 = -39 \end{bmatrix} \qquad \textbf{Checks}$$

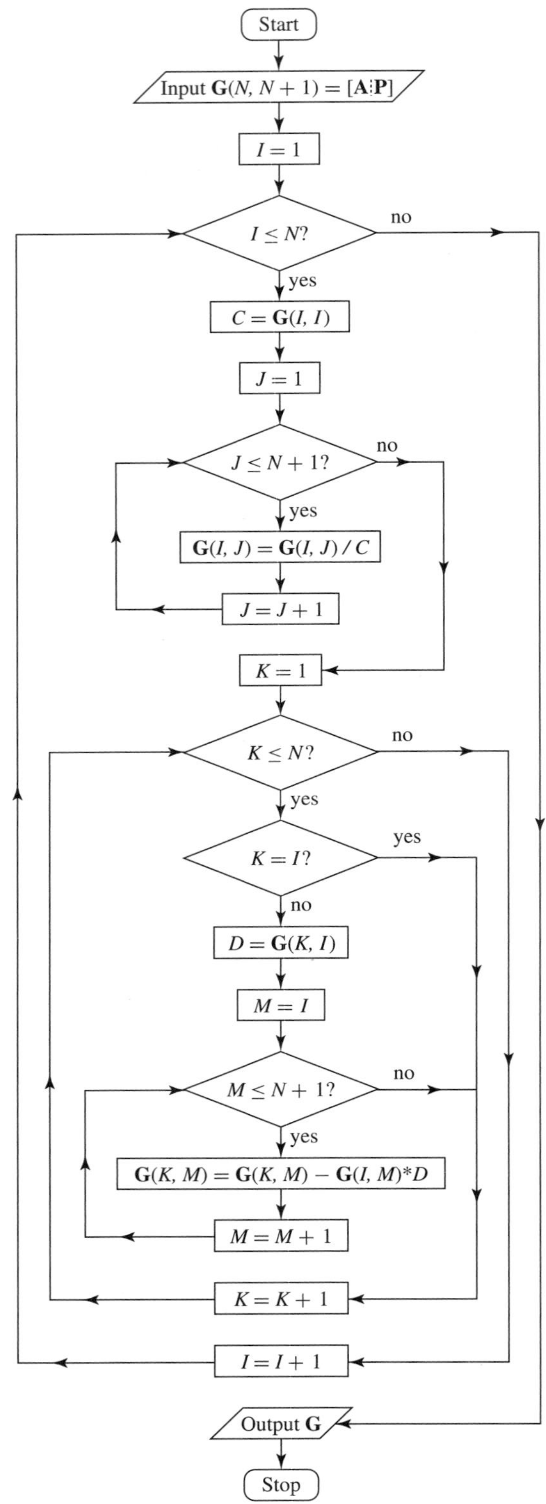

Fig. 2.2 *Flowchart for Solution of Simultaneous Equations by Gauss–Jordan Method*

The solution of large systems of simultaneous equations by the Gauss–Jordan method is usually carried out by computer, and a flowchart for programming this procedure is given in Fig. 2.2. The reader should write this program in a general form (e.g., as a subroutine), so that it can be conveniently included in the structural analysis computer programs to be developed in later chapters.

It should be noted that the Gauss–Jordan method as described in the preceding paragraphs breaks down if a diagonal element of the coefficient matrix **A** becomes zero during the elimination process. This situation can be remedied by interchanging the row of the augmented matrix containing the zero diagonal element with another row, to place a nonzero element on the diagonal; the elimination process is then continued. However, when solving the systems of equations encountered in structural analysis, the condition of a zero diagonal element should not arise; the occurrence of such a condition would indicate that the structure being analyzed is unstable [2]*.

Matrix Inversion

The procedure for determining inverses of matrices by the Gauss–Jordan method is similar to that described previously for solving simultaneous equations. The procedure involves forming an augmented matrix **G** composed of the matrix **A** that is to be inverted and a unit matrix **I** of the same order as **A**; that is,

$$\underset{n \times 2n}{\mathbf{G}} = [\underset{n \times n}{\mathbf{A}} \;\vdots\; \underset{n \times n}{\mathbf{I}}] \tag{2.24}$$

Elementary operations are then applied to the rows of the augmented matrix to reduce **A** to a unit matrix. Matrix **I**, which was initially the unit matrix, now represents the inverse matrix $\mathbf{A}^{-1}$; thus,

$$\mathbf{G} = \begin{cases} [\mathbf{A} \;\vdots\; \mathbf{I}] \\ \quad\downarrow \quad \text{elementary operations} \\ [\mathbf{I} \;\vdots\; \mathbf{A}^{-1}] \end{cases} \tag{2.25}$$

EXAMPLE 2.12 Determine the inverse of the matrix shown using the Gauss–Jordan method.

$$\mathbf{A} = \begin{bmatrix} 13 & -6 & 6 \\ -6 & 12 & -1 \\ 6 & -1 & 9 \end{bmatrix}$$

SOLUTION The augmented matrix is given by

$$\mathbf{G} = [\mathbf{A} \;\vdots\; \mathbf{I}] = \left[\begin{array}{ccc|ccc} 13 & -6 & 6 & 1 & 0 & 0 \\ -6 & 12 & -1 & 0 & 1 & 0 \\ 6 & -1 & 9 & 0 & 0 & 1 \end{array}\right] \tag{1}$$

*Numbers in brackets refer to items listed in the bibliography.

We begin the Gauss–Jordan elimination process by dividing row 1 of the augmented matrix by $G_{11} = 13$:

$$\mathbf{G} = \left[\begin{array}{ccc|ccc} 1 & -0.4615 & 0.4615 & 0.07692 & 0 & 0 \\ -6 & 12 & -1 & 0 & 1 & 0 \\ 6 & -1 & 9 & 0 & 0 & 1 \end{array}\right] \quad (2)$$

Next, we multiply row 1 by $G_{21} = -6$ and subtract it from row 2, and then multiply row 1 by $G_{31} = 6$ and subtract it from row 3. This yields

$$\mathbf{G} = \left[\begin{array}{ccc|ccc} 1 & -0.4615 & 0.4615 & 0.07692 & 0 & 0 \\ 0 & 9.231 & 1.769 & 0.4615 & 1 & 0 \\ 0 & 1.769 & 6.231 & -0.4615 & 0 & 1 \end{array}\right] \quad (3)$$

Dividing row 2 by $G_{22} = 9.231$, we obtain

$$\mathbf{G} = \left[\begin{array}{ccc|ccc} 1 & -0.4615 & 0.4615 & 0.07692 & 0 & 0 \\ 0 & 1 & 0.1916 & 0.04999 & 0.1083 & 0 \\ 0 & 1.769 & 6.231 & -0.4615 & 0 & 1 \end{array}\right] \quad (4)$$

Next, we multiply row 2 by $G_{12} = -0.4615$ and subtract it from row 1; then multiply row 2 by $G_{32} = 1.769$ and subtract it from row 3. This yields

$$\mathbf{G} = \left[\begin{array}{ccc|ccc} 1 & 0 & 0.5499 & 0.09999 & 0.04998 & 0 \\ 0 & 1 & 0.1916 & 0.04999 & 0.1083 & 0 \\ 0 & 0 & 5.892 & -0.5499 & -0.1916 & 1 \end{array}\right] \quad (5)$$

Divide row 3 by $G_{33} = 5.892$:

$$\mathbf{G} = \left[\begin{array}{ccc|ccc} 1 & 0 & 0.5499 & 0.09999 & 0.04998 & 0 \\ 0 & 1 & 0.1916 & 0.04999 & 0.1083 & 0 \\ 0 & 0 & 1 & -0.09333 & -0.03252 & 0.1697 \end{array}\right] \quad (6)$$

Multiply row 3 by $G_{13} = 0.5499$ and subtract it from row 1; then multiply row 3 by $G_{23} = 0.1916$ and subtract it from row 2 to obtain

$$\mathbf{G} = \left[\begin{array}{ccc|ccc} 1 & 0 & 0 & 0.1513 & 0.06787 & -0.09333 \\ 0 & 1 & 0 & 0.06787 & 0.1145 & -0.03252 \\ 0 & 0 & 1 & -0.09333 & -0.03252 & 0.1697 \end{array}\right] \quad (7)$$

Thus, the inverse of the given matrix $\mathbf{A}$ is

$$\mathbf{A}^{-1} = \begin{bmatrix} 0.1513 & 0.06787 & -0.09333 \\ 0.06787 & 0.1145 & -0.03252 \\ -0.09333 & -0.03252 & 0.1697 \end{bmatrix} \quad \textbf{Ans}$$

Finally, we check our computations by using the relationship $\mathbf{AA}^{-1} = \mathbf{I}$:

$$\mathbf{AA}^{-1} = \begin{bmatrix} 13 & -6 & 6 \\ -6 & 12 & -1 \\ 6 & -1 & 9 \end{bmatrix} \begin{bmatrix} 0.1513 & 0.06787 & -0.09333 \\ 0.06787 & 0.1145 & -0.03252 \\ -0.09333 & -0.03252 & 0.1697 \end{bmatrix}$$

$$= \begin{bmatrix} 0.9997 & 0.0002 & 0 \\ 0 & 0.9993 & 0 \\ 0 & 0 & 0.9998 \end{bmatrix} \approx \mathbf{I} \quad \textbf{Checks}$$

SUMMARY

In this chapter, we discussed the basic concepts of matrix algebra that are necessary for formulating the matrix methods of structural analysis:

1. A matrix is defined as a rectangular array of quantities (elements) arranged in rows and columns. The size of a matrix is measured by its number of rows and columns, and is referred to as its order.

2. Two matrices are considered to be equal if they are of the same order, and if their corresponding elements are identical.

3. Two matrices of the same order can be added (or subtracted) by adding (or subtracting) their corresponding elements.

4. The matrix multiplication $\mathbf{AB} = \mathbf{C}$ is defined only if the number of columns of the first matrix $\mathbf{A}$ equals the number of rows of the second matrix $\mathbf{B}$. Any element C_{ij} of the product matrix $\mathbf{C}$ can be evaluated by using the relationship

$$C_{ij} = \sum_{k=1}^{m} A_{ik} B_{kj} \tag{2.6}$$

where m is the number of columns of $\mathbf{A}$, or the number of rows of $\mathbf{B}$. Matrix multiplication is generally not commutative; that is, $\mathbf{AB} \neq \mathbf{BA}$.

5. The transpose of a matrix is obtained by interchanging its corresponding rows and columns. If $\mathbf{C}$ is a symmetric matrix, then $\mathbf{C}^T = \mathbf{C}$. Another useful property of matrix transposition is that

$$(\mathbf{AB})^T = \mathbf{B}^T\mathbf{A}^T \tag{2.15}$$

6. A matrix can be differentiated (or integrated) by differentiating (or integrating) each of its elements.

7. The inverse of a square matrix $\mathbf{A}$ is defined as a matrix $\mathbf{A}^{-1}$ which satisfies the relationship:

$$\mathbf{AA}^{-1} = \mathbf{A}^{-1}\mathbf{A} = \mathbf{I} \tag{2.17}$$

8. If the inverse of a matrix equals its transpose, the matrix is called an orthogonal matrix.

9. The Gauss–Jordan method of solving simultaneous equations essentially involves successively eliminating each unknown from all but one of the equations of the system by performing the following operations: dividing an equation by a scalar; and multiplying an equation by a scalar and subtracting the resulting equation from another equation. These elementary operations are applied repeatedly until a system with each equation containing only one unknown is obtained.

PROBLEMS

Section 2.3

2.1 Determine the matrices $\mathbf{C} = \mathbf{A} + \mathbf{B}$ and $\mathbf{D} = \mathbf{A} - \mathbf{B}$ if

$$\mathbf{A} = \begin{bmatrix} 3 & 8 & -1 \\ 8 & -7 & -4 \\ -1 & -4 & 5 \end{bmatrix} \qquad \mathbf{B} = \begin{bmatrix} 5 & 9 & -2 \\ -9 & 6 & 3 \\ 2 & -3 & -4 \end{bmatrix}$$

2.2 Determine the matrices $\mathbf{C} = 2\mathbf{A} + \mathbf{B}$ and $\mathbf{D} = \mathbf{A} - 3\mathbf{B}$ if

$$\mathbf{A} = \begin{bmatrix} 8 & -6 & -3 \\ 1 & -2 & 0 \\ -6 & 5 & -1 \\ -2 & 8 & 0 \end{bmatrix} \qquad \mathbf{B} = \begin{bmatrix} 3 & 2 & -3 \\ -4 & 3 & 0 \\ 2 & -8 & 6 \\ -1 & 4 & -7 \end{bmatrix}$$

2.3 Determine the products $\mathbf{C} = \mathbf{AB}$ and $\mathbf{D} = \mathbf{BA}$ if

$$\mathbf{A} = [4 \quad -6 \quad 2] \qquad \mathbf{B} = \begin{bmatrix} 3 \\ 1 \\ -5 \end{bmatrix}$$

2.4 Determine the products $\mathbf{C} = \mathbf{AB}$ and $\mathbf{D} = \mathbf{BA}$ if

$$\mathbf{A} = \begin{bmatrix} 4 & 6 \\ -7 & -5 \\ 1 & -9 \\ -3 & 11 \end{bmatrix} \qquad \mathbf{B} = \begin{bmatrix} -1 & 3 & -5 & 2 \\ -13 & -4 & 7 & 6 \end{bmatrix}$$

2.5 Determine the products $\mathbf{C} = \mathbf{AB}$ and $\mathbf{D} = \mathbf{BA}$ if

$$\mathbf{A} = \begin{bmatrix} 4 & -6 & 1 \\ -6 & 5 & 7 \\ 1 & 7 & 8 \end{bmatrix} \qquad \mathbf{B} = \begin{bmatrix} 3 & 5 & 0 \\ 5 & 7 & -2 \\ 0 & -2 & 9 \end{bmatrix}$$

2.6 Determine the products $\mathbf{C} = \mathbf{AB}$ if

$$\mathbf{A} = \begin{bmatrix} 12 & -11 & 10 \\ 0 & 2 & -4 \\ -7 & 9 & 8 \\ 6 & 15 & -5 \end{bmatrix} \qquad \mathbf{B} = \begin{bmatrix} 13 & -1 & 5 \\ 16 & -9 & 0 \\ -3 & 20 & -7 \end{bmatrix}$$

2.7 Develop a computer program to determine the matrix product $\mathbf{C} = \mathbf{AB}$ of two conformable matrices $\mathbf{A}$ and $\mathbf{B}$ of any order. Check the program by solving Problems 2.4–2.6 and comparing the computer-generated results to those determined by hand calculations.

2.8 Show that $(\mathbf{AB})^T = \mathbf{B}^T\mathbf{A}^T$ by using the following matrices

$$\mathbf{A} = \begin{bmatrix} 21 & 10 & 16 \\ -15 & 11 & 0 \\ 13 & 20 & -9 \\ 7 & -17 & 14 \end{bmatrix} \qquad \mathbf{B} = \begin{bmatrix} 7 & -4 \\ -1 & 9 \\ 3 & -6 \end{bmatrix}$$

2.9 Show that $(\mathbf{ABC})^T = \mathbf{C}^T\mathbf{B}^T\mathbf{A}^T$ by using the following matrices

$$\mathbf{A} = \begin{bmatrix} -9 & 0 \\ 13 & 20 \\ 8 & -3 \\ -11 & -5 \end{bmatrix} \qquad \mathbf{B} = \begin{bmatrix} 15 & -1 & -4 \\ 6 & 16 & 9 \end{bmatrix}$$

$$\mathbf{C} = \begin{bmatrix} -7 & 10 & 6 & 0 \\ -1 & 2 & -8 & -2 \\ 16 & 12 & 2 & 8 \end{bmatrix}$$

2.10 Determine the matrix triple product $\mathbf{C} = \mathbf{B}^T\mathbf{AB}$ if

$$\mathbf{A} = \begin{bmatrix} 40 & -10 & -25 \\ -10 & 15 & 12 \\ -25 & 12 & 30 \end{bmatrix} \qquad \mathbf{B} = \begin{bmatrix} 5 & 7 & -3 \\ -7 & 8 & 4 \\ 3 & -4 & 9 \end{bmatrix}$$

2.11 Determine the matrix triple product $\mathbf{C} = \mathbf{B}^T\mathbf{AB}$ if

$$\mathbf{A} = \begin{bmatrix} 300 & -100 \\ -100 & 200 \end{bmatrix}$$

$$\mathbf{B} = \begin{bmatrix} 0.6 & 0.8 & -0.6 & -0.8 \\ -0.8 & 0.6 & 0.8 & -0.6 \end{bmatrix}$$

2.12 Develop a computer program to determine the matrix triple product $\mathbf{C} = \mathbf{B}^T\mathbf{AB}$, where $\mathbf{A}$ is a square matrix of any order. Check the program by solving Problems 2.10 and 2.11 and comparing the results to those determined by hand calculations.

2.13 Determine the derivative $d\mathbf{A}/dx$ if

$$\mathbf{A} = \begin{bmatrix} -2x^2 & 3\sin x & -7x \\ 3\sin x & \cos^2 x & -3x^3 \\ -7x & -3x^3 & 3\sin^2 x \end{bmatrix}$$

2.14 Determine the derivative $d(\mathbf{A} + \mathbf{B})/dx$ if

$$\mathbf{A} = \begin{bmatrix} -3x & 5 \\ 4x^2 & -x^3 \\ -7 & 5x \\ 2x^3 & -x^2 \end{bmatrix} \qquad \mathbf{B} = \begin{bmatrix} 2x^2 & -x \\ -12x & 8 \\ 2x^3 & -3x^2 \\ -1 & 6x \end{bmatrix}$$

2.15 Determine the derivative $d(\mathbf{AB})/dx$ if

$$\mathbf{A} = \begin{bmatrix} 4x & 2 & -5x^2 \\ 2 & -3x^3 & -x \\ -5x^2 & -x & 7 \end{bmatrix} \qquad \mathbf{B} = \begin{bmatrix} -5x^3 & -x \\ 6 & -3x^2 \\ 2x^2 & 4x \end{bmatrix}$$

2.16 Determine the partial derivatives $\partial\mathbf{A}/\partial x$, $\partial\mathbf{A}/\partial y$, and $\partial\mathbf{A}/\partial z$, if

$$\mathbf{A} = \begin{bmatrix} x^2 & -y^2 & 2z^2 \\ -y^2 & 3xy & -yz \\ 2z^2 & -yz & 4xz \end{bmatrix}$$

2.17 Calculate the integral $\int_0^L \mathbf{A}\,dx$ if

$$\mathbf{A} = \begin{bmatrix} -5 & -3x^2 \\ 4x & -x^3 \\ 2x^4 & 6 \\ 5x^2 & -x \end{bmatrix}$$

2.18 Calculate the integral $\int_0^L \mathbf{A}\,dx$ if

$$\mathbf{A} = \begin{bmatrix} 2x & -\sin x & 2\cos^2 x \\ -\sin x & 5 & -4x^3 \\ 2\cos^2 x & -4x^3 & (1-x^2) \end{bmatrix}$$

2.19 Calculate the integral $\int_0^L \mathbf{AB}\,dx$ if

$$\mathbf{A} = \begin{bmatrix} -x^3 & 2x^2 & 3 \\ 2x & -x^2 & 2x^3 \end{bmatrix} \qquad \mathbf{B} = \begin{bmatrix} -2x & x^2 \\ 5 & -2x \\ 3x^3 & -3 \end{bmatrix}$$

2.20 Determine whether the matrices **A** and **B** given below are orthogonal matrices.

$$\mathbf{A} = \begin{bmatrix} -0.28 & -0.96 & 0 & 0 \\ 0.96 & -0.28 & 0 & 0 \\ 0 & 0 & -0.28 & -0.96 \\ 0 & 0 & 0.96 & -0.28 \end{bmatrix}$$

$$\mathbf{B} = \begin{bmatrix} -0.28 & 0.96 & 0 & 0 \\ 0.96 & -0.28 & 0 & 0 \\ 0 & 0 & -0.28 & 0.96 \\ 0 & 0 & 0.96 & -0.28 \end{bmatrix}$$

Section 2.4

2.21 through 2.25 Solve the following systems of simultaneous equations by the Gauss–Jordan method.

2.21

$$\begin{aligned} 2x_1 - 3x_2 + x_3 &= -18 \\ -9x_1 + 5x_2 + 3x_3 &= 18 \\ 4x_1 + 7x_2 - 8x_3 &= 53 \end{aligned}$$

2.22

$$\begin{aligned} 20x_1 - 9x_2 + 15x_3 &= 354 \\ -9x_1 + 16x_2 - 5x_3 &= -275 \\ 15x_1 - 5x_2 + 18x_3 &= 307 \end{aligned}$$

2.23

$$\begin{aligned} 4x_1 - 2x_2 + 3x_3 &= 37.2 \\ 3x_1 + 5x_2 - x_3 &= -7.2 \\ x_1 - 4x_2 + 2x_3 &= 30.3 \end{aligned}$$

2.24

$$\begin{aligned} 6x_1 + 15x_2 - 24x_3 + 40x_4 &= 190.9 \\ 15x_1 + 9x_2 - 13x_3 &= 69.8 \\ -24x_1 - 13x_2 + 8x_3 - 11x_4 &= -96.3 \\ 40x_1 - 11x_3 + 5x_4 &= 119.35 \end{aligned}$$

2.25

$$\begin{aligned} 2x_1 - 5x_2 + 8x_3 + 11x_4 &= 39 \\ 10x_1 + 7x_2 + 4x_3 - x_4 &= 127 \\ -3x_1 + 9x_2 + 5x_3 - 6x_4 &= 58 \\ x_1 - 4x_2 - 2x_3 + 9x_4 &= -14 \end{aligned}$$

2.26 Develop a computer program to solve a system of simultaneous equations of any size by the Gauss–Jordan method. Check the program by solving Problems 2.21 through 2.25 and comparing the computer-generated results to those determined by hand calculations.

2.27 through 2.30 Determine the inverse of the matrices shown by the Gauss–Jordan method.

2.27 $\mathbf{A} = \begin{bmatrix} 5 & 3 & -4 \\ 3 & 8 & -2 \\ -4 & -2 & 7 \end{bmatrix}$

2.28 $\mathbf{A} = \begin{bmatrix} 6 & -4 & 1 \\ -1 & 9 & 3 \\ 4 & 2 & 5 \end{bmatrix}$

2.29 $\mathbf{A} = \begin{bmatrix} 7 & -6 & 3 & -2 \\ -6 & 4 & -1 & 5 \\ 3 & -1 & 8 & 9 \\ -2 & 5 & 9 & 2 \end{bmatrix}$

2.30 $\mathbf{A} = \begin{bmatrix} 5 & -7 & -3 & 11 \\ 10 & -6 & -13 & 2 \\ -1 & 12 & 8 & -4 \\ -9 & 7 & -5 & 6 \end{bmatrix}$

3 PLANE TRUSSES

Goethals Bridge, a Cantilever Truss Bridge between Staten Island, NY, and Elizabeth, NJ.

(Photo courtesy of Port Authority of New York and New Jersey)

A *plane truss* is defined as *a two-dimensional framework of straight prismatic members connected at their ends by frictionless hinged joints, and subjected to loads and reactions that act only at the joints and lie in the plane of the structure.* The members of a plane truss are subjected to axial compressive or tensile forces only.

The objective of this chapter is to develop the analysis of plane trusses based on the matrix stiffness method. This method of analysis is general, in the sense that it can be applied to statically determinate, as well as indeterminate, plane trusses of any size and shape. We begin the chapter with the definitions of the global and local coordinate systems to be used in the analysis. The concept of "degrees of freedom" is introduced in Section 3.2; and the member force–displacement relations are established in the local coordinate system, using the equilibrium equations and the principles of *mechanics of materials,* in Section 3.3. The finite-element formulation of member stiffness relations using the principle of virtual work is presented in Section 3.4; and transformation of member forces and displacements from a local to a global coordinate system, and vice versa, is considered in Section 3.5. Member stiffness relations in the global coordinate system are derived in Section 3.6; the formulation of the stiffness relations for the entire truss, by combining the member stiffness relations, is discussed in Section 3.7; and a step-by-step procedure for the analysis of plane trusses subjected to joint loads is developed in Section 3.8.

3.1 GLOBAL AND LOCAL COORDINATE SYSTEMS

In the matrix stiffness method, two types of coordinate systems are employed to specify the structural and loading data and to establish the necessary force–displacement relations. These are referred to as the *global* (or *structural*) and the *local* (or *member*) coordinate systems.

Global Coordinate System

The overall geometry and the load–deformation relationships for an entire structure are described with reference to a Cartesian or rectangular global coordinate system.

> *The global coordinate system used in this text is a right-handed XYZ coordinate system with the plane structure lying in the XY plane.*

When analyzing a plane (two-dimensional) structure, the origin of the global *XY* coordinate system can be located at any point in the plane of the structure, with the *X* and *Y* axes oriented in any mutually perpendicular directions in the structure's plane. However, it is usually convenient to locate the origin at a lower left joint of the structure, with the X and Y axes oriented in the horizontal (positive to the right) and vertical (positive upward) directions, respectively, so that the X and Y coordinates of most of the joints are positive.

Consider, for example, the truss shown in Fig. 3.1(a), which is composed of six members and four joints. Figure 3.1(b) shows the analytical model of the truss as represented by a line diagram, on which all the joints and members are identified by numbers that have been assigned arbitrarily. The global coordinate system chosen for analysis is usually drawn on the line diagram of the structure as shown in Fig. 3.1(b). Note that the origin of the global XY coordinate system is located at joint 1.

Local Coordinate System

Since it is convenient to derive the basic member force–displacement relationships in terms of the forces and displacements in the directions along and perpendicular to members, a local coordinate system is defined for each member of the structure.

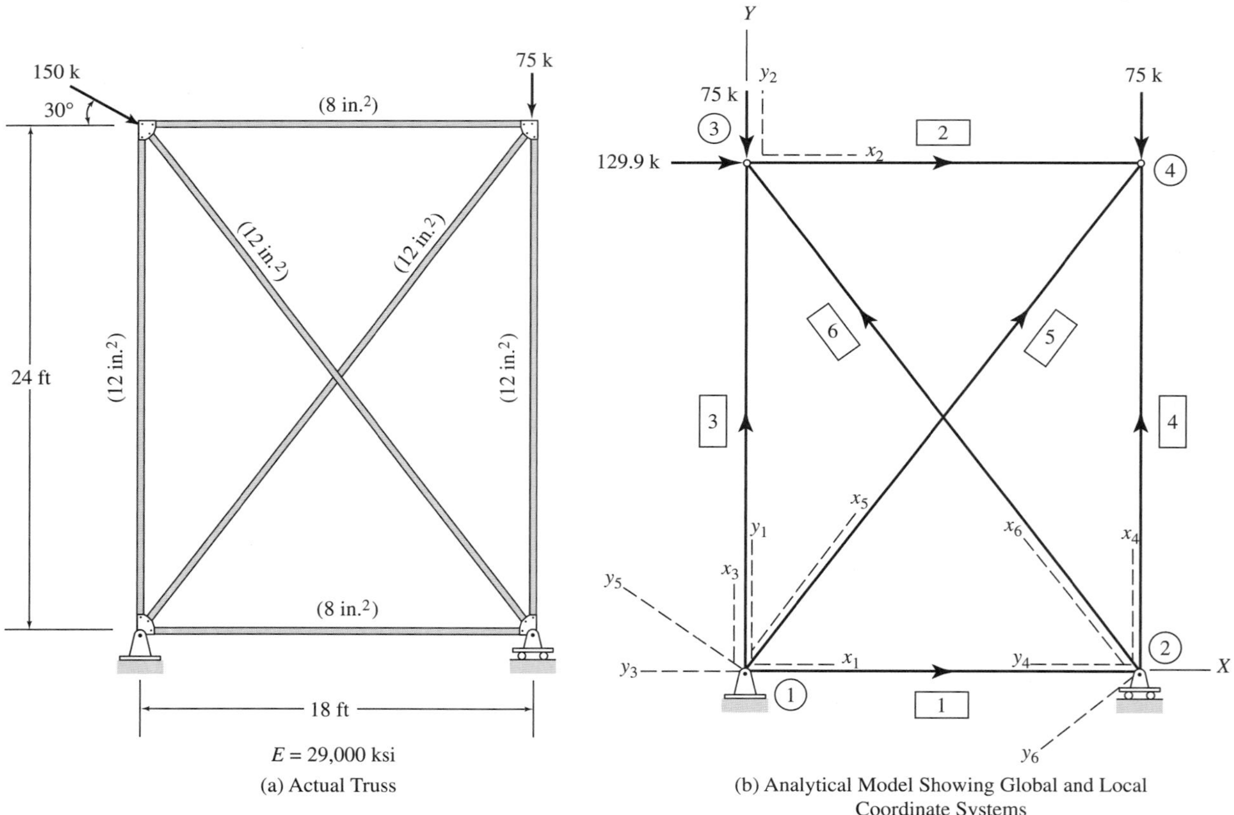

Fig. 3.1

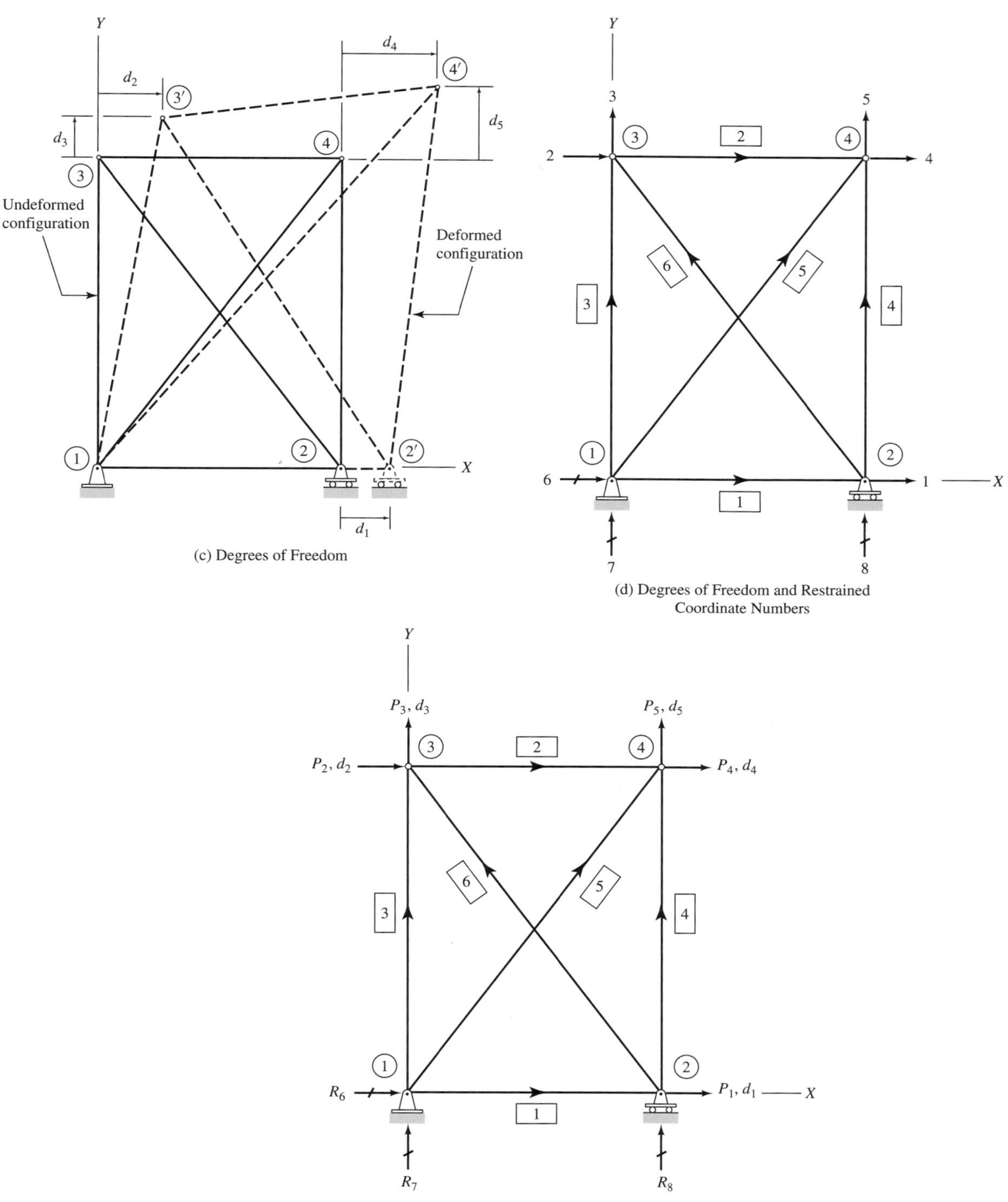

Fig. 3.1 *(continued)*

> *The origin of the local xyz coordinate system for a member may be arbitrarily located at one of the ends of the member in its undeformed state, with the x axis directed along the member's centroidal axis in the undeformed state. The positive direction of the y axis is defined so that the coordinate system is right-handed, with the local z axis pointing in the positive direction of the global Z axis.*

On the line diagram of the structure, the positive direction of the x axis for each member is indicated by drawing an arrow along each member, as shown in Fig. 3.1(b). For example, this figure shows the origin of the local coordinate system for member 1 located at its end connected to joint 1, with the x_1 axis directed from joint 1 to joint 2; the origin of the local coordinate system for member 4 located at its end connected to joint 2, with the x_4 axis directed from joint 2 to joint 4, etcetera. The joint to which the member end with the origin of the local coordinate system is connected is termed the *beginning joint* for the member, and the joint adjacent to the opposite end of the member is referred to as the *end joint*. For example, in Fig. 3.1(b), member 1 begins at joint 1 and ends at joint 2, member 4 begins at joint 2 and ends at joint 4, and so on. Once the local x axis is specified for a member, its y axis can be established by applying the right-hand rule. The local y axes thus obtained for all six members of the truss are depicted in Fig. 3.1(b). It can be seen that, for each member, if we curl the fingers of our right hand from the direction of the x axis toward the direction of the corresponding y axis, then the extended thumb points out of the plane of the page, which is the positive direction of the global Z axis.

3.2 DEGREES OF FREEDOM

The *degrees of freedom* of a structure, in general, are defined as *the independent joint displacements (translations and rotations) that are necessary to specify the deformed shape of the structure when subjected to an arbitrary loading.* Since the joints of trusses are assumed to be frictionless hinges, they are not subjected to moments and, therefore, their rotations are zero. Thus, only joint translations must be considered in establishing the degrees of freedom of trusses.

Consider again the plane truss of Fig. 3.1(a). The deformed shape of the truss, for an arbitrary loading, is depicted in Fig. 3.1(c) using an exaggerated scale. From this figure, we can see that joint 1, which is attached to the hinged support, cannot translate in any direction; therefore, it has no degrees of freedom. Because joint 2 is attached to the roller support, it can translate in the X direction, but not in the Y direction. Thus, joint 2 has only one degree of freedom, which is designated d_1 in the figure. As joint 3 is not attached to a support, two displacements (namely, the translations d_2 and d_3 in the X and Y directions, respectively) are needed to completely specify its deformed position $3'$. Thus, joint 3 has two degrees of freedom. Similarly, joint 4, which is also a *free* joint, has two degrees of freedom, designated d_4 and d_5. Thus, the entire truss has a total of five degrees of freedom. As shown in Fig. 3.1(c), the joint displacements are defined relative to the global coordinate system, and are considered to be positive when in the positive directions of the X and Y axes. Note that all the joint displacements are shown in the positive sense in Fig. 3.1(c). The five joint displacement of the truss can be collectively written in matrix form as

$$\mathbf{d} = \begin{bmatrix} d_1 \\ d_2 \\ d_3 \\ d_4 \\ d_5 \end{bmatrix}$$

in which **d** is called the *joint displacement vector,* with the number of rows equal to the number of degrees of freedom of the structure.

It is important to note that the five joint displacements d_1 through d_5 are necessary and sufficient to uniquely define the deformed shape of the truss under any arbitrary loading condition. Furthermore, the five joint displacements are independent, in the sense that each displacement can be varied arbitrarily and independently of the others.

As the foregoing example illustrates, the degrees of freedom of all types of framed structures, in general, are the same as the actual joint displacements. Thus, the number of degrees of freedom of a framed structure can be determined by subtracting the number of joint displacements restrained by supports from the total number of joint displacements of the unsupported structure; that is,

$$\begin{pmatrix} \text{number of} \\ \text{degrees of} \\ \text{freedom} \end{pmatrix} = \begin{pmatrix} \text{number of joint} \\ \text{displacements of} \\ \text{the unsupported} \\ \text{structure} \end{pmatrix} - \begin{pmatrix} \text{number of joint} \\ \text{displacements} \\ \text{restrained by} \\ \text{supports} \end{pmatrix} \tag{3.1}$$

As the number of displacements of an unsupported structure equals the product of the number of degrees of freedom of a free joint of the structure and the total number of joints of the structure, we can express Eq. (3.1) as

$$NDOF = NCJT\,(NJ) - NR \tag{3.2}$$

in which $NDOF$ represents the number of degrees of freedom of the structure (sometimes referred to as the *degree of kinematic indeterminacy* of the structure); $NCJT$ represents the number of degrees of freedom of a free joint (also called the number of structure coordinates per joint); NJ is the number of joints; and NR denotes the number of joint displacements restrained by supports.

Since a free joint of a plane truss has two degrees of freedom, which are translations in the X and Y directions, we can specialize Eq. (3.2) for the case of plane trusses:

$$\left.\begin{aligned} NCJT &= 2 \\ NDOF &= 2(NJ) - NR \end{aligned}\right\} \quad \text{for plane trusses} \tag{3.3}$$

Let us apply Eq. (3.3) to the truss of Fig. 3.1(a). The truss has four joints (i.e., $NJ = 4$), and the hinged support at joint 1 restrains two joint displacements, namely, the translations of joint 1 in the X and Y directions; whereas the roller support at joint 2 restrains one joint displacement, which is the translation of joint 2 in the Y direction. Thus, the total number of joint displacements that are restrained by all supports of the truss is 3 (i.e., $NR = 3$). Substituting the numerical values of NJ and NR into Eq. (3.3), we obtain

$$NDOF = 2(4) - 3 = 5$$

which is the same as the number of degrees of freedom of the truss obtained previously.

The degrees of freedom (or joint displacements) of a structure are also termed the structure's *free coordinates;* the joint displacements restrained by supports are commonly called the *restrained coordinates* of the structure. The free and restrained coordinates are referred to collectively as simply the *structure coordinates*. It should be noted that each structure coordinate represents an unknown quantity to be determined by the analysis, with a free coordinate representing an unknown joint displacement, and a restrained coordinate representing an unknown support reaction. Realizing that $NCJT$ (i.e., the number of structure coordinates per joint) equals the number of unknown joint displacements and/or support reactions per joint of the structure, the total number of unknown joint displacements and reactions for a structure can be expressed as

$$\begin{pmatrix} \text{number of unknown} \\ \text{joint displacements} \\ \text{and support reactions} \end{pmatrix} = NDOF + NR = NCJT(NJ)$$

Numbering of Degrees of Freedom and Restrained Coordinates

When analyzing a structure, it is not necessary to draw the structure's deformed shape, as shown in Fig. 3.1(c), to identify its degrees of freedom. Instead, the degrees of freedom can be directly specified on the line diagram of the structure by assigning numbers to the arrows drawn at the joints in the directions of the joint displacements, as shown in Fig. 3.1(d). The restrained coordinates are identified in a similar manner. However, the arrows representing the restrained coordinates are usually drawn with a slash ($\not\rightarrow$) to distinguish them from the arrows identifying the degrees of freedom.

The degrees of freedom of a plane truss are numbered starting at the lowest-numbered joint that has a degree of freedom, and proceeding sequentially to the highest-numbered joint. In the case of more than one degree of freedom at a joint, the translation in the X direction is numbered first, followed by the translation in the Y direction. The first degree of freedom is assigned the number one, and the last degree of freedom is assigned a number equal to $NDOF$.

Once all the degrees of freedom of the structure have been numbered, we number the restrained coordinates in a similar manner, but begin with a number equal to $NDOF + 1$. We start at the lowest-numbered joint that is attached to a support, and proceed sequentially to the highest-numbered joint. In the case of more than one restrained coordinate at a joint, the coordinate in the

X direction is numbered first, followed by the coordinate in the Y direction. Note that this procedure will always result in the last restrained coordinate of the structure being assigned a number equal to $2(NJ)$.

The degrees of freedom and the restrained coordinates of the truss in Fig. 3.1(d) have been numbered using the foregoing procedure. We start numbering the degrees of freedom by examining joint 1. Since the displacements of joint 1 in both the X and Y directions are restrained, this joint does not have any degrees of freedom; therefore, at this point, we do not assign any numbers to the two arrows representing its restrained coordinates, and move on to the next joint. Focusing our attention on joint 2, we realize that this joint is free to displace in the X direction, but not in the Y direction. Therefore, we assign the number 1 to the horizontal arrow indicating that the X displacement of joint 2 will be denoted by d_1. Note that, at this point, we do not assign any number to the vertical arrow at joint 2, and change our focus to the next joint. Joint 3 is free to displace in both the X and Y directions; we number the X displacement first by assigning the number 2 to the horizontal arrow, and then number the Y displacement by assigning the number 3 to the vertical arrow. This indicates that the X and Y displacements of joint 3 will be denoted by d_2 and d_3, respectively. Next, we focus our attention on joint 4, which is also free to displace in both the X and Y directions; we assign numbers 4 and 5, respectively, to its displacements in the X and Y directions, as shown in Fig. 3.1(d). Again, the arrow that is numbered 4 indicates the location and direction of the joint displacement d_4; the arrow numbered 5 shows the location and direction of d_5.

Having numbered all the degrees of freedom of the truss, we now return to joint 1, and start numbering the restrained coordinates of the structure. As previously discussed, joint 1 has two restrained coordinates; we first assign the number 6 (i.e., $NDOF + 1 = 5 + 1 = 6$) to the X coordinate (horizontal arrow), and then assign the number 7 to the Y coordinate (vertical arrow). Finally, we consider joint 2, and assign the number 8 to the vertical arrow representing the restrained coordinate in the Y direction at that joint. We realize that the displacements corresponding to the restrained coordinates 6 through 8 are zero (i.e., $d_6 = d_7 = d_8 = 0$). However, we use these restrained coordinate numbers to specify the reactions at supports of the structure, as discussed subsequently in this section.

Joint Load Vector

External loads applied to the joints of trusses are specified as force components in the global X and Y directions. These load components are considered positive when acting in the positive directions of the X and Y axes, and vice versa. Any loads initially given in inclined directions are resolved into their X and Y components, before proceeding with an analysis. For example, the 150 k inclined load acting on a joint of the truss in Fig. 3.1(a) is resolved into its rectangular components as

$$\text{load component in } X \text{ direction} = 150 \cos 30^\circ = 129.9 \text{ k}\rightarrow$$

$$\text{load component in } Y \text{ direction} = 150 \sin 30^\circ = 75 \text{ k}\downarrow$$

These components are applied at joint 3 of the line diagram of the truss shown in Fig. 3.1(b).

In general, a load can be applied to a structure at the location and in the direction of each of its degrees of freedom. For example, a five-degree-of-freedom truss can be subjected to a maximum of five loads, P_1 through P_5, as shown in Fig. 3.1(e). As indicated there, the numbers assigned to the degrees of freedom are also used to identify the joint loads. In other words, a load corresponding to a degree of freedom d_i is denoted by the symbol P_i. The five joint loads of the truss can be collectively written in matrix form as

$$\mathbf{P} = \begin{bmatrix} P_1 \\ P_2 \\ P_3 \\ P_4 \\ P_5 \end{bmatrix} = \begin{bmatrix} 0 \\ 129.9 \\ -75 \\ 0 \\ -75 \end{bmatrix} \text{ k}$$

in which $\mathbf{P}$ is called the *joint load vector* of the truss. The numerical form of $\mathbf{P}$ is obtained by comparing Figs. 3.1(b) and 3.1(e). This comparison shows that: $P_1 = 0$; $P_2 = 129.9$ k; $P_3 = -75$ k; $P_4 = 0$; and $P_5 = -75$ k. The negative signs assigned to the magnitudes of P_3 and P_5 indicate that these loads act in the negative Y (i.e., downward) direction. The numerical values of P_1 through P_5 are then stored in the appropriate rows of the joint load vector $\mathbf{P}$, as shown in the foregoing equation. It should be noted that the number of rows of $\mathbf{P}$ equals the number of degrees of freedom ($NDOF$) of the structure.

Reaction Vector

A support that prevents the translation of a joint of a structure in a particular direction exerts a reaction force on the joint in that direction. Thus, when a truss is subjected to external loads, a reaction force component can develop at the location and in the direction of each of its restrained coordinates. For example, a truss with three restrained coordinates can develop up to three reactions, as shown in Fig. 3.1(e). As indicated there, the numbers assigned to the restrained coordinates are used to identify the support reactions. In other words, a reaction corresponding to an ith restrained coordinate is denoted by the symbol R_i. The three support reactions of the truss can be collectively expressed in matrix form as

$$\mathbf{R} = \begin{bmatrix} R_6 \\ R_7 \\ R_8 \end{bmatrix}$$

in which $\mathbf{R}$ is referred to as the *reaction vector* of the structure. Note that the number of rows of $\mathbf{R}$ equals the number of restrained coordinates (NR) of the structure.

The procedure presented in this section for numerically identifying the degrees of freedom, joint loads, and reactions of a structure considerably simplifies the task of programming an analysis on a computer, as will become apparent in Chapter 4.

EXAMPLE 3.1 Identify numerically the degrees of freedom and restrained coordinates of the tower truss shown in Fig. 3.2(a). Also, form the joint load vector **P** for the truss.

SOLUTION The truss has nine degrees of freedom, which are identified by the numbers 1 through 9 in Fig. 3.2(c). The five restrained coordinates of the truss are identified by the numbers 10 through 14 in the same figure. **Ans**

By comparing Figs. 3.2(b) and (c), we express the joint load vector as

$$\mathbf{P} = \begin{bmatrix} 20 \\ 0 \\ 0 \\ 20 \\ 0 \\ 0 \\ -35 \\ 10 \\ -20 \end{bmatrix} \text{k}$$ **Ans**

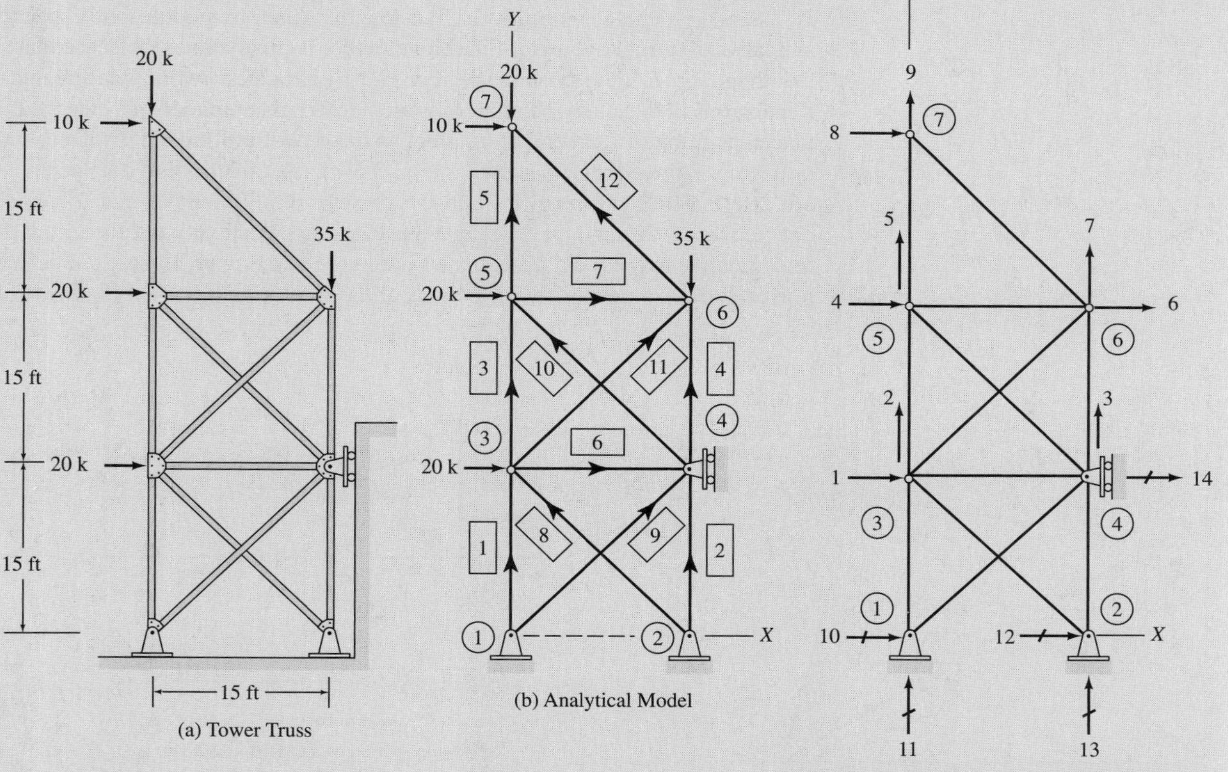

(a) Tower Truss

(b) Analytical Model

(c) Degrees of Freedom and Restrained Coordinates ($NDOF = 9$, $NR = 5$)

Fig. 3.2

3.3 MEMBER STIFFNESS RELATIONS IN THE LOCAL COORDINATE SYSTEM

In the stiffness method of analysis, the joint displacements, **d**, of a structure due to an external loading, **P**, are determined by solving a system of simultaneous equations, expressed in the form

$$\mathbf{P} = \mathbf{S}\mathbf{d} \tag{3.4}$$

in which **S** is called the *structure stiffness matrix*. It will be shown subsequently that the stiffness matrix for the entire structure, **S**, is formed by assembling the stiffness matrices for its individual members. *The stiffness matrix for a member expresses the forces at the ends of the member as functions of the displacements of those ends.* In this section, we derive the stiffness matrix for the members of plane trusses in the local coordinate system.

To establish the member stiffness relations, let us focus our attention on an arbitrary member m of the plane truss shown in Fig. 3.3(a). When the truss is subjected to external loads, m deforms and internal forces are induced at its ends. The initial and displaced positions of m are shown in Fig. 3.3(b), where L, E, and A denote, respectively, the length, Young's modulus of elasticity, and the cross-sectional area of m. The member is prismatic in the sense that its axial rigidity, EA, is constant. As Fig. 3.3(b) indicates, two displacements—translations in the x and y directions—are needed to completely specify the displaced position of each end of m. Thus, m has a total of four end displacements or degrees of freedom. As shown in Fig. 3.3(b), the member end displacements are denoted by u_1 through u_4, and the corresponding member end forces are denoted by Q_1 through Q_4. Note that these end displacements and forces are defined relative to the local coordinate system of the member, and are considered positive when in the positive directions of the local x and y axes. As indicated in

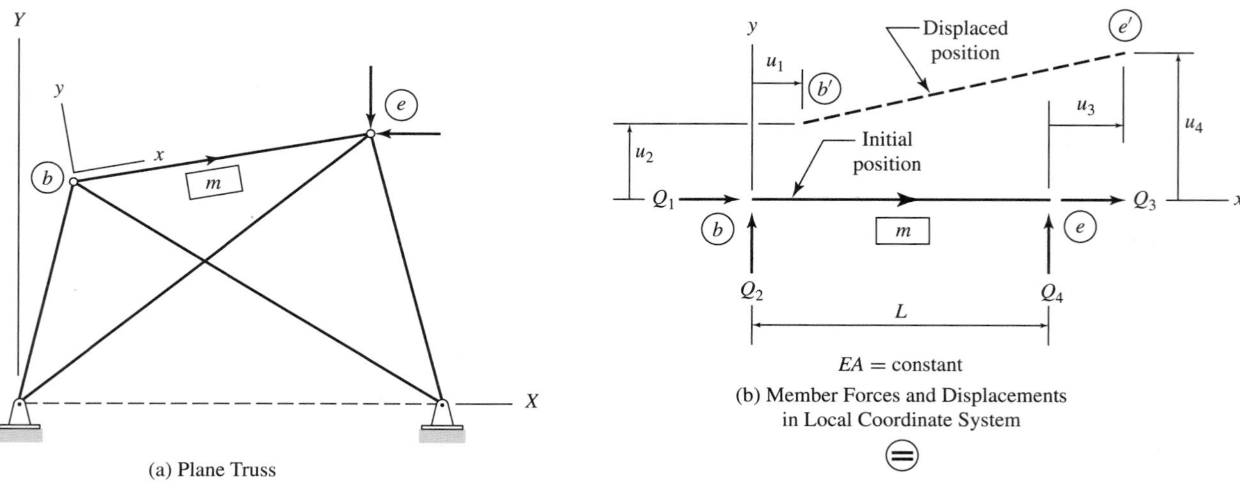

Fig. 3.3

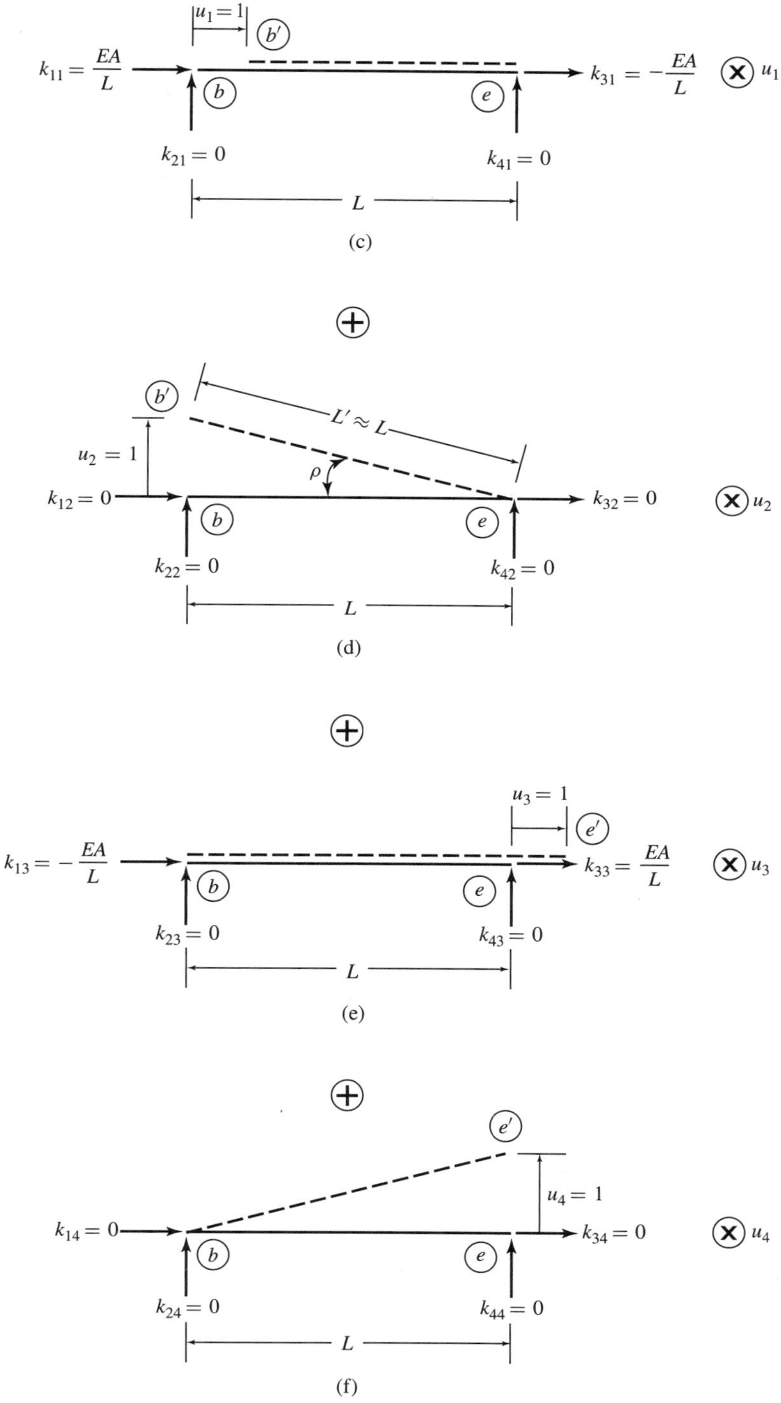

Fig. 3.3 *(continued)*

Fig. 3.3(b), the numbering scheme used to identify the member end displacements and forces is as follows:

> *Member end displacements and forces are numbered by beginning at the end of the member designated "b", where the origin of the local coordinate system is located, with the translation and force in the x direction numbered first, followed by the translation and force in the y direction. The displacements and forces at the opposite end of the member, designated "e," are then numbered in the same sequential order.*

It should be remembered that our objective here is to determine the relationships between member end forces and end displacements. Such relationships can be conveniently established by subjecting the member, separately, to each of the four end displacements as shown in Figs. 3.3(c) through (f); and by expressing the total member end forces as the algebraic sums of the end forces required to cause the individual end displacements. Thus, from Figs. 3.3(b) through (f), we can see that

$$Q_1 = k_{11}u_1 + k_{12}u_2 + k_{13}u_3 + k_{14}u_4 \tag{3.5a}$$

$$Q_2 = k_{21}u_1 + k_{22}u_2 + k_{23}u_3 + k_{24}u_4 \tag{3.5b}$$

$$Q_3 = k_{31}u_1 + k_{32}u_2 + k_{33}u_3 + k_{34}u_4 \tag{3.5c}$$

$$Q_4 = k_{41}u_1 + k_{42}u_2 + k_{43}u_3 + k_{44}u_4 \tag{3.5d}$$

in which k_{ij} *represents the force at the location and in the direction of* Q_i *required, along with other end forces, to cause a unit value of displacement* u_j, *while all other end displacements are zero*. These forces per unit displacement are called *stiffness coefficients*. It should be noted that a double-subscript notation is used for stiffness coefficients, with the first subscript identifying the force and the second subscript identifying the displacement.

By using the definition of matrix multiplication, Eqs. (3.5) can be expressed in matrix form as

$$\begin{bmatrix} Q_1 \\ Q_2 \\ Q_3 \\ Q_4 \end{bmatrix} = \begin{bmatrix} k_{11} & k_{12} & k_{13} & k_{14} \\ k_{21} & k_{22} & k_{23} & k_{24} \\ k_{31} & k_{32} & k_{33} & k_{34} \\ k_{41} & k_{42} & k_{43} & k_{44} \end{bmatrix} \begin{bmatrix} u_1 \\ u_2 \\ u_3 \\ u_4 \end{bmatrix} \tag{3.6}$$

or, symbolically, as

$$\boxed{\mathbf{Q} = \mathbf{k}\mathbf{u}} \tag{3.7}$$

in which **Q** and **u** are the member end force and member end displacement vectors, respectively, in the local coordinate system; and **k** is called the *member stiffness matrix in the local coordinate system*.

The stiffness coefficients k_{ij} can be evaluated by subjecting the member, separately, to unit values of each of the four end displacements. The member end forces required to cause the individual unit displacements are then determined by applying the equations of equilibrium, and by using the principles of *mechanics of materials*. The member end forces thus obtained represent the stiffness coefficients for the member.

Let us determine the stiffness coefficients corresponding to a unit value of the displacement u_1 at end b of m, as shown in Fig. 3.3(c). Note that all other displacements of m are zero (i.e., $u_2 = u_3 = u_4 = 0$). Since m is in equilibrium, the end forces k_{11}, k_{21}, k_{31}, and k_{41} acting on it must satisfy the three equilibrium equations: $\sum F_x = 0$, $\sum F_y = 0$, and $\sum M = 0$. Applying the equations of equilibrium, we write

$$+ \rightarrow \sum F_x = 0 \qquad k_{11} + k_{31} = 0$$

$$k_{31} = -k_{11} \tag{3.8}$$

$$+ \uparrow \sum F_y = 0 \qquad k_{21} + k_{41} = 0 \tag{3.9}$$

$$+ \circlearrowleft \sum M_e = 0 \qquad -k_{21}(L) = 0$$

Since L is not zero, k_{21} must be zero; that is

$$k_{21} = 0 \tag{3.10}$$

By substituting Eq. (3.10) into Eq. (3.9), we obtain

$$k_{41} = 0 \tag{3.11}$$

Equations (3.8), (3.10), and (3.11) indicate that m is in equilibrium under the action of two axial forces, of equal magnitude but with opposite senses, applied at its ends. Furthermore, since the displacement $u_1 = 1$ results in the shortening of the member's length, the two axial forces causing this displacement must be compressive; that is, k_{11} must act in the positive direction of the local x axis, and k_{31} (with a magnitude equal to k_{11}) must act in the negative direction of the x axis.

To relate the axial force k_{11} to the unit axial deformation ($u_1 = 1$) of m, we use the principles of the *mechanics of materials*. Recall that in a prismatic member subjected to axial tension or compression, the normal stress σ is given by

$$\sigma = \frac{\text{axial force}}{\text{cross-sectional area}} = \frac{k_{11}}{A} \tag{3.12}$$

and the normal strain, ε, is expressed as

$$\varepsilon = \frac{\text{change in length}}{\text{original length}} = \frac{1}{L} \tag{3.13}$$

For linearly elastic materials, the stress–strain relationship is given by Hooke's law as

$$\sigma = E\varepsilon \tag{3.14}$$

Substitution of Eqs. (3.12) and (3.13) into Eq. (3.14) yields

$$\frac{k_{11}}{A} = E\left(\frac{1}{L}\right)$$

from which we obtain the expression for the stiffness coefficient k_{11},

$$k_{11} = \frac{EA}{L} \tag{3.15}$$

The expression for k_{31} can now be obtained from Eq. (3.8) as

$$k_{31} = -k_{11} = -\frac{EA}{L} \tag{3.16}$$

in which the negative sign indicates that this force acts in the negative x direction. Figure 3.3(c) shows the expressions for the four stiffness coefficients required to cause the end displacement $u_1 = 1$ of m.

By using a similar approach, it can be shown that the stiffness coefficients required to cause the axial displacement $u_3 = 1$ at end e of m are as follows (Fig. 3.3e).

$$k_{13} = -\frac{EA}{L} \qquad k_{23} = 0 \qquad k_{33} = \frac{EA}{L} \qquad k_{43} = 0 \tag{3.17}$$

The deformed shape of m due to a unit value of displacement u_2, while all other displacements are zero, is shown in Fig. 3.3(d). Applying the equilibrium equations, we write

$$+\rightarrow \sum F_x = 0 \qquad k_{12} + k_{32} = 0$$

$$k_{32} = -k_{12} \tag{3.18}$$

$$+\uparrow \sum F_y = 0 \qquad k_{22} + k_{42} = 0 \tag{3.19}$$

$$+\circlearrowleft \sum M_e = 0 \qquad -k_{22}(L) = 0$$

from which we obtain

$$k_{22} = 0 \tag{3.20}$$

Substitution of Eq. (3.20) into Eq. (3.19) yields

$$k_{42} = 0 \tag{3.21}$$

Thus, the forces k_{22} and k_{42}, which act perpendicular to the longitudinal axis of m, are both zero.

As for the axial forces k_{12} and k_{32}, Eq. (3.18) indicates that they must be of equal magnitude but with opposite senses. From Fig. 3.3(d), we can see that the deformed length of the member, L', can be expressed in terms of its undeformed length L as

$$L' = \frac{L}{\cos\rho} \tag{3.22}$$

in which the angle ρ denotes the rotation of the member due to the end displacement $u_2 = 1$. Since the displacements are assumed to be small, $\cos\rho \approx 1$ and Eq. (3.22) reduces to

$$L' \approx L \tag{3.23}$$

which can be rewritten as

$$L' - L \approx 0 \tag{3.24}$$

As Eq. (3.24) indicates, the change in the length of m (or its axial deformation) is negligibly small and, therefore, no axial forces develop at the ends of m; that is,

$$k_{12} = k_{32} = 0 \tag{3.25}$$

Thus, as shown in Fig. 3.3(d), no end forces are required to produce the displacement $u_2 = 1$ of m.

Similarly, the stiffness coefficients required to cause the small end displacement $u_4 = 1$, in the direction perpendicular to the longitudinal axis of m, are also all zero, as shown in Fig. 3.3(f). Thus,

$$k_{14} = k_{24} = k_{34} = k_{44} = 0 \tag{3.26}$$

By substituting the foregoing values of the stiffness coefficients into Eq. (3.6), we obtain the following stiffness matrix for the members of plane trusses in their local coordinate systems.

$$\mathbf{k} = \begin{bmatrix} \frac{EA}{L} & 0 & -\frac{EA}{L} & 0 \\ 0 & 0 & 0 & 0 \\ -\frac{EA}{L} & 0 & \frac{EA}{L} & 0 \\ 0 & 0 & 0 & 0 \end{bmatrix} = \frac{EA}{L} \begin{bmatrix} 1 & 0 & -1 & 0 \\ 0 & 0 & 0 & 0 \\ -1 & 0 & 1 & 0 \\ 0 & 0 & 0 & 0 \end{bmatrix} \tag{3.27}$$

Note that the ith column of the member stiffness matrix $\mathbf{k}$ consists of the end forces required to cause a unit value of the end displacement u_i, while all other displacements are zero. For example, the third column of $\mathbf{k}$ consists of the four end forces required to cause the displacement $u_3 = 1$, as shown in Fig. 3.3(e), and so on. The units of the stiffness coefficients are expressed in terms of force divided by length (e.g., k/in or kN/m). When evaluating a stiffness matrix for analysis, it is important to use a consistent set of units. For example, if we wish to use the units of kips and feet, then the modulus of elasticity (E) must be expressed in k/ft^2, area of cross-section (A) in ft^2, and the member length (L) in ft.

From Eq. (3.27), we can see that the stiffness matrix $\mathbf{k}$ is symmetric; that is, $k_{ij} = k_{ji}$. As shown in Section 3.4, the stiffness matrices for linear elastic structures are always symmetric.

EXAMPLE 3.2 Determine the local stiffness matrices for the members of the truss shown in Fig. 3.4.

SOLUTION **Members 1 and 2** $E = 29{,}000$ ksi, $A = 8$ in.^{2}, $L = 18$ ft $= 216$ in.

$$\frac{EA}{L} = \frac{29{,}000(8)}{216} = 1{,}074.1 \text{ k/in.}$$

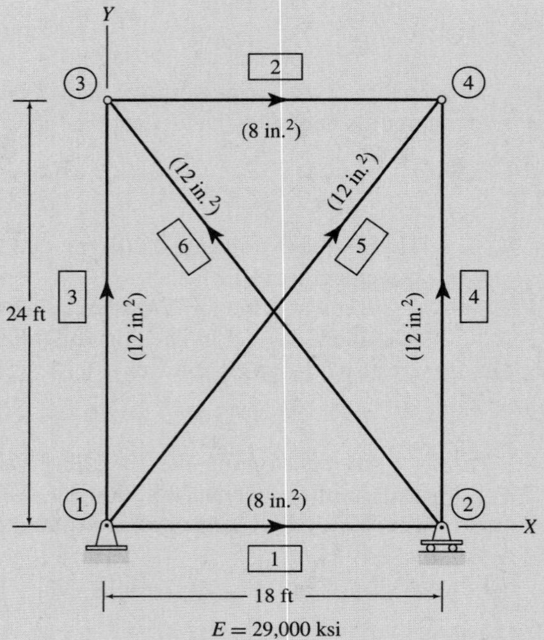

Fig. 3.4

Substitution into Eq. (3.27) yields

$$\mathbf{k}_1 = \mathbf{k}_2 = \begin{bmatrix} 1{,}074.1 & 0 & -1{,}074.1 & 0 \\ 0 & 0 & 0 & 0 \\ -1{,}074.1 & 0 & 1{,}074.1 & 0 \\ 0 & 0 & 0 & 0 \end{bmatrix} \text{ k/in.}$$ **Ans**

Members 3 and 4 $E = 29{,}000$ ksi, $A = 12$ in.^{2}, $L = 24$ ft $= 288$ in.

$$\frac{EA}{L} = \frac{29{,}000(12)}{288} = 1{,}208.3 \text{ k/in.}$$

Thus, from Eq. (3.27),

$$\mathbf{k}_3 = \mathbf{k}_4 = \begin{bmatrix} 1{,}208.3 & 0 & -1{,}208.3 & 0 \\ 0 & 0 & 0 & 0 \\ -1{,}208.3 & 0 & 1{,}208.3 & 0 \\ 0 & 0 & 0 & 0 \end{bmatrix} \text{ k/in.}$$ **Ans**

Members 5 and 6 $E = 29{,}000$ ksi, $A = 12$ in.^{2},

$$L = \sqrt{(18)^2 + (24)^2} = 30 \text{ ft} = 360 \text{ in.}$$

$$\frac{EA}{L} = \frac{29{,}000(12)}{360} = 966.67 \text{ k/in.}$$

Thus,

$$\mathbf{k}_5 = \mathbf{k}_6 = \begin{bmatrix} 966.67 & 0 & -966.67 & 0 \\ 0 & 0 & 0 & 0 \\ -966.67 & 0 & 966.67 & 0 \\ 0 & 0 & 0 & 0 \end{bmatrix} \text{ k/in.}$$ **Ans**

EXAMPLE 3.3 The displaced position of member 8 of the truss in Fig. 3.5(a) is given in Fig. 3.5(b). Calculate the axial force in this member.

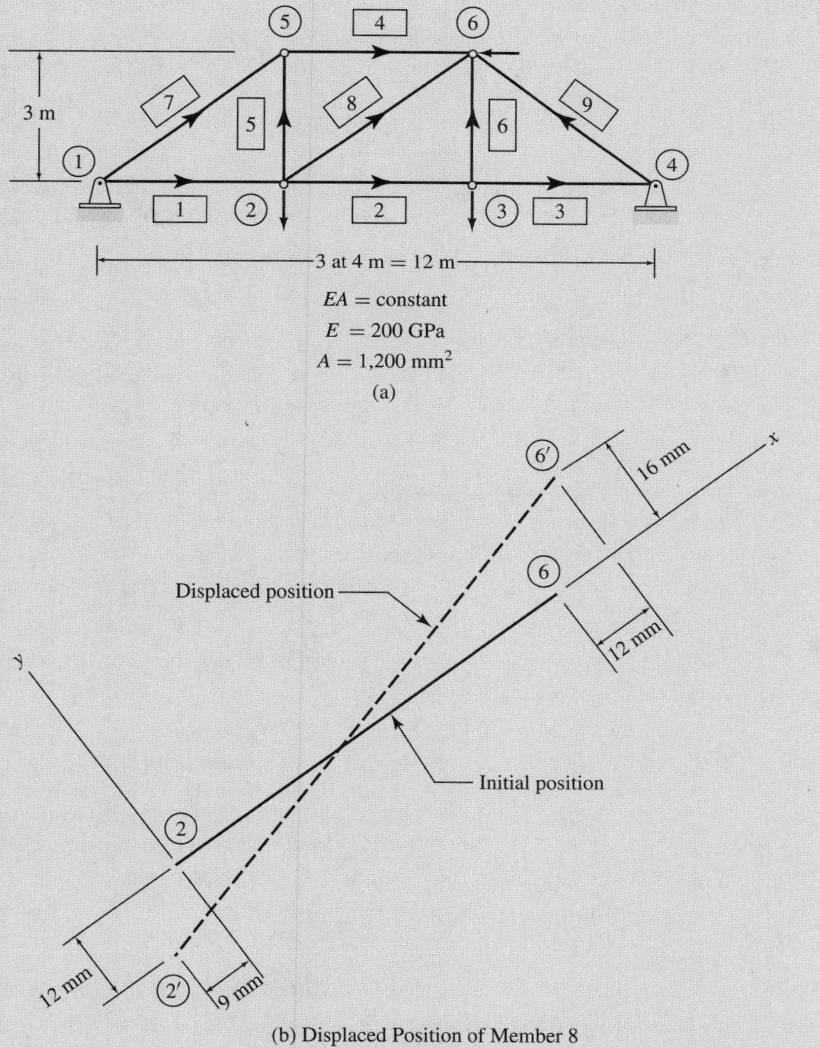

Fig. 3.5

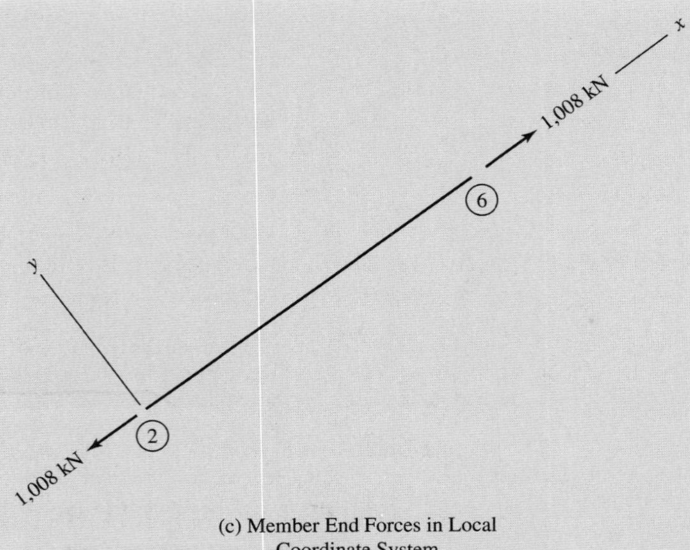

(c) Member End Forces in Local Coordinate System

Fig. 3.5 *(continued)*

SOLUTION *Member Stiffness Matrix in the Local Coordinate System:* From Fig. 3.5(a), we can see that $E = 200$ GPa $= 200(10^6)$ kN/m^2; $A = 1{,}200$ mm$^2 = 0.0012$ m^2; and $L = \sqrt{(4)^2 + (3)^2} = 5$ m. Thus,

$$\frac{EA}{L} = \frac{200(10^6)(0.0012)}{5} = 48{,}000 \text{ kN/m}$$

From Eq. (3.27), we obtain

$$\mathbf{k}_8 = 48{,}000 \begin{bmatrix} 1 & 0 & -1 & 0 \\ 0 & 0 & 0 & 0 \\ -1 & 0 & 1 & 0 \\ 0 & 0 & 0 & 0 \end{bmatrix} \text{ kN/m}$$

Member End Displacements in the Local Coordinate System: From Fig. 3.5(b), we can see that the beginning end, 2, of the member displaces 9 mm in the negative x direction and 12 mm in the negative y direction. Thus, $u_1 = -9$ mm $= -0.009$ m and $u_2 = -12$ mm $= -0.012$ m. Similarly, the opposite end, 6, of the member displaces 12 mm and 16 mm, respectively, in the x and y directions; that is, $u_3 = 12$ mm $= 0.012$ m and $u_4 = 16$ mm $= 0.016$ m. Thus, the member end displacement vector in the local coordinate system is given by

$$\mathbf{u}_8 = \begin{bmatrix} -0.009 \\ -0.012 \\ 0.012 \\ 0.016 \end{bmatrix} \text{ m}$$

Member End Forces in the Local Coordinate System: We calculate the member end forces by applying Eq. (3.7). Thus,

$$\mathbf{Q} = \mathbf{ku}$$

$$\mathbf{Q}_8 = \begin{bmatrix} Q_1 \\ Q_2 \\ Q_3 \\ Q_4 \end{bmatrix} = 48{,}000 \begin{bmatrix} 1 & 0 & -1 & 0 \\ 0 & 0 & 0 & 0 \\ -1 & 0 & 1 & 0 \\ 0 & 0 & 0 & 0 \end{bmatrix} \begin{bmatrix} -0.009 \\ -0.012 \\ 0.012 \\ 0.016 \end{bmatrix} = \begin{bmatrix} -1{,}008 \\ 0 \\ 1{,}008 \\ 0 \end{bmatrix} \text{kN}$$

The member end forces, **Q**, are depicted on the free-body diagram of the member in Fig. 3.5(c), from which we can see that, since the end force Q_1 is negative, but Q_3 is positive, member 8 is subjected to a tensile axial force, Q_a, of magnitude 1,008 kN; that is,

$$Q_{a8} = 1{,}008 \text{ kN (T)}$$ **Ans**

3.4 FINITE-ELEMENT FORMULATION USING VIRTUAL WORK*

In this section, we present an alternate formulation of the member stiffness matrix **k** in the local coordinate system. This approach, which is commonly used in the finite-element method, essentially involves expressing the strains and stresses at points within the member in terms of its end displacements **u**, and applying the principle of virtual work for deformable bodies as delineated by Eq. (1.16) in Section 1.6. Before proceeding with the derivation of **k**, let us rewrite Eq. (1.16) in a more convenient matrix form as

$$\delta W_e = \int_V \delta\varepsilon^T \sigma \, dV \tag{3.28}$$

in which δW_e denotes virtual external work; V represents member volume; and $\delta\varepsilon$ and σ denote, respectively, the virtual strain and real stress vectors, which for a general three-dimensional stress condition can be expressed as follows (see Fig. 1.16).

$$\delta\varepsilon = \begin{bmatrix} \delta\varepsilon_x \\ \delta\varepsilon_y \\ \delta\varepsilon_z \\ \delta\gamma_{xy} \\ \delta\gamma_{yz} \\ \delta\gamma_{zx} \end{bmatrix} \qquad \sigma = \begin{bmatrix} \sigma_x \\ \sigma_y \\ \sigma_z \\ \tau_{xy} \\ \tau_{yz} \\ \tau_{zx} \end{bmatrix} \tag{3.29}$$

Displacement Functions

In the finite-element method, member stiffness relations are based on *assumed* variations of displacements within members. Such displacement variations are

*This section can be omitted without loss of continuity.

referred to as the *displacement* or *interpolation functions.* A displacement function describes *the variation of a displacement component along the centroidal axis of a member in terms of its end displacements.*

Consider a prismatic member of a plane truss, subjected to end displacements u_1 through u_4, as shown in Fig. 3.6(a). Since the member displaces in both the x and y directions, we need to define two displacement functions,

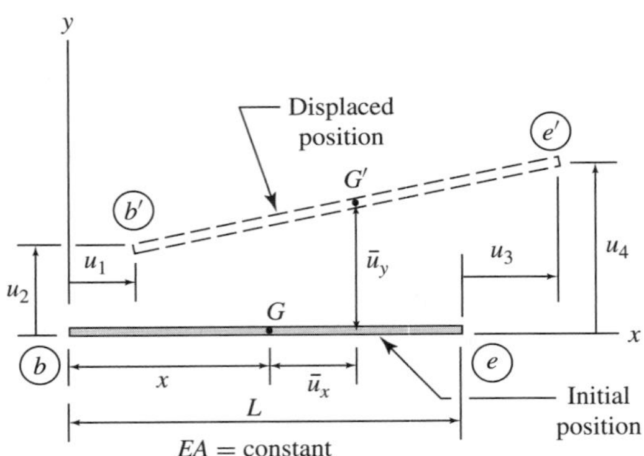

(a) Member Displacements in Local Coordinate System

=

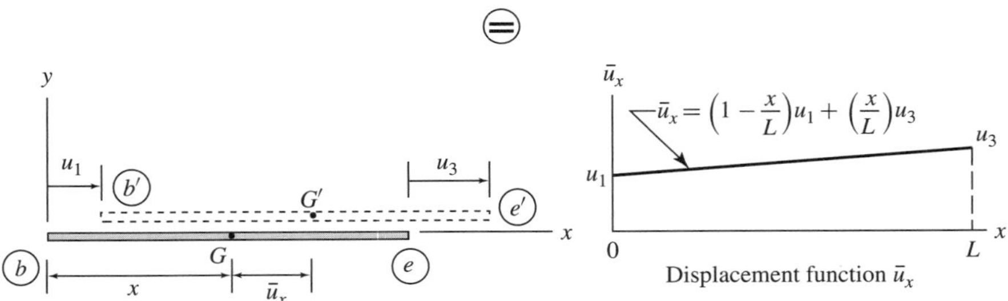

(b) Member Displacements in x Direction

+

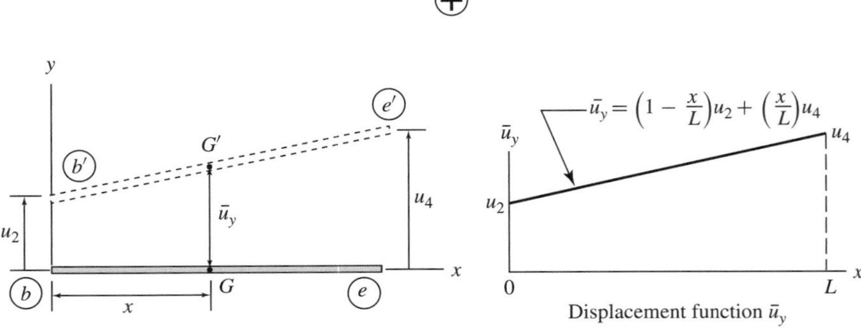

(c) Member Displacements in y Direction

Fig. 3.6

$\bar{u}_x$ and $\bar{u}_y$, for the displacements in the x and y directions, respectively. In Fig. 3.6(a), the displacement functions $\bar{u}_x$ and $\bar{u}_y$ are depicted as the displacements of an arbitrary point G located on the member's centroidal axis at a distance x from end b (left end).

The total displacement of the member (due to u_1 through u_4) can be decomposed into the displacements in the x and y directions, as shown in Figs. 3.6(b) and (c), respectively. Note that Fig. 3.6(b) shows the member subjected to the two end displacements, u_1 and u_3, in the x direction (with $u_2 = u_4 = 0$); Fig. 3.6(c) depicts the displacement of the member due to the two end displacements, u_2 and u_4, in the y direction (with $u_1 = u_3 = 0$).

The displacement functions assumed in the finite-element method are usually in the form of *complete* polynomials,

$$\bar{u}(x) = \sum_{i=0}^{n} a_i x^i \qquad \text{with } a_i \neq 0 \tag{3.30}$$

in which n is the degree of the polynomial. The polynomial used for a particular displacement function should be of such a degree that all of its coefficients can be evaluated by applying the available boundary conditions; that is,

$$n = n_{\text{bc}} - 1 \tag{3.31}$$

with n_{bc} = number of boundary conditions.

Thus, the displacement function $\bar{u}_x$ for the truss member (Fig. 3.6b) is assumed in the form of a linear polynomial as

$$\bar{u}_x = a_0 + a_1 x \tag{3.32}$$

in which a_0 and a_1 are the constants that can be determined by applying the following two boundary conditions:

$$\text{at } x = 0 \qquad \bar{u}_x = u_1$$

$$\text{at } x = L \qquad \bar{u}_x = u_3$$

By applying the first boundary condition—that is, by setting $x = 0$ and $\bar{u}_x = u_1$ in Eq. (3.32)—we obtain

$$a_0 = u_1 \tag{3.33}$$

Next, by using the second boundary condition—that is, by setting $x = L$ and $\bar{u}_x = u_3$—we obtain

$$u_3 = u_1 + a_1 L$$

from which follows

$$a_1 = \frac{u_3 - u_1}{L} \tag{3.34}$$

By substituting Eqs. (3.33) and (3.34) into Eq. (3.32), we obtain the expression for $\bar{u}_x$,

$$\bar{u}_x = u_1 + \left(\frac{u_3 - u_1}{L}\right) x$$

or

$$\bar{u}_x = \left(1 - \frac{x}{L}\right)u_1 + \left(\frac{x}{L}\right)u_3 \tag{3.35}$$

The displacement function $\bar{u}_y$, for the member displacement in the y direction (Fig. 3.6(c)), can be determined in a similar manner; that is, using a linear polynomial and evaluating its coefficients by applying the boundary conditions. The result is

$$\bar{u}_y = \left(1 - \frac{x}{L}\right)u_2 + \left(\frac{x}{L}\right)u_4 \tag{3.36}$$

The plots of the displacement functions $\bar{u}_x$ and $\bar{u}_y$ are shown in Figs. 3.6(b) and (c), respectively.

It is important to realize that the displacement functions as given by Eqs. (3.35) and (3.36) have been *assumed,* as is usually done in the finite-element method. There is no guarantee that an assumed displacement function defines the actual displacements of the member, except at its ends. In general, the displacement functions used in the finite-element method only approximate the actual displacements within members (or elements), because they represent approximate solutions of the underlying differential equations. For this reason, the finite-element method is generally considered to be an approximate method of analysis. However, for the prismatic members of framed structures, the displacement functions in the form of complete polynomials do happen to describe exactly the actual member displacements and, therefore, such functions yield exact member stiffness matrices for prismatic members.

From Fig. 3.6(c), we observe that the graph of the displacement function $\bar{u}_y$ exactly matches the displaced shape of the member's centroidal axis due to the end displacements u_2 and u_4. As this displaced shape defines the actual displacements in the y direction of all points along the member's length, we can conclude that the function $\bar{u}_y$, as given by Eq. (3.36), is exact.

To demonstrate that Eq. (3.35) describes the actual displacements in the x direction of all points along the truss member's centroidal axis, consider again the member subjected to end displacements, u_1 and u_3, in the x direction as shown in Fig. 3.7(a). Since the member is subjected to forces only at its ends, the axial force, Q_a, is constant throughout the member's length. Thus, the axial stress, σ, at point G of the member is

$$\sigma = \frac{Q_a}{A} \tag{3.37}$$

in which A represents the cross-sectional area of the member at point G. Note that the axial stress is distributed uniformly over the cross-sectional area A. By substituting the linear stress–strain relationship $\varepsilon = \sigma/E$ into Eq. (3.37), we obtain the strain at point G as

$$\varepsilon = \frac{Q_a}{EA} = \text{constant} = a_1 \tag{3.38}$$

in which a_1 is a constant. As this equation indicates, since the member is prismatic (i.e., EA = constant), the axial strain is constant throughout the member length.

To relate the strain ε to the displacement $\bar{u}_x$, we focus our attention on the differential element GH of length dx (Fig. 3.7(a)). The undeformed and deformed positions of the element are shown in Fig. 3.7(b), in which $\bar{u}_x$ and $\bar{u}_x + d\bar{u}_x$ denote, respectively, the displacements of the ends G and H of the element in the x direction. From this figure, we can see that

$$\begin{aligned}\text{deformed length of element} &= dx + (\bar{u}_x + d\bar{u}_x) - \bar{u}_x \\ &= dx + d\bar{u}_x\end{aligned}$$

Therefore, the strain in the element is given by

$$\varepsilon = \frac{\text{deformed length} - \text{initial length}}{\text{initial length}} = \frac{(dx + d\bar{u}_x) - dx}{dx}$$

or

$$\boxed{\varepsilon = \frac{d\bar{u}_x}{dx}} \tag{3.39}$$

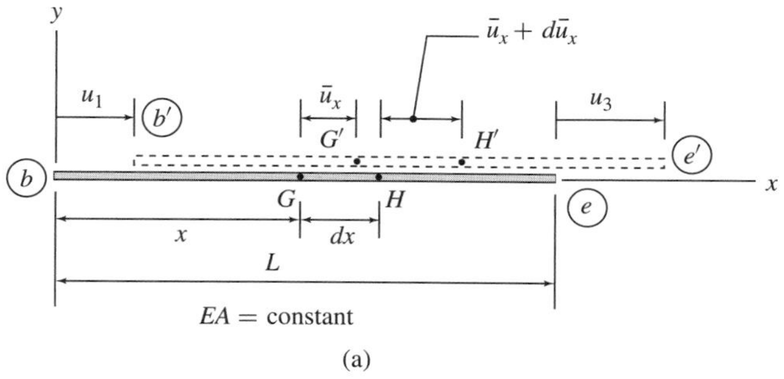

(a)

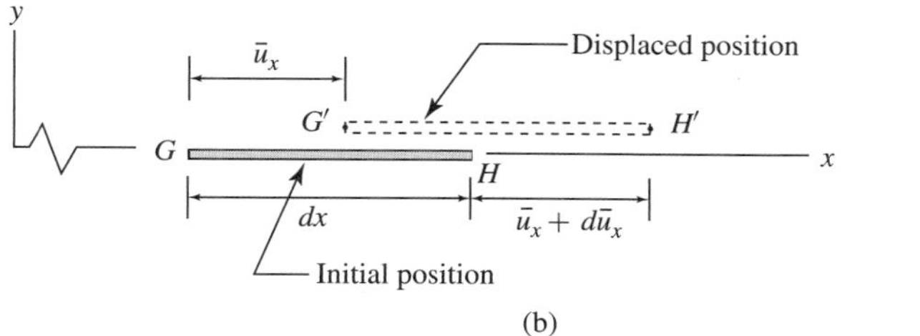

(b)

Fig. 3.7

By equating the two expressions for strain as given by Eqs. (3.38) and (3.39), we obtain

$$\frac{d\bar{u}_x}{dx} = a_1 \tag{3.40}$$

which can be rewritten as

$$d\bar{u}_x = a_1 dx \tag{3.41}$$

By integrating Eq. (3.41), we obtain

$$\bar{u}_x = a_1 x + a_0 \tag{3.42}$$

in which a_0 is the constant of integration. Note that Eq. (3.42), obtained by integrating the actual strain–displacement relationship, indicates that the linear polynomial form assumed for $\bar{u}_x$ in Eq. (3.32) was indeed correct. Furthermore, if we evaluate the two constants in Eq. (3.42) by applying the boundary conditions, we obtain an equation which is identical to Eq. (3.35), indicating that our assumed displacement function $\bar{u}_x$ (as given by Eq. (3.35)) does indeed describe the actual member displacements in the x direction.

Shape Functions

The displacement functions, as given by Eqs. (3.35) and (3.36), can be expressed alternatively as

$$\bar{u}_x = N_1 u_1 + N_3 u_3 \tag{3.43a}$$

$$\bar{u}_y = N_2 u_2 + N_4 u_4 \tag{3.43b}$$

with

$$N_1 = N_2 = 1 - \frac{x}{L} \tag{3.44a}$$

$$N_3 = N_4 = \frac{x}{L} \tag{3.44b}$$

in which N_i (with $i = 1{,}4$) are called the *shape functions*. The plots of the four shape functions for a plane truss member are given in Fig. 3.8. We can see from this figure that *a shape function N_i describes the displacement variation along a member's centroidal axis due to a unit value of the end displacement u_i, while all other end displacements are zero.*

Equations (3.43) can be written in matrix form as

$$\begin{bmatrix} \bar{u}_x \\ \bar{u}_y \end{bmatrix} = \begin{bmatrix} N_1 & 0 & N_3 & 0 \\ 0 & N_2 & 0 & N_4 \end{bmatrix} \begin{bmatrix} u_1 \\ u_2 \\ u_3 \\ u_4 \end{bmatrix} \tag{3.45}$$

or, symbolically, as

$$\boxed{\bar{\mathbf{u}} = \mathbf{N}\mathbf{u}} \tag{3.46}$$

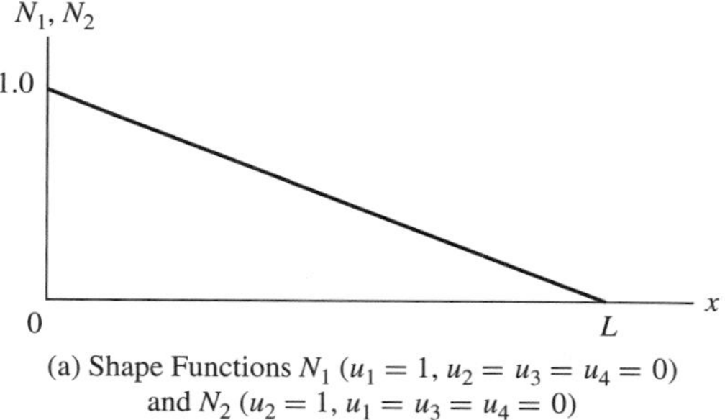

(a) Shape Functions N_1 ($u_1 = 1, u_2 = u_3 = u_4 = 0$) and N_2 ($u_2 = 1, u_1 = u_3 = u_4 = 0$)

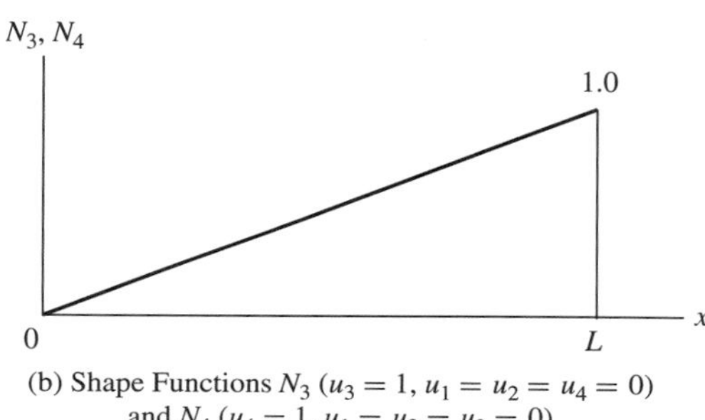

(b) Shape Functions N_3 ($u_3 = 1, u_1 = u_2 = u_4 = 0$) and N_4 ($u_4 = 1, u_1 = u_2 = u_3 = 0$)

Fig. 3.8 *Shape Functions for Plane Truss Member*

in which $\bar{\mathbf{u}}$ is the *member displacement function vector,* and **N** is called the *member shape function matrix.*

Strain–Displacement Relationship

As discussed previously, the relationship between the axial strain, ε, and the displacement, $\bar{u}_x$, is given by $\varepsilon = d\bar{u}_x/dx$ (see Eq. (3.39)). This strain–displacement relationship can be expressed in matrix form as

$$\varepsilon = \begin{bmatrix} \dfrac{d}{dx} & 0 \end{bmatrix} \begin{bmatrix} \bar{u}_x \\ \bar{u}_y \end{bmatrix} = \mathbf{D}\bar{\mathbf{u}} \tag{3.47}$$

in which **D** is known as the *differential operator matrix*. To relate the strain, ε, to the member end displacements, **u**, we substitute Eq. (3.46) into Eq. (3.47) to obtain

$$\varepsilon = \mathbf{D}(\mathbf{N}\mathbf{u}) \tag{3.48}$$

Since the end displacement vector **u** does not depend on x, it can be treated as a constant in the differentiation indicated by Eq. (3.48). In other words, the differentiation applies to **N**, but not to **u**. Thus, Eq. (3.48) can be rewritten as

$$\boxed{\varepsilon = (\mathbf{D}\mathbf{N})\mathbf{u} = \mathbf{B}\mathbf{u}} \tag{3.49}$$

in which, $\mathbf{B} = \mathbf{DN}$ is called the *member strain–displacement matrix*. To determine $\mathbf{B}$, we write

$$\mathbf{B} = \mathbf{DN} = \begin{bmatrix} \dfrac{d}{dx} & 0 \end{bmatrix} \begin{bmatrix} N_1 & 0 & N_3 & 0 \\ 0 & N_2 & 0 & N_4 \end{bmatrix}$$

By multiplying matrices $\mathbf{D}$ and $\mathbf{N}$,

$$\mathbf{B} = \begin{bmatrix} \dfrac{dN_1}{dx} & 0 & \dfrac{dN_3}{dx} & 0 \end{bmatrix}$$

Next, we substitute the expressions for N_1 and N_3 from Eqs. (3.44) into the preceding equation to obtain

$$\mathbf{B} = \begin{bmatrix} \dfrac{d}{dx}\left(1 - \dfrac{x}{L}\right) & 0 & \dfrac{d}{dx}\left(\dfrac{x}{L}\right) & 0 \end{bmatrix}$$

Finally, by performing the necessary differentiations, we determine the strain–displacement matrix $\mathbf{B}$:

$$\mathbf{B} = \begin{bmatrix} -\dfrac{1}{L} & 0 & \dfrac{1}{L} & 0 \end{bmatrix} = \frac{1}{L}\,[-1 \quad 0 \quad 1 \quad 0] \qquad \textbf{(3.50)}$$

Stress–Displacement Relationship

The relationship between member axial stress and member end displacements can now be established by substituting Eq. (3.49) into the stress–strain relationship, $\sigma = E\varepsilon$. Thus,

$$\boxed{\sigma = E\mathbf{Bu}} \qquad \textbf{(3.51)}$$

Member Stiffness Matrix, k

With both member strain and stress expressed in terms of end displacements, we can now establish the relationship between member end forces $\mathbf{Q}$ and end displacements $\mathbf{u}$, by applying the principle of virtual work for deformable bodies. Consider an arbitrary member of a plane truss that is in equilibrium under the action of end forces Q_1 through Q_4. Assume that the member is given small virtual end displacements δu_1 through δu_4, as shown in Fig. 3.9. The virtual

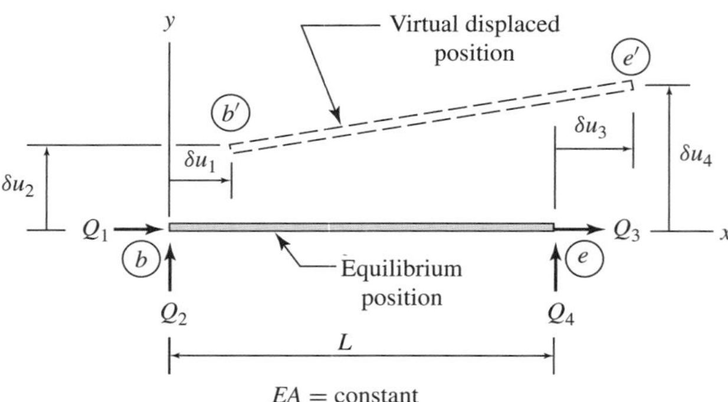

Fig. 3.9

external work done by the real member end forces Q_1 through Q_4 as they move through the corresponding virtual end displacements δu_1 through δu_4 is

$$\delta W_e = Q_1\,\delta u_1 + Q_2\,\delta u_2 + Q_3\,\delta u_3 + Q_4\,\delta u_4$$

which can be expressed in matrix form as

$$\delta W_e = [\,\delta u_1 \quad \delta u_2 \quad \delta u_3 \quad \delta u_4\,]\begin{bmatrix} Q_1 \\ Q_2 \\ Q_3 \\ Q_4 \end{bmatrix}$$

or

$$\delta W_e = \delta \mathbf{u}^T \mathbf{Q} \tag{3.52}$$

By substituting Eq. (3.52) into the expression for the principle of virtual work for deformable bodies, as given by Eq. (3.28), we obtain

$$\delta \mathbf{u}^T \mathbf{Q} = \int_V \delta \varepsilon^T \sigma\, dV \tag{3.53}$$

Recall that the right-hand side of Eq. (3.53) represents the virtual strain energy stored in the member. Substitution of Eqs. (3.49) and (3.51) into Eq. (3.53) yields

$$\delta \mathbf{u}^T \mathbf{Q} = \int_V (\mathbf{B}\,\delta\mathbf{u})^T E \mathbf{B}\, dV \mathbf{u}$$

Since $(\mathbf{B}\,\delta\mathbf{u})^T = \delta\mathbf{u}^T\mathbf{B}^T$, we can write the preceding equation as

$$\delta \mathbf{u}^T \mathbf{Q} = \delta \mathbf{u}^T \int_V \mathbf{B}^T E \mathbf{B}\, dV \mathbf{u}$$

or

$$\delta \mathbf{u}^T \left(\mathbf{Q} - \int_V \mathbf{B}^T E \mathbf{B}\, dV \mathbf{u}\right) = 0$$

Since $\delta\mathbf{u}^T$ can be arbitrarily chosen and is not zero, the quantity in the parentheses must be zero. Thus,

$$\mathbf{Q} = \left(\int_V \mathbf{B}^T E \mathbf{B}\, dV\right)\mathbf{u} = \mathbf{k}\mathbf{u} \tag{3.54}$$

in which

$$\boxed{\mathbf{k} = \int_V \mathbf{B}^T E \mathbf{B}\, dV} \tag{3.55}$$

is the member stiffness matrix in the local coordinate system. To determine the explicit form of **k**, we substitute Eq. (3.50) for **B** into Eq. (3.55) to obtain

$$\mathbf{k} = \frac{E}{L^2}\begin{bmatrix} -1 \\ 0 \\ 1 \\ 0 \end{bmatrix}[-1 \quad 0 \quad 1 \quad 0]\int_V dV = \frac{E}{L^2}\begin{bmatrix} 1 & 0 & -1 & 0 \\ 0 & 0 & 0 & 0 \\ -1 & 0 & 1 & 0 \\ 0 & 0 & 0 & 0 \end{bmatrix}\int_V dV$$

Since $\int_V dV = V = AL$, the member stiffness matrix, **k**, becomes

$$\mathbf{k} = \frac{EA}{L}\begin{bmatrix} 1 & 0 & -1 & 0 \\ 0 & 0 & 0 & 0 \\ -1 & 0 & 1 & 0 \\ 0 & 0 & 0 & 0 \end{bmatrix}$$

Note that the preceding expression for **k** is identical to that derived in Section 3.3 (Eq. (3.27)) using the direct equilibrium approach.

Symmetry of the Member Stiffness Matrix

The expression for the stiffness matrix **k** as given by Eq. (3.55) is general, in the sense that the stiffness matrices for members of other types of framed structures, as well as for elements of surface structures and solids, can also be expressed in the integral form of this equation. We can deduce from Eq. (3.55) that *for linear elastic structures, the member stiffness matrices are symmetric.*

Transposing both sides of Eq. (3.55), we write

$$\mathbf{k}^T = \int_V (\mathbf{B}^T\mathbf{E}\mathbf{B})^T dV$$

Now, recall from Section 2.3 that the transpose of a product of matrices equals the product of the transposed matrices in reverse order; that is, $(\mathbf{ABC})^T = \mathbf{C}^T\mathbf{B}^T\mathbf{A}^T$. Thus, the preceding equation becomes

$$\mathbf{k}^T = \int_V \mathbf{B}^T\mathbf{E}^T(\mathbf{B}^T)^T dV$$

For linear elastic structures, **E** is either a scalar (in the case of framed structures), or a symmetric matrix (for surface structures and solids). Therefore, $\mathbf{E}^T = \mathbf{E}$. Furthermore, by realizing that $(\mathbf{B}^T)^T = \mathbf{B}$, we can express the preceding equation as

$$\mathbf{k}^T = \int_V \mathbf{B}^T\mathbf{E}\mathbf{B}\,dV \tag{3.56}$$

Finally, a comparison of Eqs. (3.55) and (3.56) yields

$$\boxed{\mathbf{k}^T = \mathbf{k}} \tag{3.57}$$

which shows that **k** is a symmetric matrix.

3.5 COORDINATE TRANSFORMATIONS

When members of a structure are oriented in different directions, it becomes necessary to transform the stiffness relations for each member from its local coordinate system to a single global coordinate system selected for the entire structure. The member stiffness relations as expressed in the global coordinate system are then combined to establish the stiffness relations for the whole structure. In this section, we consider the transformation of member end forces and end displacements from local to global coordinate systems, and vice versa, for members of plane trusses. The transformation of the stiffness matrices is discussed in Section 3.6.

Transformation from Global to Local Coordinate Systems

Consider an arbitrary member m of a plane truss (Fig. 3.10(a)). As shown in this figure, the orientation of m relative to the global XY coordinate system is defined by an angle θ, measured counterclockwise from the positive direction of the global X axis to the positive direction of the local x axis. Recall that the stiffness matrix $\mathbf{k}$ derived in the preceding sections relates member end forces $\mathbf{Q}$ and end displacement $\mathbf{u}$ described with reference to the local xy coordinate system of the member, as shown in Fig. 3.10(b).

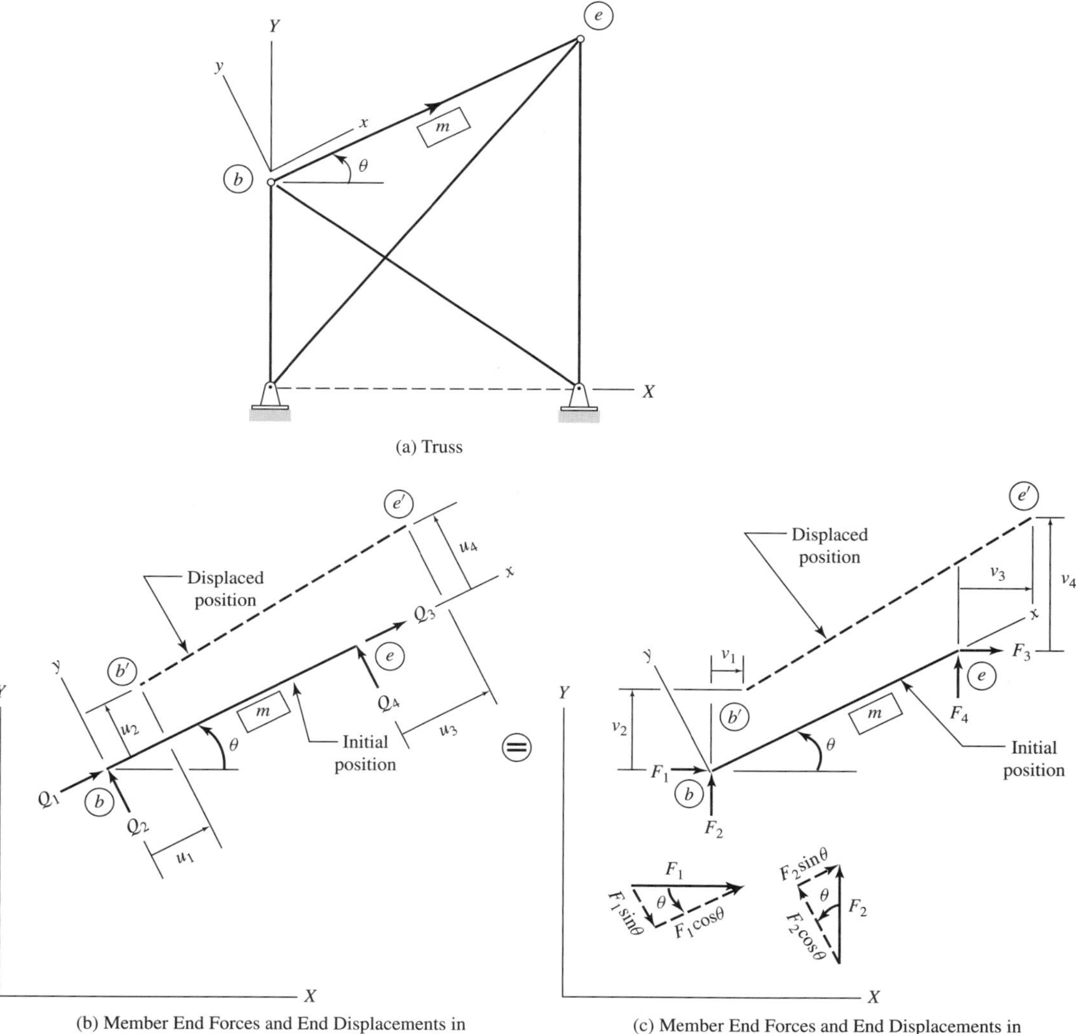

(b) Member End Forces and End Displacements in the Local Coordinate System

(c) Member End Forces and End Displacements in the Global Coordinate System

Fig. 3.10

Now, suppose that the member end forces and end displacements are specified with reference to the global XY coordinate system (Fig. 3.10(c)), and we wish to determine the equivalent system of end forces and end displacements, in the local xy coordinates, which has the same effect on m. As indicated in Fig. 3.10(c), the member end forces in the global coordinate system are denoted by F_1 through F_4, and the corresponding end displacements are denoted by v_1 through v_4. These global member end forces and end displacements are numbered beginning at member end b, with the force and translation in the X direction numbered first, followed by the force and translation in the Y direction. The forces and displacements at the member's opposite end e are then numbered in the same sequential order.

By comparing Figs. 3.10(b) and (c), we observe that at end b of m, the local force Q_1 must be equal to the algebraic sum of the components of the global forces F_1 and F_2 in the direction of the local x axis; that is,

$$Q_1 = F_1 \cos\theta + F_2 \sin\theta \tag{3.58a}$$

Similarly, the local force Q_2 equals the algebraic sum of the components of F_1 and F_2 in the direction of the local y axis. Thus,

$$Q_2 = -F_1 \sin\theta + F_2 \cos\theta \tag{3.58b}$$

By using a similar reasoning at end e, we express the local forces in terms of the global forces as

$$Q_3 = F_3 \cos\theta + F_4 \sin\theta \tag{3.58c}$$

$$Q_4 = -F_3 \sin\theta + F_4 \cos\theta \tag{3.58d}$$

Equations 3.58(a) through (d) can be written in matrix form as

$$\begin{bmatrix} Q_1 \\ Q_2 \\ Q_3 \\ Q_4 \end{bmatrix} = \begin{bmatrix} \cos\theta & \sin\theta & 0 & 0 \\ -\sin\theta & \cos\theta & 0 & 0 \\ 0 & 0 & \cos\theta & \sin\theta \\ 0 & 0 & -\sin\theta & \cos\theta \end{bmatrix} \begin{bmatrix} F_1 \\ F_2 \\ F_3 \\ F_4 \end{bmatrix} \tag{3.59}$$

or, symbolically, as

$$\mathbf{Q} = \mathbf{T}\mathbf{F} \tag{3.60}$$

with

$$\mathbf{T} = \begin{bmatrix} \cos\theta & \sin\theta & 0 & 0 \\ -\sin\theta & \cos\theta & 0 & 0 \\ 0 & 0 & \cos\theta & \sin\theta \\ 0 & 0 & -\sin\theta & \cos\theta \end{bmatrix} \tag{3.61}$$

in which $\mathbf{T}$ is referred to as the *transformation matrix*. The direction cosines of the member, necessary for the evaluation of $\mathbf{T}$, can be conveniently determined by using the following relationships:

$$\cos\theta = \frac{X_e - X_b}{L} = \frac{X_e - X_b}{\sqrt{(X_e - X_b)^2 + (Y_e - Y_b)^2}} \tag{3.62a}$$

$$\sin\theta = \frac{Y_e - Y_b}{L} = \frac{Y_e - Y_b}{\sqrt{(X_e - X_b)^2 + (Y_e - Y_b)^2}} \tag{3.62b}$$

in which X_b and Y_b denote the global coordinates of the beginning joint b for the member, and X_e and Y_e represent the global coordinates of the end joint e.

The member end displacements, like end forces, are vectors, which are defined in the same directions as the corresponding forces. Therefore, the transformation matrix **T** (Eq. (3.61)), developed for transforming end forces, can also be used to transform member end displacements from the global to local coordinate system; that is,

$$\boxed{\mathbf{u} = \mathbf{T}\mathbf{v}} \tag{3.63}$$

Transformation from Local to Global Coordinate Systems

Next, let us consider the transformation of member end forces and end displacements from local to global coordinate systems. A comparison of Figs. 3.10(b) and (c) indicates that at end b of m, the global force F_1 must be equal to the algebraic sum of the components of the local forces Q_1 and Q_2 in the direction of the global X axis; that is,

$$F_1 = Q_1 \cos\theta - Q_2 \sin\theta \tag{3.64a}$$

In a similar manner, the global force F_2 equals the algebraic sum of the components of Q_1 and Q_2 in the direction of the global Y axis. Thus,

$$F_2 = Q_1 \sin\theta + Q_2 \cos\theta \tag{3.64b}$$

By using a similar reasoning at end e, we express the global forces in terms of the local forces as

$$F_3 = Q_3 \cos\theta - Q_4 \sin\theta \tag{3.64c}$$

$$F_4 = Q_3 \sin\theta + Q_4 \cos\theta \tag{3.64d}$$

We can write Eqs. 3.64(a) through (d) in matrix form as

$$\begin{bmatrix} F_1 \\ F_2 \\ F_3 \\ F_4 \end{bmatrix} = \begin{bmatrix} \cos\theta & -\sin\theta & 0 & 0 \\ \sin\theta & \cos\theta & 0 & 0 \\ 0 & 0 & \cos\theta & -\sin\theta \\ 0 & 0 & \sin\theta & \cos\theta \end{bmatrix} \begin{bmatrix} Q_1 \\ Q_2 \\ Q_3 \\ Q_4 \end{bmatrix} \tag{3.65}$$

By comparing Eqs. (3.59) and (3.65), we observe that the transformation matrix in Eq. (3.65), which transforms the forces from the local to the global coordinate system, is the transpose of the transformation matrix **T** in Eq. (3.59), which transforms the forces from the global to the local coordinate system. Therefore, Eq. (3.65) can be expressed as

$$\boxed{\mathbf{F} = \mathbf{T}^T \mathbf{Q}} \tag{3.66}$$

Furthermore, a comparison of Eqs. (3.60) and (3.66) indicates that the inverse of the transformation matrix must be equal to its transpose; that is,

$$\mathbf{T}^{-1} = \mathbf{T}^T \tag{3.67}$$

which indicates that the transformation matrix **T** is orthogonal.

As discussed previously, because the member end displacements are also vectors, which are defined in the same directions as the corresponding forces, the matrix $\mathbf{T}^T$ also defines the transformation of member end displacements from the local to the global coordinate system; that is,

$$\mathbf{v} = \mathbf{T}^T\mathbf{u} \tag{3.68}$$

EXAMPLE 3.4 Determine the transformation matrices for the members of the truss shown in Fig. 3.11.

SOLUTION **Member 1** From Fig. 3.11, we can see that joint 1 is the beginning joint and joint 2 is the end joint for member 1. By applying Eqs. (3.62), we determine

$$\cos\theta = \frac{X_2 - X_1}{L} = \frac{6-0}{6} = 1$$

$$\sin\theta = \frac{Y_2 - Y_1}{L} = \frac{0-0}{6} = 0$$

The transformation matrix for member 1 can now be obtained by using Eq. (3.61)

$$\mathbf{T}_1 = \begin{bmatrix} 1 & 0 & 0 & 0 \\ 0 & 1 & 0 & 0 \\ 0 & 0 & 1 & 0 \\ 0 & 0 & 0 & 1 \end{bmatrix} = \mathbf{I}$$ **Ans**

As the preceding result indicates, for any member with the positive directions of its local x and y axes oriented in the positive directions of the global X and Y axes, respectively, the transformation matrix always equals a unit matrix, $\mathbf{I}$.

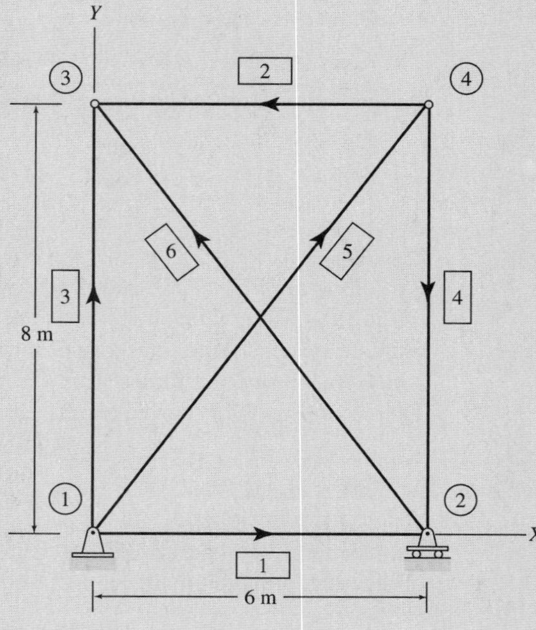

Fig. 3.11

Member 2

$$\cos\theta = \frac{X_3 - X_4}{L} = \frac{0-6}{6} = -1$$

$$\sin\theta = \frac{Y_3 - Y_4}{L} = \frac{8-8}{6} = 0$$

Thus, from Eq. (3.61)

$$\mathbf{T}_2 = \begin{bmatrix} -1 & 0 & 0 & 0 \\ 0 & -1 & 0 & 0 \\ 0 & 0 & -1 & 0 \\ 0 & 0 & 0 & -1 \end{bmatrix}$$ **Ans**

Member 3

$$\cos\theta = \frac{X_3 - X_1}{L} = \frac{0-0}{8} = 0$$

$$\sin\theta = \frac{Y_3 - Y_1}{L} = \frac{8-0}{8} = 1$$

Thus,

$$\mathbf{T}_3 = \begin{bmatrix} 0 & 1 & 0 & 0 \\ -1 & 0 & 0 & 0 \\ 0 & 0 & 0 & 1 \\ 0 & 0 & -1 & 0 \end{bmatrix}$$ **Ans**

Member 4

$$\cos\theta = \frac{X_2 - X_4}{L} = \frac{6-6}{8} = 0$$

$$\sin\theta = \frac{Y_2 - Y_4}{L} = \frac{0-8}{8} = -1$$

Thus,

$$\mathbf{T}_4 = \begin{bmatrix} 0 & -1 & 0 & 0 \\ 1 & 0 & 0 & 0 \\ 0 & 0 & 0 & -1 \\ 0 & 0 & 1 & 0 \end{bmatrix}$$ **Ans**

Member 5

$$L = \sqrt{(X_4 - X_1)^2 + (Y_4 - Y_1)^2} = \sqrt{(6-0)^2 + (8-0)^2} = 10 \text{ m}$$

$$\cos\theta = \frac{X_4 - X_1}{L} = \frac{6-0}{10} = 0.6$$

$$\sin\theta = \frac{Y_4 - Y_1}{L} = \frac{8-0}{10} = 0.8$$

$$\mathbf{T}_5 = \begin{bmatrix} 0.6 & 0.8 & 0 & 0 \\ -0.8 & 0.6 & 0 & 0 \\ 0 & 0 & 0.6 & 0.8 \\ 0 & 0 & -0.8 & 0.6 \end{bmatrix}$$ **Ans**

Member 6

$$L = \sqrt{(X_3 - X_2)^2 + (Y_3 - Y_2)^2} = \sqrt{(0-6)^2 + (8-0)^2} = 10 \text{ m}$$

$$\cos\theta = \frac{X_3 - X_2}{L} = \frac{0-6}{10} = -0.6$$

$$\sin\theta = \frac{Y_3 - Y_2}{L} = \frac{8-0}{10} = 0.8$$

$$\mathbf{T}_6 = \begin{bmatrix} -0.6 & 0.8 & 0 & 0 \\ -0.8 & -0.6 & 0 & 0 \\ 0 & 0 & -0.6 & 0.8 \\ 0 & 0 & -0.8 & -0.6 \end{bmatrix}$$ **Ans**

EXAMPLE 3.5 For the truss shown in Fig. 3.12(a), the end displacements of member 2 in the global coordinate system are (Fig. 3.12(b)):

$$\mathbf{v}_2 = \begin{bmatrix} 0.75 \\ 0 \\ 1.5 \\ -2 \end{bmatrix} \text{ in.}$$

Calculate the end forces for this member in the global coordinate system. Is the member in equilibrium under these forces?

SOLUTION *Member Stiffness Matrix in the Local Coordinate System:* $E = 10{,}000$ ksi, $A = 9$ in.^{2}, $L = \sqrt{(9)^2 + (12)^2} = 15$ ft $= 180$ in.

$$\frac{EA}{L} = \frac{10{,}000(9)}{180} = 500 \text{ k/in.}$$

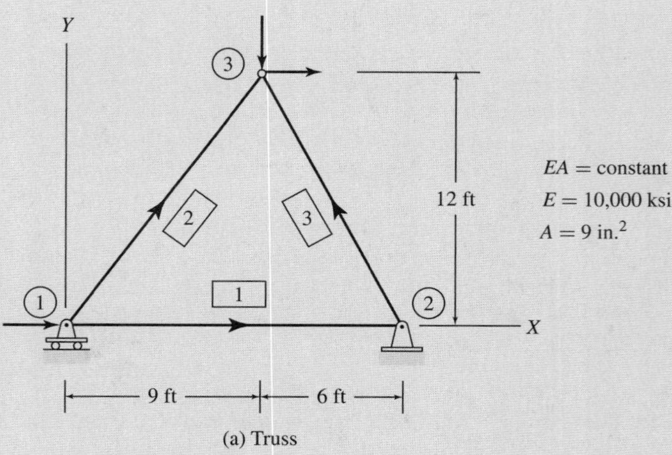

(a) Truss

Fig. 3.12

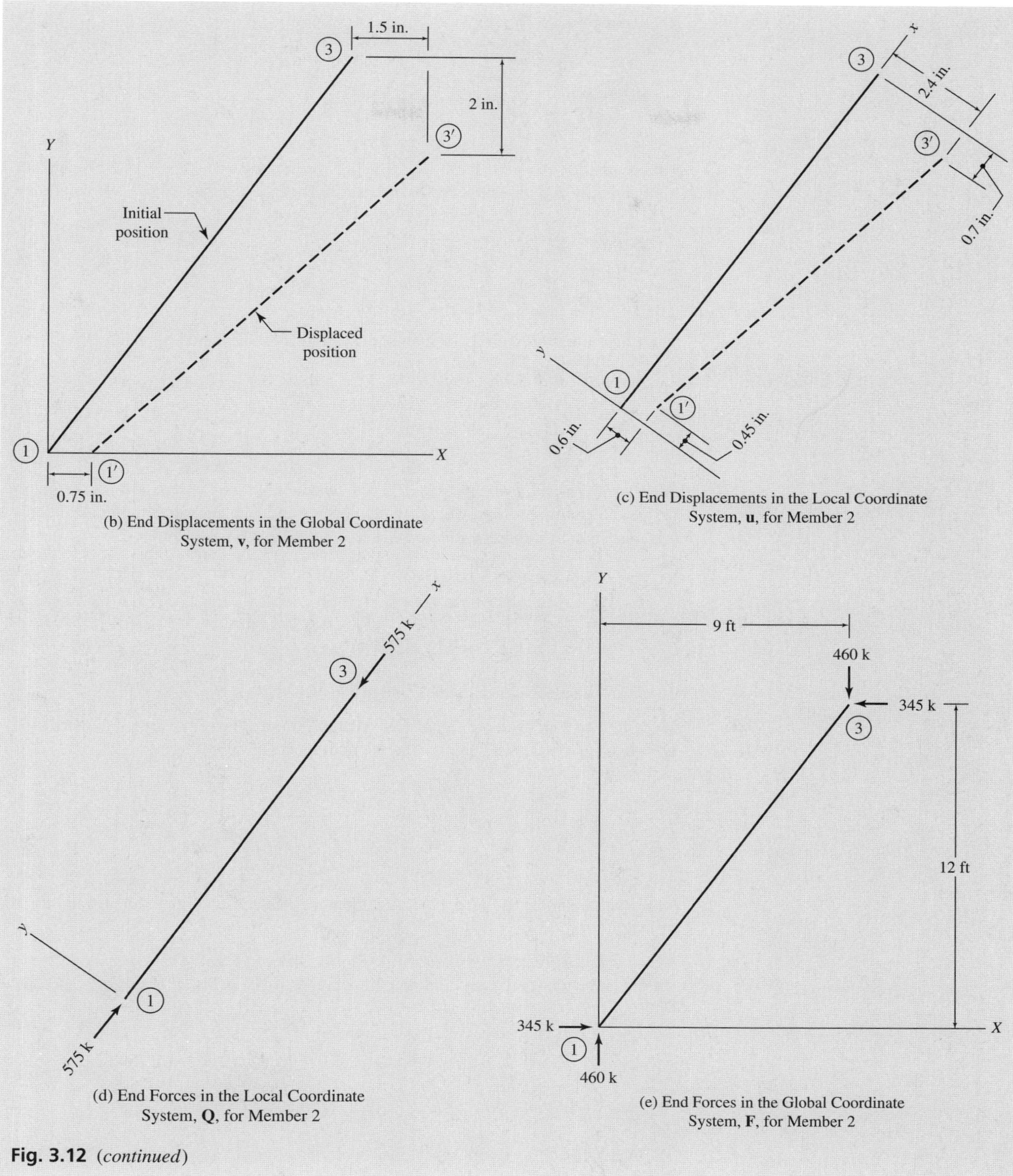

(b) End Displacements in the Global Coordinate System, **v**, for Member 2

(c) End Displacements in the Local Coordinate System, **u**, for Member 2

(d) End Forces in the Local Coordinate System, **Q**, for Member 2

(e) End Forces in the Global Coordinate System, **F**, for Member 2

Fig. 3.12 *(continued)*

Thus, from Eq. (3.27),

$$\mathbf{k}_2 = 500 \begin{bmatrix} 1 & 0 & -1 & 0 \\ 0 & 0 & 0 & 0 \\ -1 & 0 & 1 & 0 \\ 0 & 0 & 0 & 0 \end{bmatrix} \text{ k/in.}$$

Transformation Matrix: From Fig. 3.12(a), we can see that joint 1 is the beginning joint and joint 3 is the end joint for member 2. By applying Eqs. (3.62), we determine

$$\cos\theta = \frac{X_3 - X_1}{L} = \frac{9 - 0}{15} = 0.6$$

$$\sin\theta = \frac{Y_3 - Y_1}{L} = \frac{12 - 0}{15} = 0.8$$

The transformation matrix for member 2 can now be evaluated by using Eq. (3.61):

$$\mathbf{T}_2 = \begin{bmatrix} 0.6 & 0.8 & 0 & 0 \\ -0.8 & 0.6 & 0 & 0 \\ 0 & 0 & 0.6 & 0.8 \\ 0 & 0 & -0.8 & 0.6 \end{bmatrix}$$

Member End Displacements in the Local Coordinate System: To determine the member global end forces, first we calculate member end displacements in the local coordinate system by using the relationship $\mathbf{u} = \mathbf{Tv}$ (Eq. (3.63)). Thus,

$$\mathbf{u}_2 = \begin{bmatrix} u_1 \\ u_2 \\ u_3 \\ u_4 \end{bmatrix} = \begin{bmatrix} 0.6 & 0.8 & 0 & 0 \\ -0.8 & 0.6 & 0 & 0 \\ 0 & 0 & 0.6 & 0.8 \\ 0 & 0 & -0.8 & 0.6 \end{bmatrix} \begin{bmatrix} 0.75 \\ 0 \\ 1.5 \\ -2 \end{bmatrix} = \begin{bmatrix} 0.45 \\ -0.6 \\ -0.7 \\ -2.4 \end{bmatrix} \text{ in.}$$

These end displacements are depicted in Fig. 3.12(c).

Member End Forces in the Local Coordinate System: Next, by using the expression $\mathbf{Q} = \mathbf{ku}$ (Eq. (3.7)), we compute the member local end forces as

$$\mathbf{Q}_2 = \begin{bmatrix} Q_1 \\ Q_2 \\ Q_3 \\ Q_4 \end{bmatrix} = 500 \begin{bmatrix} 1 & 0 & -1 & 0 \\ 0 & 0 & 0 & 0 \\ -1 & 0 & 1 & 0 \\ 0 & 0 & 0 & 0 \end{bmatrix} \begin{bmatrix} 0.45 \\ -0.6 \\ -0.7 \\ -2.4 \end{bmatrix} = \begin{bmatrix} 575 \\ 0 \\ -575 \\ 0 \end{bmatrix} \text{ k}$$

Note that, as shown in Fig. 3.12(d), the member is in compression with an axial force of magnitude 575 k.

Member End Forces in the Global Coordinate System: Finally, we determine the desired member end forces by applying the relationship $\mathbf{F} = \mathbf{T}^T\mathbf{Q}$ as given in Eq. (3.66). Thus,

$$\mathbf{F}_2 = \begin{bmatrix} F_1 \\ F_2 \\ F_3 \\ F_4 \end{bmatrix} = \begin{bmatrix} 0.6 & -0.8 & 0 & 0 \\ 0.8 & 0.6 & 0 & 0 \\ 0 & 0 & 0.6 & -0.8 \\ 0 & 0 & 0.8 & 0.6 \end{bmatrix} \begin{bmatrix} 575 \\ 0 \\ -575 \\ 0 \end{bmatrix} = \begin{bmatrix} 345 \\ 460 \\ -345 \\ -460 \end{bmatrix} \text{ k}$$ **Ans**

The member end forces in the global coordinate system are shown in Fig. 3.12(e).

Equilibrium Check: To check whether or not the member is in equilibrium, we apply the three equations of equilibrium, as follows.

$$+\rightarrow \sum F_X = 0 \qquad 345 - 345 = 0$$ **Checks**

$$+\uparrow \sum F_Y = 0 \qquad 460 - 460 = 0$$ **Checks**

$$+\circlearrowleft \sum M_{①} = 0 \qquad 345(12) - 460(9) = 0$$ **Checks**

Therefore, the member is in equilibrium. **Ans**

3.6 MEMBER STIFFNESS RELATIONS IN THE GLOBAL COORDINATE SYSTEM

By using the member stiffness relations in the local coordinate system from Sections 3.3 and 3.4, and the transformation relations from Section 3.5, we can now establish the stiffness relations for members in the global coordinate system.

First, we substitute the local stiffness relations $\mathbf{Q} = \mathbf{ku}$ (Eq. (3.7)) into the force transformation relations $\mathbf{F} = \mathbf{T}^T\mathbf{Q}$ (Eq. (3.66)) to obtain

$$\mathbf{F} = \mathbf{T}^T\mathbf{Q} = \mathbf{T}^T\mathbf{ku} \tag{3.69}$$

Then, by substituting the displacement transformation relations $\mathbf{u} = \mathbf{Tv}$ (Eq. (3.63)) into Eq. (3.69), we determine that the desired relationship between the member end forces $\mathbf{F}$ and end displacements $\mathbf{v}$, in the global coordinate system, is

$$\mathbf{F} = \mathbf{T}^T\mathbf{kTv} \tag{3.70}$$

Equation (3.70) can be conveniently expressed as

$$\boxed{\mathbf{F} = \mathbf{Kv}} \tag{3.71}$$

with

$$\boxed{\mathbf{K} = \mathbf{T}^T\mathbf{kT}} \tag{3.72}$$

in which the matrix $\mathbf{K}$ is called the *member stiffness matrix in the global coordinate system.* The explicit form of $\mathbf{K}$ can be determined by substituting Eqs. (3.27) and (3.61) into Eq. (3.72), as

$$\mathbf{K} = \begin{bmatrix} \cos\theta & -\sin\theta & 0 & 0 \\ \sin\theta & \cos\theta & 0 & 0 \\ 0 & 0 & \cos\theta & -\sin\theta \\ 0 & 0 & \sin\theta & \cos\theta \end{bmatrix} \frac{EA}{L} \begin{bmatrix} 1 & 0 & -1 & 0 \\ 0 & 0 & 0 & 0 \\ -1 & 0 & 1 & 0 \\ 0 & 0 & 0 & 0 \end{bmatrix} \begin{bmatrix} \cos\theta & \sin\theta & 0 & 0 \\ -\sin\theta & \cos\theta & 0 & 0 \\ 0 & 0 & \cos\theta & \sin\theta \\ 0 & 0 & -\sin\theta & \cos\theta \end{bmatrix}$$

Performing the matrix multiplications, we obtain

$$\mathbf{K} = \frac{EA}{L}\begin{bmatrix} \cos^2\theta & \cos\theta\sin\theta & -\cos^2\theta & -\cos\theta\sin\theta \\ \cos\theta\sin\theta & \sin^2\theta & -\cos\theta\sin\theta & -\sin^2\theta \\ -\cos^2\theta & -\cos\theta\sin\theta & \cos^2\theta & \cos\theta\sin\theta \\ -\cos\theta\sin\theta & -\sin^2\theta & \cos\theta\sin\theta & \sin^2\theta \end{bmatrix} \tag{3.73}$$

Note that, like the member local stiffness matrix, *the member global stiffness matrix,* **K**, *is symmetric*. The physical interpretation of the member global stiffness matrix **K** is similar to that of the member local stiffness matrix; that is, *a stiffness coefficient K_{ij} represents the force at the location and in the direction of F_i required, along with other end forces, to cause a unit value of displacement v_j, while all other end displacements are zero*. Thus, the *j*th column of matrix **K** consists of the end forces in the global coordinate system required to cause a unit value of the end displacement v_j, while all other end displacements are zero.

As the preceding interpretation indicates, the member global stiffness matrix **K** can alternately be derived by subjecting an inclined truss member, separately, to unit values of each of the four end displacements in the global coordinate system as shown in Fig. 3.13, and by evaluating the end forces in the global coordinate system required to cause the individual unit displacements. Let us verify the expression for **K** given in Eq. (3.73), using this alternative approach. Consider a prismatic plane truss member inclined at an angle θ relative to the global X axis, as shown in Fig. 3.13(a). When end b of the member is given a unit displacement $v_1 = 1$, while the other end displacements are held at zero, the member shortens and an axial compressive force develops in it. In the case of small displacements (as assumed herein), the axial deformation u_a of the member due to v_1 is equal to the component of $v_1 = 1$ in the undeformed direction of the member; that is (Fig. 3.13(a)),

$$u_a = v_1 \cos\theta = 1 \cos\theta = \cos\theta$$

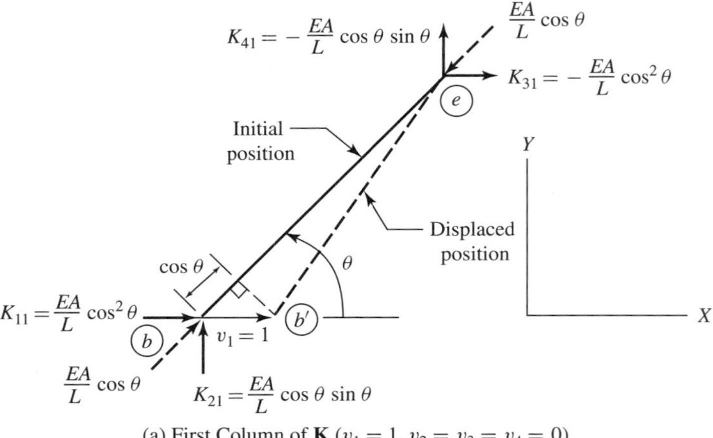

(a) First Column of **K** ($v_1 = 1$, $v_2 = v_3 = v_4 = 0$)

Fig. 3.13

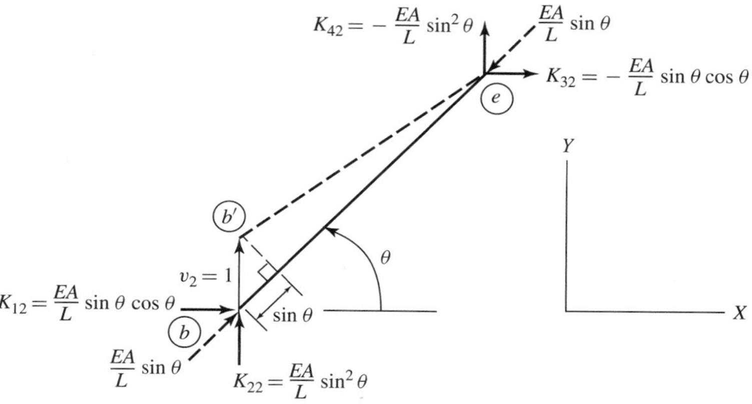

(b) Second Column of **K** ($v_2 = 1$, $v_1 = v_3 = v_4 = 0$)

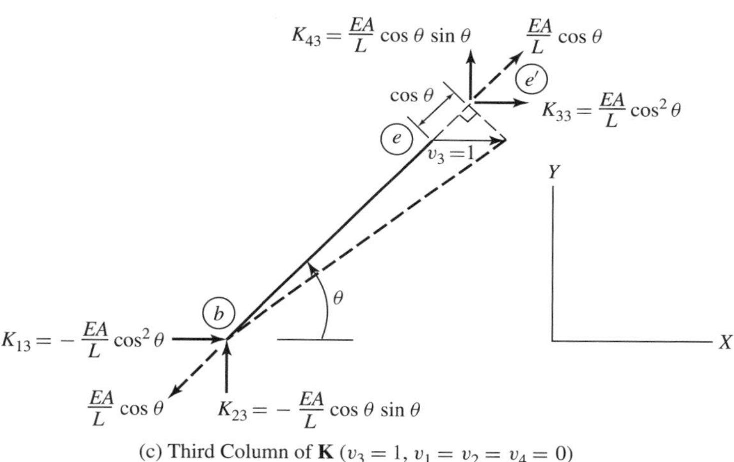

(c) Third Column of **K** ($v_3 = 1$, $v_1 = v_2 = v_4 = 0$)

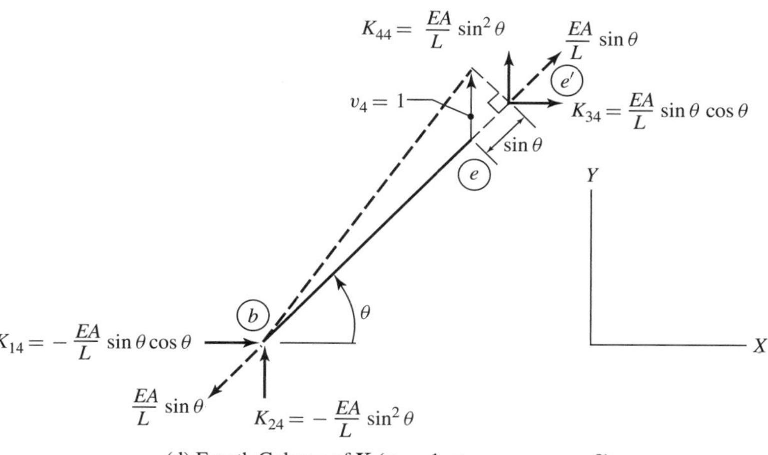

(d) Fourth Column of **K** ($v_4 = 1$, $v_1 = v_2 = v_3 = 0$)

Fig. 3.13 *(continued)*

The axial compressive force Q_a in the member caused by the axial deformation u_a can be expressed as

$$Q_a = \left(\frac{EA}{L}\right) u_a = \left(\frac{EA}{L}\right) \cos\theta$$

From Fig. 3.13(a), we can see that the stiffness coefficients must be equal to the components of the member axial force Q_a in the directions of the global X and Y axes. Thus, at end b,

$$K_{11} = Q_a \cos\theta = \left(\frac{EA}{L}\right) \cos^2\theta \qquad \textbf{(3.74a)}$$

$$K_{21} = Q_a \sin\theta = \left(\frac{EA}{L}\right) \cos\theta \sin\theta \qquad \textbf{(3.74b)}$$

Similarly, at end e,

$$K_{31} = -Q_a \cos\theta = -\left(\frac{EA}{L}\right) \cos^2\theta \qquad \textbf{(3.74c)}$$

$$K_{41} = -Q_a \sin\theta = -\left(\frac{EA}{L}\right) \cos\theta \sin\theta \qquad \textbf{(3.74d)}$$

in which the negative signs for K_{31} and K_{41} indicate that these forces act in the negative directions of the X and Y axes, respectively. Note that the member must be in equilibrium under the action of the four end forces, K_{11}, K_{21}, K_{31}, and K_{41}. Also, note that the expressions for these stiffness coefficients (Eqs. (3.74)) are identical to those given in the first column of the **K** matrix in Eq. (3.73).

The stiffness coefficients corresponding to the unit values of the remaining end displacements v_2, v_3, and v_4 can be evaluated in a similar manner, and are given in Figs. 3.13 (b) through (d), respectively. As expected, these stiffness coefficients are the same as those previously obtained by transforming the stiffness relations from the local to the global coordinate system (Eq. (3.73)).

EXAMPLE 3.6 Solve Example 3.5 by using the member stiffness relationship in the global coordinate system, $\mathbf{F} = \mathbf{K}\mathbf{v}$.

SOLUTION *Member Stiffness Matrix in the Global Coordinate System:* It was shown in Example 3.5 that for member 2,

$$\frac{EA}{L} = 500 \text{ k/in.}, \cos\theta = 0.6, \sin\theta = 0.8$$

Thus, from Eq. (3.73):

$$\mathbf{K}_2 = \begin{bmatrix} 180 & 240 & -180 & -240 \\ 240 & 320 & -240 & -320 \\ -180 & -240 & 180 & 240 \\ -240 & -320 & 240 & 320 \end{bmatrix} \text{ k/in.}$$

Member End Forces in the Global Coordinate System: By applying the relationship $\mathbf{F} = \mathbf{Kv}$ as given in Eq. (3.71), we obtain

$$\mathbf{F} = \begin{bmatrix} F_1 \\ F_2 \\ F_3 \\ F_4 \end{bmatrix} = \begin{bmatrix} 180 & 240 & -180 & -240 \\ 240 & 320 & -240 & -320 \\ -180 & -240 & 180 & 240 \\ -240 & -320 & 240 & 320 \end{bmatrix} \begin{bmatrix} 0.75 \\ 0 \\ 1.5 \\ -2 \end{bmatrix} = \begin{bmatrix} 345 \\ 460 \\ -345 \\ -460 \end{bmatrix} \text{k}$$

Ans

Equilibrium check: See Example 3.5.

3.7 STRUCTURE STIFFNESS RELATIONS

Having determined the member force–displacement relationships in the global coordinate system, we are now ready to establish the stiffness relations for the entire structure. The structure stiffness relations express the external loads **P** acting at the joints of the structure, as functions of the joint displacements **d**. Such relationships can be established as follows:

1. The joint loads **P** are first expressed in terms of the member end forces in the global coordinate system, **F**, by applying the equations of equilibrium for the joints of the structure.
2. The joint displacements **d** are then related to the member end displacements in the global coordinate system, **v**, by using the compatibility conditions that the displacements of the member ends must be the same as the corresponding joint displacements.
3. Next, the compatibility equations are substituted into the member force–displacement relations, $\mathbf{F} = \mathbf{Kv}$, to express the member global end forces **F** in terms of the joint displacements **d**. The **F**–**d** relations thus obtained are then substituted into the joint equilibrium equations to establish the desired structure stiffness relationships between the joint loads **P** and the joint displacements **d**.

Consider, for example, an arbitrary plane truss as shown in Fig. 3.14(a). The analytical model of the truss is given in Fig. 3.14(b), which indicates that the structure has two degrees of freedom, d_1 and d_2. The joint loads corresponding to these degrees of freedom are designated P_1 and P_2, respectively. The global end forces **F** and end displacements **v** for the three members of the truss are shown in Fig. 3.14(c), in which the superscript (i) denotes the member number. Note that for members 1 and 3, the bottom joints (i.e., joints 2 and 4, respectively) have been defined as the beginning joints; whereas, for member 2, the top joint 1 is the beginning joint. As stated previously, our objective is to express the joint loads **P** as functions of the joint displacement **d**.

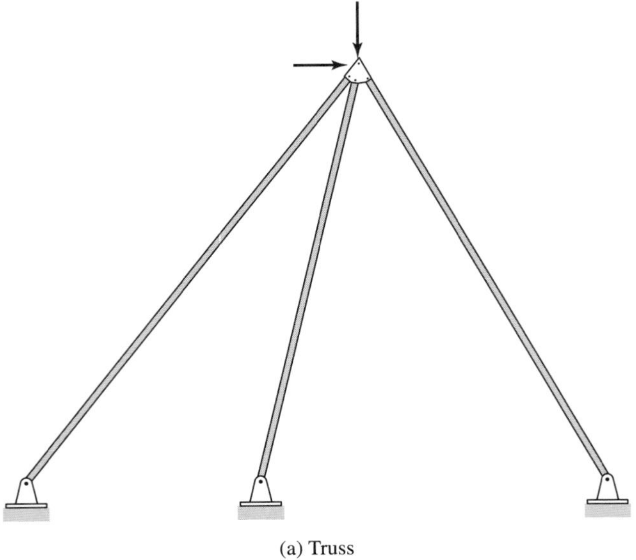

(a) Truss

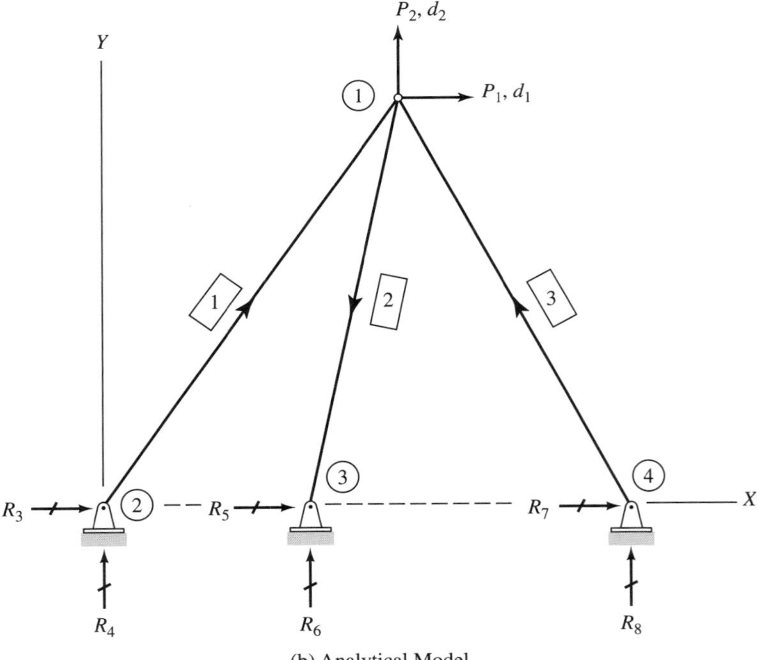

(b) Analytical Model

Fig. 3.14

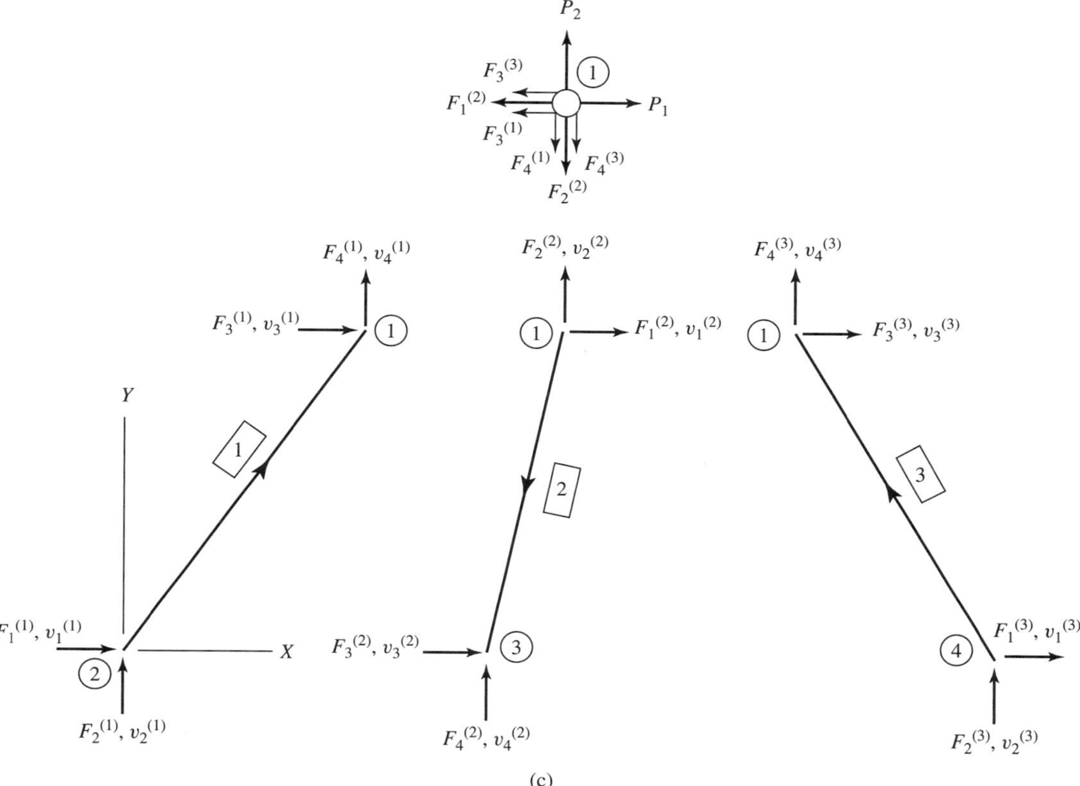

Fig. 3.14 *(continued)*

Equilibrium Equations

To relate the external joint loads **P** to the internal member end forces **F**, we apply the two equations of equilibrium, $\sum F_X = 0$ and $\sum F_Y = 0$, to the free body of joint 1 shown in Fig. 3.14(c). This yields the equilibrium equations,

$$P_1 = F_3^{(1)} + F_1^{(2)} + F_3^{(3)} \qquad \textbf{(3.75a)}$$

$$P_2 = F_4^{(1)} + F_2^{(2)} + F_4^{(3)} \qquad \textbf{(3.75b)}$$

Compatibility Equations

By comparing Figs. 3.14(b) and (c), we observe that since the lower end 2 of member 1 is connected to the hinged support 2, which cannot translate in any direction, the two displacements of end 2 of the member must be zero. Similarly, since end 1 of this member is connected to joint 1, the displacements of end 1 must be the same as the displacements of joint 1. Thus, the compatibility conditions for member 1 are

$$v_1^{(1)} = v_2^{(1)} = 0 \qquad v_3^{(1)} = d_1 \qquad v_4^{(1)} = d_2 \qquad \textbf{(3.76)}$$

In a similar manner, the compatibility conditions for members 2 and 3, respectively, are found to be

$$v_1^{(2)} = d_1 \qquad v_2^{(2)} = d_2 \qquad v_3^{(2)} = v_4^{(2)} = 0 \tag{3.77}$$

$$v_1^{(3)} = v_2^{(3)} = 0 \qquad v_3^{(3)} = d_1 \qquad v_4^{(3)} = d_2 \tag{3.78}$$

Member Stiffness Relations

Of the two types of relationships established thus far, the equilibrium equations (Eqs. (3.75)) express joint loads in terms of member end forces, whereas the compatibility equations (Eqs. (3.76) through (3.78)) relate joint displacements to member end displacements. Now, we will link the two types of relationships by employing the member stiffness relationship in the global coordinate system derived in the preceding section.

We can write the member global stiffness relation $\mathbf{F} = \mathbf{Kv}$ (Eq. (3.71)) in expanded form for member 1 as

$$\begin{bmatrix} F_1^{(1)} \\ F_2^{(1)} \\ F_3^{(1)} \\ F_4^{(1)} \end{bmatrix} = \begin{bmatrix} K_{11}^{(1)} & K_{12}^{(1)} & K_{13}^{(1)} & K_{14}^{(1)} \\ K_{21}^{(1)} & K_{22}^{(1)} & K_{23}^{(1)} & K_{24}^{(1)} \\ K_{31}^{(1)} & K_{32}^{(1)} & K_{33}^{(1)} & K_{34}^{(1)} \\ K_{41}^{(1)} & K_{42}^{(1)} & K_{43}^{(1)} & K_{44}^{(1)} \end{bmatrix} \begin{bmatrix} v_1^{(1)} \\ v_2^{(1)} \\ v_3^{(1)} \\ v_4^{(1)} \end{bmatrix} \tag{3.79}$$

from which we obtain the expressions for forces at end 1 of the member:

$$F_3^{(1)} = K_{31}^{(1)} v_1^{(1)} + K_{32}^{(1)} v_2^{(1)} + K_{33}^{(1)} v_3^{(1)} + K_{34}^{(1)} v_4^{(1)} \tag{3.80a}$$

$$F_4^{(1)} = K_{41}^{(1)} v_1^{(1)} + K_{42}^{(1)} v_2^{(1)} + K_{43}^{(1)} v_3^{(1)} + K_{44}^{(1)} v_4^{(1)} \tag{3.80b}$$

In a similar manner, we write the stiffness relations for member 2 as

$$\begin{bmatrix} F_1^{(2)} \\ F_2^{(2)} \\ F_3^{(2)} \\ F_4^{(2)} \end{bmatrix} = \begin{bmatrix} K_{11}^{(2)} & K_{12}^{(2)} & K_{13}^{(2)} & K_{14}^{(2)} \\ K_{21}^{(2)} & K_{22}^{(2)} & K_{23}^{(2)} & K_{24}^{(2)} \\ K_{31}^{(2)} & K_{32}^{(2)} & K_{33}^{(2)} & K_{34}^{(2)} \\ K_{41}^{(2)} & K_{42}^{(2)} & K_{43}^{(2)} & K_{44}^{(2)} \end{bmatrix} \begin{bmatrix} v_1^{(2)} \\ v_2^{(2)} \\ v_3^{(2)} \\ v_4^{(2)} \end{bmatrix} \tag{3.81}$$

from which we obtain the forces at end 1 of the member:

$$F_1^{(2)} = K_{11}^{(2)} v_1^{(2)} + K_{12}^{(2)} v_2^{(2)} + K_{13}^{(2)} v_3^{(2)} + K_{14}^{(2)} v_4^{(2)} \tag{3.82a}$$

$$F_2^{(2)} = K_{21}^{(2)} v_1^{(2)} + K_{22}^{(2)} v_2^{(2)} + K_{23}^{(2)} v_3^{(2)} + K_{24}^{(2)} v_4^{(2)} \tag{3.82b}$$

Similarly, for member 3, the stiffness relations are written as

$$\begin{bmatrix} F_1^{(3)} \\ F_2^{(3)} \\ F_3^{(3)} \\ F_4^{(3)} \end{bmatrix} = \begin{bmatrix} K_{11}^{(3)} & K_{12}^{(3)} & K_{13}^{(3)} & K_{14}^{(3)} \\ K_{21}^{(3)} & K_{22}^{(3)} & K_{23}^{(3)} & K_{24}^{(3)} \\ K_{31}^{(3)} & K_{32}^{(3)} & K_{33}^{(3)} & K_{34}^{(3)} \\ K_{41}^{(3)} & K_{42}^{(3)} & K_{43}^{(3)} & K_{44}^{(3)} \end{bmatrix} \begin{bmatrix} v_1^{(3)} \\ v_2^{(3)} \\ v_3^{(3)} \\ v_4^{(3)} \end{bmatrix} \tag{3.83}$$

and the forces at end 1 of the member are given by

$$F_3^{(3)} = K_{31}^{(3)} v_1^{(3)} + K_{32}^{(3)} v_2^{(3)} + K_{33}^{(3)} v_3^{(3)} + K_{34}^{(3)} v_4^{(3)} \tag{3.84a}$$

$$F_4^{(3)} = K_{41}^{(3)} v_1^{(3)} + K_{42}^{(3)} v_2^{(3)} + K_{43}^{(3)} v_3^{(3)} + K_{44}^{(3)} v_4^{(3)} \tag{3.84b}$$

Note that Eqs. (3.80), (3.82), and (3.84) express the six member end forces that appear in the joint equilibrium equations (Eqs. (3.75)), in terms of member end displacements.

To relate the joint displacements **d** to the member end forces **F**, we substitute the compatibility equations into the foregoing member force–displacement relations. Thus, by substituting the compatibility equations for member 1 (Eqs. (3.76)) into its force–displacement relations as given by Eqs. (3.80), we express the member end forces $\mathbf{F}^{(1)}$ in terms of the joint displacements **d** as

$$F_3^{(1)} = K_{33}^{(1)} d_1 + K_{34}^{(1)} d_2 \tag{3.85a}$$

$$F_4^{(1)} = K_{43}^{(1)} d_1 + K_{44}^{(1)} d_2 \tag{3.85b}$$

In a similar manner, for member 2, by substituting Eqs. (3.77) into Eqs. (3.82), we obtain

$$F_1^{(2)} = K_{11}^{(2)} d_1 + K_{12}^{(2)} d_2 \tag{3.86a}$$

$$F_2^{(2)} = K_{21}^{(2)} d_1 + K_{22}^{(2)} d_2 \tag{3.86b}$$

Similarly, for member 3, substitution of Eqs. (3.78) into Eqs. (3.84) yields

$$F_3^{(3)} = K_{33}^{(3)} d_1 + K_{34}^{(3)} d_2 \tag{3.87a}$$

$$F_4^{(3)} = K_{43}^{(3)} d_1 + K_{44}^{(3)} d_2 \tag{3.87b}$$

Structure Stiffness Relations

Finally, by substituting Eqs. (3.85) through (3.87) into the joint equilibrium equations (Eqs. (3.75)), we establish the desired relationships between the joint loads **P** and the joint displacements **d** of the truss:

$$P_1 = \left(K_{33}^{(1)} + K_{11}^{(2)} + K_{33}^{(3)}\right) d_1 + \left(K_{34}^{(1)} + K_{12}^{(2)} + K_{34}^{(3)}\right) d_2 \tag{3.88a}$$

$$P_2 = \left(K_{43}^{(1)} + K_{21}^{(2)} + K_{43}^{(3)}\right) d_1 + \left(K_{44}^{(1)} + K_{22}^{(2)} + K_{44}^{(3)}\right) d_2 \tag{3.88b}$$

Equations (3.88) can be conveniently expressed in condensed matrix form as

$$\boxed{\mathbf{P} = \mathbf{S}\mathbf{d}} \tag{3.89}$$

in which

$$\mathbf{S} = \begin{bmatrix} K_{33}^{(1)} + K_{11}^{(2)} + K_{33}^{(3)} & K_{34}^{(1)} + K_{12}^{(2)} + K_{34}^{(3)} \\ K_{43}^{(1)} + K_{21}^{(2)} + K_{43}^{(3)} & K_{44}^{(1)} + K_{22}^{(2)} + K_{44}^{(3)} \end{bmatrix} \tag{3.90}$$

The matrix **S**, which is a square matrix with the number of rows and columns equal to the degrees of freedom (*NDOF*), is called the *structure stiffness matrix.* The preceding method of determining the structure stiffness relationships by combining the member stiffness relations is commonly referred to as the *direct stiffness method* [48].

Like member stiffness matrices, *structure stiffness matrices of linear elastic structures are always symmetric*. Note that in Eq. (3.90) the two off-diagonal elements of **S** are equal to each other, because $K_{34}^{(1)} = K_{43}^{(1)}$, $K_{12}^{(2)} = K_{21}^{(2)}$, and $K_{34}^{(3)} = K_{43}^{(3)}$; thereby making **S** a symmetric matrix.

Physical Interpretation of Structure Stiffness Matrix

The structure stiffness matrix **S** can be interpreted in a manner analogous to the member stiffness matrix. *A structure stiffness coefficient S_{ij} represents the force at the location and in the direction of P_i required, along with other joint forces, to cause a unit value of the displacement d_j, while all other joint displacements are zero.* Thus, the *j*th column of the structure stiffness matrix **S** consists of the joint loads required, at the locations and in the directions of all the degrees of freedom of the structure, to cause a unit value of the displacement d_j while all other displacements are zero. This interpretation of the structure stiffness matrix indicates that such a matrix can, alternatively, be determined by subjecting the structure, separately, to unit values of each of its joint displacements, and by evaluating the joint loads required to cause the individual displacements.

To illustrate this approach, consider again the three-member truss of Fig. 3.14. To determine its structure stiffness matrix **S**, we subject the truss to the joint displacements $d_1 = 1$ (with $d_2 = 0$), and $d_2 = 1$ (with $d_1 = 0$), as shown in Figs. 3.15(a) and (b), respectively. As depicted in Fig. 3.15(a), the stiffness coefficients S_{11} and S_{21} (elements of the first column of **S**) represent the horizontal and vertical forces at joint 1 required to cause a unit displacement of the joint in the horizontal direction ($d_1 = 1$), while holding it in place vertically ($d_2 = 0$). The unit horizontal displacement of joint 1 induces unit displacements, in the same direction, at the top ends of the three members connected to the joint. The member stiffness coefficients (or end forces) necessary to cause these unit end displacements of the individual members are shown in Fig. 3.15(a). Note that these stiffness coefficients are labeled in accordance with the notation for member end forces adopted in Section 3.5. (Also, recall

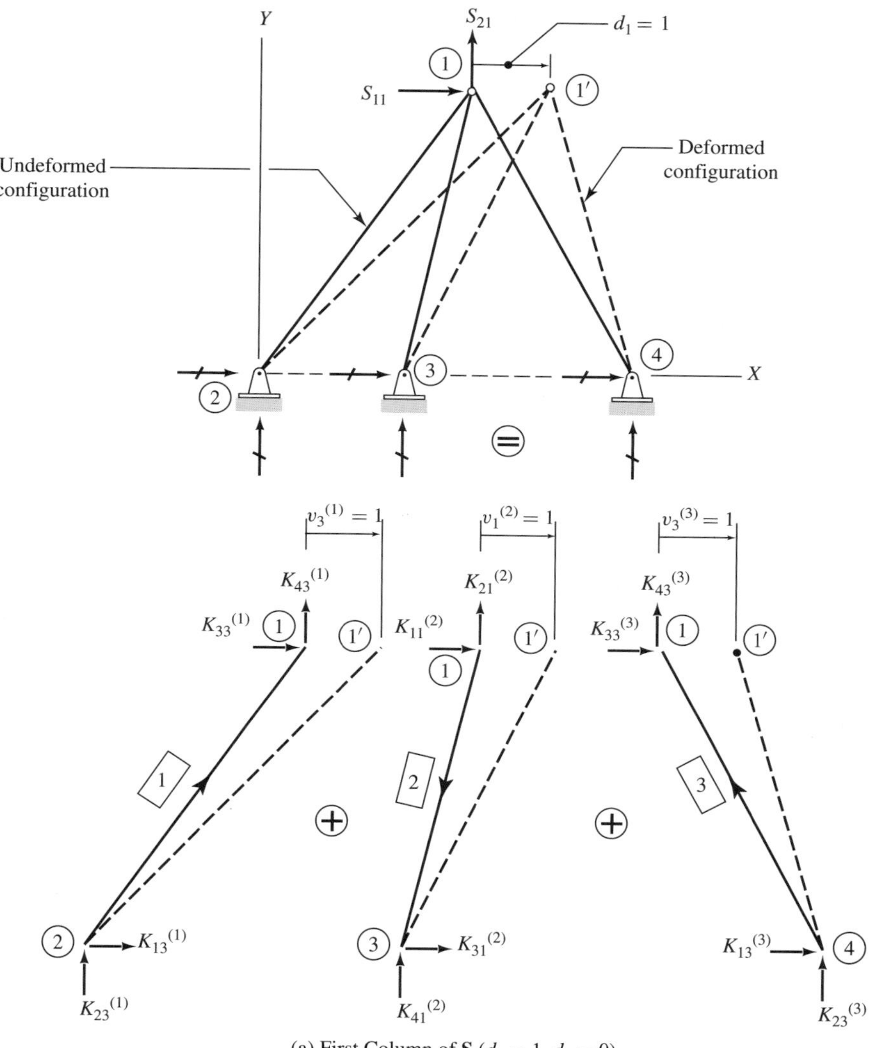

(a) First Column of **S** ($d_1 = 1$, $d_2 = 0$)

Fig. 3.15

that the explicit expressions for member stiffness coefficients, in terms of E, A, L, and θ of a member, were derived in Section 3.6.)

From Fig. 3.15(a), we realize that the total horizontal force S_{11} at joint 1, required to cause the joint displacement $d_1 = 1$ (with $d_2 = 0$), must be equal to the algebraic sum of the horizontal forces at the top ends of the three members connected to the joint; that is,

$$S_{11} = K_{33}^{(1)} + K_{11}^{(2)} + K_{33}^{(3)} \tag{3.91a}$$

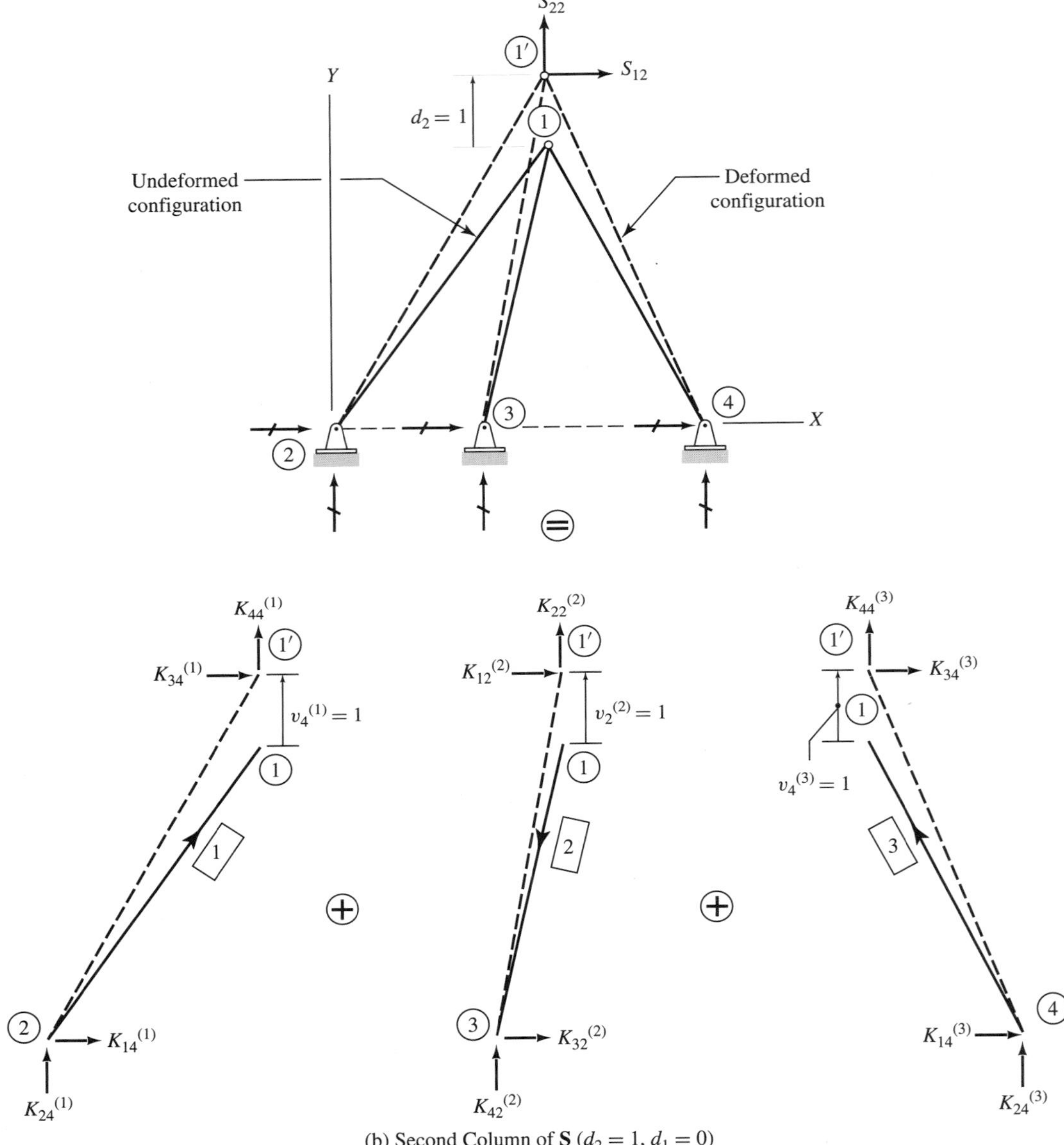

(b) Second Column of **S** ($d_2 = 1$, $d_1 = 0$)

Fig. 3.15 *(continued)*

Similarly, the total vertical force S_{21} at joint 1 must be equal to the algebraic sum of the vertical forces at the top ends of all the members connected to the joint. Thus (Fig. 3.15a),

$$S_{21} = K_{43}^{(1)} + K_{21}^{(2)} + K_{43}^{(3)} \tag{3.91b}$$

Note that the expressions for S_{11} and S_{21}, as given in Eqs. 3.91(a) and (b), are identical to those listed in the first column of the **S** matrix in Eq. (3.90).

The stiffness coefficients in the second column of the **S** matrix can be determined in a similar manner. As depicted in Fig. 3.15(b), the structure stiffness coefficients S_{12} and S_{22} represent the horizontal and vertical forces at joint 1 required to cause a unit displacement of the joint in the vertical direction (d_2), while holding it in place horizontally ($d_1 = 0$). The joint displacement $d_2 = 1$ induces unit vertical displacements at the top ends of the three members; these, in turn, cause the forces (member stiffness coefficients) to develop at the ends of the members. From Fig. 3.15(b), we can see that the stiffness coefficient S_{12} of joint 1, in the horizontal direction, must be equal to the algebraic sum of the member stiffness coefficients, in the same direction, at the top ends of all the members connected to the joint; that is,

$$S_{12} = K_{34}^{(1)} + K_{12}^{(2)} + K_{34}^{(3)} \tag{3.91c}$$

Similarly, the structure stiffness coefficient S_{22}, in the vertical direction, equals the algebraic sum of the vertical member stiffness coefficients at the top ends of the three members connected to joint 1. Thus (Fig. 3.15b),

$$S_{22} = K_{44}^{(1)} + K_{22}^{(2)} + K_{44}^{(3)} \tag{3.91d}$$

Again, the expressions for S_{12} and S_{22}, as given in Eqs. 3.91(c) and (d), are the same as those listed in the second column of the **S** matrix in Eq. (3.90).

Assembly of the Structure Stiffness Matrix Using Member Code Numbers

In the preceding paragraphs of this section, we have studied two procedures for determining the structure stiffness matrix **S**. Although a study of the foregoing procedures is essential for developing an understanding of the concept of the stiffness of multiple-degrees-of-freedom structures, these procedures cannot be implemented easily on computers and, therefore, are seldom used in practice.

From Eqs. (3.91), we observe that the structure stiffness coefficient of a joint in a direction equals the algebraic sum of the member stiffness coefficients, in that direction, at all the member ends connected to the joint. This fact indicates that the structure stiffness matrix **S** can be formulated directly by adding the elements of the member stiffness matrices into their proper positions in the structure matrix. This technique of directly forming a structure stiffness matrix by assembling the elements of the member global stiffness matrices can be programmed conveniently on computers. The technique was introduced by S. S. Tezcan in 1963 [44], and is sometimes referred to as the *code number technique*.

To illustrate this technique, consider again the three-member truss of Fig. 3.14. The analytical model of the truss is redrawn in Fig. 3.16(a), which shows that the structure has two degrees of freedom (numbered 1 and 2), and six restrained coordinates (numbered from 3 to 8). The stiffness matrices in the global coordinate system for members 1, 2, and 3 of the truss are designated $\mathbf{K}_1$, $\mathbf{K}_2$, and $\mathbf{K}_3$, respectively (Fig. 3.16(c)). Our objective is to form the structure stiffness matrix **S** by assembling the elements of $\mathbf{K}_1$, $\mathbf{K}_2$, and $\mathbf{K}_3$.

To determine the positions of the elements of a member matrix **K** in the structure matrix **S**, we identify the number of the structure's degree of freedom or restrained coordinate, at the location and in the direction of each of the

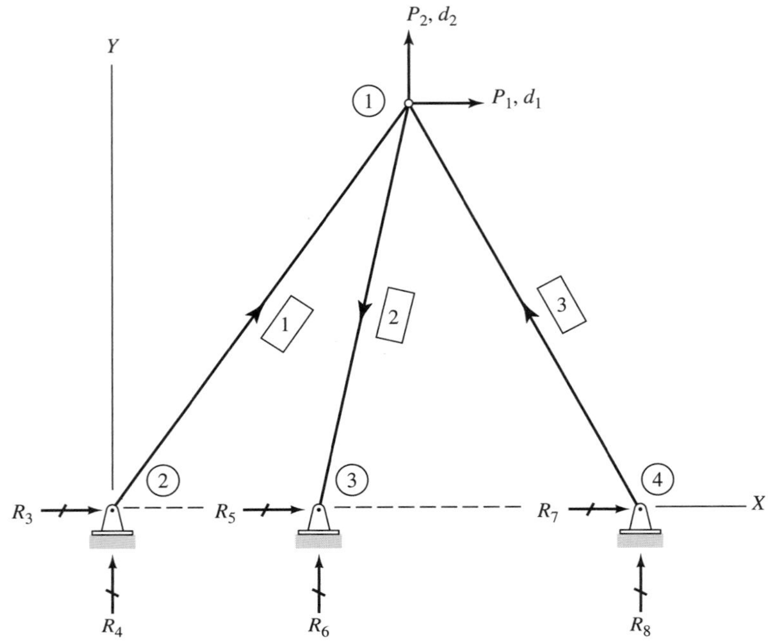

Two degrees of freedom (1 and 2);
six restrained coordinates (3 through 8)

(a) Analytical Model

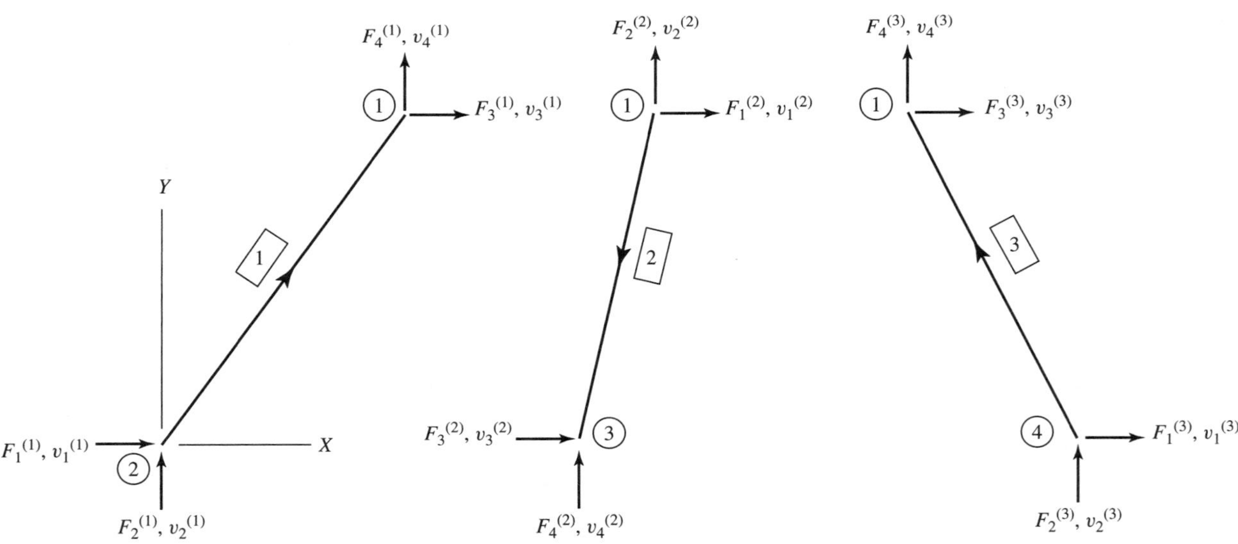

(b) Member End Forces and End Displacements in the Global Coordinate System

Fig. 3.16

$$\mathbf{K}_1 = \begin{bmatrix} K_{11}^{(1)} & K_{12}^{(1)} & K_{13}^{(1)} & K_{14}^{(1)} \\ K_{21}^{(1)} & K_{22}^{(1)} & K_{23}^{(1)} & K_{24}^{(1)} \\ K_{31}^{(1)} & K_{32}^{(1)} & K_{33}^{(1)} & K_{34}^{(1)} \\ K_{41}^{(1)} & K_{42}^{(1)} & K_{43}^{(1)} & K_{44}^{(1)} \end{bmatrix} \begin{matrix} 3 \\ 4 \\ 1 \\ 2 \end{matrix} \quad (\text{columns } 3\ 4\ 1\ 2)$$

$$\mathbf{K}_2 = \begin{bmatrix} K_{11}^{(2)} & K_{12}^{(2)} & K_{13}^{(2)} & K_{14}^{(2)} \\ K_{21}^{(2)} & K_{22}^{(2)} & K_{23}^{(2)} & K_{24}^{(2)} \\ K_{31}^{(2)} & K_{32}^{(2)} & K_{33}^{(2)} & K_{34}^{(2)} \\ K_{41}^{(2)} & K_{42}^{(2)} & K_{43}^{(2)} & K_{44}^{(2)} \end{bmatrix} \begin{matrix} 1 \\ 2 \\ 5 \\ 6 \end{matrix} \quad (\text{columns } 1\ 2\ 5\ 6)$$

$$\mathbf{K}_3 = \begin{bmatrix} K_{11}^{(3)} & K_{12}^{(3)} & K_{13}^{(3)} & K_{14}^{(3)} \\ K_{21}^{(3)} & K_{22}^{(3)} & K_{23}^{(3)} & K_{24}^{(3)} \\ K_{31}^{(3)} & K_{32}^{(3)} & K_{33}^{(3)} & K_{34}^{(3)} \\ K_{41}^{(3)} & K_{42}^{(3)} & K_{43}^{(3)} & K_{44}^{(3)} \end{bmatrix} \begin{matrix} 7 \\ 8 \\ 1 \\ 2 \end{matrix} \quad (\text{columns } 7\ 8\ 1\ 2)$$

$$\mathbf{S} = \begin{bmatrix} K_{33}^{(1)} + K_{11}^{(2)} + K_{33}^{(3)} & K_{34}^{(1)} + K_{12}^{(2)} + K_{34}^{(3)} \\ K_{43}^{(1)} + K_{21}^{(2)} + K_{43}^{(3)} & K_{44}^{(1)} + K_{22}^{(2)} + K_{44}^{(3)} \end{bmatrix} \begin{matrix} 1 \\ 2 \end{matrix} \quad (\text{columns } 1\ 2)$$

(c) Assembling of Structure Stiffness Matrix **S**

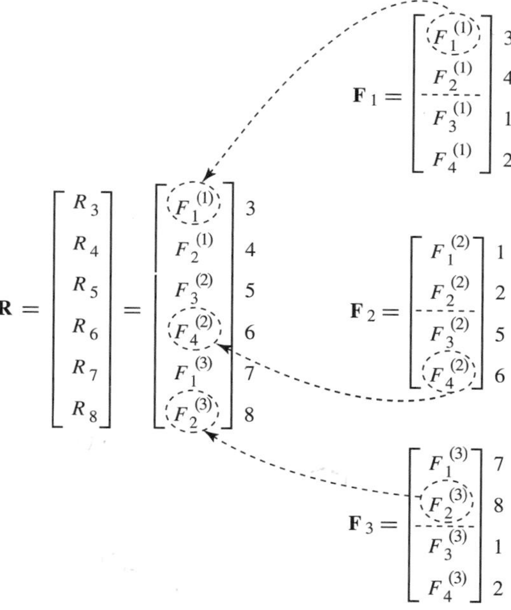

(d) Assembly of Support Reaction Vector **R**

Fig. 3.16 *(continued)*

member's global end displacements, **v**. Such structure degrees of freedom and restrained coordinate numbers for a member, when arranged in the same order as the member's end displacements, are referred to as the member's *code numbers*. In accordance with the notation for member end displacements adopted in Section 3.5, the first two end displacements, v_1 and v_2, are always specified in the X and Y directions, respectively, at the beginning of the member; and the last two end displacements, v_3 and v_4, are always in the X and Y directions, respectively, at the end of the member. Therefore, the first two code numbers for a member are always the numbers of the structure degrees of freedom and/or restrained coordinates in the X and Y directions, respectively, at the beginning joint for the member; and the third and fourth member code numbers are always the numbers of the structure degrees of freedom and/or restrained coordinates in the X and Y directions, respectively, at the end joint for the member.

From Fig. 3.16(a), we can see that for member 1 of the truss, the beginning and the end joints are 2 and 1, respectively. At the beginning joint 2, the restrained coordinate numbers are 3 and 4 in the X and Y directions, respectively; whereas, at the end joint 1, the structure degree of freedom numbers, in the X and Y directions, are 1 and 2, respectively. Thus, the code numbers for member 1 are 3, 4, 1, 2. Similarly, since the beginning and end joints for member 2 are 1 and 3, respectively, the code numbers for this member are 1, 2, 5, 6. In a similar manner, the code numbers for member 3 are found to be 7, 8, 1, 2. The code numbers for the three members of the truss can be verified by comparing the member global end displacements shown in Fig. 3.16(b) with the structure degrees of freedom and restrained coordinates given in Fig. 3.16(a).

The code numbers for a member define the compatibility equations for the member. For example, the code numbers 3, 4, 1, 2 imply the following compatibility equations for member 1:

$$v_1^{(1)} = d_3 \qquad v_2^{(1)} = d_4 \qquad v_3^{(1)} = d_1 \qquad v_4^{(1)} = d_2$$

Since the displacements corresponding to the restrained coordinates 3 and 4 are zero (i.e., $d_3 = d_4 = 0$), the compatibility equations for member 1 become

$$v_1^{(1)} = v_2^{(1)} = 0 \qquad v_3^{(1)} = d_1 \qquad v_4^{(1)} = d_2$$

which are identical to those given in Eqs. (3.76).

The member code numbers can also be used to formulate the joint equilibrium equations for a structure (such as those given in Eqs. (3.75)). The equilibrium equation corresponding to an ith degree of freedom (or restrained coordinate) can be obtained by equating the joint load P_i (or the reaction R_i) to the algebraic sum of the member end forces, with the code number i, of all the members of the structure. For example, to obtain the equilibrium equations for the truss of Fig. 3.16(a), we write the code numbers for its three members by the side of their respective end force vectors, as

$$\mathbf{F}_1 = \begin{bmatrix} F_1^{(1)} \\ F_2^{(1)} \\ F_3^{(1)} \\ F_4^{(1)} \end{bmatrix} \begin{matrix} 3 \\ 4 \\ 1 \\ 2 \end{matrix} \qquad \mathbf{F}_2 = \begin{bmatrix} F_1^{(2)} \\ F_2^{(2)} \\ F_3^{(2)} \\ F_4^{(2)} \end{bmatrix} \begin{matrix} 1 \\ 2 \\ 5 \\ 6 \end{matrix} \qquad \mathbf{F}_3 = \begin{bmatrix} F_1^{(3)} \\ F_2^{(3)} \\ F_3^{(3)} \\ F_4^{(3)} \end{bmatrix} \begin{matrix} 7 \\ 8 \\ 1 \\ 2 \end{matrix} \tag{3.92}$$

From Eq. (3.92), we can see that the member end forces with the code number 1 are: $F_3^{(1)}$ of member 1, $F_1^{(2)}$ of member 2, and $F_3^{(3)}$ of member 3. Thus, the equilibrium equation corresponding to degree of freedom 1 is given by

$$P_1 = F_3^{(1)} + F_1^{(2)} + F_3^{(3)}$$

which is identical to Eq. 3.75(a). Similarly, the equilibrium equation corresponding to degree of freedom 2 can be obtained by equating P_2 to the sum of the end forces, with code number 2, of the three members. Thus, from Eq. (3.92)

$$P_2 = F_4^{(1)} + F_2^{(2)} + F_4^{(3)}$$

which is the same as Eq. (3.75(b)).

To establish the structure stiffness matrix **S**, we write the code numbers of each member on the right side and at the top of its stiffness matrix **K**, as shown in Fig. 3.16(c). These code numbers now define the positions of the elements of the member stiffness matrices in the structure stiffness matrix **S**. In other words, the code numbers on the right side of a matrix **K** represent the row numbers of the **S** matrix, and the code numbers at the top represent the column numbers of **S**. Furthermore, since the number of rows and columns of **S** equal the number of degrees of freedom (*NDOF*) of the structure, only those elements of a **K** matrix with both row and column code numbers less than or equal to *NDOF* belong in **S**. For example, since the truss of Fig. 3.16(a) has two degrees of freedom, only the bottom-right quarters of the member matrices $\mathbf{K}_1$ and $\mathbf{K}_3$, and the top-left quarter of $\mathbf{K}_2$, belong in **S** (see Fig. 3.16(c)).

The structure stiffness matrix **S** is established by algebraically adding the pertinent elements of the **K** matrices of all the members, in their proper positions, in the **S** matrix. For example, to assemble **S** for the truss of Fig. 3.16(a), we start by storing the pertinent elements of $\mathbf{K}_1$ in **S** (see Fig. 3.16(c)). Thus, the element $K_{33}^{(1)}$ is stored in row 1 and column 1 of **S**, the element $K_{43}^{(1)}$ is stored in row 2 and column 1 of **S**, the element $K_{34}^{(1)}$ is stored in row 1 and column 2 of **S** (see Fig. 3.16(c)), and the element $K_{44}^{(1)}$ is stored in row 2 and column 2 of **S**. Note that only those elements of $\mathbf{K}_1$ whose row and column code numbers are either 1 or 2 are stored in **S**. The same procedure is then repeated for members 2 and 3. When two or more member stiffness coefficients are stored in the same position in **S**, then the coefficients must be algebraically added. The completed structure stiffness matrix **S** for the truss is shown in Fig. 3.16(c). Note that this matrix is identical to the one determined previously by substituting the member compatibility equations and stiffness relations into the joint equilibrium equations (Eq. (3.90)).

Once **S** has been determined, the structure stiffness relations, $\mathbf{P} = \mathbf{Sd}$ (Eq. (3.89)), which now represent a system of simultaneous linear algebraic equations, can be solved for the unknown joint displacements **d**. With **d** known, the end displacements **v** for each member can be obtained by applying the compatibility equations defined by its code numbers; then the corresponding end displacements **u** and end forces **Q** and **F** can be computed by using the member's transformation and stiffness relations. Finally, the support reactions **R** can be determined from the member end forces **F**, by considering the equilibrium of the support joints in the directions of the restrained coordinates, as discussed in the following paragraphs.

Assembly of the Support Reaction Vector Using Member Code Numbers

The support reactions **R** of a structure can be expressed in terms of the member global end forces **F**, using the equilibrium requirement that the reaction in a direction at a joint must be equal to the algebraic sum of all the forces, in that direction, at all the member ends connected to the joint. Because the code numbers of a member specify the locations and directions of its global end forces with respect to the structure's degrees of freedom and/or restrained coordinates, the reaction corresponding to a restrained coordinate can be evaluated by algebraically summing those elements of the **F** vectors of all the members whose code numbers are the same as the restrained coordinate.

As the foregoing discussion suggests, the reaction vector **R** can be assembled from the member end force vectors **F**, using a procedure similar to that for forming the structure stiffness matrix. To determine the reactions, we write the restrained coordinate numbers ($NDOF + 1$ through $2(NJ)$) on the right side of vector **R**, as shown in Fig. 3.16(d). Next, the code numbers of each member are written on the right side of its end force vector **F** (Fig. 3.16(d)). Any member code number that is greater than the number of degrees of freedom of the structure (*NDOF*) now represents the restrained coordinate number of the row of **R** in which the corresponding member force is to be stored. The reaction vector **R** is obtained by algebraically adding the pertinent elements of the **F** vectors of all the members in their proper positions in **R**.

For example, to assemble **R** for the truss of Fig. 3.16(a), we begin by storing the pertinent elements of $\mathbf{F}_1$ in **R**. Thus, as shown in Fig. 3.16(d), the element $F_1^{(1)}$ with code number 3 is stored in row 1 of **R**, which has the restrained coordinate number 3 by its side. Similarly, the element $F_2^{(1)}$ (with code number 4) is stored in row 2 (with restrained coordinate number 4) of **R**. Note that only those elements of $\mathbf{F}_1$ whose code numbers are greater than 2 ($= NDOF$) are stored in **R**. The same procedure is then repeated for members 2 and 3. The completed support reaction vector **R** for the truss is shown in Fig. 3.16(d).

EXAMPLE 3.7

Determine the structure stiffness matrix for the truss shown in Fig. 3.17(a).

SOLUTION

Analytical Model: The analytical model of the truss is shown in Fig. 3.17(b). The structure has three degrees of freedom—the translation in the *X* direction of joint 1, and the translations in the *X* and *Y* directions of joint 4. These degrees of freedom are identified by numbers 1 through 3; and the five restrained coordinates of the truss are identified by numbers 4 through 8, as shown in Fig. 3.17(b).

Structure Stiffness Matrix: To generate the 3×3 structure stiffness matrix **S**, we will determine, for each member, the global stiffness matrix **K** and store its pertinent elements in their proper positions in **S** by using the member's code numbers.

Member 1 $L = 6$ m, $\cos\theta = 1$, $\sin\theta = 0$

$$\frac{EA}{L} = \frac{70(10^6)\,(0.0015)}{6} = 17{,}500 \text{ kN/m}$$

The member stiffness matrix in global coordinates can now be evaluated by using Eq. (3.73).

$$\mathbf{K}_1 = \begin{matrix} \scriptstyle 7 & \scriptstyle 8 & \scriptstyle 2 & \scriptstyle 3 \\ \end{matrix} \\ \left[\begin{array}{cc|cc} 17{,}500 & 0 & -17{,}500 & 0 \\ 0 & 0 & 0 & 0 \\ \hline -17{,}500 & 0 & 17{,}500 & 0 \\ 0 & 0 & 0 & 0 \end{array}\right] \begin{matrix} 7 \\ 8 \\ 2 \\ 3 \end{matrix} \text{ kN/m} \qquad (1)$$

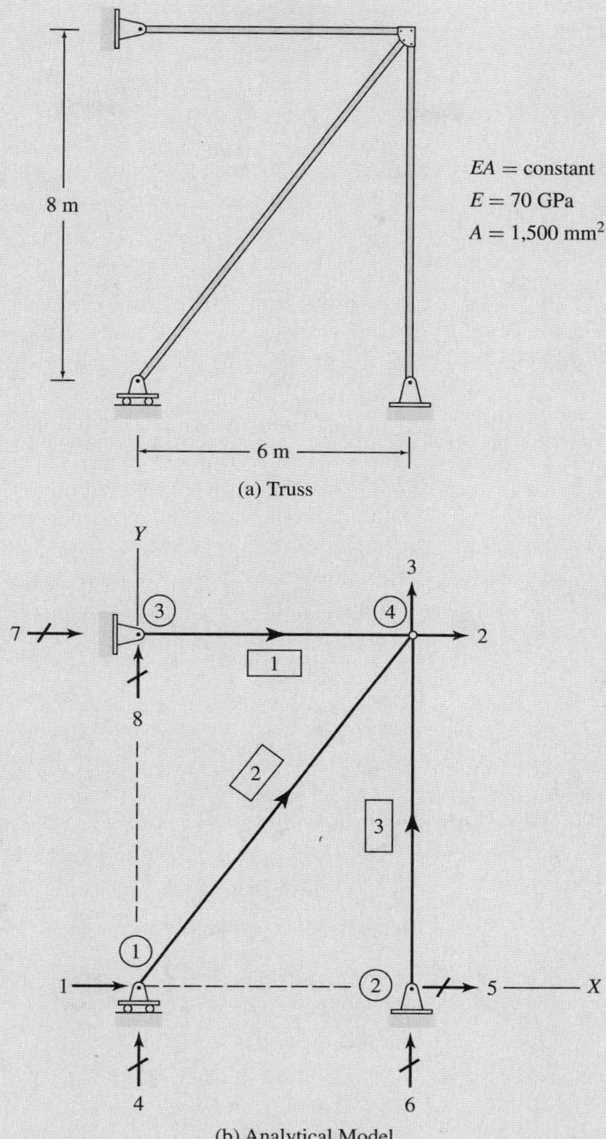

Fig. 3.17

From Fig. 3.17(b), we observe that joint 3 has been selected as the beginning joint, and joint 4 as the end joint, for member 1. Thus, the code numbers for this member are 7, 8, 2, 3. These numbers are written on the right side and at the top of $\mathbf{K}_1$ (see Eq. (1)) to indicate the rows and columns, respectively, of the structure stiffness matrix $\mathbf{S}$, where the elements of $\mathbf{K}_1$ must be stored. Note that the elements of $\mathbf{K}_1$ that correspond to the restrained coordinate numbers 7 and 8 are simply disregarded. Thus, the element in row 3 and column 3 of $\mathbf{K}_1$ is stored in row 2 and column 2 of **S**, as

$$\mathbf{S} = \begin{matrix} 1 & 2 & 3 & \\ \end{matrix} \begin{bmatrix} 0 & 0 & 0 \\ 0 & 17{,}500 & 0 \\ 0 & 0 & 0 \end{bmatrix} \begin{matrix} 1 \\ 2 \\ 3 \end{matrix} \tag{2}$$

Member 2 As shown in Fig. 3.17(b), joint 1 is the beginning joint, and joint 4 is the end joint, for member 2. By applying Eqs. (3.62), we determine

$$L = \sqrt{(X_4 - X_1)^2 + (Y_4 - Y_1)^2} = \sqrt{(6-0)^2 + (8-0)^2} = 10 \text{ m}$$

$$\cos\theta = \frac{X_4 - X_1}{L} = \frac{6-0}{10} = 0.6$$

$$\sin\theta = \frac{Y_4 - Y_1}{L} = \frac{8-0}{10} = 0.8$$

$$\frac{EA}{L} = \frac{70(10^6)(0.0015)}{10} = 10{,}500 \text{ kN/m}$$

By using the expression for **K** given in Eq. (3.73), we obtain

$$\mathbf{K}_2 = \begin{matrix} 1 & 4 & 2 & 3 \end{matrix} \begin{bmatrix} 3{,}780 & 5{,}040 & -3{,}780 & -5{,}040 \\ 5{,}040 & 6{,}720 & -5{,}040 & -6{,}720 \\ -3{,}780 & -5{,}040 & 3{,}780 & 5{,}040 \\ -5{,}040 & -6{,}720 & 5{,}040 & 6{,}720 \end{bmatrix} \begin{matrix} 1 \\ 4 \\ 2 \\ 3 \end{matrix} \text{ kN/m} \tag{3}$$

From Fig. 3.17(b), we can see that the code numbers for this member are 1, 4, 2, 3. These numbers are used to add the pertinent elements of $\mathbf{K}_2$ in their proper positions in **S**, as given in Eq. (2). Thus, **S** now becomes

$$\mathbf{S} = \begin{matrix} 1 & 2 & 3 \end{matrix} \begin{bmatrix} 3{,}780 & -3{,}780 & -5{,}040 \\ -3{,}780 & 17{,}500 + 3{,}780 & 5{,}040 \\ -5{,}040 & 5{,}040 & 6{,}720 \end{bmatrix} \begin{matrix} 1 \\ 2 \\ 3 \end{matrix} \tag{4}$$

Member 3 $L = 8$ m, $\cos\theta = 0$, $\sin\theta = 1$

$$\frac{EA}{L} = \frac{70(10^6)(0.0015)}{8} = 13{,}125 \text{ kN/m}$$

By using Eq. (3.73),

$$\mathbf{K}_3 = \begin{matrix} 5 & 6 & 2 & 3 \end{matrix} \begin{bmatrix} 0 & 0 & 0 & 0 \\ 0 & 13{,}125 & 0 & -13{,}125 \\ 0 & 0 & 0 & 0 \\ 0 & -13{,}125 & 0 & 13{,}125 \end{bmatrix} \begin{matrix} 5 \\ 6 \\ 2 \\ 3 \end{matrix} \text{ kN/m} \tag{5}$$

The code numbers for this member are 5, 6, 2, 3. By using these code numbers, the pertinent elements of $\mathbf{K}_3$ are added in **S** (as given in Eq. (4)), yielding

$$\mathbf{S} = \begin{matrix} 1 & 2 & 3 \end{matrix} \begin{bmatrix} 3{,}780 & -3{,}780 & -5{,}040 \\ -3{,}780 & 17{,}500 + 3{,}780 & 5{,}040 \\ -5{,}040 & 5{,}040 & 6{,}720 + 13{,}125 \end{bmatrix} \begin{matrix} 1 \\ 2 \\ 3 \end{matrix} \text{ kN/m}$$

Since the stiffnesses of all three members of the truss have now been stored in **S**, the structure stiffness matrix for the given truss is

$$\mathbf{S} = \begin{bmatrix} 3,780 & -3,780 & -5,040 \\ -3,780 & 21,280 & 5,040 \\ -5,040 & 5,040 & 19,845 \end{bmatrix} \begin{matrix} 1 \\ 2 \\ 3 \end{matrix} \text{ kN/m}$$

(columns numbered 1, 2, 3) **Ans**

Note that the structure stiffness matrix **S**, obtained by assembling the stiffness coefficients of the three members, is symmetric.

3.8 PROCEDURE FOR ANALYSIS

Based on the discussion presented in the previous sections, the following step-by-step procedure can be developed for the analysis of plane trusses subjected to joint loads.

1. Prepare an analytical model of the truss as follows.
   a. Draw a line diagram of the structure, on which each joint and member is identified by a number.
   b. Establish a global *XY* coordinate system, with the *X* and *Y* axes oriented in the horizontal (positive to the right) and vertical (positive upward) directions, respectively. It is usually convenient to locate the origin of the global coordinate system at a lower left joint of the structure, so that the *X* and *Y* coordinates of most of the joints are positive.
   c. For each member, establish a local *xy* coordinate system by selecting one of the joints at its ends as the beginning joint and the other as the end joint. On the structure's line diagram, indicate the positive direction of the local *x* axis for each member by drawing an arrow along the member pointing toward its end joint. For horizontal members, the coordinate transformations can be avoided by selecting the joint at the member's left end as the beginning joint.
   d. Identify the degrees of freedom (or joint displacements) and the restrained coordinates of the structure. These quantities are specified on the line diagram by assigning numbers to the arrows drawn at the joints in the *X* and *Y* directions. The degrees of freedom are numbered first, starting at the lowest-numbered joint and proceeding sequentially to the highest. In the case of more than one degree of freedom at a joint, the *X*-displacement is numbered first, followed by the *Y*-displacement. After all the degrees of freedom have been numbered, the restrained coordinates are numbered, beginning with a number equal to $NDOF + 1$. Starting at the lowest-numbered joint and proceeding sequentially to the highest, all of the restrained coordinates of the structure are numbered. In the case of more than one restrained coordinate at a joint, the *X*-coordinate is numbered first, followed by the *Y*-coordinate.
2. Evaluate the structure stiffness matrix **S**. The number of rows and columns of **S** must be equal to the degrees of freedom (*NDOF*) of the

structure. For each member of the truss, perform the following operations.

a. Calculate its length and direction cosines. (The expressions for $\cos\theta$ and $\sin\theta$ are given in Eqs. (3.62).)
b. Compute the member stiffness matrix in the global coordinate system, **K**, using Eq. (3.73).
c. Identify its code numbers, and store the pertinent elements of **K** in their proper positions in **S**, using the procedure described in Section 3.7.

The complete structure stiffness matrix, obtained by assembling the stiffness coefficients of all the members of the truss, must be a symmetric matrix.

3. Form the $NDOF \times 1$ joint load vector **P**.
4. Determine the joint displacements **d**. Substitute **P** and **S** into the structure stiffness relations, $\mathbf{P} = \mathbf{Sd}$ (Eq. (3.89)), and solve the resulting system of simultaneous equations for the unknown joint displacements **d**. To check that the solution of simultaneous equations has been carried out correctly, substitute the numerical values of **d** back into the structure stiffness relations, $\mathbf{P} = \mathbf{Sd}$. If the solution is correct, then the stiffness relations should be satisfied. Note that joint displacements are considered positive when in the positive directions of the global X and Y axes; similarly, the displacements are negative in the negative directions.
5. Compute member end displacements and end forces, and support reactions. For each member of the truss, do the following.
   a. Obtain member end displacements in the global coordinate system, **v**, from the joint displacements, **d**, using the member's code numbers.
   b. Calculate the member's transformation matrix **T** by using Eq. (3.61), and determine member end displacements in the local coordinate system, **u**, using the transformation relationship $\mathbf{u} = \mathbf{Tv}$ (Eq. (3.63)). For horizontal members with local x axis positive to the right (i.e., in the same direction as the global X axis), member end displacements in the global and local coordinate systems are the same; that is, $\mathbf{u} = \mathbf{v}$. Member axial deformation, u_a, if desired, can be obtained from the relationship $u_a = u_1 - u_3$, in which u_1 and u_3 are the first and third elements, respectively, of vector **u**. A positive value of u_a indicates shortening (or contraction) of the member in the axial direction, and a negative value indicates elongation.
   c. Determine the member stiffness matrix in the local coordinate system, **k**, using Eq. (3.27); then calculate member end forces in the local coordinate system by using the stiffness relationship $\mathbf{Q} = \mathbf{ku}$ (Eq. (3.7)). The member axial force, Q_a, equals the first element, Q_1, of the vector **Q** (i.e., $Q_a = Q_1$); a positive value of Q_a indicates that the axial force is compressive, and a negative value indicates that the axial force is tensile.
   d. Compute member end forces in the global coordinate system, **F**, by using the transformation relationship $\mathbf{F} = \mathbf{T}^T\mathbf{Q}$ (Eq. (3.66)). For horizontal members with the local x axis positive to the right, the member

end forces in the local and global coordinate systems are the same; that is, $\mathbf{F} = \mathbf{Q}$.

e. By using member code numbers, store the pertinent elements of **F** in their proper positions in the support reaction vector **R**, as discussed in Section 3.7.

6. To check the calculation of member end forces and support reactions, apply the three equations of equilibrium ($\sum F_X = 0$, $\sum F_Y = 0$, and $\sum M = 0$) to the free body of the entire truss. If the calculations have been carried out correctly, then the equilibrium equations should be satisfied.

Instead of following steps 5c and d, the member end forces can be determined alternatively by first evaluating the global forces **F**, using the global stiffness relationship $\mathbf{F} = \mathbf{Kv}$ (Eq. (3.71)), and then obtaining the local forces **Q** from the transformation relationship $\mathbf{Q} = \mathbf{TF}$ (Eq. (3.60)).

EXAMPLE 3.8

Determine the joint displacements, member axial forces, and support reactions for the truss shown in Fig. 3.18(a) by the matrix stiffness method.

SOLUTION

Analytical Model: The analytical model of the truss is shown in Fig. 3.18(b). The truss has two degrees of freedom, which are the translations of joint 1 in the X and Y directions. These are numbered as 1 and 2, respectively. The six restrained coordinates of the truss are identified by numbers 3 through 8.

Structure Stiffness Matrix:

Member 1 As shown in Fig. 3.18(b), we have selected joint 2 as the beginning joint, and joint 1 as the end joint, for member 1. By applying Eqs. (3.62), we determine

$$L = \sqrt{(X_1 - X_2)^2 + (Y_1 - Y_2)^2} = \sqrt{(12 - 0)^2 + (16 - 0)^2} = 20 \text{ ft}$$

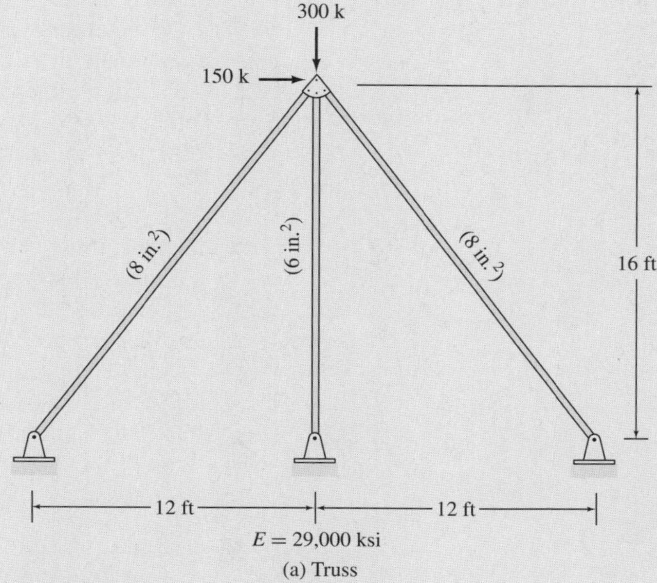

(a) Truss

Fig. 3.18

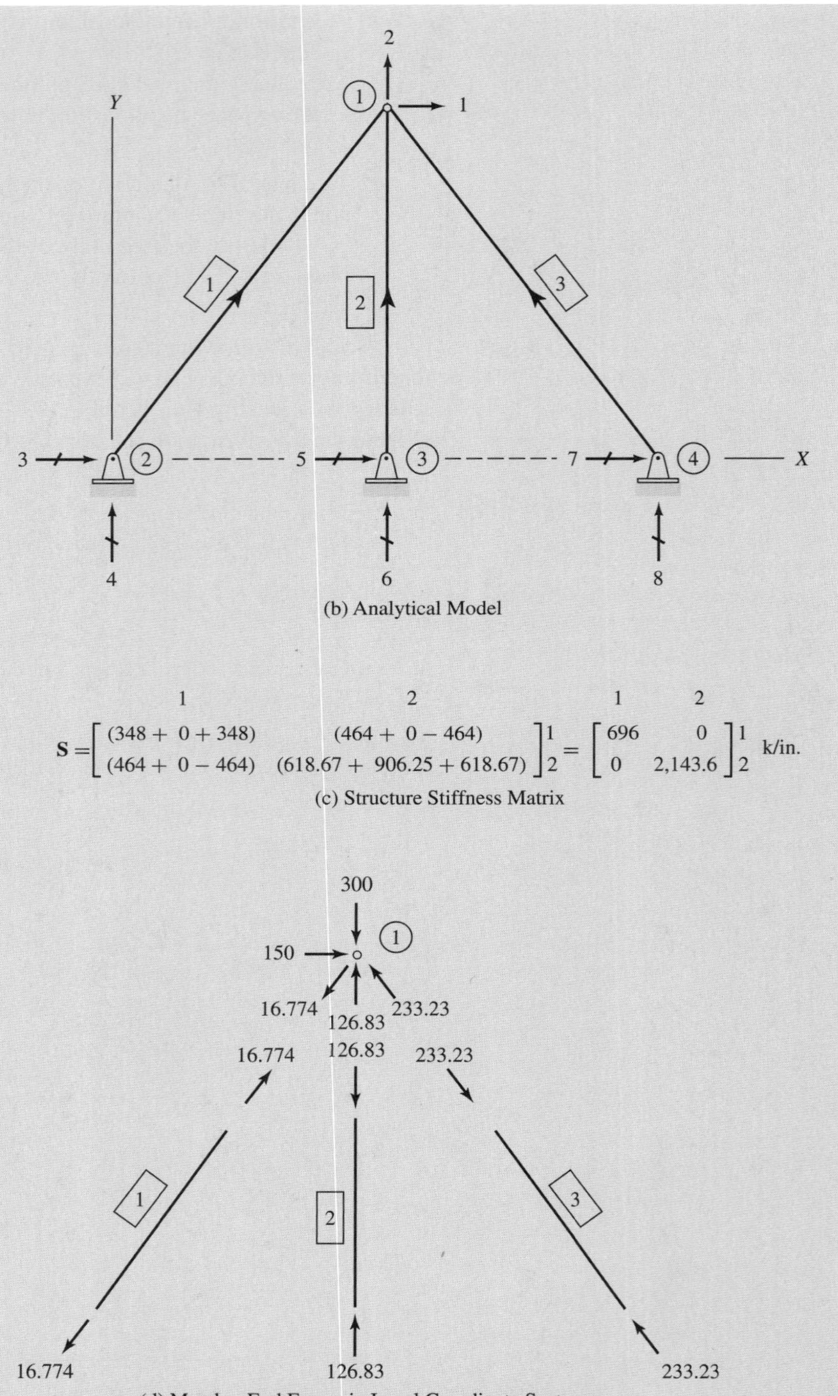

(b) Analytical Model

$$\mathbf{S} = \begin{bmatrix} (348 + 0 + 348) & (464 + 0 - 464) \\ (464 + 0 - 464) & (618.67 + 906.25 + 618.67) \end{bmatrix} = \begin{bmatrix} 696 & 0 \\ 0 & 2{,}143.6 \end{bmatrix} \text{ k/in.}$$

(c) Structure Stiffness Matrix

(d) Member End Forces in Local Coordinate Systems

Fig. 3.18 *(continued)*

$$\mathbf{R} = \begin{bmatrix} -10.064 \\ -13.419 \\ 0 \\ 126.83 \\ -139.94 \\ 186.58 \end{bmatrix} \begin{matrix} 3 \\ 4 \\ 5 \\ 6 \\ 7 \\ 8 \end{matrix} \text{ k}$$

(e) Support Reaction Vector

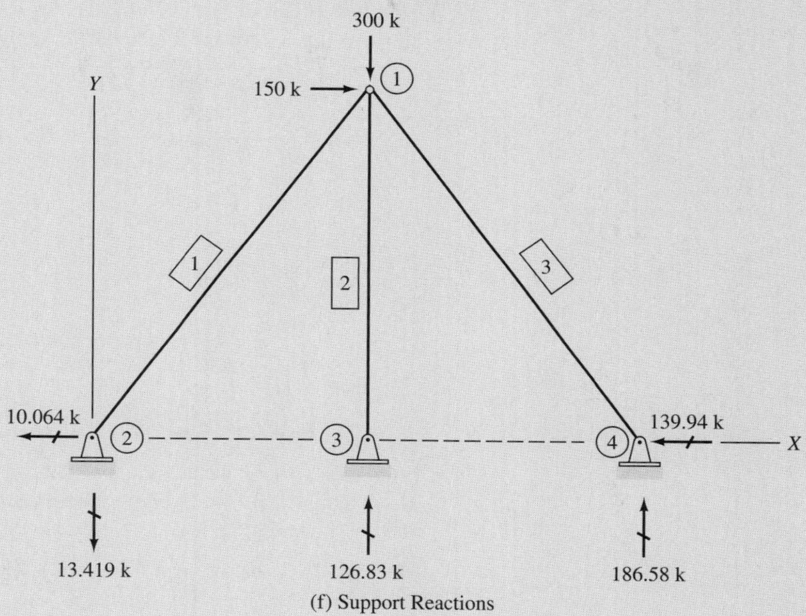

(f) Support Reactions

Fig. 3.18 *(continued)*

$$\cos\theta = \frac{X_1 - X_2}{L} = \frac{12 - 0}{20} = 0.6$$

$$\sin\theta = \frac{Y_1 - Y_2}{L} = \frac{16 - 0}{20} = 0.8$$

Using the units of kips and inches, we evaluate the member's global stiffness matrix (Eq. (3.73)) as

$$\mathbf{K}_1 = \frac{(29{,}000)(8)}{(20)(12)} \begin{bmatrix} 0.36 & 0.48 & -0.36 & -0.48 \\ 0.48 & 0.64 & -0.48 & -0.64 \\ -0.36 & -0.48 & 0.36 & 0.48 \\ -0.48 & -0.64 & 0.48 & 0.64 \end{bmatrix}$$

or

$$\mathbf{K}_1 = \begin{matrix} & \begin{matrix} 3 & \quad 4 & \quad\quad 1 & \quad\quad 2 \end{matrix} \\ & \begin{bmatrix} 348 & 464 & -348 & -464 \\ 464 & 618.67 & -464 & -618.67 \\ -348 & -464 & 348 & 464 \\ -464 & -618.67 & 464 & 618.67 \end{bmatrix} \begin{matrix} 3 \\ 4 \\ 1 \\ 2 \end{matrix} \end{matrix} \text{ k/in.} \qquad \textbf{(1)}$$

From Fig. 3.18(b), we observe that the code numbers for member 1 are 3, 4, 1, 2. These numbers are written on the right side and at the top of $\mathbf{K}_1$ (see Eq. (1)) to indicate the rows and columns, respectively, of the structure stiffness matrix **S**, in which the elements of $\mathbf{K}_1$ must be stored. Note that the elements of $\mathbf{K}_1$, which correspond to the restrained coordinate numbers 3 and 4, are simply ignored. Thus, the element in row 3 and column 3 of $\mathbf{K}_1$ is stored in row 1 and column 1 of **S**, as shown in Fig. 3.18(c); and the element in row 4 and column 3 of $\mathbf{K}_1$ is stored in row 2 and column 1 of **S**. The remaining elements of $\mathbf{K}_1$ are stored in **S** in a similar manner, as shown in Fig. 3.18(c).

Member 2 From Fig. 3.18(b), we can see that joint 3 is the beginning joint, and joint 1 is the end joint, for member 2. Applying Eqs. (3.62), we write

$$L = \sqrt{(X_1 - X_3)^2 + (Y_1 - Y_3)^2} = \sqrt{(12-12)^2 + (16-0)^2} = 16 \text{ ft}$$

$$\cos\theta = \frac{X_1 - X_3}{L} = \frac{12-12}{16} = 0$$

$$\sin\theta = \frac{Y_1 - Y_3}{L} = \frac{16-0}{16} = 1$$

Thus, using Eq. (3.73),

$$\mathbf{K}_2 = \begin{array}{c} \begin{matrix} 5 & 6 & 1 & 2 \end{matrix} \\ \left[\begin{array}{cc|cc} 0 & 0 & 0 & 0 \\ 0 & 906.25 & 0 & -906.25 \\ \hline 0 & 0 & 0 & 0 \\ 0 & -906.25 & 0 & 906.25 \end{array}\right] \begin{matrix} 5 \\ 6 \\ 1 \\ 2 \end{matrix} \end{array} \text{ k/in.}$$

From Fig. 3.18(b), we can see that the code numbers for this member are 5, 6, 1, 2. These numbers are used to store the pertinent elements of $\mathbf{K}_2$ in their proper positions in **S**, as shown in Fig. 3.18(c).

Member 3 It can be seen from Fig. 3.18(b) that joint 4 is the beginning joint, and joint 1 is the end joint, for member 3. Thus,

$$L = \sqrt{(X_1 - X_4)^2 + (Y_1 - Y_4)^2} = \sqrt{(12-24)^2 + (16-0)^2} = 20 \text{ ft}$$

$$\cos\theta = \frac{X_1 - X_4}{L} = \frac{12-24}{20} = -0.6$$

$$\sin\theta = \frac{Y_1 - Y_4}{L} = \frac{16-0}{20} = 0.8$$

Using Eq. (3.73),

$$\mathbf{K}_3 = \begin{array}{c} \begin{matrix} 7 & 8 & 1 & 2 \end{matrix} \\ \left[\begin{array}{cc|cc} 348 & -464 & -348 & 464 \\ -464 & 618.67 & 464 & -618.67 \\ \hline -348 & 464 & 348 & -464 \\ 464 & -618.67 & -464 & 618.67 \end{array}\right] \begin{matrix} 7 \\ 8 \\ 1 \\ 2 \end{matrix} \end{array} \text{ k/in.}$$

The code numbers for this member are 7, 8, 1, 2. Using these numbers, the pertinent elements of $\mathbf{K}_3$ are stored in **S**, as shown in Fig. 3.18(c).

The complete structure stiffness matrix **S**, obtained by assembling the stiffness coefficients of the three members of the truss, is given in Fig. 3.18(c). Note that **S** is symmetric.

Joint Load Vector: By comparing Figs. 3.18(a) and (b), we realize that

$$P_1 = 150 \text{ k} \qquad P_2 = -300 \text{ k}$$

Thus, the joint load vector is

$$\mathbf{P} = \begin{bmatrix} 150 \\ -300 \end{bmatrix} \text{k} \tag{2}$$

Joint Displacements: By substituting **P** and **S** into the structure stiffness relationship given by Eq. (3.89), we write

$$\mathbf{P} = \mathbf{Sd}$$

$$\begin{bmatrix} 150 \\ -300 \end{bmatrix} = \begin{bmatrix} 696 & 0 \\ 0 & 2{,}143.6 \end{bmatrix} \begin{bmatrix} d_1 \\ d_2 \end{bmatrix}$$

Solving these equations, we determine the joint displacements:

$$d_1 = 0.21552 \text{ in.} \qquad d_2 = -0.13995 \text{ in.}$$

or

$$\mathbf{d} = \begin{bmatrix} 0.21552 \\ -0.13995 \end{bmatrix} \text{in.}$$ **Ans**

To check that the solution of equations has been carried out correctly, we substitute the numerical values of joint displacements back into the structure stiffness relationship to obtain

$$\mathbf{P} = \mathbf{Sd} = \begin{bmatrix} 696 & 0 \\ 0 & 2{,}143.6 \end{bmatrix} \begin{bmatrix} 0.21552 \\ -0.13995 \end{bmatrix} = \begin{bmatrix} 150 \\ -300 \end{bmatrix}$$ **Checks**

which is the same as the load vector **P** given in Eq. (2), thereby indicating that the calculated joint displacements do indeed satisfy the structure stiffness relations.

Member End Displacements and End Forces:

Member 1 The member end displacements in the global coordinate system can be obtained simply by comparing the member's global degree of freedom numbers with its code numbers, as follows:

$$\mathbf{v}_1 = \begin{bmatrix} v_1 \\ v_2 \\ v_3 \\ v_4 \end{bmatrix} \begin{matrix} 3 \\ 4 \\ 1 \\ 2 \end{matrix} = \begin{bmatrix} 0 \\ 0 \\ d_1 \\ d_2 \end{bmatrix} = \begin{bmatrix} 0 \\ 0 \\ 0.21552 \\ -0.13995 \end{bmatrix} \text{in.} \tag{3}$$

Note that the code numbers for the member (3, 4, 1, 2) are written on the right side of **v**, as shown in Eq. (3). Because the code numbers corresponding to v_1 and v_2 are the restrained coordinate numbers 3 and 4, this indicates that $v_1 = v_2 = 0$. Similarly, the code numbers 1 and 2 corresponding to v_3 and v_4, respectively, indicate that $v_3 = d_1$ and $v_4 = d_2$. It should be clear that these compatibility equations could have been established alternatively by a simple visual inspection of the line diagram of the structure depicted in Fig. 3.18(b). However, as will be shown in Chapter 4, the use of the member code numbers enables us to conveniently program this procedure on a computer.

To determine the member end displacements in the local coordinate system, we first evaluate its transformation matrix **T** as defined in Eq. (3.61):

$$\mathbf{T}_1 = \begin{bmatrix} 0.6 & 0.8 & 0 & 0 \\ -0.8 & 0.6 & 0 & 0 \\ 0 & 0 & 0.6 & 0.8 \\ 0 & 0 & -0.8 & 0.6 \end{bmatrix}$$

The member local end displacements can now be calculated, using the relationship $\mathbf{u} = \mathbf{Tv}$ (Eq. (3.63)), as

$$\mathbf{u}_1 = \begin{bmatrix} u_1 \\ u_2 \\ u_3 \\ u_4 \end{bmatrix} = \begin{bmatrix} 0.6 & 0.8 & 0 & 0 \\ -0.8 & 0.6 & 0 & 0 \\ 0 & 0 & 0.6 & 0.8 \\ 0 & 0 & -0.8 & 0.6 \end{bmatrix} \begin{bmatrix} 0 \\ 0 \\ 0.21552 \\ -0.13995 \end{bmatrix} = \begin{bmatrix} 0 \\ 0 \\ 0.017352 \\ -0.25639 \end{bmatrix} \text{ in.}$$

Before we can calculate the member end forces in the local coordinate system, we need to determine its local stiffness matrix $\mathbf{k}$, using Eq. (3.27):

$$\mathbf{k}_1 = \begin{bmatrix} 966.67 & 0 & -966.67 & 0 \\ 0 & 0 & 0 & 0 \\ -966.67 & 0 & 966.67 & 0 \\ 0 & 0 & 0 & 0 \end{bmatrix} \text{ k/in.}$$

Now, using Eq. (3.7), we compute the member local end forces as

$$\mathbf{Q} = \mathbf{ku}$$

$$\mathbf{Q}_1 = \begin{bmatrix} Q_1 \\ Q_2 \\ Q_3 \\ Q_4 \end{bmatrix} = \begin{bmatrix} 966.67 & 0 & -966.67 & 0 \\ 0 & 0 & 0 & 0 \\ -966.67 & 0 & 966.67 & 0 \\ 0 & 0 & 0 & 0 \end{bmatrix} \begin{bmatrix} 0 \\ 0 \\ 0.017352 \\ -0.25639 \end{bmatrix} = \begin{bmatrix} -16.774 \\ 0 \\ 16.774 \\ 0 \end{bmatrix} \text{ k}$$

The member axial force is equal to the first element of the vector $\mathbf{Q}_1$; that is,

$$Q_{a1} = -16.774 \text{ k}$$

in which the negative sign indicates that the axial force is tensile, or

$$Q_{a1} = 16.774 \text{ k (T)}$$ **Ans**

This member axial force can be verified by visually examining the free-body diagram of the member subjected to the local end forces, as shown in Fig. 3.18(d).

By applying Eq. (3.66), we determine the member end forces in the global coordinate system:

$$\mathbf{F} = \mathbf{T}^T\mathbf{Q}$$

$$\mathbf{F}_1 = \begin{bmatrix} F_1 \\ F_2 \\ F_3 \\ F_4 \end{bmatrix} = \begin{bmatrix} 0.6 & -0.8 & 0 & 0 \\ 0.8 & 0.6 & 0 & 0 \\ 0 & 0 & 0.6 & -0.8 \\ 0 & 0 & 0.8 & 0.6 \end{bmatrix} \begin{bmatrix} -16.774 \\ 0 \\ 16.774 \\ 0 \end{bmatrix} = \begin{bmatrix} -10.064 \\ -13.419 \\ 10.064 \\ 13.419 \end{bmatrix} \begin{matrix} 3 \\ 4 \\ 1 \\ 2 \end{matrix} \text{ k} \tag{4}$$

Next, we write the member code numbers (3, 4, 1, 2) on the right side of $\mathbf{F}_1$ (see Eq. (4)), and store the pertinent elements of $\mathbf{F}_1$ in their proper positions in the reaction vector $\mathbf{R}$ by matching the code numbers (on the side of $\mathbf{F}_1$) to the restrained coordinate numbers written on the right side of $\mathbf{R}$ (see Fig. 3.18(e)). Thus, the element in row 1 of $\mathbf{F}_1$ (with code number 3) is stored in row 1 of $\mathbf{R}$ (with restrained coordinate number 3); and the element in row 2 of $\mathbf{F}_1$ (with code number 4) is stored in row 2 of $\mathbf{R}$ (with restrained coordinate number 4), as shown in Fig. 3.18(e). Note that the elements in rows 3 and 4 of $\mathbf{F}_1$, with code numbers corresponding to degrees of freedom 1 and 2 of the structure, are simply disregarded.

Member 2 The member end displacements in the global coordinate system are given by

$$\mathbf{v}_2 = \begin{bmatrix} v_1 \\ v_2 \\ v_3 \\ v_4 \end{bmatrix} \begin{matrix} 5 \\ 6 \\ 1 \\ 2 \end{matrix} = \begin{bmatrix} 0 \\ 0 \\ d_1 \\ d_2 \end{bmatrix} = \begin{bmatrix} 0 \\ 0 \\ 0.21552 \\ -0.13995 \end{bmatrix} \text{ in.}$$

The member end displacements in the local coordinate system can now be determined by using the relationship $\mathbf{u} = \mathbf{Tv}$ (Eq. (3.63)), with $\mathbf{T}$ as defined in Eq. (3.61):

$$\mathbf{u}_2 = \begin{bmatrix} u_1 \\ u_2 \\ u_3 \\ u_4 \end{bmatrix} = \begin{bmatrix} 0 & 1 & 0 & 0 \\ -1 & 0 & 0 & 0 \\ 0 & 0 & 0 & 1 \\ 0 & 0 & -1 & 0 \end{bmatrix} \begin{bmatrix} 0 \\ 0 \\ 0.21552 \\ -0.13995 \end{bmatrix} = \begin{bmatrix} 0 \\ 0 \\ -0.13995 \\ -0.21552 \end{bmatrix} \text{ in.}$$

Using Eq. (3.7), we compute member end forces in the local coordinate system:

$$\mathbf{Q} = \mathbf{ku}$$

$$\mathbf{Q}_2 = \begin{bmatrix} Q_1 \\ Q_2 \\ Q_3 \\ Q_4 \end{bmatrix} = 906.25 \begin{bmatrix} 1 & 0 & -1 & 0 \\ 0 & 0 & 0 & 0 \\ -1 & 0 & 1 & 0 \\ 0 & 0 & 0 & 0 \end{bmatrix} \begin{bmatrix} 0 \\ 0 \\ -0.13995 \\ -0.21552 \end{bmatrix} = \begin{bmatrix} 126.83 \\ 0 \\ -126.83 \\ 0 \end{bmatrix} \text{ k}$$

from which we obtain the member axial force (see also Fig. 3.18(d)):

$$Q_{a2} = 126.83 \text{ k (C)}$$ **Ans**

Using the relationship $\mathbf{F} = \mathbf{T}^T\mathbf{Q}$ (Eq. (3.66)), we calculate the member global end forces to be

$$\mathbf{F}_2 = \begin{bmatrix} F_1 \\ F_2 \\ F_3 \\ F_4 \end{bmatrix} = \begin{bmatrix} 0 & -1 & 0 & 0 \\ 1 & 0 & 0 & 0 \\ 0 & 0 & 0 & -1 \\ 0 & 0 & 1 & 0 \end{bmatrix} \begin{bmatrix} 126.83 \\ 0 \\ -126.83 \\ 0 \end{bmatrix} = \begin{bmatrix} 0 \\ 126.83 \\ \hdashline 0 \\ -126.83 \end{bmatrix} \begin{matrix} 5 \\ 6 \\ 1 \\ 2 \end{matrix} \text{ k}$$

The pertinent elements of $\mathbf{F}_2$ are now stored in their proper positions in the reaction vector $\mathbf{R}$, by using member code numbers (5, 6, 1, 2), as shown in Fig. 3.18(e).

Member 3 The member global end displacements are

$$\mathbf{v}_3 = \begin{bmatrix} v_1 \\ v_2 \\ v_3 \\ v_4 \end{bmatrix} \begin{matrix} 7 \\ 8 \\ 1 \\ 2 \end{matrix} = \begin{bmatrix} 0 \\ 0 \\ d_1 \\ d_2 \end{bmatrix} = \begin{bmatrix} 0 \\ 0 \\ 0.21552 \\ -0.13995 \end{bmatrix} \text{ in.}$$

As in the case of members 1 and 2, we can determine the end forces $\mathbf{Q}_3$ and $\mathbf{F}_3$ for member 3 by using the relationships $\mathbf{u} = \mathbf{Tv}$, $\mathbf{Q} = \mathbf{ku}$, and $\mathbf{F} = \mathbf{T}^T\mathbf{Q}$, in sequence. However, such member forces can also be obtained by applying sequentially the global stiffness relationship $\mathbf{F} = \mathbf{Kv}$ (Eq. (3.71)) and the transformation relation $\mathbf{Q} = \mathbf{TF}$ (Eq. (3.60)). Let us apply this alternative approach to determine the end forces for member 3.

Applying Eq. (3.71), we compute the member end forces in the global coordinate system:

$$\mathbf{F} = \mathbf{K}\mathbf{v}$$

$$\mathbf{F}_3 = \begin{bmatrix} F_1 \\ F_2 \\ F_3 \\ F_4 \end{bmatrix} = \begin{bmatrix} 348 & -464 & -348 & 464 \\ -464 & 618.67 & 464 & -618.67 \\ -348 & 464 & 348 & -464 \\ 464 & -618.67 & -464 & 618.67 \end{bmatrix} \begin{bmatrix} 0 \\ 0 \\ 0.21552 \\ -0.13995 \end{bmatrix}$$

$$= \begin{bmatrix} -139.94 \\ 186.58 \\ \hdashline 139.94 \\ -186.58 \end{bmatrix} \begin{matrix} 7 \\ 8 \\ 1 \\ 2 \end{matrix} \text{ k}$$

Using the member code numbers (7, 8, 1, 2), the pertinent elements of $\mathbf{F}_3$ are stored in the reaction vector $\mathbf{R}$, as shown in Fig. 3.18(e).

The member end forces in the local coordinate system can now be obtained by using the transformation relationship $\mathbf{Q} = \mathbf{TF}$ (Eq. (3.60)), with $\mathbf{T}$ as defined in Eq. (3.61).

$$\mathbf{Q}_3 = \begin{bmatrix} Q_1 \\ Q_2 \\ Q_3 \\ Q_4 \end{bmatrix} = \begin{bmatrix} -0.6 & 0.8 & 0 & 0 \\ -0.8 & -0.6 & 0 & 0 \\ 0 & 0 & -0.6 & 0.8 \\ 0 & 0 & -0.8 & -0.6 \end{bmatrix} \begin{bmatrix} -139.94 \\ 186.58 \\ 139.94 \\ -186.58 \end{bmatrix} = \begin{bmatrix} 233.23 \\ 0 \\ -233.23 \\ 0 \end{bmatrix} \text{ k}$$

from which the member axial force is found to be (see also Fig. 3.18(d))

$$Q_{a3} = 233.23 \text{ k (C)}$$ **Ans**

Support Reactions: The completed reaction vector $\mathbf{R}$ is shown in Fig. 3.18(e), and the support reactions are depicted on a line diagram of the truss in Fig. 3.18(f). **Ans**

Equilibrium Check: Applying the equations of equilibrium to the free body of the entire truss (Fig. 3.18(f)), we obtain

$$+\rightarrow \sum F_X = 0 \qquad 150 - 10.064 - 139.94 = -0.004 \approx 0$$ **Checks**

$$+\uparrow \sum F_Y = 0 \qquad -300 - 13.419 + 126.83 + 186.58 = -0.009 \approx 0$$ **Checks**

$$+\circlearrowleft \sum M_{①} = 0 \qquad -10.064(16) + 13.419(12) - 139.94(16) + 186.58(12) = -0.076 \text{ k-ft} \approx 0$$ **Checks**

EXAMPLE 3.9

Determine the joint displacements, member axial forces, and support reactions for the truss shown in Fig. 3.19(a), using the matrix stiffness method.

SOLUTION

Analytical Model: From the analytical model of the truss shown in Fig. 3.19(b), we observe that the structure has three degrees of freedom (numbered 1, 2, and 3), and five restrained coordinates (numbered 4 through 8). Note that for horizontal member 2, the left end joint 3 is chosen as the beginning joint, so that the positive directions of local axes are the same as the global axes. Thus, no coordinate transformations are necessary for this member; that is, the member stiffness relations in the local and global coordinate systems are the same.

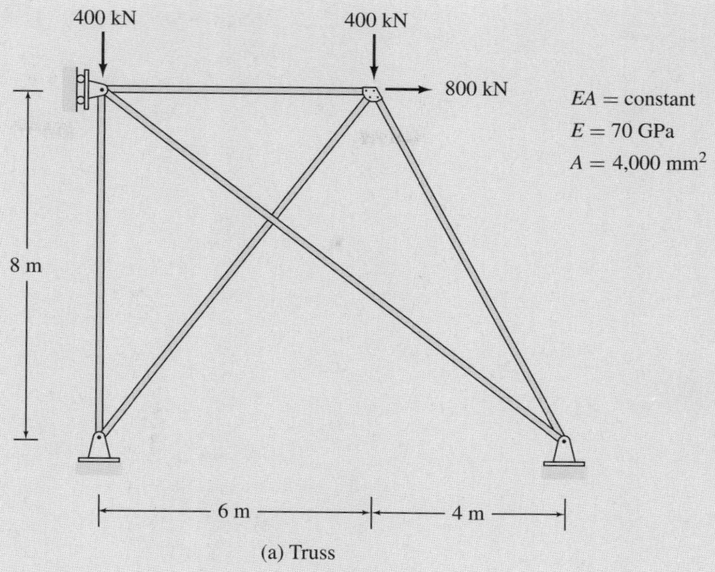

(a) Truss

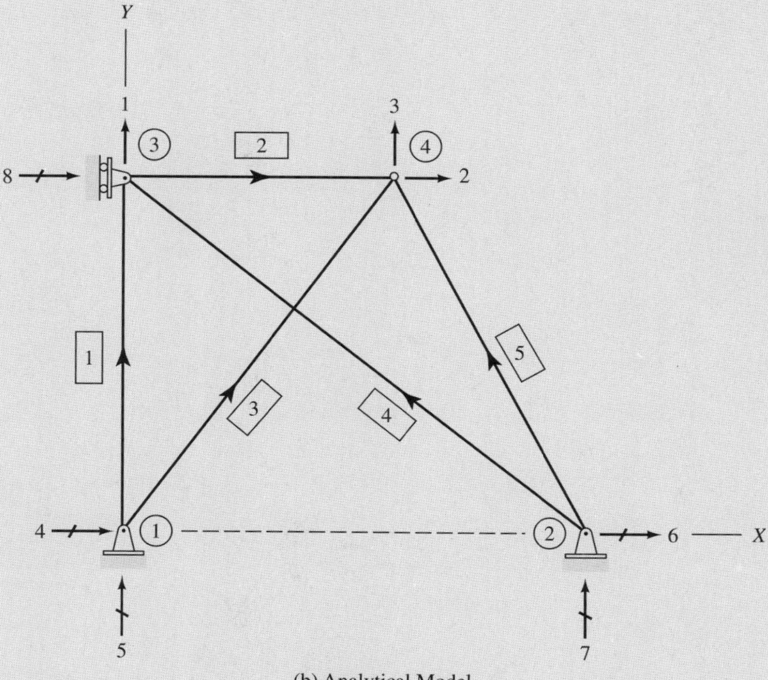

(b) Analytical Model

Fig. 3.19

$$\mathbf{S} = \begin{bmatrix} 35{,}000 + 8{,}533 & 0 & 0 \\ 0 & 46{,}667 + 10{,}080 + 6{,}260.9 & 13{,}440 - 12{,}522 \\ 0 & 13{,}440 - 12{,}522 & 17{,}920 + 25{,}043 \end{bmatrix} \begin{matrix} 1 \\ 2 \\ 3 \end{matrix}$$

(column numbers 1, 2, 3)

$$= \begin{bmatrix} 43{,}533 & 0 & 0 \\ 0 & 63{,}008 & 918 \\ 0 & 918 & 42{,}963 \end{bmatrix} \begin{matrix} 1 \\ 2 \\ 3 \end{matrix} \text{ kN/m}$$

(column numbers 1, 2, 3)

(c) Structure Stiffness Matrix

$$\mathbf{R} = \begin{bmatrix} -0.57994 \\ 321.59 - 0.77325 \\ -98.008 - 200.38 \\ 78.407 + 400.76 \\ -599.06 + 98.008 \end{bmatrix} \begin{matrix} 4 \\ 5 \\ 6 \\ 7 \\ 8 \end{matrix} = \begin{bmatrix} -0.57994 \\ 320.82 \\ -298.39 \\ 479.17 \\ -501.05 \end{bmatrix} \begin{matrix} 4 \\ 5 \\ 6 \\ 7 \\ 8 \end{matrix} \text{ kN}$$

(d) Support Reaction Vector

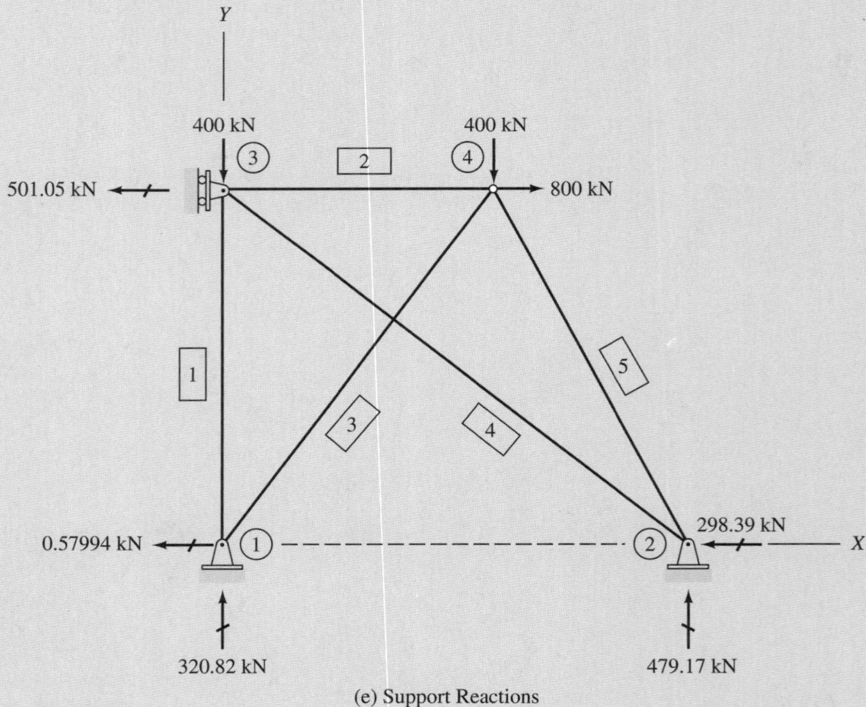

(e) Support Reactions

Fig. 3.19 *(continued)*

Structure Stiffness Matrix:

Member 1 Using Eqs. (3.62), we write

$$L = \sqrt{(X_3 - X_1)^2 + (Y_3 - Y_1)^2} = \sqrt{(0-0)^2 + (8-0)^2} = 8\text{ m}$$

$$\cos\theta = \frac{X_3 - X_1}{L} = \frac{0-0}{8} = 0$$

$$\sin\theta = \frac{Y_3 - Y_1}{L} = \frac{8-0}{8} = 1$$

Using the units of kN and meters, we obtain the member global stiffness matrix (Eq. (3.73)):

$$\mathbf{K}_1 = \frac{70(10^6)(0.004)}{8}\begin{bmatrix} 0 & 0 & 0 & 0 \\ 0 & 1 & 0 & -1 \\ 0 & 0 & 0 & 0 \\ 0 & -1 & 0 & 1 \end{bmatrix} = \begin{array}{c} \begin{matrix} 4 & \quad 5 & \quad 8 & \quad 1 \end{matrix} \\ \begin{bmatrix} 0 & 0 & 0 & 0 \\ 0 & 35{,}000 & 0 & -35{,}000 \\ 0 & 0 & 0 & 0 \\ 0 & -35{,}000 & 0 & 35{,}000 \end{bmatrix} \end{array} \begin{matrix} 4 \\ 5 \\ 8 \\ 1 \end{matrix} \text{ kN/m}$$

From Fig. 3.19(b), we observe that the code numbers for member 1 are 4, 5, 8, 1. These numbers are written on the right side and at the top of $\mathbf{K}_1$, and the pertinent elements of $\mathbf{K}_1$ are stored in their proper positions in the structure stiffness matrix $\mathbf{S}$, as shown in Fig. 3.19(c).

Member 2 As discussed, no coordinate transformations are needed for this horizontal member; that is, $\mathbf{T}_2 = \mathbf{I}$, and $\mathbf{K}_2 = \mathbf{k}_2$. Substituting $E = 70(10^6)$ kN/m^2, $A = 0.004$ m^2, and $L = 6$ m into Eq. (3.27), we obtain

$$\mathbf{K}_2 = \mathbf{k}_2 = \begin{array}{c} \begin{matrix} 8 & \quad 1 & \quad 2 & \quad 3 \end{matrix} \\ \begin{bmatrix} 46{,}667 & 0 & -46{,}667 & 0 \\ 0 & 0 & 0 & 0 \\ -46{,}667 & 0 & 46{,}667 & 0 \\ 0 & 0 & 0 & 0 \end{bmatrix} \end{array} \begin{matrix} 8 \\ 1 \\ 2 \\ 3 \end{matrix} \text{ kN/m}$$

From Fig. 3.19(b), we can see that the code numbers for member 2 are 8, 1, 2, 3. These numbers are used to store the appropriate elements of $\mathbf{K}_2$ in $\mathbf{S}$, as shown in Fig. 3.19(c).

Member 3

$$L = \sqrt{(X_4 - X_1)^2 + (Y_4 - Y_1)^2} = \sqrt{(6-0)^2 + (8-10)^2} = 10\text{ m}$$

$$\cos\theta = \frac{X_4 - X_1}{L} = \frac{6-0}{10} = 0.6$$

$$\sin\theta = \frac{Y_4 - Y_1}{L} = \frac{8-0}{10} = 0.8$$

$$\mathbf{K}_3 = \begin{array}{c} \begin{matrix} 4 & \quad 5 & \quad 2 & \quad 3 \end{matrix} \\ \begin{bmatrix} 10{,}080 & 13{,}440 & -10{,}080 & -13{,}440 \\ 13{,}440 & 17{,}920 & -13{,}440 & -17{,}920 \\ -10{,}080 & -13{,}440 & 10{,}080 & 13{,}440 \\ -13{,}440 & -17{,}920 & 13{,}440 & 17{,}920 \end{bmatrix} \end{array} \begin{matrix} 4 \\ 5 \\ 2 \\ 3 \end{matrix} \text{ kN/m}$$

Using the code numbers (4, 5, 2, 3) of member 3, the relevant elements of $\mathbf{K}_3$ are stored in $\mathbf{S}$, as shown in Fig. 3.19(c).

Member 4

$$L = \sqrt{(X_3 - X_2)^2 + (Y_3 - Y_2)^2} = \sqrt{(0-10)^2 + (8-0)^2} = 12.806 \text{ m}$$

$$\cos\theta = \frac{X_3 - X_2}{L} = \frac{0-10}{12.806} = -0.78088$$

$$\sin\theta = \frac{Y_3 - Y_2}{L} = \frac{8-0}{12.806} = 0.62471$$

$$\mathbf{K}_4 = \begin{array}{c} \begin{matrix} 6 & \quad 7 & \quad 8 & \quad 1 \end{matrix} \\ \left[\begin{array}{rrr:r} 13{,}333 & -10{,}666 & -13{,}333 & 10{,}666 \\ -10{,}666 & 8{,}533 & 10{,}666 & -8{,}533 \\ -13{,}333 & 10{,}666 & 13{,}333 & -10{,}666 \\ \hdashline 10{,}666 & -8{,}533 & -10{,}666 & 8{,}533 \end{array}\right] \end{array} \begin{array}{c} \\ 6 \\ 7 \\ 8 \\ 1 \end{array} \text{ kN/m}$$

The member code numbers are 6, 7, 8, 1. Thus, the element in row 4 and column 4 of $\mathbf{K}_4$ is stored in row 1 and column 1 of $\mathbf{S}$, as shown in Fig. 3.19(c).

Member 5

$$L = \sqrt{(X_4 - X_2)^2 + (Y_4 - Y_2)^2} = \sqrt{(6-10)^2 + (8-0)^2} = 8.9443 \text{ m}$$

$$\cos\theta = \frac{X_4 - X_2}{L} = \frac{6-10}{8.9443} = -0.44721$$

$$\sin\theta = \frac{Y_4 - Y_2}{L} = \frac{8-0}{8.9443} = 0.89442$$

$$\mathbf{K}_5 = \begin{array}{c} \begin{matrix} 6 & \quad 7 & \quad 2 & \quad 3 \end{matrix} \\ \left[\begin{array}{rr:rr} 6{,}260.9 & -12{,}522 & -6{,}260.9 & 12{,}522 \\ -12{,}522 & 25{,}043 & 12{,}522 & -25{,}043 \\ \hdashline -6{,}260.9 & 12{,}522 & 6{,}260.9 & -12{,}522 \\ 12{,}522 & -25{,}043 & -12{,}522 & 25{,}043 \end{array}\right] \end{array} \begin{array}{c} \\ 6 \\ 7 \\ 2 \\ 3 \end{array} \text{ kN/m}$$

The code numbers for member 5 are 6, 7, 2, 3. These numbers are used to store the pertinent elements of $\mathbf{K}_5$ in $\mathbf{S}$.

The completed structure stiffness matrix $\mathbf{S}$ is given in Fig. 3.19(c).

Joint Load Vector: By comparing Figs. 3.19(a) and (b), we obtain

$$\mathbf{P} = \begin{bmatrix} -400 \\ 800 \\ -400 \end{bmatrix} \text{ kN}$$

Joint Displacements: The structure stiffness relationship (Eq. (3.89)) can now be written as

$$\mathbf{P} = \mathbf{Sd}$$

$$\begin{bmatrix} -400 \\ 800 \\ -400 \end{bmatrix} = \begin{bmatrix} 43{,}533 & 0 & 0 \\ 0 & 63{,}008 & 918 \\ 0 & 918 & 42{,}963 \end{bmatrix} \begin{bmatrix} d_1 \\ d_2 \\ d_3 \end{bmatrix}$$

Solving these equations simultaneously, we determine the joint displacements.

$$\mathbf{d} = \begin{bmatrix} -0.0091884 \\ 0.012837 \\ -0.0095846 \end{bmatrix} \text{ m} = \begin{bmatrix} -9.1884 \\ 12.837 \\ -9.5846 \end{bmatrix} \text{ mm}$$

Ans

To check our solution, the numerical values of **d** are back-substituted into the structure stiffness relation $\mathbf{P} = \mathbf{Sd}$ to obtain

$$\mathbf{P} = \mathbf{Sd} = \begin{bmatrix} 43{,}533 & 0 & 0 \\ 0 & 63{,}008 & 918 \\ 0 & 918 & 42{,}963 \end{bmatrix} \begin{bmatrix} -0.0091884 \\ 0.012837 \\ -0.0095846 \end{bmatrix} = \begin{bmatrix} -400 \\ 800.04 \approx 800 \\ -400 \end{bmatrix}$$

Checks

Member End Displacements and End Forces:

Member 1 The global end displacements of member 1 are obtained by comparing its global degree-of-freedom numbers with its code numbers. Thus,

$$\mathbf{v}_1 = \begin{bmatrix} v_1 \\ v_2 \\ v_3 \\ v_4 \end{bmatrix} \begin{matrix} 4 \\ 5 \\ 8 \\ 1 \end{matrix} = \begin{bmatrix} 0 \\ 0 \\ 0 \\ d_1 \end{bmatrix} = \begin{bmatrix} 0 \\ 0 \\ 0 \\ -0.0091884 \end{bmatrix} \text{ m}$$

To determine its local end displacements, we apply the relationship $\mathbf{u} = \mathbf{Tv}$ (Eq. (3.63)), with **T** as given in Eq. (3.61):

$$\mathbf{u}_1 = \begin{bmatrix} u_1 \\ u_2 \\ u_3 \\ u_4 \end{bmatrix} = \begin{bmatrix} 0 & 1 & 0 & 0 \\ -1 & 0 & 0 & 0 \\ 0 & 0 & 0 & 1 \\ 0 & 0 & -1 & 0 \end{bmatrix} \begin{bmatrix} 0 \\ 0 \\ 0 \\ -0.0091884 \end{bmatrix} = \begin{bmatrix} 0 \\ 0 \\ -0.0091884 \\ 0 \end{bmatrix} \text{ m}$$

Next, we compute the end forces in the local coordinate system by using the relationship $\mathbf{Q} = \mathbf{ku}$ (Eq. (3.7)), with **k** as defined in Eq. (3.27). Thus,

$$\mathbf{Q}_1 = \begin{bmatrix} Q_1 \\ Q_2 \\ Q_3 \\ Q_4 \end{bmatrix} = 35{,}000 \begin{bmatrix} 1 & 0 & -1 & 0 \\ 0 & 0 & 0 & 0 \\ -1 & 0 & 1 & 0 \\ 0 & 0 & 0 & 0 \end{bmatrix} \begin{bmatrix} 0 \\ 0 \\ -0.0091884 \\ 0 \end{bmatrix} = \begin{bmatrix} 321.59 \\ 0 \\ -321.59 \\ 0 \end{bmatrix} \text{ kN}$$

Therefore, the member axial force, which equals the first element of the vector $\mathbf{Q}_1$, is

$$Q_{a1} = 321.59 \text{ kN (C)}$$

Ans

The global end forces can now be obtained by using the relationship $\mathbf{F} = \mathbf{T}^T\mathbf{Q}$ (Eq. (3.66)):

$$\mathbf{F}_1 = \begin{bmatrix} F_1 \\ F_2 \\ F_3 \\ F_4 \end{bmatrix} = \begin{bmatrix} 0 & -1 & 0 & 0 \\ 1 & 0 & 0 & 0 \\ 0 & 0 & 0 & -1 \\ 0 & 0 & 1 & 0 \end{bmatrix} \begin{bmatrix} 321.59 \\ 0 \\ -321.59 \\ 0 \end{bmatrix} = \begin{bmatrix} 0 \\ 321.59 \\ 0 \\ -321.59 \end{bmatrix} \begin{matrix} 4 \\ 5 \\ 8 \\ 1 \end{matrix} \text{ kN}$$

Using the code numbers (4, 5, 8, 1), the elements of $\mathbf{F}_1$ corresponding to the restrained coordinates (4 through 8) are stored in their proper positions in **R**, as shown in Fig. 3.19(d).

Member 2

$$\mathbf{u}_2 = \mathbf{v}_2 = \begin{bmatrix} v_1 \\ v_2 \\ v_3 \\ v_4 \end{bmatrix} \begin{matrix} 8 \\ 1 \\ 2 \\ 3 \end{matrix} = \begin{bmatrix} 0 \\ d_1 \\ d_2 \\ d_3 \end{bmatrix} = \begin{bmatrix} 0 \\ -0.0091884 \\ 0.012837 \\ -0.0095846 \end{bmatrix} \text{ m}$$

Using the relationship $\mathbf{Q} = \mathbf{ku}$ (Eq. (3.7)), we determine the member end forces:

$$\mathbf{F}_2 = \mathbf{Q}_2 = 46{,}667 \begin{bmatrix} 1 & 0 & -1 & 0 \\ 0 & 0 & 0 & 0 \\ -1 & 0 & 1 & 0 \\ 0 & 0 & 0 & 0 \end{bmatrix} \begin{bmatrix} 0 \\ -0.0091884 \\ 0.012837 \\ -0.0095846 \end{bmatrix} = \begin{bmatrix} -599.06 \\ 0 \\ 599.06 \\ 0 \end{bmatrix} \begin{matrix} 8 \\ 1 \\ 2 \\ 3 \end{matrix} \text{ kN}$$

from which the member axial force is found.

$$Q_{a2} = -599.06 \text{ kN} = 599.06 \text{ kN (T)} \qquad \textbf{Ans}$$

The element in the first row of $\mathbf{F}_2$ (with code number 8) is stored in the fifth row of $\mathbf{R}$ (with restrained coordinate number 8), as shown in Fig. 3.19(d).

Member 3

$$\mathbf{v}_3 = \begin{bmatrix} 0 \\ 0 \\ 0.012837 \\ -0.0095846 \end{bmatrix} \begin{matrix} 4 \\ 5 \\ 2 \\ 3 \end{matrix} \text{ m}$$

Using Eq. (3.63),

$$\mathbf{u} = \mathbf{Tv}$$

$$\mathbf{u}_3 = \begin{bmatrix} 0.6 & 0.8 & 0 & 0 \\ -0.8 & 0.6 & 0 & 0 \\ 0 & 0 & 0.6 & 0.8 \\ 0 & 0 & -0.8 & 0.6 \end{bmatrix} \begin{bmatrix} 0 \\ 0 \\ 0.012837 \\ -0.0095846 \end{bmatrix} = \begin{bmatrix} 0 \\ 0 \\ 0.00003452 \\ -0.01602 \end{bmatrix} \text{ m}$$

Applying Eq. (3.7),

$$\mathbf{Q} = \mathbf{ku}$$

$$\mathbf{Q}_3 = 28{,}000 \begin{bmatrix} 1 & 0 & -1 & 0 \\ 0 & 0 & 0 & 0 \\ -1 & 0 & 1 & 0 \\ 0 & 0 & 0 & 0 \end{bmatrix} \begin{bmatrix} 0 \\ 0 \\ 0.00003452 \\ -0.01602 \end{bmatrix} = \begin{bmatrix} -0.96656 \\ 0 \\ 0.96656 \\ 0 \end{bmatrix} \text{ kN}$$

from which,

$$Q_{a3} = -0.96656 \text{ kN} = 0.96656 \text{ kN (T)} \qquad \textbf{Ans}$$

From Eq. (3.66), we obtain

$$\mathbf{F} = \mathbf{T}^T\mathbf{Q}$$

$$\mathbf{F}_3 = \begin{bmatrix} 0.6 & -0.8 & 0 & 0 \\ 0.8 & 0.6 & 0 & 0 \\ 0 & 0 & 0.6 & -0.8 \\ 0 & 0 & 0.8 & 0.6 \end{bmatrix} \begin{bmatrix} -0.96656 \\ 0 \\ 0.96656 \\ 0 \end{bmatrix} = \begin{bmatrix} -0.57994 \\ -0.77325 \\ 0.57994 \\ 0.77325 \end{bmatrix} \begin{matrix} 4 \\ 5 \\ 2 \\ 3 \end{matrix} \text{ kN}$$

The pertinent elements of $\mathbf{F}_3$ are stored in $\mathbf{R}$, using the member code numbers (4, 5, 2, 3), as shown in Fig. 3.19(d).

Member 4

$$\mathbf{v}_4 = \begin{bmatrix} 0 \\ 0 \\ 0 \\ -0.0091884 \end{bmatrix} \begin{matrix} 6 \\ 7 \\ 8 \\ 1 \end{matrix} \quad \text{m}$$

$$\mathbf{u} = \mathbf{T}\mathbf{v}$$

$$\mathbf{u}_4 = \begin{bmatrix} -0.78088 & 0.62471 & 0 & 0 \\ -0.62471 & -0.78088 & 0 & 0 \\ 0 & 0 & -0.78088 & 0.62471 \\ 0 & 0 & -0.62471 & -0.78088 \end{bmatrix} \begin{bmatrix} 0 \\ 0 \\ 0 \\ -0.0091884 \end{bmatrix}$$

$$= \begin{bmatrix} 0 \\ 0 \\ -0.0057401 \\ 0.007175 \end{bmatrix} \text{m}$$

$$\mathbf{Q} = \mathbf{k}\mathbf{u}$$

$$\mathbf{Q}_4 = 21{,}865 \begin{bmatrix} 1 & 0 & -1 & 0 \\ 0 & 0 & 0 & 0 \\ -1 & 0 & 1 & 0 \\ 0 & 0 & 0 & 0 \end{bmatrix} \begin{bmatrix} 0 \\ 0 \\ -0.0057401 \\ 0.007175 \end{bmatrix} = \begin{bmatrix} 125.51 \\ 0 \\ -125.51 \\ 0 \end{bmatrix} \text{kN}$$

from which,

$$Q_{a4} = 125.51 \text{ kN (C)} \qquad \textbf{Ans}$$

$$\mathbf{F} = \mathbf{T}^T\mathbf{Q}$$

$$\mathbf{F}_4 = \begin{bmatrix} -0.78088 & -0.62471 & 0 & 0 \\ 0.62471 & -0.78088 & 0 & 0 \\ 0 & 0 & -0.78088 & -0.62471 \\ 0 & 0 & 0.62471 & -0.78088 \end{bmatrix} \begin{bmatrix} 125.51 \\ 0 \\ -125.51 \\ 0 \end{bmatrix}$$

$$= \begin{bmatrix} -98.008 \\ 78.407 \\ 98.008 \\ -78.407 \end{bmatrix} \begin{matrix} 6 \\ 7 \\ 8 \\ 1 \end{matrix} \quad \text{kN}$$

The relevant elements of $\mathbf{F}_4$ are stored in $\mathbf{R}$, as shown in Fig. 3.19(d).

Member 5

$$\mathbf{v}_5 = \begin{bmatrix} 0 \\ 0 \\ 0.012837 \\ -0.0095846 \end{bmatrix} \begin{matrix} 6 \\ 7 \\ 2 \\ 3 \end{matrix} \quad \text{m}$$

$$\mathbf{u} = \mathbf{T}\mathbf{v}$$

$$\mathbf{u}_5 = \begin{bmatrix} -0.44721 & 0.89442 & 0 & 0 \\ -0.89442 & -0.44721 & 0 & 0 \\ 0 & 0 & -0.44721 & 0.89442 \\ 0 & 0 & -0.89442 & -0.44721 \end{bmatrix} \begin{bmatrix} 0 \\ 0 \\ 0.012837 \\ -0.0095846 \end{bmatrix}$$

$$= \begin{bmatrix} 0 \\ 0 \\ -0.014313 \\ -0.0071953 \end{bmatrix} \text{m}$$

$$\mathbf{Q} = \mathbf{k}\mathbf{u}$$

$$\mathbf{Q}_5 = 31{,}305 \begin{bmatrix} 1 & 0 & -1 & 0 \\ 0 & 0 & 0 & 0 \\ -1 & 0 & 1 & 0 \\ 0 & 0 & 0 & 0 \end{bmatrix} \begin{bmatrix} 0 \\ 0 \\ -0.014313 \\ -0.0071953 \end{bmatrix} = \begin{bmatrix} 448.07 \\ 0 \\ -448.07 \\ 0 \end{bmatrix} \text{kN}$$

Thus,

$$Q_{a5} = 448.07 \text{ kN (C)}$$ **Ans**

$$\mathbf{F} = \mathbf{T}^T\mathbf{Q}$$

$$\mathbf{F}_5 = \begin{bmatrix} -0.44721 & -0.89442 & 0 & 0 \\ 0.89442 & -0.44721 & 0 & 0 \\ 0 & 0 & -0.44721 & -0.89442 \\ 0 & 0 & 0.89442 & -0.44721 \end{bmatrix} \begin{bmatrix} 448.07 \\ 0 \\ -448.07 \\ 0 \end{bmatrix}$$

$$= \begin{bmatrix} -200.38 \\ 400.76 \\ 200.38 \\ -400.76 \end{bmatrix} \begin{matrix} 6 \\ 7 \\ 2 \\ 3 \end{matrix} \text{kN}$$

The pertinent elements of $\mathbf{F}_5$ are stored in $\mathbf{R}$, as shown in Fig. 3.19(d).

Support Reactions: The completed reaction vector $\mathbf{R}$ is given in Fig. 3.19(d), and the support reactions are shown on a line diagram of the structure in Fig. 3.19(e). **Ans**

Equilibrium Check: Considering the equilibrium of the entire truss, we write (Fig. 3.19(e)),

$$+ \rightarrow \sum F_X = 0 \quad -0.57994 - 298.39 - 501.05 + 800 = -0.02 \text{ kN} \approx 0$$ **Checks**

$$+ \uparrow \sum F_Y = 0 \quad 320.82 + 479.17 - 400 - 400 = -0.01 \text{ kN} \approx 0$$ **Checks**

$$+ \circlearrowleft \sum M_{①} = 0 \quad 479.17(10) + 501.05(8) - 800(8) - 400(6) = 0.1 \text{ kN} \cdot \text{m} \approx 0$$ **Checks**

SUMMARY

In this chapter, we have studied the basic concepts of the analysis of plane trusses based on the matrix stiffness method. A block diagram that summarizes the various steps involved in this analysis is presented in Fig. 3.20.

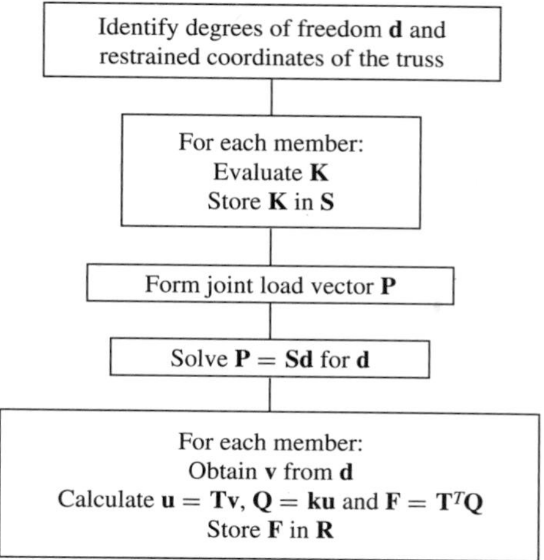
Identify degrees of freedom **d** and restrained coordinates of the truss
For each member:
Evaluate **K**
Store **K** in **S**
Form joint load vector **P**
Solve **P** = **Sd** for **d**
For each member:
Obtain **v** from **d**
Calculate **u** = **Tv**, **Q** = **ku** and **F** = $\mathbf{T}^T\mathbf{Q}$
Store **F** in **R**

Fig. 3.20

PROBLEMS

Section 3.2

3.1 through 3.3 Identify by numbers the degrees of freedom and restrained coordinates of the trusses shown in Figs. P3.1–P3.3. Also, form the joint load vector **P** for the trusses.

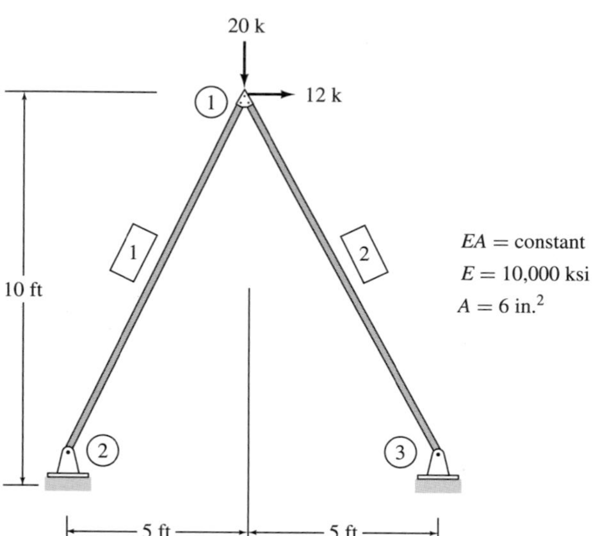
20 k
12 k
10 ft
5 ft
5 ft
EA = constant
E = 10,000 ksi
A = 6 in.²

Fig. P3.1, P3.17

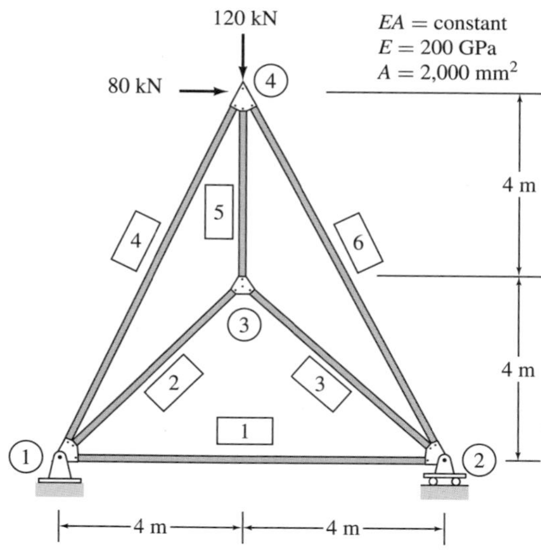
120 kN
80 kN
EA = constant
E = 200 GPa
A = 2,000 mm²
4 m
4 m
4 m
4 m

Fig. P3.2, P3.23

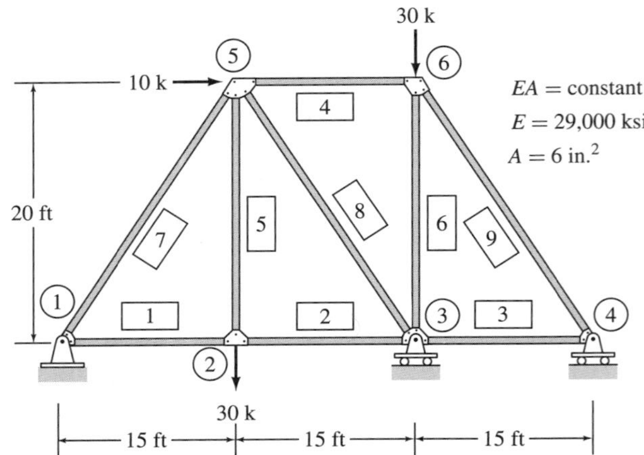

Fig. P3.3, P3.25

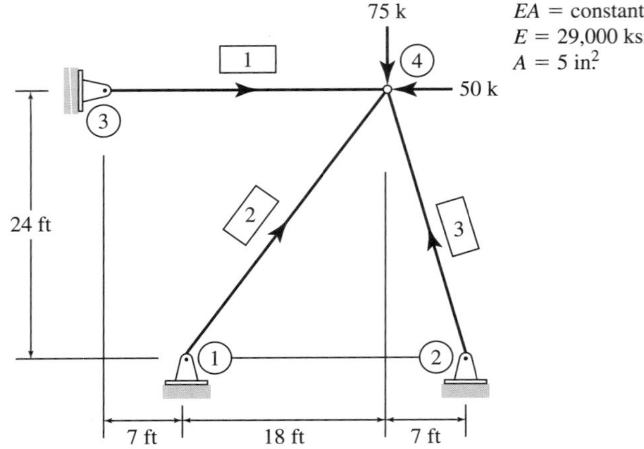

Fig. P3.5, P3.9, P3.15, P3.19

Section 3.3

3.4 and 3.5 Determine the local stiffness matrix **k** for each member of the trusses shown in Figs. P3.4 and P3.5.

3.6 If end displacements in the local coordinate system for member 5 of the truss shown in Fig. P3.6 are

$$\mathbf{u}_5 = \begin{bmatrix} -0.5 \\ 0.5 \\ 0.75 \\ 1.25 \end{bmatrix} \text{ in.}$$

calculate the axial force in the member.

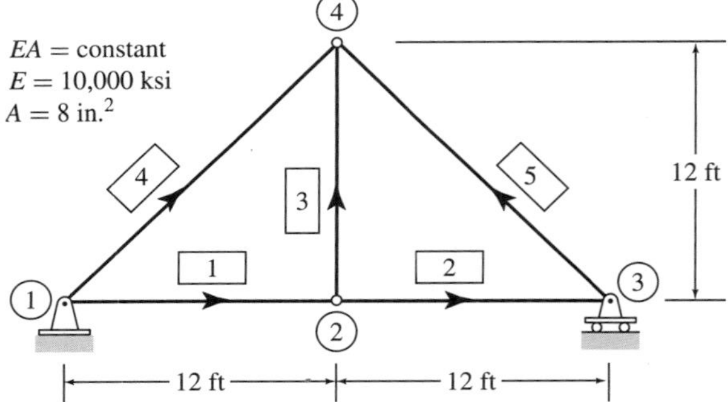

Fig. P3.6, P3.10, P3.12

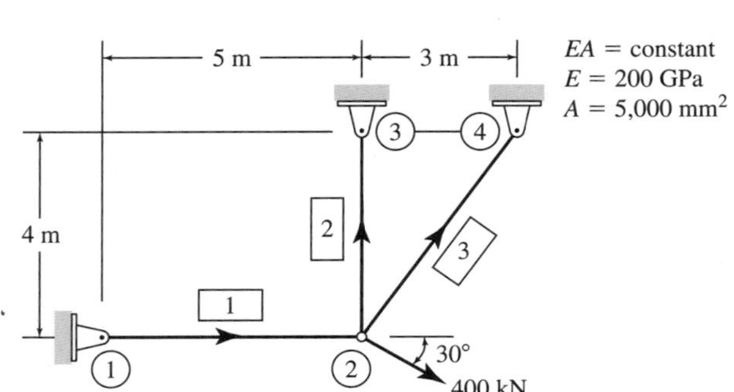

Fig. P3.4, P3.8, P3.14, P3.18

3.7 If end displacements in the local coordinate system for member 9 of the truss shown in Fig. P3.7 are

$$\mathbf{u}_9 = \begin{bmatrix} 17.6 \\ 3.2 \\ 33 \\ 6 \end{bmatrix} \text{ mm}$$

calculate the axial force in the member.

Section 3.5

3.8 and 3.9 Determine the transformation matrix **T** for each member of the trusses shown in Figs. P3.8 and P3.9.

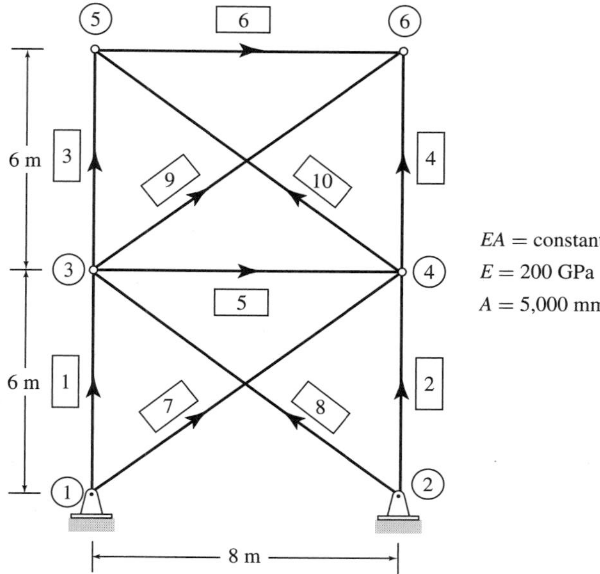

Fig. P3.7, P3.11, P3.13

3.10 If the end displacements in the global coordinate system for member 5 of the truss shown in Fig. P3.10 are

$$\mathbf{v}_5 = \begin{bmatrix} 0.5 \\ 0 \\ 0.25 \\ -1 \end{bmatrix} \text{ in.}$$

calculate the end forces for the member in the global coordinate system. Is the member in equilibrium under these forces?

3.11 If the end displacements in the global coordinate system for member 9 of the truss shown in Fig. P3.11 are

$$\mathbf{v}_9 = \begin{bmatrix} 16 \\ -8 \\ 30 \\ -15 \end{bmatrix} \text{ mm}$$

calculate the end forces for the member in the global coordinate system. Is the member in equilibrium under these forces?

Section 3.6

3.12 Solve Problem 3.10, using the member stiffness relationship in the global coordinate system, $\mathbf{F} = \mathbf{K}\mathbf{v}$.

3.13 Solve Problem 3.11, using the member stiffness relationship in the global coordinate system, $\mathbf{F} = \mathbf{K}\mathbf{v}$.

Section 3.7

3.14 and 3.15 Determine the structure stiffness matrices **S** for the trusses shown in Figs. P3.14 and P3.15.

Section 3.8

3.16 through 3.25 Determine the joint displacements, member axial forces, and support reactions for the trusses shown in Figs. P3.16 through P3.25, using the matrix stiffness method. Check the hand-calculated results by using the computer program on the publisher's website for this book (www.cengage.com/engineering), or by using any other general purpose structural analysis program available.

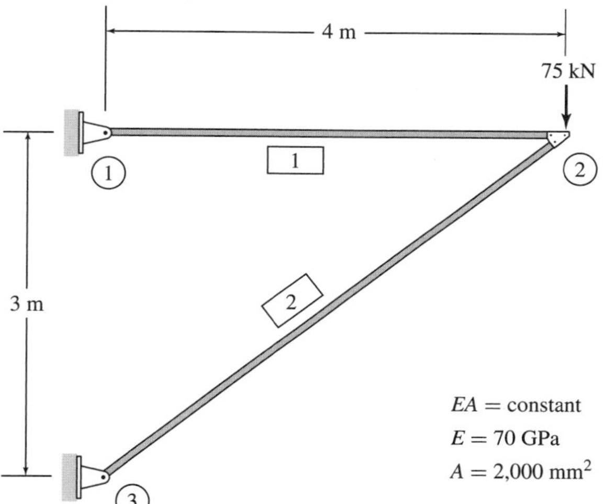

Fig. P3.16

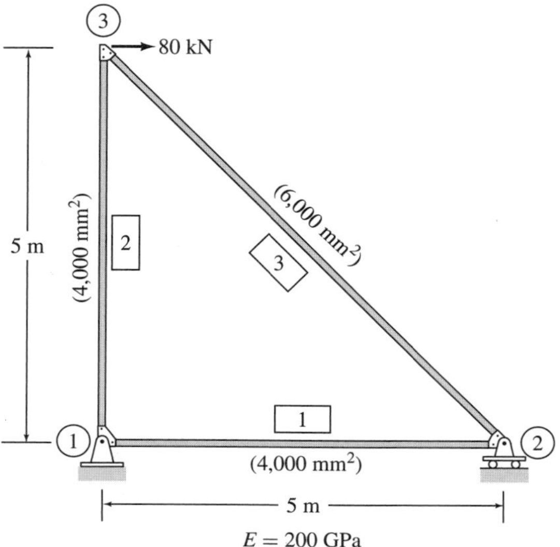

Fig. P3.20

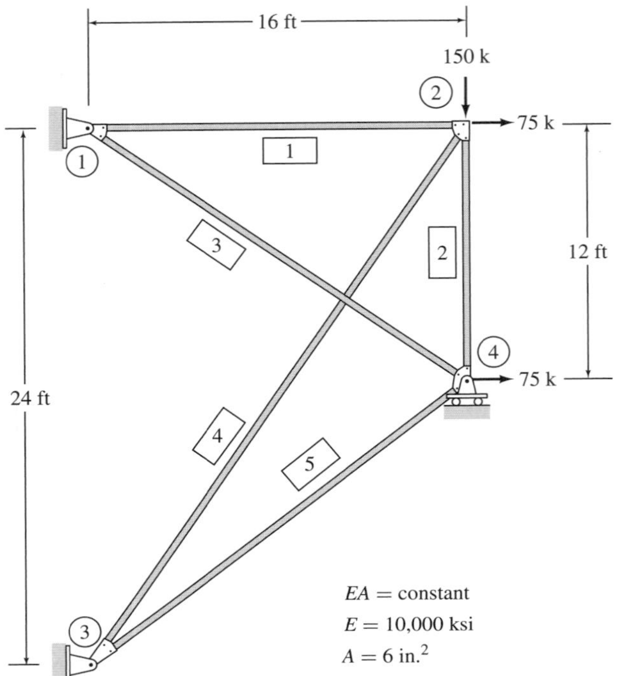

Fig. P3.21

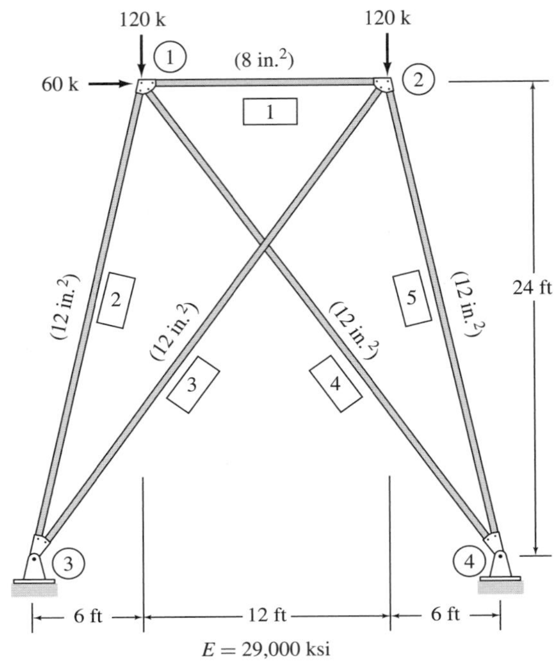

Fig. P3.22

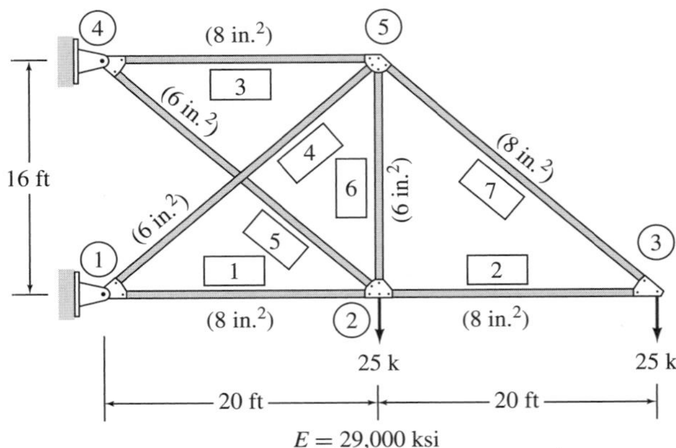

Fig. P3.24

3.26 and 3.27 Using a structural analysis computer program, determine the joint displacements, member axial forces, and support reactions for the Fink roof truss and the Baltimore bridge truss shown in Figs. P3.26 and P3.27, respectively. Verify the computer-generated results by manually checking the equilibrium equations for the entire truss, and for its joints numbered 5,10 and 15.

3.28 and 3.29 Using a structural analysis computer program, determine the largest value of the load parameter P that can be applied to the trusses shown in Figs. P3.28 and P3.29 without causing yielding and buckling of any of the members.

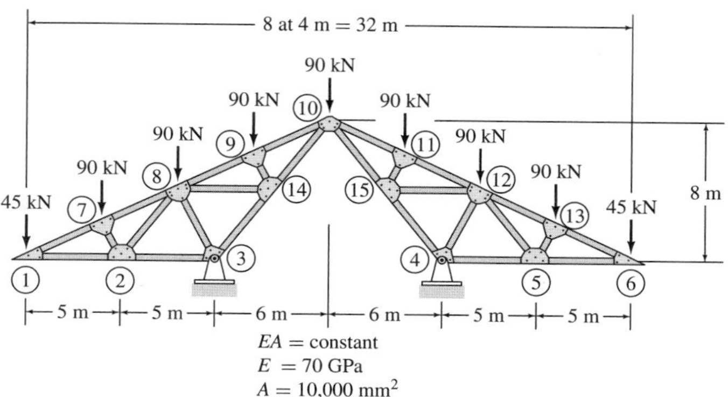

Fig. P3.26

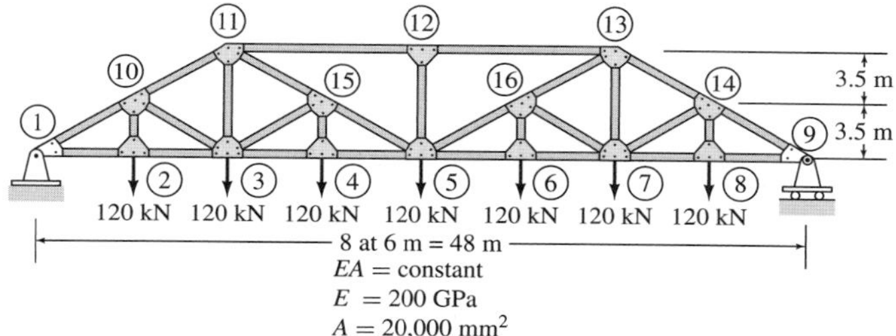

Fig. P3.27

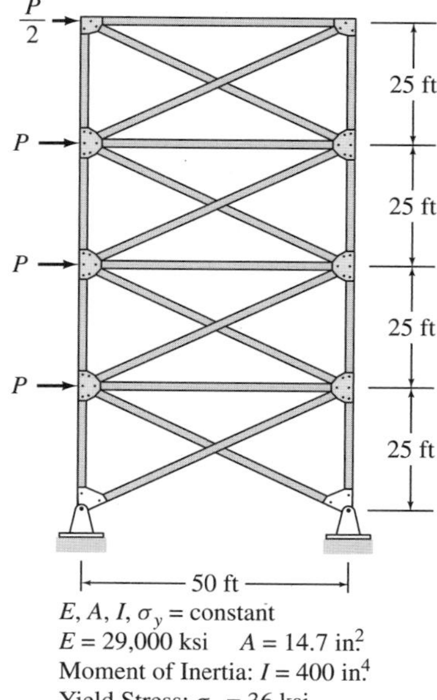

Fig. P3.28

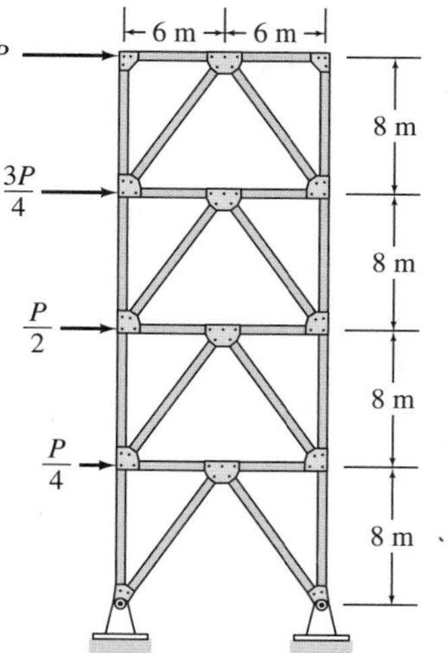

Fig. P3.29

4 COMPUTER PROGRAM FOR ANALYSIS OF PLANE TRUSSES

Truss Bridge
(Capricornis Photographic Inc. / Shutterstock)

In the previous chapter, we studied the basic principles of the analysis of plane trusses by the matrix stiffness method. In this chapter, we consider the computer implementation of the foregoing method of analysis. Our objective is to develop a general computer program that can be used to analyze any statically determinate or indeterminate plane truss, of any arbitrary configuration, subjected to any system of joint loads.

From a programming viewpoint, it is generally convenient to divide a structural analysis program into two parts or modules: (a) *input module,* and (b) *analysis module* (Fig. 4.1). The input module reads, and stores into the computer's memory, the structural and loading data necessary for the analysis; the analysis module uses the input data to perform the analysis, and communicates the results back to the user via an output device, such as a printer or a monitor. The development of a relatively simple input module is presented in Section 4.1; in the following five sections (4.2 through 4.6), we consider programming of the five analysis steps discussed in Chapter 3 (Fig. 3.20). The topics covered in these sections are as follows: assignment of the degree-of-freedom and restrained coordinate numbers for plane trusses (Section 4.2); generation of the structure stiffness matrix by assembling the elements of the member stiffness matrices (Section 4.3); formation of the joint load vector (Section 4.4); solution of the structure stiffness equations to obtain joint displacements (Section 4.5); and, finally, evaluation of the member axial forces and support reactions (Section 4.6).

The entire programming process is described by means of detailed flowcharts, so that readers can write this computer program in any programming language. It is important to realize that the programming process presented in this chapter represents only one of many ways in which the matrix stiffness method of analysis can be implemented on computers. Readers are strongly encouraged to conceive, and attempt, alternative strategies that can make the computer implementation (and/or application of the method) more efficient. One such strategy, which takes advantage of the banded form of the structure stiffness matrix, will be discussed in Chapter 9.

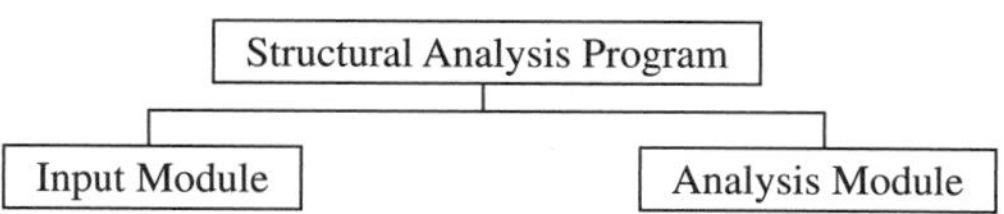

Fig. 4.1

4.1 DATA INPUT

In this section, we focus our attention on the input module of our computer program. As stated previously, *the input module of a structural analysis program reads the structural and loading data necessary for analysis from a file or another type of input device, and stores it in the computer's memory so that it can be processed conveniently by the program for structural analysis.*

When structural analysis is carried out by hand calculations (e.g., as in Chapter 3), the information needed for the analysis is obtained by visually inspecting the analytical model of the structure (as represented by the line diagram). In computerized structural analysis, however, all of the data necessary for analysis must be specified in the form of numbers, and must be organized in the computer's memory in the form of matrices (arrays), in such a way that it can be used for analysis without any reference to a visual image (or line diagram) of the structure. This data in numerical form must completely and uniquely define the analytical model of the structure. In other words, a person with no knowledge of the actual structure or its analytical model should be able to reconstruct the visual analytical model of the structure, using only the numerical data and the knowledge of how this data is organized.

The input data necessary for the analysis of plane trusses can be divided into the following six categories:

- joint data
- support data
- material property data
- cross-sectional property data
- member data
- load data

In the following, we discuss procedures for inputting data belonging to each of the foregoing categories, using the truss of Fig. 4.2(a) as an example. The analytical model of this truss is depicted in Fig. 4.2(b). Note that all the information in this figure is given in units of kips and inches. This is because we plan to design a computer program that can work with any *consistent* set of units. Thus, all the data must be converted into a consistent set of units before being input into the program.

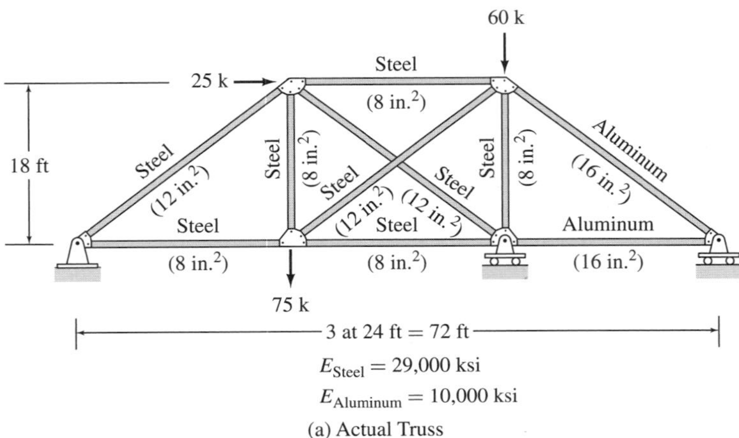

(a) Actual Truss

Fig. 4.2

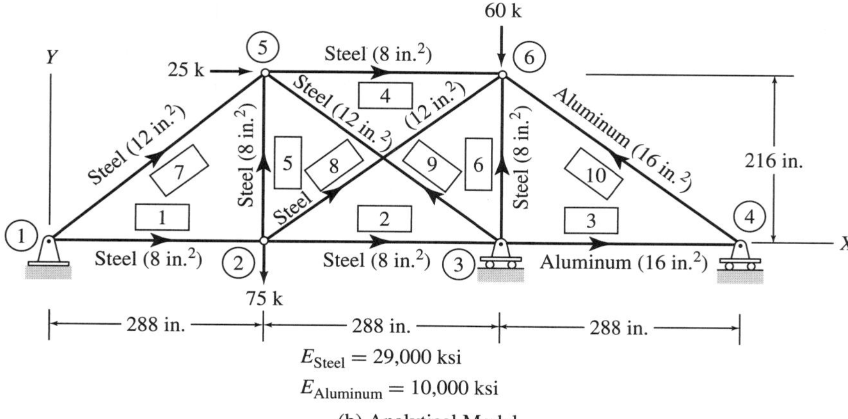

$E_{\text{Steel}} = 29{,}000$ ksi

$E_{\text{Aluminum}} = 10{,}000$ ksi

(b) Analytical Model

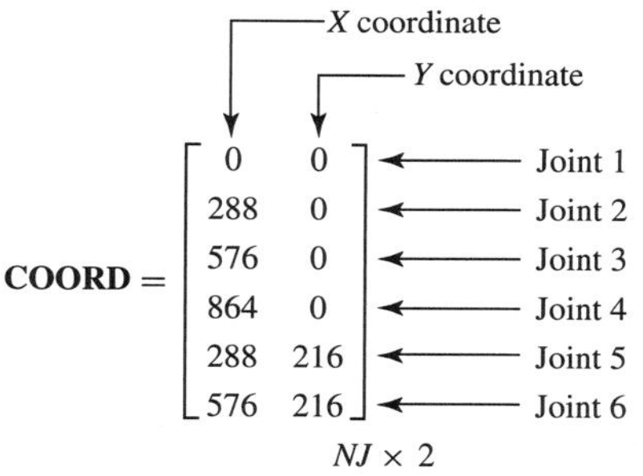

(c) Joint Coordinate Matrix

$$\mathbf{CP} = \begin{bmatrix} 8 \\ 12 \\ 16 \end{bmatrix} \begin{matrix} \leftarrow \text{Cross-section type no. 1} \\ \leftarrow \text{Cross-section type no. 2} \\ \leftarrow \text{Cross-section type no. 3} \end{matrix}$$

$NCP \times 1$

(f) Cross-sectional Property Vector

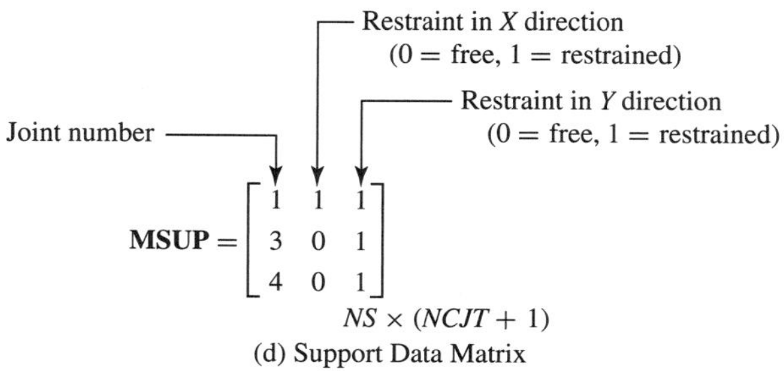

(d) Support Data Matrix

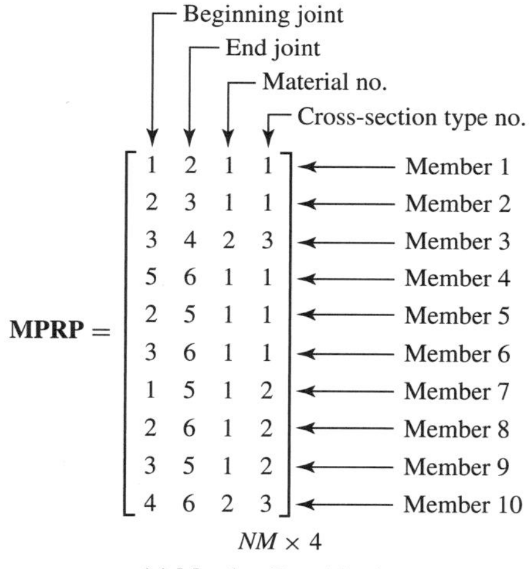

(g) Member Data Matrix

$$\mathbf{EM} = \begin{bmatrix} 29000 \\ 10000 \end{bmatrix} \begin{matrix} \leftarrow \text{Material no. 1} \\ \leftarrow \text{Material no. 2} \end{matrix}$$

$NMP \times 1$

(e) Elastic Modulus Vector

Joint number; Force in X direction; Force in Y direction

$$\mathbf{JP} = \begin{bmatrix} 2 \\ 5 \\ 6 \end{bmatrix} \qquad \mathbf{PJ} = \begin{bmatrix} 0 & -75 \\ 25 & 0 \\ 0 & -60 \end{bmatrix}$$

$NJL \times 1$ $\qquad$ $NJL \times NCJT$

(h) Load Data Matrices

Fig. 4.2 *(continued)*

Joint Data

The joint data consists of: (a) the total number of joints (NJ) of the truss, and (b) the global (X and Y) coordinates of each joint. The relative positions of the joints of the truss are specified by means of the global (X and Y) coordinates of the joints. These joint coordinates are usually stored in the computer's memory in the form of a matrix, so that they can be accessed easily by the computer program for analysis. In our program, we store the joint coordinates in a matrix **COORD** of the order $NJ \times 2$ (Fig. 4.2(c)). The matrix, which is referred to as the *joint coordinate matrix,* has two columns, and its number of rows equals the total number of joints (NJ) of the structure. The X and Y coordinates of a joint i are stored in the first and second columns, respectively, of the ith row of the matrix **COORD**. Thus, for the truss of Fig. 4.2(b) (which has six joints), the joint coordinate matrix is a 6×2 matrix, as shown in Fig. 4.2(c). Note that the joint coordinates are stored in the sequential order of joint numbers. Thus, by comparing Figs. 4.2(b) and (c), we can see that the X and Y coordinates of joint 1 (i.e., 0 and 0) are stored in the first and second columns, respectively, of the first row of **COORD**. Similarly, the X and Y coordinates of joint 5 (288 and 216) are stored in the first and second columns, respectively, of the fifth row of **COORD**, and so on.

A flowchart for programming the reading and storing of the joint data for plane trusses is given in Fig. 4.3(a). As shown there, the program first reads the value of the integer variable NJ, which represents the total number of joints of the truss. Then, using a *Do Loop* command, the X and Y coordinates of each joint are read, and stored in the first and second columns, respectively, of the matrix **COORD**. The *Do Loop* starts with joint number 1 and ends with joint

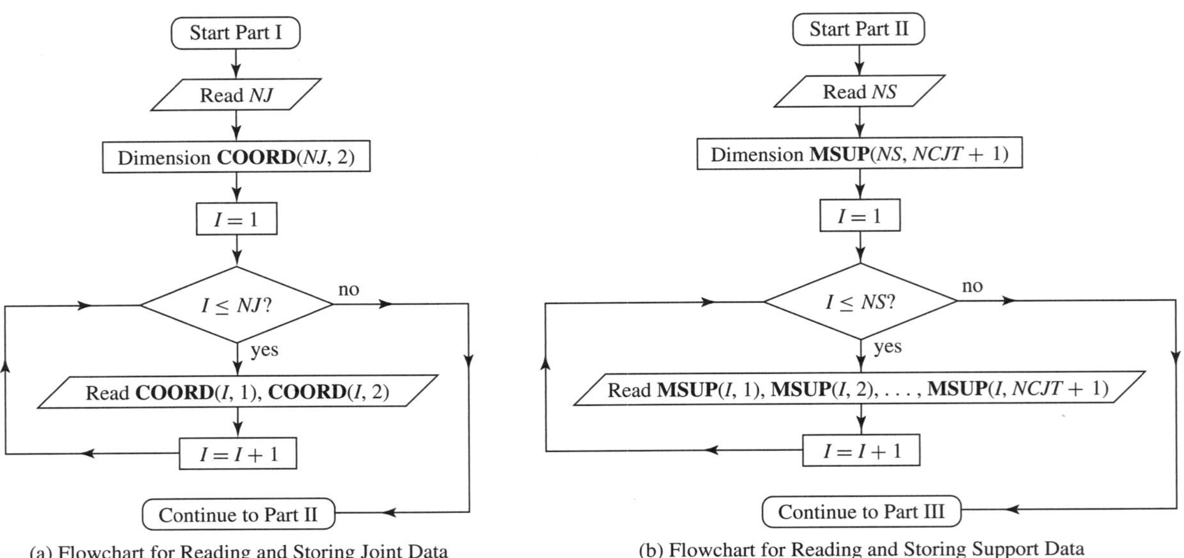

(a) Flowchart for Reading and Storing Joint Data

(b) Flowchart for Reading and Storing Support Data

Fig. 4.3

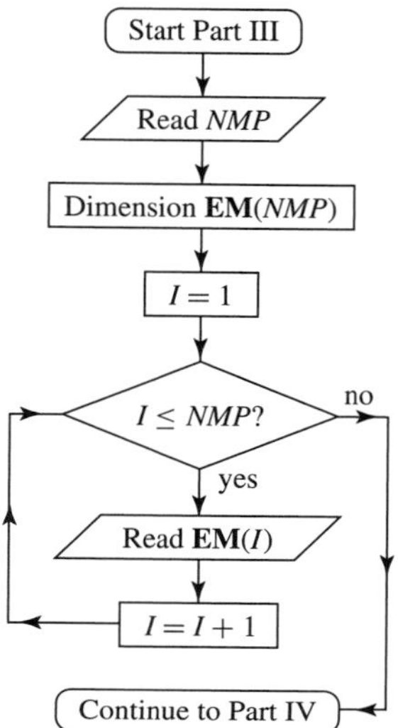

(c) Flowchart for Reading and Storing Material Property Data

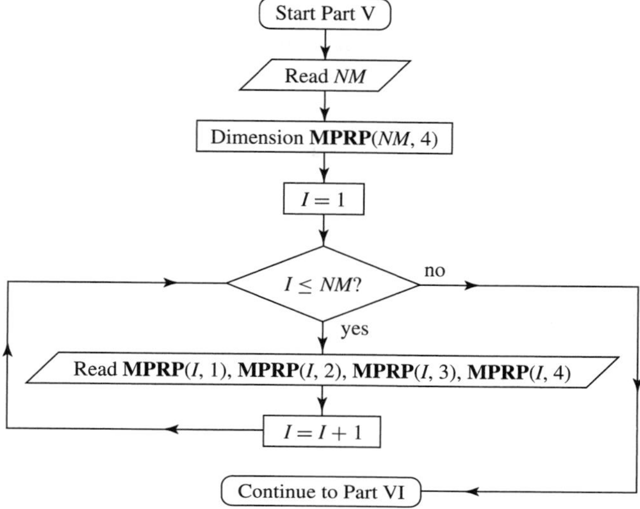

(e) Flowchart for Reading and Storing Member Data

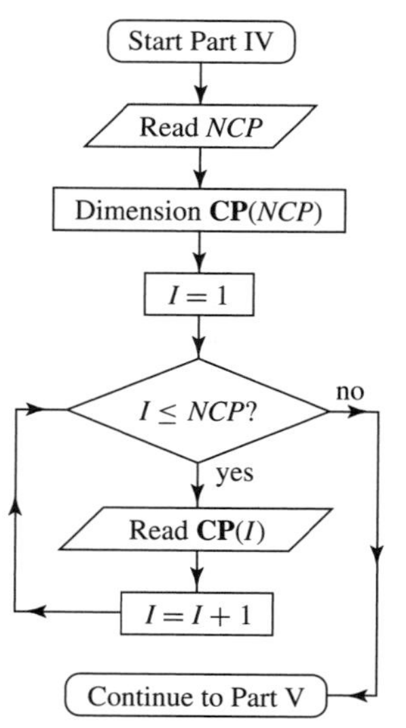

(d) Flowchart for Reading and Storing Cross-sectional Property Data

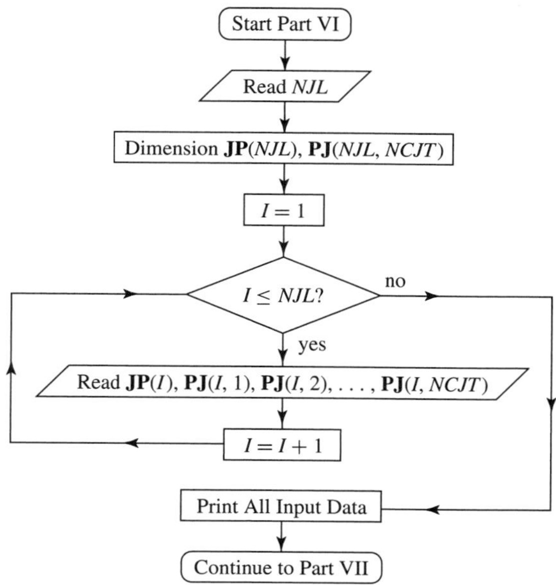

(f) Flowchart for Reading and Storing Load Data

Fig. 4.3 (*continued*)

number *NJ*. It should be noted that, depending upon the type of programming language and/or compiler being used, some additional statements (such as variable type declaration and formatted read/write statements) may be needed to implement the foregoing program. (It is assumed herein that the reader has a working knowledge of a programming language.)

The input data to be read by the computer program is either entered interactively by the user (responding to prompts on the screen), or is supplied in the form of a data file. The former approach is used in the computer software which can be downloaded from the publisher's website for this book. However, the latter approach is recommended for beginning programmers, because it is straightforward and requires significantly less programming. As an example, the input data file (in free-format) for the truss of Fig. 4.2(b) is given in Fig. 4.4. Note that the first line

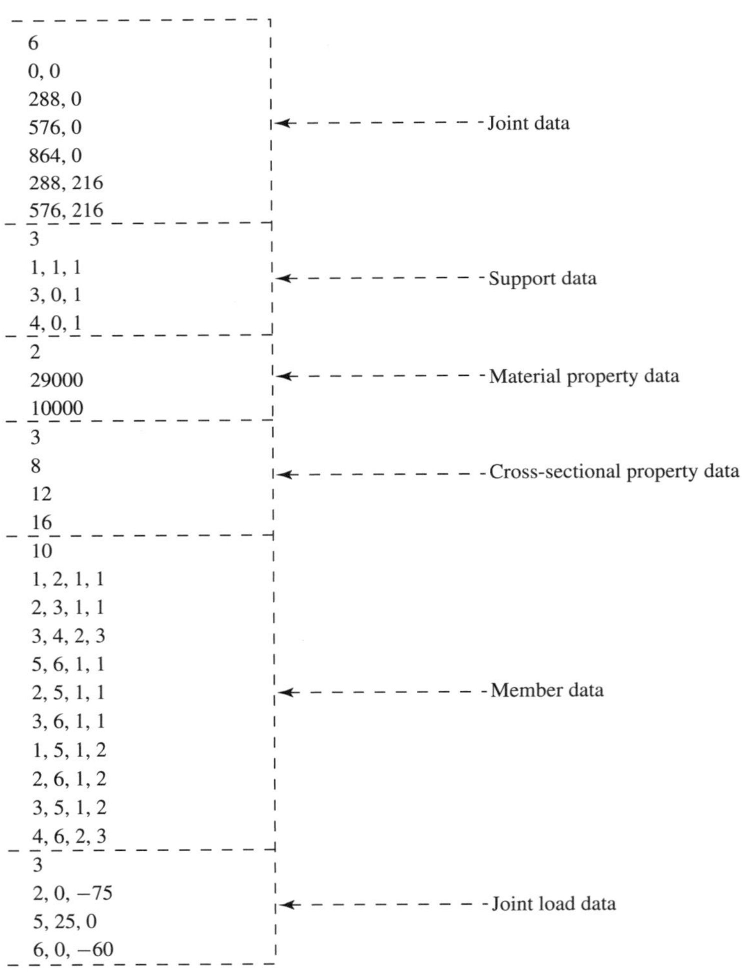

Fig. 4.4 *An Example of an Input Data File*

of this data file contains the total number of joints of the truss (i.e., 6); the next six lines contain the X and Y coordinates of joints 1 through 6, respectively.

Support Data

The support data consists of (a) the number of joints that are attached to supports (NS); and (b) the joint number, and the directions of restraints, for each support joint. Since there can be at most two restrained coordinates at a joint of a plane truss (i.e., $NCJT = 2$), the restraints at a support joint of such a structure can be conveniently specified by using a two-digit code in which each digit is either a 0 or a 1. The first digit of the code represents the restraint condition at the joint in the global X direction; it is 0 if the joint is free to translate in the X direction, or it is 1 if the joint is restrained in the X direction. Similarly, the second digit of the code represents the restraint condition at the joint in the global Y direction; a 0 indicates that the joint is unrestrained in the Y direction, and a 1 indicates that it is restrained. The restraint codes for the various types of supports for plane trusses are given in Fig. 4.5. (The special case of inclined roller supports will be considered in Chapter 9.)

Considering again the example truss of Fig. 4.2(b), we can see that joint 1 is attached to a hinged support that prevents it from translating in any direction.

Type of Support		Restraint Code
Free joint (no support)	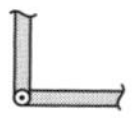	0, 0
Roller with horizontal reaction	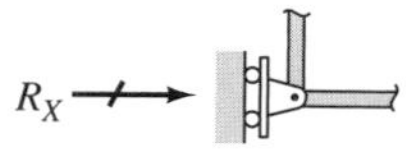 R_X	1, 0
Roller with vertical reaction	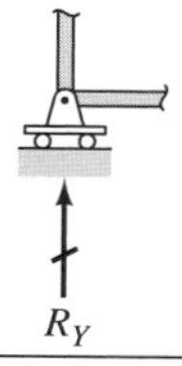 R_Y	0, 1
Hinge	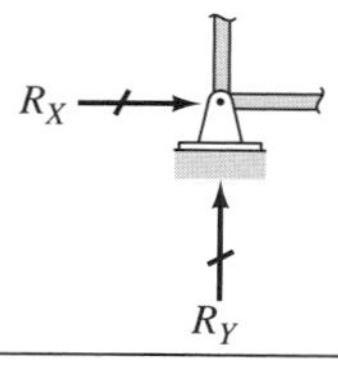 R_X R_Y	1, 1

Fig. 4.5 *Restraint Codes for Plane Trusses*

Thus, the restraint code for joint 1 is 1,1 indicating that this joint is restrained from translating in both the X and Y directions. Similarly, the restraint codes for joints 3 and 4, which are attached to roller supports, are 0,1 because these joints are free to translate in the horizontal (X) direction, but are restrained by the rollers from translating in the vertical (Y) direction. The restraint codes of the remaining joints of the truss, which are free to translate in any direction, can be considered to be 0,0. However, it is not necessary to input codes for free joints, because the computer program considers every joint to be free, unless it is identified as a support joint.

The support data can be stored in the computer's memory in the form of an integer matrix **MSUP** of order $NS \times (NCJT + 1)$ (Fig. 4.2(d)). For plane trusses, because $NCJT = 2$, the *support data matrix* **MSUP** consists of three columns, with the number of rows equal to the number of support joints (NS). In each row of **MSUP**, the support joint number is stored in the first column, and the first and second digits of the corresponding restraint code are stored in the second and third columns, respectively. Thus, for the truss of Fig. 4.2(b), which has three support joints, the support data matrix is a 3×3 matrix, as shown in Fig. 4.2(d). Note that in the first row of **MSUP** the support joint number 1 is stored in the first column, and the first and second digits of the restraint code for this joint (i.e., 1 and 1) are stored in the second and third columns, respectively. Similarly, the second row of **MSUP** consists of the support joint number 3 in the first column, and the two digits of the corresponding restraint code (i.e., 0 and 1) in the second and third columns, respectively, and so on.

A flowchart for programming the reading and storing of the support data is given in Fig. 4.3(b), in which, as noted previously, the integer variable $NCJT$ denotes the number of structure coordinates per joint. Like this flowchart, many parts of the computer program presented in this chapter are given in a general form in terms of the variable $NCJT$, so that they can be conveniently incorporated into computer programs for analyzing other types of framed structures (e.g., beams and plane frames), which are considered in subsequent chapters. For example, as discussed in this section, by setting $NCJT = 2$, the flowchart of Fig. 4.3(b) can be used to input support data for plane trusses; whereas, as discussed subsequently in Chapter 6, the same flowchart can be used to input support data for plane frames, provided $NCJT$ is set equal to three.

An example of how the support data for a plane truss may appear in an input data file is given in Fig. 4.4.

Material Property Data

The material property data involves (a) the number of materials used in the structure (NMP), and (b) the modulus of elasticity (E) of each material. The elastic moduli are stored by the program in an *elastic modulus vector* **EM**. The number of rows of **EM** equals the number of materials (NMP), with the elastic modulus of material i stored in the ith row of the vector (Fig. 4.2(e)).

Consider, for example, the truss of Fig. 4.2(b). The truss is composed of two materials; namely, steel and aluminum. We arbitrarily select the steel ($E = 29{,}000$ ksi) to be material number 1, and the aluminum ($E = 10{,}000$ ksi) to be material number 2. Thus, the elastic modulus vector, **EM**, of the truss consists of two rows, as shown in Fig. 4.2(e); the elastic modulus of material number 1 (i.e., 29,000) is stored in the first row of **EM**, and the elastic modulus of material number 2 (i.e., 10,000) is stored in the second row.

Figure 4.3(c) shows a flowchart for programming the reading and storing of the material property data; Fig. 4.4 illustrates how this type of data may appear in an input data file.

Cross-Sectional Property Data

The cross-sectional property data consists of (a) the number of different cross-section types used for the truss members (NCP); and (b) the cross-sectional area (A) for each cross-section type. The cross-sectional areas are stored by the program in a *cross-sectional property vector* **CP**. The number of rows of **CP** equals the number of cross-section types (NCP), with the area of cross-section i stored in the ith row of the vector (Fig. 4.2(f)).

For example, three types of member cross-sections are used for the truss of Fig. 4.2(b). We arbitrarily assign the numbers 1, 2, and 3 to the cross-sections with areas of 8, 12, and 16 in.^{2}, respectively. Thus, the cross-sectional property vector, **CP**, consists of three rows; areas of cross-section types 1, 2, and 3 are stored in rows 1, 2, and 3, respectively, as shown in Fig. 4.2(f).

A flowchart for reading and storing the cross-sectional property data into computer memory is given in Fig. 4.3(d); Fig. 4.4 shows an example of an input data file containing this type of data.

Member Data

The member data consists of (a) the total number of members (NM) of the truss; and (b) for each member, the beginning joint number, the end joint number, the material number, and the cross-section type number.

The member data can be stored in computer memory in the form of an integer *member data matrix,* **MPRP**, of order $NM \times 4$ (Fig. 4.2(g)). The information corresponding to a member i is stored in the ith row of **MPRP**; its beginning and end joint numbers are stored in the first and second columns, respectively, and the material and cross-section numbers are stored in the third and fourth columns, respectively.

For example, since the truss of Fig. 4.2(b) has 10 members, its member data matrix is a 10×4 matrix, as shown in Fig. 4.2(g). From Fig. 4.2(b), we can see that the beginning and end joints for member 1 are 1 and 2, respectively; the material and cross-section numbers for this member are 1 and 1, respectively. Thus, the numbers 1, 2, 1, and 1 are stored in columns 1 through 4, respectively, of the first row of **MPRP**, as shown in Fig. 4.2(g). Similarly, we see from Fig. 4.2(b) that the beginning joint, end joint, material, and cross-section numbers for member 3 are 3, 4, 2, and 3, respectively, and they are stored, respectively, in columns 1 through 4 of row 3 of **MPRP**, and so on.

Figure 4.3(e) shows a flowchart for programming the reading and storing of the member data. An example of how member data may appear in an input data file is given in Fig. 4.4.

Load Data

The load data involves (a) the number of joints that are subjected to external loads (*NJL*); and (b) the joint number, and the magnitudes of the force components in the global *X* and *Y* directions, for each loaded joint. The numbers of the loaded joints are stored in an integer vector **JP** of order $NJL \times 1$; the corresponding load components in the *X* and *Y* directions are stored in the first and second columns, respectively, of a real matrix **PJ** of order $NJL \times NCJT$, with $NCJT = 2$ for plane trusses (see Fig. 4.2(h)). Thus, for the example truss of Fig. 4.2(a), which has three joints (2, 5, and 6) that are subjected to loads, the *load data matrices,* **JP** and **PJ**, are of orders 3×1 and 3×2, respectively, as shown in Fig. 4.2(h). The first row of **JP** contains joint number 2; the loads in the *X* and *Y* directions at this joint (i.e., 0 and -75 k) are stored in the first and second columns, respectively, of the same row of **PJ**. The information about joints 5 and 6 is then stored in a similar manner in the second and third rows, respectively, of **JP** and **PJ**, as shown in the figure.

A flowchart for programming the reading and storing of the load data is given in Fig. 4.3(f), in which *NCJT* must be set equal to 2 for plane trusses. Figure 4.4 shows the load data for the example truss in an input file.

It is important to recognize that the numerical data stored in the various matrices in Figs. 4.2(c) through (h) *completely* and *uniquely* defines the analytical model of the example truss, without any need to refer to the line diagram of the structure (Fig. 4.2(b)).

After all the input data has been read and stored in computer memory, it is considered a good practice to print this data directly from the matrices in the computer memory (or view it on the screen), so that its validity can be verified (Fig. 4.3(f)). An example of such a printout, showing the input data for the example truss of Fig. 4.2, is given in Fig. 4.6.

```
          **********************************
          *       Computer Software        *
          *              for               *
          *  MATRIX ANALYSIS OF STRUCTURES  *
          *         Second Edition         *
          *               by               *
          *         Aslam Kassimali        *
          **********************************

              =======================
              General Structural Data
              =======================

Project Title: Figure 4-2
Structure Type : Plane Truss
Number of Joints : 6
Number of Members : 10
Number of Material Property Sets (E) : 2
Number of Cross-Sectional Property Sets : 3
```

Fig. 4.6 *A Sample Printout of Input Data*

Joint Coordinates

Joint No.	X Coordinate	Y Coordinate
1	0.0000E+00	0.0000E+00
2	2.8800E+02	0.0000E+00
3	5.7600E+02	0.0000E+00
4	8.6400E+02	0.0000E+00
5	2.8800E+02	2.1600E+02
6	5.7600E+02	2.1600E+02

Supports

Joint No.	X Restraint	Y Restraint
1	Yes	Yes
3	No	Yes
4	No	Yes

Material Properties

Material No.	Modulus of Elasticity (E)	Co-efficient of Thermal Expansion
1	2.9000E+04	0.0000E+00
2	1.0000E+04	0.0000E+00

Cross-Sectional Properties

Property No.	Area (A)
1	8.0000E+00
2	1.2000E+01
3	1.6000E+01

Member Data

Member No.	Beginning Joint	End Joint	Material No.	Cross-Sectional Property No.
1	1	2	1	1
2	2	3	1	1
3	3	4	2	3
4	5	6	1	1
5	2	5	1	1
6	3	6	1	1
7	1	5	1	2
8	2	6	1	2
9	3	5	1	2
10	4	6	2	3

Joint Loads

Joint No.	X Force	Y Force
2	0.0000E+00	-7.5000E+01
5	2.5000E+01	0.0000E+00
6	0.0000E+00	-6.0000E+01

************* End of Input Data *************

Fig. 4.6 *(continued)*

4.2 ASSIGNMENT OF STRUCTURE COORDINATE NUMBERS

Having completed the input module, we are now ready to develop the analysis module of our computer program. *The analysis module of a structural analysis program uses the input data stored in computer memory to calculate the desired response characteristics of the structure, and communicates these results to the user through an output device, such as a printer or a monitor.*

As discussed in Section 3.8, the first step of the analysis involves specification of the structure's degrees of freedom and restrained coordinates, which are collectively referred to as, simply, the *structure coordinates*. Recall that when the analysis was carried out by hand calculations (in Chapter 3), the structure coordinate numbers were written next to the arrows, in the global X and Y directions, drawn at the joints. In computerized analysis, however, these numbers must be organized in computer memory in the form of a matrix or a vector. In our program, the structure coordinate numbers are stored in an integer vector **NSC**, with the number of rows equal to the number of structure coordinates per joint ($NCJT$) times the number of joints of the structure (NJ). For plane trusses, because $NCJT = 2$, the number of rows of **NSC** equals twice the number of joints of the truss (i.e., $2NJ$). The structure coordinate numbers are arranged in **NSC** in the sequential order of joint numbers, with the number for the X coordinate at a joint followed by the number for its Y coordinate. In other words, the numbers for the X and Y structure coordinates at a joint i are stored in rows $(i - 1)2 + 1$ and $(i - 1)2 + 2$, respectively, of **NSC**. For example, the line diagram of the truss of Fig. 4.2(a) is depicted in Fig. 4.7(a) with its degrees of freedom and restrained coordinates indicated, and the corresponding 12×1 **NSC** vector is given in Fig. 4.7(b).

The procedure for assigning the structure coordinate (i.e., degrees of freedom and restrained coordinate) numbers was discussed in detail in Section 3.2.

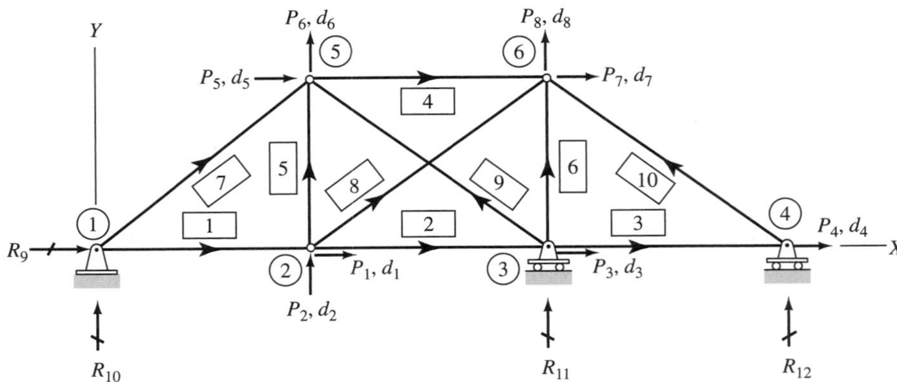

(a) Line Diagram Showing Degrees of Freedom and Restrained Coordinates

Fig. 4.7

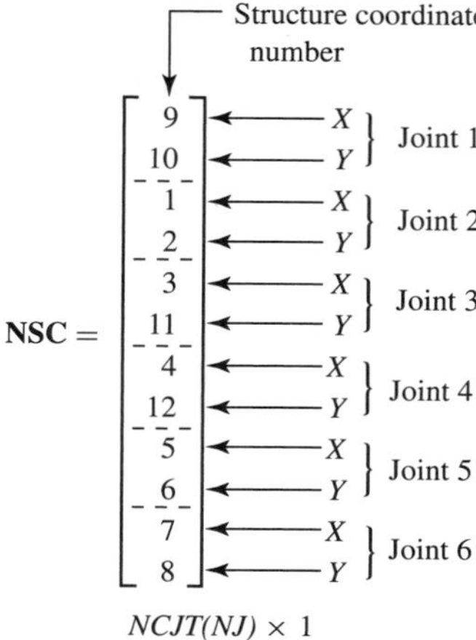

(b) Structure Coordinate Number Vector

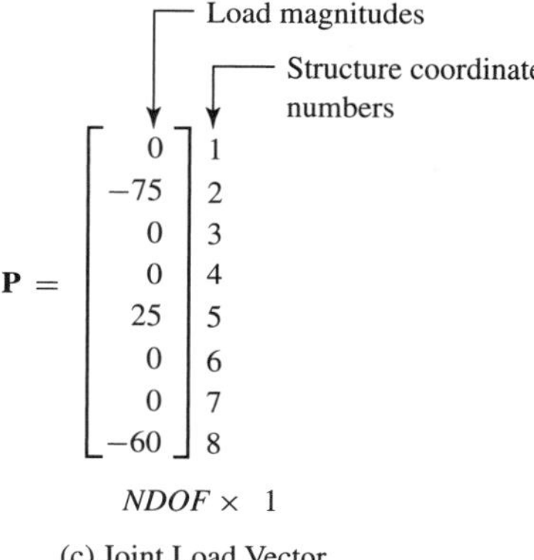

(c) Joint Load Vector

Fig. 4.7 *(continued)*

This procedure can be conveniently programmed using the flowcharts given in Fig. 4.8 on the next page. Figure 4.8(a) describes a program for determining the number of degrees of freedom and the number of restrained coordinates of the structure. (Note again that $NCJT = 2$ for plane trusses.) The program first determines the number of restrained coordinates (NR) by simply counting the number of 1s in the second and third columns of the support data matrix **MSUP**. Recall from our discussion of restraint codes in Section 4.1 that each 1 in the second or third column of **MSUP** represents a restraint (in either the X or Y direction) at a joint of the structure. With NR known, the number of degrees of freedom ($NDOF$) is evaluated from the following relationship (Eq. (3.3)).

$$NDOF = 2(NJ) - NR$$

For example, since the **MSUP** matrix for the example truss, given in Fig. 4.2(d), contains four 1s in its second and third columns, the number of

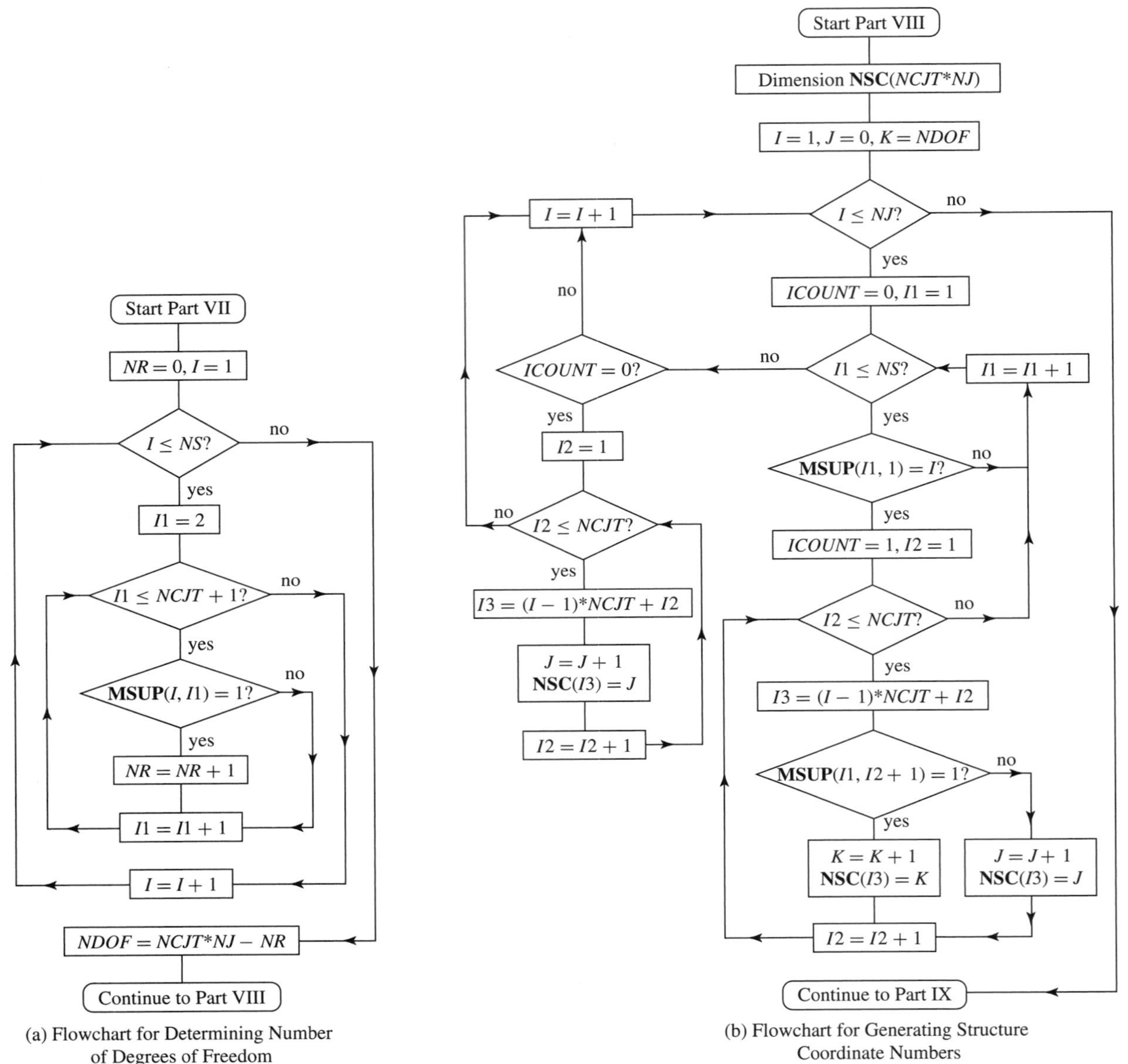

(a) Flowchart for Determining Number of Degrees of Freedom

(b) Flowchart for Generating Structure Coordinate Numbers

Fig. 4.8

restrained coordinates of the truss is four. Furthermore, since the truss has six joints (Fig. 4.2(c)), its number of degrees of freedom equals

$$NDOF = 2(6) - 4 = 8$$

Once the number of degrees of freedom ($NDOF$) has been determined, the program generates the structure coordinate number vector **NSC**, as shown by the flowchart in Fig. 4.8(b). Again, *NCJT* should be set equal to 2 for plane trusses. As this flowchart indicates, the program uses two integer counters, J and K, to keep track of the degrees-of-freedom and restrained coordinate numbers, respectively. The initial value of J is set equal to 0, whereas the initial value of K is set equal to *NDOF*.

The structure coordinates are numbered, one joint at a time, starting at joint 1 and proceeding sequentially to the joint number *NJ*. First, the number of the joint under consideration, I, is compared with the numbers in the first column of the support matrix **MSUP** to determine whether or not I is a support joint. If a match is found between I and one of the numbers in the first column of **MSUP**, then the counter *ICOUNT* is set equal to 1; otherwise, the value of *ICOUNT* remains 0 as initially assigned.

If joint I is not a support joint (i.e., $ICOUNT = 0$), then the number for its degree of freedom in the X direction is obtained by increasing the degrees-of-freedom counter J by 1 (i.e., $J = J + 1$), and this value of J is stored in row number $(I - 1)2 + 1$ of the **NSC** vector. Next, the value of J is again increased by 1 (i.e., $J = J + 1$) to obtain the number for the degree of freedom of joint I in the Y direction, and the new value of J is stored in row $(I - 1)2 + 2$ of the **NSC** vector.

If joint I is found to be a support joint (i.e., $ICOUNT = 1$), then the second column of the corresponding row of **MSUP** is checked to determine whether joint I is restrained in the X direction. If the joint is restrained in the X direction, then the number for its X-restrained coordinate is obtained by increasing the restrained coordinate counter K by 1 (i.e., $K = K + 1$), and this value of K is stored in row $(I - 1)2 + 1$ of the **NSC** vector. However, if the joint is not restrained in the X direction, then the degrees-of-freedom counter J is increased by 1, and its value (instead of that of K) is stored in row $(I - 1)2 + 1$ of the **NSC** vector. Next, the restraint condition in the Y direction at joint I is determined by checking the third column of the corresponding row of **MSUP**. If the joint is found to be restrained, then the counter K is increased by 1; otherwise, the counter J is increased by the same amount. The new value of either K or J is then stored in row $(I - 1)2 + 2$ of the **NSC** vector.

The computer program repeats the foregoing procedure for each joint of the structure to complete the structure coordinate number vector, **NSC**. As shown in Fig. 4.8(b), this part of the program (for generating structure coordinate numbers) can be conveniently coded using *Do Loop* or *For–Next* types of programming statements.

4.3 GENERATION OF THE STRUCTURE STIFFNESS MATRIX

The structure coordinate number vector **NSC**, defined in the preceding section, can be used to conveniently determine the member code numbers needed to establish the structure stiffness matrix **S**, without any reference to the visual image of the structure (e.g., the line diagram). The code numbers at the beginning of a member of a general framed structure are stored in the following rows of the **NSC** vector:

$$\left.\begin{aligned}
\textbf{NSC}\text{ row for the first code number} &= (JB-1)NCJT+1\\
\textbf{NSC}\text{ row for the second code number} &= (JB-1)NCJT+2\\
&\vdots\\
\textbf{NSC}\text{ row for the }NCJT\text{th code number} &= (JB-1)NCJT+NCJT
\end{aligned}\right\} \qquad \textbf{(4.1a)}$$

in which JB is the beginning joint of the member. Similarly, the code numbers at the end of the member, connected to joint JE, can be obtained from the following rows of the **NSC**:

$$\left.\begin{aligned}
\textbf{NSC}\text{ row for the first code number} &= (JE-1)NCJT+1\\
\textbf{NSC}\text{ row for the second code number} &= (JE-1)NCJT+2\\
&\vdots\\
\textbf{NSC}\text{ row for the }NCJT\text{th code number} &= (JE-1)NCJT+NCJT
\end{aligned}\right\} \qquad \textbf{(4.1b)}$$

Suppose, for example, that we wish to determine the code numbers for member 9 of the truss of Fig. 4.2(b). First, from the member data matrix **MPRP** of this truss (Fig. 4.2(g)), we obtain the beginning and end joints for this member as 3 and 5, respectively. (This information is obtained from row 9, columns 1 and 2, respectively, of **MPRP**.) Next, we determine the row numbers of the **NSC** vector in which the structure coordinate numbers for joints 3 and 5 are stored. Thus, at the beginning of the member (Eq. (4.1a) with $NCJT = 2$),

$$\begin{aligned}
\textbf{NSC}\text{ row for the first code number} &= (3-1)2+1=5\\
\textbf{NSC}\text{ row for the second code number} &= (3-1)2+2=6
\end{aligned}$$

Similarly, at the end of the member (Eq. (4.1b)),

$$\begin{aligned}
\textbf{NSC}\text{ row for the first code number} &= (5-1)2+1=9\\
\textbf{NSC}\text{ row for the second code number} &= (5-1)2+2=10
\end{aligned}$$

The foregoing calculations indicate that the code numbers for member 9 are stored in rows 5, 6, 9, and 10 of the **NSC** vector. Thus, from the appropriate rows of the **NSC** vector of the truss given in Fig. 4.7(b), we obtain the member's code numbers to be 3, 11, 5, 6. A visual check of the truss's line diagram in Fig. 4.7(a) indicates that these code numbers are indeed correct.

The procedure for forming the structure stiffness matrix **S** by assembling the elements of the member global stiffness matrices **K** was discussed in detail in Sections 3.7 and 3.8. A flowchart for programming this procedure is presented in Fig. 4.9, in which $NCJT$ should be set equal to 2 for plane trusses. As indicated by the flowchart, this part of our computer program begins by initializing all the elements of the **S** matrix to 0. The assembly of the structure stiffness matrix is then carried out using a *Do Loop,* in which the following operations are performed for each member of the structure.

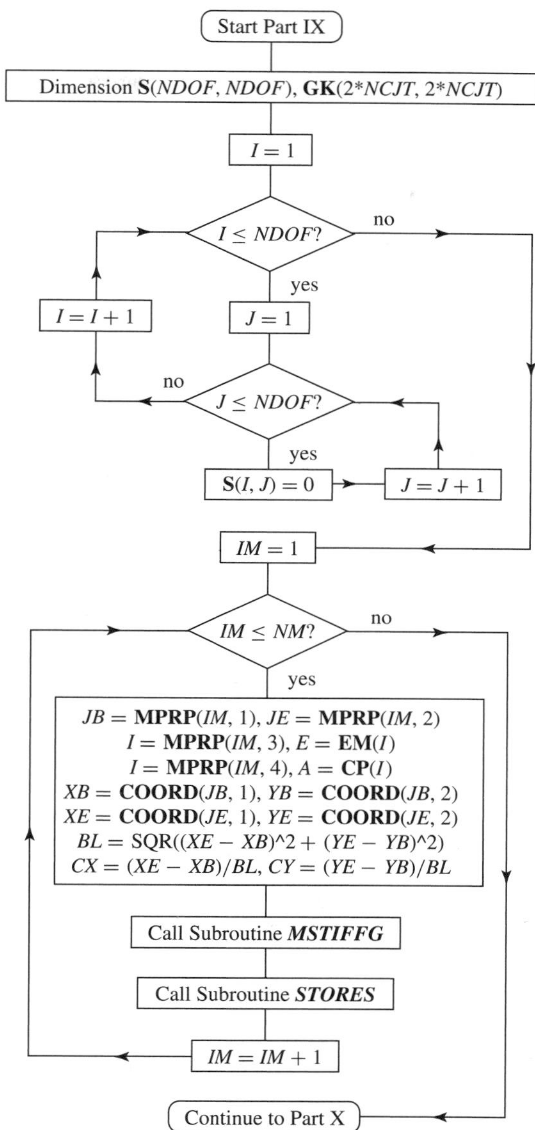

Fig. 4.9 *Flowchart for Generating Structure Stiffness Matrix for Plane Trusses*

1. *Evaluation of member properties.* For the member under consideration, *IM*, the program reads the beginning joint number, *JB*, and the end joint number, *JE*, from the first and second columns, respectively, of the member data matrix **MPRP**. Next, the material property number is read from the third column of **MPRP**, and the corresponding value of the modulus of elasticity, *E*, is obtained from the elastic modulus vector **EM**. The program then reads the number of the member cross-section type from the fourth column of **MPRP**,

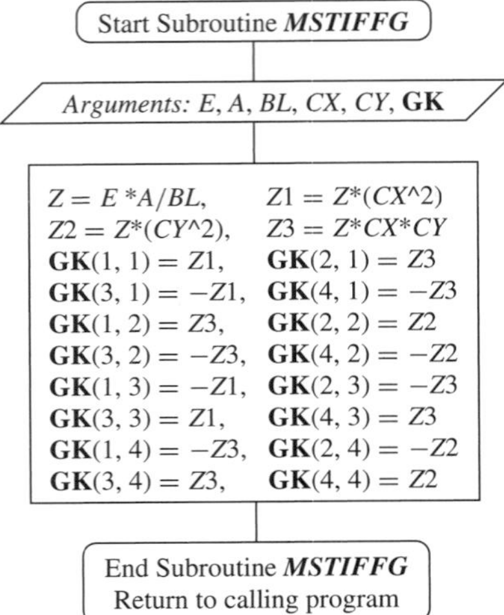

Fig. 4.10 *Flowchart of Subroutine **MSTIFFG** for Determining Member Global Stiffness Matrix for Plane Trusses*

and obtains the corresponding value of the cross-sectional area, A, from the cross-sectional property vector **CP**. Finally, the X and Y coordinates of the beginning joint JB and the end joint JE are obtained from the joint coordinate matrix **COORD**, and the member's length, BL, and its direction cosines, CX $(= \cos\theta)$ and CY $(= \sin\theta)$, are calculated using Eqs. (3.62).

2. *Determination of member global stiffness matrix* **GK** $(= \mathbf{K})$ *by subroutine* ***MSTIFFG***. After the necessary properties of the member under consideration, IM, have been evaluated, the program calls on the subroutine ***MSTIFFG*** to form the member stiffness matrix in the global coordinate system. (A flowchart of this subroutine is shown in Fig. 4.10.) Note that in the computer program, the member global stiffness matrix is named **GK** (instead of **K**) to indicate that it is a real (not an integer) matrix. As the flowchart in Fig. 4.10 indicates, the subroutine simply calculates the values of the various stiffness coefficients, and stores them into appropriate elements of the **GK** matrix, in accordance with Eq. (3.73).

3. *Storage of the elements of member global stiffness matrix* **GK** *into structure stiffness matrix* **S** *by subroutine* ***STORES***. Once the matrix **GK** has been determined for the member under consideration, IM, the program (Fig. 4.9) calls the subroutine ***STORES*** to store the pertinent elements of **GK** in their proper positions in the structure stiffness matrix **S**. A flowchart of this subroutine, which essentially consists of two nested *Do Loops,* is given in Fig. 4.11. As this flowchart indicates, the outer *Do Loop* performs the following operations sequentially for each row of the **GK** matrix, starting with row 1 and ending with row $2(NCJT)$: (a) the member code number $N1$ corresponding to the row under consideration, I, is obtained from the **NSC** vector using the procedure discussed previously in this

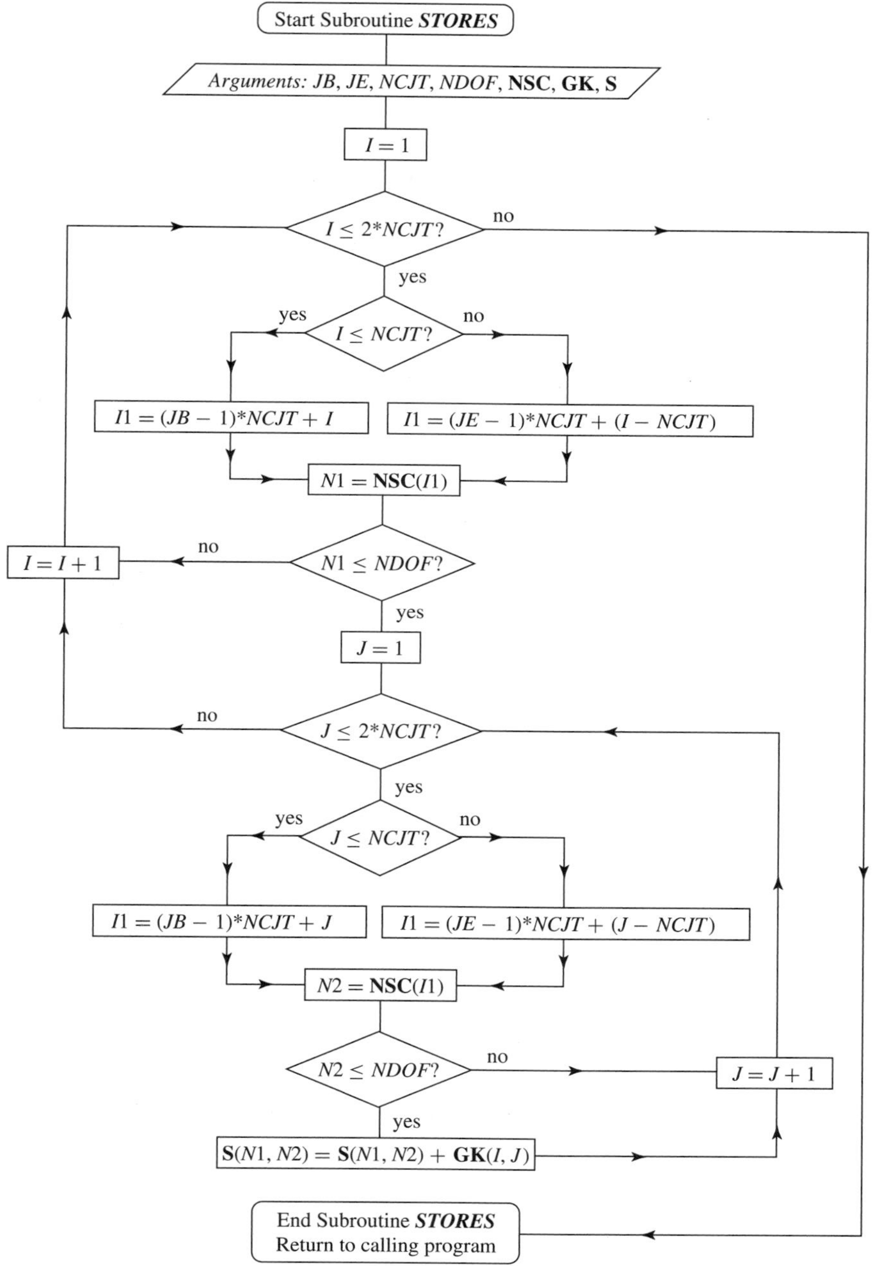

Fig. 4.11 *Flowchart of Subroutine* ***STORES*** *for Storing Member Global Stiffness Matrix in Structure Stiffness Matrix*

section; (b) if $N1$ is less than or equal to $NDOF$, then the inner *Do Loop* is activated; otherwise, the inner loop is skipped; and (c) the row number I is increased by 1, and steps (a) through (c) are repeated. The inner loop, activated from the outer loop, performs the following operations sequentially for each column of the **GK** matrix, starting with column 1 and ending with column 2($NCJT$): (a) the member code number $N2$ corresponding to the column under consideration, J, is obtained from the **NSC** vector; (b) if $N2$ is less than or equal to $NDOF$, then the value of the element in the Ith row and Jth column of **GK** is added to the value of the element in the $N1$th row and $N2$th column of **S**; otherwise, no action is taken; and (c) the column number J is increased by 1, and steps (a) through (c) are repeated. The inner loop ends when its steps (a) and (b) have been applied to all the columns of **GK**; the program control is then returned to step (c) of the outer loop. The subroutine ***STORES*** ends when steps (a) and (b) of the outer loop have been applied to all the rows of the **GK** matrix, thereby storing all the pertinent elements of the global stiffness matrix of the member under consideration, IM, in their proper positions in the structure stiffness matrix **S**.

Refocusing our attention on Fig. 4.9, we can see that formation of the structure stiffness matrix is complete when the three operations, described in the foregoing paragraphs, have been performed for each member of the structure.

4.4 FORMATION OF THE JOINT LOAD VECTOR

In this section, we consider the programming of the next analysis step, which involves formation of the joint load vector **P**. A flowchart for programming this process is shown in Fig. 4.12. Again, when analyzing plane trusses, the value of $NCJT$ should be set equal to 2 in the program. It is seen from the figure that this part of our computer program begins by initializing each element of **P** to 0. The program then generates the load vector **P** by performing the following operations for each row of the load data vector **JP**, starting with row 1 and proceeding sequentially to row NJL:

1. For the row under consideration, I, the number of the loaded joint $I1$ is read from the **JP** vector.

2. The number of the X structure coordinate, N, at joint $I1$ is obtained from row $I2 = (I1 - 1)2 + 1$ of the **NSC** vector. If $N \le NDOF$, then the value of the element in the Ith row and the first column of the load data matrix **PJ** (i.e., the X load component) is added to the Nth row of the load vector **P**; otherwise, no action is taken.

3. The **NSC** row number $I2$ is increased by 1 (i.e., $I2 = I2 + 1$), and the structure coordinate number, N, of the Y coordinate is read from the **NSC**. If $N \le NDOF$, then the value of the element in the Ith row and the second column of **PJ** (i.e., the Y load component) is added to the Nth row of **P**; otherwise, no action is taken.

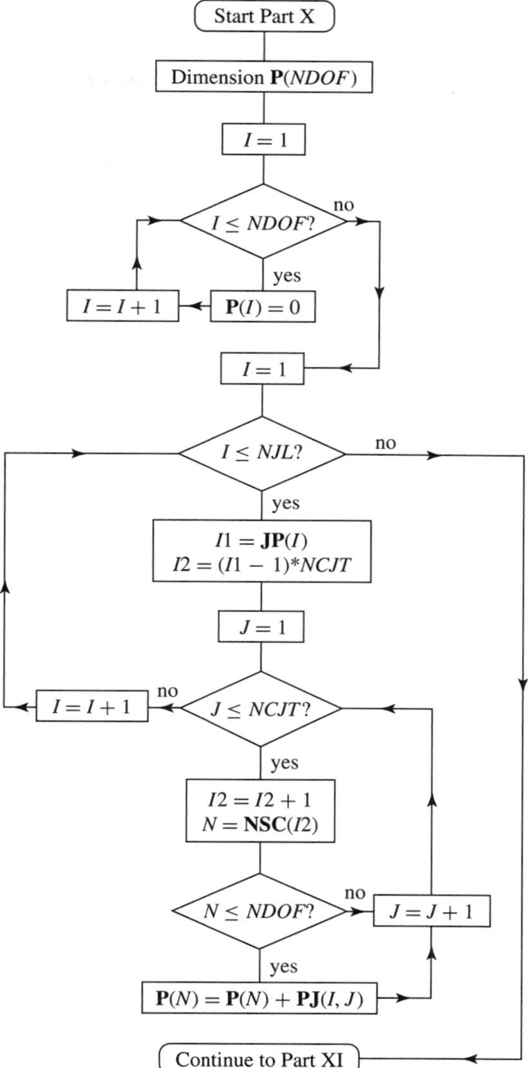

Fig. 4.12 *Flowchart for Forming Joint Load Vector*

The foregoing operations are repeated for each loaded joint of the structure to complete the joint load vector **P**.

To illustrate this procedure, let us form the joint load vector **P** for the example truss of Fig. 4.2(a) without referring to its visual image or line diagram (i.e., using only the input data matrices and the **NSC** vector). Recall that in Section 4.2, using the **MSUP** matrix, we determined that the number of degrees of freedom of this structure equals 8. Thus, the joint load vector **P** for the truss must be of order 8×1.

We begin generating **P** by focusing our attention on row 1 (i.e., $I = 1$) of the load data vector **JP** (Fig. 4.2(h)), from which we determine the number of the first loaded joint, $I1$, to be 2. We then determine the row of the **NSC** in which the number of the X structure coordinate at joint 2 is stored, using the following relationship:

$$I2 = (2 - 1)2 + 1 = 3$$

From row 3 of the **NSC** vector given in Fig. 4.7(b), we read the number of the structure coordinate under consideration as 1 (i.e., $N = 1$). This indicates that the force component in the first row and first column of the load data matrix **PJ** (i.e., the X component of the load acting at joint 2) must be stored in the first row of **P**; that is, **P**(1) = 0. Next, we increase $I2$ by 1 (i.e., $I2 = 4$) and, from row 4 of the **NSC**, we find the number of the Y structure coordinate at the joint to be 2 (i.e., $N = 2$). This indicates that the load component in the first row and second column of **PJ** is to be stored in the second row of **P**; that is, **P**(2) = −75.

Having stored the loads acting at joint 2 in the load vector **P**, we now focus our attention on the second row of **JP** (i.e., $I = 2$), and read the number of the next loaded joint, $I1$, as 5. We then determine the **NSC** row where the number of the X structure coordinate at joint 5 is stored as

$$I2 = (5 - 1)2 + 1 = 9$$

From row 9 of the **NSC** (Fig. 4.7(b)), we find the number of the structure coordinate under consideration to be 5 (i.e., $N = 5$). Thus, the force component in row 2 and column 1 of **PJ** must be stored in row 5 of **P**; or **P**(5) = 25. Next, we increase $I2$ by 1 to 10, and from row 10 of the **NSC** read the structure coordinate number, N, as 6. Thus, the load component in the second row and second column of **PJ** is stored in the sixth row of **P**; or **P**(6) = 0.

Finally, by repeating the foregoing procedure for row 3 of **JP**, we store the X and Y force components at joint 6 in rows 7 and 8, respectively, of **P**. The completed joint load vector **P** thus obtained is shown in Fig. 4.7(c).

4.5 SOLUTION FOR JOINT DISPLACEMENTS

Having programmed the generation of the structure stiffness matrix **S** and the joint load vector **P**, we now proceed to the next part of our computer program, which calculates the joint displacements, **d**, by solving the structure stiffness relationship, **Sd** = **P** (Eq. (3.89)). A flowchart for programming this analysis step is depicted in Fig. 4.13. The program solves the system of simultaneous equations, representing the stiffness relationship, **Sd** = **P**, using the Gauss–Jordan elimination method discussed in Section 2.4.

It should be recognized that the program for the calculation of joint displacements, as presented in Fig. 4.13, involves essentially the same operations as the program for the solution of simultaneous equations given in Fig. 2.2. However, in the previous program (Fig. 2.2), the elementary operations were

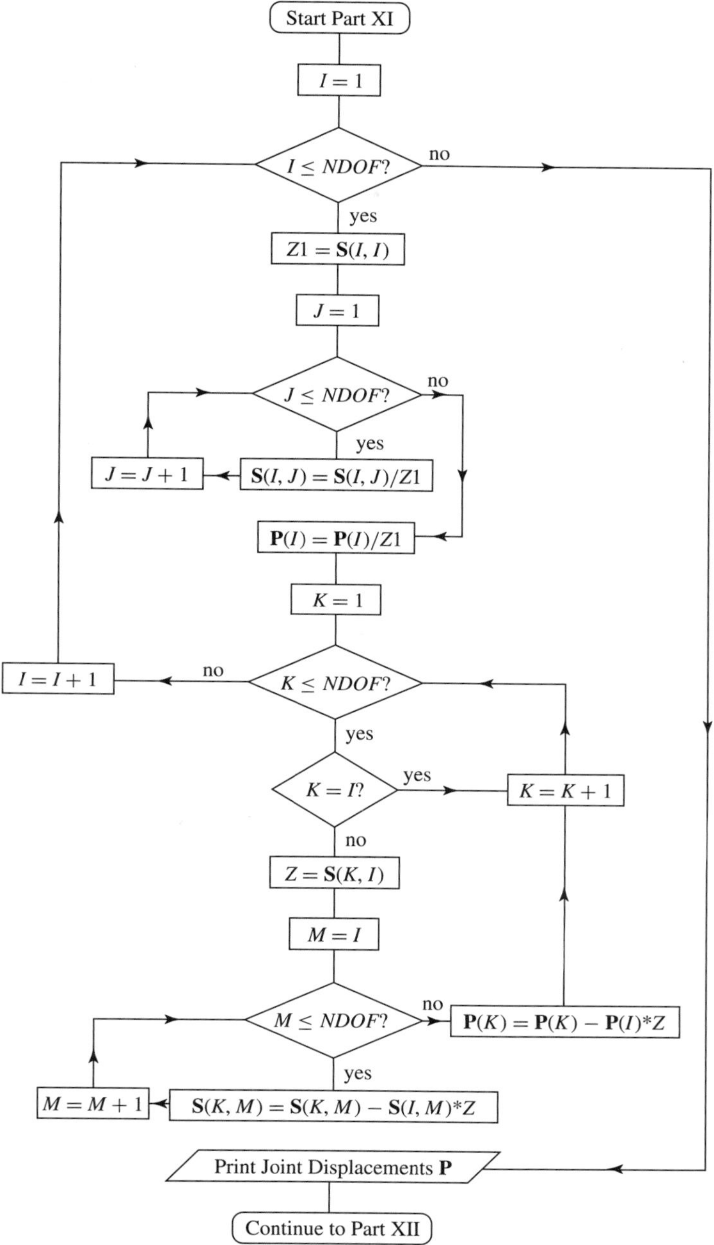

Fig. 4.13 *Flowchart for Calculation of Joint Displacements by Gauss–Jordan Method*

applied to an augmented matrix; in the present program (Fig. 4.13), to save space in computer memory, no augmented matrix is formed, and the elementary operations are applied directly to the structure stiffness matrix **S** and the joint load vector **P**. Thus, at the end of the Gauss–Jordan elimination process, the **S** matrix is reduced to a unit matrix, and the **P** vector contains values of the joint displacements. In the rest of our computer program, therefore, **P** (instead of **d**) is considered to be the joint displacement vector. The joint displacements thus obtained can be communicated to the user through a printout or on the screen.

4.6 CALCULATION OF MEMBER FORCES AND SUPPORT REACTIONS

In this section, we consider programming of the final analysis step, which involves calculation of the member forces and support reactions. A flowchart for programming this analysis step is presented in Fig. 4.14, with $NCJT = 2$ for plane trusses. As shown there, this part of our computer program begins by initializing each element of the reaction vector, **R**, to 0. The member forces and support reactions are then determined by performing the following operations for each member of the structure, via a *Do Loop*.

1. *Evaluation of member properties.* For the member under consideration, *IM*, the program reads the beginning joint number *JB*, the end joint number *JE*, the modulus of elasticity *E*, the cross-sectional area *A*, and the *X* and *Y* coordinates of the beginning and end joints. It then calculates the member length, *BL*, and direction cosines, *CX* ($= \cos\theta$) and *CY* ($= \sin\theta$), using Eqs. (3.62).

2. *Evaluation of member global end displacements* **V** (= **v**) *by subroutine **MDISPG**.* After the properties of the member under consideration, *IM*, have been calculated, the computer program calls subroutine ***MDISPG***, to obtain the member end displacements in the global coordinate system. A flowchart of this subroutine is given in Fig. 4.15. As this flowchart indicates, after initializing **V** to 0, the subroutine reads, in order, for each of the member end displacements, V_I, the number of the corresponding structure coordinate, *N*, at joint *JB* or *JE*, from the **NSC** vector. If the structure coordinate number *N*, corresponding to an end displacement V_I, is found to be less than or equal to *NDOF*, then the value of the element in the *N*th row of the joint-displacement vector **P** (= **d**) is stored in the *I*th row of the member displacement vector **V**.

3. *Determination of member transformation matrix* **T** *by subroutine **MTRANS***. After the global end-displacement vector **V** for the member under consideration, *IM*, has been evaluated, the main program (Fig. 4.14) calls on the subroutine ***MTRANS*** to form the member transformation matrix **T**. A flowchart of this subroutine is shown in Fig. 4.16. As this figure indicates, the subroutine first initializes **T** to 0, and then simply stores the values of the direction

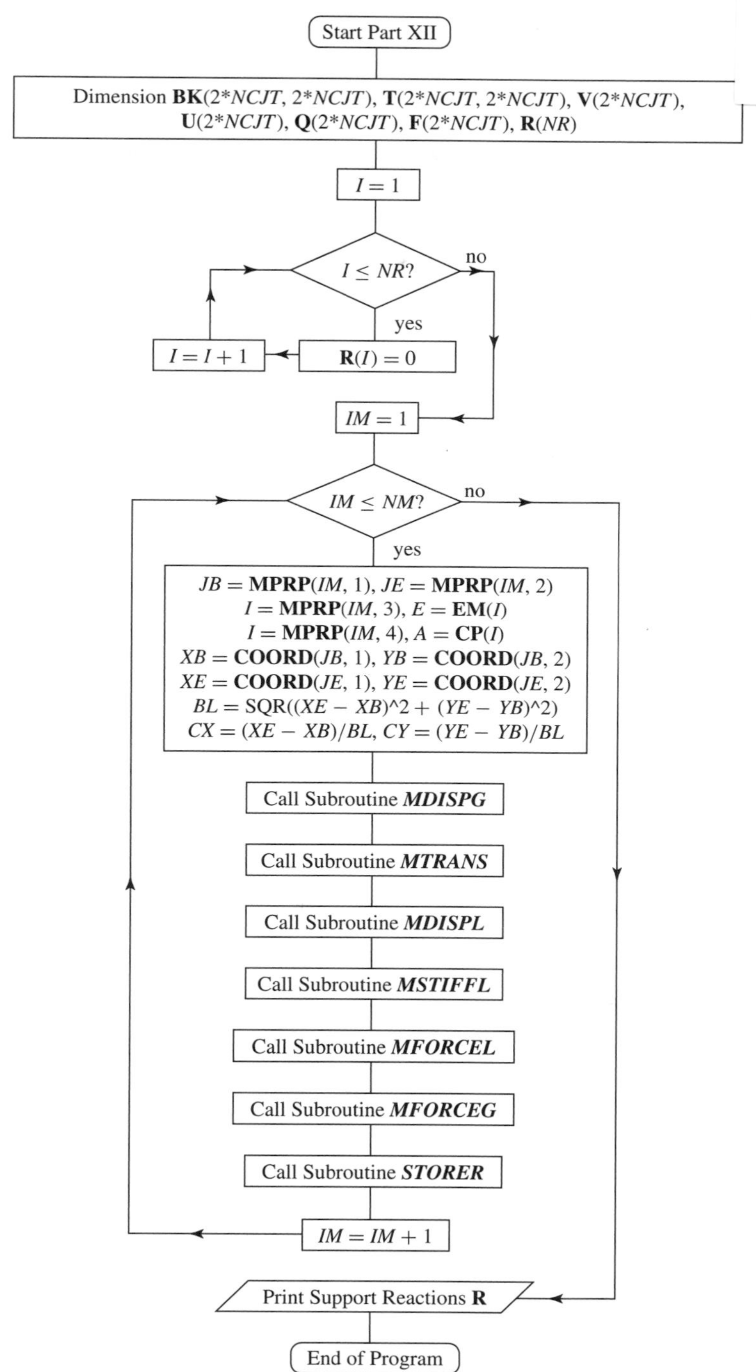

Fig. 4.14 *Flowchart for Determination of Member Forces and Support Reactions for Plane Trusses*

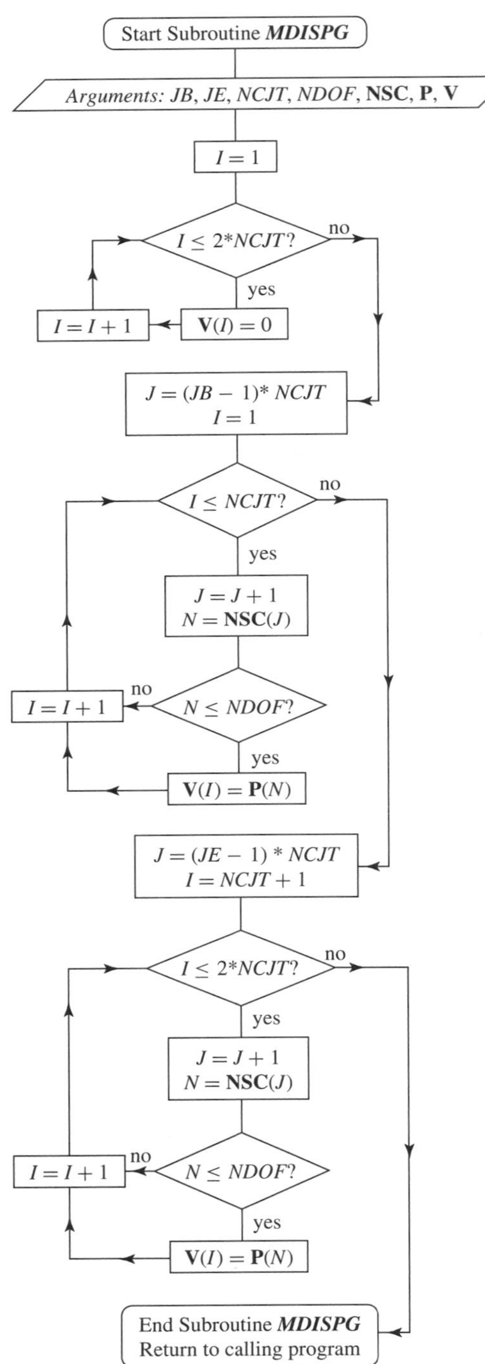

Fig. 4.15 *Flowchart of Subroutine* ***MDISPG*** *for Determining Member Global Displacement Vector*

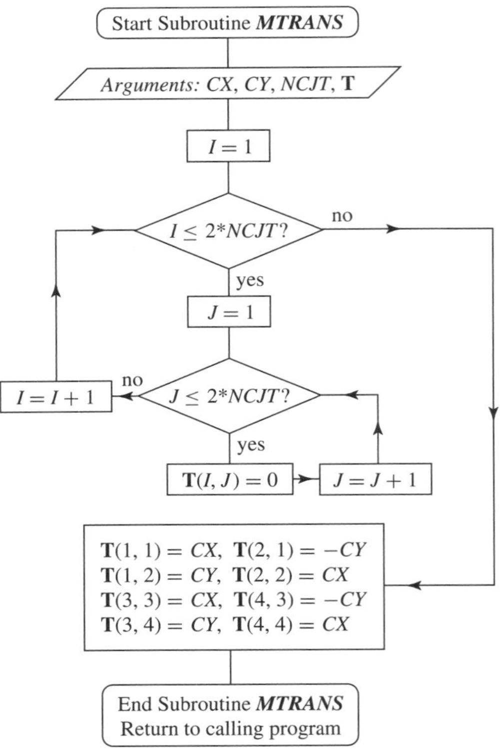

Fig. 4.16 *Flowchart of Subroutine* ***MTRANS*** *for Determining Member Transformation Matrix for Plane Trusses*

cosines *CX* and *CY*, with appropriate plus or minus signs, into various elements of **T** in accordance with Eq. (3.61).

4. *Calculation of member local end displacements* **U** (= **u**) *by subroutine* ***MDISPL***. Next, as shown in Fig. 4.14, the program calls subroutine ***MDISPL*** to obtain the local end displacements of the member under consideration, *IM*. From the flowchart given in Fig. 4.17, we can see that after initializing **U** to 0, this subroutine calculates the member local end-displacement vector by applying the relationship **U** = **TV** (Eq. (3.63)). The procedure for multiplying matrices was discussed in Section 2.3, and subroutine ***MDISPL*** (Fig. 4.17) uses essentially the same operations as the program for matrix multiplication given in Fig. 2.1.

5. *Determination of member local stiffness matrix* **BK** (= **k**) *by subroutine* ***MSTIFFL***. After the local end displacements of the member under

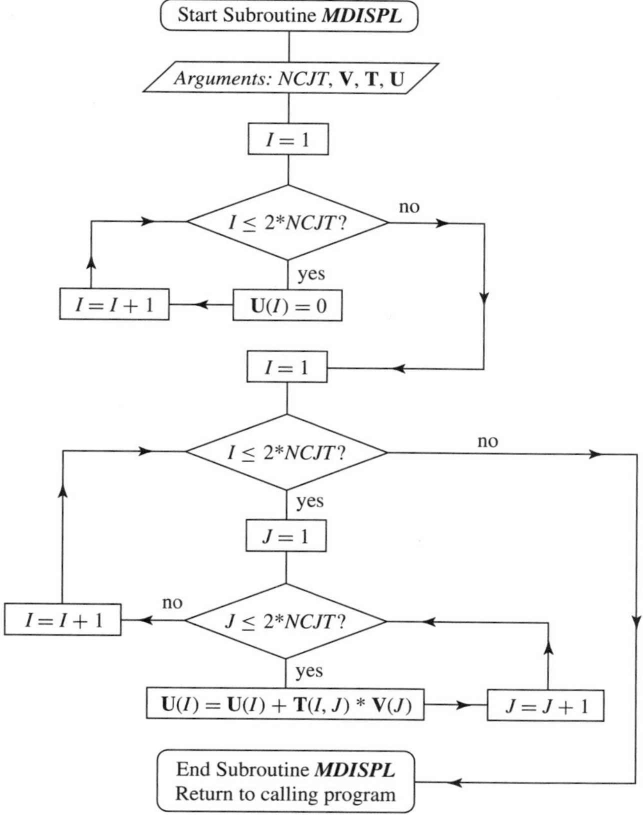

Fig. 4.17 *Flowchart of Subroutine **MDISPL** for Determining Member Local Displacement Vector*

consideration, *IM*, have been evaluated, the program calls subroutine ***MSTIFFL*** to form the member stiffness matrix in the local coordinate system. A flowchart of this subroutine is shown in Fig. 4.18, in which the member local stiffness matrix is identified by the name **BK** (instead of **k**) to indicate that it is a real matrix. As this figure indicates, the subroutine, after initializing **BK** to 0, simply calculates the values of the various stiffness coefficients and stores them in appropriate elements of **BK**, in accordance with Eq. (3.27).

6. *Evaluation of member local end forces* **Q** *by subroutine* ***MFORCEL***. As shown in Fig. 4.14, the program then calls subroutine ***MFORCEL*** to obtain the local end forces of the member under consideration, *IM*. From the flowchart depicted in Fig. 4.19, we can see that, after initializing **Q** to 0, this

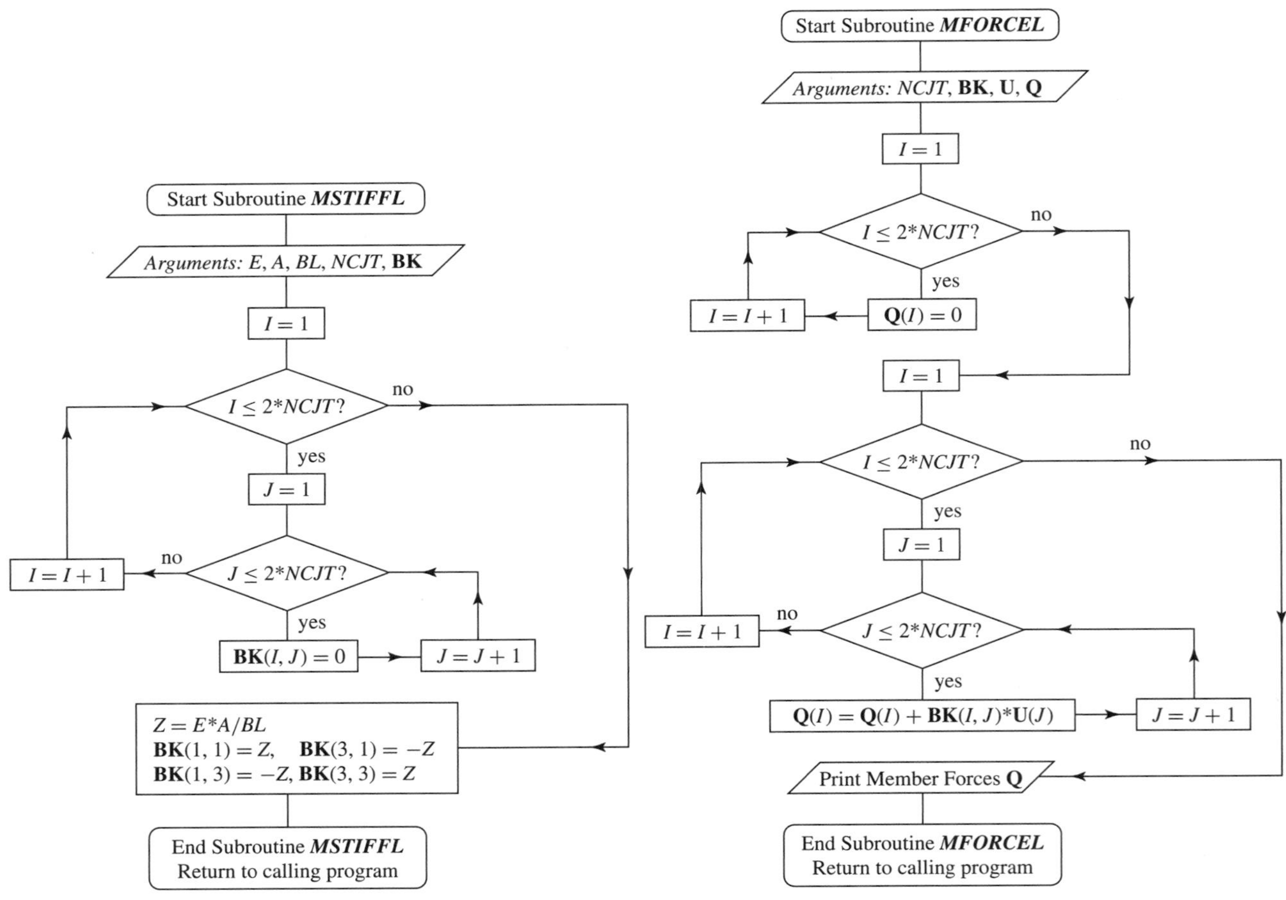

Fig. 4.18 *Flowchart of Subroutine* ***MSTIFFL*** *for Determining Member Local Stiffness Matrix for Plane Trusses*

Fig. 4.19 *Flowchart of Subroutine* ***MFORCEL*** *for Determining Member Local Force Vector*

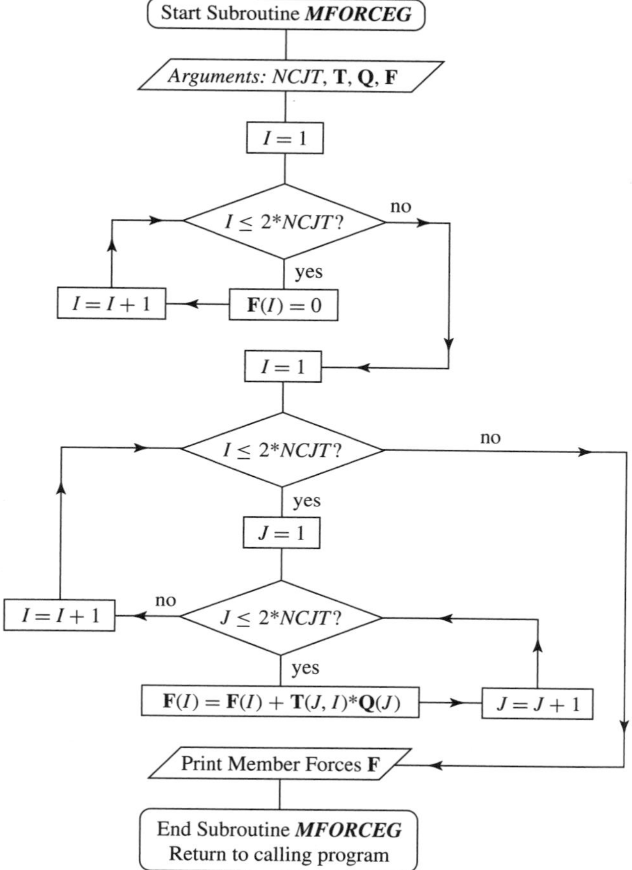

Fig. 4.20 *Flowchart of Subroutine* ***MFORCEG*** *for Determining Member Global Force Vector*

subroutine calculates the member local end forces using the relationship $\mathbf{Q} = \mathbf{BK}\,\mathbf{U}$ (Eq. (3.7)). The $\mathbf{Q}$ vector thus obtained is then printed or displayed on the screen.

7. *Calculation of member global end forces* **F** *by subroutine* ***MFORCEG****.* After the local end forces of the member under consideration, *IM*, have been evaluated, the computer program calls subroutine ***MFORCEG*** to calculate the member end forces in the global coordinate system. A flowchart of this subroutine is given in Fig. 4.20. From the figure, we can see that after initializing **F** to 0, the subroutine calculates the global end forces by applying the relationship $\mathbf{F} = \mathbf{T}^T\mathbf{Q}$ (Eq. (3.66)); these forces are then communicated to the user through a printer or on the screen.

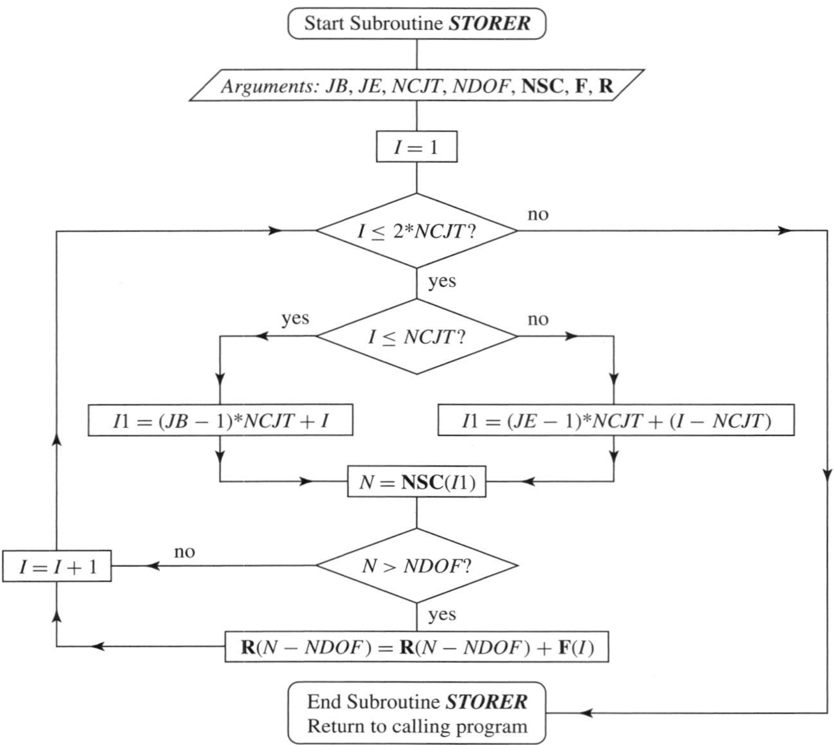

Fig. 4.21 *Flowchart of Subroutine* ***STORER*** *for Storing Member Global Forces in Support Reaction Vector*

8. *Storage of the elements of member global force vector* **F** *in reaction vector* **R** *by subroutine* ***STORER***. Once the global force vector **F** has been determined for the member under consideration, *IM*, the program (Fig. 4.14) calls subroutine ***STORER*** to store the pertinent elements of **F** in their proper positions in the support reaction vector **R**. A flowchart of this subroutine, which essentially consists of a *Do Loop,* is given in Fig. 4.21. As shown in this flowchart, the subroutine reads, in order, for each of the member's forces, F_I, the number of the corresponding structure coordinate, *N*, from the **NSC** vector. If $N > NDOF$, then the value of F_I is added to the $(N - NDOF)$th row of the reaction vector **R**.

Returning our attention to Fig. 4.14, we can see that the formation of the reaction vector **R** is completed when the foregoing eight operations have been performed for each member of the structure. The support reactions thus obtained can then be communicated to the user via a printer or on the screen. A sample printout is given in Fig. 4.22, showing the results of the analysis for the example truss of Fig. 4.2.

```
****************************************************
*              Results of Analysis                 *
****************************************************

                 =======================
                 Joint Displacements
                 =======================

Joint No.        X Translation        Y Translation
---------        -------------        -------------
    1             0.0000E+00           0.0000E+00
    2             7.4568E-02          -2.0253E-01
    3             1.1362E-01           0.0000E+00
    4             1.0487E-01           0.0000E+00
    5             5.7823E-02          -1.5268E-01
    6             2.8344E-02          -7.9235E-02

                 =======================
                 Member Axial Forces
                 =======================

          Member          Axial Force (Qa)
          ------          ----------------
             1            6.0069E+01 (T)
             2            3.1459E+01 (T)
             3            4.8629E+00 (C)
             4            2.3747E+01 (C)
             5            5.3543E+01 (T)
             6            8.5105E+01 (C)
             7            4.3836E+01 (C)
             8            3.5762E+01 (T)
             9            4.5402E+01 (C)
            10            6.0787E+00 (T)

                 =====================
                 Support Reactions
                 =====================

  Joint No.           X Force             Y Force
  ---------           -------             -------
      1             -2.5000E+01          2.6301E+01
      3              0.0000E+00          1.1235E+02
      4              0.0000E+00         -3.6472E+00

**************** End of Analysis ****************
```

Fig. 4.22 *A Sample Printout of Analysis Results*

SUMMARY

In this chapter, we have developed a general computer program for the analysis of plane trusses subjected to joint loads. The general program consists of a main program, which is subdivided into twelve parts, and nine subroutines. Brief descriptions of the various parts of the main program, and the subroutines, are provided in Table 4.1 for quick reference.

Table 4.1 *Computer Program for Analysis of Plane Trusses*

Main program part	Description
I	Reads and stores joint data (Fig. 4.3(a))
II	Reads and stores support data (Fig. 4.3(b))
III	Reads and stores material properties (Fig. 4.3(c))
IV	Reads and stores cross-sectional properties (Fig. 4.3(d))
V	Reads and stores member data (Fig. 4.3(e))
VI	Reads and stores joint loads (Fig. 4.3(f))
VII	Determines the number of degrees of freedom *NDOF* of the structure (Fig. 4.8(a))
VIII	Forms the structure coordinate number vector **NSC** (Fig. 4.8(b))
IX	Generates the structure stiffness matrix **S** (Fig. 4.9); subroutines called: ***MSTIFFG*** and ***STORES***
X	Forms the joint load vector **P** (Fig. 4.12)
XI	Calculates the structure's joint displacements by solving the stiffness relationship, **Sd** = **P**, using the Gauss–Jordan elimination method. The vector **P** now contains joint displacements (Fig. 4.13).
XII	Determines the member end force vectors **Q** and **F**, and the support reaction vector **R** (Fig. 4.14); subroutines called: ***MDISPG, MTRANS, MDISPL, MSTIFFL, MFORCEL, MFORCEG,*** and ***STORER***

Subroutine	Description
MDISPG	Determines the member global displacement vector **V** from the joint displacement vector **P** (Fig. 4.15)
MDISPL	Calculates the member local displacement vector $\mathbf{U} = \mathbf{TV}$ (Fig. 4.17)
MFORCEG	Determines the member global force vector $\mathbf{F} = \mathbf{T}^T\mathbf{Q}$ (Fig. 4.20)
MFORCEL	Evaluates the member local force vector $\mathbf{Q} = \mathbf{BK}\ \mathbf{U}$ (Fig. 4.19)
MSTIFFG	Forms the member global stiffness matrix **GK** (Fig. 4.10)
MSTIFFL	Forms the member local stiffness matrix **BK** (Fig. 4.18)
MTRANS	Forms the member transformation matrix **T** (Fig. 4.16)
STORER	Stores the pertinent elements of the member global force vector **F** in the reaction vector **R** (Fig. 4.21)
STORES	Stores the pertinent elements of the member global stiffness matrix **GK** in the structure stiffness matrix **S** (Fig. 4.11)

PROBLEMS

The objective of the following problems is to develop, incrementally, a computer program for the analysis of plane trusses; while testing each program increment for correctness, as it is being developed. The reader is strongly encouraged to manually solve as many of the problems (3.16 through 3.25) as possible, so that these hand-calculation results can be used to check the correctness of the various parts of the computer program.

Section 4.1

4.1 Develop an input module of a computer program for the analysis of plane trusses, which can perform the following operations:

a. read from a data file, or computer screen, all the necessary input data;
b. store the input data in computer memory in the form of scalars, vectors, and/or matrices, as appropriate; and
c. print the input data from computer memory.

Check the program for correctness by inputting data for the trusses of Problems 3.16 through 3.25, and by carefully examining the printouts of the input data to ensure that all data have been correctly read and stored.

Section 4.2

4.2 Extend the program developed in Problem 4.1, so that it can perform the following additional operations:

a. determining the number of degrees of freedom (*NDOF*) of the structure;
b. forming the structure coordinate number vector **NSC**; and
c. printing out the *NDOF* and **NSC**.

To check the program for correctness, use it to determine the *NDOF* and **NSC** for the trusses of Problems 3.16 through 3.25, and compare the computer-generated results to those obtained by hand calculations.

Section 4.3

4.3 Extend the program of Problem 4.2 to generate, and print, the structure stiffness matrix **S**. Use the program to generate the structure stiffness matrices for the trusses of Problems 3.16 through 3.25, and compare the computer-generated **S** matrices to those obtained by hand calculations.

Section 4.4

4.4 Extend the program developed in Problem 4.3 to form, and print, the joint load vector **P**. Apply the program to the trusses of Problems 3.16 through 3.25, and compare the computer-generated **P** vectors to those obtained by hand calculations.

Section 4.5

4.5 Extend the program of Problem 4.4 so that it can: (a) calculate the structure's joint displacements by solving the stiffness relationship, $\mathbf{Sd} = \mathbf{P}$, using the Gauss–Jordan elimination method; and (b) print the joint displacements. Using the program, determine the joint displacements for the trusses of Problems 3.16 through 3.25, and compare the computer-generated results to those obtained by hand calculations.

Section 4.6

4.6 Extend the program developed in Problem 4.5 so that it can determine and print: (a) the local end forces, **Q**, for each member of the truss; and (b) the support reaction vector **R**. Use the program to analyze the trusses of Problems 3.16 through 3.25, and compare the computer-generated results to those obtained by hand calculations.

5 BEAMS

A Continuous Beam Bridge
(Photo courtesy of Bethlehem Steel Corporation)

The term "beam" is used herein to refer to *a long straight structure, which is supported and loaded in such a way that all the external forces and couples (including reactions) acting on it lie in a plane of symmetry of its cross-section, with all the forces perpendicular to its centroidal axis.* Under the action of external loads, beams are subjected only to bending moments and shear forces (but no axial forces).

In this chapter, we study the basic concepts of the analysis of beams by the matrix stiffness method, and develop a computer program for the analysis of beams based on the matrix stiffness formulation. As we proceed through the chapter, the reader will notice that, although the member stiffness relations for beams differ from those for plane trusses, the overall format of the method of analysis remains essentially the same—and many of the analysis steps developed in Chapter 3 for the case of plane trusses can be directly applied to beams. Therefore, the computer program developed in Chapter 4 for the analysis of plane trusses can be modified with relative ease for the analysis of beams.

We begin by discussing the preparation of analytical models of beams in Section 5.1, where the global and local coordinate systems and the degrees of freedom of beams are defined. Next, we derive the member stiffness relations in the local coordinate system in Section 5.2; and present the finite-element formulation of the member stiffness matrix, via the principle of virtual work, in Section 5.3. The derivation of the member fixed-end forces, due to external loads applied to members, is considered in Section 5.4; and the formation of the stiffness relations for the entire beam, by combining the member stiffness relations, is discussed in Section 5.5. The procedure for forming the structure fixed-joint force vectors, and the concept of equivalent joint loads, are introduced in Section 5.6; and a step-by-step procedure for the analysis of beams is presented in Section 5.7. Finally, a computer program for the analysis of beams is developed in Section 5.8.

5.1 ANALYTICAL MODEL

For analysis by the matrix stiffness method, *the continuous beam is modeled as a series of straight prismatic members connected at their ends to joints, so that the unknown external reactions act only at the joints*. Consider, for example, the two-span continuous beam shown in Fig. 5.1(a). Although the structure actually consists of a single continuous beam between the two fixed supports at the ends, for the purpose of analysis it is considered to be composed of three members (1, 2, and 3), rigidly connected at four joints (1 through 4), as shown in Fig. 5.1(b). Note that joint 2 has been introduced in the analytical model so that the vertical reaction at the roller support acts on a joint (instead of on a member), and joint 3 is used to subdivide the right span of the beam into two members, each with constant flexural rigidity (EI) along its length. This division of the beam into members and joints is necessary because the formulation of the stiffness method requires that the unknown external reactions act only at the joints (i.e., all the member loads be known in advance of analysis), and the member stiffness relationships used in the analysis (to be derived in the following sections) are valid for prismatic members only.

It is important to realize that because joints 1 through 4 (Fig. 5.1(b)) are modeled as rigid joints (i.e., the corresponding ends of the adjacent members are rigidly connected to the joints), they satisfy the continuity and restraint conditions of the actual structure (Fig. 5.1(a)). In other words, since the left end of member 1 and the right end of member 3 of the analytical model are rigidly

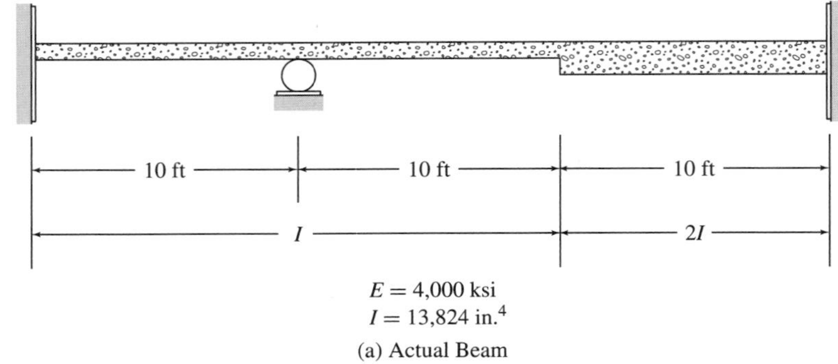

(a) Actual Beam

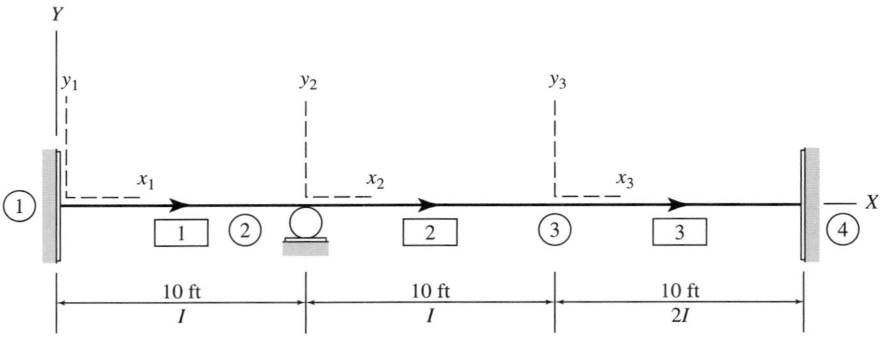

(b) Analytical Model Showing Global and Local Coordinate Systems

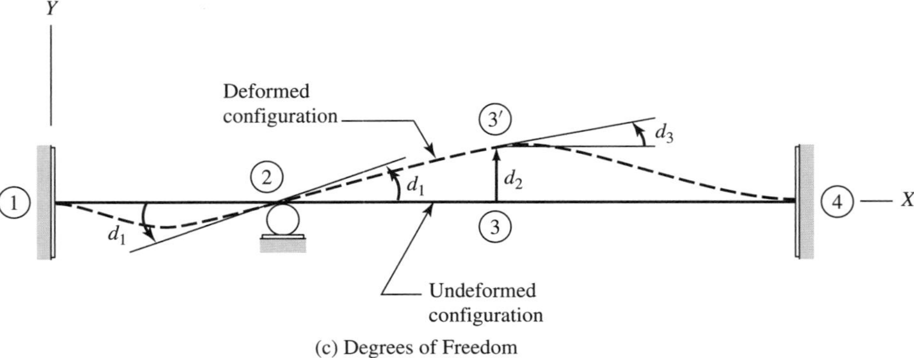

(c) Degrees of Freedom

Fig. 5.1

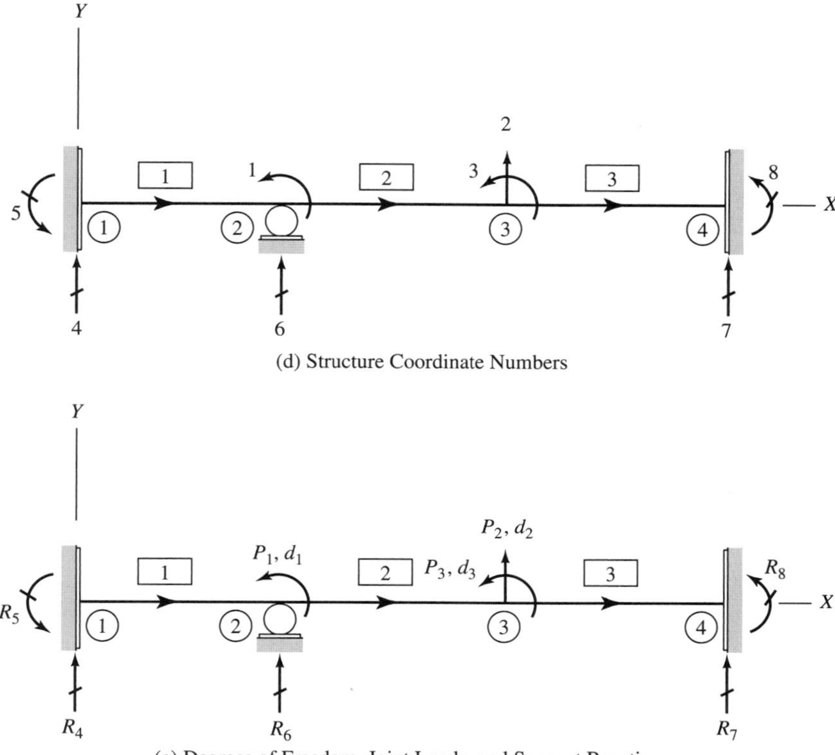

(d) Structure Coordinate Numbers

(e) Degrees of Freedom, Joint Loads, and Support Reactions

Fig. 5.1 *(continued)*

connected to joints 1 and 4, respectively, which are in turn attached to the fixed supports, the displacements and rotations at the exterior ends of the members are 0; thereby satisfying the restraint conditions of the actual beam at the two ends. Similarly, as the right end of member 1 and the left end of member 2 (Fig. 5.1(b)) are connected to the rigid joint 2, which is attached to a roller support, the displacements at the foregoing ends of members 1 and 2 are 0, and the rotations at the two ends are equal. This indicates that the analytical model satisfies the restraint and continuity conditions of the actual beam at the location of joint 2. Finally, the right end of member 2 and the left end of member 3 (Fig. 5.1(b)) are rigidly connected to joint 3, to ensure that the continuity of both the displacement and the rotation is maintained at the location of joint 3 in the analytical model.

Global and Local Coordinate Systems

As discussed in Chapter 3, the overall geometry, as well as the loads and displacements (including rotations) at the joints of a structure are described with reference to a Cartesian global (*XYZ*) coordinate system. The particular orientation of the global coordinate system, used in this chapter, is as follows.

> *The global coordinate system used for the analysis of beams is a right-handed XYZ coordinate system, with the X axis oriented in the horizontal (positive to the right) direction, and coinciding with the centroidal axis of the beam in the undeformed state. The Y axis is oriented in the vertical (positive upward) direction, with all the external loads and reactions of the beam lying in the XY plane.*

Although not necessary, it is usually convenient to locate the origin of the global *XY* coordinate system at the leftmost joint of the beam, as shown in Fig. 5.1(b), so that the *X* coordinates of all the joints are positive. As will become apparent in Section 5.8, this definition of the global coordinate system simplifies the computer programming of beam analysis, because only one (*X*) coordinate is needed to specify the location of each joint of the structure.

As in the case of plane trusses (Chapter 3), a local (right-handed, *xyz*) coordinate system is defined for each member of the beam, to establish the relationships between member end forces and end displacements, in terms of member loads. Note that *the terms forces (or loads) and displacements are used in this text in the general sense to include moments and rotations, respectively.* The local coordinate system is defined as follows.

> *The origin of the local xyz coordinate system for a member is located at the left end (beginning) of the member in its undeformed state, with the x axis directed along its centroidal axis in the undeformed state, and the y axis oriented in the vertical (positive upward) direction.*

The local coordinate systems for the three members of the example continuous beam are depicted in Fig. 5.1(b). As this figure indicates, the local coordinate system of each member is oriented so that the positive directions of the local *x* and *y* axes are the same as the positive directions of the global *X* and *Y* axes, respectively.

The selection of the global and local coordinate systems, as specified in this section, considerably simplifies the analysis of continuous beams by eliminating the need for transformation of member end forces, end displacements, and stiffnesses, from the local to the global coordinate system and vice-versa.

Degrees of Freedom

The degrees of freedom (or free coordinates) of a beam are simply its unknown joint displacements (translations and rotations). Since the axial deformations of the beam are neglected, the translations of its joints in the global *X* direction are 0. Therefore, a joint of a beam can have up to two degrees of freedom, namely, a translation in the global *Y* direction (i.e., in the direction perpendicular to the beam's centroidal axis) and a rotation (about the global *Z* axis). Thus, the number of structure coordinates (i.e., free and/or restrained coordinates) at a joint of a beam equals 2, or $NCJT = 2$.

Let us consider the analytical model of the continuous beam as given in Fig. 5.1(b). The deformed shape of the beam, due to an arbitrary loading, is depicted in Fig. 5.1(c) using an exaggerated scale. From this figure, we can see that joint 1, which is attached to the fixed support, can neither translate nor rotate; therefore, it does not have any degrees of freedom. Since joint 2 of the beam is attached to the roller support, it can rotate, but not translate. Thus, joint 2 has only one degree of freedom, which is designated d_1 in the figure. As joint 3 is not attached to a support, two displacements—the translation d_2 in the Y direction, and the rotation d_3 about the Z axis—are needed to completely specify its deformed position $3'$. Thus, joint 3 has two degrees of freedom. Finally, joint 4, which is attached to the fixed support, can neither translate nor rotate; therefore, it does not have any degrees of freedom. Thus, the entire beam has a total of three degrees of freedom.

As indicated in Fig. 5.1(c), joint translations are considered positive when vertically upward, and joint rotations are considered positive when counterclockwise. All the joint displacements in Fig. 5.1(c) are shown in the positive sense. The $NDOF \times 1$ joint displacement vector $\mathbf{d}$ for the beam is written as

$$\mathbf{d} = \begin{bmatrix} d_1 \\ d_2 \\ d_3 \end{bmatrix}$$

Since the number of structure coordinates per joint equals 2 (i.e., $NCJT = 2$), the number of degrees of freedom, $NDOF$, of a beam can be obtained from Eq. (3.2) as

$$\left.\begin{array}{l} NCJT = 2 \\ NDOF = 2(NJ) - NR \end{array}\right\} \quad \text{for beams} \tag{5.1}$$

in which, as in the case of plane trusses, NJ represents the number of joints of the beam, and NR denotes the number of joint displacements restrained by supports (or the number of restrained coordinates). Let us apply Eq. (5.1) to the analytical model of the beam in Fig. 5.1(b). The beam has four joints (i.e., $NJ = 4$); two joints, 1 and 4, are attached to the fixed supports that together restrain four joint displacements (namely, the translations in the Y direction and the rotations of joints 1 and 4). Furthermore, the roller support at joint 2 restrains one joint displacement, which is the translation of joint 2 in the Y direction. Thus, the total number of joint displacements that are restrained by all supports of the beam is 5 (i.e., $NR = 5$). Substitution of the numerical values of NJ and NR into Eq. (5.1) yields

$$NDOF = 2(4) - 5 = 3$$

which is the same as the number of degrees of freedom of the beam obtained previously. As in the case of plane trusses, the free and restrained coordinates of a beam are collectively referred to simply as the structure coordinates.

When analyzing a beam, it is not necessary to draw its deformed shape, as shown in Fig. 5.1(c), to identify the degrees of freedom. Instead, all the structure coordinates (i.e., degrees of freedom and restrained coordinates) are usually directly specified on the beam's line diagram by assigning numbers to the arrows drawn at the joints in the directions of the joint displacements, as shown in Fig. 5.1(d). In this figure, a slash (/) has been added to the arrows corresponding to the restrained coordinates to distinguish them from those representing the degrees of freedom.

The procedure for assigning numbers to the structure coordinates of beams is similar to that for the case of plane trusses, discussed in detail in Section 3.2. The degrees of freedom are numbered first, starting at the lowest-numbered joint, that has a degree of freedom, and proceeding sequentially to the highest-numbered joint. If a joint has two degrees of freedom, then the translation in the *Y* direction is numbered first, followed by the rotation. The first degree of freedom is assigned the number 1, and the last degree of freedom is assigned a number equal to *NDOF*.

After all the degrees of freedom of the beam have been numbered, its restrained coordinates are numbered beginning with a number equal to $NDOF + 1$. Starting at the lowest-numbered joint that is attached to a support, and proceeding sequentially to the highest-numbered joint, all of the restrained coordinates of the beam are numbered. If a joint has two restrained coordinates, then the coordinate in the *Y* direction (corresponding to the reaction force) is numbered first, followed by the rotation coordinate (corresponding to the reaction couple). The number assigned to the last restrained coordinate of the beam is always $2(NJ)$. The structure coordinate numbers for the example beam, obtained by applying the foregoing procedure, are given in Fig. 5.1(d).

Joint Load and Reaction Vectors

Unlike plane trusses, which are subjected only to joint loads, the external loads on beams may be applied at the joints as well as on the members. The external loads (i.e., forces and couples or moments) applied at the joints of a structure are referred to as the *joint loads,* whereas the external loads acting between the ends of the members of the structure are termed the *member loads*. In this section, we focus our attention only on the joint loads, with the member loads considered in subsequent sections. As discussed in Section 3.2, an external joint load can, in general, be applied to the beam at the location and in the direction of each of its degrees of freedom. For example, the beam of Fig. 5.1(b), with three degrees of freedom, can be subjected to a maximum of three joint loads, P_1 through P_3, as shown in Fig. 5.1(e). As indicated there, a load corresponding to a degree of freedom d_i is denoted symbolically by P_i. The 3×1 joint load vector $\mathbf{P}$ for the beam is written in the form

$$\mathbf{P} = \underset{NDOF \times 1}{\begin{bmatrix} P_1 \\ P_2 \\ P_3 \end{bmatrix}}$$

As for the support reactions, when a beam is subjected to external joint and/or member loads, a reaction (force or moment) can develop at the location and in the direction of each of its restrained coordinates. For example, the beam of Fig. 5.1(b), which has five restrained coordinates, can develop up to five reactions, as shown in Fig. 5.1(e). As indicated in this figure, the reaction corresponding to the ith restrained coordinate is denoted symbolically by R_i. The 5×1 reaction vector $\mathbf{R}$ for the beam is expressed as

$$\mathbf{R} = \underset{NR \times 1}{\begin{bmatrix} R_4 \\ R_5 \\ R_6 \\ R_7 \\ R_8 \end{bmatrix}}$$

EXAMPLE 5.1 Identify by numbers the degrees of freedom and restrained coordinates of the continuous beam with a cantilever overhang shown in Fig. 5.2(a). Also, form the beam's joint load vector **P**.

SOLUTION The beam has four degrees of freedom, which are identified by numbers 1 through 4 in Fig. 5.2(b). The four restrained coordinates of the beam are identified by numbers 5 through 8 in the same figure. **Ans**

By comparing Figs. 5.2(a) and (b), we can see that $P_1 = -50$ k-ft; $P_2 = 0$; $P_3 = -20$ k; and $P_4 = 0$. The negative signs assigned to the magnitudes of P_1 and P_3 indicate that these loads act in the clockwise and downward directions, respectively. Thus, the joint load vector can be expressed in the units of kips and feet, as

$$\mathbf{P} = \begin{bmatrix} -50 \\ 0 \\ -20 \\ 0 \end{bmatrix}$$ **Ans**

Alternative Approach: The analysis of beams with cantilever overhangs can be considerably expedited by realizing that the cantilever portions are statically determinate (in the sense that the shear and moment at a cantilever's end can be evaluated directly by applying the equilibrium equations to the free-body of the cantilever portion). Therefore, the cantilever portions can be removed from the beam, and only the remaining indeterminate part needs to be analyzed by the stiffness method. However, the end moments and the end forces exerted by the cantilevers on the remaining indeterminate part of the structure must be included in the stiffness analysis, as illustrated in the following paragraphs.

Since the beam of Fig. 5.2(a) has a cantilever member CD, we separate this statically determinate member from the rest of the beam, as shown in Fig. 5.2(c). The force S_{CD} and the moment M_{CD} at end C of the cantilever are then calculated by applying the equilibrium equations, as follows.

$$+\uparrow \sum F_Y = 0 \qquad S_{CD} - 20 = 0 \qquad S_{CD} = 20 \text{ k} \uparrow$$

$$+\circlearrowleft \sum M_C = 0 \qquad M_{CD} - 20(10) = 0 \qquad M_{CD} = 200 \text{ k-ft} \circlearrowleft$$

Next, the moment M_{CD} is applied as a joint load, in the clockwise (opposite) direction, at joint C of the indeterminate part AC of the beam, as shown in Fig. 5.2(c). Note that

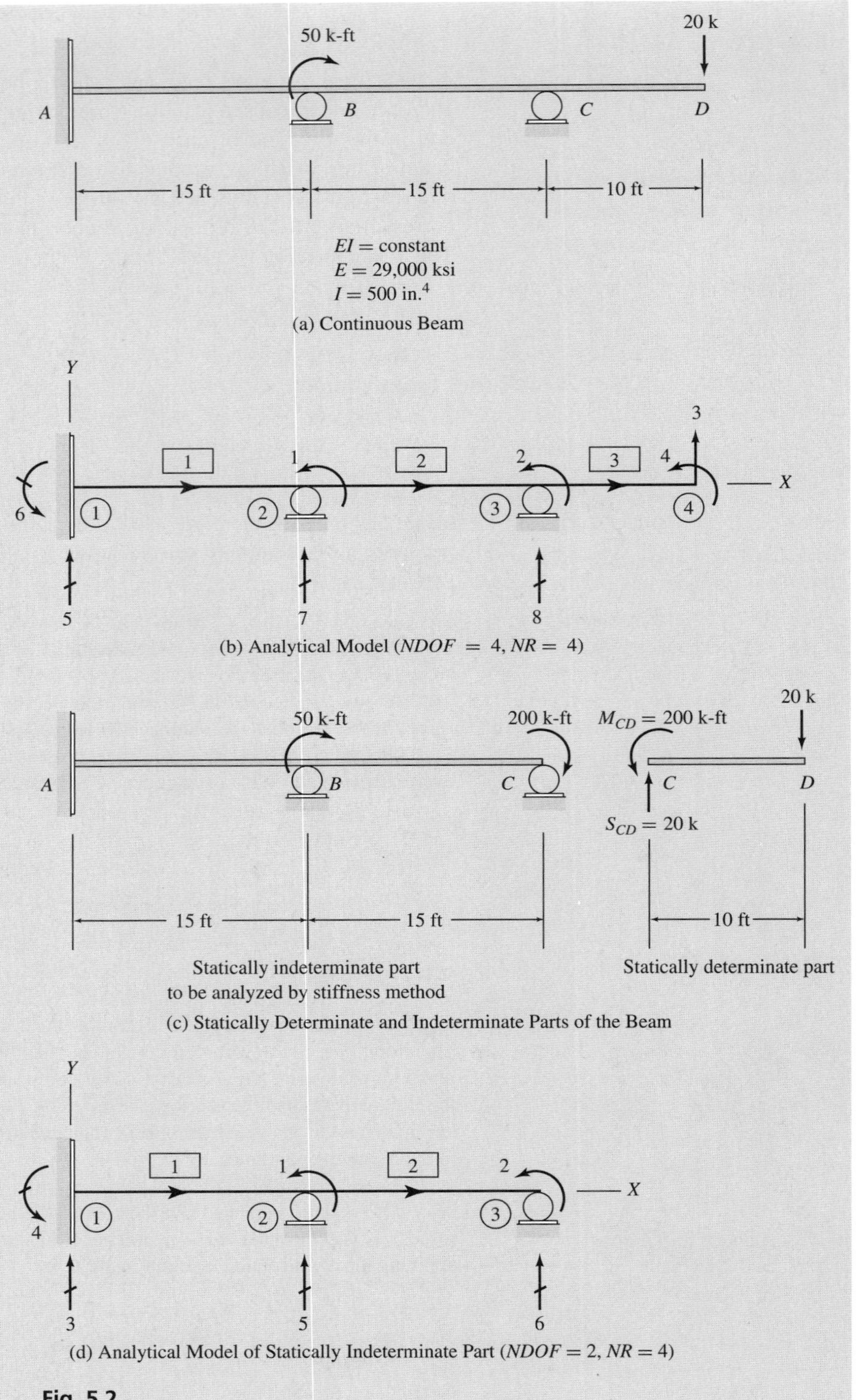
50 k-ft
20 k
A
B
C
D
15 ft
15 ft
10 ft
EI = constant
E = 29,000 ksi
I = 500 in.4
(a) Continuous Beam
Y
X
1
2
3
4
5
6
7
8
(b) Analytical Model (NDOF = 4, NR = 4)
50 k-ft
200 k-ft
MCD = 200 k-ft
20 k
SCD = 20 k
15 ft
15 ft
10 ft
Statically indeterminate part
to be analyzed by stiffness method
Statically determinate part
(c) Statically Determinate and Indeterminate Parts of the Beam
(d) Analytical Model of Statically Indeterminate Part (NDOF = 2, NR = 4)

Fig. 5.2

the end force S_{CD} (= 20 k) need not be considered in the analysis of the indeterminate part because its only effect is to increase the reaction at support C by 20 k.

The analytical model of the indeterminate part of the beam is drawn in Fig. 5.2(d). Note that the number of degrees of freedom has now been reduced to only two, identified by numbers 1 and 2 in the figure. The number of restrained coordinates remains at four, and these coordinates are identified by numbers 3 through 6 in Fig. 5.2(d). By comparing the indeterminate part of the beam in Fig. 5.2(c) to its analytical model in Fig. 5.2(d), we obtain the joint load vector as

$$\mathbf{P} = \begin{bmatrix} -50 \\ -200 \end{bmatrix} \text{ k-ft}$$ **Ans**

Once the analytical model of Fig. 5.2(d) has been analyzed by the stiffness method, the reaction force R_6 must be adjusted (i.e., increased by 20 k) to account for the end force S_{CD} being exerted by the cantilever CD on support C.

5.2 MEMBER STIFFNESS RELATIONS

When a beam is subjected to external loads, internal moments and shears generally develop at the ends of its individual members. *The equations expressing the forces (including moments) at the end of a member as functions of the displacements (including rotations) of its ends, in terms of the external loads applied to the member, are referred to as the member stiffness relations.* Such member stiffness relations are necessary for establishing the stiffness relations for the entire beam, as discussed in Section 5.5. In this section, we derive the stiffness relations for the members of beams.

To develop the member stiffness relations, we focus our attention on an arbitrary prismatic member m of the continuous beam shown in Fig. 5.3(a). When the beam is subjected to external loads, member m deforms and internal shear forces and moments are induced at its ends. The initial and displaced positions of m are depicted in Fig. 5.3(b), in which L, E, and I denote the length, Young's modulus of elasticity, and moment of inertia, respectively, of the member. It can be seen from this figure that two displacements—translation in the y direction and rotation about the z axis—are necessary to completely specify

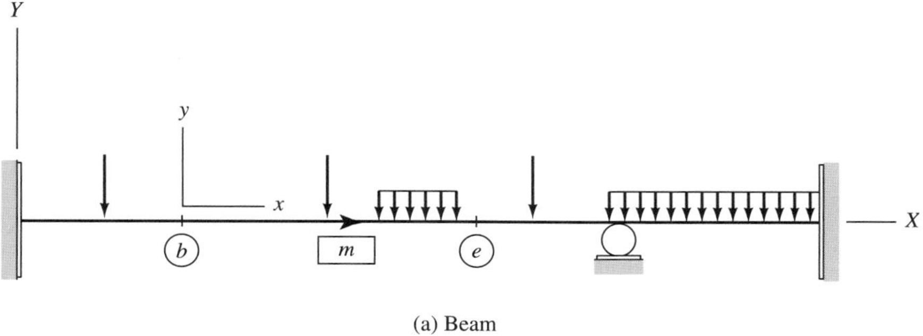

(a) Beam

Fig. 5.3

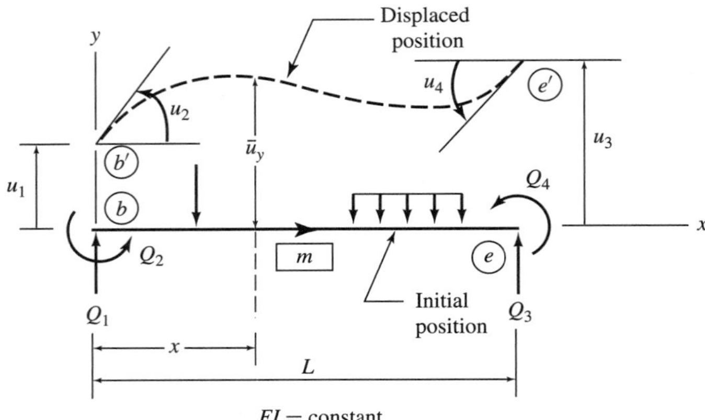

(b) Member Forces and Displacements in the Local Coordinate System

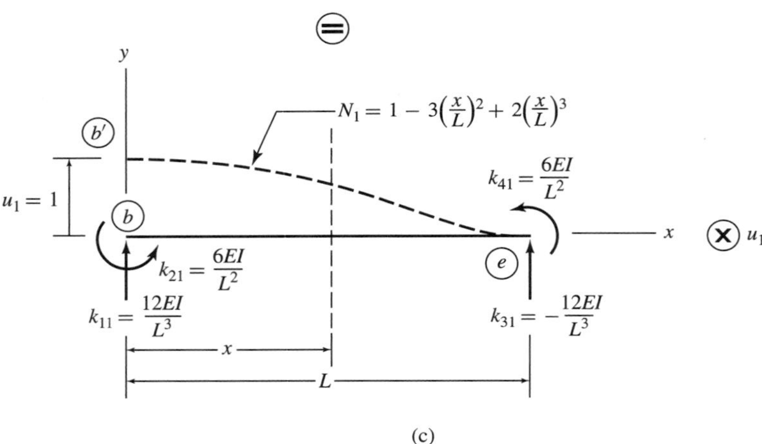

(c)

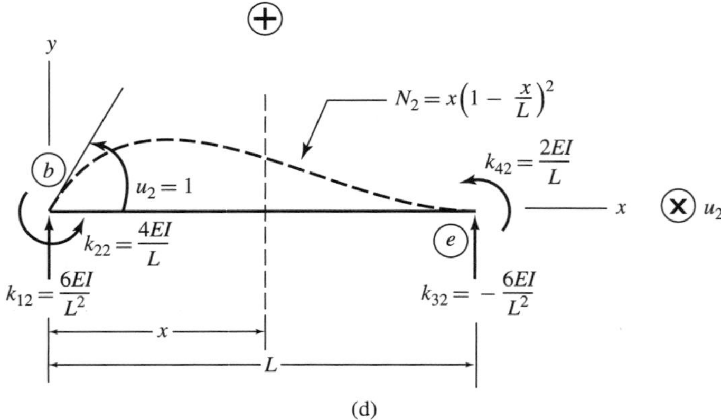

(d)

Fig. 5.3 *(continued)*

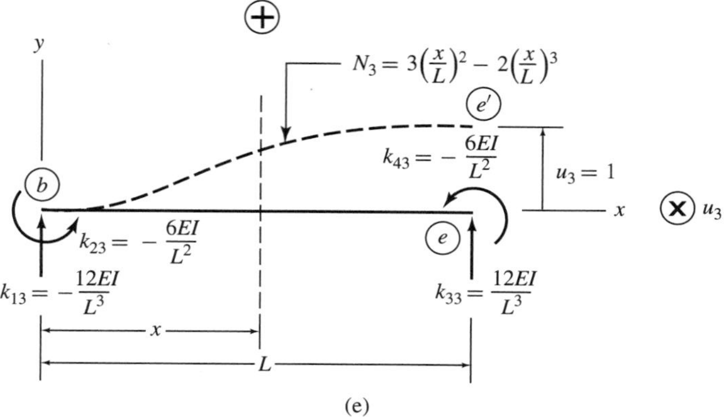

(e)

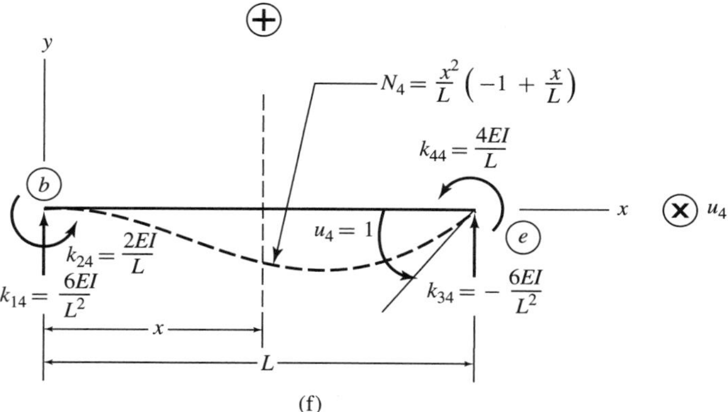

(f)

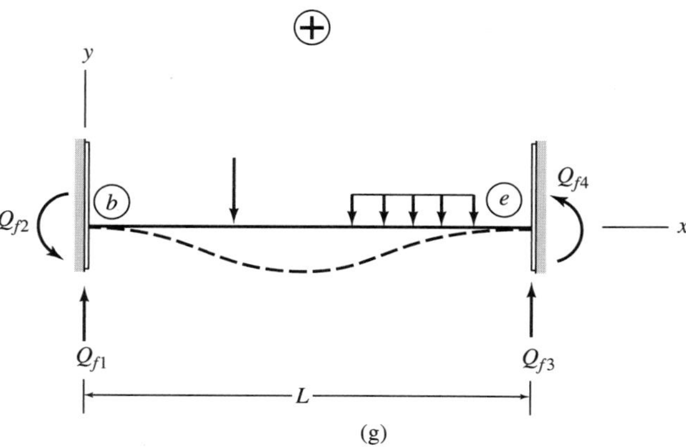

(g)

Fig. 5.3 *(continued)*

the displaced position of each end of the member. Thus, the member has a total of four end displacements or degrees of freedom. As Fig. 5.3(b) indicates, the member end displacements (including rotations) are denoted by u_1 through u_4, and the corresponding end forces (including moments) are denoted by Q_1 through Q_4. Note that the member end translations and forces are considered positive when vertically upward (i.e., in the positive direction of the local y axis), and the end rotations and moments are considered positive when counterclockwise. The numbering scheme used for identifying the member end displacements and forces is similar to that used previously for plane trusses in Chapter 3. As indicated in Fig. 5.3(b), *the member end displacements and forces are numbered by beginning at the left end b of the member, which is the origin of the local coordinate system, with the vertical translation and force numbered first, followed by the rotation and moment. The displacements and forces at the opposite end e of the member are then numbered in the same sequential order.*

The relationships between member end forces and end displacements can be conveniently established by subjecting the member, separately, to each of the four end displacements and external loads, as shown in Figs. 5.3(c) through (g); and by expressing the total member end forces as the algebraic sums of the end forces required to cause the individual end displacements and the forces caused by the external loads acting on the member with no end displacements. Thus, from Figs. 5.3(b) through (g), we can see that

$$Q_1 = k_{11}u_1 + k_{12}u_2 + k_{13}u_3 + k_{14}u_4 + Q_{f1} \tag{5.2a}$$

$$Q_2 = k_{21}u_1 + k_{22}u_2 + k_{23}u_3 + k_{24}u_4 + Q_{f2} \tag{5.2b}$$

$$Q_3 = k_{31}u_1 + k_{32}u_2 + k_{33}u_3 + k_{34}u_4 + Q_{f3} \tag{5.2c}$$

$$Q_4 = k_{41}u_1 + k_{42}u_2 + k_{43}u_3 + k_{44}u_4 + Q_{f4} \tag{5.2d}$$

in which, as defined in Chapter 3, *a stiffness coefficient k_{ij} represents the force at the location and in the direction of Q_i required, along with other end forces, to cause a unit value of displacement u_j, while all other end displacements are* 0, *and the member is not subjected to any external loading between its ends.* The last terms, Q_{fi} (with $i = 1$ to 4), on the right sides of Eqs. (5.2), represent *the forces that would develop at the member ends, due to external loads, if both ends of the member were fixed against translations and rotations* (see Fig. 5.3(g)). These forces are commonly referred to as the *member fixed-end forces* due to external loads. Equations (5.2) can be written in matrix form as

$$\begin{bmatrix} Q_1 \\ Q_2 \\ Q_3 \\ Q_4 \end{bmatrix} = \begin{bmatrix} k_{11} & k_{12} & k_{13} & k_{14} \\ k_{21} & k_{22} & k_{23} & k_{24} \\ k_{31} & k_{32} & k_{33} & k_{34} \\ k_{41} & k_{42} & k_{43} & k_{44} \end{bmatrix} \begin{bmatrix} u_1 \\ u_2 \\ u_3 \\ u_4 \end{bmatrix} + \begin{bmatrix} Q_{f1} \\ Q_{f2} \\ Q_{f3} \\ Q_{f4} \end{bmatrix} \tag{5.3}$$

or, symbolically, as

$$\boxed{\mathbf{Q} = \mathbf{k}\mathbf{u} + \mathbf{Q}_f} \tag{5.4}$$

in which **Q** and **u** represent the member end force and member end displacement vectors, respectively, in the local coordinate system; **k** is the member stiffness matrix in the local coordinate system; and $\mathbf{Q}_f$ is called the *member fixed-end force vector in the local coordinate system.*

In the rest of this and the following section, we focus our attention on the derivation of the member stiffness matrix **k**. The fixed-end force vector $\mathbf{Q}_f$ is considered in detail in Section 5.4.

Derivation of Member Stiffness Matrix k

Various classical methods of structural analysis, such as the *method of consistent deformations* and the *slope-deflection equations,* can be used to determine the expressions for the stiffness coefficients k_{ij} in terms of member length and its flexural rigidity, EI. In the following, however, we derive such stiffness expressions by directly integrating the differential equation for beam deflection. This direct integration approach is not only relatively simple and straightforward, but it also yields member shape functions as a part of the solution. The shape functions are often used to establish the member mass matrices for the dynamic analysis of beams [34]; they also provide insight into the finite-element formulation of beam analysis (considered in the next section).

It may be recalled from a previous course on *mechanics of materials* that the differential equation for small-deflection bending of a beam, composed of linearly elastic homogenous material and loaded in a plane of symmetry of its cross-section, can be expressed as

$$\boxed{\frac{d^2\bar{u}_y}{dx^2} = \frac{M}{EI}} \tag{5.5}$$

in which $\bar{u}_y$ represents the deflection of the beam's centroidal axis (which coincides with the neutral axis) in the y direction, at a distance x from the origin of the xy coordinate system as shown in Fig. 5.3(b); and M denotes the bending moment at the beam section at the same location, x. It is important to realize that the bending moment M is considered positive in accordance with the *beam sign convention*, which can be stated as follows (see Fig. 5.4).

> *The bending moment at a section of a beam is considered positive when the external force or couple tends to bend the beam concave upward (in the positive y direction), causing compression in the fibers above (in the positive y direction), and tension in the fibers below (in the negative y direction), the neutral axis of the beam at the section.*

To obtain the expressions for the coefficients k_{i1} ($i = 1$ through 4) in the first column of the member stiffness matrix **k** (Eq. (5.3)), we subject the

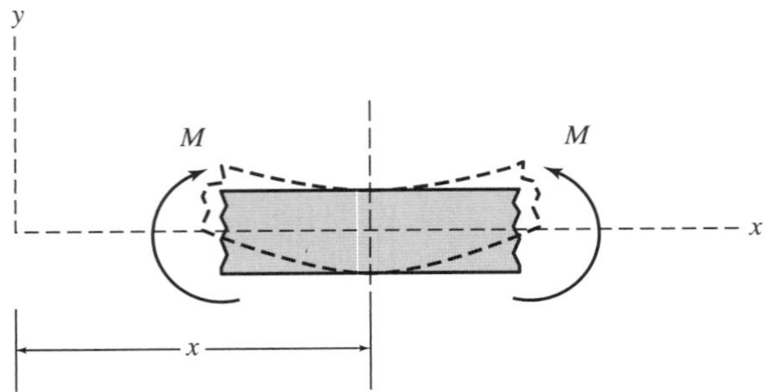

Fig. 5.4 *Beam Sign Convention*

member to a unit value of the end displacement u_1 at end b, as shown in Fig. 5.3(c). Note that all other end displacements of the member are 0 (i.e., $u_2 = u_3 = u_4 = 0$), and the member is in equilibrium under the action of two end moments k_{21} and k_{41}, and two end shears k_{11} and k_{31}. To determine the equation for bending moment for the member, we pass a section at a distance x from end b, as shown in Fig. 5.3(c). Considering the free body to the left of this section, we obtain the bending moment M at the section as

$$M = -k_{21} + k_{11}x \tag{5.6}$$

Note that the bending moment due to the couple k_{21} is negative, in accordance with the *beam sign convention,* because of its tendency to bend the member concave downward, causing tension in the fibers above and compression in the fibers below the neutral axis. The bending moment $k_{11}x$ due to the end shear k_{11} is positive, however, in accordance with the *beam sign convention.*

Substitution of Eq. (5.6) into Eq. (5.5) yields

$$\frac{d^2\bar{u}_y}{dx^2} = \frac{1}{EI}(-k_{21} + k_{11}x) \tag{5.7}$$

in which the flexural rigidity EI of the member is constant because the member is assumed to be prismatic. The equation for the slope θ of the member can be determined by integrating Eq. (5.7) as

$$\theta = \frac{d\bar{u}_y}{dx} = \frac{1}{EI}\left(-k_{21}x + k_{11}\frac{x^2}{2}\right) + C_1 \tag{5.8}$$

in which C_1 denotes a constant of integration. By integrating Eq. (5.8), we obtain the equation for deflection as

$$\bar{u}_y = \frac{1}{EI}\left(-k_{21}\frac{x^2}{2} + k_{11}\frac{x^3}{6}\right) + C_1x + C_2 \tag{5.9}$$

in which C_2 is another constant of integration. The four unknowns in Eqs. (5.8) and (5.9)—that is, two constants of integration C_1 and C_2, and two stiffness coefficients k_{11} and k_{21}—can now be evaluated by applying the following four boundary conditions.

$$\begin{array}{llll} \text{At end } b, & x = 0, & \theta = 0 \\ & x = 0, & \bar{u}_y = 1 \\ \text{At end } e, & x = L, & \theta = 0 \\ & x = L, & \bar{u}_y = 0 \end{array}$$

By applying the first boundary condition—that is, by setting $x = 0$ and $\theta = 0$ in Eq. (5.8)—we obtain $C_1 = 0$. Next, by using the second boundary condition—that is, by setting $x = 0$ and $\bar{u}_y = 1$ in Eq. (5.9)—we obtain $C_2 = 1$. Thus, the equations for the slope and deflection of the member become

$$\theta = \frac{1}{EI}\left(-k_{21}x + k_{11}\frac{x^2}{2}\right) \tag{5.10}$$

$$\bar{u}_y = \frac{1}{EI}\left(-k_{21}\frac{x^2}{2} + k_{11}\frac{x^3}{6}\right) + 1 \tag{5.11}$$

We now apply the third boundary condition—that is, we set $x = L$ and $\theta = 0$ in Eq. (5.10)—to obtain

$$0 = \frac{1}{EI}\left(-k_{21}L + k_{11}\frac{L^2}{2}\right)$$

from which

$$k_{21} = k_{11}\frac{L}{2} \tag{5.12}$$

Next, we use the last boundary condition—that is, we set $x = L$ and $\bar{u}_y = 0$ in Eq. (5.11)—to obtain

$$0 = \frac{1}{EI}\left(-k_{21}\frac{L^2}{2} + k_{11}\frac{L^3}{6}\right) + 1$$

from which

$$k_{21} = \frac{2EI}{L^2} + k_{11}\frac{L}{3} \tag{5.13}$$

By substituting Eq. (5.12) into Eq. (5.13), we determine the expression for the stiffness coefficient k_{11}:

$$\boxed{k_{11} = \frac{12EI}{L^3}} \tag{5.14}$$

and the substitution of Eq. (5.14) into Eq. (5.12) yields

$$\boxed{k_{21} = \frac{6EI}{L^2}} \tag{5.15}$$

The remaining two stiffness coefficients, k_{31} and k_{41}, can now be determined by applying the equations of equilibrium to the free body of the member shown in Fig. 5.3(c). Thus,

$$+\uparrow \sum F_y = 0 \qquad \frac{12EI}{L^3} + k_{31} = 0$$

$$\boxed{k_{31} = -\frac{12EI}{L^3}} \tag{5.16}$$

$$+\circlearrowleft \sum M_e = 0 \qquad \frac{6EI}{L^2} - \frac{12EI}{L^3}(L) + k_{41} = 0$$

$$\boxed{k_{41} = \frac{6EI}{L^2}} \tag{5.17}$$

To determine the deflected shape of the member, we substitute the expressions for k_{11} (Eq. (5.14)) and k_{21} (Eq. (5.15)) into Eq. (5.11). This yields

$$\bar{u}_y = 1 - 3\left(\frac{x}{L}\right)^2 + 2\left(\frac{x}{L}\right)^3 \tag{5.18}$$

Since the foregoing equation describes the variation of $\bar{u}_y$ (i.e., the y displacement) along the member's length due to a unit value of the end displacement u_1, while all other end displacements are zero, it represents the member shape function N_1; that is,

$$\boxed{N_1 = 1 - 3\left(\frac{x}{L}\right)^2 + 2\left(\frac{x}{L}\right)^3} \tag{5.19}$$

The expressions for coefficients k_{i2} ($i = 1$ through 4) in the second column of the member stiffness matrix **k** (Eq. (5.3)) can be evaluated in a similar manner. We subject the member to a unit value of the end displacement u_2 at end b, as shown in Fig. 5.3(d). Note that all other member end displacements are 0 (i.e., $u_1 = u_3 = u_4 = 0$), and the member is in equilibrium under the action of two end moments k_{22} and k_{42}, and two end shears k_{12} and k_{32}. The equation for bending moment at a distance x from end b of the member can be written as

$$M = -k_{22} + k_{12}x \tag{5.20}$$

By substituting Eq. (5.20) into the differential equation for beam deflection (Eq. (5.5)), we obtain

$$\frac{d^2\bar{u}_y}{dx^2} = \frac{1}{EI}(-k_{22} + k_{12}x) \tag{5.21}$$

By integrating Eq. (5.21) twice, we obtain the equations for the slope and deflection of the member as

$$\theta = \frac{d\bar{u}_y}{dx} = \frac{1}{EI}\left(-k_{22}x + k_{12}\frac{x^2}{2}\right) + C_1 \tag{5.22}$$

$$\bar{u}_y = \frac{1}{EI}\left(-k_{22}\frac{x^2}{2} + k_{12}\frac{x^3}{6}\right) + C_1 x + C_2 \tag{5.23}$$

The four unknowns, C_1, C_2, k_{12} and k_{22}, in Eqs. (5.22) and (5.23) can now be evaluated by applying the boundary conditions, as follows.

At end b,	$x = 0,$	$\theta = 1$
	$x = 0,$	$\bar{u}_y = 0$
At end e,	$x = L,$	$\theta = 0$
	$x = L,$	$\bar{u}_y = 0$

Application of the first boundary condition (i.e., $\theta = 1$ at $x = 0$) yields $C_1 = 1$; using the second boundary condition (i.e., $\bar{u}_y = 0$ at $x = 0$), we obtain $C_2 = 0$. By applying the third boundary condition (i.e., $\theta = 0$ at $x = L$), we obtain

$$0 = \frac{1}{EI}\left(-k_{22}L + k_{12}\frac{L^2}{2}\right) + 1$$

from which

$$k_{22} = \frac{EI}{L} + k_{12}\frac{L}{2} \tag{5.24}$$

and application of the last boundary condition (i.e., $\bar{u}_y = 0$ at $x = L$) yields

$$0 = \frac{1}{EI}\left(-k_{22}\frac{L^2}{2} + k_{12}\frac{L^3}{6}\right) + L$$

from which

$$k_{22} = \frac{2EI}{L} + k_{12}\frac{L}{3} \tag{5.25}$$

By substituting Eq. (5.24) into Eq. (5.25), we obtain the expression for the stiffness coefficient k_{12}:

$$\boxed{k_{12} = \frac{6EI}{L^2}} \tag{5.26}$$

and by substituting Eq. (5.26) into either Eq. (5.24) or Eq. (5.25), we obtain

$$\boxed{k_{22} = \frac{4EI}{L}} \tag{5.27}$$

To determine the two remaining stiffness coefficients, k_{32} and k_{42}, we apply the equilibrium equations to the free body of the member shown in Fig. 5.3(d):

$$+\uparrow \sum F_y = 0 \qquad \frac{6EI}{L^2} + k_{32} = 0$$

$$\boxed{k_{32} = -\frac{6EI}{L^2}} \tag{5.28}$$

$$+\circlearrowleft \sum M_e = 0 \qquad \frac{4EI}{L} - \frac{6EI}{L^2}(L) + k_{42} = 0$$

$$\boxed{k_{42} = \frac{2EI}{L}} \tag{5.29}$$

The shape function (i.e., deflected shape) of the member, due to a unit end displacement u_2, can now be obtained by substituting the expressions for k_{12} (Eq. (5.26)) and k_{22} (Eq. (5.27)) into Eq. (5.23), with $C_1 = 1$ and $C_2 = 0$. Thus,

$$\boxed{N_2 = x\left(1 - \frac{x}{L}\right)^2} \tag{5.30}$$

Next, we subject the member to a unit value of the end displacement u_3 at end e, as shown in Fig. 5.3(e), to determine the coefficients k_{i3} ($i = 1$ through 4) in the third column of the member stiffness matrix **k**. The bending moment at a distance x from end b of the member is given by

$$M = -k_{23} + k_{13}x \tag{5.31}$$

Substitution of Eq. (5.31) into the beam deflection differential equation (Eq. (5.5)) yields

$$\frac{d^2\bar{u}_y}{dx^2} = \frac{1}{EI}(-k_{23} + k_{13}x) \tag{5.32}$$

By integrating Eq. (5.32) twice, we obtain

$$\theta = \frac{d\bar{u}_y}{dx} = \frac{1}{EI}\left(-k_{23}x + k_{13}\frac{x^2}{2}\right) + C_1 \tag{5.33}$$

$$\bar{u}_y = \frac{1}{EI}\left(-k_{23}\frac{x^2}{2} + k_{13}\frac{x^3}{6}\right) + C_1x + C_2 \tag{5.34}$$

The four unknowns, C_1, C_2, k_{13} and k_{23}, in Eqs. (5.33) and (5.34) are evaluated using the boundary conditions, as follows.

$$\begin{array}{lll} \text{At end } b, & x = 0, & \theta = 0 \\ & x = 0, & \bar{u}_y = 0 \\ \text{At end } e, & x = L, & \theta = 0 \\ & x = L, & \bar{u}_y = 1 \end{array}$$

Using the first two boundary conditions, we obtain $C_1 = C_2 = 0$. Application of the third boundary condition yields

$$0 = \frac{1}{EI}\left(-k_{23}L + k_{13}\frac{L^2}{2}\right)$$

from which

$$k_{23} = k_{13}\frac{L}{2} \tag{5.35}$$

and, using the last boundary condition, we obtain

$$1 = \frac{1}{EI}\left(-k_{23}\frac{L^2}{2} + k_{13}\frac{L^3}{6}\right)$$

from which

$$k_{23} = -\frac{2EI}{L^2} + k_{13}\frac{L}{3} \tag{5.36}$$

By substituting Eq. (5.35) into Eq. (5.36), we determine the stiffness coefficient k_{13} to be

$$\boxed{k_{13} = -\frac{12EI}{L^3}} \tag{5.37}$$

and the substitution of Eq. (5.37) into Eq. (5.35) yields

$$\boxed{k_{23} = -\frac{6EI}{L^2}} \tag{5.38}$$

The two remaining stiffness coefficients, k_{33} and k_{43}, are determined by considering the equilibrium of the free body of the member (Fig. 5.3(e)):

$$+\uparrow \sum F_y = 0 \qquad -\frac{12EI}{L^3} + k_{33} = 0$$

$$\boxed{k_{33} = \frac{12EI}{L^3}} \tag{5.39}$$

$$+\circlearrowleft \sum M_e = 0 \qquad -\frac{6EI}{L^2} + \frac{12EI}{L^3}(L) + k_{43} = 0$$

$$\boxed{k_{43} = -\frac{6EI}{L^2}} \tag{5.40}$$

and the shape function N_3 for the member is obtained by substituting Eqs. (5.37) and (5.38) into Eq. (5.34) with $C_1 = C_2 = 0$. Thus,

$$\boxed{N_3 = 3\left(\frac{x}{L}\right)^2 - 2\left(\frac{x}{L}\right)^3} \tag{5.41}$$

To determine the stiffness coefficients k_{i4} ($i = 1$ through 4) in the last (fourth) column of **k**, we subject the member to a unit value of the end displacement u_4 at end e, as shown in Fig. 5.3(f). The bending moment in the member is given by

$$M = -k_{24} + k_{14}x \tag{5.42}$$

Substitution of Eq. (5.42) into Eq. (5.5) yields

$$\frac{d^2\bar{u}_y}{dx^2} = \frac{1}{EI}(-k_{24} + k_{14}x) \tag{5.43}$$

By integrating Eq. (5.43) twice, we obtain

$$\theta = \frac{d\bar{u}_y}{dx} = \frac{1}{EI}\left(-k_{24}x + k_{14}\frac{x^2}{2}\right) + C_1 \tag{5.44}$$

$$\bar{u}_y = \frac{1}{EI}\left(-k_{24}\frac{x^2}{2} + k_{14}\frac{x^3}{6}\right) + C_1x + C_2 \tag{5.45}$$

To evaluate the four unknowns, C_1, C_2, k_{14} and k_{24}, in Eqs. (5.44) and (5.45), we use the boundary conditions, as follows.

At end b, $\quad x = 0, \quad \theta = 0$
$\qquad\qquad x = 0, \quad \bar{u}_y = 0$

At end e, $\quad x = L, \quad \theta = 1$
$\qquad\qquad x = L, \quad \bar{u}_y = 0$

Application of the first two boundary conditions yields $C_1 = C_2 = 0$. Using the third boundary condition, we obtain

$$1 = \frac{1}{EI}\left(-k_{24}L + k_{14}\frac{L^2}{2}\right)$$

or

$$k_{24} = -\frac{EI}{L} + k_{14}\frac{L}{2} \tag{5.46}$$

and the use of the fourth boundary condition yields

$$0 = \frac{1}{EI}\left(-k_{24}\frac{L^2}{2} + k_{14}\frac{L^3}{6}\right)$$

from which

$$k_{24} = k_{14}\frac{L}{3} \tag{5.47}$$

By substituting Eq. (5.47) into Eq. (5.46), we obtain the stiffness coefficient k_{14}:

$$\boxed{k_{14} = \frac{6EI}{L^2}} \tag{5.48}$$

and by substituting Eq. (5.48) into Eq. (5.47), we obtain

$$\boxed{k_{24} = \frac{2EI}{L}} \tag{5.49}$$

Next, we determine the remaining stiffness coefficients by considering the equilibrium of the free body of the member (Fig. 5.3(f)):

$$+\uparrow \sum F_y = 0 \qquad \frac{6EI}{L^2} + k_{34} = 0$$

$$\boxed{k_{34} = -\frac{6EI}{L^2}} \tag{5.50}$$

$$+\circlearrowleft \sum M_e = 0 \qquad \frac{2EI}{L} - \frac{6EI}{L^2}(L) + k_{44} = 0$$

$$\boxed{k_{44} = \frac{4EI}{L}} \tag{5.51}$$

To obtain the shape function N_4 of the beam, we substitute Eqs. (5.48) and (5.49) into Eq. (5.45), yielding

$$\boxed{N_4 = \frac{x^2}{L}\left(-1 + \frac{x}{L}\right)} \tag{5.52}$$

Finally, by substituting the expressions for the stiffness coefficients (Eqs. (5.14–5.17), (5.26–5.29), (5.37–5.40), and (5.48–5.51)), into the matrix form of **k** given in Eq. (5.3), we obtain the following local stiffness matrix for the members of beams.

$$\mathbf{k} = \frac{EI}{L^3}\begin{bmatrix} 12 & 6L & -12 & 6L \\ 6L & 4L^2 & -6L & 2L^2 \\ -12 & -6L & 12 & -6L \\ 6L & 2L^2 & -6L & 4L^2 \end{bmatrix} \tag{5.53}$$

Note that the stiffness matrix **k** is symmetric; that is, $k_{ij} = k_{ji}$.

EXAMPLE 5.2

Determine the stiffness matrices for the members of the beam shown in Fig. 5.5.

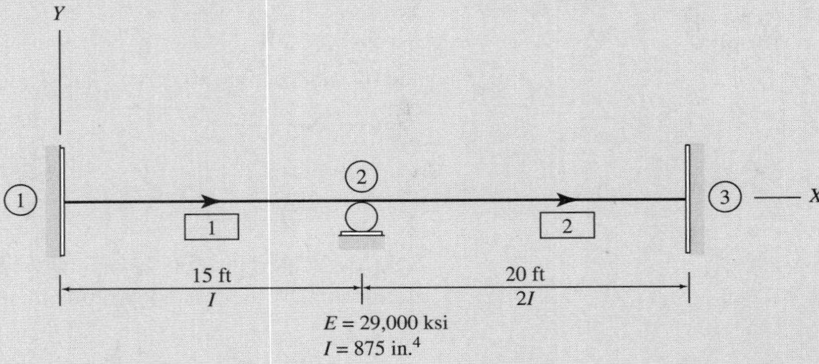

Fig. 5.5

SOLUTION

Member 1 $E = 29{,}000$ ksi, $I = 875$ in.^{4}, $L = 15$ ft $= 180$ in.

$$\frac{EI}{L^3} = \frac{29{,}000\,(875)}{(180)^3} = 4.351 \text{ k/in.}$$

Substitution in Eq. (5.53) yields

$$\mathbf{k}_1 = \begin{bmatrix} 52.212 & 4{,}699.1 & -52.212 & 4{,}699.1 \\ 4{,}699.1 & 563{,}889 & -4{,}699.1 & 281{,}944 \\ -52.212 & -4{,}699.1 & 52.212 & -4{,}699.1 \\ 4{,}699.1 & 281{,}944 & -4{,}699.1 & 563{,}889 \end{bmatrix}$$ **Ans**

Member 2 $E = 29{,}000$ ksi, $I = 1{,}750$ in.^{4}, $L = 20$ ft $= 240$ in.

$$\frac{EI}{L^3} = \frac{29{,}000\,(1{,}750)}{(240)^3} = 3.6712 \text{ k/in.}$$

Thus, from Eq. (5.53)

$$\mathbf{k}_2 = \begin{bmatrix} 44.054 & 5{,}286.5 & -44.054 & 5{,}286.5 \\ 5{,}286.5 & 845{,}833 & -5{,}286.5 & 422{,}917 \\ -44.054 & -5{,}286.5 & 44.054 & -5{,}286.5 \\ 5{,}286.5 & 422{,}917 & -5{,}286.5 & 845{,}833 \end{bmatrix}$$ **Ans**

5.3 FINITE-ELEMENT FORMULATION USING VIRTUAL WORK*

The member stiffness matrix **k**, as given by Eq. (5.53), is usually derived in the finite-element method by applying the principle of virtual work. The formulation involves essentially the same general steps that were outlined in Section 3.4 for the case of the members of plane trusses.

Displacement Function

Consider a prismatic member of a beam, subjected to end displacements u_1 through u_4, as shown in Fig. 5.6. Since the member displaces only in the y direction, only one displacement function $\bar{u}_y$ needs to be defined. In Fig. 5.6, the displacement function $\bar{u}_y$ is depicted as the displacement of an arbitrary point G located on the member's centroidal axis (which coincides with the neutral axis) at a distance x from the end b.

As discussed in Section 3.4, in the finite-element method, a displacement function is usually *assumed* in the form of a complete polynomial of such a degree that all of its coefficients can be evaluated from the available boundary conditions of the member. From Fig. 5.6, we realize that the boundary conditions for the member under consideration are as follows.

At end b, $\quad x = 0, \quad \bar{u}_y = u_1 \quad$ **(5.54a)**

$$x = 0, \qquad \theta = \frac{d\bar{u}_y}{dx} = u_2 \tag{5.54b}$$

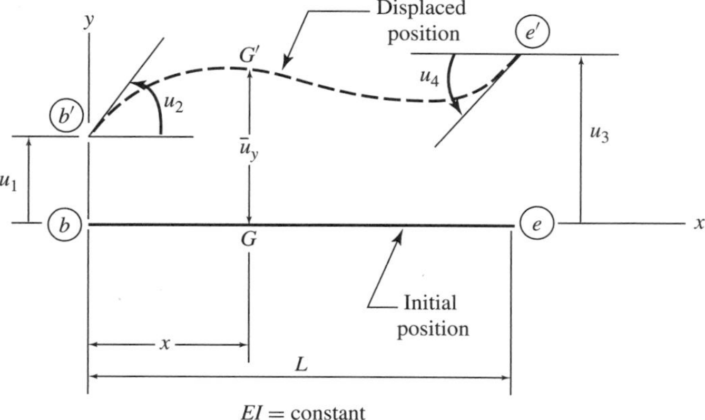

Fig. 5.6

*This section can be omitted without loss of continuity.

At end e, $\quad x = L, \quad \bar{u}_y = u_3$ **(5.54c)**

$$x = L, \quad \theta = \frac{d\bar{u}_y}{dx} = u_4 \tag{5.54d}$$

Since there are four boundary conditions, we can use a cubic polynomial (with four coefficients) for the displacement function $\bar{u}_y$, as

$$\bar{u}_y = a_0 + a_1 x + a_2 x^2 + a_3 x^3 \tag{5.55}$$

in which a_0 through a_3 are the constants to be determined by applying the four boundary conditions specified in Eqs. (5.54). By differentiating Eq. (5.55) with respect to x, we obtain the equation for the slope of the member as

$$\theta = \frac{d\bar{u}_y}{dx} = a_1 + 2a_2 x + 3a_3 x^2 \tag{5.56}$$

Now, we apply the first boundary condition (Eq. (5.54a)) by setting $x = 0$ and $\bar{u}_y = u_1$ in Eq. (5.55). This yields

$$a_0 = u_1 \tag{5.57}$$

Similarly, using the second boundary condition—that is, by setting $x = 0$ and $\theta = u_2$ in Eq. (5.56)—we obtain

$$a_1 = u_2 \tag{5.58}$$

Next, we apply the third boundary condition, setting $x = L$ and $\bar{u}_y = u_3$ in Eq. (5.55). This yields

$$u_3 = a_0 + a_1 L + a_2 L^2 + a_3 L^3 \tag{5.59}$$

By substituting $a_0 = u_1$ (Eq. (5.57)) and $a_1 = u_2$ (Eq. (5.58)) into Eq. (5.59), we obtain

$$a_3 = \frac{1}{L^3}\left(-u_1 - u_2 L + u_3 - a_2 L^2\right) \tag{5.60}$$

To apply the fourth boundary condition (Eq. (5.54d)), we set $x = L$ and $\theta = u_4$ in Eq. (5.56). This yields

$$u_4 = a_1 + 2a_2 L + 3a_3 L^2 \tag{5.61}$$

By substituting Eqs. (5.57), (5.58), and (5.60) into Eq. (5.61), and solving the resulting equation for a_2, we obtain

$$a_2 = \frac{1}{L^2}(-3u_1 - 2u_2 L + 3u_3 - u_4 L) \tag{5.62}$$

and the backsubstitution of Eq. (5.62) into Eq. (5.60) yields

$$a_3 = \frac{1}{L^3}(2u_1 + u_2 L - 2u_3 + u_4 L) \tag{5.63}$$

Finally, by substituting Eqs. (5.57), (5.58), (5.62), and (5.63) into Eq. (5.55), we obtain the following expression for the displacement function $\bar{u}_y$, in terms of the end displacements u_1 through u_4.

$$\bar{u}_y = \left[1 - 3\left(\frac{x}{L}\right)^2 + 2\left(\frac{x}{L}\right)^3\right] u_1 + \left[x\left(1 - \frac{x}{L}\right)^2\right] u_2 + \left[3\left(\frac{x}{L}\right)^2 - 2\left(\frac{x}{L}\right)^3\right] u_3 + \left[\frac{x^2}{L}\left(-1 + \frac{x}{L}\right)\right] u_4 \tag{5.64}$$

Shape Functions

The displacement function $\bar{u}_y$, as given by Eq. (5.64), can alternatively be written as

$$\bar{u}_y = N_1 u_1 + N_2 u_2 + N_3 u_3 + N_4 u_4 \tag{5.65}$$

with

$$N_1 = 1 - 3\left(\frac{x}{L}\right)^2 + 2\left(\frac{x}{L}\right)^3 \tag{5.66a}$$

$$N_2 = x\left(1 - \frac{x}{L}\right)^2 \tag{5.66b}$$

$$N_3 = 3\left(\frac{x}{L}\right)^2 - 2\left(\frac{x}{L}\right)^3 \tag{5.66c}$$

$$N_4 = \frac{x^2}{L}\left(-1 + \frac{x}{L}\right) \tag{5.66d}$$

in which N_i ($i = 1$ through 4) are the member shape functions. A comparison of Eqs. (5.66a) through (5.66d) with Eqs. (5.19), (5.30), (5.41), and (5.52), respectively, indicates that the shape functions determined herein by assuming a cubic displacement function are identical to those obtained in Section 5.2 by exactly solving the differential equation for bending of beams. This is because a cubic polynomial represents the actual (or exact) solution of the governing differential equation (Eq. (5.5)), provided that the member is prismatic and it is not subjected to any external loading.

Equation (5.65) can be written in matrix form as

$$\bar{u}_y = [\, N_1 \quad N_2 \quad N_3 \quad N_4 \,] \begin{bmatrix} u_1 \\ u_2 \\ u_3 \\ u_4 \end{bmatrix} \tag{5.67}$$

or, symbolically, as

$$\bar{u}_y = \mathbf{N}\mathbf{u} \tag{5.68}$$

in which **N** is the member shape-function matrix.

Strain–Displacement Relationship

We recall from *mechanics of materials* that the normal (longitudinal) strain ε in a fiber of a member, located at a distance y above the neutral axis, can be expressed in terms of the displacement $\bar{u}_y$ of the member's neutral axis, by the relationship

$$\varepsilon = -y\frac{d^2\bar{u}_y}{dx^2} \tag{5.69}$$

in which the minus sign indicates that the tensile strain is considered positive. By substituting Eq. (5.68) into Eq. (5.69), we write

$$\varepsilon = -y\frac{d^2}{dx^2}(\mathbf{Nu}) \tag{5.70}$$

Since the end-displacement vector $\mathbf{u}$ is not a function of x, it can be treated as a constant for the purpose of differentiation. Thus, Eq. (5.70) can be expressed as

$$\boxed{\varepsilon = \left(-y\frac{d^2\mathbf{N}}{dx^2}\right)\mathbf{u} = \mathbf{Bu}} \tag{5.71}$$

To determine the member strain-displacement matrix $\mathbf{B}$, we write

$$\mathbf{B} = -y\frac{d^2\mathbf{N}}{dx^2} = -y\begin{bmatrix}\dfrac{d^2N_1}{dx^2} & \dfrac{d^2N_2}{dx^2} & \dfrac{d^2N_3}{dx^2} & \dfrac{d^2N_4}{dx^2}\end{bmatrix} \tag{5.72}$$

By differentiating twice the equations for the shape functions as given by Eqs. (5.66), and substituting the resulting expressions into Eq. (5.72), we obtain

$$\mathbf{B} = -\frac{y}{L^2}\left[6\left(-1+2\frac{x}{L}\right) \quad 2L\left(-2+3\frac{x}{L}\right) \quad 6\left(1-2\frac{x}{L}\right) \quad 2L\left(-1+3\frac{x}{L}\right)\right] \tag{5.73}$$

Stress–Displacement Relationship

To establish the relationship between the member normal stress and the end displacements, we substitute Eq. (5.71) into the stress–strain relation $\sigma = E\varepsilon$. This yields

$$\boxed{\sigma = E\mathbf{Bu}} \tag{5.74}$$

Member Stiffness Matrix, k

With both the member strain and stress expressed in terms of end displacements, we can now establish the member stiffness matrix **k** by applying the principle of virtual work for deformable bodies. Consider an arbitrary member of a beam in equilibrium under the action of end forces Q_1 through Q_4, as shown in Fig. 5.7. Note that the member is not subjected to any external loading between its ends; therefore, the fixed-end forces $\mathbf{Q}_f$ are 0.

Now, assume that the member is given small virtual end displacements δu_1 through δu_4, as shown in Fig. 5.7. The virtual external work done by the real member end forces Q_1 through Q_4 as they move through the corresponding virtual end displacements δu_1 through δu_4 is

$$\delta W_e = Q_1 \delta u_1 + Q_2 \delta u_2 + Q_3 \delta u_3 + Q_4 \delta u_4$$

which can be written in matrix form as

$$\delta W_e = \delta \mathbf{u}^T \mathbf{Q} \tag{5.75}$$

Substitution of Eq. (5.75) into the expression for the principle of virtual work for deformable bodies as given in Eq. (3.28) in Section 3.4, yields

$$\delta \mathbf{u}^T \mathbf{Q} = \int_V \delta \varepsilon^T \sigma \, dV \tag{5.76}$$

in which the right-hand side represents the virtual strain energy stored in the member. By substituting Eqs. (5.71) and (5.74) into Eq. (5.76), we obtain

$$\delta \mathbf{u}^T \mathbf{Q} = \int_V (\mathbf{B} \, \delta \mathbf{u})^T E \mathbf{B} \, dV \, \mathbf{u}$$

Since $(\mathbf{B} \, \delta \mathbf{u})^T = \delta \mathbf{u}^T \mathbf{B}^T$, the foregoing equation becomes

$$\delta \mathbf{u}^T \mathbf{Q} = \delta \mathbf{u}^T \int_V \mathbf{B}^T E \mathbf{B} \, dV \, \mathbf{u}$$

or

$$\delta \mathbf{u}^T \left(\mathbf{Q} - \int_V \mathbf{B}^T E \mathbf{B} \, dV \, \mathbf{u} \right) = 0$$

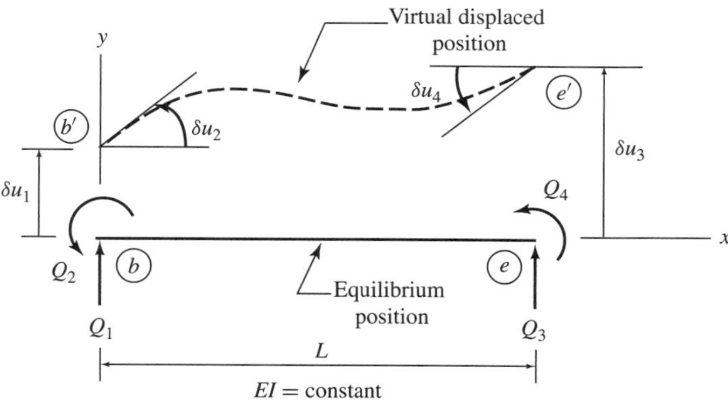

Fig. 5.7

As $\delta\mathbf{u}^T$ may be arbitrarily chosen and is not 0, the quantity in the parentheses must be 0; thus,

$$\mathbf{Q} = \left(\int_V \mathbf{B}^T E \mathbf{B} \, dV \right) \mathbf{u} = \mathbf{k}\mathbf{u} \tag{5.77}$$

with

$$\boxed{\mathbf{k} = \int_V \mathbf{B}^T E \mathbf{B} \, dV} \tag{5.78}$$

Note that the foregoing general form of **k** for beam members is the same as that obtained in Section 3.4 for the members of plane trusses (Eq. (3.55)). To explicitly determine the member stiffness matrix **k**, we substitute Eq. (5.73) for **B** into Eq. (5.78). This yields

$$\mathbf{k} = \frac{E}{L^4} \int_V y^2 \begin{bmatrix} 6\left(-1 + 2\frac{x}{L}\right) \\ 2L\left(-2 + 3\frac{x}{L}\right) \\ 6\left(1 - 2\frac{x}{L}\right) \\ 2L\left(-1 + 3\frac{x}{L}\right) \end{bmatrix} \left[6\left(-1 + 2\frac{x}{L}\right) \quad 2L\left(-2 + 3\frac{x}{L}\right) \quad 6\left(1 - 2\frac{x}{L}\right) \quad 2L\left(-1 + 3\frac{x}{L}\right) \right] dV \tag{5.79}$$

By substituting $dV = (dA)\,dx$ into Eq. (5.79), and realizing that $\int_A y^2 \, dA = I$, we obtain

$$\mathbf{k} = \frac{EI}{L^4} \int_0^L$$
$$\times \begin{bmatrix} 36\left(-1+2\frac{x}{L}\right)^2 & 12L\left(-2+3\frac{x}{L}\right)\left(-1+2\frac{x}{L}\right) & -36\left(-1+2\frac{x}{L}\right)^2 & 12L\left(-1+3\frac{x}{L}\right)\left(-1+2\frac{x}{L}\right) \\ 12L\left(-2+3\frac{x}{L}\right)\left(-1+2\frac{x}{L}\right) & 4L^2\left(-2+3\frac{x}{L}\right)^2 & 12L\left(-2+3\frac{x}{L}\right)\left(1-2\frac{x}{L}\right) & 4L^2\left(-2+3\frac{x}{L}\right)\left(-1+3\frac{x}{L}\right) \\ -36\left(-1+2\frac{x}{L}\right)^2 & 12L\left(-2+3\frac{x}{L}\right)\left(1-2\frac{x}{L}\right) & 36\left(-1+2\frac{x}{L}\right)^2 & 12L\left(-1+3\frac{x}{L}\right)\left(1-2\frac{x}{L}\right) \\ 12L\left(-1+3\frac{x}{L}\right)\left(-1+2\frac{x}{L}\right) & 4L^2\left(-2+3\frac{x}{L}\right)\left(-1+3\frac{x}{L}\right) & 12L\left(-1+3\frac{x}{L}\right)\left(1-2\frac{x}{L}\right) & 4L^2\left(-1+3\frac{x}{L}\right)^2 \end{bmatrix} dx \tag{5.80}$$

which, upon integration, becomes

$$\mathbf{k} = \frac{EI}{L^3}\begin{bmatrix} 12 & 6L & -12 & 6L \\ 6L & 4L^2 & -6L & 2L^2 \\ -12 & -6L & 12 & -6L \\ 6L & 2L^2 & -6L & 4L^2 \end{bmatrix}$$

Note that the foregoing expression for $\mathbf{k}$ is identical to that derived in Section 5.2 (Eq. (5.53)) by directly integrating the differential equation for beam deflection and applying the equilibrium equations.

5.4 MEMBER FIXED-END FORCES DUE TO LOADS

It was shown in Section 5.2 that the stiffness relationships for a member of a beam can be written in matrix form (see Eq. (5.4)) as

$$\mathbf{Q} = \mathbf{ku} + \mathbf{Q}_f$$

As the foregoing relationship indicates, the total forces $\mathbf{Q}$ that can develop at the ends of a member can be expressed as the sum of the forces $\mathbf{ku}$ due to the end displacements $\mathbf{u}$, and the fixed-end forces $\mathbf{Q}_f$ that would develop at the member ends due to external loads if both member ends were fixed against translations and rotations.

In this section, we consider the derivation of the expressions for fixed-end forces due to external loads applied to the members of beams. To illustrate the procedure, consider a fixed member subjected to a concentrated load W, as shown in Fig. 5.8(a). As indicated in this figure, the fixed-end moments at the member ends b and e are denoted by FM_b and FM_e, respectively, whereas FS_b and FS_e denote the fixed-end shears at member ends b and e, respectively. Our objective is to determine expressions for the fixed-end moments and shears in terms of the magnitude and location of the load W; we will use the direct integration approach, along with the equations of equilibrium, for this purpose.

As the concentrated load W acts at point A of the member (Fig. 5.8(a)), the bending moment M cannot be expressed as a single continuous function of x over the entire length of the member. Therefore, we divide the member into two segments, bA and Ae; and we determine the following equations for bending moment in segments bA and Ae, respectively:

$$0 \leq x \leq l_1 \qquad M = -FM_b + FS_b x \tag{5.81}$$

$$l_1 \leq x \leq L \qquad M = -FM_b + FS_b x - W(x - l_1) \tag{5.82}$$

By substituting Eqs. (5.81) and (5.82) into the differential equation for beam deflection (Eq. (5.5)), we obtain, respectively,

$$0 \leq x \leq l_1 \qquad \frac{d^2\bar{u}_y}{dx^2} = \frac{1}{EI}(-FM_b + FS_b x) \tag{5.83}$$

$$l_1 \leq x \leq L \qquad \frac{d^2\bar{u}_y}{dx^2} = \frac{1}{EI}[-FM_b + FS_b x - W(x - l_1)] \tag{5.84}$$

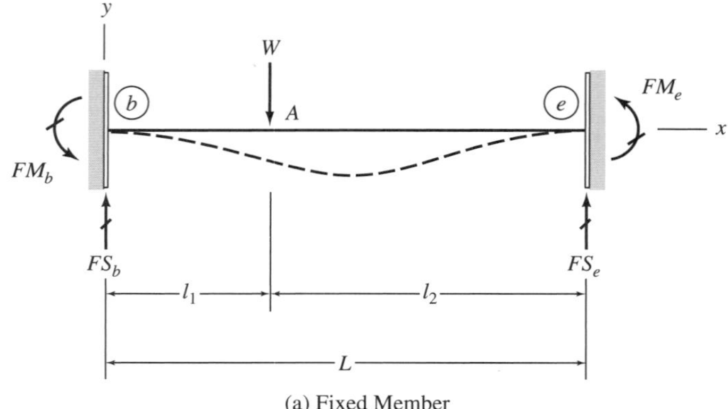

(a) Fixed Member

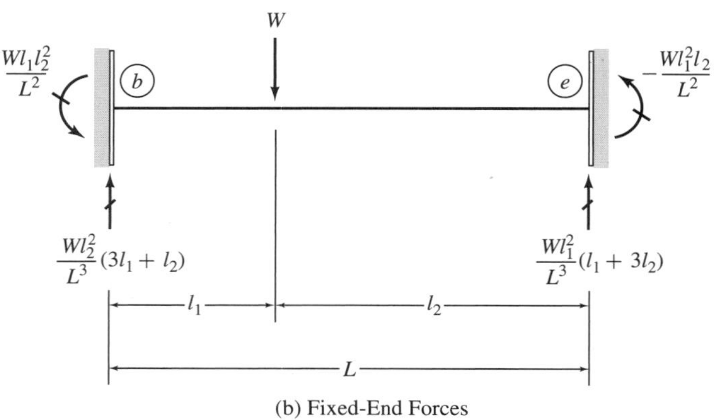

(b) Fixed-End Forces

Fig. 5.8

By integrating Eq. (5.83) twice, we obtain the equations for the slope and deflection in segment bA of the member:

$$0 \le x \le l_1 \qquad \theta = \frac{d\bar{u}_y}{dx} = \frac{1}{EI}\left(-FM_b x + FS_b\frac{x^2}{2}\right) + C_1 \tag{5.85}$$

$$0 \le x \le l_1 \qquad \bar{u}_y = \frac{1}{EI}\left(-FM_b\frac{x^2}{2} + FS_b\frac{x^3}{6}\right) + C_1x + C_2 \tag{5.86}$$

Similarly, by integrating Eq. (5.84) twice, we obtain the equations for the slope and deflection in the segment Ae:

$$l_1 \le x \le L \qquad \theta = \frac{1}{EI}\left[-FM_b x + FS_b\frac{x^2}{2} - \frac{Wx}{2}(x - 2l_1)\right] + C_3 \tag{5.87}$$

$$l_1 \le x \le L \qquad \bar{u}_y = \frac{1}{EI}\left[-FM_b\frac{x^2}{2} + FS_b\frac{x^3}{6} - \frac{Wx^2}{6}(x - 3l_1)\right] + C_3x + C_4 \tag{5.88}$$

Equations (5.85) through (5.88) indicate that the equations for the slope and deflection of the member contain a total of six unknowns; that is, four constants of integration, C_1 through C_4, and two fixed-end forces, FM_b and FS_b. These six unknowns can be evaluated by applying four boundary conditions (i.e., the slopes and deflections at the two fixed ends, b and e, must be 0), and two continuity conditions requiring that the slope and the deflection of the member's elastic curve be continuous at point A. By applying the two boundary conditions for the fixed end b (i.e., at $x = 0, \theta = \bar{u}_y = 0$) to Eqs. (5.85) and (5.86), we obtain

$$C_1 = C_2 = 0 \tag{5.89}$$

Next, to evaluate the constant C_3, we use the condition that the slope must be continuous at point A. This condition requires that the two slope equations (Eqs. (5.85) and (5.87)) yield the same slope θ_A at $x = l_1$. By setting $x = l_1$ in Eqs. (5.85) and (5.87), and equating the resulting expressions, we obtain

$$\frac{1}{EI}\left(-FM_b l_1 + FS_b \frac{l_1^2}{2}\right) = \frac{1}{EI}\left(-FM_b l_1 + FS_b \frac{l_1^2}{2} + \frac{Wl_1^2}{2}\right) + C_3$$

By solving for C_3, we determine that

$$C_3 = -\frac{Wl_1^2}{2EI} \tag{5.90}$$

In a similar manner, we evaluate the constant C_4 by applying the condition of continuity of deflection at point A. By setting $x = l_1$ in the two deflection equations (Eqs. (5.86) and (5.88)), and equating the resulting expressions, we obtain

$$\frac{1}{EI}\left(-FM_b \frac{l_1^2}{2} + FS_b \frac{l_1^3}{6}\right)$$
$$= \frac{1}{EI}\left(-FM_b \frac{l_1^2}{2} + FS_b \frac{l_1^3}{6} + \frac{Wl_1^3}{3}\right) - \frac{Wl_1^3}{2EI} + C_4$$

from which

$$C_4 = \frac{Wl_1^3}{6EI} \tag{5.91}$$

With the four constants of integration known, we can now evaluate the two remaining unknowns, FS_b and FM_b, by applying the boundary conditions that the slope and deflection at the fixed end e must be 0 (i.e., at $x = L, \theta = \bar{u}_y = 0$). By setting $x = L$ in Eqs. (5.87) and (5.88), with $\theta = 0$ in Eq. (5.87) and $\bar{u}_y = 0$ in Eq. (5.88), we obtain

$$\frac{1}{EI}\left[-FM_b L + FS_b \frac{L^2}{2} - \frac{WL}{2}(L - 2l_1)\right] - \frac{Wl_1^2}{2EI} = 0 \tag{5.92}$$

$$\frac{1}{EI}\left[-FM_b \frac{L^2}{2} + FS_b \frac{L^3}{6} - \frac{WL^2}{6}(L - 3l_1)\right] - \frac{Wl_1^2 L}{2EI} + \frac{Wl_1^3}{6EI} = 0 \tag{5.93}$$

To solve Eqs. (5.92) and (5.93) for FS_b and FM_b, we rewrite Eq. (5.92) to express FM_b in terms of FS_b as

$$FM_b = FS_b\frac{L}{2} - \frac{W}{2}(L - 2l_1) - \frac{Wl_1^2}{2L} \tag{5.94}$$

By substituting Eq. (5.94) into Eq. (5.93), and solving the resulting equation for FS_b, we obtain

$$FS_b = \frac{W}{L^3}\left(L^3 - 3l_1^2L + 2l_1^3\right)$$

Substitution of $L = l_1 + l_2$ into the numerator of the foregoing equation yields the expression for the fixed-end shear FS_b as

$$FS_b = \frac{Wl_2^2}{L^3}(3l_1 + l_2) \tag{5.95}$$

By back substituting Eq. (5.95) into Eq. (5.94), we obtain the expression for the fixed-end moment as

$$FM_b = \frac{Wl_1l_2^2}{L^2} \tag{5.96}$$

Finally, the fixed-end forces, FS_e and FM_e, at the member end e, can be determined by applying the equations of equilibrium to the free body of the member (Fig. 5.8(a)). Thus,

$$+\uparrow \sum F_y = 0 \qquad \frac{Wl_2^2}{L^3}(3l_1 + l_2) - W + FS_e = 0$$

By substituting $L = l_1 + l_2$ into the numerator and solving for FS_e, we obtain

$$FS_e = \frac{Wl_1^2}{L^3}(l_1 + 3l_2) \tag{5.97}$$

and

$$+\circlearrowleft \sum M_e = 0 \qquad \frac{Wl_1l_2^2}{L^2} - \frac{Wl_2^2}{L^3}(3l_1 + l_2)L + Wl_2 + FM_e = 0$$

$$FM_e = -\frac{Wl_1^2l_2}{L^2} \tag{5.98}$$

in which the negative answer for FM_e indicates that its actual sense is clockwise for the loading condition under consideration. Figure 5.8(b) depicts the four fixed-end forces that develop in a member of a beam subjected to a single concentrated load.

The expressions for fixed-end forces due to other types of loading conditions can be derived by using the direct integration approach as illustrated here, or by employing another classical method, such as the *method of consistent deformations*. The expressions for fixed-end forces due to some common types of member loads are given inside the front cover of this book for convenient reference.

Member Fixed-End Force Vector $\mathbf{Q}_f$

Once the fixed-end forces for a member have been evaluated, its fixed-end force vector $\mathbf{Q}_f$ can be generated by storing the fixed-end forces in their proper positions in a 4×1 vector. In accordance with the scheme for numbering member end forces adopted in Section 5.2, the fixed-end shear FS_b and the fixed-end moment FM_b, at the left end b of the member, must be stored in the first and second rows, respectively, of the $\mathbf{Q}_f$ vector; the fixed-end shear FS_e and the fixed-end moment FM_e, at the opposite member end e, are stored in the third and fourth rows, respectively, of the $\mathbf{Q}_f$ vector. Thus, the fixed-end force vector for a member of a beam (Fig 5.8(a)) is expressed as

$$\mathbf{Q}_f = \begin{bmatrix} Q_{f1} \\ Q_{f2} \\ Q_{f3} \\ Q_{f4} \end{bmatrix} = \begin{bmatrix} FS_b \\ FM_b \\ FS_e \\ FM_e \end{bmatrix} \tag{5.99}$$

When storing numerical values or fixed-end force expressions in $\mathbf{Q}_f$, the appropriate sign convention for member end forces must be followed. In accordance with the sign convention adopted in Section 5.2, the fixed-end shears are considered positive when upward (i.e., in the positive direction of the local y axis); the fixed-end moments are considered positive when counterclockwise. For example, the fixed-end force vector for the beam member shown in Fig. 5.8(b) is given by

$$\mathbf{Q}_f = \begin{bmatrix} \dfrac{Wl_2^2}{L^3}(3l_1 + l_2) \\ \dfrac{Wl_1l_2^2}{L^2} \\ \dfrac{Wl_1^2}{L^3}(l_1 + 3l_2) \\ -\dfrac{Wl_1^2l_2}{L^2} \end{bmatrix}$$

EXAMPLE 5.3 Determine the fixed-end force vectors for the members of the two-span continuous beam shown in Fig. 5.9. Use the fixed-end force equations given inside the front cover.

SOLUTION **Member 1** By substituting $w = 2$ k/ft, $L = 30$ ft, and $l_1 = l_2 = 0$ into the fixed-end force expressions given for loading type 3, we obtain

$$FS_b = FS_e = \frac{2(30)}{2} = 30 \text{ k}$$

$$FM_b = \frac{2(30)^2}{12} = 150 \text{ k-ft}$$

$$FM_e = -\frac{2(30)^2}{12} = -150 \text{ k-ft}$$

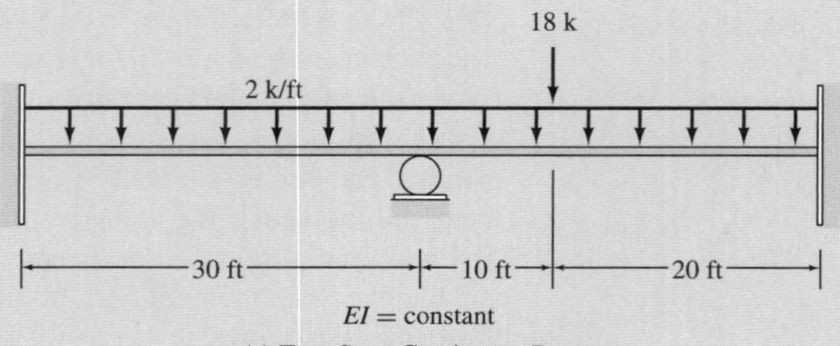

(a) Two-Span Continuous Beam

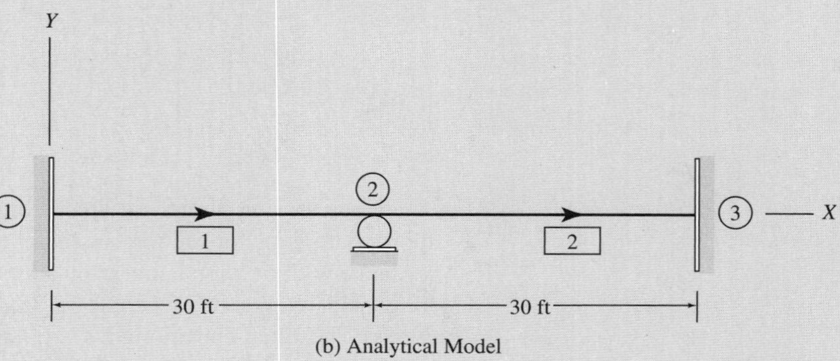

(b) Analytical Model

Fig. 5.9

By substituting these values of fixed-end forces into Eq. (5.99), we obtain the fixed-end force vector for member 1:

$$\mathbf{Q}_{f1} = \begin{bmatrix} 30 \\ 150 \\ 30 \\ -150 \end{bmatrix}$$ **Ans**

Member 2 From Fig. 5.9(a), we can see that this member is subjected to two different loadings—a concentrated load $W = 18$ k with $l_1 = 10$ ft, $l_2 = 20$ ft, and $L = 30$ ft (load type 1), and a uniformly distributed load $w = 2$ k/ft with $l_1 = l_2 = 0$ and $L = 30$ ft (load type 3). The fixed-end forces for such a member, due to the combined effect of several loads, can be conveniently determined by superimposing (algebraically adding) the fixed-end forces due to each of the loads acting individually on the member. By using superposition, we determine the fixed-end forces for member 2 to be

$$FS_b = \frac{18(20)^2}{(30)^3}[3(10) + (20)] + \frac{2\,(30)}{2} = 43.333 \text{ k}$$

$$FM_b = \frac{18\,(10)\,(20)^2}{(30)^2} + \frac{2\,(30)^2}{12} = 230 \text{ k-ft}$$

$$FS_e = \frac{18(10)^2}{(30)^3}[10 + 3(20)] + \frac{2\,(30)}{2} = 34.667 \text{ k}$$

$$FM_e = -\frac{18\,(10)^2\,(20)}{(30)^2} - \frac{2\,(30)^2}{12} = -190 \text{ k-ft}$$

Thus, the fixed-end force vector for member 2 is

$$\mathbf{Q}_{f2} = \begin{bmatrix} 43.333 \\ 230 \\ 34.667 \\ -190 \end{bmatrix}$$ **Ans**

5.5 STRUCTURE STIFFNESS RELATIONS

The procedure for establishing the structure stiffness relations for beams is essentially the same as that for plane trusses discussed in Section 3.7. The procedure, called the *direct stiffness method,* involves: (a) expressing the joint loads $\mathbf{P}$ in terms of the member end forces $\mathbf{Q}$ by applying the joint equilibrium equations; (b) relating the joint displacements $\mathbf{d}$ to the member end displacements $\mathbf{u}$ by using the compatibility conditions that the member end displacements and rotations must be the same as the corresponding joint displacements and rotations; and (c) linking the joint displacements $\mathbf{d}$ to the joint loads $\mathbf{P}$ by means of the member force-displacement relations $\mathbf{Q} = \mathbf{ku} + \mathbf{Q}_f$.

Consider, for example, an arbitrary beam subjected to joint and member loads, as depicted in Fig. 5.10(a). The structure has three degrees of freedom, d_1 through d_3, as shown in Fig. 5.10(b). Our objective is to establish the structure stiffness relationships, which express the external loads as functions of the joint displacements $\mathbf{d}$. The member end forces $\mathbf{Q}$ and end displacements $\mathbf{u}$ for the three members of the beam are given in Fig. 5.10(c), in which the superscript (i) denotes the member number.

By applying the equations of equilibrium $\sum F_Y = 0$ and $\sum M = 0$ to the free body of joint 2, and the equilibrium equation $\sum M = 0$ to the free body of joint 3, we obtain the following relationships between the external joint loads $\mathbf{P}$ and the internal member end forces $\mathbf{Q}$.

$$P_1 = Q_3^{(1)} + Q_1^{(2)} \tag{5.100a}$$

$$P_2 = Q_4^{(1)} + Q_2^{(2)} \tag{5.100b}$$

$$P_3 = Q_4^{(2)} + Q_2^{(3)} \tag{5.100c}$$

Next, we determine compatibility conditions for the three members of the beam. Since the left end 1 of member 1 is connected to fixed support 1 (Fig. 5.10(b)), which can neither translate nor rotate, the displacements $u_1^{(1)}$ and $u_2^{(1)}$ of end 1 of the member (Fig. 5.10(c)) must be 0. Similarly, since end 2 of this member is connected to joint 2, the displacements $u_3^{(1)}$ and $u_4^{(1)}$ of end 2 must be the same as the displacements d_1 and d_2, respectively, of joint 2. Thus, the compatibility equations for member 1 are:

$$u_1^{(1)} = u_2^{(1)} = 0 \qquad u_3^{(1)} = d_1 \qquad u_4^{(1)} = d_2 \tag{5.101}$$

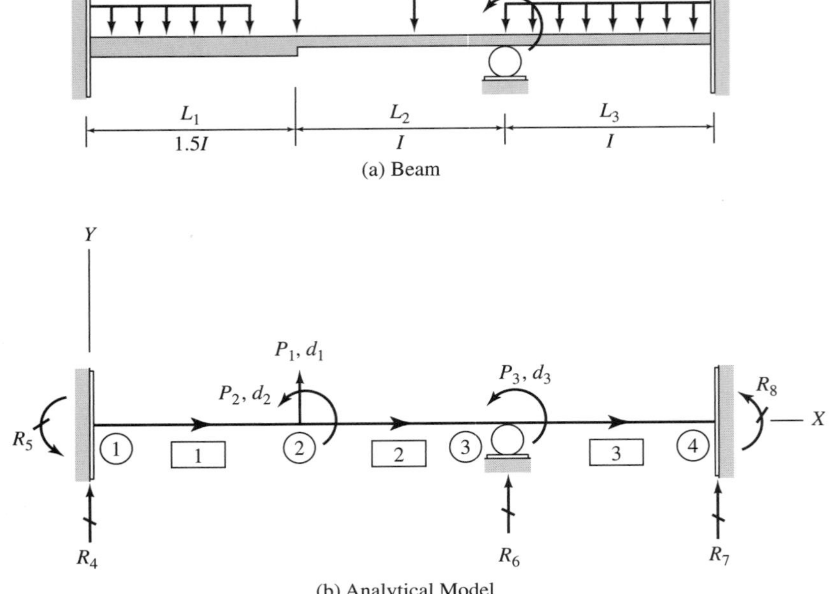

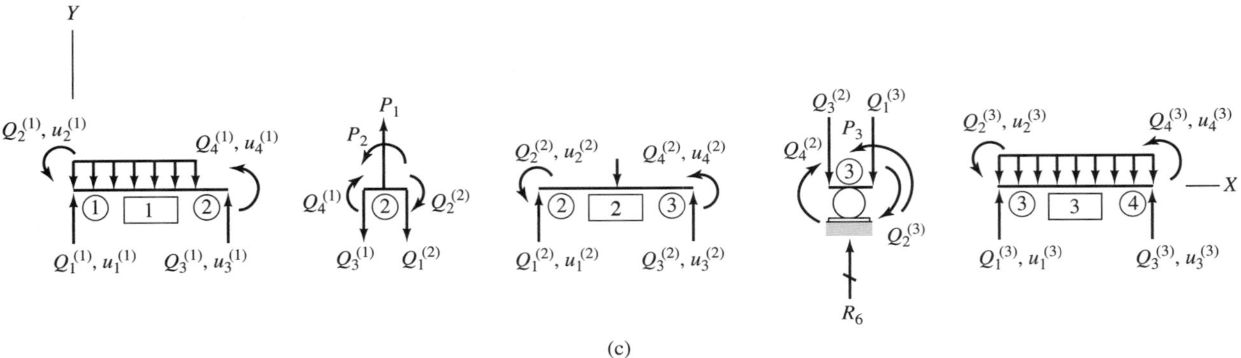

Fig. 5.10

In a similar manner, the compatibility equations for members 2 and 3, respectively, are given by

$$u_1^{(2)} = d_1 \qquad u_2^{(2)} = d_2 \qquad u_3^{(2)} = 0 \qquad u_4^{(2)} = d_3 \tag{5.102}$$

$$u_1^{(3)} = 0 \qquad u_2^{(3)} = d_3 \qquad u_3^{(3)} = u_4^{(3)} = 0 \tag{5.103}$$

The link between the joint equilibrium equations (Eqs. (5.100)) and the compatibility conditions (Eqs. (5.101) through (5.103)) is provided by the member stiffness relationship $\mathbf{Q} = \mathbf{ku} + \mathbf{Q}_f$ (Eq. (5.4)). To express the six member end forces that appear in Eqs. (5.100) in terms of the member end displacements, we will use the expanded form of the member stiffness relationship given in Eqs. (5.2). Thus, the end forces $Q_3^{(1)}$ and $Q_4^{(1)}$, of member 1, can be expressed in terms of the member end displacements as

$$Q_3^{(1)} = k_{31}^{(1)}u_1^{(1)} + k_{32}^{(1)}u_2^{(1)} + k_{33}^{(1)}u_3^{(1)} + k_{34}^{(1)}u_4^{(1)} + Q_{f3}^{(1)} \quad \textbf{(5.104a)}$$

$$Q_4^{(1)} = k_{41}^{(1)}u_1^{(1)} + k_{42}^{(1)}u_2^{(1)} + k_{43}^{(1)}u_3^{(1)} + k_{44}^{(1)}u_4^{(1)} + Q_{f4}^{(1)} \quad \textbf{(5.104b)}$$

Similarly, the end forces $Q_1^{(2)}$, $Q_2^{(2)}$, and $Q_4^{(2)}$, of member 2, are written as

$$Q_1^{(2)} = k_{11}^{(2)}u_1^{(2)} + k_{12}^{(2)}u_2^{(2)} + k_{13}^{(2)}u_3^{(2)} + k_{14}^{(2)}u_4^{(2)} + Q_{f1}^{(2)} \quad \textbf{(5.105a)}$$

$$Q_2^{(2)} = k_{21}^{(2)}u_1^{(2)} + k_{22}^{(2)}u_2^{(2)} + k_{23}^{(2)}u_3^{(2)} + k_{24}^{(2)}u_4^{(2)} + Q_{f2}^{(2)} \quad \textbf{(5.105b)}$$

$$Q_4^{(2)} = k_{41}^{(2)}u_1^{(2)} + k_{42}^{(2)}u_2^{(2)} + k_{43}^{(2)}u_3^{(2)} + k_{44}^{(2)}u_4^{(2)} + Q_{f4}^{(2)} \quad \textbf{(5.105c)}$$

and the end force $Q_2^{(3)}$, of member 3, is expressed as

$$Q_2^{(3)} = k_{21}^{(3)}u_1^{(3)} + k_{22}^{(3)}u_2^{(3)} + k_{23}^{(3)}u_3^{(3)} + k_{24}^{(3)}u_4^{(3)} + Q_{f2}^{(3)} \quad \textbf{(5.106)}$$

Next, we relate the joint displacements **d** to the member end forces **Q** by substituting the compatibility equations, Eqs. (5.101), (5.102), and (5.103), into the member force-displacement relations given by Eqs. (5.104), (5.105), and (5.106), respectively. Thus,

$$Q_3^{(1)} = k_{33}^{(1)}d_1 + k_{34}^{(1)}d_2 + Q_{f3}^{(1)} \quad \textbf{(5.107a)}$$

$$Q_4^{(1)} = k_{43}^{(1)}d_1 + k_{44}^{(1)}d_2 + Q_{f4}^{(1)} \quad \textbf{(5.107b)}$$

$$Q_1^{(2)} = k_{11}^{(2)}d_1 + k_{12}^{(2)}d_2 + k_{14}^{(2)}d_3 + Q_{f1}^{(2)} \quad \textbf{(5.107c)}$$

$$Q_2^{(2)} = k_{21}^{(2)}d_1 + k_{22}^{(2)}d_2 + k_{24}^{(2)}d_3 + Q_{f2}^{(2)} \quad \textbf{(5.107d)}$$

$$Q_4^{(2)} = k_{41}^{(2)}d_1 + k_{42}^{(2)}d_2 + k_{44}^{(2)}d_3 + Q_{f4}^{(2)} \quad \textbf{(5.107e)}$$

$$Q_2^{(3)} = k_{22}^{(3)}d_3 + Q_{f2}^{(3)} \quad \textbf{(5.107f)}$$

Finally, by substituting Eqs. (5.107) into the joint equilibrium equations (Eqs. (5.100)), we establish the desired structure stiffness relationships as

$$P_1 = \left(k_{33}^{(1)} + k_{11}^{(2)}\right)d_1 + \left(k_{34}^{(1)} + k_{12}^{(2)}\right)d_2 + k_{14}^{(2)}d_3 + \left(Q_{f3}^{(1)} + Q_{f1}^{(2)}\right) \quad \textbf{(5.108a)}$$

$$P_2 = \left(k_{43}^{(1)} + k_{21}^{(2)}\right)d_1 + \left(k_{44}^{(1)} + k_{22}^{(2)}\right)d_2 + k_{24}^{(2)}d_3 + \left(Q_{f4}^{(1)} + Q_{f2}^{(2)}\right) \quad \textbf{(5.108b)}$$

$$P_3 = k_{41}^{(2)}d_1 + k_{42}^{(2)}d_2 + \left(k_{44}^{(2)} + k_{22}^{(3)}\right)d_3 + \left(Q_{f4}^{(2)} + Q_{f2}^{(3)}\right) \quad \textbf{(5.108c)}$$

Equations (5.108) can be conveniently expressed in matrix form as

$$\mathbf{P} = \mathbf{S}\mathbf{d} + \mathbf{P}_f$$

or

$$\boxed{\mathbf{P} - \mathbf{P}_f = \mathbf{S}\mathbf{d}} \quad \textbf{(5.109)}$$

in which

$$\mathbf{S} = \begin{bmatrix} k_{33}^{(1)} + k_{11}^{(2)} & k_{34}^{(1)} + k_{12}^{(2)} & k_{14}^{(2)} \\ k_{43}^{(1)} + k_{21}^{(2)} & k_{44}^{(1)} + k_{22}^{(2)} & k_{24}^{(2)} \\ k_{41}^{(2)} & k_{42}^{(2)} & k_{44}^{(2)} + k_{22}^{(3)} \end{bmatrix} \qquad \textbf{(5.110)}$$

is the $NDOF \times NDOF$ structure stiffness matrix for the beam of Fig. 5.10(b), and

$$\mathbf{P}_f = \begin{bmatrix} Q_{f3}^{(1)} + Q_{f1}^{(2)} \\ Q_{f4}^{(1)} + Q_{f2}^{(2)} \\ Q_{f4}^{(2)} + Q_{f2}^{(3)} \end{bmatrix} \qquad \textbf{(5.111)}$$

is the $NDOF \times 1$ *structure fixed-joint force vector*. The structure fixed-joint force vectors are further discussed in the following section. In the rest of this section, we focus our attention on the structure stiffness matrices.

By examining Eq. (5.110), we realize that the structure stiffness matrix **S** of the beam of Fig. 5.10(b) is symmetric, because of the symmetric nature of the member stiffness matrices (i.e., $k_{ij} = k_{ji}$). (The structure stiffness matrices of all linear elastic structures are always symmetric.) As discussed in Chapter 3, a structure stiffness coefficient S_{ij} represents the force at the location and in the direction of P_i required, along with other joint forces, to cause a unit value of the displacement d_j, while all other joint displacements are 0, and the structure is not subjected to any external loads. We can use this definition to verify the **S** matrix (Eq. (5.110)) for the beam of Fig. 5.10.

In Figs. 5.11(a) through (c), the beam is subjected to the unit values of the three joint displacements d_1 through d_3, respectively. As depicted in Fig. 5.11(a),

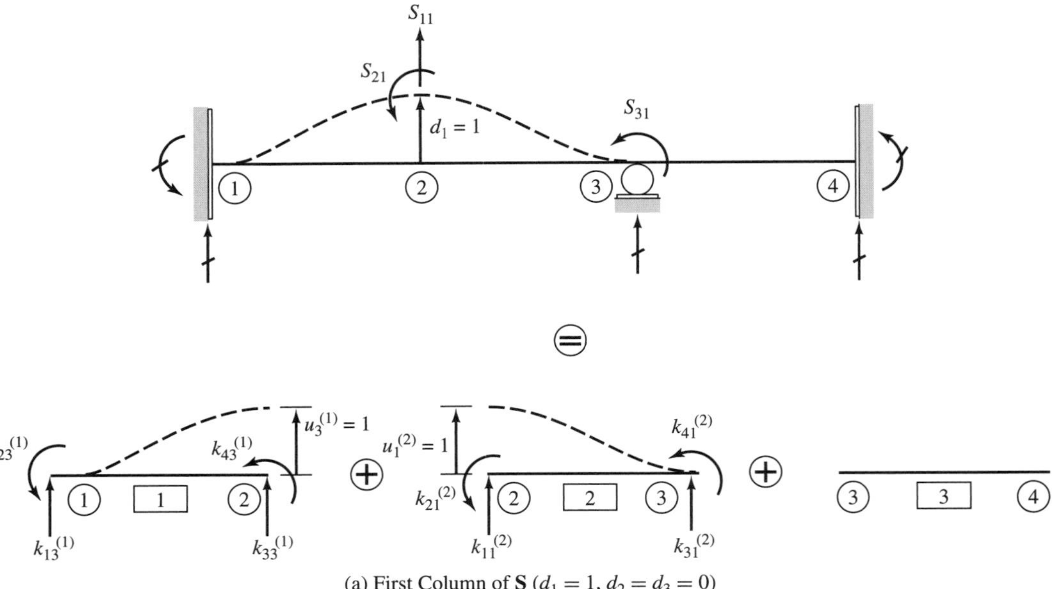

(a) First Column of **S** ($d_1 = 1, d_2 = d_3 = 0$)

Fig. 5.11

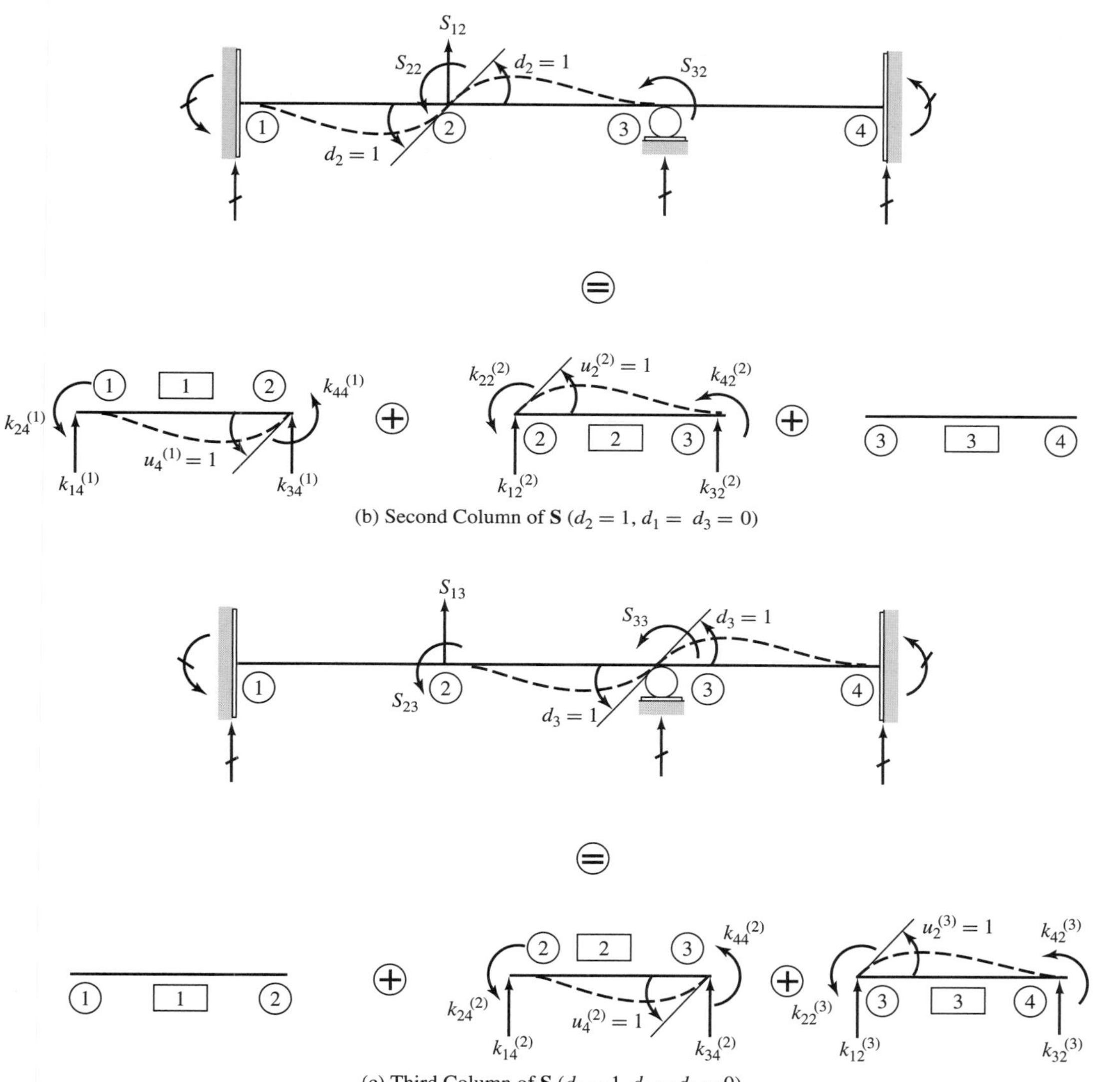

(b) Second Column of **S** ($d_2 = 1, d_1 = d_3 = 0$)

(c) Third Column of **S** ($d_3 = 1, d_1 = d_2 = 0$)

Fig. 5.11 *(continued)*

the joint displacement $d_1 = 1$ (with $d_2 = d_3 = 0$) induces unit displacements $u_3^{(1)} = 1$ at the right end of member 1 and $u_1^{(2)} = 1$ at the left end of member 2, while member 3 is not subjected to any displacements. The member stiffness coefficients (or end forces) necessary to cause the foregoing end displacements of the individual members are also shown in Fig. 5.11(a). (Recall that we derived the explicit expressions for member stiffness coefficients, in terms of E, I, and L of the member, in Section 5.2.) From the figure, we can see that the total vertical joint force S_{11} at joint 2, required to cause the joint displacement $d_1 = 1$ (with $d_2 = d_3 = 0$), must be equal to the algebraic sum of the vertical forces at the two member ends connected to this joint; that is,

$$S_{11} = k_{33}^{(1)} + k_{11}^{(2)} \tag{5.112a}$$

Similarly, the total joint moment S_{21} at joint 2 must be equal to the algebraic sum of the moments at the ends of members 1 and 2 connected to joint 2. Thus, (Fig. 5.11(a)),

$$S_{21} = k_{43}^{(1)} + k_{21}^{(2)} \tag{5.112b}$$

and the total joint moment S_{31} at joint 3 must equal the algebraic sum of the moments at the two member ends connected to the joint; that is,

$$S_{31} = k_{41}^{(2)} \tag{5.112c}$$

Note that the foregoing expressions for s_{i1} ($i = 1$ through 3) are identical to those listed in the first column of **S** in Eq. (5.110).

The second column of **S** can be verified in a similar manner. From Fig. 5.11(b), we can see that the joint rotation $d_2 = 1$ (with $d_1 = d_3 = 0$) induces unit rotations $u_4^{(1)} = 1$ at the right end of member 1, and $u_2^{(2)} = 1$ at the left end of member 2. The member stiffness coefficients associated with these end displacements are also shown in the figure. By comparing the joint forces with the member end forces, we obtain the expressions for the structure stiffness coefficients as

$$S_{12} = k_{34}^{(1)} + k_{12}^{(2)} \tag{5.112d}$$

$$S_{22} = k_{44}^{(1)} + k_{22}^{(2)} \tag{5.112e}$$

$$S_{32} = k_{42}^{(2)} \tag{5.112f}$$

which are the same as those in the second column of **S** in Eq. (5.110).

Similarly, by subjecting the beam to a unit rotation $d_3 = 1$ (with $d_1 = d_2 = 0$), as shown in Fig. 5.11(c), we obtain

$$S_{13} = k_{14}^{(2)} \tag{5.112g}$$

$$S_{23} = k_{24}^{(2)} \tag{5.112h}$$

$$S_{33} = k_{44}^{(2)} + k_{22}^{(3)} \tag{5.112i}$$

The foregoing structure stiffness coefficients are identical to those listed in the third column of **S** in Eq. (5.110).

Assembly of the Structure Stiffness Matrix Using Member Code Numbers

Although the procedures discussed thus far for formulating **S** provide clearer insight into the basic concept of the structure stiffness matrix, it is more convenient from a computer programming viewpoint to directly form the structure stiffness matrix **S** by assembling the elements of the member stiffness matrices **k**. This technique, which is sometimes referred to as the *code number technique,* was described in detail in Section 3.7 for the case of plane trusses. The technique essentially involves storing the pertinent elements of the stiffness matrix **k** for each member of the beam, in the structure stiffness matrix **S**, by using the member code numbers. The code numbers for a member are simply the structure coordinate numbers at the location and in the direction of each of the member end displacements **u**, arranged in the order of the end displacements.

To illustrate this technique, consider again the three-member beam of Fig. 5.10. The analytical model of the beam is redrawn in Fig. 5.12(a), which shows its three degrees of freedom (numbered from 1 to 3) and five restrained coordinates (numbered from 4 to 8). In accordance with the notation for member end displacements adopted in Section 5.2, the first two end displacements of a member, u_1 and u_2, are always the vertical translation and rotation, respectively, at the left end (or beginning) of the member, whereas the last two end displacements, u_3 and u_4, are always the vertical translation and rotation, respectively, at the right end (or end) of the member. Thus, the first two code numbers for a member are always the structure coordinate numbers

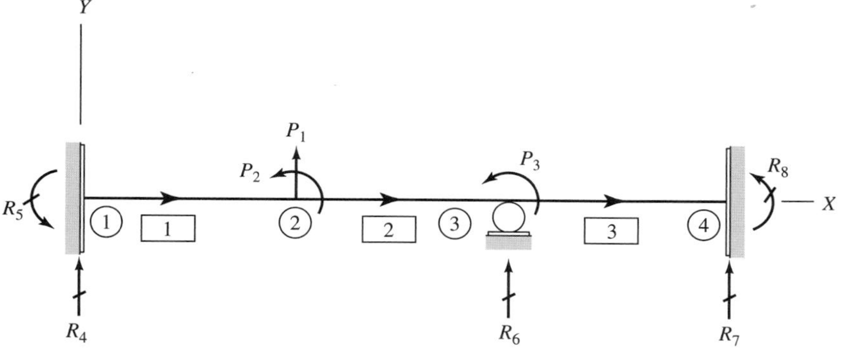

Three degrees of freedom (1 through 3);
five restrained coordinates (4 through 8)

(a) Analytical Model

$$\mathbf{k}_1 = \begin{array}{c} \begin{matrix} 4 & \quad 5 & \quad 1 & \quad 2 \end{matrix} \\ \left[\begin{matrix} k_{11}^{(1)} & k_{12}^{(1)} & k_{13}^{(1)} & k_{14}^{(1)} \\ k_{21}^{(1)} & k_{22}^{(1)} & k_{23}^{(1)} & k_{24}^{(1)} \\ k_{31}^{(1)} & k_{32}^{(1)} & k_{33}^{(1)} & k_{34}^{(1)} \\ k_{41}^{(1)} & k_{42}^{(1)} & k_{43}^{(1)} & k_{44}^{(1)} \end{matrix}\right] \begin{matrix} 4 \\ 5 \\ 1 \\ 2 \end{matrix} \end{array}
\qquad
\mathbf{k}_2 = \begin{array}{c} \begin{matrix} 1 & \quad 2 & \quad 6 & \quad 3 \end{matrix} \\ \left[\begin{matrix} k_{11}^{(2)} & k_{12}^{(2)} & k_{13}^{(2)} & k_{14}^{(2)} \\ k_{21}^{(2)} & k_{22}^{(2)} & k_{23}^{(2)} & k_{24}^{(2)} \\ k_{31}^{(2)} & k_{32}^{(2)} & k_{33}^{(2)} & k_{34}^{(2)} \\ k_{41}^{(2)} & k_{42}^{(2)} & k_{43}^{(2)} & k_{44}^{(2)} \end{matrix}\right] \begin{matrix} 1 \\ 2 \\ 6 \\ 3 \end{matrix} \end{array}
\qquad
\mathbf{k}_3 = \begin{array}{c} \begin{matrix} 6 & \quad 3 & \quad 7 & \quad 8 \end{matrix} \\ \left[\begin{matrix} k_{11}^{(3)} & k_{12}^{(3)} & k_{13}^{(3)} & k_{14}^{(3)} \\ k_{21}^{(3)} & k_{22}^{(3)} & k_{23}^{(3)} & k_{24}^{(3)} \\ k_{31}^{(3)} & k_{32}^{(3)} & k_{33}^{(3)} & k_{34}^{(3)} \\ k_{41}^{(3)} & k_{42}^{(3)} & k_{43}^{(3)} & k_{44}^{(3)} \end{matrix}\right] \begin{matrix} 6 \\ 3 \\ 7 \\ 8 \end{matrix} \end{array}$$

$$\mathbf{S} = \begin{array}{c} \begin{matrix} 1 & \qquad\qquad 2 & \qquad\qquad 3 \end{matrix} \\ \left[\begin{matrix} k_{33}^{(1)} + k_{11}^{(2)} & k_{34}^{(1)} + k_{12}^{(2)} & k_{14}^{(2)} \\ k_{43}^{(1)} + k_{21}^{(2)} & k_{44}^{(1)} + k_{22}^{(2)} & k_{24}^{(2)} \\ k_{41}^{(2)} & k_{42}^{(2)} & k_{44}^{(2)} + k_{22}^{(3)} \end{matrix}\right] \begin{matrix} 1 \\ 2 \\ 3 \end{matrix} \end{array}$$

(b) Assembling of Structure Stiffness Matrix **S**

Fig. 5.12

corresponding to the vertical translation and rotation, respectively, of the beginning joint; and the third and fourth member code numbers are always the structure coordinate numbers corresponding to the vertical translation and rotation, respectively, of the end joint.

From Fig. 5.12(a) we can see that, for member 1 of the beam, the beginning joint is 1 with restrained coordinates 4 and 5, and the end joint is 2 with degrees of freedom 1 and 2. Thus, the code numbers for member 1 are 4, 5, 1, 2. Similarly, the code numbers for member 2, for which the beginning and end joints are 2 and 3, respectively, are 1, 2, 6, 3. In a similar manner, the code numbers for member 3 are found to be 6, 3, 7, 8.

To establish the structure stiffness matrix **S**, we write the code numbers of each member on the right side and at the top of its stiffness matrix $\mathbf{k}_i$ ($i = 1, 2,$ or 3), as shown in Fig. 5.12(b). These code numbers now define the positions of the elements of the member stiffness matrices in the structure stiffness matrix **S**. In other words, the code numbers on the right side of a **k** matrix represent the row numbers of **S**; and the code numbers at the top represent the column numbers of **S**. Furthermore, since the number of rows and columns of **S** equal the number of degrees of freedom (*NDOF*) of the beam, only those elements of a **k** matrix for which both the row and the column code numbers are less than or equal to *NDOF* belong in the structure stiffness matrix **S**. The structure stiffness matrix **S** is obtained by algebraically adding the pertinent elements of the **k** matrices of all the members in their proper positions in the **S** matrix.

To assemble the **S** matrix for the beam of Fig. 5.12(a), we start by storing the pertinent elements of the stiffness matrix of member 1, $\mathbf{k}_1$, in the **S** matrix (see Fig. 5.12(b)). Thus, the element $k_{33}^{(1)}$ is stored in row 1 and column 1 of **S**, the element $k_{43}^{(1)}$ is stored in row 2 and column 1 of **S**, the element $k_{34}^{(1)}$ is stored in row 1 and column 2 of **S**, and the element $k_{44}^{(1)}$ is stored in row 2 and column 2 of **S**. It should be noted that since the beam has three degrees of freedom, only those elements of $\mathbf{k}_1$ whose row and column code numbers both are less than or equal to 3 are stored in **S**. The same procedure is then used to store the pertinent elements of $\mathbf{k}_2$ and $\mathbf{k}_3$, of members 2 and 3, respectively, in the **S** matrix. Note that when two or more member stiffness coefficients are stored in the same element of **S**, then the coefficients must be algebraically added. The completed structure stiffness matrix **S** for the beam is shown in Fig. 5.12(b), and is identical to the one derived previously (Eq. (5.110)) by substituting the member compatibility and stiffness relations into the joint equilibrium equations.

EXAMPLE 5.4 Determine the structure stiffness matrix for the three-span continuous beam shown in Fig. 5.13(a).

SOLUTION *Analytical Model:* The analytical model of the structure is shown in Fig. 5.13(b). The beam has two degrees of freedom—the rotations of joints 2 and 3—which are identified by the structure coordinate numbers 1 and 2 in the figure.

Structure Stiffness Matrix: To generate the 2×2 structure stiffness matrix **S**, we will determine, for each member, the stiffness matrix **k** and store its pertinent elements in their proper positions in **S** by using the member code numbers.

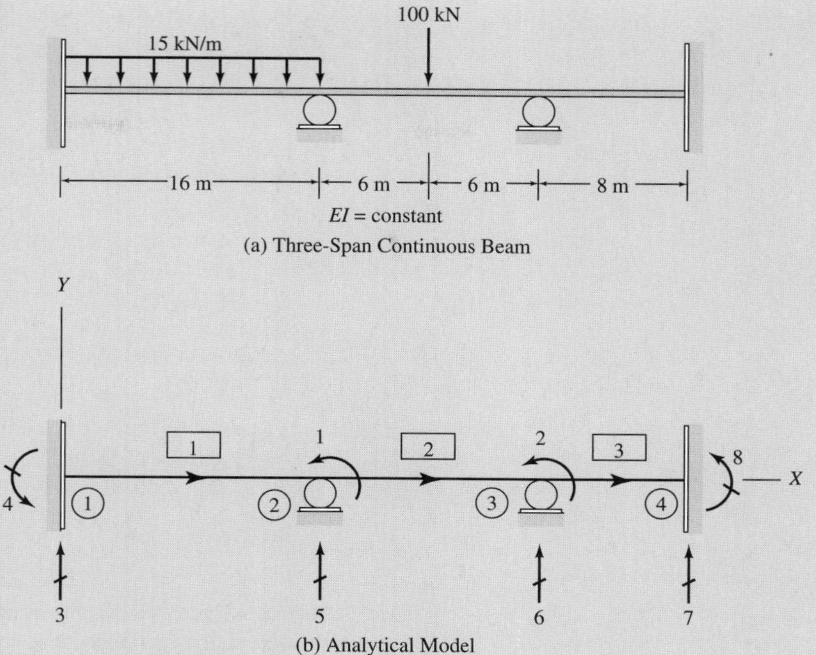

(a) Three-Span Continuous Beam

(b) Analytical Model

Fig. 5.13

Member 1 By substituting $L = 16$ m into Eq. (5.53), we obtain

$$\mathbf{k}_1 = EI \begin{matrix} 3 & 4 & 5 & 1 & \\ \left[\begin{matrix} 0.0029297 & 0.023438 & -0.0029297 & 0.023438 \\ 0.023438 & 0.25 & -0.023438 & 0.125 \\ -0.0029297 & -0.023438 & 0.0029297 & -0.023438 \\ 0.023438 & 0.125 & -0.023438 & \boxed{0.25} \end{matrix}\right] & \begin{matrix} 3 \\ 4 \\ 5 \\ 1 \end{matrix} \end{matrix} \tag{1}$$

From Fig. 5.13(b), we observe that the code numbers for this member are 3, 4, 5, 1. These numbers are written on the right side and at the top of $\mathbf{k}_1$ in Eq. (1), to indicate the rows and columns, respectively, of the structure stiffness matrix **S**, where the elements of $\mathbf{k}_1$ are to be stored. Thus, the element in row 4 and column 4 of $\mathbf{k}_1$ is stored in row 1 and column 1 of **S**, as

$$\mathbf{S} = EI \begin{matrix} 1 & 2 & \\ \left[\begin{matrix} 0.25 & 0 \\ 0 & 0 \end{matrix}\right] & \begin{matrix} 1 \\ 2 \end{matrix} \end{matrix} \tag{2}$$

Note that the elements of $\mathbf{k}_1$ corresponding to the restrained coordinate numbers 3, 4, and 5 are disregarded.

Member 2 $L = 12$ m. By using Eq. (5.53),

$$\mathbf{k}_2 = EI \begin{matrix} 5 & 1 & 6 & 2 & \\ \left[\begin{matrix} 0.0069444 & 0.041667 & -0.0069444 & 0.041667 \\ 0.041667 & \boxed{0.33333} & -0.041667 & \boxed{0.16667} \\ -0.0069444 & -0.041667 & 0.0069444 & -0.041667 \\ 0.041667 & \boxed{0.16667} & -0.041667 & \boxed{0.33333} \end{matrix}\right] & \begin{matrix} 5 \\ 1 \\ 6 \\ 2 \end{matrix} \end{matrix} \tag{3}$$

From Fig. 5.13(b), we can see that the code numbers for this member are 5, 1, 6, 2. These numbers are used to add the pertinent elements of $\mathbf{k}_2$ in their proper positions in the structure stiffness matrix $\mathbf{S}$ given in Eq. (2), which now becomes

$$\mathbf{S} = EI \begin{array}{c} \begin{array}{cc} \quad 1 \quad & \quad\quad 2 \quad \end{array} \\ \left[\begin{array}{cc} 0.25+0.33333 & 0.16667 \\ 0.16667 & 0.33333 \end{array}\right] \end{array} \begin{array}{c} \\ 1 \\ 2 \end{array} \tag{4}$$

Member 3 $L = 8$ m. Thus,

$$\mathbf{k}_3 = EI \begin{array}{c} \begin{array}{cccc} \quad 6 \quad & \quad 2 \quad & \quad 7 \quad & \quad 8 \quad \end{array} \\ \left[\begin{array}{cccc} 0.023438 & 0.09375 & -0.023438 & 0.09375 \\ 0.09375 & \boxed{0.5} & -0.09375 & 0.25 \\ -0.023438 & -0.09375 & 0.023438 & -0.09375 \\ 0.09375 & 0.25 & -0.09375 & 0.5 \end{array}\right] \end{array} \begin{array}{c} \\ 6 \\ 2 \\ 7 \\ 8 \end{array} \tag{5}$$

The code numbers for this member are 6, 2, 7, 8. Thus, the element in row 2 and column 2 of $\mathbf{k}_3$ is added in row 2 and column 2 of $\mathbf{S}$ in Eq. (4), as

$$\mathbf{S} = EI \begin{array}{c} \begin{array}{cc} \quad 1 \quad & \quad\quad 2 \quad \end{array} \\ \left[\begin{array}{cc} 0.25+0.33333 & 0.16667 \\ 0.16667 & 0.33333+0.5 \end{array}\right] \end{array} \begin{array}{c} \\ 1 \\ 2 \end{array}$$

Since the stiffnesses of all three members of the beam have now been stored in $\mathbf{S}$, the structure stiffness matrix for the given beam is

$$\mathbf{S} = EI \begin{array}{c} \begin{array}{cc} \quad 1 \quad & \quad 2 \quad \end{array} \\ \left[\begin{array}{cc} 0.58333 & 0.16667 \\ 0.16667 & 0.83333 \end{array}\right] \end{array} \begin{array}{c} \\ 1 \\ 2 \end{array} \qquad \textbf{Ans}$$

Note that the structure stiffness matrix is symmetric.

5.6 STRUCTURE FIXED-JOINT FORCES AND EQUIVALENT JOINT LOADS

As discussed in the preceding section, the force–displacement relationships for an entire structure can be expressed in matrix form (see Eq. (5.109)) as

$$\mathbf{P} - \mathbf{P}_f = \mathbf{S}\mathbf{d}$$

in which $\mathbf{P}_f$ represents the structure fixed-joint force vector. It was also shown in the preceding section that by using the basic equations of equilibrium, compatibility, and member stiffness, the structure fixed-joint forces $\mathbf{P}_f$ can be expressed in terms of the member fixed-end forces $\mathbf{Q}_f$ (see Eq. (5.111)). In this section, we consider the physical interpretation of the structure fixed-joint forces; and discuss the formation of the $\mathbf{P}_f$ vector, by assembling the elements of the member $\mathbf{Q}_f$ vectors, using the member code numbers.

The concept of the structure fixed-joint forces $\mathbf{P}_f$ is analogous to that of the member fixed-end forces $\mathbf{Q}_f$. *The structure fixed-joint forces represent the reaction forces (and/or moments) that would develop at the locations and in the directions of the structure's degrees of freedom, due to the external member loads, if all the joints of the structure were fixed against translations and rotations.*

To develop some insight into the concept of structure fixed-joint forces, let us reconsider the beam of Fig. 5.10. The beam, subjected only to the member loads, is redrawn in Fig. 5.14(a), with its analytical model depicted in Fig. 5.14(b). Now, assume that joint 2, which is free to translate and rotate, is restrained against these displacements by an imaginary restraint (or clamp) applied to it, as shown in Fig. 5.14(c). Similarly, joint 3, which is free to rotate, is also restrained against rotation by means of an imaginary restraint (or clamp). When external loads are applied to the members of this hypothetical

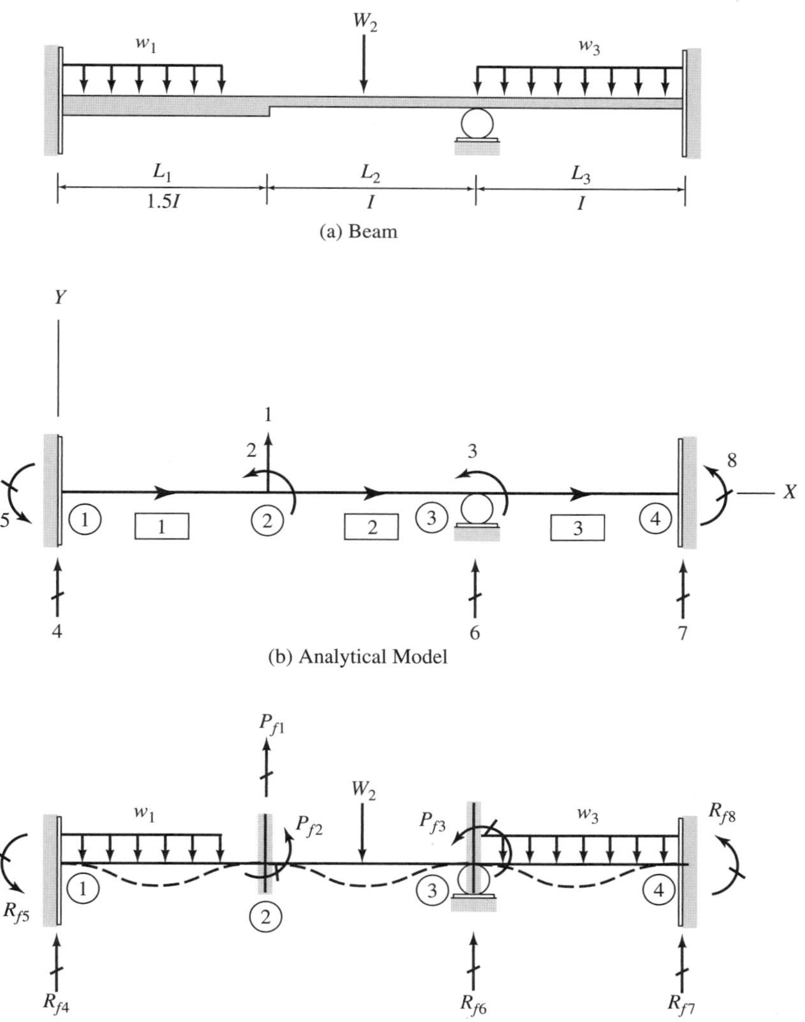

Fig. 5.14

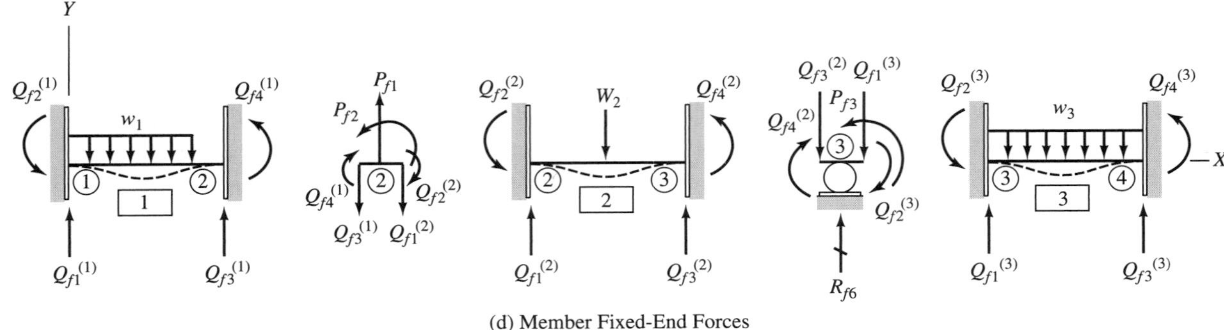

(d) Member Fixed-End Forces

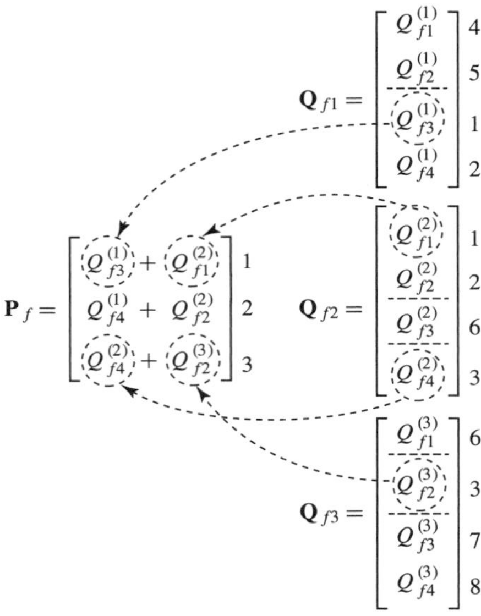

$$\mathbf{Q}_{f1} = \begin{bmatrix} Q_{f1}^{(1)} \\ Q_{f2}^{(1)} \\ Q_{f3}^{(1)} \\ Q_{f4}^{(1)} \end{bmatrix} \begin{matrix} 4 \\ 5 \\ 1 \\ 2 \end{matrix} \qquad \mathbf{Q}_{f2} = \begin{bmatrix} Q_{f1}^{(2)} \\ Q_{f2}^{(2)} \\ Q_{f3}^{(2)} \\ Q_{f4}^{(2)} \end{bmatrix} \begin{matrix} 1 \\ 2 \\ 6 \\ 3 \end{matrix} \qquad \mathbf{Q}_{f3} = \begin{bmatrix} Q_{f1}^{(3)} \\ Q_{f2}^{(3)} \\ Q_{f3}^{(3)} \\ Q_{f4}^{(3)} \end{bmatrix} \begin{matrix} 6 \\ 3 \\ 7 \\ 8 \end{matrix}$$

$$\mathbf{P}_f = \begin{bmatrix} Q_{f3}^{(1)} + Q_{f1}^{(2)} \\ Q_{f4}^{(1)} + Q_{f2}^{(2)} \\ Q_{f4}^{(2)} + Q_{f2}^{(3)} \end{bmatrix} \begin{matrix} 1 \\ 2 \\ 3 \end{matrix}$$

(e) Assembly of Structure Fixed-Joint Force Vector $\mathbf{P}_f$

Fig. 5.14 *(continued)*

completely fixed structure, reaction forces and moments develop at each of its joints. Note that, in Fig. 5.14(c), the reactions due to the imaginary restraints are denoted symbolically by P_{fi} (i = 1 through 3), whereas the reactions at the actual supports are denoted by R_{fi} (i = 4 through 8). The imaginary reactions P_{f1}, P_{f2}, and P_{f3}, which are at the locations and in the directions of the structure's three degrees of freedom 1, 2, and 3, respectively, are considered the structure fixed-joint forces due to member loads. Thus, the structure fixed-joint force vector, $\mathbf{P}_f$, for the beam, can be written as

$$\mathbf{P}_f = \begin{bmatrix} P_{f1} \\ P_{f2} \\ P_{f3} \end{bmatrix} \tag{5.113}$$

To relate the structure fixed-joint forces $\mathbf{P}_f$ to the member fixed-end forces $\mathbf{Q}_f$, we draw the free-body diagrams of the members and the interior joints of the hypothetical fixed beam, as shown in Fig. 5.14(d). In this figure, the superscript (i) denotes the member number. Note that, because all the joints of the beam are completely restrained, the member ends, which are rigidly connected to the joints, are also fixed against any displacements. Therefore, only the fixed-end forces due to member loads, $\mathbf{Q}_f$, can develop at the ends of the three members of the beam. By applying the equations of equilibrium $\sum F_Y = 0$ and $\sum M = 0$ to the free body of joint 2, and the equilibrium equation $\sum M = 0$ to the free body of joint 3, we obtain the following relationships between the structure fixed-joint forces and the member fixed-end forces.

$$P_{f1} = Q_{f3}^{(1)} + Q_{f1}^{(2)}$$

$$P_{f2} = Q_{f4}^{(1)} + Q_{f2}^{(2)}$$

$$P_{f3} = Q_{f4}^{(2)} + Q_{f2}^{(3)}$$

which can be expressed in vector form as

$$\mathbf{P}_f = \begin{bmatrix} P_{f1} \\ P_{f2} \\ P_{f3} \end{bmatrix} = \begin{bmatrix} Q_{f3}^{(1)} + Q_{f1}^{(2)} \\ Q_{f4}^{(1)} + Q_{f2}^{(2)} \\ Q_{f4}^{(2)} + Q_{f2}^{(3)} \end{bmatrix}$$

Note that the foregoing $\mathbf{P}_f$ vector is identical to that determined for the example beam in the preceding section (Eq. (5.111)).

Assembly of Structure Fixed-Joint Force Vector Using Member Code Numbers

The structure fixed-joint force vector $\mathbf{P}_f$ can be conveniently assembled from the member fixed-end force vectors $\mathbf{Q}_f$, using the member code number technique. The technique is similar to that for forming the structure stiffness matrix $\mathbf{S}$, described in the preceding section. Essentially, the procedure involves storing the pertinent elements of the fixed-end force vector $\mathbf{Q}_f$ for each member of the beam in their proper positions in the structure fixed-joint force vector $\mathbf{P}_f$, using the member code numbers.

The foregoing procedure is illustrated for the example beam in Fig. 5.14(e). As shown there, the code numbers of each member are written on the right side of its fixed-end force vector $\mathbf{Q}_f$. Any member code number that is less than or equal to the number of degrees of freedom of the structure ($NDOF$), now identifies the row of $\mathbf{P}_f$ in which the corresponding member force is to be stored. Thus, as shown in Fig. 5.14(e), the third and fourth elements of $\mathbf{Q}_{f1}$, with code numbers 1 and 2, respectively, are stored in the first and second rows of $\mathbf{P}_f$. The same procedure is then repeated for members 2 and 3. Note that the completed $\mathbf{P}_f$ vector for the beam is identical to that obtained previously (Eq. (5.111)).

Equivalent Joint Loads

The negatives of the structure fixed-joint forces (i.e., $-\mathbf{P}_f$) are commonly known as the *equivalent joint loads*. This is because the structure fixed-joint forces, when applied to a structure with their directions reversed, cause the same joint displacements as the actual member loads.

The validity of the foregoing interpretation can be shown easily using the principle of superposition (Section 1.7), as illustrated in Fig. 5.15. Figure 5.15(a) shows a continuous beam subjected to arbitrary member loads. (This beam was considered previously, and its analytical model is given in Fig. 5.14(b).) In Fig. 5.15(b), joints 2 and 3 of the beam are fixed by imaginary restraints so that the translation and rotation of joint 2, and the rotation of joint 3, are 0. This hypothetical completely fixed beam is then subjected to member loads, causing the structure fixed-joint forces P_{f1}, P_{f2}, and P_{f3} to develop at the imaginary restraints, as shown in Fig. 5.15(b). Lastly, as shown in Fig. 5.15(c),

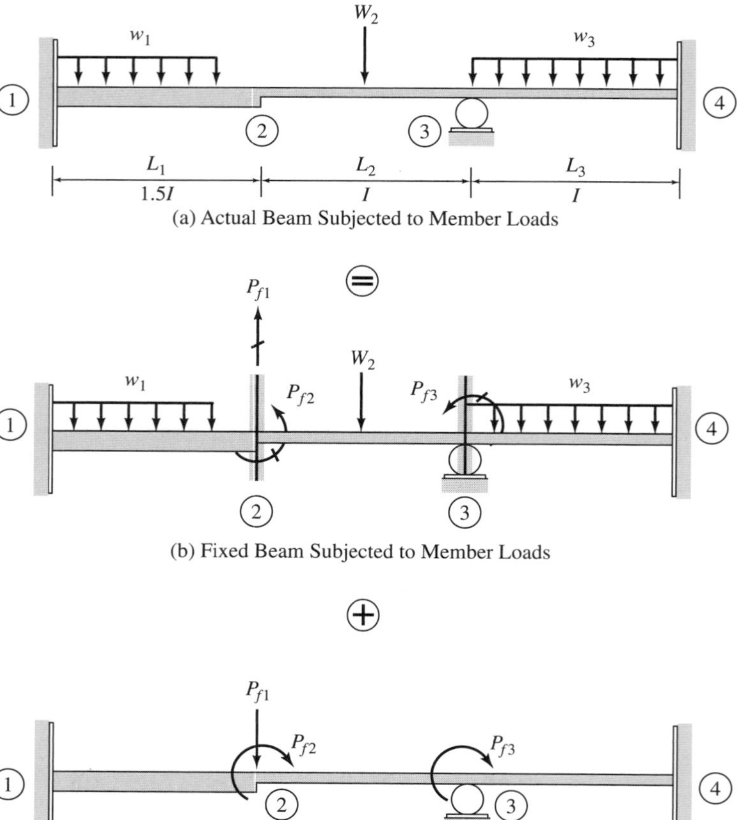

Fig. 5.15

the actual beam is subjected to external loads at its joints, which are equal in magnitude to P_{f1}, P_{f2}, and P_{f3}, but are reversed in direction.

By comparing Figs. 5.15(a), (b), and (c), we realize that the actual loading applied to the beam in Fig. 5.15(a) equals the algebraic sum of the loadings given in Figs. 5.15(b) and (c), because the reactive forces P_{f1}, P_{f2}, and P_{f3} in Fig. 5.15(b) cancel the corresponding applied loads in Fig. 5.15(c). Thus, according to the principle of superposition, any joint displacement of the actual beam due to the member loads (Fig. 5.15(a)) must equal the algebraic sum of the corresponding joint displacement of the fixed beam due to the member loads (Fig. 5.15(b)), and the corresponding joint displacement of the actual beam subjected to no member loads, but to the fixed-joint forces with their directions reversed. However, the joint displacements of the fixed beam (Fig. 5.15(b)) are 0. Therefore, the joint displacements of the beam due to the member loads (Fig. 5.15(a)) must be equal to the corresponding joint displacements of the beam due to the negatives of the fixed-joint forces (Fig. 5.15(c)). In other words, the negatives of the structure fixed-joint forces do indeed cause the same joint displacements of the beam as the actual member loads; and in that sense, such forces can be considered to be the equivalent joint loads. It is important to realize that this equivalency between the negative fixed-joint forces and the member loads is valid only for joint displacements, and cannot be generalized to member end forces and reactions, because such forces are generally not 0 in fixed structures subjected to member loads.

Based on the foregoing discussion of equivalent joint loads, we can define the *equivalent joint load vector* $\mathbf{P}_e$ for a structure as simply the negative of its fixed-joint force vector $\mathbf{P}_f$; that is,

$$\boxed{\mathbf{P}_e = -\mathbf{P}_f} \qquad \textbf{(5.114)}$$

An alternative form of the structure stiffness relations, in terms of the equivalent joint loads, can now be obtained by substituting Eq. (5.114) into Eq. (5.109). This yields

$$\boxed{\mathbf{P} + \mathbf{P}_e = \mathbf{S}\mathbf{d}} \qquad \textbf{(5.115)}$$

Once $\mathbf{S}$, $\mathbf{P}_f$ (or $\mathbf{P}_e$), and $\mathbf{P}$ have been evaluated, the structure stiffness relations (Eq. (5.109) or Eq. (5.115)), which now represent a system of simultaneous linear equations, can be solved for the unknown joint displacements $\mathbf{d}$. With $\mathbf{d}$ known, the end displacements $\mathbf{u}$ for each member can be determined by applying the compatibility equations defined by its code numbers; then the corresponding end forces $\mathbf{Q}$ can be computed by applying the member stiffness relations. Finally, the support reactions $\mathbf{R}$ are determined from the member end forces $\mathbf{Q}$, by considering the equilibrium of the support joints in the directions of the restrained coordinates via member code numbers, as discussed in Chapter 3 for the case of plane trusses.

EXAMPLE 5.5 Determine the fixed-joint force vector and the equivalent joint load vector for the propped-cantilever beam shown in Fig. 5.16(a).

SOLUTION *Analytical Model:* See Fig. 5.16(b).

Structure Fixed-Joint Force Vector: To generate the 3 × 1 structure fixed-joint force vector $\mathbf{P}_f$, we will, for each member: (a) determine the fixed-end force vector $\mathbf{Q}_f$, using the fixed-end force equations given inside the front cover; and (b) store the pertinent elements of $\mathbf{Q}_f$ in their proper positions in $\mathbf{P}_f$, using the member code numbers.

Member 1 By substituting $w = 30$ kN/m, $L = 9$ m, and $l_1 = l_2 = 0$ into the fixed-end force expressions for loading type 3, we obtain

$$FS_b = FS_e = \frac{30(9)}{2} = 135 \text{ kN}$$

$$FM_b = \frac{30(9)^2}{12} = 202.5 \text{ kN} \cdot \text{m}$$

$$FM_e = -\frac{30(9)^2}{12} = -202.5 \text{ kN} \cdot \text{m}$$

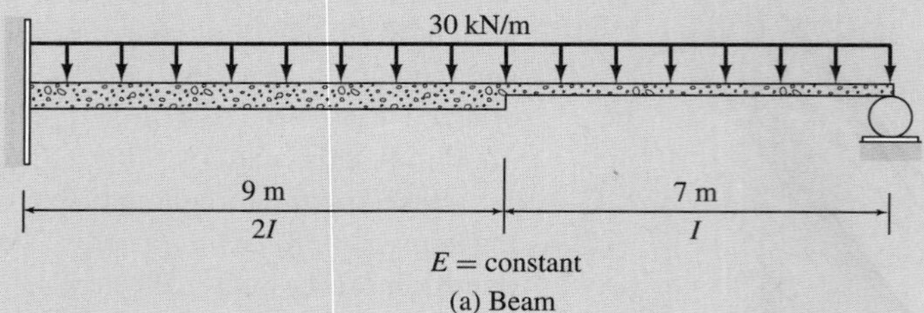

(a) Beam

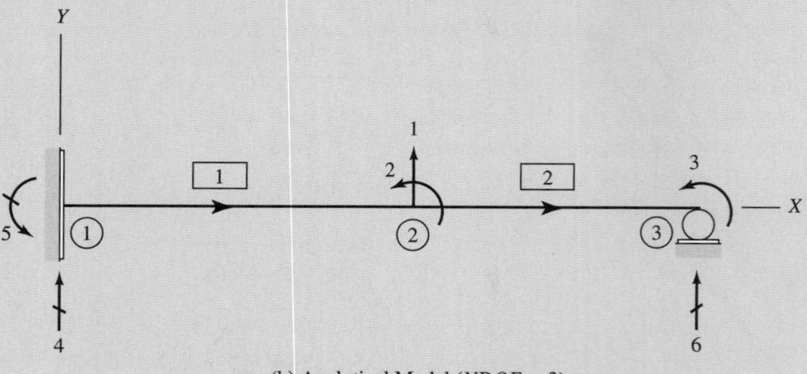

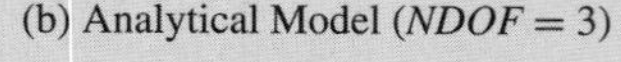
(b) Analytical Model ($NDOF = 3$)

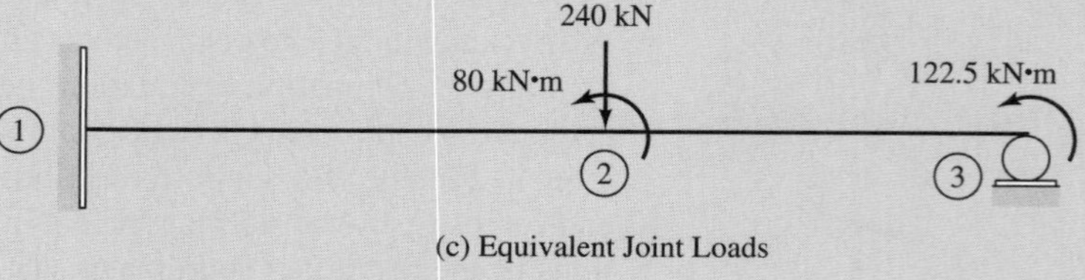

(c) Equivalent Joint Loads

Fig. 5.16

Thus, the fixed-end force vector for member 1 is given by

$$\mathbf{Q}_{f1} = \begin{bmatrix} 135 \\ 202.5 \\ \hdashline 135 \\ -202.5 \end{bmatrix} \begin{matrix} 4 \\ 5 \\ 1 \\ 2 \end{matrix} \qquad \textbf{(1)}$$

From Fig. 5.16(b), we can see that the code numbers for member 1 are 4, 5, 1, 2. These numbers are written on the right side of $\mathbf{Q}_{f1}$ in Eq. (1) to indicate the rows of the structure fixed-joint vector $\mathbf{P}_f$, where the elements of $\mathbf{Q}_{f1}$ are to be stored. Thus, the elements in the third and fourth rows of $\mathbf{Q}_{f1}$ are stored in rows 1 and 2, respectively, of $\mathbf{P}_f$, as

$$\mathbf{P}_f = \begin{bmatrix} 135 \\ -202.5 \\ 0 \end{bmatrix} \begin{matrix} 1 \\ 2 \\ 3 \end{matrix} \qquad \textbf{(2)}$$

Note that the elements of $\mathbf{Q}_{f1}$ corresponding to the restrained coordinate numbers 4 and 5 are disregarded.

Member 2 By substituting $w = 30$ kN/m, $L = 7$ m, and $l_1 = l_2 = 0$ into the fixed-end force expressions for loading type 3, we obtain

$$FS_b = FS_e = \frac{30(7)}{2} = 105 \text{ kN}$$

$$FM_b = \frac{30(7)^2}{12} = 122.5 \text{ kN} \cdot \text{m}$$

$$FM_e = -\frac{30(7)^2}{12} = -122.5 \text{ kN} \cdot \text{m}$$

Thus,

$$\mathbf{Q}_{f2} = \begin{bmatrix} 105 \\ 122.5 \\ \hdashline 105 \\ \hdashline -122.5 \end{bmatrix} \begin{matrix} 1 \\ 2 \\ 6 \\ 3 \end{matrix} \qquad \textbf{(3)}$$

From Fig. 5.16(b), we observe that the code numbers for this member are 1, 2, 6, 3. These numbers are used to add the pertinent elements of $\mathbf{Q}_{f2}$ in their proper positions in $\mathbf{P}_f$ given in Eq. (2), which now becomes

$$\mathbf{P}_f = \begin{bmatrix} 135 + 105 \\ -202.5 + 122.5 \\ -122.5 \end{bmatrix} \begin{matrix} 1 \\ 2 \\ 3 \end{matrix}$$

Since the fixed-end forces for both members of the beam have now been stored in $\mathbf{P}_f$, the structure fixed-joint force vector for the given beam is

$$\mathbf{P}_f = \begin{bmatrix} 240 \\ -80 \\ -122.5 \end{bmatrix} \begin{matrix} 1 \\ 2 \\ 3 \end{matrix} \qquad \textbf{Ans}$$

Equivalent Joint Load Vector: By using Eq. (5.114), we obtain

$$\mathbf{P}_e = -\mathbf{P}_f = \begin{bmatrix} -240 \\ 80 \\ 122.5 \end{bmatrix} \begin{matrix} 1 \\ 2 \\ 3 \end{matrix} \qquad \textbf{Ans}$$

These equivalent joint loads, when applied to the beam as shown in Fig. 5.16(c), cause the same joint displacements as the actual 30 kN/m uniformly distributed load given in Fig. 5.16(a).

5.7 PROCEDURE FOR ANALYSIS

Based on the concepts presented in the previous sections, we can develop the following step-by-step procedure for the analysis of beams by the matrix stiffness method. The reader should note that the overall format of this procedure is essentially the same as the procedure for analysis of plane trusses presented in Chapter 3.

1. Prepare an analytical model of the beam, as follows.
   a. Draw a line diagram of the beam, and identify each joint and member by a number. The origin of the global *XY* coordinate system is usually located at the farthest left joint, with the *X* and *Y* axes oriented in the horizontal (positive to the right) and vertical (positive upward) directions, respectively. For each member, establish a local *xy* coordinate system, with the origin at the left end (beginning) of the member, and the *x* and *y* axes oriented in the horizontal (positive to the right) and vertical (positive upward) directions, respectively.
   b. Number the degrees of freedom and restrained coordinates of the beam, as discussed in Section 5.1.
2. Evaluate the structure stiffness matrix **S** and fixed-joint force vector $\mathbf{P}_f$. The number of rows and columns of **S** must be equal to the number of degrees of freedom (*NDOF*) of the beam; the number of rows of $\mathbf{P}_f$ must equal *NDOF*. For each member of the structure, perform the following operations.
   a. Compute the member stiffness matrix **k** (Eq. (5.53)).
   b. If the member is subjected to external loads, then evaluate its fixed-end force vector $\mathbf{Q}_f$, using the expressions for fixed-end forces given inside the front cover.
   c. Identify the member code numbers and store the pertinent elements of **k** and $\mathbf{Q}_f$ in their proper positions in the structure stiffness matrix **S**, and the fixed-joint force vector $\mathbf{P}_f$, respectively. The complete structure stiffness matrix **S**, obtained by assembling the stiffness coefficients of all the members of the beam, must be symmetric.
3. If the beam is subjected to joint loads, then form the $NDOF \times 1$ joint load vector **P**.
4. Determine the joint displacements **d**. Substitute **P**, $\mathbf{P}_f$, and **S** into the structure stiffness relations, $\mathbf{P} - \mathbf{P}_f = \mathbf{Sd}$ (Eq. (5.109)), and solve the resulting system of simultaneous equations for the unknown joint displacements **d**. To check that the simultaneous equations have been solved correctly, substitute the numerical values of the joint displacements **d** back into the structure stiffness relations, $\mathbf{P} - \mathbf{P}_f = \mathbf{Sd}$.

If the solution is correct, then the stiffness relations should be satisfied. It should be noted that joint translations are considered positive when in the positive direction of the Y axis, and joint rotations are considered positive when counterclockwise.

5. Compute member end displacements and end forces, and support reactions. For each member of the beam, do the following.

   a. Obtain member end displacements $\mathbf{u}$ from the joint displacements $\mathbf{d}$, using the member code numbers.

   b. Compute member end forces, using the relationship $\mathbf{Q} = \mathbf{ku} + \mathbf{Q}_f$ (Eq. (5.4)).

   c. Using the member code numbers, store the pertinent elements of $\mathbf{Q}$ in their proper positions in the support reaction vector $\mathbf{R}$ (as discussed in Chapter 3).

6. Check the calculation of member end forces and support reactions by applying the equations of equilibrium, $\sum F_Y = 0$ and $\sum M = 0$, to the free body of the entire beam. If the calculations have been carried out correctly, then the equilibrium equations should be satisfied.

EXAMPLE 5.6 Determine the joint displacements, member end forces, and support reactions for the three-span continuous beam shown in Fig. 5.17(a), using the matrix stiffness method.

SOLUTION *Analytical Model:* See Fig. 5.17(b). The beam has two degrees of freedom—the rotations of joints 2 and 3—which are numbered 1 and 2, respectively. The six restrained coordinates of the beam are numbered 3 through 8.

Structure Stiffness Matrix and Fixed-Joint Force Vector:

Member 1 By substituting $E = 29{,}000$ ksi, $I = 510$ in.^{4}, and $L = 240$ in. into Eq. (5.53), we obtain

$$\mathbf{k}_1 = \begin{array}{c} \begin{array}{cccc} 3 & 4 & 5 & 1 \end{array} \\ \left[\begin{array}{cccc} 12.839 & 1{,}540.6 & -12.839 & 1{,}540.6 \\ 1{,}540.6 & 246{,}500 & -1{,}540.6 & 123{,}250 \\ -12.839 & -1{,}540.6 & 12.839 & -1{,}540.6 \\ 1{,}540.6 & 123{,}250 & -1{,}540.6 & 246{,}500 \end{array}\right] \begin{array}{c} 3 \\ 4 \\ 5 \\ 1 \end{array} \end{array} \quad (1)$$

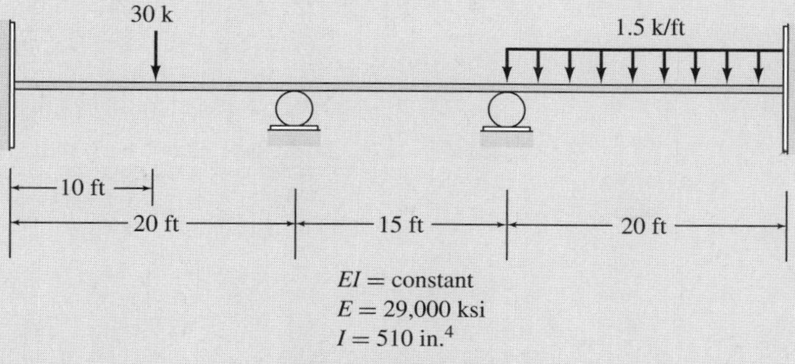

(a) Three-Span Continuous Beam

Fig. 5.17

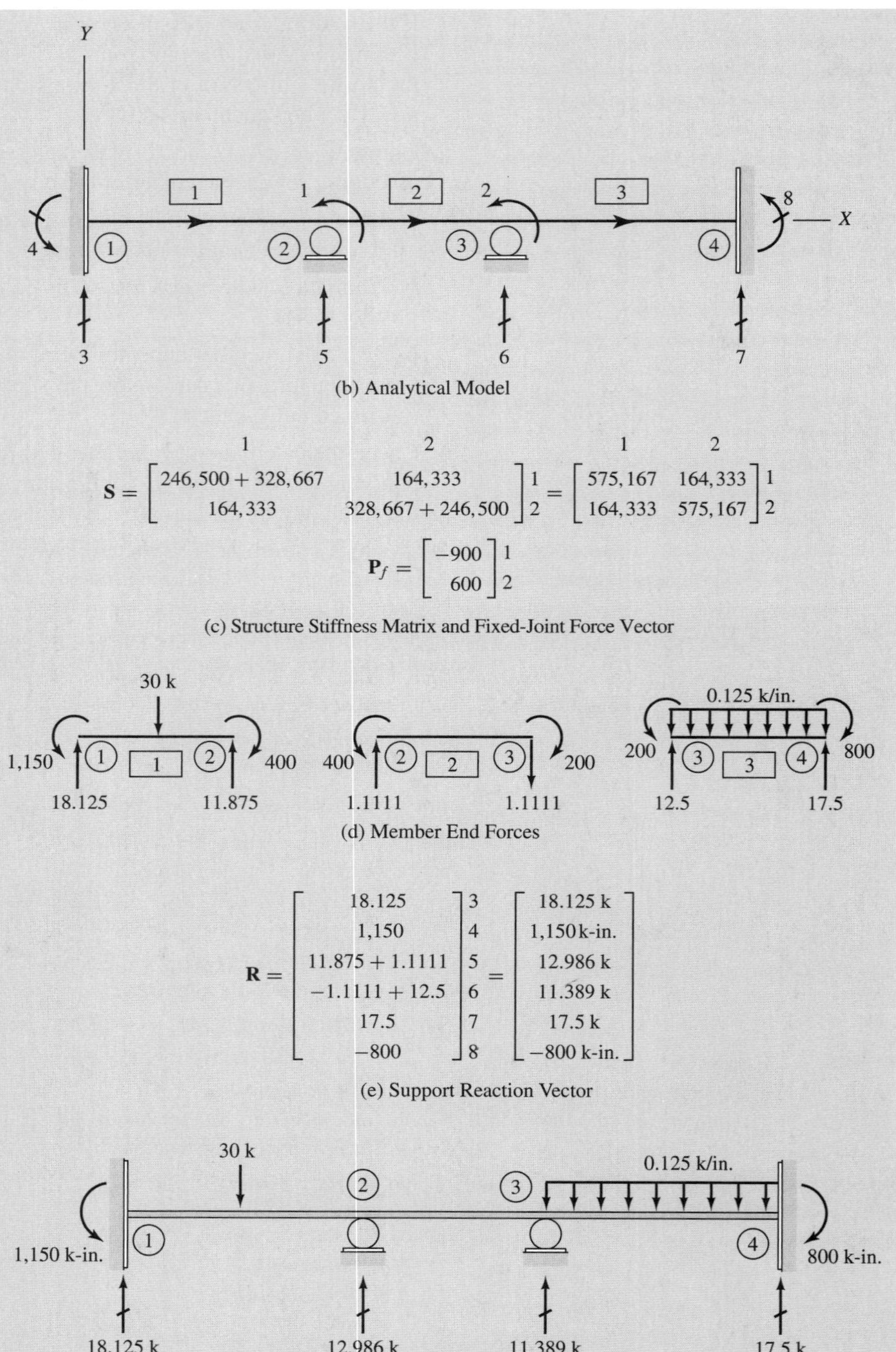

(b) Analytical Model

$$\mathbf{S} = \begin{bmatrix} 246,500 + 328,667 & 164,333 \\ 164,333 & 328,667 + 246,500 \end{bmatrix} \begin{matrix} 1 \\ 2 \end{matrix} = \begin{bmatrix} 575,167 & 164,333 \\ 164,333 & 575,167 \end{bmatrix} \begin{matrix} 1 \\ 2 \end{matrix}$$

$$\mathbf{P}_f = \begin{bmatrix} -900 \\ 600 \end{bmatrix} \begin{matrix} 1 \\ 2 \end{matrix}$$

(c) Structure Stiffness Matrix and Fixed-Joint Force Vector

(d) Member End Forces

$$\mathbf{R} = \begin{bmatrix} 18.125 \\ 1,150 \\ 11.875 + 1.1111 \\ -1.1111 + 12.5 \\ 17.5 \\ -800 \end{bmatrix} \begin{matrix} 3 \\ 4 \\ 5 \\ 6 \\ 7 \\ 8 \end{matrix} = \begin{bmatrix} 18.125 \text{ k} \\ 1,150 \text{ k-in.} \\ 12.986 \text{ k} \\ 11.389 \text{ k} \\ 17.5 \text{ k} \\ -800 \text{ k-in.} \end{bmatrix}$$

(e) Support Reaction Vector

(f) Support Reactions

Fig. 5.17 *(continued)*

Using the fixed-end force equations given inside the front cover, we evaluate the fixed-end forces due to the 30 k concentrated load as

$$FS_b = \frac{30(120)^2}{(240)^3}[3(120) + 120] = 15 \text{ k}$$

$$FM_b = \frac{30(120)(120)^2}{(240)^2} = 900 \text{ k-in.}$$

$$FS_e = \frac{30(120)^2}{(240)^3}[120 + 3(120)] = 15 \text{ k}$$

$$FM_e = -\frac{30(120)^2(120)}{(240)^2} = -900 \text{ k-in.}$$

Thus, the fixed-end force vector for member 1 is

$$\mathbf{Q}_{f1} = \begin{bmatrix} 15 \\ 900 \\ 15 \\ \hline -900 \end{bmatrix} \begin{matrix} 3 \\ 4 \\ 5 \\ 1 \end{matrix} \tag{2}$$

From Fig. 5.17(b), we observe that the code numbers for member 1 are 3, 4, 5, 1. Using these code numbers, the pertinent elements of $\mathbf{k}_1$ and $\mathbf{Q}_{f1}$ are stored in their proper positions in the 2 × 2 structure stiffness matrix $\mathbf{S}$ and the 2 × 1 structure fixed-joint force vector $\mathbf{P}_f$, respectively, as shown in Fig. 5.17(c).

Member 2 $E = 29{,}000$ ksi, $I = 510$ in.^{4}, and $L = 180$ in.

$$\mathbf{k}_2 = \begin{matrix} 5 & 1 & 6 & 2 \end{matrix} \\ \begin{bmatrix} 30.432 & 2{,}738.9 & -30.432 & 2{,}738.9 \\ 2{,}738.9 & \boxed{328{,}667} & -2{,}738.9 & \boxed{164{,}333} \\ -30.432 & -2{,}738.9 & 30.432 & -2{,}738.9 \\ 2{,}738.9 & \boxed{164{,}333} & -2{,}738.9 & \boxed{328{,}667} \end{bmatrix} \begin{matrix} 5 \\ 1 \\ 6 \\ 2 \end{matrix} \tag{3}$$

Since this member is not subjected to any external loads, its fixed-end force vector is 0; that is,

$$\mathbf{Q}_{f2} = \mathbf{0} \tag{4}$$

Using the code numbers 5, 1, 6, 2 for this member (see Fig. 5.17(b)), the relevant elements of $\mathbf{k}_2$ are stored into $\mathbf{S}$, as shown in Fig. 5.17(c).

Member 3 $E = 29{,}000$ ksi, $I = 510$ in.^{4}, and $L = 240$ in.

$$\mathbf{k}_3 = \begin{matrix} 6 & 2 & 7 & 8 \end{matrix} \\ \begin{bmatrix} 12.839 & 1{,}540.6 & -12.839 & 1{,}540.6 \\ 1{,}540.6 & \boxed{246{,}500} & -1{,}540.6 & 123{,}250 \\ -12.839 & -1{,}540.6 & 12.839 & -1{,}540.6 \\ 1{,}540.6 & 123{,}250 & -1{,}540.6 & 246{,}500 \end{bmatrix} \begin{matrix} 6 \\ 2 \\ 7 \\ 8 \end{matrix} \tag{5}$$

The fixed-end forces due to the 0.125 k/in. (=1.5 k/ft) uniformly distributed load are

$$FS_b = \frac{0.125(240)}{2} = 15 \text{ k}$$

$$FM_b = \frac{0.125(240)^2}{12} = 600 \text{ k-in.}$$

$$FS_e = \frac{0.125(240)}{2} = 15 \text{ k}$$

$$FM_e = -\frac{0.125(240)^2}{12} = -600 \text{ k-in.}$$

Thus,

$$\mathbf{Q}_{f3} = \begin{bmatrix} 15 \\ 600 \\ 15 \\ -600 \end{bmatrix} \begin{matrix} 6 \\ 2 \\ 7 \\ 8 \end{matrix} \tag{6}$$

The relevant elements of $\mathbf{k}_3$ and $\mathbf{Q}_{f3}$ are stored in **S** and $\mathbf{P}_f$, respectively, using the member code numbers 6, 2, 7, 8.

The completed structure stiffness matrix **S** and structure fixed-joint force vector $\mathbf{P}_f$ are given in Fig. 5.17(c). Note that the **S** matrix is symmetric.

Joint Load Vector: Since no external loads (i.e., moments) are applied to the beam at joints 2 and 3, the joint load vector is 0; that is,

$$\mathbf{P} = \mathbf{0}$$

Joint Displacements: By substituting the numerical values of **P**, $\mathbf{P}_f$, and **S** into Eq. (5.109), we write the stiffness relations for the entire continuous beam as

$$\mathbf{P} - \mathbf{P}_f = \mathbf{Sd}$$

$$\begin{bmatrix} 0 \\ 0 \end{bmatrix} - \begin{bmatrix} -900 \\ 600 \end{bmatrix} = \begin{bmatrix} 575{,}167 & 164{,}333 \\ 164{,}333 & 575{,}167 \end{bmatrix} \begin{bmatrix} d_1 \\ d_2 \end{bmatrix}$$

or

$$\begin{bmatrix} 900 \\ -600 \end{bmatrix} = \begin{bmatrix} 575{,}167 & 164{,}333 \\ 164{,}333 & 575{,}167 \end{bmatrix} \begin{bmatrix} d_1 \\ d_2 \end{bmatrix}$$

By solving these equations simultaneously, we determine the joint displacements to be

$$\mathbf{d} = \begin{bmatrix} 2.0284 \\ -1.6227 \end{bmatrix} \times 10^{-3} \text{ rad}$$ **Ans**

To check the foregoing solution, we substitute the numerical values of **d** back into the structure stiffness relationship to obtain

$$\mathbf{P} - \mathbf{P}_f = \mathbf{Sd} = \begin{bmatrix} 575{,}167 & 164{,}333 \\ 164{,}333 & 575{,}167 \end{bmatrix} \begin{bmatrix} 2.0284 \\ -1.6227 \end{bmatrix} \times 10^{-3} = \begin{bmatrix} 900.01 \\ -599.99 \end{bmatrix}$$

Checks

Member End Displacements and End Forces:

Member 1 The member end displacements **u** can be obtained simply by comparing the member's degree of freedom numbers with its code numbers, as follows:

$$\mathbf{u}_1 = \begin{bmatrix} u_1 \\ u_2 \\ u_3 \\ u_4 \end{bmatrix} \begin{matrix} 3 \\ 4 \\ 5 \\ 1 \end{matrix} = \begin{bmatrix} 0 \\ 0 \\ 0 \\ d_1 \end{bmatrix} = \begin{bmatrix} 0 \\ 0 \\ 0 \\ 2.0284 \end{bmatrix} \times 10^{-3} \tag{7}$$

Note that the member code numbers (3, 4, 5, 1), when written on a side of **u** as shown in Eq. (7), define the compatibility equations for the member. Since the code numbers corresponding to u_1, u_2, and u_3 are the restrained coordinate numbers 3, 4, and 5, respectively, this indicates that $u_1 = u_2 = u_3 = 0$. Similarly, the code number 1 corresponding to u_4 indicates that $u_4 = d_1$. The foregoing compatibility equations can be easily verified by a visual inspection of the beam's line diagram, given in Fig. 5.17(b).

The member end forces can now be calculated, using the member stiffness relationship $\mathbf{Q} = \mathbf{ku} + \mathbf{Q}_f$ (Eq. (5.4)). Using $\mathbf{k}_1$ and $\mathbf{Q}_{f1}$ from Eqs. (1) and (2), respectively, we write

$$\mathbf{Q}_1 = \begin{bmatrix} 12.839 & 1{,}540.6 & -12.839 & 1{,}540.6 \\ 1{,}540.6 & 246{,}500 & -1{,}540.6 & 123{,}250 \\ -12.839 & -1{,}540.6 & 12.839 & -1{,}540.6 \\ 1{,}540.6 & 123{,}250 & -1{,}540.6 & 246{,}500 \end{bmatrix} \begin{bmatrix} 0 \\ 0 \\ 0 \\ 2.0284 \end{bmatrix} \times 10^{-3}$$

$$+ \begin{bmatrix} 15 \\ 900 \\ 15 \\ -900 \end{bmatrix} = \begin{bmatrix} 18.125 \text{ k} \\ 1{,}150 \text{ k-in.} \\ 11.875 \text{ k} \\ -400 \text{ k-in.} \end{bmatrix} \begin{matrix} 3 \\ 4 \\ 5 \\ 1 \end{matrix} \qquad (8) \quad \textsf{Ans}$$

The end forces for member 1 are shown in Fig. 5.17(d). We can check our calculation of end forces by applying the equilibrium equations, $\sum F_y = 0$ and $\sum M = 0$, to the free body of member 1 to ensure that it is in equilibrium. Thus,

$$+\uparrow \sum F_y = 0 \qquad 18.125 - 30 + 11.875 = 0 \qquad \textsf{Checks}$$

$$+\circlearrowleft \sum M_{①} = 0 \qquad 1{,}150 - 30(120) - 400 + 11.875(240) = 0 \qquad \textsf{Checks}$$

Next, to generate the support reaction vector **R**, we write the member code numbers (3, 4, 5, 1) on the right side of $\mathbf{Q}_1$, as shown in Eq. (8), and store the pertinent elements of $\mathbf{Q}_1$ in their proper positions in **R** by matching the code numbers on the side of $\mathbf{Q}_1$ to the restrained coordinate numbers on the right side of **R** (see Fig. 5.17(e)).

Member 2 The member end displacements are given by

$$\mathbf{u}_2 = \begin{bmatrix} u_1 \\ u_2 \\ u_3 \\ u_4 \end{bmatrix} \begin{matrix} 5 \\ 1 \\ 6 \\ 2 \end{matrix} = \begin{bmatrix} 0 \\ d_1 \\ 0 \\ d_2 \end{bmatrix} = \begin{bmatrix} 0 \\ 2.0284 \\ 0 \\ -1.6227 \end{bmatrix} \times 10^{-3}$$

By using $\mathbf{k}_2$ from Eq. (3) and $\mathbf{Q}_{f2} = \mathbf{0}$, we compute member end forces as

$$\mathbf{Q} = \mathbf{ku} + \mathbf{Q}_f$$

$$\mathbf{Q}_2 = \begin{bmatrix} 30.432 & 2{,}738.9 & -30.432 & 2{,}738.9 \\ 2{,}738.9 & 328{,}667 & -2{,}738.9 & 164{,}333 \\ -30.432 & -2{,}738.9 & 30.432 & -2{,}738.9 \\ 2{,}738.9 & 164{,}333 & -2{,}738.9 & 328{,}667 \end{bmatrix} \begin{bmatrix} 0 \\ 2.0284 \\ 0 \\ -1.6227 \end{bmatrix} \times 10^{-3}$$

$$= \begin{bmatrix} 1.1111 \text{ k} \\ 400 \text{ k-in.} \\ -1.1111 \text{ k} \\ -200 \text{ k-in.} \end{bmatrix} \begin{matrix} 5 \\ 1 \\ 6 \\ 2 \end{matrix} \qquad \textsf{Ans}$$

The foregoing member end forces are shown in Fig. 5.17(d). To check our calculations, we apply the equations of equilibrium to the free body of member 2 as

$$+\uparrow \sum F_y = 0 \qquad 1.1111 - 1.1111 = 0 \qquad \textsf{Checks}$$

$$+\circlearrowleft \sum M_{②} = 0 \qquad 400 - 200 - 1.1111(180) = 0.002 \approx 0 \qquad \textsf{Checks}$$

Next, we store the pertinent elements of $\mathbf{Q}_2$ in their proper positions in the reaction vector **R**, using the member code numbers (5, 1, 6, 2), as shown in Fig. 5.17(e).

Member 3

$$\mathbf{u}_3 = \begin{bmatrix} u_1 \\ u_2 \\ u_3 \\ u_4 \end{bmatrix} \begin{matrix} 6 \\ 2 \\ 7 \\ 8 \end{matrix} = \begin{bmatrix} 0 \\ d_2 \\ 0 \\ 0 \end{bmatrix} = \begin{bmatrix} 0 \\ -1.6227 \\ 0 \\ 0 \end{bmatrix} \times 10^{-3}$$

By substituting $\mathbf{k}_3$ and $\mathbf{Q}_{f3}$ from Eqs. (5) and (6), respectively, into the member stiffness relationship $\mathbf{Q} = \mathbf{ku} + \mathbf{Q}_f$, we determine the end forces for member 3 to be

$$\mathbf{Q}_3 = \begin{bmatrix} 12.839 & 1{,}540.6 & -12.839 & 1{,}540.6 \\ 1{,}540.6 & 246{,}500 & -1{,}540.6 & 123{,}250 \\ -12.839 & -1{,}540.6 & 12.839 & -1{,}540.6 \\ 1{,}540.6 & 123{,}250 & -1{,}540.6 & 246{,}500 \end{bmatrix} \begin{bmatrix} 0 \\ -1.6227 \\ 0 \\ 0 \end{bmatrix} \times 10^{-3}$$

$$+ \begin{bmatrix} 15 \\ 600 \\ 15 \\ -600 \end{bmatrix} = \begin{bmatrix} 12.5 \text{ k} \\ 200 \text{ k-in.} \\ 17.5 \text{ k} \\ -800 \text{ k-in.} \end{bmatrix} \begin{matrix} 6 \\ 2 \\ 7 \\ 8 \end{matrix}$$ **Ans**

These member end forces are shown in Fig. 5.17(d). To check our calculations, we apply the equilibrium equations:

$+\uparrow \sum F_y = 0 \qquad 12.5 - 0.125(240) + 17.5 = 0$ **Checks**

$+\circlearrowleft \sum M_{③} = 0 \qquad 200 - 0.125(240)(120) - 800 + 17.5(240) = 0$ **Checks**

Next, by using the code numbers (6, 2, 7, 8) for member 3, we store the relevant elements of $\mathbf{Q}_3$ in their proper positions in $\mathbf{R}$.

Support Reactions: The completed reaction vector $\mathbf{R}$ is shown in Fig. 5.17(e), and the support reactions are depicted on a line diagram of the structure in Fig. 5.17(f). **Ans**

Equilibrium Check: Finally, applying the equilibrium equations to the free body of the entire beam (Fig. 5.17(f)), we write

$+\uparrow \sum F_y = 0$

$18.125 - 30 + 12.986 + 11.389 - 0.125(240) + 17.5 = 0$ **Checks**

$+\circlearrowleft \sum M_{①} = 0$

$1{,}150 - 30(120) + 12.986(240) + 11.389(420) - 0.125(240)(540)$

$+ 17.5(660) - 800 = 0.02 \approx 0$ **Checks**

EXAMPLE 5.7

Determine the joint displacements, member end forces, and support reactions for the beam shown in Fig. 5.18(a), using the matrix stiffness method.

SOLUTION

Analytical Model: See Fig. 5.18(b). The beam has four degrees of freedom (numbered 1 through 4) and four restrained coordinates (numbered 5 through 8).

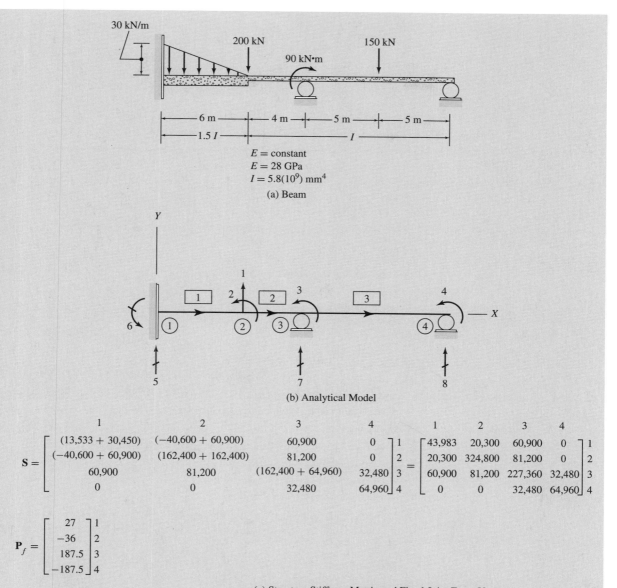

30 kN/m

200 kN

90 kN•m

150 kN

6 m — 4 m — 5 m — 5 m

1.5 *I* — *I*

E = constant

E = 28 GPa

$I = 5.8(10^9)\ \text{mm}^4$

(a) Beam

Y

1 2 3 4 5 6 7 8

[1] [2] [3]

(1) (2) (3) (4)

X

(b) Analytical Model

$$\mathbf{S} = \begin{bmatrix} (13{,}533 + 30{,}450) & (-40{,}600 + 60{,}900) & 60{,}900 & 0 \\ (-40{,}600 + 60{,}900) & (162{,}400 + 162{,}400) & 81{,}200 & 0 \\ 60{,}900 & 81{,}200 & (162{,}400 + 64{,}960) & 32{,}480 \\ 0 & 0 & 32{,}480 & 64{,}960 \end{bmatrix} = \begin{bmatrix} 43{,}983 & 20{,}300 & 60{,}900 & 0 \\ 20{,}300 & 324{,}800 & 81{,}200 & 0 \\ 60{,}900 & 81{,}200 & 227{,}360 & 32{,}480 \\ 0 & 0 & 32{,}480 & 64{,}960 \end{bmatrix}$$

(Rows and columns numbered 1, 2, 3, 4.)

$$\mathbf{P}_f = \begin{bmatrix} 27 \\ -36 \\ 187.5 \\ -187.5 \end{bmatrix} \begin{matrix} 1 \\ 2 \\ 3 \\ 4 \end{matrix}$$

(c) Structure Stiffness Matrix and Fixed-Joint Force Vector

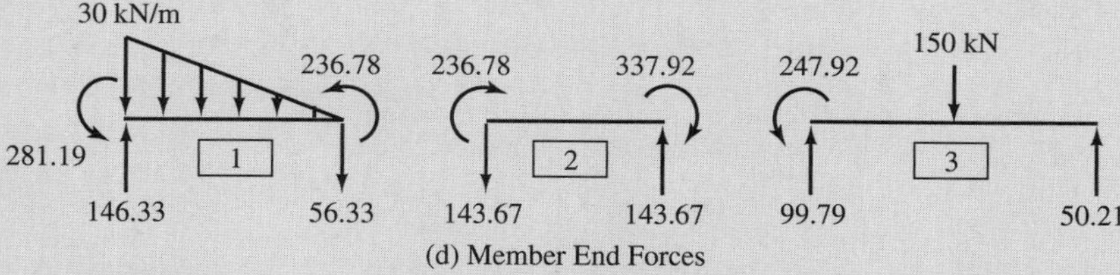

(d) Member End Forces

Fig. 5.18

$$\mathbf{R} = \begin{bmatrix} 146.33 \\ 281.19 \\ 143.67 + 99.79 \\ 50.21 \end{bmatrix} \begin{matrix} 5 \\ 6 \\ 7 \\ 8 \end{matrix} = \begin{bmatrix} 146.33 \text{ kN} \\ 281.19 \text{ kN} \cdot \text{m} \\ 243.46 \text{ kN} \\ 50.21 \text{ kN} \end{bmatrix}$$

(e) Support Reaction Vector

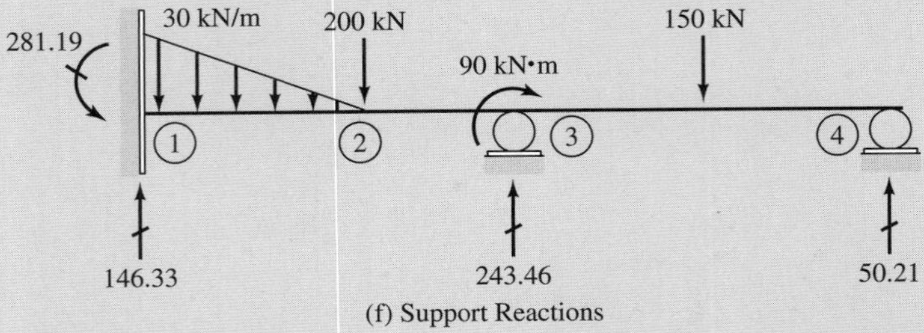

(f) Support Reactions

Fig. 5.18 *(continued)*

Structure Stiffness Matrix and Fixed-Joint Force Vector:

Member 1 By substituting $E = 28(10^6)$ kN/m², $I = 8{,}700(10^{-6})$ m⁴, and $L = 6$ m into Eq. (5.53), we write

$$\mathbf{k}_1 = \begin{matrix} 5 & 6 & 1 & 2 \\ \left[\begin{matrix} 13{,}533 \\ 40{,}600 \\ -13{,}533 \\ 40{,}600 \end{matrix}\right. & \begin{matrix} 40{,}600 \\ 162{,}400 \\ -40{,}600 \\ 81{,}200 \end{matrix} & \begin{matrix} -13{,}533 \\ -40{,}600 \\ 13{,}533 \\ -40{,}600 \end{matrix} & \left.\begin{matrix} 40{,}600 \\ 81{,}200 \\ -40{,}600 \\ 162{,}400 \end{matrix}\right] \end{matrix} \begin{matrix} \\ 5 \\ 6 \\ 1 \\ 2 \end{matrix}$$

Using the fixed-end force expressions given inside the front cover, we obtain $FS_b = 63$ kN, $FM_b = 54$ kN · m, $FS_e = 27$ kN, and $FM_e = -36$ kN · m. Thus,

$$\mathbf{Q}_{f1} = \begin{bmatrix} 63 \\ 54 \\ 27 \\ -36 \end{bmatrix} \begin{matrix} 5 \\ 6 \\ 1 \\ 2 \end{matrix}$$

Using the code numbers (5, 6, 1, 2) for member 1, we store the pertinent elements of $\mathbf{k}_1$ and $\mathbf{Q}_{f1}$ in their proper positions in the $\mathbf{S}$ matrix and the $\mathbf{P}_f$ vector, respectively, as shown in Fig. 5.18(c).

Member 2 $E = 28(10^6)$ kN/m², $I = 5{,}800(10^{-6})$ m⁴, and $L = 4$ m. Thus,

$$\mathbf{k}_2 = \begin{matrix} 1 & 2 & 7 & 3 \\ \left[\begin{matrix} 30{,}450 \\ 60{,}900 \\ -30{,}450 \\ 60{,}900 \end{matrix}\right. & \begin{matrix} 60{,}900 \\ 162{,}400 \\ -60{,}900 \\ 81{,}200 \end{matrix} & \begin{matrix} -30{,}450 \\ -60{,}900 \\ 30{,}450 \\ -60{,}900 \end{matrix} & \left.\begin{matrix} 60{,}900 \\ 81{,}200 \\ -60{,}900 \\ 162{,}400 \end{matrix}\right] \end{matrix} \begin{matrix} \\ 1 \\ 2 \\ 7 \\ 3 \end{matrix}$$

As this member is not subjected to any loads,

$$\mathbf{Q}_{f2} = \mathbf{0}$$

Using the member code numbers 1, 2, 7, 3, the relevant elements of $\mathbf{k}_2$ are stored in $\mathbf{S}$ in Fig. 5.18(c).

Member 3 $E = 28(10^6)$ kN/m^2, $I = 5{,}800(10^{-6})$ m^4, and $L = 10$ m.

$$\mathbf{k}_3 = \begin{array}{c} \\ \left[\begin{array}{cccc} 1{,}948.8 & 9{,}744 & -1{,}948.8 & 9{,}744 \\ 9{,}744 & 64{,}960 & -9{,}744 & 32{,}480 \\ -1{,}948.8 & -9{,}744 & 1{,}948.8 & -9{,}744 \\ 9{,}744 & 32{,}480 & -9{,}744 & 64{,}960 \end{array}\right] \end{array} \begin{array}{c} 7 \\ 3 \\ 8 \\ 4 \end{array}$$

(column code numbers: 7, 3, 8, 4)

The fixed-end forces are determined to be $FS_b = 75$ kN, $FM_b = 187.5$ kN · m, $FS_e = 75$ kN, and $FM_e = -187.5$ kN · m. Thus,

$$\mathbf{Q}_{f3} = \left[\begin{array}{c} 75 \\ 187.5 \\ 75 \\ -187.5 \end{array}\right] \begin{array}{c} 7 \\ 3 \\ 8 \\ 4 \end{array}$$

The relevant elements of $\mathbf{k}_3$ and $\mathbf{Q}_{f3}$ are stored in $\mathbf{S}$ and $\mathbf{P}_f$ respectively, using the member code numbers 7, 3, 8, 4. The completed structure stiffness matrix $\mathbf{S}$ and structure fixed-joint force vector $\mathbf{P}_f$ are given in Fig. 5.18(c).

Joint Load Vector: By comparing Figs. 5.18(a) and (b), we realize that $P_1 = -200$ kN, $P_2 = 0$, $P_3 = -90$ kN · m, and $P_4 = 0$. Thus, the joint load vector can be expressed as

$$\mathbf{P} = \left[\begin{array}{c} -200 \\ 0 \\ -90 \\ 0 \end{array}\right]$$

Joint Displacements: The stiffness relations for the entire beam can be expressed as

$$\mathbf{P} - \mathbf{P}_f = \mathbf{S}\mathbf{d}$$

By substituting the numerical values of $\mathbf{P}$, $\mathbf{P}_f$, and $\mathbf{S}$, we obtain

$$\left[\begin{array}{c} -200 \\ 0 \\ -90 \\ 0 \end{array}\right] - \left[\begin{array}{c} 27 \\ -36 \\ 187.5 \\ -187.5 \end{array}\right] = \left[\begin{array}{cccc} 43{,}983 & 20{,}300 & 60{,}900 & 0 \\ 20{,}300 & 324{,}800 & 81{,}200 & 0 \\ 60{,}900 & 81{,}200 & 227{,}360 & 32{,}480 \\ 0 & 0 & 32{,}480 & 64{,}960 \end{array}\right] \left[\begin{array}{c} d_1 \\ d_2 \\ d_3 \\ d_4 \end{array}\right]$$

or

$$\left[\begin{array}{c} -227 \\ 36 \\ -277.5 \\ 187.5 \end{array}\right] = \left[\begin{array}{cccc} 43{,}983 & 20{,}300 & 60{,}900 & 0 \\ 20{,}300 & 324{,}800 & 81{,}200 & 0 \\ 60{,}900 & 81{,}200 & 227{,}360 & 32{,}480 \\ 0 & 0 & 32{,}480 & 64{,}960 \end{array}\right] \left[\begin{array}{c} d_1 \\ d_2 \\ d_3 \\ d_4 \end{array}\right]$$

By solving the foregoing system of simultaneous equations, we determine the joint displacements to be

$$\mathbf{d} = \left[\begin{array}{l} -4.4729 \text{ m} \\ 0.56143 \text{ rad} \\ -0.68415 \text{ rad} \\ 3.2285 \text{ rad} \end{array}\right] \times 10^{-3}$$ **Ans**

Back substitution of the foregoing numerical values of $\mathbf{d}$ into the structure stiffness relationship $\mathbf{P} - \mathbf{P}_f = \mathbf{S}\mathbf{d}$ indicates that the solution of the simultaneous equations has indeed been carried out correctly.

Member End Displacements and End Forces:

Member 1

$$\mathbf{u}_1 = \begin{bmatrix} u_1 \\ u_2 \\ u_3 \\ u_4 \end{bmatrix} \begin{matrix} 5 \\ 6 \\ 1 \\ 2 \end{matrix} = \begin{bmatrix} 0 \\ 0 \\ d_1 \\ d_2 \end{bmatrix} = \begin{bmatrix} 0 \\ 0 \\ -4.4729 \\ 0.56143 \end{bmatrix} \times 10^{-3}$$

$$\mathbf{Q}_1 = \mathbf{k}_1\mathbf{u}_1 + \mathbf{Q}_{f1} = \begin{bmatrix} 146.33 \text{ kN} \\ 281.19 \text{ kN} \cdot \text{m} \\ -56.33 \text{ kN} \\ 236.78 \text{ kN} \cdot \text{m} \end{bmatrix} \begin{matrix} 5 \\ 6 \\ 1 \\ 2 \end{matrix}$$ **Ans**

Member 2

$$\mathbf{u}_2 = \begin{bmatrix} -4.4729 \\ 0.56143 \\ 0 \\ -0.68415 \end{bmatrix} \begin{matrix} 1 \\ 2 \\ 7 \\ 3 \end{matrix} \times 10^{-3}; \quad \mathbf{Q}_2 = \mathbf{k}_2\mathbf{u}_2 + \mathbf{Q}_{f2} = \begin{bmatrix} -143.67 \text{ kN} \\ -236.78 \text{ kN} \cdot \text{m} \\ 143.67 \text{ kN} \\ -337.92 \text{ kN} \cdot \text{m} \end{bmatrix} \begin{matrix} 1 \\ 2 \\ 7 \\ 3 \end{matrix}$$ **Ans**

Member 3

$$\mathbf{u}_3 = \begin{bmatrix} 0 \\ -0.68415 \\ 0 \\ 3.2285 \end{bmatrix} \begin{matrix} 7 \\ 3 \\ 8 \\ 4 \end{matrix} \times 10^{-3}; \quad \mathbf{Q}_3 = \mathbf{k}_3\mathbf{u}_3 + \mathbf{Q}_{f3} = \begin{bmatrix} 99.79 \text{ kN} \\ 247.92 \text{ kN} \cdot \text{m} \\ 50.21 \text{ kN} \\ 0 \end{bmatrix} \begin{matrix} 7 \\ 3 \\ 8 \\ 4 \end{matrix}$$ **Ans**

The member end forces are shown in Fig. 5.18(d).

Support Reactions: The reaction vector **R**, as assembled from the appropriate elements of the member end-force vectors, is given in Fig. 5.18(e). Also, Fig. 5.18(f) depicts the support reactions on a line diagram of the structure. **Ans**

Equilibrium Check: The equilibrium equations check.

5.8 COMPUTER PROGRAM

In this section, we consider computer implementation of the procedure for the analysis of beams presented in this chapter. Because of the similarity in the methods for the analysis of beams and plane trusses, the overall format of the program for beam analysis remains the same as that for the analysis of plane trusses developed in Chapter 4. Therefore, many parts of the plane truss program can be copied and used, without any modifications, in the program for beam analysis. In the following, we discuss the development of an input module and consider programming of the analysis steps for beams.

Input Module

Joint Data The joint data consists of (a) the total number of joints (NJ) of the beam, and (b) the global X coordinate of each joint. (Recall that the global

XY coordinate system must be oriented so that the *X* axis coincides with the beam's centroidal axis.) The joint coordinates are stored in computer memory in the form of a *joint coordinate vector* **COORD** of the order $NJ \times 1$. Consider, for example, the continuous beam shown in Fig. 5.19(a), with its analytical model given in Fig. 5.19(b). As the beam has four joints, its **COORD** vector has four rows, with the *X* coordinate of a joint *i* stored in the *i*th row, as shown in Fig. 5.19(c). A flowchart for programming the reading

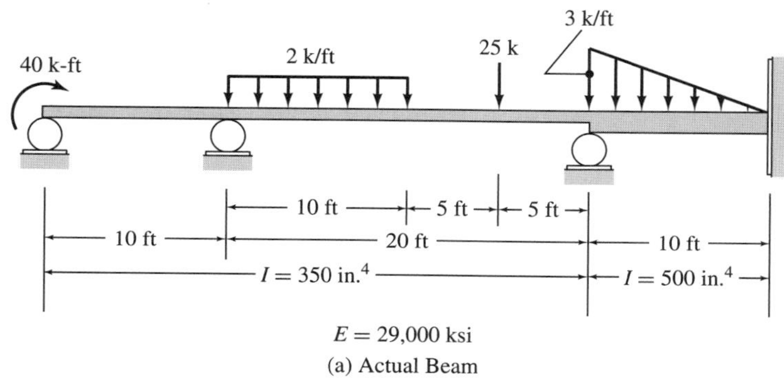

(a) Actual Beam

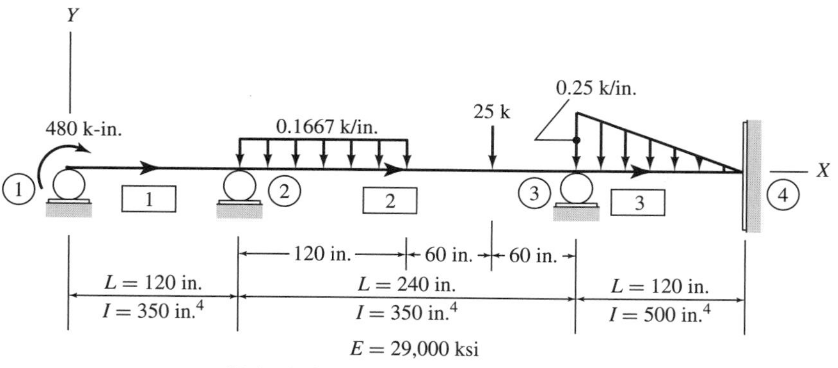

(b) Analytical Model (Units: Kips, Inches)

X Coordinate

$$\mathbf{COORD} = \begin{bmatrix} 0 \\ 120 \\ 360 \\ 480 \end{bmatrix} \begin{matrix} \leftarrow \text{Joint 1} \\ \leftarrow \text{Joint 2} \\ \leftarrow \text{Joint 3} \\ \leftarrow \text{Joint 4} \end{matrix}$$

$NJ \times 1$

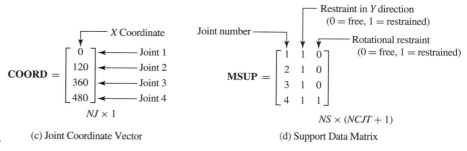

(c) Joint Coordinate Vector

Joint number

Restraint in *Y* direction (0 = free, 1 = restrained)

Rotational restraint (0 = free, 1 = restrained)

$$\mathbf{MSUP} = \begin{bmatrix} 1 & 1 & 0 \\ 2 & 1 & 0 \\ 3 & 1 & 0 \\ 4 & 1 & 1 \end{bmatrix}$$

$NS \times (NCJT + 1)$

(d) Support Data Matrix

Fig. 5.19

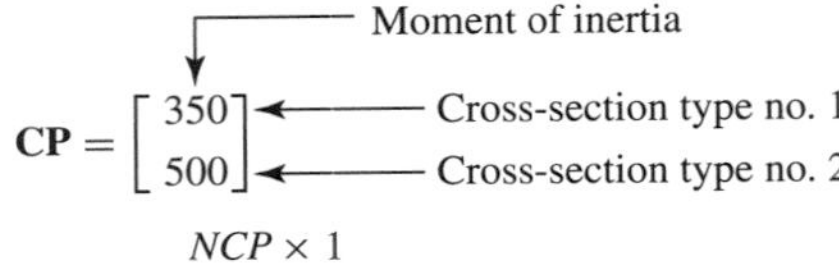

(e) Elastic Modulus Vector

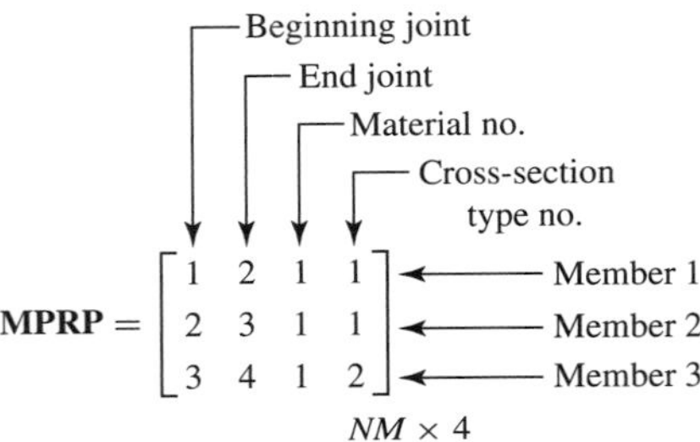

(f) Cross-Sectional Property Vector

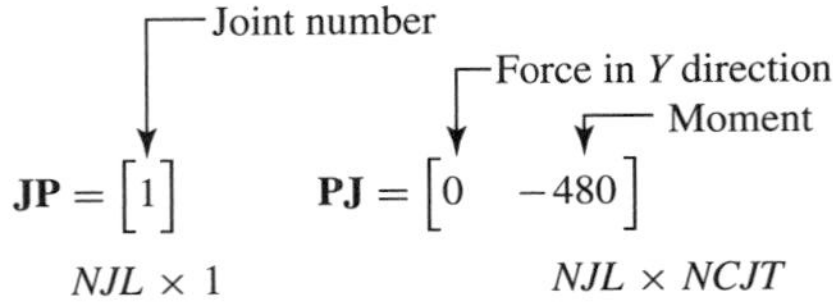

(g) Member Data Matrix

(h) Joint Load Data Matrices

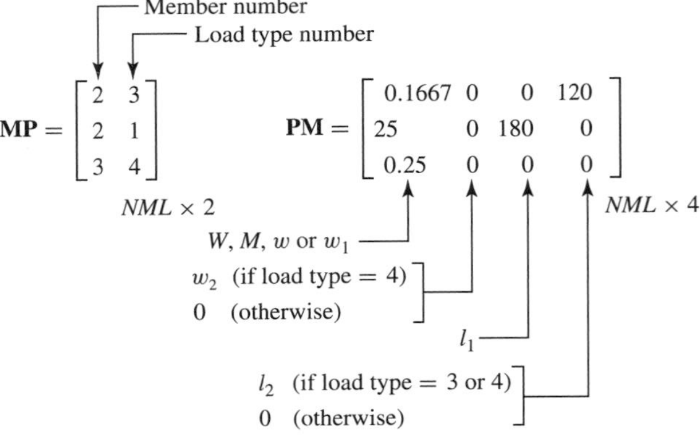

(i) Member Load Data Matrices

Fig. 5.19 *(continued)*

and storing of the joint data for beams is given in Fig. 5.20(a), and an example of the input data file for the beam of Fig. 5.19(b) is shown in Fig. 5.21 on page 229. Note that the first line of this data file contains the total number of joints of the beam (i.e., 4), with the next four lines containing the X coordinates of joints 1 through 4, respectively.

Support Data The support data consists of (a) the number of joints that are attached to supports (NS), and (b) the joint number, and the restraint code, for each support joint. Since the number of structure coordinates per joint of a beam equals 2 (i.e., $NCJT = 2$), a two-digit code is used to specify the restraints at a support joint. The first digit of the code represents the restraint condition at the joint in the global Y direction; it equals 0 if the joint is free to translate in the Y direction, or it equals 1 if the joint is restrained in the Y direction. Similarly, the second digit of the code represents the rotational restraint condition at the joint; a 0 indicates that the joint is free to rotate, and a 1 indicates that it is restrained against rotation. The restraint codes for the various types of supports for beams are given in Fig. 5.22 on page 230. Since the joints 1, 2, and 3 of the example beam (Fig. 5.19(b)) are attached to roller supports, their restraint codes are 1,0, indicating that these joints are restrained from translating in the Y direction, but are free to rotate. Similarly, the restraint code for joint 4, which is attached to a fixed support, is 1,1, because this joint can neither translate nor rotate. The support data for beams is stored in computer

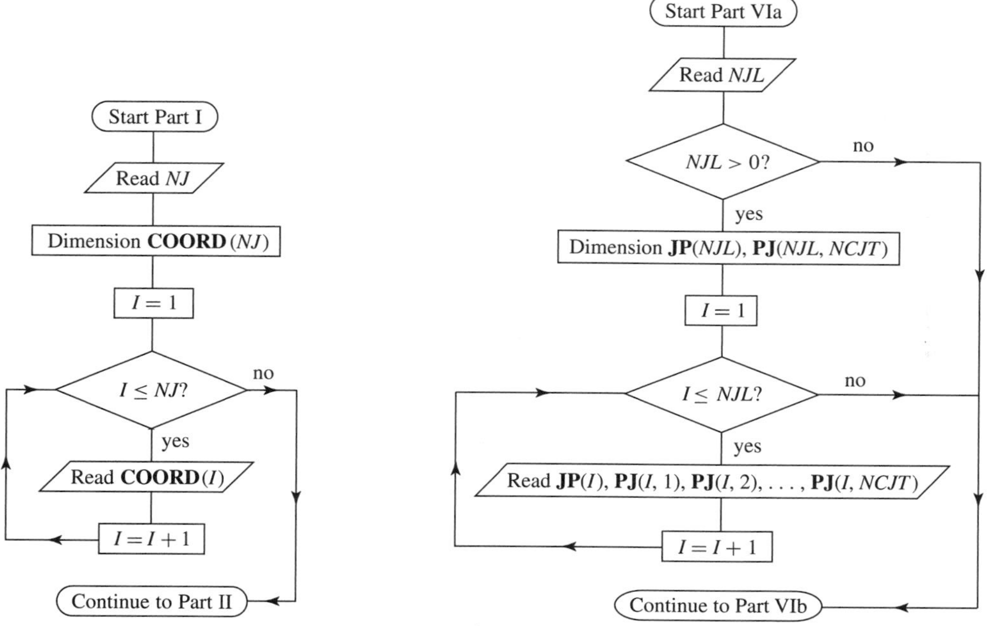

(a) Flowchart for Reading and Storing Joint Data for Beams

(b) Flowchart for Reading and Storing Joint Load Data

Fig. 5.20

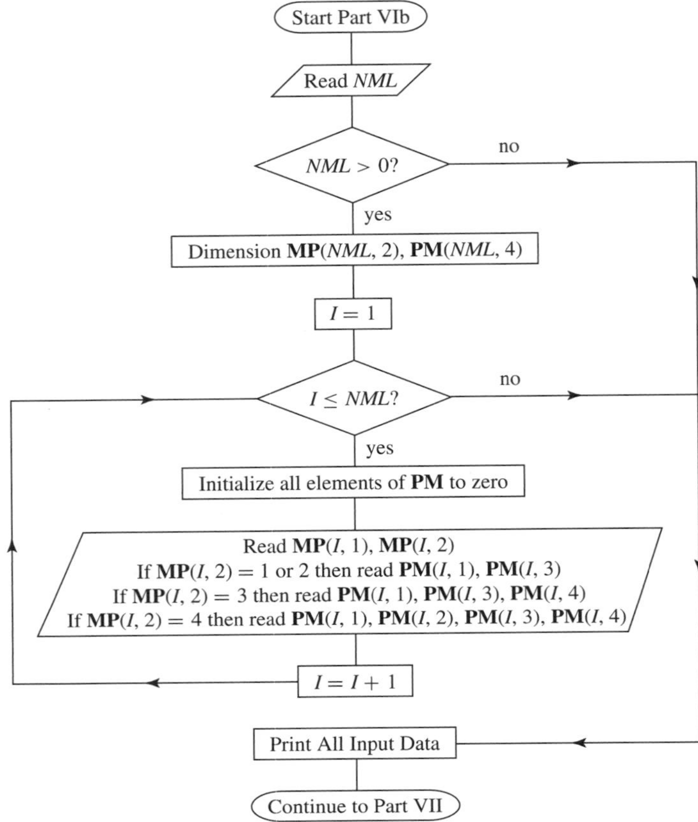

(c) Flowchart for Reading and Storing Member Load Data

Fig. 5.20 *(continued)*

memory as an integer matrix **MSUP** of order $NS \times (NCJT + 1)$, as discussed in Chapter 4 for the case of plane trusses. Thus, for the beam of Fig. 5.19(b), which has four support joints, **MSUP** is a 4×3 matrix, as shown in Fig. 5.19(d). The computer code developed previously for Part II of the plane truss analysis program (see flowchart in Fig. 4.3(b)) can be copied and used in the beam analysis program for reading the support data, and storing it in computer memory. An example of how the support data for beams may appear in an input data file is given in Fig. 5.21.

Material Property Data The procedure for inputting material property data for beams is identical to that for the case of plane trusses, as described in Chapter 4. Thus, the computer code written for Part III of the plane truss program (see flowchart in Fig. 4.3(c)) can be used in the beam analysis program for inputting the material property data. The elastic modulus vector for the example beam of Fig. 5.19(b) is given in Fig. 5.19(e); Fig. 5.21 illustrates how this type of data may appear in an input data file.

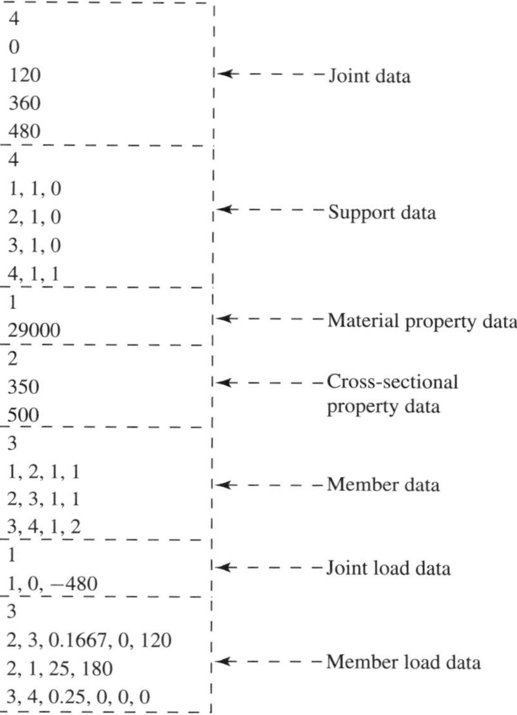

Fig. 5.21 *An Example of an Input Data File*

Cross-Sectional Property Data The cross-sectional property data consists of (a) the number of different cross-section types used for the members of the beam (NCP), and (b) the moment of inertia (I) for each cross-section type. The moments of inertia are stored in computer memory in a cross-sectional property vector **CP** of order $NCP \times 1$, with the moment of inertia of cross-section i stored in the ith row of **CP**. For example, two types of member cross-sections are used for the beam of Fig. 5.19(b). We arbitrarily assign the numbers 1 and 2 to the cross-sections with the moments of inertia of 350 and 500 in.^{4}, respectively. Thus, the **CP** vector consists of two rows, with the moments of inertia of cross-section types 1 and 2 stored in rows 1 and 2, respectively, as shown in Fig. 5.19(f). The computer code developed in Part IV of the plane truss program (see flowchart in Fig. 4.3(d)) can be used for inputting cross-sectional property data for beams. An example of how this type of data may appear in an input data file is given in Fig. 5.21.

Member Data As in the case of plane trusses, the member data for beams consists of (a) the total number of members (NM) of the beam, and (b) for each member: the beginning joint number, the end joint number, the material number, and the cross-section type number. This member data is organized in

Type of Support		Restraint Code
Free joint (no support)		0, 0
Roller or hinge	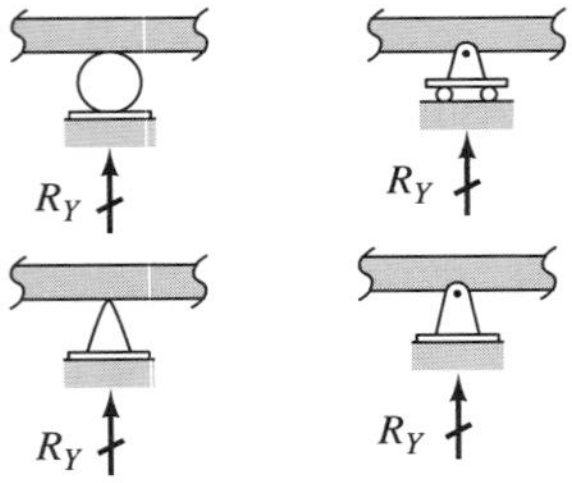 R_Y R_Y R_Y R_Y	1, 0
Support which prevents rotation, but not translation in Y direction; for example, a collar on a smooth shaft	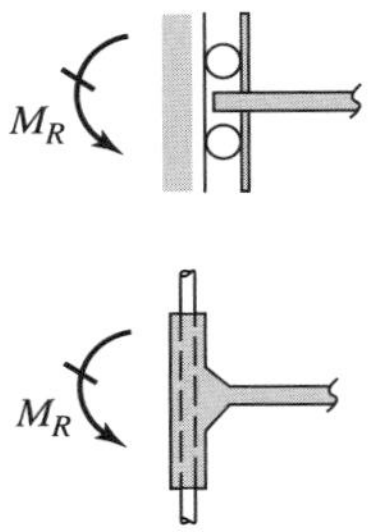 M_R M_R	0, 1
Fixed	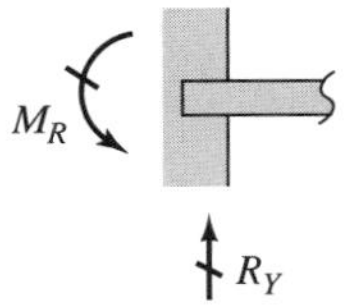 M_R R_Y	1, 1

Fig. 5.22 *Restraint Codes for Beams*

computer memory in the form of an integer member data matrix, **MPRP**, of order $NM \times 4$, as discussed in Chapter 4. The computer code for Part V of the plane truss program (see flowchart in Fig. 4.3(e)) can be used for inputting member data for beams. The **MPRP** matrix for the example beam is shown in Fig. 5.19(g), with the corresponding input data file given in Fig. 5.21.

Joint Load Data The joint load data involves (a) the number of joints that are subjected to external loads (NJL), and (b) for each loaded joint, the joint number, and the magnitudes of the force in the global Y direction and the couple. As in the case of plane trusses (Chapter 4), the numbers of the loaded joints are stored in an integer vector **JP** of order $NJL \times 1$, with the corresponding force and couple being stored in the first and second columns, respectively, of a real matrix **PJ** of order $NJL \times NCJT$ (with $NCJT = 2$ for beams). The joint load matrices for the example beam of Fig. 5.19(b) are shown in Fig. 5.19(h). A flowchart for programming the input of joint load data is given in Fig. 5.20(b); and Fig. 5.21 shows the joint load data for the example beam in an input file that can be read by the program.

Member Load Data The member load data consists of (a) the total number of loads applied to the members of the beam (*NML*), and (b) for each member load: the member number, the load type, and the magnitude(s) and location(s) of the load. The four common types of member loads for beams are depicted as load types 1 through 4 inside the front cover of this book, along with the expressions for the corresponding member fixed-end forces. The total number of member loads, *NML*, represents the sum of the different loads acting on the individual members of the structure. From Fig. 5.19(b), we can see that member 1 of the example beam is not subjected to any loads, whereas member 2 is subjected to two loads—namely, a uniformly distributed load (type 3) and a concentrated load (type 1). Also, member 3 of the beam is subjected to one load—a linearly varying load (type 4). Thus, the beam is subjected to a total of three member loads; that is, $NML = 3$. For each member load, the member number and the load type are stored in the first and second columns, respectively, of an integer matrix **MP** of order $NML \times 2$, with the corresponding load magnitude(s) and location(s) being stored in a real matrix **PM** of order $NML \times 4$. With reference to the load types depicted inside the front cover: when the load type is 1 or 2, the magnitude of W or M is stored in the first column, and the distance l_1 is stored in the third column, of the **PM** matrix, with the elements of the second and fourth columns of **PM** left blank (or set equal to 0). In the case of load type 3, the magnitude of w is stored in the first column, and distances l_1 and l_2 are stored in the third and fourth columns, respectively, of the **PM** matrix, with the second column element left blank. When the load type is 4, the magnitudes of w_1 and w_2 are stored in the first and second columns, respectively, and the distances l_1 and l_2 are stored in the third and fourth columns, respectively, of the **PM** matrix. For example, as the beam of Fig. 5.19(b) is subjected to three member loads, its *member load-data matrices,* **MP** and **PM**, are of the orders 3×2 and 3×4, respectively, as shown in Fig. 5.19(i). The first rows of these matrices contain information about the first member load, which is arbitrarily chosen to be the uniformly distributed load acting on member 2. Thus, the first row of **MP** contains the member number, 2, and the load type, 3, stored in the first and second columns; and the first row of **PM** contains $w = 0.1667$ in column 1, 0 in column 2, $l_1 = 0$ in column 3, and $l_2 = 120$ in column 4. The information about the second member load—the concentrated load acting on member 2—is then stored in the second rows of **MP** and **PM**; with the member number 2 and the load type 1 stored in the first and second columns of **MP**, and $W = 25$ and $l_1 = 180$ stored in the first and third columns of **PM**. Similarly, the third member load—the linearly varying load on member 3—is defined in the third rows of **MP** and **PM**; with the member number 3 and the load type 4 stored in the first and second columns of **MP**, and $w_1 = 0.25$, $w_2 = 0$, $l_1 = 0$, and $l_2 = 0$ stored in columns 1 through 4, respectively, of **PM**, as shown in Fig. 5.19(i). It is important to realize that the member fixed-end force expressions given inside the front cover are based on the sign convention that the member loads W, w, w_1, and w_2 are positive when acting downward (i.e., in the negative direction of the member y axis), and the couple M is positive when clockwise. A flowchart for programming the input of member load data is given in Fig. 5.20(c); Fig. 5.21 shows the member load data in an input file that can be read by the program.

An example of a computer printout of the input data for the beam of Fig. 5.19 is given in Fig. 5.23.

```
**********************************
*        Computer Software       *
*              for               *
*  MATRIX ANALYSIS OF STRUCTURES *
*         Second Edition         *
*               by               *
*         Aslam Kassimali        *
**********************************

        =======================
        General Structural Data
        =======================

Project Title: Figure 5-19
Structure Type: Beam
Number of Joints: 4
Number of Members: 3
Number of Material Property Sets (E): 1
Number of Cross-Sectional Property Sets: 2

             =================
             Joint Coordinates
             =================

       Joint No.        X Coordinate
           1             0.0000E+00
           2             1.2000E+02
           3             3.6000E+02
           4             4.8000E+02

                ========
                Supports
                ========

Joint No.     Y Restraint     Rotational Restraint
    1             Yes                   No
    2             Yes                   No
    3             Yes                   No
    4             Yes                   Yes
```

Fig. 5.23 *A Sample Printout of Input Data*

```
Material Properties
```

Material No.	Modulus of Elasticity (E)	Co-efficient of Thermal Expansion
1	2.9000E+04	0.0000E+00

```
Cross-Sectional Properties
```

Property No.	Moment of Inertia
1	3.5000E+02
2	5.0000E+02

```
Member Data
```

Member No.	Beginning Joint	End Joint	Material No.	Cross-Sectional Property No.
1	1	2	1	1
2	2	3	1	1
3	3	4	1	2

```
Joint Loads
```

Joint No.	Y Force	Moment
1	0.0000E+00	-4.8000E+02

```
Member Loads
```

Member No.	Load Type	Load Magnitude (W or M) or Intensity (w or w1)	Load Intensity w2	Distance l1	Distance l2
2	Conc.	2.500E+1	---	1.80E+2	----
2	Uniform	1.667E-1	---	0.00E+0	1.20E+2
3	Linear	2.500E-1	0.000E+0	0.00E+0	0.00E+0

```
************** End of Input Data **************
```

Fig. 5.23 *(continued)*

Analysis Module

Assignment of Structure Coordinate Numbers The process of programming the determination, for beams, of the number of degrees of freedom, *NDOF*, and the formation of the structure coordinate number vector, **NSC**, is identical to that for plane trusses. Thus, Parts VII and VIII of the plane truss program (as described by flowcharts in Figs. 4.8(a) and (b)) can be copied and used without any modifications in the program for the analysis of beams.

Generation of the Structure Stiffness Matrix and the Equivalent Joint Load Vector A flowchart for programming this part of our computer program is presented in Fig. 5.24. As the flowchart indicates, this part of the program

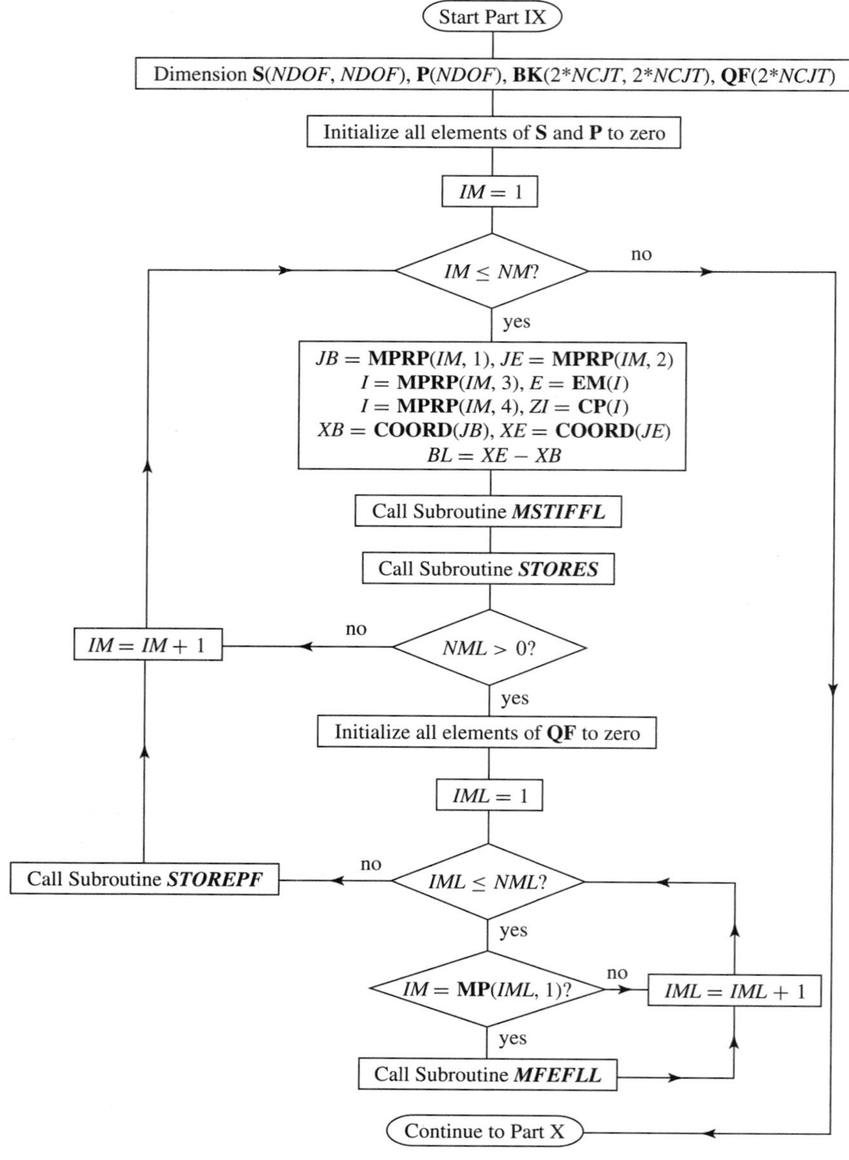

Fig. 5.24 *Flowchart for Generating Structure Stiffness Matrix and Equivalent Joint Load Vector for Beams*

begins by initializing all the elements of the **S** matrix, and a structure load vector **P** of order *NDOF* $\times$ 1, to 0. The assembly of the structure stiffness matrix, and the equivalent joint load vector due to member loads, is then carried out by using a *Do Loop,* in which the following operations are performed for each member of the beam: (a) For the member under consideration, *IM*, the program reads the modulus of elasticity E and the moment of inertia *ZI*, and calculates the member length *BL*. (b) Next, the program calls the subroutine ***MSTIFFL*** to form the member stiffness matrix **BK** (= **k**). As the flowchart in Fig. 5.25 indicates, this subroutine simply calculates the values of the various elements of the **BK** matrix, in accordance with Eq. (5.53). (c) The program then calls the subroutine ***STORES*** to store the pertinent elements of **BK** in their proper positions in the structure stiffness matrix **S**. A flowchart of this subroutine is given in Fig. 5.26 on the next page. By comparing the flowchart of the present ***STORES*** subroutine (Fig. 5.26) with that of the ***STORES*** subroutine of the plane truss program in Fig. 4.11, we can see that the two subroutines are identical, except that the present subroutine stores the elements of the member local stiffness matrix **BK** (instead of the global stiffness matrix **GK**) in **S**. (d) Returning our attention to Fig. 5.24, we can see that after the ***STORES*** subroutine has been executed, the program checks the first column of the member load data matrix **MP** to determine whether the member under consideration, *IM*, is subjected to any loads. If the member is subjected to loads, then the subroutine ***MFEFLL*** is called to form the member fixed-end force vector **QF** (= $\mathbf{Q}_f$). As the flowchart in Fig. 5.27 on page 237 indicates, this subroutine calculates the values of the member fixed-end forces, for load types 1 through 4, using the equations given inside

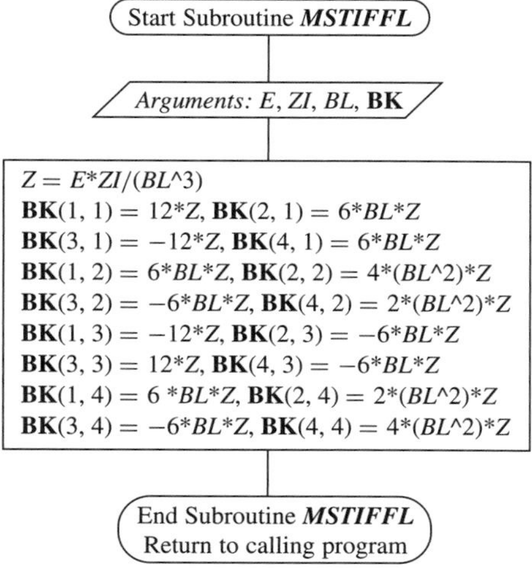

Fig. 5.25 *Flowchart of Subroutine* ***MSTIFFL*** *for Determining Member Stiffness Matrix for Beams*

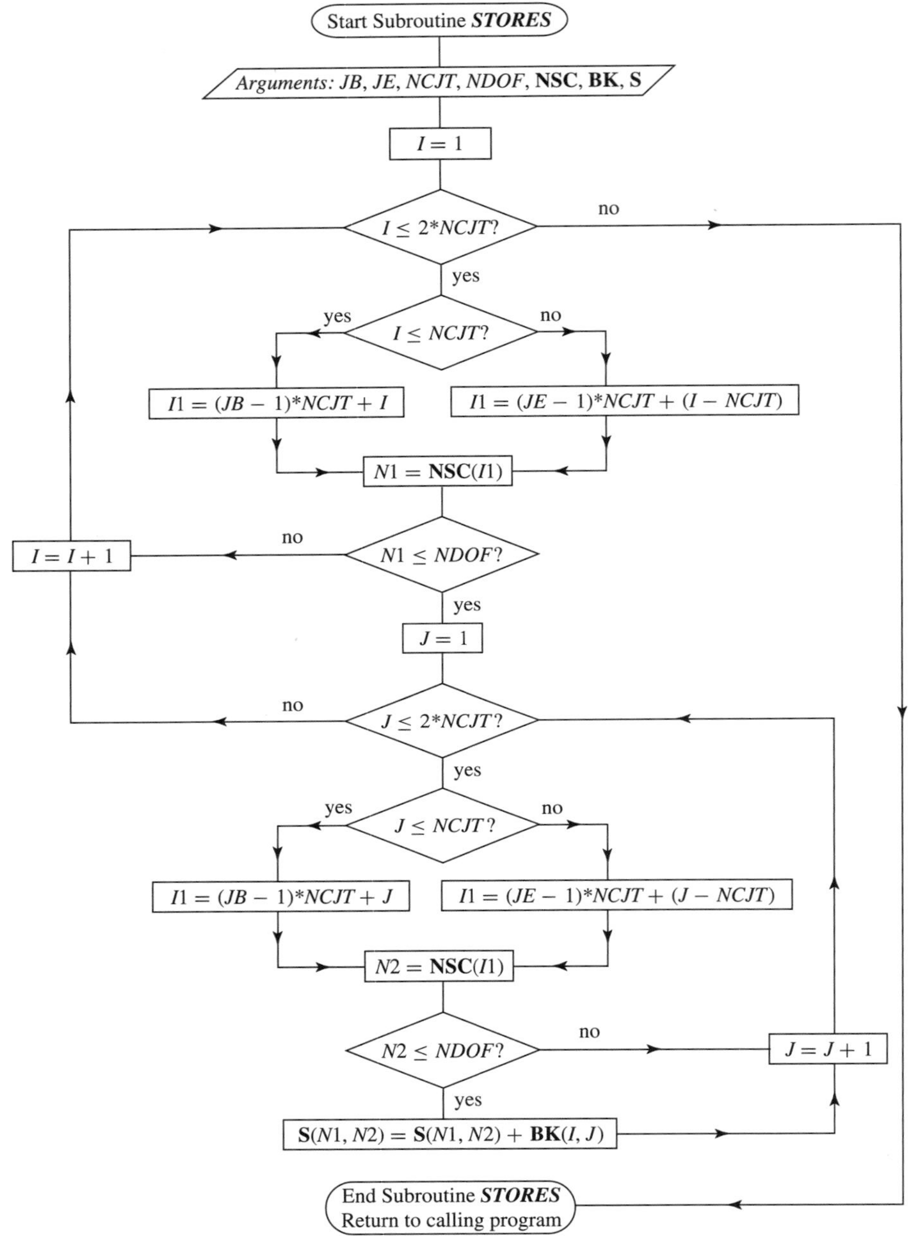

Fig. 5.26 *Flowchart of Subroutine* ***STORES*** *for Storing Member Stiffness Matrix in Structure Stiffness Matrix for Beams*

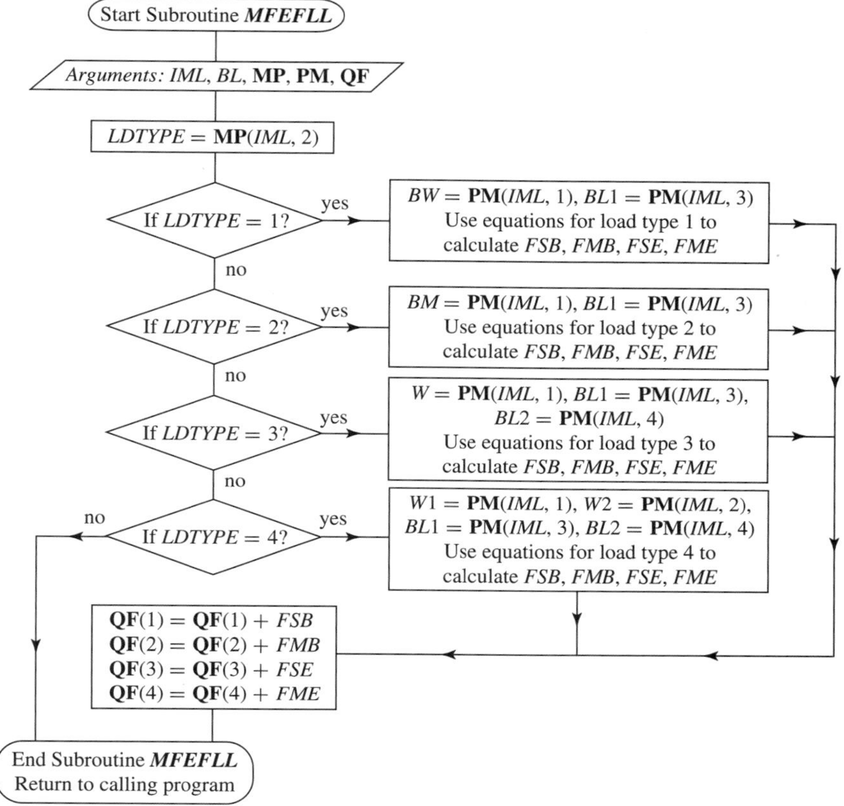

Fig. 5.27 *Flowchart of Subroutine* ***MFEFLL*** *for Determining Member Fixed-End Force Vector for Beams*

the front cover. (e) The program then calls the subroutine ***STOREPF*** to store the *negative* values of the pertinent elements of **QF** in their proper positions in the load vector **P**. A flowchart of this subroutine, which essentially consists of a *Do Loop,* is given in Fig. 5.28 on the next page. As shown in this flowchart, the subroutine reads, in order, for each of the member fixed-end forces, QF_I, the number of the corresponding structure coordinate, $N1$, from the **NSC** vector. If $N1$ is less than or equal to *NDOF*, then the value of QF_I is subtracted from the $N1$th row of the load vector **P**. From Fig. 5.24, we can see that when the foregoing operations have been performed for each member of the beam, the structure stiffness matrix **S** is completed, and the structure load vector **P** equals the equivalent joint load vector $\mathbf{P}_e$, or the negative of the structure fixed-joint force vector $\mathbf{P}_f$ (i.e., $\mathbf{P} = \mathbf{P}_e = -\mathbf{P}_f$).

Storage of the Joint Loads into the Structure Load Vector In this part of our computer program, the joint loads are added to the structure load vector **P**.

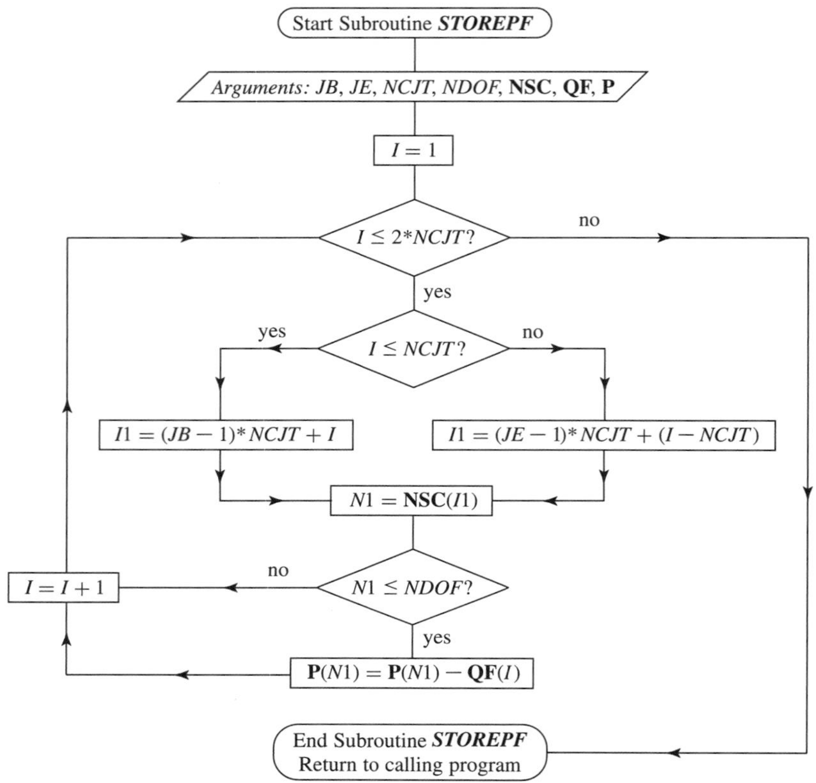

Fig. 5.28 *Flowchart of Subroutine **STOREPF** for Storing Member Fixed-End Force Vector in Structure Load Vector for Beams*

A flowchart for programming this process is shown in Fig. 5.29. This flowchart is the same as the previous flowchart (Fig. 4.12) for forming the joint load vector for plane trusses, except that the load vector **P** is not initialized to 0 in this part of the program (as it was previously), because it now contains the equivalent joint loads due to member loads.

Solution for Joint Displacements In this part, the program solves the system of simultaneous equations representing the beam's stiffness relationship, **Sd** = **P**, using Gauss–Jordan elimination. The programming of this process has been discussed previously (see the flowchart in Fig. 4.13), and it may be recalled that, upon completion of the Gauss–Jordan elimination process, the vector **P** contains the values of the joint displacements **d**. The computer code developed in Chapter 4 for Part XI of the plane truss program can be transported, without any alteration, into the beam analysis program for the calculation of joint displacements.

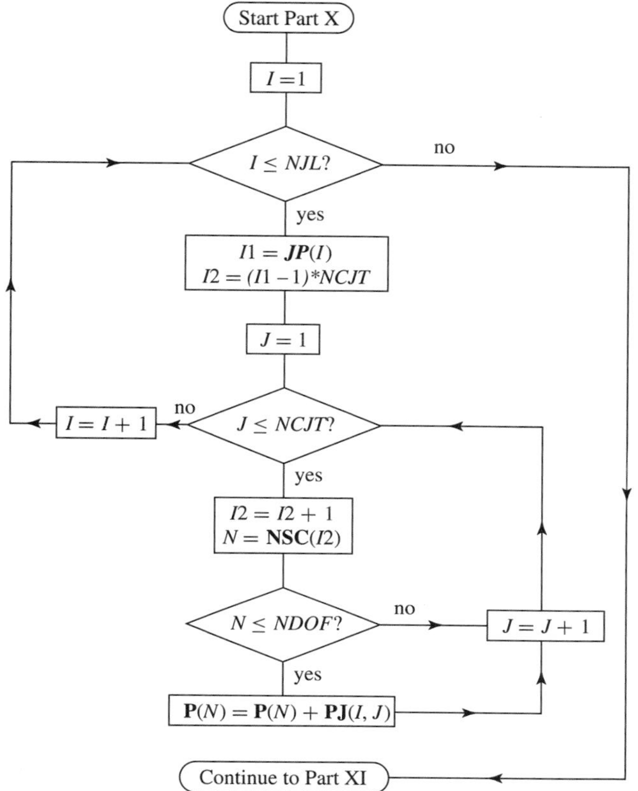

Fig. 5.29 *Flowchart for Storing Joint Loads in Structure Load Vector*

Calculation of Member Forces and Support Reactions The last part of our program involves the calculation of member forces and support reactions. A flowchart for programming this process is given in Fig. 5.30 on the next page. As this figure indicates, after initializing the reaction vector **R** to 0, the program uses a *Do Loop* to perform the following operations for each member of the beam: (a) For the member under consideration, *IM*, the program reads the modulus of elasticity *E* and the moment of inertia *ZI*, and calculates the member length *BL*. (b) Next, the program calls the subroutine ***MDISPL*** to obtain the member end displacements **U** (= **u**) from the joint displacements **P** (= **d**), using the member code numbers, as depicted by the flowchart in Fig. 5.31 on page 241. (c) The program then calls the subroutine ***MSTIFFL*** (Fig. 5.25) to form the member stiffness matrix **BK** (= **k**). (d) Returning our attention to Fig. 5.30, we can see that the program then initializes the **QF** vector to 0, and checks the first column of the member load matrix **MP** to determine if the member *IM* is subjected to any loads. If the member is subjected to loads, then the subroutine ***MFEFLL*** (Fig. 5.27) is used to form the fixed-end force vector **QF**. (e) Next, the program

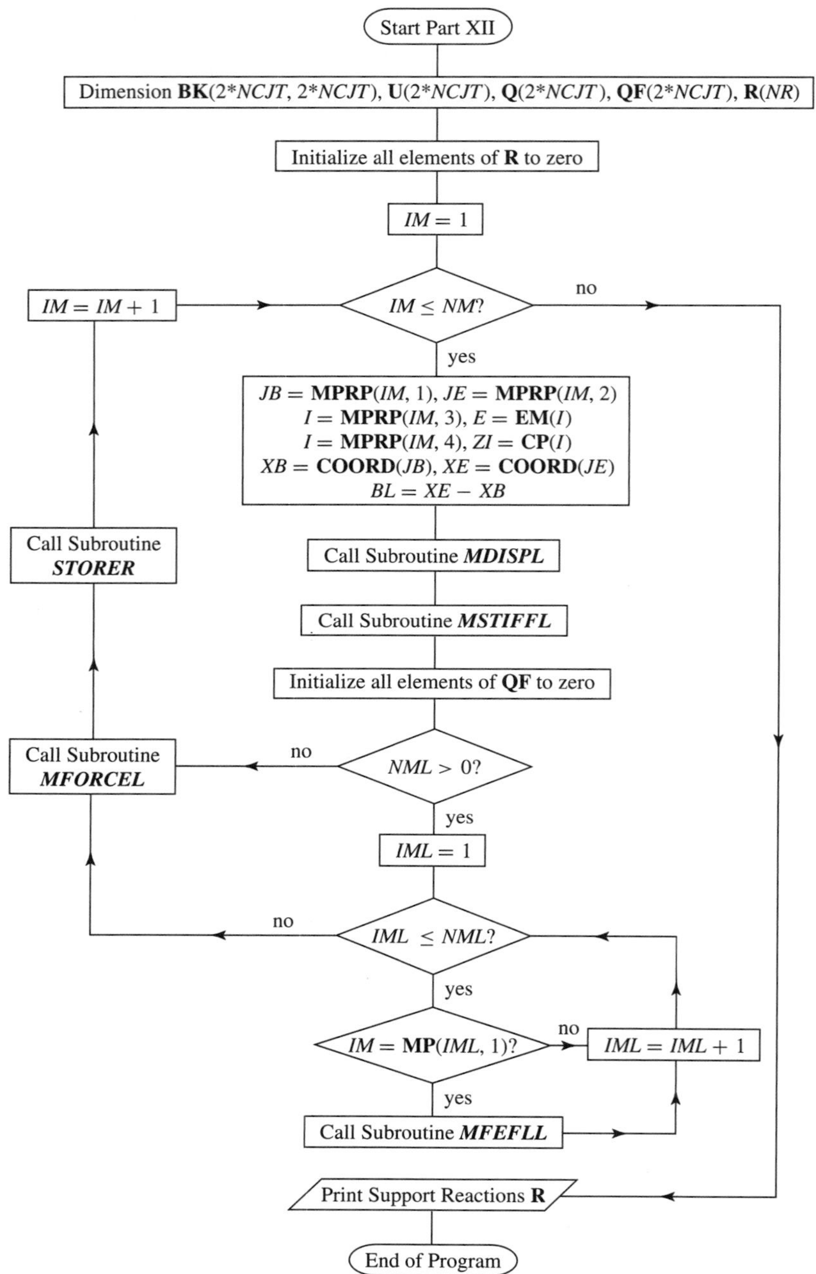

Fig. 5.30 *Flowchart for Determination of Member Forces and Support Reactions for Beams*

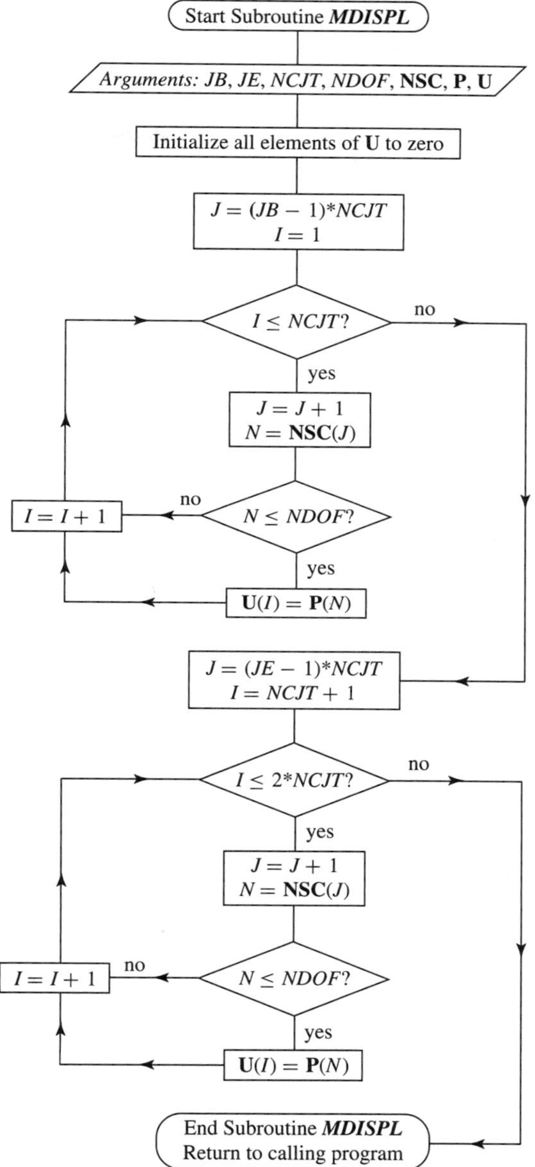

Fig. 5.31 *Flowchart of Subroutine **MDISPL** for Determining Member Displacement Vector for Beams*

calls the subroutine ***MFORCEL*** to evaluate the member end forces **Q**, using the relationship $\mathbf{Q} = \mathbf{BK}\ \mathbf{U} + \mathbf{QF}$ (i.e., $\mathbf{Q} = \mathbf{ku} + \mathbf{Q}_f$, see Eq. (5.4)). A flowchart of this subroutine is shown in Fig. 5.32. (f) The program then stores the pertinent elements of **Q** in the support reaction vector **R**, using the subroutine ***STORER***. The present ***STORER*** subroutine, whose flowchart is given in

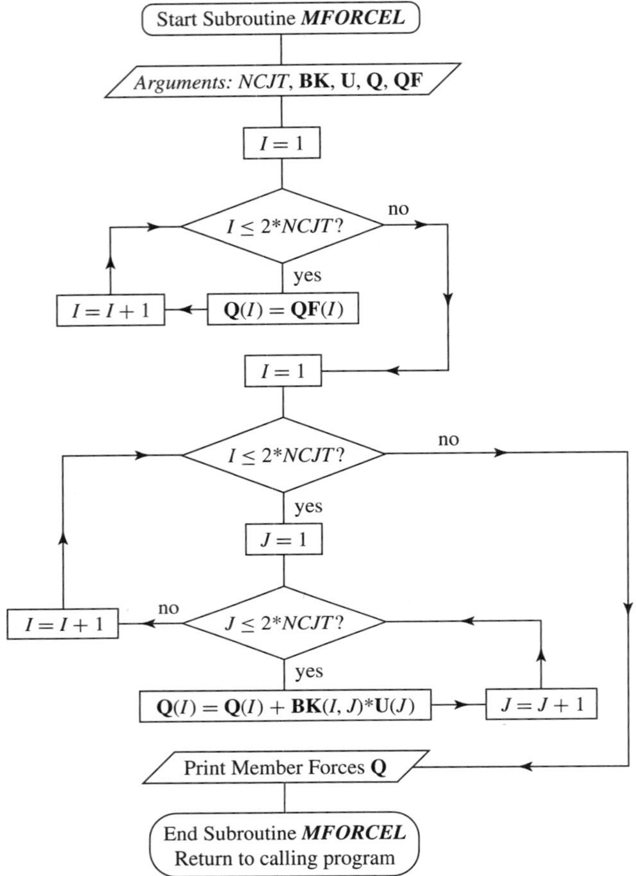

Fig. 5.32 *Flowchart of Subroutine **MFORCEL** for Determining Member Local Force Vector*

Fig. 5.33, is identical to the ***STORER*** subroutine of the plane truss program (Fig. 4.21), except that the present subroutine stores the elements of the member local force vector **Q** (instead of the global force vector **F**) in **R**. A sample computer printout, showing the results of the analysis of the example beam of Fig. 5.19, is given in Fig. 5.34.

Finally, the entire program for the analysis of beams is summarized in Table 5.1. As shown in this table, the program consists of a main program, divided into twelve parts, and seven subroutines. Brief descriptions of the various parts and subroutines of the program are also provided in Table 5.1 for quick reference. It should be noted that seven parts of the main program can be obtained from the plane truss computer program developed in Chapter 4. Furthermore, the computer code for many of the remaining parts of the main program, as well as the subroutines, can be conveniently developed by modifying the computer code written previously for the corresponding part or subroutine of the plane truss program.

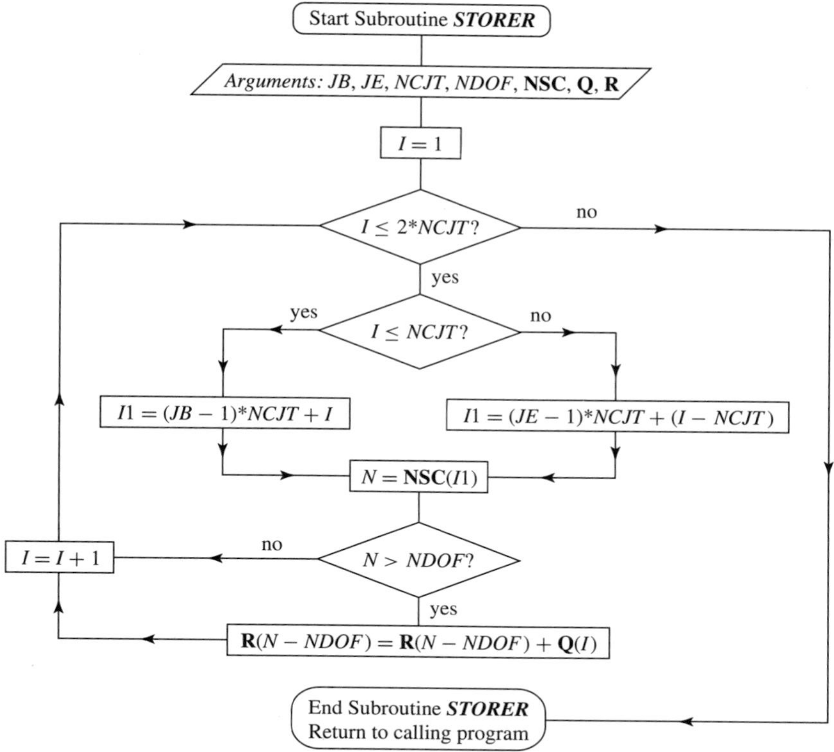

Fig. 5.33 *Flowchart of Subroutine **STORER** for Storing Member Forces in Support Reaction Vector for Beams*

```
****************************************************
*                Results of Analysis               *
****************************************************

                 =====================
                 Joint Displacements
                 =====================

 Joint No.        Y Translation        Rotation (Rad)
 ---------        -------------        --------------
     1             0.0000E+00           -5.5719E-04
     2             0.0000E+00           -1.7231E-03
     3             0.0000E+00            1.6238E-03
     4             0.0000E+00            0.0000E+00
```

Fig. 5.34 *A Sample Printout of Analysis Results*

```
Member End Forces in Local Coordinates

Member       Joint      Shear Force       Moment
  1            1        -9.6435E+00    -4.8000E+02
               2         9.6435E+00    -6.7722E+02

  2            2         2.0055E+01     6.7722E+02
               3         2.4949E+01    -9.6485E+02

  3            3         2.0311E+01     9.6485E+02
               4        -5.3106E+00     2.7242E+02

                 Support Reactions

Joint No.             Y Force            Moment
    1              -9.6435E+00        0.0000E+00
    2               2.9698E+01        0.0000E+00
    3               4.5260E+01        0.0000E+00
    4              -5.3106E+00        2.7242E+02
**************** End of Analysis ****************
```

Fig. 5.34 *(continued)*

Table 5.1 *Computer Program for Analysis of Beams*

Main program part	Description
I	Reads and stores joint data (Fig. 5.20(a))
II	Reads and stores support data (Fig. 4.3(b))
III	Reads and stores material properties (Fig. 4.3(c))
IV	Reads and stores cross-sectional properties (Fig. 4.3(d))
V	Reads and stores member data (Fig. 4.3(e))
VIa	Reads and stores joint loads (Fig. 5.20(b))
VIb	Reads and stores member loads (Fig. 5.20(c))
VII	Determines the number of degrees of freedom *NDOF* of the structure (Fig. 4.8(a))
VIII	Forms the structure coordinate number vector **NSC** (Fig. 4.8(b))
IX	Generates the structure stiffness matrix **S** and the structure load vector $\mathbf{P} = \mathbf{P}_e = -\mathbf{P}_f$ due to member loads (Fig. 5.24) Subroutines called: ***MSTIFFL***, ***STORES***, ***MFEFLL***, and ***STOREPF***
X	Stores joint loads in the structure load vector **P** (Fig. 5.29)
XI	Calculates the structure joint displacements by solving the stiffness relationship, $\mathbf{Sd} = \mathbf{P}$, using Gauss–Jordan elimination. The vector **P** now contains joint displacements (Fig. 4.13).

(continued)

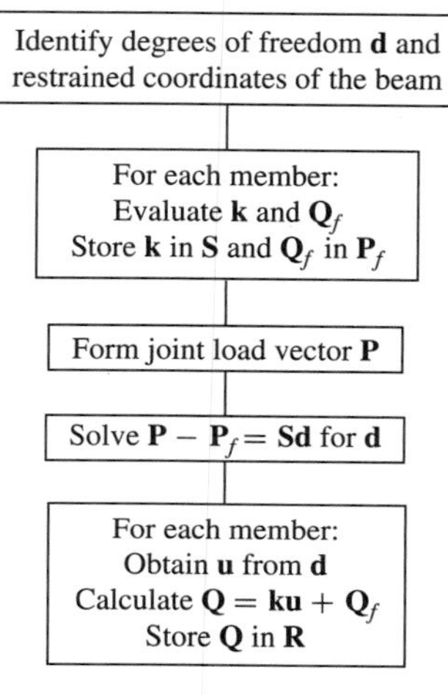

Identify degrees of freedom **d** and restrained coordinates of the beam
For each member: Evaluate **k** and $\mathbf{Q}_f$ Store **k** in **S** and $\mathbf{Q}_f$ in $\mathbf{P}_f$
Form joint load vector **P**
Solve $\mathbf{P} - \mathbf{P}_f = \mathbf{Sd}$ for **d**
For each member: Obtain **u** from **d** Calculate $\mathbf{Q} = \mathbf{ku} + \mathbf{Q}_f$ Store **Q** in **R**

Fig. 5.35

Table 5.1 *(continued)*

Main program part	Description
XII	Determines the member end force vector **Q**, and the support reaction vector **R** (Fig. 5.30). Subroutines called: ***MDISPL***, ***MSTIFFL***, ***MFEFLL***, ***MFORCEL***, and ***STORER***
Subroutine	**Description**
MDISPL	Determines the member displacement vector **U** from the joint displacement vector **P** (Fig. 5.31)
MFEFLL	Calculates the member fixed-end force vector **QF** (Fig. 5.27)
MFORCEL	Evaluates the member local force vector **Q** = **BK U** + **QF** (Fig. 5.32)
MSTIFFL	Forms the member stiffness matrix **BK** (Fig. 5.25)
STOREPF	Stores the negative values of the pertinent elements of the member fixed-end force vector **QF** in the structure load vector **P** (Fig. 5.28)
STORER	Stores the pertinent elements of the member force vector **Q** in the reaction vector **R** (Fig. 5.33)
STORES	Stores the pertinent elements of the member stiffness matrix **BK** in the structure stiffness matrix **S** (Fig. 5.26)

SUMMARY

In this chapter, we have developed the matrix stiffness method for the analysis of beams. A block diagram summarizing the various steps of the analysis is presented in Fig. 5.35.

PROBLEMS

Section 5.1

5.1 through 5.4 Identify by numbers the degrees of freedom and restrained coordinates of the beams shown in Figs. P5.1 through P5.4. Also, form the joint load vector **P** for the beams.

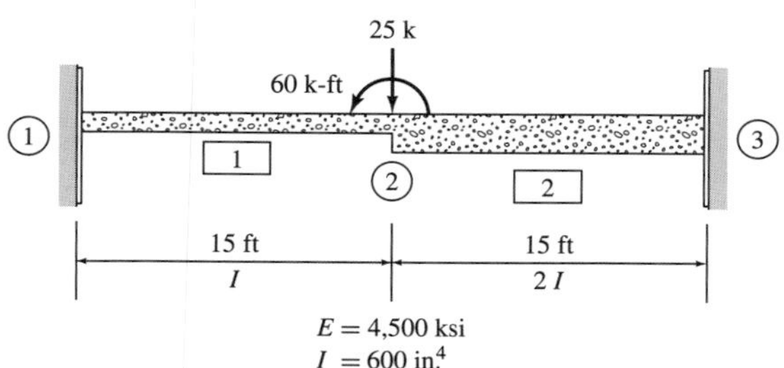

Fig. P5.1, P5.5, P5.19, P5.27

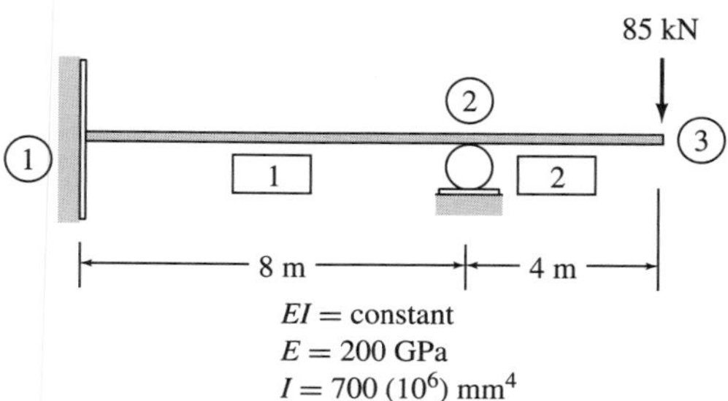

Fig. P5.2, P5.28

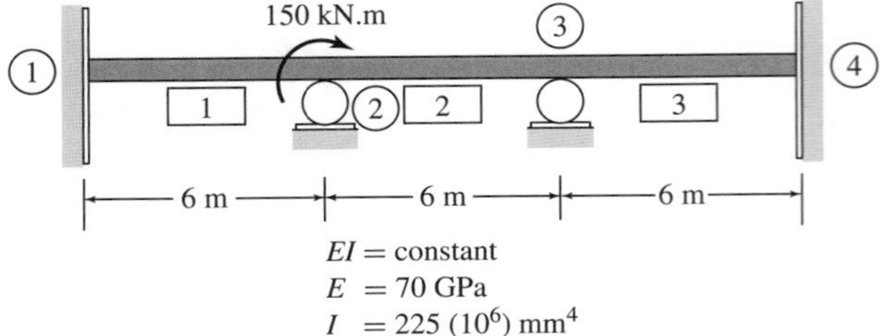

Fig. P5.3, P5.6, P5.20, P5.29

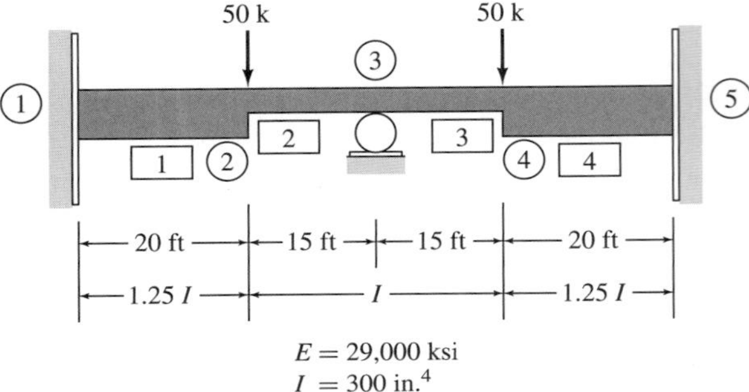

Fig. P5.4, P5.7, P5.21, P5.35

Section 5.2

5.5 through 5.8 Determine the stiffness matrices for the members of the beams shown in Figs. P5.5 through P5.8.

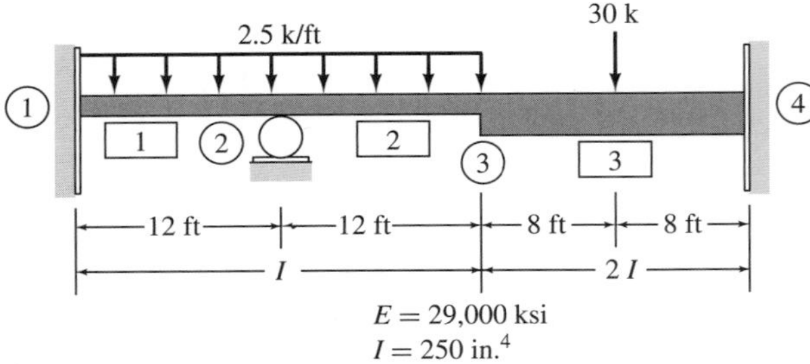

Fig. P5.8, P5.22, P5.23, P5.30

5.9 If the end displacements of member 1 of the beam shown in Fig. P5.9 are

$$\mathbf{u}_1 = \begin{bmatrix} 0 \\ 0 \\ -0.6667 \text{ in.} \\ -0.006667 \text{ rad} \end{bmatrix}$$

calculate the end forces for the member. Is the member in equilibrium under these forces?

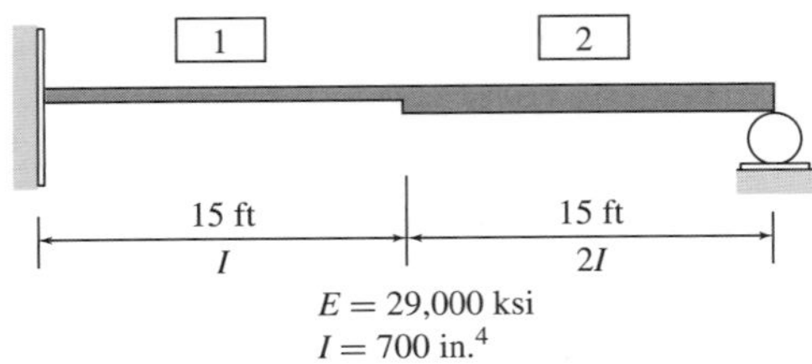

Fig. P5.9

5.10 If the end displacements of member 2 of the beam shown in Fig. P5.10 are

$$\mathbf{u}_2 = \begin{bmatrix} 0 \\ 0.08581 \text{ rad} \\ 0 \\ -0.08075 \text{ rad} \end{bmatrix}$$

calculate the end forces for the member. Is the member in equilibrium under these forces?

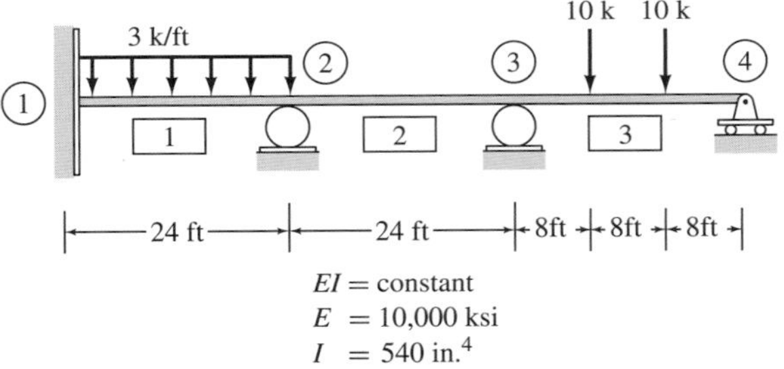

Fig. P5.10, P5.17, P5.24, P5.31

Section 5.4

5.11 through 5.14 Using the direct integration approach, derive the equations of fixed-end forces due to the member loads shown in Figs. P5.11 through P5.14. Check the results, using the fixed-end force expressions given inside the front cover.

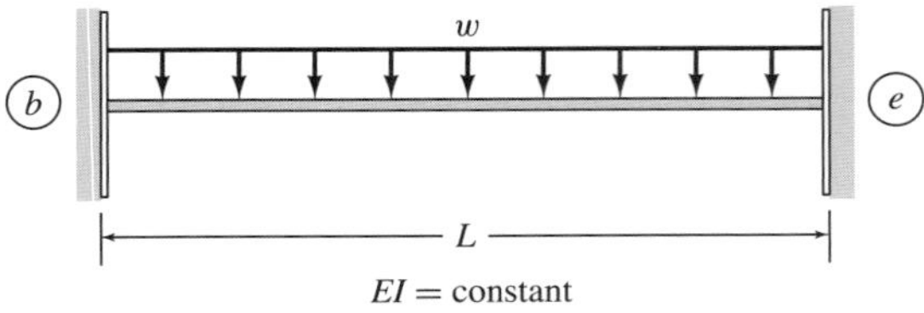

Fig. P5.11

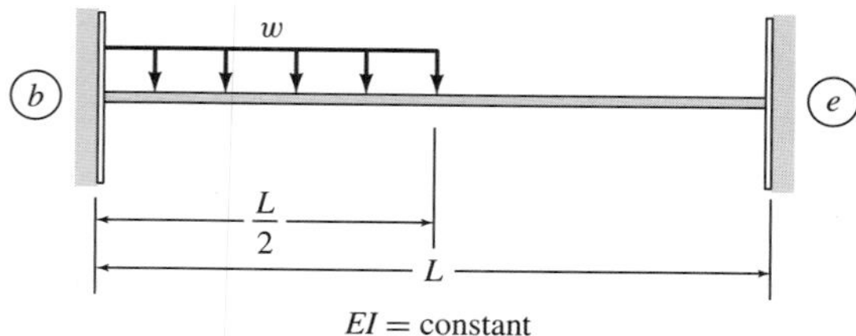

Fig. P5.12

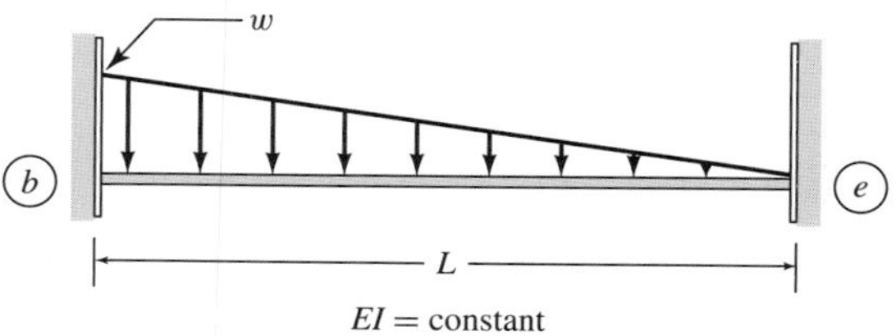

Fig. P5.13

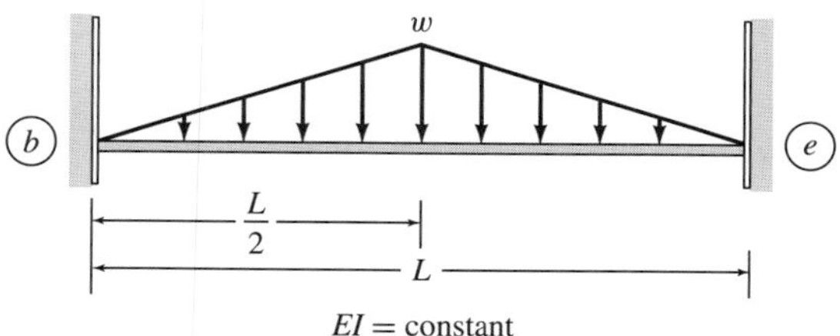

Fig. P5.14

5.15 and 5.16 Determine the fixed-end force vectors for the members of the beams shown in Figs. P5.15 and P5.16. Use the fixed-end force equations given inside the front cover.

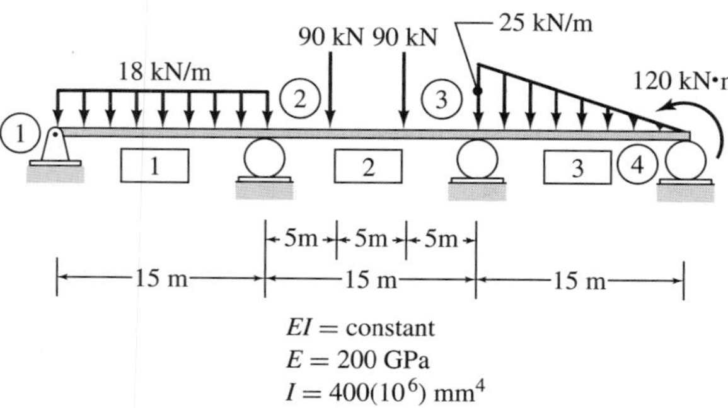

Fig. P5.15, P5.25, P5.33

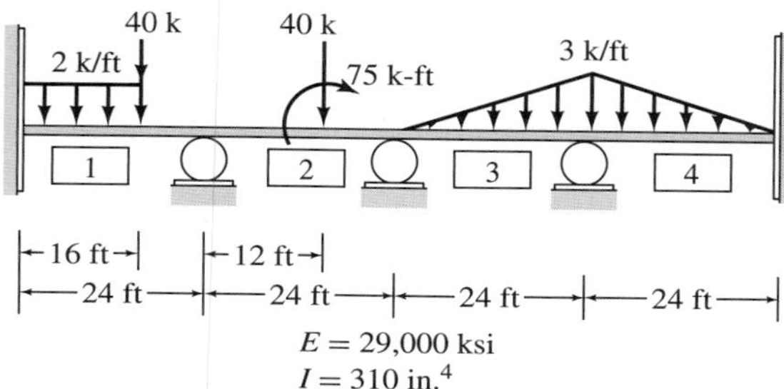

Fig. P5.16, P5.26, P5.32

5.17 If the end displacements of member 1 of the beam shown in Fig. P5.17 are

$$\mathbf{u}_1 = \begin{bmatrix} 0 \\ 0 \\ 0 \\ 0.08581 \text{ rad} \end{bmatrix}$$

calculate the end forces for the member. Is the member in equilibrium under these forces?

5.18 If the end displacements of member 2 of the beam shown in Fig. P5.18 are

$$\mathbf{u}_2 = \begin{bmatrix} -0.02532 \text{ m} \\ -0.00434 \text{ rad} \\ -0.02532 \text{ m} \\ 0.00434 \text{ rad} \end{bmatrix}$$

calculate the end forces for the member. Is the member in equilibrium under these forces?

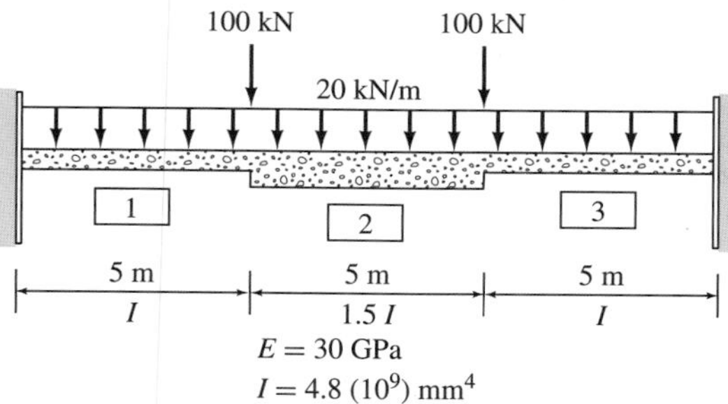

Fig. P5.18, P5.34

Section 5.5

5.19 through 5.22 Determine the structure stiffness matrices **S** for the beams shown in Figs. P5.19 through P5.22.

Section 5.6

5.23 through 5.26 Determine the fixed-joint force vectors and the equivalent joint load vectors for the beams shown in Figs. P5.23 through P5.26.

Section 5.7

5.27 through 5.35 Determine the joint displacements, member end forces, and support reactions for the beams shown in Figs. P5.27 through P5.35, using the matrix stiffness method. Check the hand-calculated results by using the computer program provided with this book, the publisher's website for this book (www.cengage.com/engineering), or by using any other general purpose structural analysis program available.

Section 5.8

5.36 Develop a general computer program for the analysis of beams by the matrix stiffness method. Use the program to analyze the beams of Problems 5.27 through 5.35, and compare the computer-generated results to those obtained by hand calculations.

6 PLANE FRAMES

Beekman Tower, New York
(Estormiz, Wikimedia Commons)

A plane frame is defined as *a two-dimensional assemblage of straight members connected together by rigid and/or hinged connections, and subjected to loads and reactions that lie in the plane of the structure.* Under the action of external loads, the members of a plane frame may be subjected to axial forces like the members of plane trusses, as well as bending moments and shears like the members of beams. Therefore, the stiffness relations for plane frame members can be conveniently obtained by combining the stiffness relations for plane truss and beam members.

The members of frames are usually connected by rigid connections, although hinged connections are sometimes used. In this chapter, we develop the analysis of rigidly connected plane frames based on the matrix stiffness method. The modifications in the method of analysis necessary to account for the presence of any hinged connections in the frame are considered in Chapter 7.

We begin, in Section 6.1, with a discussion of the process of developing an analytical model of the frame. We establish the force–displacement relations for the members of plane frames in their local coordinate systems in Section 6.2, where we also consider derivation of the member fixed-end axial forces due to external loads applied to the members. The transformation of member forces and displacements from a local to a global coordinate system, and vice versa, is considered in Section 6.3; and the member stiffness relations in the global coordinate system are developed in Section 6.4. The stiffness relations for the entire frame are formulated in Section 6.5, where the process of forming the structure fixed-joint force vectors, due to member loads, is also discussed. We then develop a step-by-step procedure for the analysis of plane frames in Section 6.6; finally, in Section 6.7, we cover the computer implementation of the procedure for analysis of plane frames.

6.1 ANALYTICAL MODEL

The process of dividing plane frames into members and joints, for the purpose of analysis, is the same as that for beams (Chapter 5); that is, *a plane frame is divided into members and joints so that:* (a) *all of the members are straight and prismatic, and* (b) *all the external reactions act only at the joints.* Consider, for example, the frame shown in Fig. 6.1(a). The analytical model of the frame is depicted in Fig. 6.1(b), which shows that, for the purpose of analysis, the frame is considered to be composed of four members and five joints. Note that because the member stiffness relationships to be used in the analysis are valid for prismatic members only, the left column of the frame has been subdivided into two members, each with constant cross-sectional properties (i.e., cross-sectional area and moment of inertia) along its length.

Global and Local Coordinate Systems

The global and local coordinate systems for plane frames are established in a manner similar to that for plane trusses (Chapter 3). The global coordinate system used for plane frames is a right-handed *XYZ* coordinate system with the frame lying in the *XY* plane, as shown in Fig. 6.1(b). It is usually convenient to

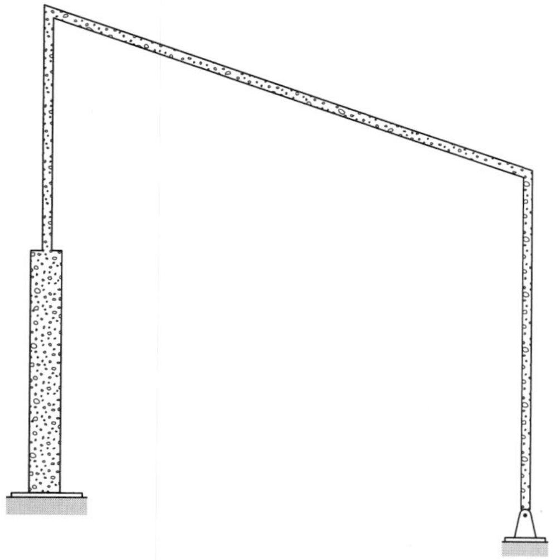

(a) Actual Frame

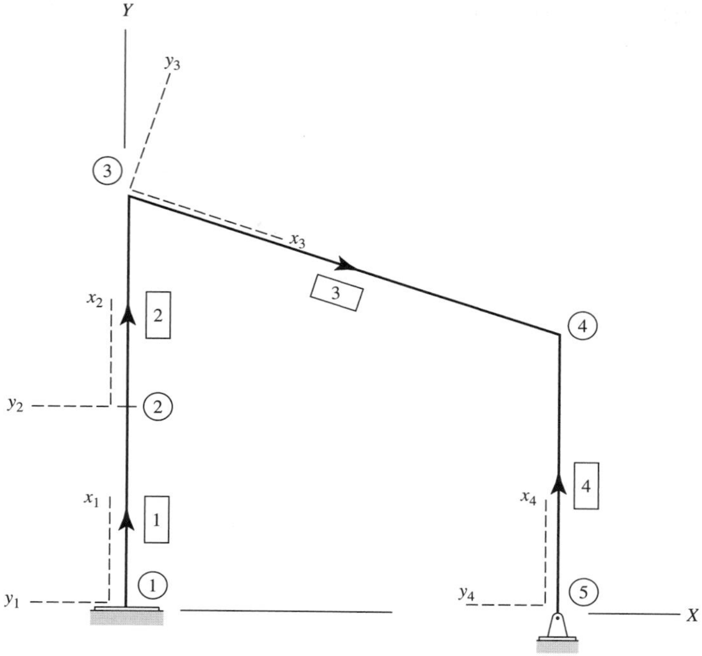

(b) Analytical Model Showing Global and Local Coordinate Systems

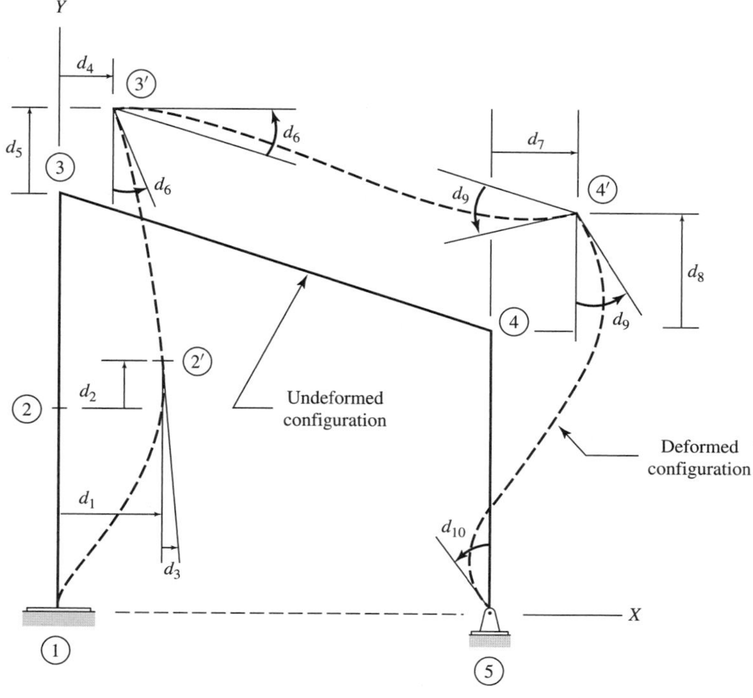

(c) Degrees of Freedom

Fig. 6.1

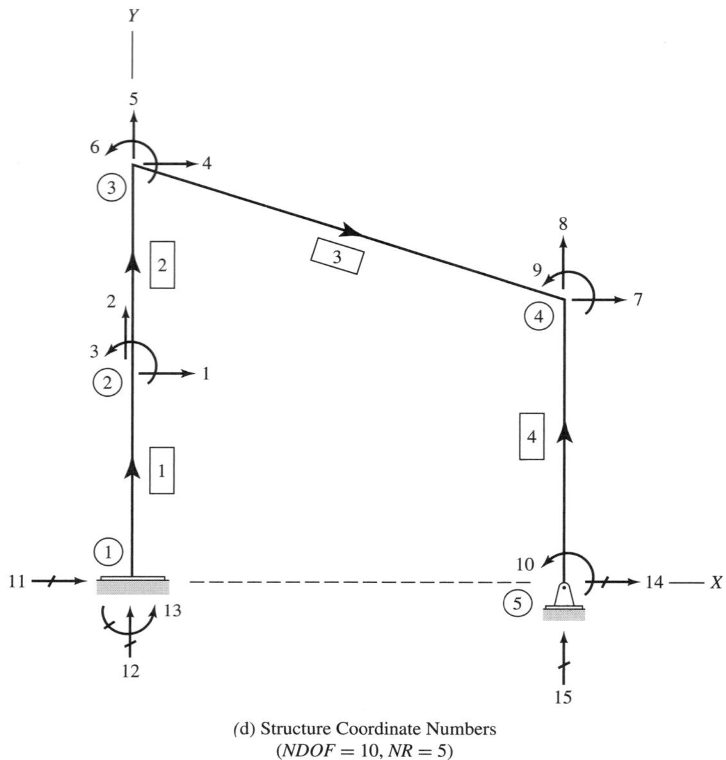

(d) Structure Coordinate Numbers
($NDOF = 10$, $NR = 5$)

Fig. 6.1 (*continued*)

locate the origin of the global coordinate system at a lower left joint of the frame with the X and Y axes oriented in the horizontal (positive to the right) and the vertical (positive upward) directions, respectively (see Fig. 6.1(b)).

For each member of the frame, a local xyz coordinate system is established, with its origin at an end of the member and the x axis directed along the member's centroidal axis in the undeformed state. The positive direction of the y axis is defined so that the local coordinate system is right-handed, with the local z axis pointing in the positive direction of the global Z axis. The member end at which the origin of the local coordinate system is located can be chosen arbitrarily, and is usually considered to be the *beginning* of the member; the opposite member end is simply referred to as the *end* of the member. The local coordinate systems selected for the four members of the example frame are depicted in Fig. 6.1(b). As indicated in this figure, the member local coordinate systems can be conveniently shown on the line diagram of the structure by drawing an arrow on each member in the positive direction of its x axis.

Degrees of Freedom and Restrained Coordinates

The degrees of freedom of a plane frame are simply the unknown displacements (translations and rotations) of its joints. Since an unsupported joint of a plane frame can translate in any direction in the *XY* plane and rotate about the *Z* axis, three displacements—the translations in the *X* and *Y* directions and the rotation about the *Z* axis—are needed to completely specify its deformed position. Thus, a free joint of a plane frame has three degrees of freedom, and three structure coordinates (i.e., free and/or restrained coordinates) need to be defined at each joint, for the purpose of analysis (i.e., $NCJT = 3$).

Let us examine the degrees of freedom of the analytical model of the example frame given in Fig. 6.1(b). The deformed shape of the frame, due to an arbitrary loading, is depicted in Fig. 6.1(c), using an exaggerated scale. From this figure, we can see that joint 1, which is attached to a fixed support, can neither translate nor rotate; therefore, it does not have any degrees of freedom. Since joint 2 is not attached to any support, it is free to translate as well as rotate, and three displacements—the translations d_1 and d_2 in the *X* and *Y* directions, respectively, and the rotation d_3—are needed to completely specify its deformed position $2'$. Thus, joint 2 has three degrees of freedom. Similarly, joints 3 and 4, which are also free joints, have three degrees of freedom each. The displacements of joint 3 are designated d_4, d_5, and d_6; the degrees of freedom of joint 4 are designated d_7, d_8, and d_9. Finally, joint 5, which is attached to a hinged support, can rotate, but it cannot translate; therefore, it has only one degree of freedom, designated d_{10}. Thus, the entire frame has a total of ten degrees of freedom. All the joint displacements are shown in Fig. 6.1(c) in the positive sense. As indicated in this figure, the joint translations are considered positive when in the positive directions of the *X* and *Y* axes and joint rotations considered positive when counterclockwise. The $NDOF \times 1$ joint displacement vector **d** for this frame is written as

$$\mathbf{d} = \underset{10 \times 1}{\begin{bmatrix} d_1 \\ d_2 \\ \vdots \\ d_9 \\ d_{10} \end{bmatrix}}$$

As discussed in Section 3.2, the number of degrees of freedom, *NDOF*, of a framed structure, in general, can be determined by subtracting the number of joint displacements restrained by supports, *NR*, from the total number of joint displacements of the unsupported structure (which equals $NCJT \times NJ$). Since *NCJT* equals 3 for plane frames, the number of degrees of freedom of such structures can be expressed as (see Eq. (3.2))

$$\left.\begin{aligned} NCJT &= 3 \\ NDOF &= 3(NJ) - NR \end{aligned}\right\} \quad \text{for plane frames} \qquad \textbf{(6.1)}$$

From Fig. 6.1(b), we can see that the example frame has five joints (i.e., $NJ = 5$); of these, joint 1 is attached to a fixed support that restrains three joint displacements, and joint 5 is attached to a hinged support that restrains two joint displacements. Thus, the total number of joint displacements that are restrained by all supports of the frame equals 5 (i.e., $NR = 5$). Substitution of $NJ = 5$ and $NR = 5$ into Eq. (6.1) yields the number of degrees of freedom of the frame:

$$NDOF = 3(5) - 5 = 10$$

which is the same as the number of degrees of freedom of the frame obtained previously.

As in the case of plane trusses and beams, the structure coordinates of a plane frame are usually specified on the frame's line diagram by assigning numbers to the arrows drawn at the joints in the directions of the joint displacements, with a slash (/) added to the arrows representing the restrained coordinates to distinguish them from the degrees of freedom, as shown in Fig. 6.1(d). The procedure for assigning numbers to the structure coordinates of a plane frame is analogous to that for plane trusses and beams. The degrees of freedom of the frame are numbered first by beginning at the lowest-numbered joint with a degree of freedom, and proceeding sequentially to the highest-numbered joint. If a joint has more than one degree of freedom, then the translation in the X direction is numbered first, followed by the translation in the Y direction, and then the rotation. The first degree of freedom is assigned the number one, and the last degree of freedom is assigned the number equal to $NDOF$. After all the degrees of freedom have been numbered, the restrained coordinates of the frame are numbered in the same manner as the degrees of freedom, but starting with the number equal to $NDOF + 1$ and ending with the number equal to $3(NJ)$. The structure coordinate numbers for the example frame, obtained by applying this procedure, are given in Fig. 6.1(d).

EXAMPLE 6.1 Identify by numbers the degrees of freedom and restrained coordinates of the frame shown in Fig. 6.2(a). Also, form the joint load vector **P** for the frame.

SOLUTION *Degrees of Freedom and Restrained Coordinates:* See Fig. 6.2(b). **Ans**

Joint Load Vector: Units are kips and feet.

$$\mathbf{P} = \begin{bmatrix} 0 \\ 0 \\ 20 \\ 0 \\ 0 \\ 0 \\ 0 \\ -75 \\ 10 \\ -11.5 \\ 0 \\ 0 \\ -11.5 \\ 0 \end{bmatrix}$$ **Ans**

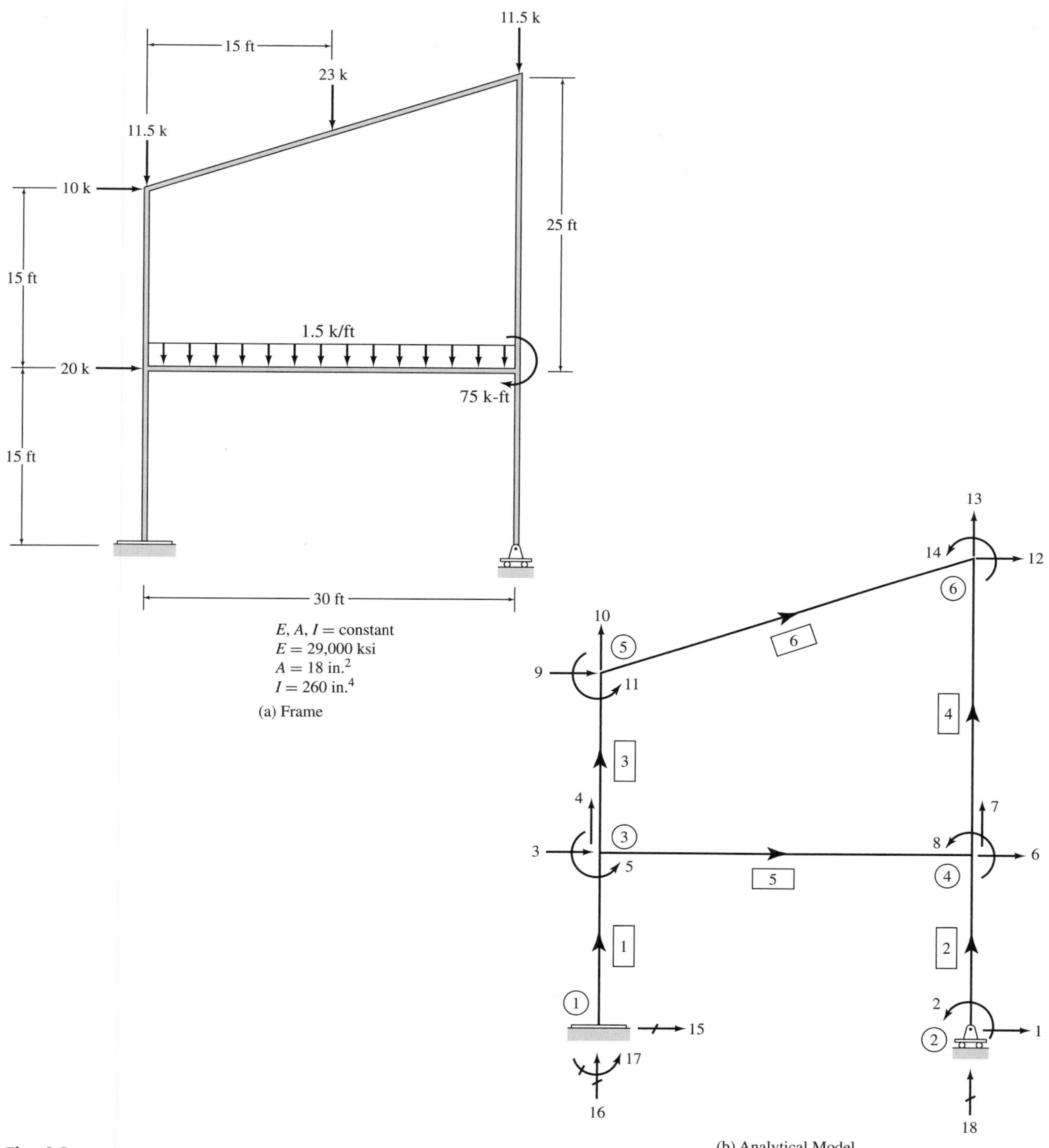

(a) Frame

(b) Analytical Model
($NDOF = 14$, $NR = 4$)

Fig. 6.2

6.2 MEMBER STIFFNESS RELATIONS IN THE LOCAL COORDINATE SYSTEM

Consider an arbitrary prismatic member m of the plane frame shown in Fig. 6.3(a). When the frame is subjected to external loads, member m deforms and internal axial forces, shears, and moments are induced at its ends. The initial and displaced positions of the member are shown in Fig. 6.3(b), from

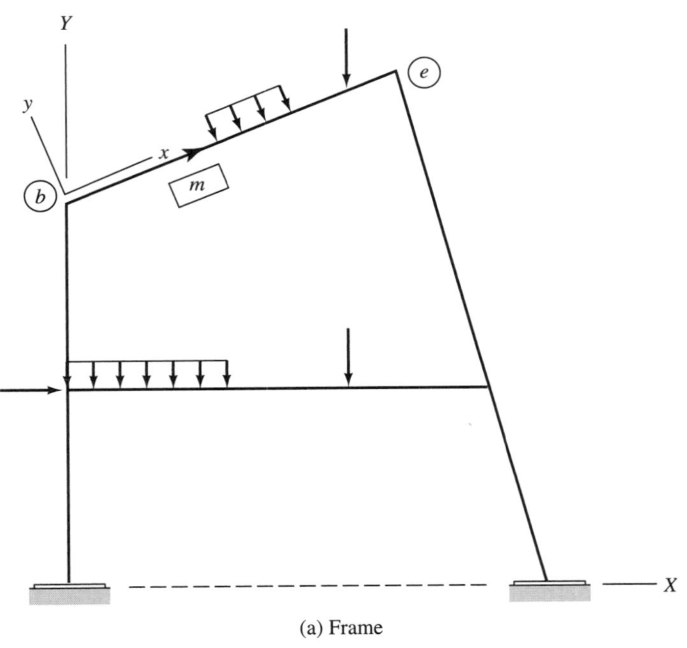

(a) Frame

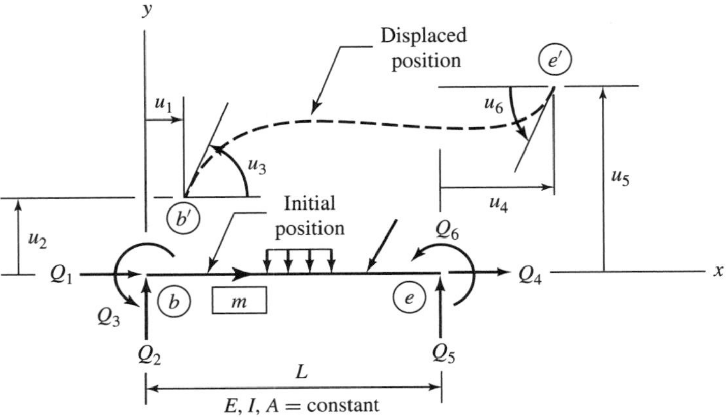

(b) Member Forces and Displacements in the Local Coordinate System

Fig. 6.3

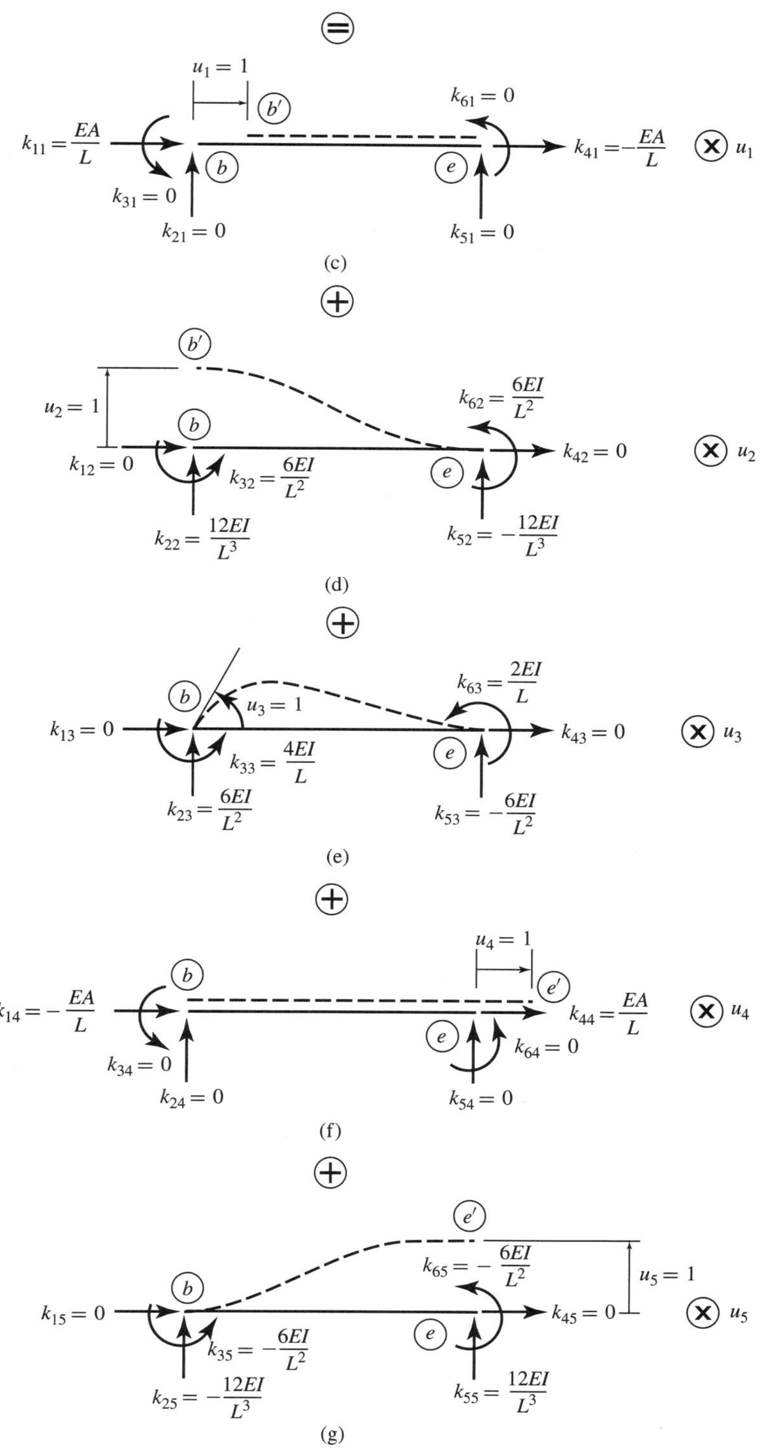

Fig. 6.3 *(continued)*

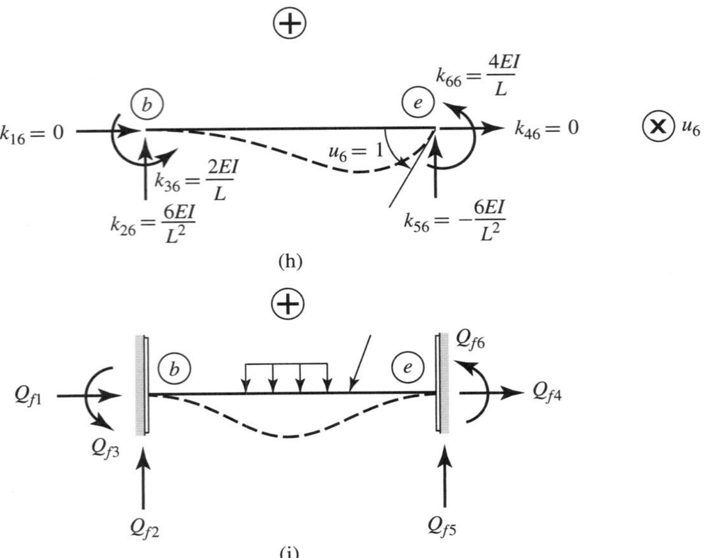

Fig. 6.3 *(continued)*

which we can see that three displacements—translations in the x and y directions and rotation about the z axis—are needed to completely specify the displaced position of each end of the member. Thus, the member has a total of six degrees of freedom. As indicated in Fig. 6.3(b), the six member end displacements are denoted by u_1 through u_6, and the corresponding member end forces are denoted by Q_1 through Q_6. Note that the member end displacements **u** and end forces **Q** are defined relative to the local coordinate system of the member, with translations and forces in the positive directions of the local x and y axes considered positive, and counterclockwise rotations and moments considered positive. As shown in Fig. 6.3(b), *a member's local end displacements and end forces are numbered by beginning at its end b, with the translation and force in the x direction numbered first, followed by the translation and force in the y direction, and then the rotation and moment. The displacements and forces at the member's opposite end e are then numbered in the same sequential order.*

The relationships between the end forces **Q** and the end displacements **u**, for the members of plane frames, can be established by essentially the same process as used previously for the case of beams (Section 5.2). The process involves subjecting the member, separately, to each of the six end displacements as shown in Fig. 6.3(c) through (h), and to the external loading with no end displacements (i.e., with both member ends completely fixed against translations and rotations), as shown in Fig. 6.3(i). The total member end forces due to the combined effect of the six end displacements, and the external loading, can now be expressed as

$$Q_i = \sum_{j=1}^{6} \left(k_{ij} u_j\right) + Q_{fi} \qquad i = 1, 2, \ldots, 6 \tag{6.2}$$

in which the stiffness coefficient k_{ij} represents the force corresponding to Q_i due to a unit value of the displacement u_j, and Q_{fi} denotes the fixed-end force corresponding to Q_i due to the external loads acting on the member. Equation (6.2) can be expressed in matrix form as

$$\begin{bmatrix} Q_1 \\ Q_2 \\ Q_3 \\ Q_4 \\ Q_5 \\ Q_6 \end{bmatrix} = \begin{bmatrix} k_{11} & k_{12} & k_{13} & k_{14} & k_{15} & k_{16} \\ k_{21} & k_{22} & k_{23} & k_{24} & k_{25} & k_{26} \\ k_{31} & k_{32} & k_{33} & k_{34} & k_{35} & k_{36} \\ k_{41} & k_{42} & k_{43} & k_{44} & k_{45} & k_{46} \\ k_{51} & k_{52} & k_{53} & k_{54} & k_{55} & k_{56} \\ k_{61} & k_{62} & k_{63} & k_{64} & k_{65} & k_{66} \end{bmatrix} \begin{bmatrix} u_1 \\ u_2 \\ u_3 \\ u_4 \\ u_5 \\ u_6 \end{bmatrix} + \begin{bmatrix} Q_{f1} \\ Q_{f2} \\ Q_{f3} \\ Q_{f4} \\ Q_{f5} \\ Q_{f6} \end{bmatrix} \tag{6.3}$$

or, symbolically, as

$$\boxed{\mathbf{Q} = \mathbf{k}\mathbf{u} + \mathbf{Q}_f} \tag{6.4}$$

in which $\mathbf{Q}$ and $\mathbf{u}$ denote the 6×1 member end-force and member end-displacement vectors, respectively, in the local coordinate system; $\mathbf{k}$ represents the 6×6 member local stiffness matrix; and $\mathbf{Q}_f$ is the 6×1 member fixed-end force vector in the local coordinate system.

Member Local Stiffness Matrix k

The explicit form of the local stiffness matrix $\mathbf{k}$ (in terms of E, A, I, and L) for the members of plane frames can be conveniently developed by using the expressions for the member stiffness coefficients of trusses and beams derived in Chapters 3 and 5, respectively.

To obtain the first column of $\mathbf{k}$, we subject the member to a unit end displacement $u_1 = 1$ (with $u_2 = u_3 = u_4 = u_5 = u_6 = 0$), as shown in Fig. 6.3(c). The expressions for the member axial forces required to cause this unit axial deformation were derived in Section 3.3, and are given in Fig. 3.3(c). By comparing Figs. 6.3(c) and 3.3(c), we obtain the stiffness coefficients for the plane frame member, due to end displacement $u_1 = 1$, as

$$k_{11} = \frac{EA}{L}, \qquad k_{41} = -\frac{EA}{L}, \qquad k_{21} = k_{31} = k_{51} = k_{61} = 0 \tag{6.5a}$$

Note that the imposition of end displacement $u_1 = 1$ does not cause the member to bend; therefore, no moments or shears develop at the ends of the member.

Similarly, the fourth column of $\mathbf{k}$ can be determined by comparing Fig. 6.3(f) to Fig. 3.3(e), which yields

$$k_{14} = -\frac{EA}{L}, \qquad k_{44} = \frac{EA}{L}, \qquad k_{24} = k_{34} = k_{54} = k_{64} = 0 \tag{6.5b}$$

To determine the second column of $\mathbf{k}$, the member is subjected to a unit end displacement $u_2 = 1$ (with $u_1 = u_3 = u_4 = u_5 = u_6 = 0$), as shown in 6.3(d).

The expressions for the member end shears and moments required to cause this deflected shape were derived in Section 5.2, and are given in Fig. 5.3(c). By comparing Figs. 6.3(d) and 5.3(c), we obtain the stiffness coefficients for the plane frame member, due to $u_2 = 1$, as

$$k_{22} = \frac{12EI}{L^3}, \qquad k_{32} = \frac{6EI}{L^2}, \qquad k_{52} = -\frac{12EI}{L^3}, \qquad k_{62} = \frac{6EI}{L^2},$$
$$k_{12} = k_{42} = 0 \tag{6.5c}$$

The third, fifth, and sixth columns of **k** can be developed in a similar manner, by comparing Figs. 6.3(e), (g), and (h) to Figs. 5.3(d), (e), and (f), respectively. This process yields

$$k_{23} = \frac{6EI}{L^2}, \qquad k_{33} = \frac{4EI}{L}, \qquad k_{53} = -\frac{6EI}{L^2}, \qquad k_{63} = \frac{2EI}{L},$$
$$k_{13} = k_{43} = 0 \tag{6.5d}$$

$$k_{25} = -\frac{12EI}{L^3}, \qquad k_{35} = -\frac{6EI}{L^2}, \qquad k_{55} = \frac{12EI}{L^3},$$
$$k_{65} = -\frac{6EI}{L^2}, \qquad k_{15} = k_{45} = 0 \tag{6.5e}$$

and

$$k_{26} = \frac{6EI}{L^2}, \qquad k_{36} = \frac{2EI}{L}, \qquad k_{56} = -\frac{6EI}{L^2}, \qquad k_{66} = \frac{4EI}{L},$$
$$k_{16} = k_{46} = 0 \tag{6.5f}$$

Finally, by substituting Eqs. (6.5) into the appropriate columns of **k** given in Eq. (6.3), we can express the local stiffness matrix for the members of plane frames as

$$\mathbf{k} = \frac{EI}{L^3}\begin{bmatrix} \frac{AL^2}{I} & 0 & 0 & -\frac{AL^2}{I} & 0 & 0 \\ 0 & 12 & 6L & 0 & -12 & 6L \\ 0 & 6L & 4L^2 & 0 & -6L & 2L^2 \\ -\frac{AL^2}{I} & 0 & 0 & \frac{AL^2}{I} & 0 & 0 \\ 0 & -12 & -6L & 0 & 12 & -6L \\ 0 & 6L & 2L^2 & 0 & -6L & 4L^2 \end{bmatrix} \tag{6.6}$$

Member Local Fixed-End Force Vector $\mathbf{Q}_f$

Unlike the members of beams, which are loaded only perpendicular to their longitudinal axes, the members of plane frames can be subjected to loads oriented in any direction in the plane of the structure. Before proceeding with the calculation of the fixed-end forces for a plane frame member, any loads acting on it in inclined directions are resolved into their components in the directions

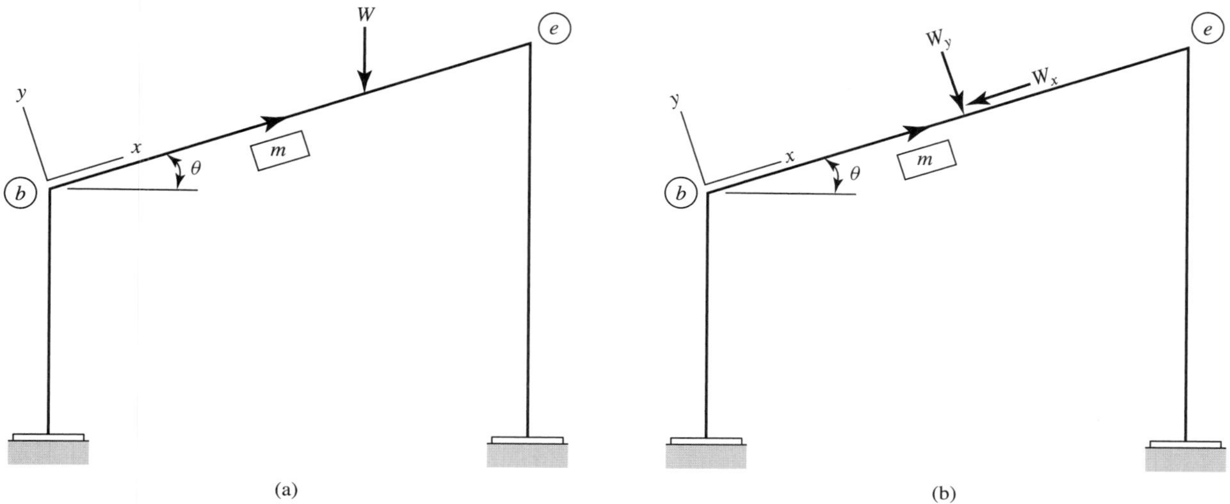

Fig. 6.4

of the local x and y axes of the member. For example, the vertical load W acting on the inclined member m of the frame of Fig. 6.4(a) is resolved into its rectangular components in the local x and y directions of the member m as

$$W_x = W \sin\theta \qquad \text{and} \qquad W_y = W \cos\theta$$

as shown in Fig. 6.4(b).

After all the loads acting on a member have been resolved into components parallel and perpendicular to the longitudinal axis of the member (i.e., in the local x and y directions, respectively), the fixed-end shears (FS_b and FS_e) and moments (FM_b and FM_e) due to the perpendicular loading and any couples can be calculated by using the fixed-end force equations for loading types 1 through 4 (given inside the front cover). The procedure for deriving these fixed-end shear and moment equations was discussed in Section 5.4.

The expressions for the member fixed-end axial forces, due to two common types of member axial loadings, are also given inside the front cover (see loading types 5 and 6). Such expressions can be conveniently determined by integrating the differential equation for the member axial deformation. This approach is illustrated in the following paragraphs, with loading type 6 taken as an example.

Consider a fixed member of a plane frame, subjected to a uniformly distributed axial load w over a part of its length, as shown in Fig. 6.5(a). As indicated there, the fixed-end axial forces at the member ends b and e are denoted by FA_b and FA_e, respectively. To develop the differential equation for axial deformation of an elastic member, we recall from Section 3.4 that the relationship between the axial strain ε_a and the axial displacement $\bar{u}_x$, of the centroidal axis of a member, is given by (see Eq. (3.39))

$$\varepsilon_a = \frac{d\bar{u}_x}{dx}$$

Substitution of this strain-displacement equation into Hooke's law yields

$$\sigma_a = E\varepsilon_a = E\frac{d\bar{u}_x}{dx}$$

in which σ_a represents the axial stress. To relate the axial displacement $\bar{u}_x$ to the axial force Q_a acting at the cross-section, we multiply both sides of the preceding equation by the cross-sectional area A to obtain

$$Q_a = \sigma_a A = EA\frac{d\bar{u}_x}{dx}$$

or

$$\boxed{\frac{d\bar{u}_x}{dx} = \frac{Q_a}{EA}} \tag{6.7}$$

Equation (6.7) represents the differential equation for axial deformation of a member composed of linearly elastic homogeneous material. In this equation, $\bar{u}_x$ denotes the displacement of the member's centroidal axis in the x direction, at a distance x from the origin b of the local xy coordinate system of the member (Fig. 6.5(a)); Q_a represents the axial force at the member cross-section at the same location, x. Furthermore, Eq. (6.7) is based on the sign convention that the axial force Q_a is considered positive when causing tension at the member cross-section. The total axial deformation of a member can be obtained by multiplying both sides of Eq. (6.7) by dx and integrating the resulting equation

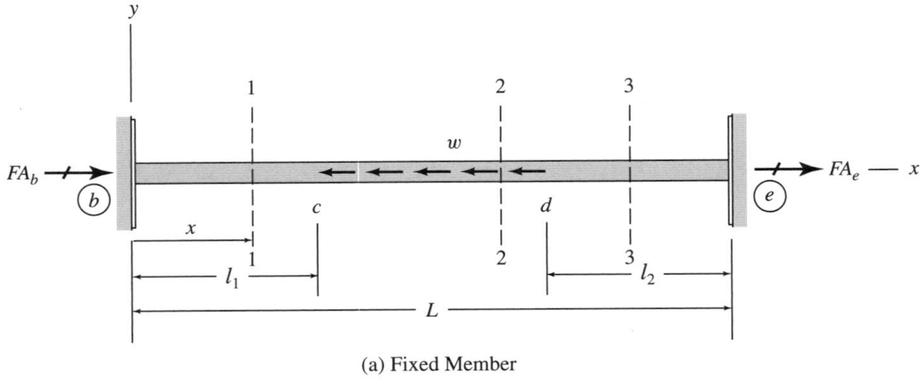

(a) Fixed Member

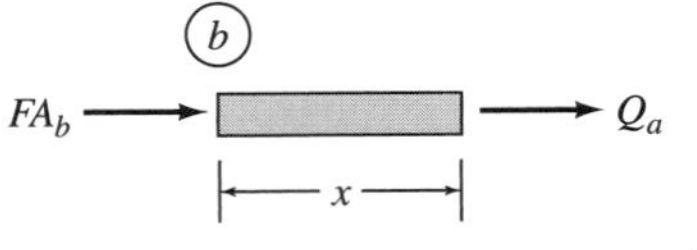

(b) Section 1–1

Fig. 6.5

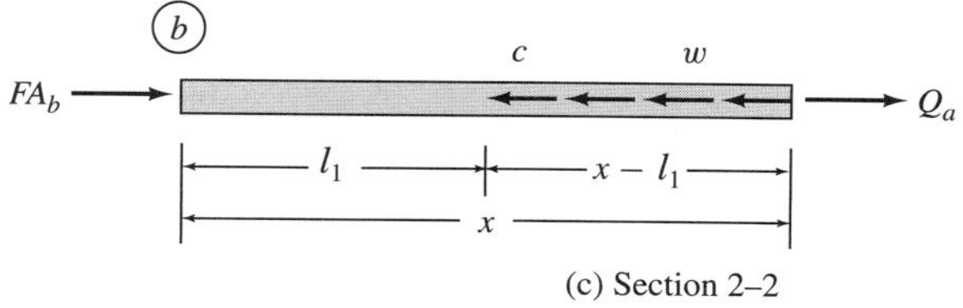

(c) Section 2–2

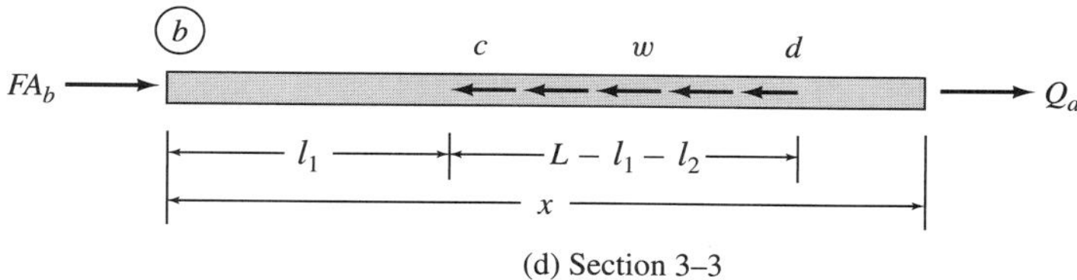

(d) Section 3–3

Fig. 6.5 *(continued)*

over the length L of the member:

$$\bar{u}_{xe}(at x = L) = \int_0^L \frac{Q_a}{EA}\, dx \tag{6.8}$$

Realizing that EA is constant for prismatic members, the axial deformation of such members can be expressed as

$$\bar{u}_{xe} = \frac{1}{EA} \int_0^L Q_a\, dx \tag{6.9}$$

To obtain the expressions for the fixed-end axial forces FA_b and FA_e for the member shown in Fig. 6.5(a), we first determine the equations for axial force Q_a in terms of one of the unknowns, FA_b. Since the uniformly distributed load w is applied over member portion cd (Fig. 6.5(a)), the axial force Q_a cannot be expressed as a single continuous function over the entire length of the member. Therefore, we divide the member into three segments, bc, cd, and de, and determine the equations for axial force in these segments by passing sections 1–1, 2–2, and 3–3, respectively, through the member, as shown in Fig. 6.5(a). By considering the equilibrium of the free body of the member to the left of section 1–1 (Fig. 6.5(b)), we determine the axial force Q_a at section 1–1 to be

$$\overset{+}{\rightarrow} \sum F_x = 0 \qquad FA_b + Q_a = 0 \qquad Q_a = -FA_b$$

Thus, the equation of the axial force in segment bc can be expressed as

$$0 \le x \le l_1 \qquad Q_a = -FA_b \tag{6.10a}$$

Similarly, by considering the free bodies of the member to the left of sections 2–2 and 3–3 (Fig. 6.5(c) and (d)), we obtain the equations of the axial force in segments cd and de, respectively, as

$$l_1 \le x \le L - l_2 \qquad Q_a = -FA_b + w(x - l_1) \tag{6.10b}$$

$$L - l_2 \le x \le L \qquad Q_a = -FA_b + w(L - l_1 - l_2) \tag{6.10c}$$

Next, by substituting Eqs. (6.10) into Eq. (6.9), we write

$$\bar{u}_{xe} = \frac{1}{EA}\left[\int_0^{l_1} -FA_b\,dx + \int_{l_1}^{L-l_2} \{-FA_b + w(x - l_1)\}\,dx + \int_{L-l_2}^{L} \{-FA_b + w(L - l_1 - l_2)\}\,dx\right]$$

By integrating and simplifying the right-hand side of the foregoing equation, we obtain the axial deformation of the member as

$$\bar{u}_{xe} = \frac{1}{EA}\left[-FA_bL + \frac{w}{2}(L - l_1 - l_2)(L - l_1 + l_2)\right] \tag{6.11}$$

The expression for FA_b can now be determined by using the compatibility condition that, because both ends b and e of the member are attached to fixed supports, the axial deformation of the member must be 0. Thus, by substituting $\bar{u}_{xe} = 0$ into Eq. (6.11), we write

$$\bar{u}_{xe} = \frac{1}{EA}\left[-FA_bL + \frac{w}{2}(L - l_1 - l_2)(L - l_1 + l_2)\right] = 0 \tag{6.12}$$

Solving Eq. (6.12) for FA_b, we obtain

$$FA_b = \frac{w}{2L}(L - l_1 - l_2)(L - l_1 + l_2) \tag{6.13}$$

With the fixed-end axial force FA_b known, we can now determine the remaining fixed-end axial force FA_e by applying the equation of equilibrium $\sum F_x = 0$ to the free body of the entire member. Thus (see Fig. 6.5(a)),

$$\overset{+}{\rightarrow} \sum F_x = 0 \qquad FA_b - w(L - l_1 - l_2) + FA_e = 0$$

Substituting Eq. (6.13) into the foregoing equation, and simplifying the result, we obtain the expression for FA_e:

$$FA_e = \frac{w}{2L}(L - l_1 - l_2)(L + l_1 - l_2) \tag{6.14}$$

The expressions for fixed-end axial forces due to other types of axial loadings can be derived in a similar manner, using the integration approach illustrated here.

Once the fixed-end axial and shear forces and moments for a member have been evaluated, its fixed-end force vector $\mathbf{Q}_f$ can be generated by storing the fixed-end forces and moments in their proper positions in a 6×1 vector, as follows.

$$\mathbf{Q}_f = \begin{bmatrix} Q_{f1} \\ Q_{f2} \\ Q_{f3} \\ Q_{f4} \\ Q_{f5} \\ Q_{f6} \end{bmatrix} = \begin{bmatrix} FA_b \\ FS_b \\ FM_b \\ FA_e \\ FS_e \\ FM_e \end{bmatrix} \tag{6.15}$$

The sign convention for member local fixed-end forces, $\mathbf{Q}_f$, is the same as that for the member end forces in the local coordinate system, $\mathbf{Q}$. Thus, the member local fixed-end axial forces and shears are considered positive when in the positive directions of the member's local x and y axes, and the local fixed-end moments are considered positive when counterclockwise. However, the member loads are commonly defined to be positive in the directions *opposite* to those for the local fixed-end forces. In other words, the member axial and perpendicular loads are considered positive when in the *negative* directions of the member's local x and y axes, respectively, and the external couples applied to the members are considered positive when clockwise. The expressions for the member fixed-end forces (including moments) given inside the front cover of this text are based on this sign convention, in which all the fixed-end forces and member loads (including couples) are shown in the positive sense.

EXAMPLE 6.2 The displaced position of member 2, of the frame of Fig. 6.6(a), is given in Fig. 6.6(b). Calculate the end forces for this member in the local coordinate system. Is the member in equilibrium under these forces?

SOLUTION *Member Local Stiffness Matrix:* From Fig. 6.6(a), we can see that, for member 2, $E = 29{,}000$ ksi, $A = 28.2$ in.^{2}, $I = 833$ in.^{4}, and $L = \sqrt{(16)^2 + (12)^2} = 20$ ft $=$ 240 in. By substituting the numerical values of E, A, I, and L into Eq. (6.6), we obtain the following local stiffness matrix for member 2, in units of kips and inches.

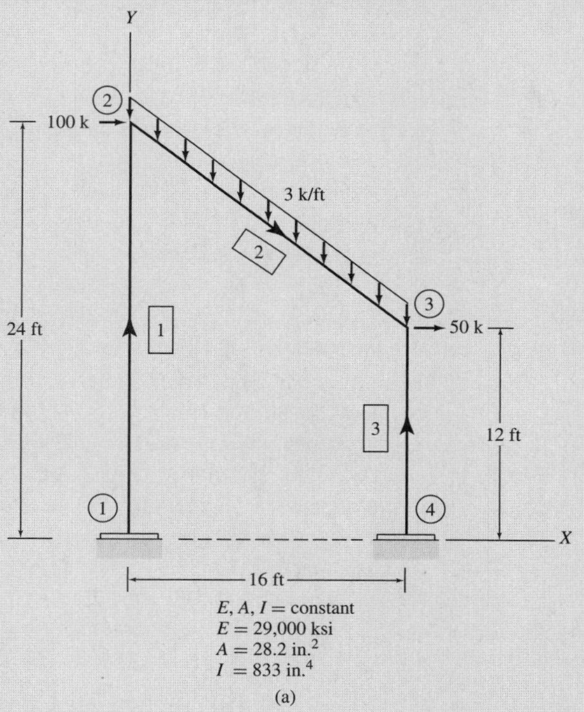

Fig. 6.6

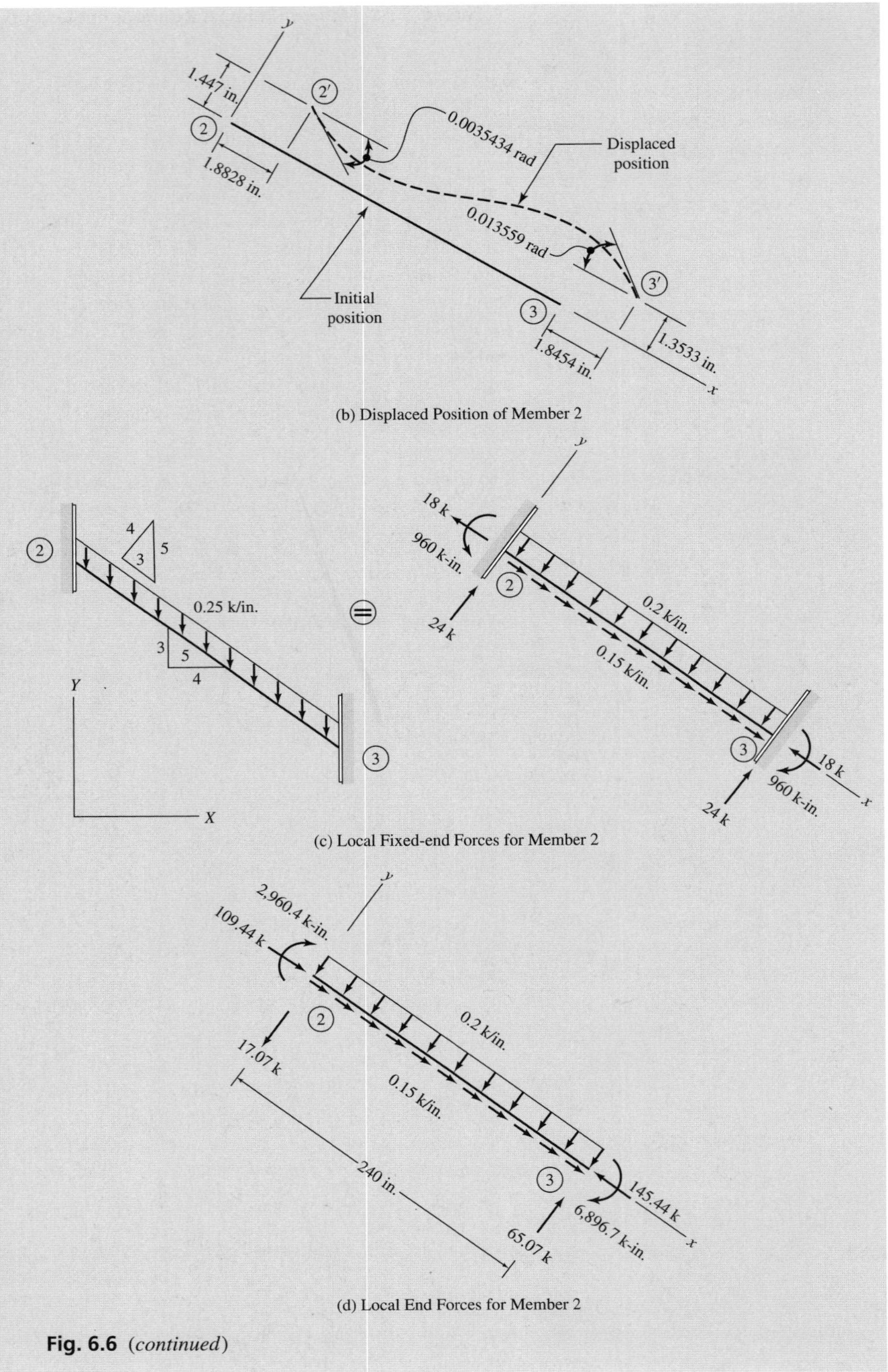

(b) Displaced Position of Member 2

(c) Local Fixed-end Forces for Member 2

(d) Local End Forces for Member 2

Fig. 6.6 (*continued*)

$$\mathbf{k}_2 = \begin{bmatrix} 3{,}407.5 & 0 & 0 & -3{,}407.5 & 0 & 0 \\ 0 & 20.97 & 2{,}516.4 & 0 & -20.97 & 2{,}516.4 \\ 0 & 2{,}516.4 & 402{,}620 & 0 & -2{,}516.4 & 201{,}310 \\ -3{,}407.5 & 0 & 0 & 3{,}407.5 & 0 & 0 \\ 0 & -20.97 & -2{,}516.4 & 0 & 20.97 & -2{,}516.4 \\ 0 & 2{,}516.4 & 201{,}310 & 0 & -2{,}516.4 & 402{,}620 \end{bmatrix} \quad (1)$$

Member Local End Displacements: See Fig. 6.6(b).

$$\mathbf{u}_2 = \begin{bmatrix} 1.8828 \\ 1.4470 \\ -0.0035434 \\ 1.8454 \\ 1.3533 \\ -0.013559 \end{bmatrix} \quad (2)$$

Note that the values of u_3 and u_6 are negative, because both member ends rotate in the clockwise direction.

Member Local Fixed-end Force Vector: As the 0.25 k/in. (= 3 k/ft) uniformly distributed load, applied to the member, acts in the vertical direction, it is necessary to resolve it into components parallel and perpendicular to the member. The components of the vertical distributed load in the local x and y directions are (see Fig. 6.6(c)):

$$w_x = -\frac{3}{5}(0.25) = -0.15 \text{ k/in.}$$

$$w_y = \frac{4}{5}(0.25) = 0.2 \text{ k/in.}$$

in which, in accordance with the sign convention for member loads discussed previously, a negative sign is assigned to the magnitude of w_x because it acts in the positive direction of the local x axis.

The local fixed-end forces can now be evaluated, using the expressions given inside the front cover. By substituting $w = -0.15$ k/in., $L = 240$ in., and $l_1 = l_2 = 0$ into the expressions for the fixed-end axial forces given for loading type 6, we obtain

$$FA_b = FA_e = \frac{-0.15(240)}{2} = -18 \text{ k}$$

Similarly, substitution of $w = 0.2$ k/in., $L = 240$ in., and $l_1 = l_2 = 0$ into the expressions for the fixed-end shears and moments given for loading type 3 yields

$$FS_b = FS_e = \frac{0.2(240)}{2} = 24 \text{ k}$$

$$FM_b = -FM_e = \frac{0.2(240)^2}{12} = 960 \text{ k-in.}$$

These fixed-end forces for member 2 are shown in Fig. 6.6(c). The local fixed-end force vector for the member is given by

$$\mathbf{Q}_{f2} = \begin{bmatrix} -18 \\ 24 \\ 960 \\ -18 \\ 24 \\ -960 \end{bmatrix} \quad (3)$$

Member Local End Forces: The local end forces for member 2 can now be determined by substituting the numerical forms of $\boldsymbol{k}_2$, $\boldsymbol{u}_2$, and $\boldsymbol{Q}_{f2}$ (Eqs. (1), (2), and (3), respectively), into Eq. (6.4), and performing the required matrix multiplication and addition. This yields

$$\mathbf{Q}_2 = \mathbf{k}_2\mathbf{u}_2 + \mathbf{Q}_{f2} = \begin{bmatrix} 109.44 \text{ k} \\ -17.07 \text{ k} \\ -2{,}960.4 \text{ k-in.} \\ -145.44 \text{ k} \\ 65.07 \text{ k} \\ -6{,}896.7 \text{ k-in.} \end{bmatrix}$$ **Ans**

These member end forces are depicted in Fig. 6.6(d).

Equilibrium Check: To check whether the member is in equilibrium, we apply the three equations of equilibrium to the free body of the member shown in Fig. 6.6(d). Thus,

$+\searrow \sum F_x = 0 \quad 109.44 + 0.15(240) - 145.44 = 0$ **Checks**

$+\nearrow \sum F_y = 0 \quad -17.07 - 0.2(240) + 65.07 = 0$ **Checks**

$+\circlearrowleft \sum M_{②} = 0 \quad -2{,}960.4 - 0.2(240)(120) - 6{,}896.7 + 65.07(240) = -0.3 \cong 0$ **Checks**

Therefore, the member is in equilibrium. **Ans**

6.3 COORDINATE TRANSFORMATIONS

Unlike beams, whose members all are oriented in the same direction, plane frames usually contain members oriented in various directions in the plane of the structure. Therefore, it becomes necessary to transform the stiffness relations of the members of a plane frame from their local coordinate systems to the global coordinate system before they can be combined to establish the stiffness relations for the entire frame. In this section, we extend the transformation relationships developed in Section 3.5 for plane truss members to include end moments and rotations, so that they can be used for the members of plane frames. The revised transformation relations thus obtained are then used in Section 6.4 to develop the member stiffness relations in the global coordinate system for plane frames.

Consider an arbitrary member m of a plane frame, as shown in Fig. 6.7(a). The orientation of the member with respect to the global XY coordinate system is defined by an angle θ, measured counterclockwise from the positive direction of the global X axis to the positive direction of the local x axis, as shown in Fig. 6.7(a). When the frame is subjected to external loads, member m deforms, and internal forces and moments develop at its ends. The displaced position of member m, due to an arbitrary loading applied to the frame, is shown in Figs. 6.7(b) and (c). In Fig. 6.7(b), the member end displacements, $\mathbf{u}$, and end forces, $\mathbf{Q}$, are measured relative to the local xy coordinate system of the member; whereas, in Fig. 6.7(c), the member end displacements, $\mathbf{v}$, and end forces, $\mathbf{F}$, are defined with respect to the global XY coordinate system of the frame. The local and global systems of member end displacements and forces are *equivalent,* in the sense that both systems cause the same translations and rotations of the member ends b and e, and produce the same state of strain and stress in the member. As shown in Fig. 6.7(c), the global member end forces, $\mathbf{F}$, and end displacements, $\mathbf{v}$, are numbered by beginning at member end b, with the force and translation in the X direction numbered first, followed by the

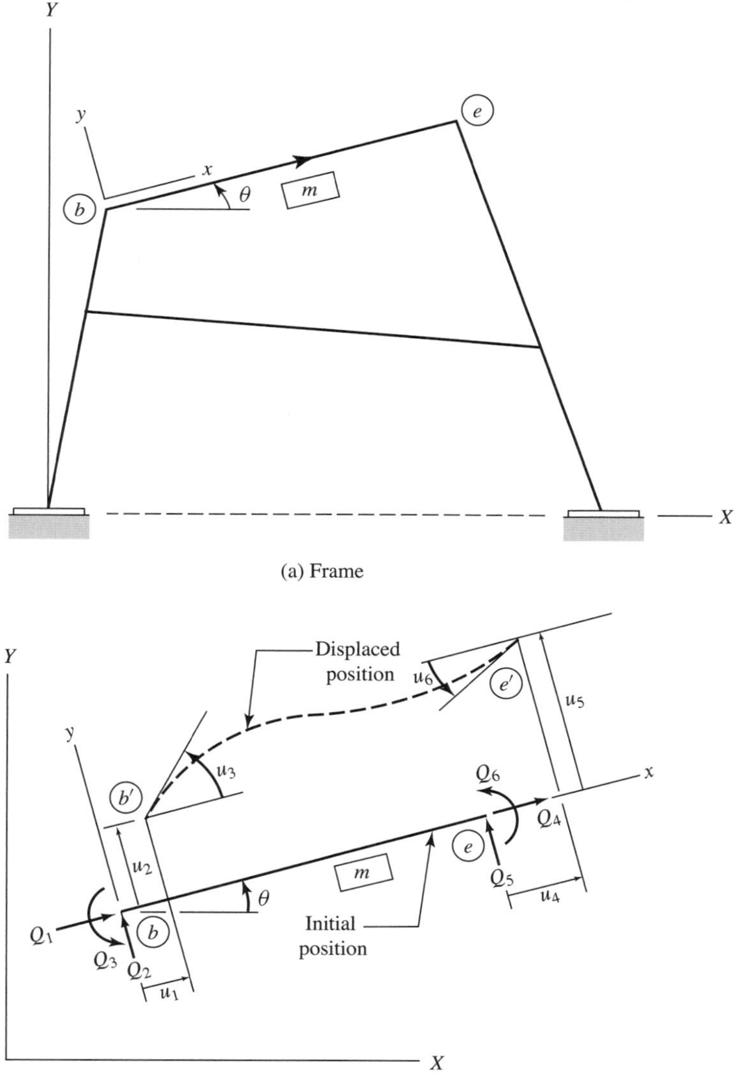

(a) Frame

(b) Member End Forces and End Displacements in the Local Coordinate System

Fig. 6.7

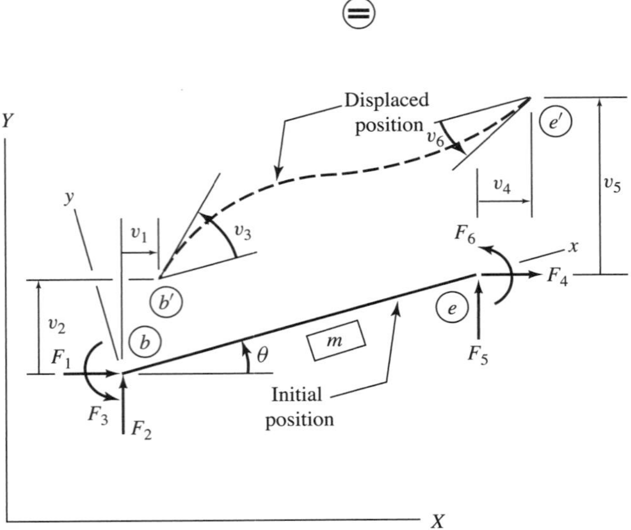

(c) Member End Forces and End Displacements in the Global Coordinate System

Fig. 6.7 *(continued)*

force and translation in the Y direction, and then the moment and rotation. The forces and displacements at the member's opposite end e are then numbered in the same sequential order.

Now, suppose that the member's global end forces and end displacements are specified, and we wish to determine the corresponding end forces and end displacements in the local coordinate system of the member. As discussed in Section 3.5, the local forces Q_1 and Q_2 must be equal to the algebraic sums of the components of the global forces F_1 and F_2 in the directions of the local x and y axes, respectively; that is,

$$Q_1 = F_1 \cos\theta + F_2 \sin\theta \tag{6.16a}$$

$$Q_2 = -F_1 \sin\theta + F_2 \cos\theta \tag{6.16b}$$

Note that Eqs. (6.16a and b) are identical to Eqs. (3.58a and b), respectively, derived previously for the case of plane truss members.

As for the relationship between the local end moment Q_3 and the global end moment F_3—because the local z axis and the global Z axis are oriented in the same direction (i.e., directed out of the plane of the page), the local moment Q_3 must be equal to the global moment F_3. Thus,

$$Q_3 = F_3 \tag{6.16c}$$

Using a similar reasoning at end e of the member, we express the local forces in terms of the global forces as

$$Q_4 = F_4 \cos\theta + F_5 \sin\theta \tag{6.16d}$$

$$Q_5 = -F_4 \sin\theta + F_5 \cos\theta \tag{6.16e}$$

$$Q_6 = F_6 \tag{6.16f}$$

We can write Eqs. (6.16a through f) in matrix form as

$$\begin{bmatrix} Q_1 \\ Q_2 \\ Q_3 \\ Q_4 \\ Q_5 \\ Q_6 \end{bmatrix} = \begin{bmatrix} \cos\theta & \sin\theta & 0 & 0 & 0 & 0 \\ -\sin\theta & \cos\theta & 0 & 0 & 0 & 0 \\ 0 & 0 & 1 & 0 & 0 & 0 \\ 0 & 0 & 0 & \cos\theta & \sin\theta & 0 \\ 0 & 0 & 0 & -\sin\theta & \cos\theta & 0 \\ 0 & 0 & 0 & 0 & 0 & 1 \end{bmatrix} \begin{bmatrix} F_1 \\ F_2 \\ F_3 \\ F_4 \\ F_5 \\ F_6 \end{bmatrix} \tag{6.17}$$

or, symbolically, as

$$\mathbf{Q} = \mathbf{TF} \tag{6.18}$$

in which the transformation matrix $\mathbf{T}$ is given by

$$\mathbf{T} = \begin{bmatrix} \cos\theta & \sin\theta & 0 & 0 & 0 & 0 \\ -\sin\theta & \cos\theta & 0 & 0 & 0 & 0 \\ 0 & 0 & 1 & 0 & 0 & 0 \\ 0 & 0 & 0 & \cos\theta & \sin\theta & 0 \\ 0 & 0 & 0 & -\sin\theta & \cos\theta & 0 \\ 0 & 0 & 0 & 0 & 0 & 1 \end{bmatrix} \tag{6.19}$$

The direction cosines ($\cos\theta$ and $\sin\theta$) of the plane frame members can be evaluated using Eqs. (3.62a and b), given in Section 3.5.

Because member end displacements, like end forces, are vectors, which are defined in the same directions as the corresponding forces, the transformation matrix $\mathbf{T}$ (Eq. (6.19)) can also be used to transform member end displacements from the global to the local coordinate system; that is,

$$\boxed{\mathbf{u} = \mathbf{Tv}} \tag{6.20}$$

Next, we consider the transformation of member end forces and end displacements from the local to the global coordinate system. Returning our attention to Figs. 6.7(b) and (c), we realize that at end b of the member, the global forces F_1 and F_2 must be equal to the algebraic sums of the components of the local forces Q_1 and Q_2 in the directions of the global X and Y axes, respectively; that is,

$$F_1 = Q_1\cos\theta - Q_2\sin\theta \tag{6.21a}$$

$$F_2 = Q_1\sin\theta + Q_2\cos\theta \tag{6.21b}$$

and, as discussed previously, the global moment F_3 equals the local moment Q_3, or

$$F_3 = Q_3 \tag{6.21c}$$

In a similar manner, the global forces at end e of the member can be expressed in terms of the local forces as

$$F_4 = Q_4\cos\theta - Q_5\sin\theta \tag{6.21d}$$

$$F_5 = Q_4\sin\theta + Q_5\cos\theta \tag{6.21e}$$

$$F_6 = Q_6 \tag{6.21f}$$

We can write Eqs. (6.21a through f) in matrix form as

$$\begin{bmatrix} F_1 \\ F_2 \\ F_3 \\ F_4 \\ F_5 \\ F_6 \end{bmatrix} = \begin{bmatrix} \cos\theta & -\sin\theta & 0 & 0 & 0 & 0 \\ \sin\theta & \cos\theta & 0 & 0 & 0 & 0 \\ 0 & 0 & 1 & 0 & 0 & 0 \\ 0 & 0 & 0 & \cos\theta & -\sin\theta & 0 \\ 0 & 0 & 0 & \sin\theta & \cos\theta & 0 \\ 0 & 0 & 0 & 0 & 0 & 1 \end{bmatrix} \begin{bmatrix} Q_1 \\ Q_2 \\ Q_3 \\ Q_4 \\ Q_5 \\ Q_6 \end{bmatrix} \quad \textbf{(6.22)}$$

By comparing Eq. (6.22) to Eq. (6.17), we realize that the transformation matrix in Eq. (6.22), which transforms the forces from the local to the global coordinate system, is the transpose of the transformation matrix **T** in Eq. (6.17), which transforms the forces from the global to the local coordinate system. Therefore, Eq. (6.22) can be written as

$$\boxed{\mathbf{F} = \mathbf{T}^T\mathbf{Q}} \quad \textbf{(6.23)}$$

Also, a comparison of Eqs. (6.18) and (6.23) indicates that the inverse of **T** equals its transpose; that is,

$$\mathbf{T}^{-1} = \mathbf{T}^T \quad \textbf{(6.24)}$$

which indicates that the transformation matrix **T** is orthogonal.

As discussed previously, because the member end displacements are also vectors defined in the directions of their corresponding forces, the matrix $\mathbf{T}^T$ also defines the transformation of member end displacements from the local to the global coordinate system; that is,

$$\mathbf{v} = \mathbf{T}^T\,\mathbf{u} \quad \textbf{(6.25)}$$

By comparing the transformation matrix **T** derived herein for plane frame members (Eq. (6.19)) with the one developed in Section 3.5 for plane truss members (Eq. (3.61)), we observe that the **T** matrix for plane trusses can be obtained by deleting the third and sixth columns and the third and sixth rows from the **T** matrix for plane frame members. This is because there are no moments and rotations induced at the ends of plane truss members, which are subjected to axial forces only.

EXAMPLE 6.3 The displaced position of member 2, of the frame of Fig. 6.8(a), is given in Fig. 6.8(b). Calculate the end displacements and end forces for this member in the global coordinate system. Is the member in equilibrium under the global end forces?

SOLUTION *Member Local End Displacements and Forces:* In Example 6.2, we obtained the local end displacement and force vectors for the member under consideration as

$$\mathbf{u}_2 = \begin{bmatrix} 1.8828 \text{ in.} \\ 1.4470 \text{ in.} \\ -0.0035434 \text{ rad} \\ 1.8454 \text{ in.} \\ 1.3533 \text{ in.} \\ -0.013559 \text{ rad} \end{bmatrix} \quad \textbf{(1)}$$

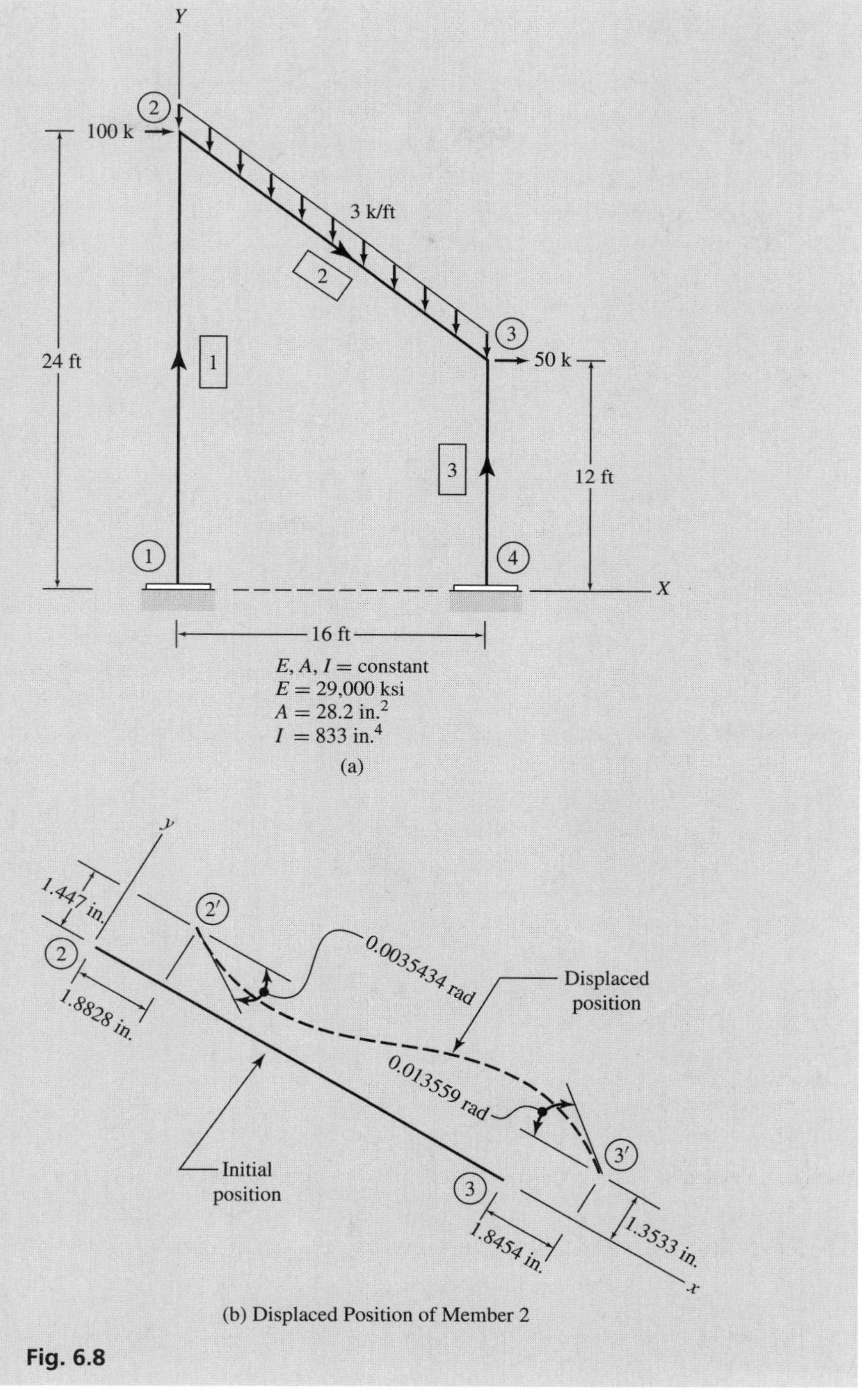

(b) Displaced Position of Member 2

Fig. 6.8

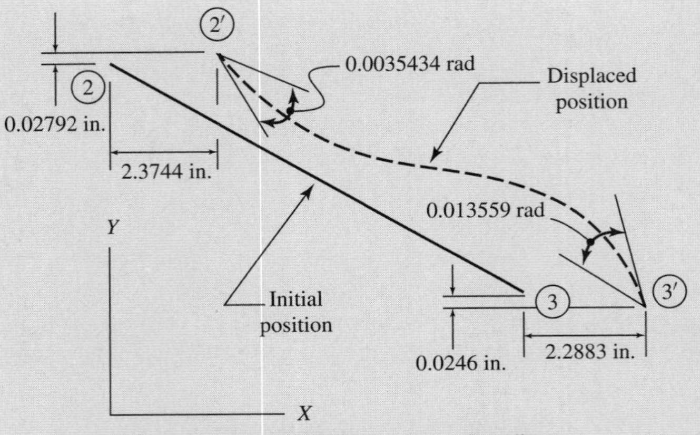

(c) End Displacements in the Global Coordinate System for Member 2

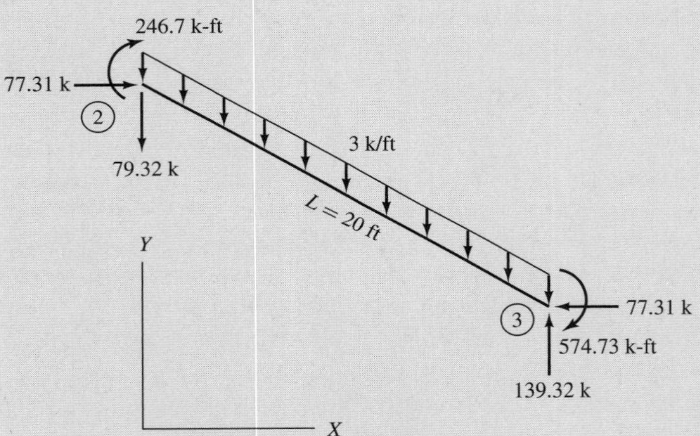

(d) End Forces in the Global Coordinate System for Member 2

Fig. 6.8 *(continued)*

and

$$\mathbf{Q}_2 = \mathbf{k}_2\mathbf{u}_2 + \mathbf{Q}_{f2} = \begin{bmatrix} 109.44 \text{ k} \\ -17.07 \text{ k} \\ -2{,}960.4 \text{ k-in.} \\ -145.44 \text{ k} \\ 65.07 \text{ k} \\ -6{,}896.7 \text{ k-in.} \end{bmatrix} \qquad \textbf{(2)}$$

Transformation Matrix: From Fig. 6.8(a), we can see that joint 2 is the beginning joint and joint 3 is the end joint for member 2. By applying Eqs. (3.62), we determine the member's direction cosines as

$$\cos\theta = \frac{X_3 - X_2}{L} = \frac{16-0}{20} = 0.8$$
$$\sin\theta = \frac{Y_3 - Y_2}{L} = \frac{12-24}{20} = -0.6$$

The transformation matrix for member 2 can now be evaluated, using Eq. (6.19).

$$\mathbf{T}_2 = \begin{bmatrix} 0.8 & -0.6 & 0 & 0 & 0 & 0 \\ 0.6 & 0.8 & 0 & 0 & 0 & 0 \\ 0 & 0 & 1 & 0 & 0 & 0 \\ 0 & 0 & 0 & 0.8 & -0.6 & 0 \\ 0 & 0 & 0 & 0.6 & 0.8 & 0 \\ 0 & 0 & 0 & 0 & 0 & 1 \end{bmatrix} \tag{3}$$

Member Global End Displacements: By substituting the transpose of $\mathbf{T}_2$ from Eq. (3), and $\mathbf{u}_2$ from Eq. (1), into Eq. (6.25), we obtain

$$\mathbf{V}_2 = \mathbf{T}_2^T\mathbf{u}_2 = \begin{bmatrix} 2.3744 \text{ in.} \\ 0.02792 \text{ in.} \\ -0.0035434 \text{ rad} \\ 2.2883 \text{ in.} \\ -0.02460 \text{ in.} \\ -0.013559 \text{ rad} \end{bmatrix}$$ **Ans**

These end displacements are depicted in Fig. 6.8(c).

Member Global End Forces: Similarly, by substituting the transpose of $\mathbf{T}_2$ from Eq. (3), and $\mathbf{Q}_2$ from Eq. (2), into Eq. (6.23), we determine the global end forces for member 2 to be

$$\mathbf{F}_2 = \mathbf{T}_2^T\mathbf{Q}_2 = \begin{bmatrix} 77.31 \text{ k} \\ -79.32 \text{ k} \\ \begin{Bmatrix} -2{,}960.4 \text{ k-in.} \\ (= -246.7 \text{ k-ft}) \end{Bmatrix} \\ -77.31 \text{ k} \\ 139.32 \text{ k} \\ \begin{Bmatrix} -6{,}896.7 \text{ k-in.} \\ (= -574.73 \text{ k-ft}) \end{Bmatrix} \end{bmatrix}$$ **Ans**

The global member end forces are shown in Fig. 6.8(d).

Equilibrium Check: See Fig. 6.8(d).

$$+\rightarrow \sum F_X = 0 \qquad 77.31 - 77.31 = 0$$ **Checks**

$$+\uparrow \sum F_Y = 0 \qquad -79.32 - 3(20) + 139.32 = 0$$ **Checks**

$$+\circlearrowleft \sum M_{②} = 0 \qquad -246.7 - 3(20)\left(\frac{16}{2}\right) - 574.73 - 77.31(12) + 139.32(16) = -0.03 \text{ k-ft} \cong 0$$ **Checks**

Therefore, the member is in equilibrium. **Ans**

6.4 MEMBER STIFFNESS RELATIONS IN THE GLOBAL COORDINATE SYSTEM

The process of establishing the stiffness relationships for plane frame members in the global coordinate system is similar to that for the members of plane trusses (Section 3.6). We first substitute the local stiffness relations $\mathbf{Q} = \mathbf{ku} + \mathbf{Q}_f$ (Eq. (6.4)) into the force transformation relations $\mathbf{F} = \mathbf{T}^T\mathbf{Q}$ (Eq. (6.23)) to obtain

$$\mathbf{F} = \mathbf{T}^T\mathbf{Q} = \mathbf{T}^T\mathbf{ku} + \mathbf{T}^T\mathbf{Q}_f \tag{6.26}$$

Then, we substitute the displacement transformation relations $\mathbf{u} = \mathbf{Tv}$ (Eq. (6.20)) into Eq. (6.26) to determine the desired relationships between the member end forces $\mathbf{F}$ and end displacements $\mathbf{v}$, in the global coordinate system:

$$\mathbf{F} = \mathbf{T}^T\mathbf{kTv} + \mathbf{T}^T\mathbf{Q}_f \tag{6.27}$$

Equation (6.27) can be conveniently expressed as

$$\boxed{\mathbf{F} = \mathbf{Kv} + \mathbf{F}_f} \tag{6.28}$$

with

$$\boxed{\mathbf{K} = \mathbf{T}^T\mathbf{kT}} \tag{6.29}$$

$$\boxed{\mathbf{F}_f = \mathbf{T}^T\mathbf{Q}_f} \tag{6.30}$$

The matrix $\mathbf{K}$ represents the member stiffness matrix in the global coordinate system; $\mathbf{F}_f$ is called the *member fixed-end force vector in the global coordinate system.*

Member Global Stiffness Matrix K

The expression of the member global stiffness matrix $\mathbf{K}$ given in Eq. (6.29), as a product of the three matrices $\mathbf{T}^T$, $\mathbf{k}$, and $\mathbf{T}$, is sometimes referred to as the *matrix triple product* form of $\mathbf{K}$. The explicit form of $\mathbf{K}$, in terms of L, E, A, I, and θ of the member, can be determined by substituting the explicit forms of the member local stiffness matrix $\mathbf{k}$ from Eq. (6.6) and the member transformation matrix $\mathbf{T}$ from Eq. (6.19) into Eq. (6.29), and by multiplying the matrices $\mathbf{T}^T$, $\mathbf{k}$, and $\mathbf{T}$, in that order. The explicit form of the member global stiffness matrix $\mathbf{K}$ thus obtained is given in Eq. (6.31).

From a computer programming viewpoint, it is usually more convenient to evaluate $\mathbf{K}$ using the numerical values of $\mathbf{k}$ and $\mathbf{T}$ in the matrix triple product given in Eq. (6.29), rather than the explicit form of $\mathbf{K}$ given in Eq. (6.31). In Section 6.7, we will develop a computer subroutine to generate $\mathbf{K}$ by multiplying the numerical forms of $\mathbf{T}^T$, $\mathbf{k}$, and $\mathbf{T}$, in sequence. The explicit form of $\mathbf{K}$ (Eq. (6.31)), however, provides insight into the physical interpretation of the member global stiffness matrix, and proves convenient for evaluating $\mathbf{K}$ by hand calculations.

$$\mathbf{K} = \frac{EI}{L^3}\begin{bmatrix} \frac{AL^2}{I}\cos^2\theta + 12\sin^2\theta & \left(\frac{AL^2}{I} - 12\right)\cos\theta\sin\theta & -6L\sin\theta & -\left(\frac{AL^2}{I}\cos^2\theta + 12\sin^2\theta\right) & -\left(\frac{AL^2}{I} - 12\right)\cos\theta\sin\theta & -6L\sin\theta \\ \left(\frac{AL^2}{I} - 12\right)\cos\theta\sin\theta & \frac{AL^2}{I}\sin^2\theta + 12\cos^2\theta & 6L\cos\theta & -\left(\frac{AL^2}{I} - 12\right)\cos\theta\sin\theta & -\left(\frac{AL^2}{I}\sin^2\theta + 12\cos^2\theta\right) & 6L\cos\theta \\ -6L\sin\theta & 6L\cos\theta & 4L^2 & 6L\sin\theta & -6L\cos\theta & 2L^2 \\ -\left(\frac{AL^2}{I}\cos^2\theta + 12\sin^2\theta\right) & -\left(\frac{AL^2}{I} - 12\right)\cos\theta\sin\theta & 6L\sin\theta & \frac{AL^2}{I}\cos^2\theta + 12\sin^2\theta & \left(\frac{AL^2}{I} - 12\right)\cos\theta\sin\theta & 6L\sin\theta \\ -\left(\frac{AL^2}{I} - 12\right)\cos\theta\sin\theta & -\left(\frac{AL^2}{I}\sin^2\theta + 12\cos^2\theta\right) & -6L\cos\theta & \left(\frac{AL^2}{I} - 12\right)\cos\theta\sin\theta & \frac{AL^2}{I}\sin^2\theta + 12\cos^2\theta & -6L\cos\theta \\ -6L\sin\theta & 6L\cos\theta & 2L^2 & 6L\sin\theta & -6L\cos\theta & 4L^2 \end{bmatrix} \tag{6.31}$$

The physical interpretation of the member global stiffness matrix $\mathbf{K}$ for plane frame members is similar to that of $\mathbf{K}$ for members of plane trusses. *A stiffness coefficient K_{ij} represents the force at the location and in the direction F_i required, along with other global end forces, to cause a unit value of displacement v_j, while all other global end displacements are 0, and the member is not subjected to any external loads between its ends.* In other words, as depicted in Figs. 6.9(a) through (f), the jth column of $\mathbf{K}$ ($j = 1$ through 6) represents the member end forces, in the global coordinate system, required to cause a unit value of the global end displacement v_j, while all other end displacements are 0, and the member is not subjected to any external loads.

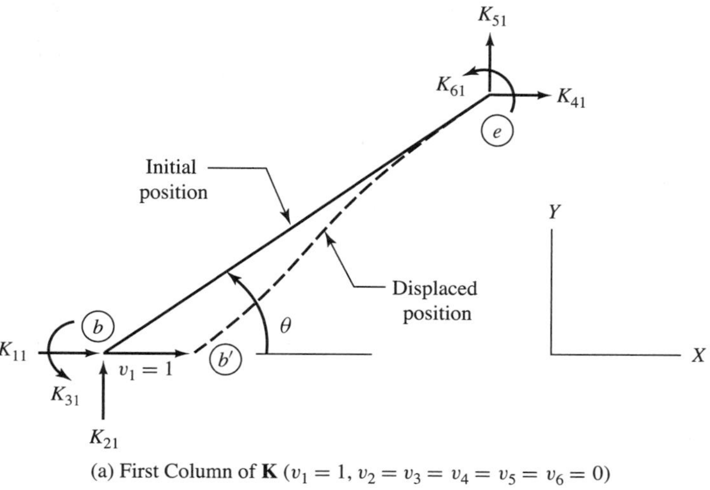

(a) First Column of $\mathbf{K}$ ($v_1 = 1$, $v_2 = v_3 = v_4 = v_5 = v_6 = 0$)

Fig. 6.9

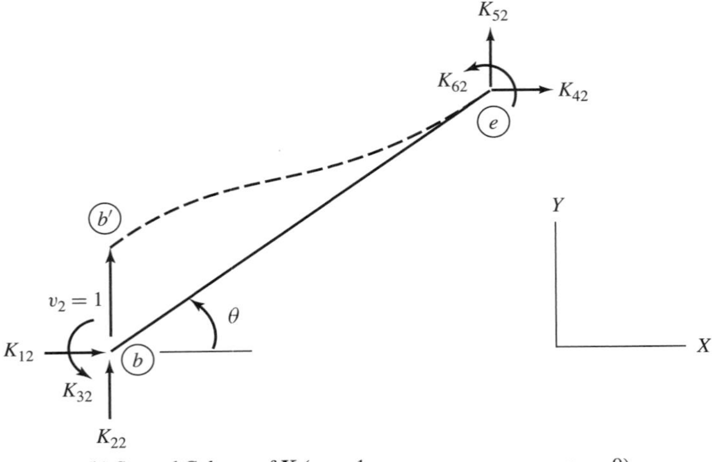

(b) Second Column of **K** ($v_2 = 1$, $v_1 = v_3 = v_4 = v_5 = v_6 = 0$)

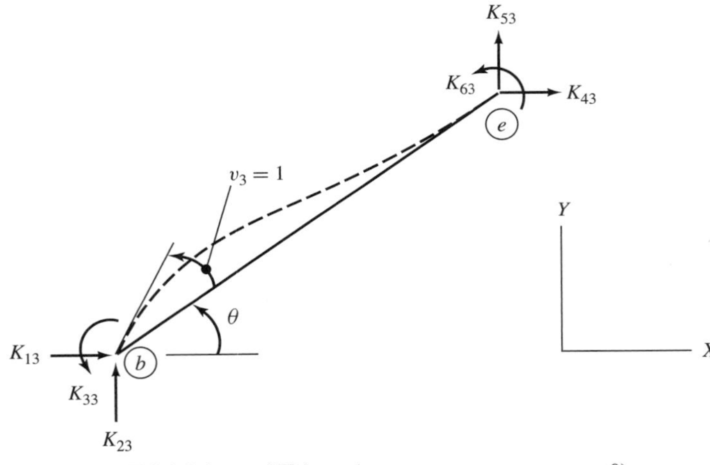

(c) Third Column of **K** ($v_3 = 1$, $v_1 = v_2 = v_4 = v_5 = v_6 = 0$)

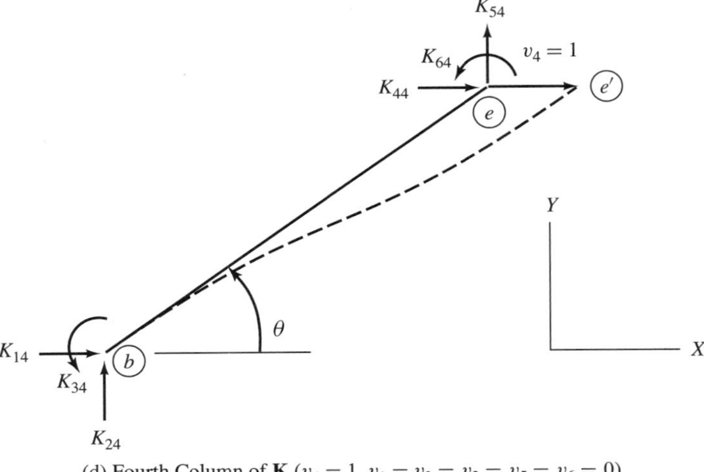

(d) Fourth Column of **K** ($v_4 = 1$, $v_1 = v_2 = v_3 = v_5 = v_6 = 0$)

Fig. 6.9 *(continued)*

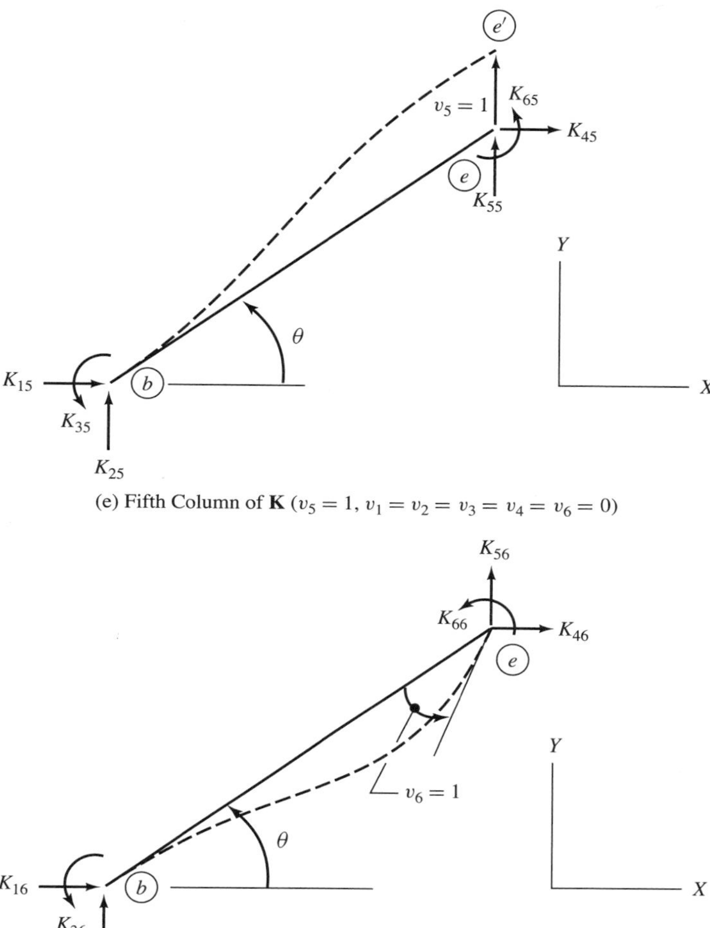

(e) Fifth Column of **K** ($v_5 = 1$, $v_1 = v_2 = v_3 = v_4 = v_6 = 0$)

(f) Sixth Column of **K** ($v_6 = 1$, $v_1 = v_2 = v_3 = v_4 = v_5 = 0$)

Fig. 6.9 *(continued)*

We can use the foregoing interpretation of the member global stiffness matrix to check the explicit form of **K** given in Eq. (6.31). For example, to determine the first column of **K**, we subject the member to a unit end displacement $v_1 = 1$, while all other end displacements are held at 0. As shown in Fig. 6.10(a), the components of this global end displacement in the directions along, and perpendicular to, the member's longitudinal axis, respectively, are

$$u_a = v_1 \cos\theta = 1 \cos\theta = \cos\theta$$
$$u_p = v_1 \sin\theta = 1 \sin\theta = \sin\theta$$

The axial compressive force in the member caused by the axial deformation u_a is shown in Fig. 6.10(b), and the member end shears and moments due to the perpendicular displacement u_p are given in Fig. 6.10(c). Note that these

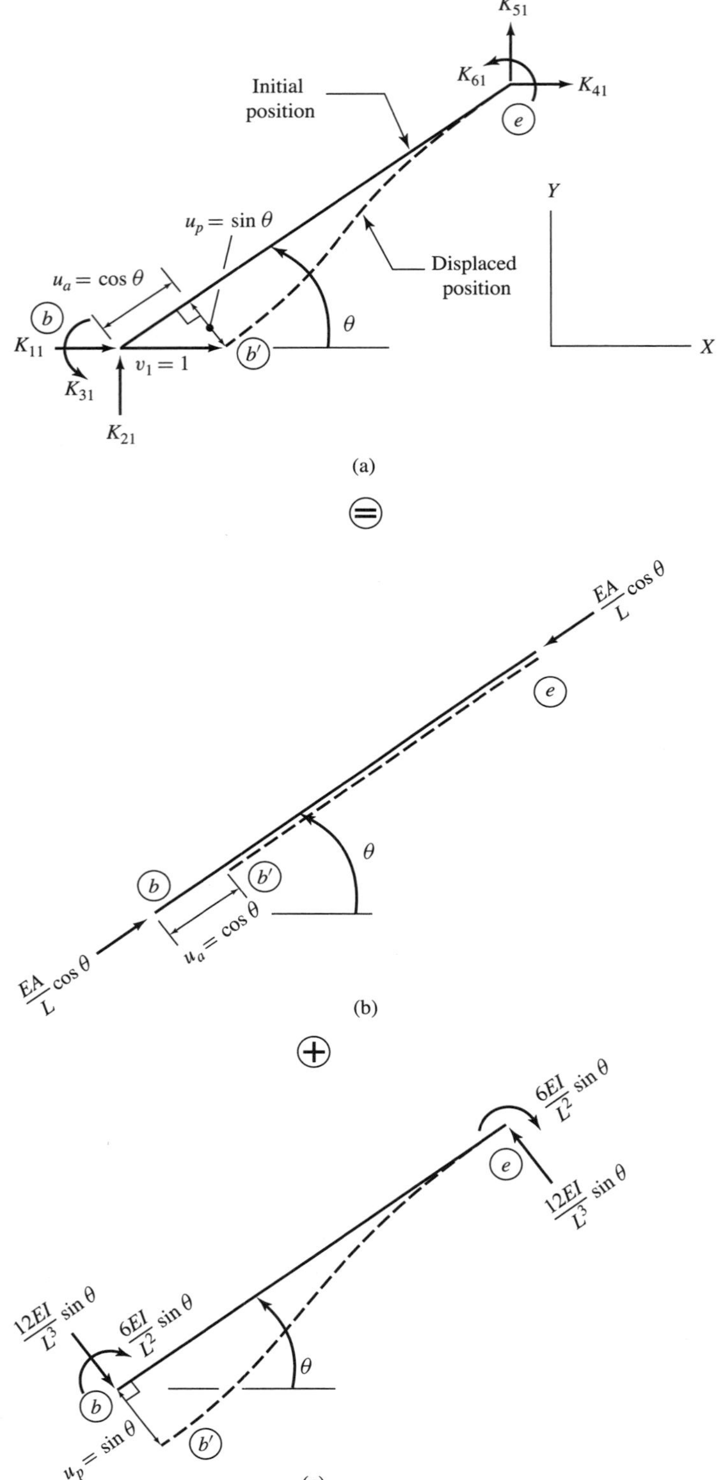

K_{51}
K_{61}
K_{41}
Initial position
e
Y
$u_p = \sin\theta$
Displaced position
$u_a = \cos\theta$
b
θ
K_{11}
$v_1 = 1$
b'
X
K_{31}
K_{21}
(a)
=
$\frac{EA}{L}\cos\theta$
e
θ
b
b'
$u_a = \cos\theta$
$\frac{EA}{L}\cos\theta$
(b)
+
$\frac{6EI}{L^2}\sin\theta$
e
$\frac{12EI}{L^3}\sin\theta$
$\frac{12EI}{L^3}\sin\theta$
$\frac{6EI}{L^2}\sin\theta$
θ
b
$u_p = \sin\theta$
b'
(c)

Fig. 6.10

member end shears and moments are obtained by multiplying the member end forces developed previously (Fig. 6.3(d)) by the negative of u_p (or by setting $u_2 = -u_p = -\sin\theta$ in Fig. 6.3(d)).

By comparing Figs. 6.10(a), (b), and (c), we realize that the global stiffness coefficients K_{11} and K_{21}, at end b of the member, must be equal to the algebraic sums in the global X and Y directions, respectively, of the member end axial force and shear at end b; that is,

$$K_{11} = \left(\frac{EA}{L}\cos\theta\right)\cos\theta + \left(\frac{12EI}{L^3}\sin\theta\right)\sin\theta$$

$$= \frac{EA}{L}\cos^2\theta + \frac{12EI}{L^3}\sin^2\theta \tag{6.32a}$$

and

$$K_{21} = \left(\frac{EA}{L}\cos\theta\right)\sin\theta - \left(\frac{12EI}{L^3}\sin\theta\right)\cos\theta$$

$$= \left(\frac{EA}{L} - \frac{12EI}{L^3}\right)\cos\theta\sin\theta \tag{6.32b}$$

Also, the global stiffness coefficient K_{31} in Fig. 6.10(a) must be equal to the member end moment in Fig. 6.10(c); that is,

$$K_{31} = -\frac{6EI}{L^2}\sin\theta \tag{6.32c}$$

Similarly, the global stiffness coefficients at end e of the member can be expressed as (see Figs. 6.10(a) through (c))

$$K_{41} = -\left(\frac{EA}{L}\cos\theta\right)\cos\theta - \left(\frac{12EI}{L^3}\sin\theta\right)\sin\theta$$

$$= -\frac{EA}{L}\cos^2\theta - \frac{12EI}{L^3}\sin^2\theta \tag{6.32d}$$

$$K_{51} = -\left(\frac{EA}{L}\cos\theta\right)\sin\theta + \left(\frac{12EI}{L^3}\sin\theta\right)\cos\theta$$

$$= -\left(\frac{EA}{L} - \frac{12EI}{L^3}\right)\cos\theta\sin\theta \tag{6.32e}$$

and

$$K_{61} = -\frac{6EI}{L^2}\sin\theta \tag{6.32f}$$

Note that the expressions for the member global stiffness coefficients, in Eqs. 6.32(a) through (f), are identical to those in the first column of the explicit form of **K** given in Eq. (6.31). The remaining columns of **K** can be verified in a similar manner.

Member Global Fixed-End Force Vector $\mathbf{F}_f$

The explicit form of the member global fixed-end force vector $\mathbf{F}_f$ can be obtained by substituting Eqs. (6.19) and (6.15) into the relationship $\mathbf{F}_f = \mathbf{T}^T\mathbf{Q}_f$

(Eq. (6.30)). This yields

$$\mathbf{F}_f = \begin{bmatrix} FA_b \cos\theta - FS_b \sin\theta \\ FA_b \sin\theta + FS_b \cos\theta \\ FM_b \\ FA_e \cos\theta - FS_e \sin\theta \\ FA_e \sin\theta + FS_e \cos\theta \\ FM_e \end{bmatrix} \tag{6.33}$$

The member global fixed-end forces $\mathbf{F}_f$, like the local fixed-end forces $\mathbf{Q}_f$, represent the forces that would develop at the member ends due to external loads, if both member ends were restrained against translations and rotations. However, the global fixed-end forces $\mathbf{F}_f$ are oriented in the global X and Y directions of the structure (Fig. 6.11(a)), whereas the local fixed-end forces $\mathbf{Q}_f$ are oriented in the local x and y directions of the member (Fig. 6.11(b)).

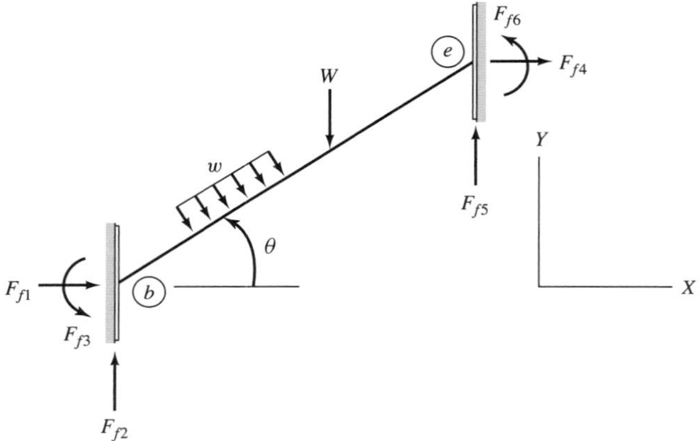

(a) Member Global Fixed-End Force Vector $\mathbf{F}_f$

=

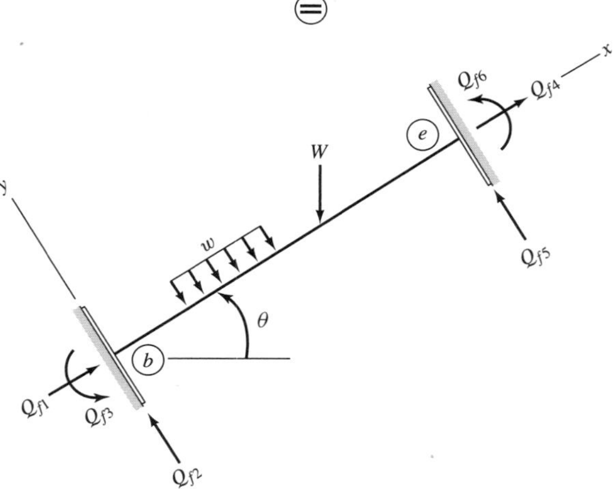

(b) Member Local Fixed-End Force Vector $\mathbf{Q}_f$

Fig. 6.11

EXAMPLE 6.4 In Example 6.3, the global end displacement vector for member 2 of the frame of Fig. 6.8 was found to be

$$\mathbf{v}_2 = \begin{bmatrix} 2.3744 \text{ in.} \\ 0.02792 \text{ in.} \\ -0.0035434 \text{ rad} \\ 2.2883 \text{ in.} \\ -0.02460 \text{ in.} \\ -0.013559 \text{ rad} \end{bmatrix}$$

Calculate the end forces for this member in the global coordinate system using the member global stiffness relationship $\mathbf{F} = \mathbf{Kv} + \mathbf{F}_f$.

SOLUTION *Member Global Stiffness Matrix:* It was shown in Example 6.3 that, for the member under consideration,

$$\cos\theta = 0.8 \qquad \text{and} \qquad \sin\theta = -0.6$$

By substituting these direction cosines, and the numerical values of $E = 29{,}000$ ksi, $A = 28.2$ in.^{2}, $I = 833$ in.^{4}, and $L = 240$ in., into Eq. (6.31), we evaluate the global stiffness matrix for member 2 as

$$\mathbf{K}_2 = \begin{bmatrix} 2{,}188.3 & -1{,}625.5 & 1{,}509.8 & -2{,}188.3 & 1{,}625.5 & 1{,}509.8 \\ -1{,}625.5 & 1{,}240.1 & 2{,}013.1 & 1{,}625.5 & -1{,}240.1 & 2{,}013.1 \\ 1{,}509.8 & 2{,}013.1 & 402{,}620 & -1{,}509.8 & -2{,}013.1 & 201{,}310 \\ -2{,}188.3 & 1{,}625.5 & -1{,}509.8 & 2{,}188.3 & -1{,}625.5 & -1{,}509.8 \\ 1{,}625.5 & -1{,}240.1 & -2{,}013.1 & -1{,}625.5 & 1{,}240.1 & -2{,}013.1 \\ 1{,}509.8 & 2{,}013.1 & 201{,}310 & -1{,}509.8 & -2{,}013.1 & 402{,}620 \end{bmatrix}$$

The matrix $\mathbf{K}_2$ can be obtained alternatively by substituting the numerical forms of $\mathbf{k}_2$ (Eq. (1) of Example 6.2) and $\mathbf{T}_2$ (Eq. (3) of Example 6.3) into the relationship $\mathbf{K} = \mathbf{T}^T\mathbf{kT}$ (Eq. (6.29)), and by evaluating the matrix triple product. The reader is encouraged to use this alternative approach to verify the foregoing $\mathbf{K}_2$ matrix.

Member Global Fixed-end Force Vector: From Example 6.2: $FA_b = FA_e = -18$ k; $FS_b = FS_e = 24$ k; and $FM_b = -FM_e = 960$ k-in. By substituting these numerical values, and $\cos\theta = 0.8$ and $\sin\theta = -0.6$, into Eq. (6.33), we obtain

$$\mathbf{F}_{f2} = \begin{bmatrix} 0 \\ 30 \\ 960 \\ 0 \\ 30 \\ -960 \end{bmatrix}$$

Again, the reader is urged to verify this $\mathbf{F}_{f2}$ vector by substituting the numerical values of $\mathbf{Q}_{f2}$ (Eq. (3) of Example 6.2) and $\mathbf{T}_2$ (Eq. (3) of Example 6.3) into the relationship $\mathbf{F}_f = \mathbf{T}^T\,\mathbf{Q}_f$ (Eq. (6.30)), and by performing the matrix multiplication.

Member Global End Forces: The global end forces for member 2 can now be determined by applying Eq. (6.28):

$$\mathbf{F}_2 = \mathbf{K}_2\mathbf{v}_2 + \mathbf{F}_{f2} = \begin{bmatrix} 77.22 \text{ k} \\ -79.25 \text{ k} \\ \begin{Bmatrix} -2{,}960.5 \text{ k-in.} \\ (= -246.7 \text{ k-ft}) \end{Bmatrix} \\ -77.22 \text{ k} \\ 139.25 \text{ k} \\ \begin{Bmatrix} -6{,}896.7 \text{ k-in.} \\ (= -574.73 \text{ k-ft}) \end{Bmatrix} \end{bmatrix} \qquad \textbf{Ans}$$

Note that this $\mathbf{F}_2$ vector is the same as the one obtained in Example 6.3 by transforming the member end forces from the local to the global coordinate system.

Equilibrium check: See Example 6.3.

6.5 STRUCTURE STIFFNESS RELATIONS

The process of establishing the structure stiffness relations for plane frames is essentially the same as that for beams (Section 5.5), except that the member global (instead of local) stiffness relations must now be used to assemble the structure stiffness matrices and the fixed-joint force vectors. Consider, for example, an arbitrary plane frame as shown in Fig. 6.12(a). As the analytical model of the frame in Fig. 6.12(b) indicates, the frame has three degrees of freedom, d_1, d_2, and d_3, with the corresponding joint loads designated P_1, P_2,

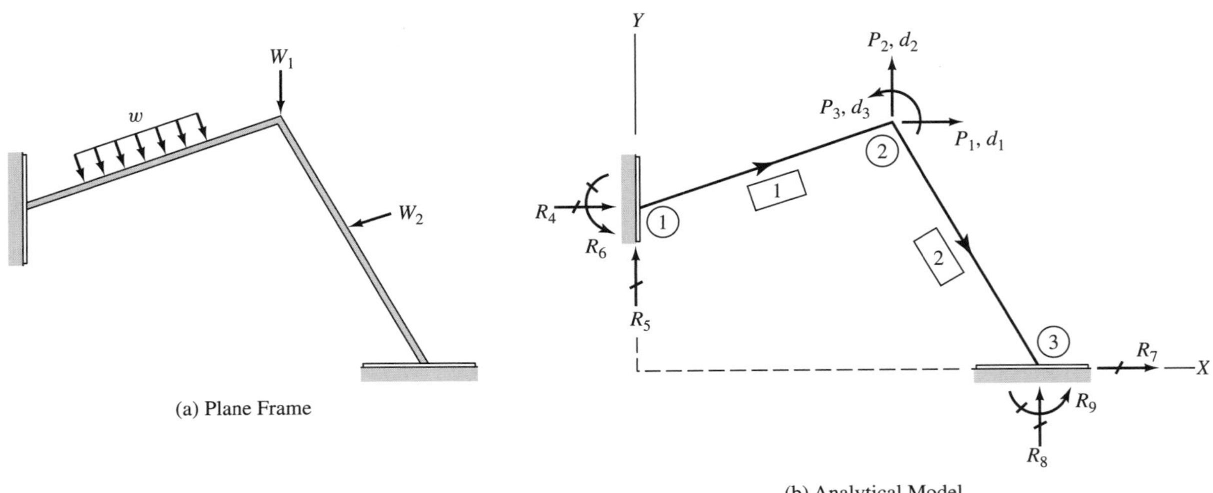

Fig. 6.12

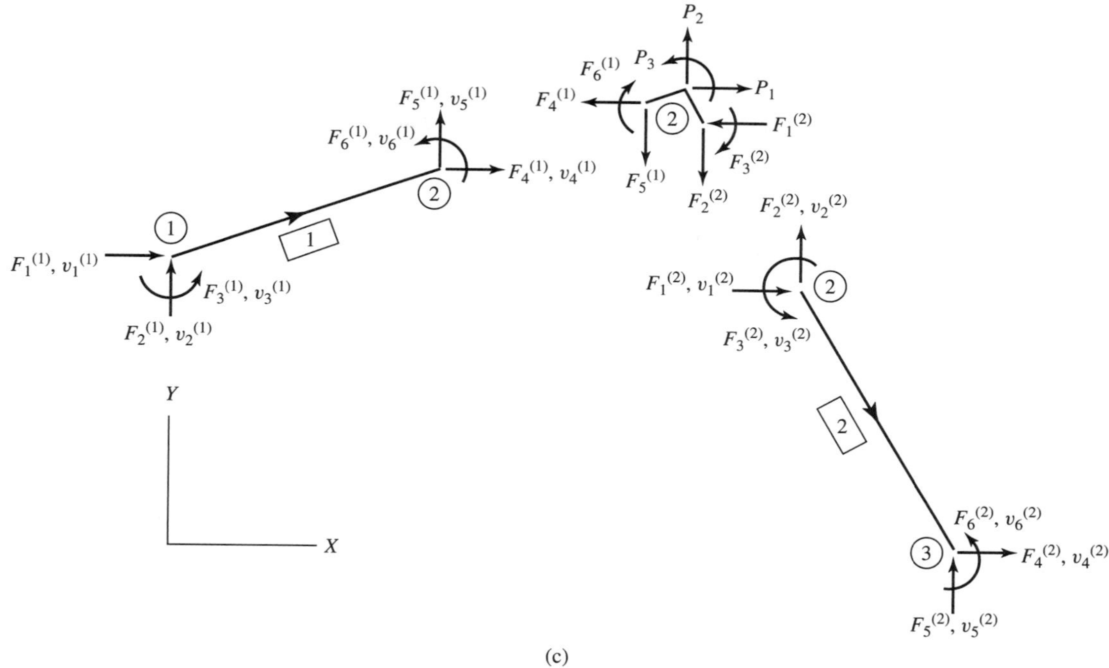

Fig. 6.12 (*continued*)

and P_3, respectively. Remember that our objective is to relate the known external joint (and member) loads to the as yet unknown joint displacements **d**.

To achieve our objective, we first relate the joint loads **P** to the member global end forces **F** by writing the joint equilibrium equations. By applying the three equations of equilibrium, $\sum F_X = 0$, $\sum F_Y = 0$, and $\sum M = 0$, to the free body of joint 2 drawn in Fig. 6.12(c), we obtain

$$P_1 = F_4^{(1)} + F_1^{(2)} \tag{6.34a}$$

$$P_2 = F_5^{(1)} + F_2^{(2)} \tag{6.34b}$$

$$P_3 = F_6^{(1)} + F_3^{(2)} \tag{6.34c}$$

in which the superscript (i) denotes the member number.

Next, we relate the joint displacements **d** to the member global end displacements **v** by applying the compatibility conditions that the member end displacements must be the same as the corresponding joint displacements. Thus, by comparing Figs. 6.12(b) and (c), we write the compatibility equations for members 1 and 2, respectively, as

$$v_1^{(1)} = v_2^{(1)} = v_3^{(1)} = 0 \qquad v_4^{(1)} = d_1 \qquad v_5^{(1)} = d_2 \qquad v_6^{(1)} = d_3 \tag{6.35}$$

$$v_1^{(2)} = d_1 \qquad v_2^{(2)} = d_2 \qquad v_3^{(2)} = d_3 \qquad v_4^{(2)} = v_5^{(2)} = v_6^{(2)} = 0 \tag{6.36}$$

With the relationships between **P** and **F**, and **v** and **d**, now established, we express the member end forces **F** that appear in the equilibrium equations (Eqs. (6.34)) in terms of the member end displacements **v**, using the member global stiffness relations $\mathbf{F} = \mathbf{K}\mathbf{v} + \mathbf{F}_f$ (Eq. (6.28)). By writing this equation in expanded form for an arbitrary member i ($i = 1$ or 2), we obtain

$$\begin{bmatrix} F_1^{(i)} \\ F_2^{(i)} \\ F_3^{(i)} \\ F_4^{(i)} \\ F_5^{(i)} \\ F_6^{(i)} \end{bmatrix} = \begin{bmatrix} K_{11}^{(i)} & K_{12}^{(i)} & K_{13}^{(i)} & K_{14}^{(i)} & K_{15}^{(i)} & K_{16}^{(i)} \\ K_{21}^{(i)} & K_{22}^{(i)} & K_{23}^{(i)} & K_{24}^{(i)} & K_{25}^{(i)} & K_{26}^{(i)} \\ K_{31}^{(i)} & K_{32}^{(i)} & K_{33}^{(i)} & K_{34}^{(i)} & K_{35}^{(i)} & K_{36}^{(i)} \\ K_{41}^{(i)} & K_{42}^{(i)} & K_{43}^{(i)} & K_{44}^{(i)} & K_{45}^{(i)} & K_{46}^{(i)} \\ K_{51}^{(i)} & K_{52}^{(i)} & K_{53}^{(i)} & K_{54}^{(i)} & K_{55}^{(i)} & K_{56}^{(i)} \\ K_{61}^{(i)} & K_{62}^{(i)} & K_{63}^{(i)} & K_{64}^{(i)} & K_{65}^{(i)} & K_{66}^{(i)} \end{bmatrix} \begin{bmatrix} v_1^{(i)} \\ v_2^{(i)} \\ v_3^{(i)} \\ v_4^{(i)} \\ v_5^{(i)} \\ v_6^{(i)} \end{bmatrix} + \begin{bmatrix} F_{f1}^{(i)} \\ F_{f2}^{(i)} \\ F_{f3}^{(i)} \\ F_{f4}^{(i)} \\ F_{f5}^{(i)} \\ F_{f6}^{(i)} \end{bmatrix} \tag{6.37}$$

From this, we determine the expressions for forces at end 2 of member 1 (i.e., $i = 1$) to be

$$F_4^{(1)} = K_{41}^{(1)} v_1^{(1)} + K_{42}^{(1)} v_2^{(1)} + K_{43}^{(1)} v_3^{(1)} + K_{44}^{(1)} v_4^{(1)} + K_{45}^{(1)} v_5^{(1)} + K_{46}^{(1)} v_6^{(1)} + F_{f4}^{(1)} \tag{6.38a}$$

$$F_5^{(1)} = K_{51}^{(1)} v_1^{(1)} + K_{52}^{(1)} v_2^{(1)} + K_{53}^{(1)} v_3^{(1)} + K_{54}^{(1)} v_4^{(1)} + K_{55}^{(1)} v_5^{(1)} + K_{56}^{(1)} v_6^{(1)} + F_{f5}^{(1)} \tag{6.38b}$$

$$F_6^{(1)} = K_{61}^{(1)} v_1^{(1)} + K_{62}^{(1)} v_2^{(1)} + K_{63}^{(1)} v_3^{(1)} + K_{64}^{(1)} v_4^{(1)} + K_{65}^{(1)} v_5^{(1)} + K_{66}^{(1)} v_6^{(1)} + F_{f6}^{(1)} \tag{6.38c}$$

Similarly, from Eq. (6.37), we determine the expressions for forces at end 2 of member 2 (i.e., $i = 2$) to be

$$F_1^{(2)} = K_{11}^{(2)} v_1^{(2)} + K_{12}^{(2)} v_2^{(2)} + K_{13}^{(2)} v_3^{(2)} + K_{14}^{(2)} v_4^{(2)} + K_{15}^{(2)} v_5^{(2)} + K_{16}^{(2)} v_6^{(2)} + F_{f1}^{(2)} \tag{6.39a}$$

$$F_2^{(2)} = K_{21}^{(2)} v_1^{(2)} + K_{22}^{(2)} v_2^{(2)} + K_{23}^{(2)} v_3^{(2)} + K_{24}^{(2)} v_4^{(2)} + K_{25}^{(2)} v_5^{(2)} + K_{26}^{(2)} v_6^{(2)} + F_{f2}^{(2)} \tag{6.39b}$$

$$F_3^{(2)} = K_{31}^{(2)} v_1^{(2)} + K_{32}^{(2)} v_2^{(2)} + K_{33}^{(2)} v_3^{(2)} + K_{34}^{(2)} v_4^{(2)} + K_{35}^{(2)} v_5^{(2)} + K_{36}^{(2)} v_6^{(2)} + F_{f3}^{(2)} \tag{6.39c}$$

By substituting the compatibility equations for members 1 and 2 (Eqs. (6.35) and (6.36)) into Eqs. (6.38) and (6.39), respectively, we obtain

$$F_4^{(1)} = K_{44}^{(1)} d_1 + K_{45}^{(1)} d_2 + K_{46}^{(1)} d_3 + F_{f4}^{(1)} \tag{6.40a}$$

$$F_5^{(1)} = K_{54}^{(1)} d_1 + K_{55}^{(1)} d_2 + K_{56}^{(1)} d_3 + F_{f5}^{(1)} \tag{6.40b}$$

$$F_6^{(1)} = K_{64}^{(1)} d_1 + K_{65}^{(1)} d_2 + K_{66}^{(1)} d_3 + F_{f6}^{(1)} \tag{6.40c}$$

$$F_1^{(2)} = K_{11}^{(2)} d_1 + K_{12}^{(2)} d_2 + K_{13}^{(2)} d_3 + F_{f1}^{(2)} \tag{6.40d}$$

$$F_2^{(2)} = K_{21}^{(2)} d_1 + K_{22}^{(2)} d_2 + K_{23}^{(2)} d_3 + F_{f2}^{(2)} \tag{6.40e}$$

$$F_3^{(2)} = K_{31}^{(2)} d_1 + K_{32}^{(2)} d_2 + K_{33}^{(2)} d_3 + F_{f3}^{(2)} \tag{6.40f}$$

Finally, by substituting Eqs. (6.40) into the joint equilibrium equations (Eqs. (6.34)), we obtain the desired structure stiffness relations for the plane frame:

$$\begin{aligned} P_1 = &\left(K_{44}^{(1)} + K_{11}^{(2)}\right) d_1 + \left(K_{45}^{(1)} + K_{12}^{(2)}\right) d_2 \\ &+ \left(K_{46}^{(1)} + K_{13}^{(2)}\right) d_3 + \left(F_{f4}^{(1)} + F_{f1}^{(2)}\right) \end{aligned} \tag{6.41a}$$

$$\begin{aligned} P_2 = &\left(K_{54}^{(1)} + K_{21}^{(2)}\right) d_1 + \left(K_{55}^{(1)} + K_{22}^{(2)}\right) d_2 \\ &+ \left(K_{56}^{(1)} + K_{23}^{(2)}\right) d_3 + \left(F_{f5}^{(1)} + F_{f2}^{(2)}\right) \end{aligned} \tag{6.41b}$$

$$\begin{aligned} P_3 = &\left(K_{64}^{(1)} + K_{31}^{(2)}\right) d_1 + \left(K_{65}^{(1)} + K_{32}^{(2)}\right) d_2 \\ &+ \left(K_{66}^{(1)} + K_{33}^{(2)}\right) d_3 + \left(F_{f6}^{(1)} + F_{f3}^{(2)}\right) \end{aligned} \tag{6.41c}$$

The foregoing equations can be symbolically expressed as

$$\mathbf{P} = \mathbf{Sd} + \mathbf{P}_f$$

or

$$\boxed{\mathbf{P} - \mathbf{P}_f = \mathbf{Sd}} \tag{6.42}$$

in which **S** represents the $NDOF \times NDOF$ structure stiffness matrix, and $\mathbf{P}_f$ is the $NDOF \times 1$ structure fixed-joint force vector, for the plane frame with

$$\mathbf{S} = \begin{bmatrix} K_{44}^{(1)} + K_{11}^{(2)} & K_{45}^{(1)} + K_{12}^{(2)} & K_{46}^{(1)} + K_{13}^{(2)} \\ K_{54}^{(1)} + K_{21}^{(2)} & K_{55}^{(1)} + K_{22}^{(2)} & K_{56}^{(1)} + K_{23}^{(2)} \\ K_{64}^{(1)} + K_{31}^{(2)} & K_{65}^{(1)} + K_{32}^{(2)} & K_{66}^{(1)} + K_{33}^{(2)} \end{bmatrix} \tag{6.43}$$

and

$$\mathbf{P}_f = \begin{bmatrix} F_{f4}^{(1)} + F_{f1}^{(2)} \\ F_{f5}^{(1)} + F_{f2}^{(2)} \\ F_{f6}^{(1)} + F_{f3}^{(2)} \end{bmatrix} \tag{6.44}$$

Structure Stiffness Matrix S

As discussed in Chapters 3 and 5, an element S_{ij} of the structure stiffness matrix **S** represents the force at the location and in the direction of P_i required, along with other joint forces, to cause a unit value of the displacement d_j, while all other joint displacements are 0, and the frame is subjected to no external loads. In other words, the jth column of **S** consists of joint forces required, at the locations and in the directions of all the degrees of freedom of the frame, to cause a unit value of the displacement d_j while all other joint displacements are 0.

We can use the foregoing interpretation to verify the **S** matrix given in Eq. (6.43) for the frame of Fig. 6.12. To obtain the first column of **S**, we subject the frame to a unit value of the joint displacement $d_1 = 1$ ($d_2 = d_3 = 0$), as shown in Fig. 6.13(a). As depicted there, this unit joint displacement induces

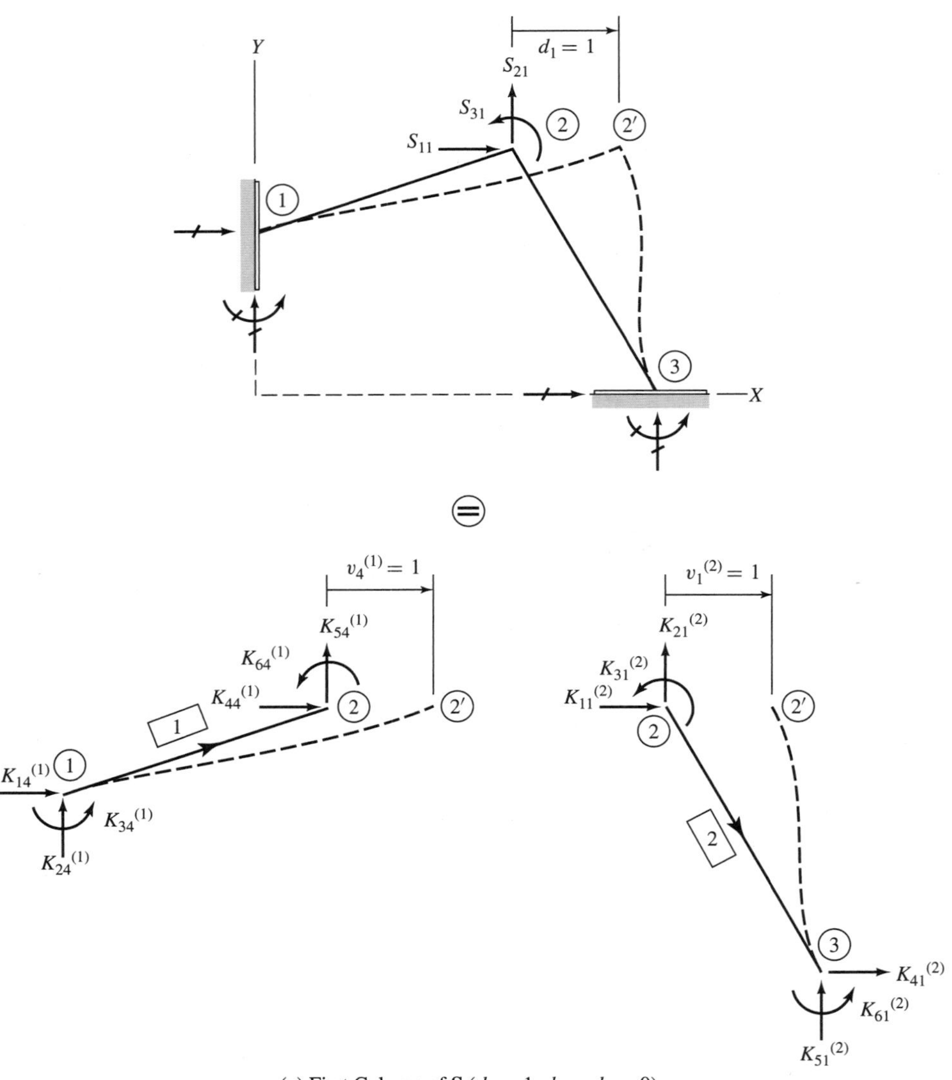

(a) First Column of **S** ($d_1 = 1$, $d_2 = d_3 = 0$)

Fig. 6.13

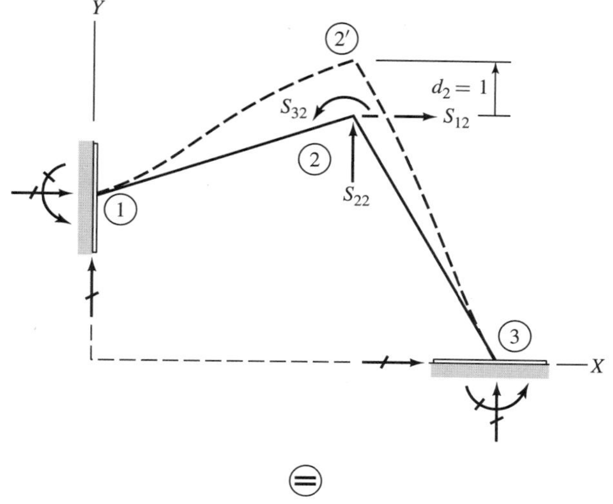

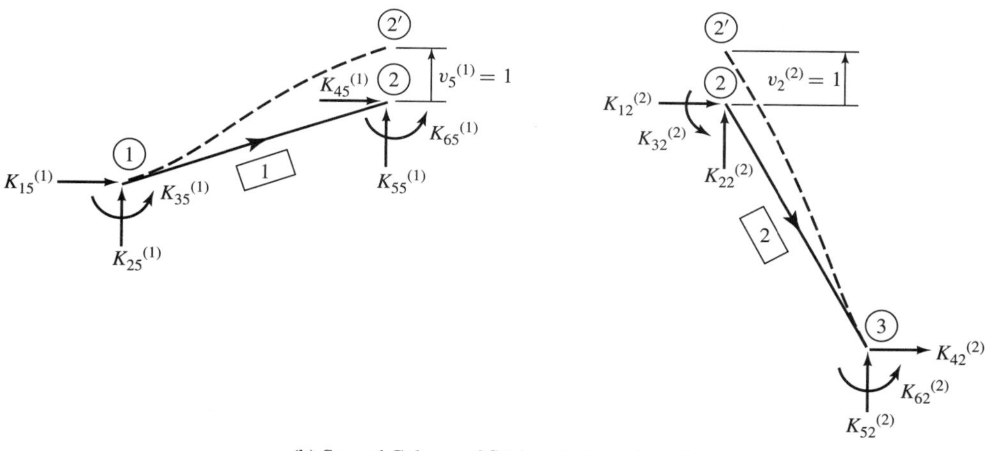

(b) Second Column of **S** ($d_2 = 1$, $d_1 = d_3 = 0$)

Fig. 6.13 (*continued*)

unit global end displacements $v_4^{(1)}$ at the end of member 1, and $v_1^{(2)}$ at the beginning of member 2. The member global stiffness coefficients, necessary to cause the foregoing end displacements, are also given in Fig. 6.13(a). From this figure, we can see that the structure stiffness coefficients (or joint forces) S_{11} and S_{21} at joint 2 must be equal to the algebraic sums of the forces in the X and Y directions, respectively, at the two member ends connected to the joint; that is,

$$S_{11} = K_{44}^{(1)} + K_{11}^{(2)} \tag{6.45a}$$

$$S_{21} = K_{54}^{(1)} + K_{21}^{(2)} \tag{6.45b}$$

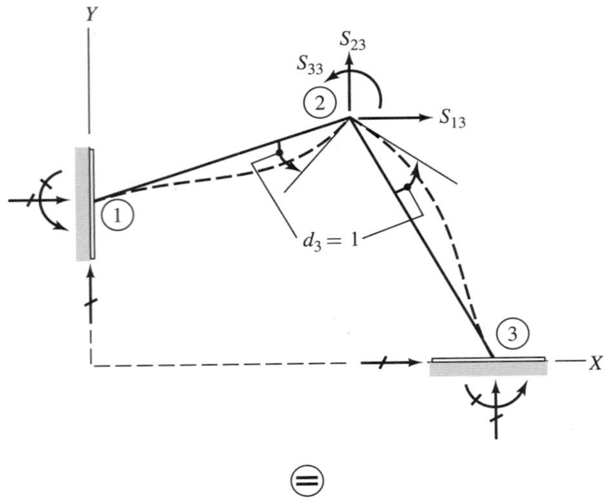

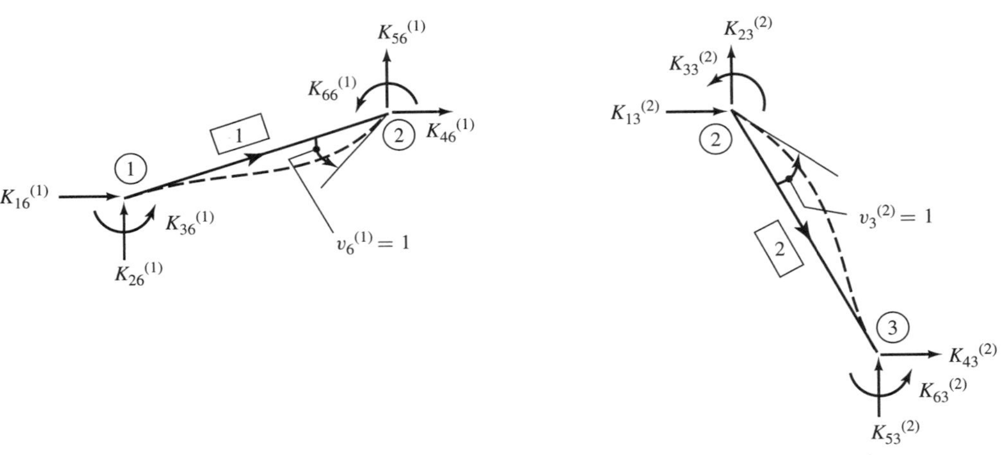

(c) Third Column of **S** ($d_3 = 1, d_1 = d_2 = 0$)

Fig. 6.13 *(continued)*

Similarly, the structure stiffness coefficient (or joint moment) S_{31} at joint 2 must be equal to the algebraic sum of the moments at the two member ends connected to the joint; thus,

$$S_{31} = K_{64}^{(1)} + K_{31}^{(2)} \tag{6.45c}$$

Note that the expressions for S_{i1} ($i = 1$ to 3) given in Eqs. (6.45) are identical to those listed in the first column of **S** in Eq. (6.43).

The second and third columns of **S** can be verified in a similar manner using Figs. 6.13(b) and (c), respectively. It should be noted that the structure stiffness matrix **S** in Eq. (6.43) is symmetric, because of the symmetry of the member global stiffness matrices (i.e., $K_{ij} = K_{ji}$). The structure stiffness matrices of linear elastic structures must always be symmetric.

Structure Fixed-Joint Force Vector $\mathbf{P}_f$ and Equivalent Joint Loads

As discussed in Chapter 5, the structure fixed-joint forces represent the reactions that would develop at the locations and in the directions of the frame's degrees of freedom, due to member loads, if all the joints of the frame were fixed against translations and rotations. This definition enables us to directly express the structure fixed-joint forces in terms of the member global fixed-end forces (instead of deriving such expressions by combining the frame's equilibrium, compatibility, and member force-displacement relations, as was done in the earlier part of this section—see Eqs. (6.34) through (6.44)).

Let us verify the $\mathbf{P}_f$ vector, given in Eq. (6.44) for the frame of Fig. 6.12, using this direct approach. The frame is redrawn in Fig. 6.14(a) with its joint 2, which is actually free to translate in the X and Y directions and rotate, now restrained against these displacements by an imaginary restraint. When this hypothetical completely fixed frame is subjected to member loads only (note that the joint load W_1 shown in Fig. 6.12(a) is not drawn in Fig. 6.14(a)), the structure fixed-joint forces (or reactions) P_{f1}, P_{f2}, and P_{f3} develop at the imaginary restraint at joint 2. As shown in Fig. 6.14(a), the structure fixed-joint force at the location and in the direction of an ith degree of freedom is denoted by P_{fi}.

To relate the structure fixed-joint forces $\mathbf{P}_f$ to the member global fixed-end forces $\mathbf{F}_f$, we draw the free-body diagrams of the two members of the hypothetical fixed frame, as shown in Fig. 6.14(b). Note that, because all the joints of the frame are restrained, the member ends, which are rigidly connected to the joints, are also fixed against any displacements. Therefore, only the fixed-end forces due to member loads, $\mathbf{F}_f$, can develop at the ends of the members.

By comparing Figs. 6.14(a) and (b), we realize that the structure fixed-joint forces P_{f1} and P_{f2} at joint 2 must be equal to the algebraic sums of the fixed-end forces in the X and Y directions, respectively, at the two member ends connected to the joint; that is,

$$P_{f1} = F_{f4}^{(1)} + F_{f1}^{(2)} \tag{6.46a}$$

$$P_{f2} = F_{f5}^{(1)} + F_{f2}^{(2)} \tag{6.46b}$$

Similarly, the structure fixed-joint moment P_{f3} at joint 2 must be equal to the algebraic sum of the fixed-end moments at the two member ends connected to the joint. Thus,

$$P_{f3} = F_{f6}^{(1)} + F_{f3}^{(2)} \tag{6.46c}$$

Note that the expressions for P_{fi} ($i = 1$ to 3) given in Eqs. (6.46) are the same as those listed in the $\mathbf{P}_f$ vector in Eq. (6.44).

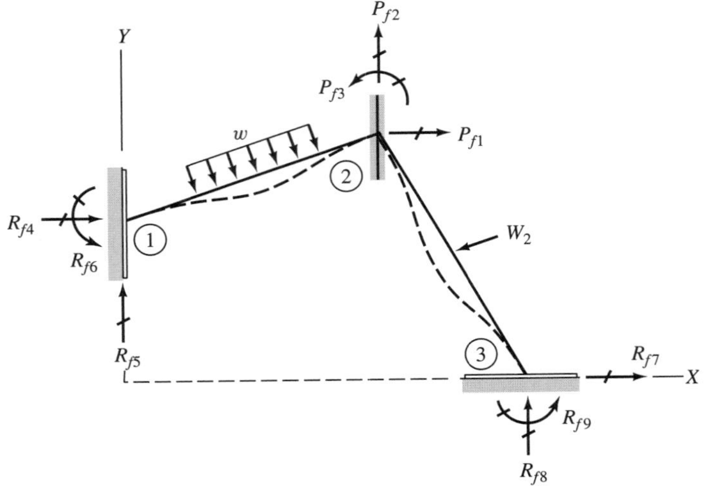

(a) Fixed Frame Subjected to Member Loads

=

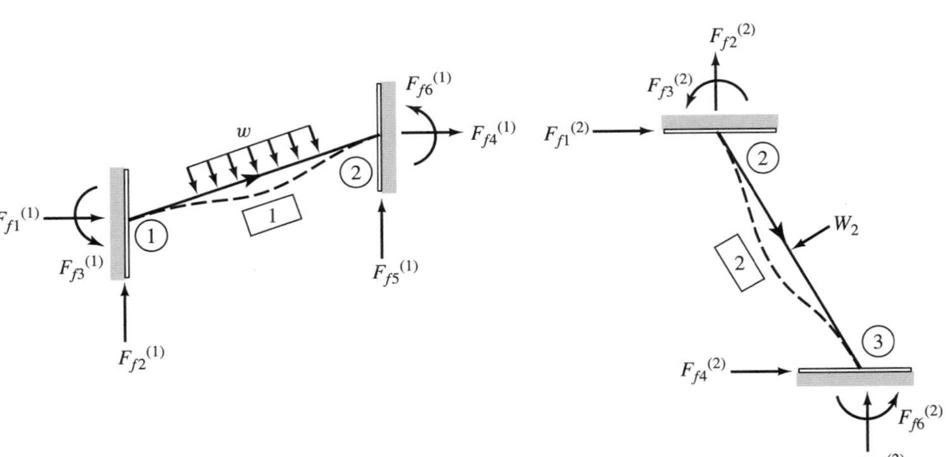

(b) Member Global Fixed-End Forces

Fig. 6.14

It may be recalled from Section 5.6 that another interpretation of the structure fixed-joint forces due to member loads is that when they are applied to the structure with their directions reversed, the fixed-joint forces cause the same joint displacements as the actual member loads. The negatives of the structure fixed-joint forces are, therefore, referred to as the equivalent joint loads. We can show the validity of this interpretation by setting the joint loads equal to 0 (i.e., $\mathbf{P} = \mathbf{0}$) in Eq. (6.42), thereby reducing the structure stiffness relationship to

$$-\mathbf{P}_f = \mathbf{S}\mathbf{d} \tag{6.47}$$

in which **d** now represents the joint displacements due to the negatives of the structure fixed-joint forces applied to the joints of the structure. However, since member loads are now the only external effects acting on the structure, the **d** vector in Eq. (6.47) must also represent the joint displacements due to member loads. Thus, we can conclude that the negatives of the structure fixed-joint forces must cause the same joint displacements as the actual member loads.

The validity of this interpretation can also be demonstrated using the principle of superposition. Figure 6.15(a) shows the two-member frame considered previously (Fig. 6.12), subjected to arbitrary member loads w and W_2. In Fig. 6.15(b), joint 2 of the frame is fixed by an imaginary restraint so that, when the fixed frame is subjected to member loads, the structure fixed-joint forces P_{f1}, P_{f2}, and P_{f3} develop at the imaginary restraint. Lastly, in Fig. 6.15(c), the actual frame is subjected to joint loads, which are equal in magnitude to the structure fixed-joint forces P_{f1}, P_{f2}, and P_{f3}, but reversed in direction.

By comparing Figs. 6.15(a) through (c), we realize that the actual loading applied to the actual frame in Fig. 6.15(a) equals the algebraic sum of the loadings in Figs. 6.15(b) and (c), because the reactions P_{f1}, P_{f2}, and P_{f3} in Fig. 6.15(b) cancel the corresponding applied loads in Fig. 6.15(c). Thus, in accordance with the superposition principle, any joint displacement of the actual frame due to the member loads (Fig. 6.15(a)) must be equal to the algebraic sum of the corresponding joint displacement of the fixed frame due to the member loads (Fig. 6.15(b)) and the corresponding joint displacement of the actual frame subjected to no member loads, but to the structure fixed-joint forces with their directions reversed. However, since the joint displacements of the fixed frame (Fig. 6.15(b)) are 0, the joint displacements of the frame due to the member loads (Fig. 6.15(a)) must be equal to the corresponding joint displacements of the frame due to the negatives of the fixed-joint forces (Fig. 6.15(c)). Thus, the negatives of the structure fixed-joint forces can be considered to be equivalent to member loads in terms of joint displacements. It should be noted that this equivalency is valid only for joint displacements, and it cannot be generalized to member end forces and support reactions.

Assembly of S and $\mathbf{P}_f$, Using Member Code Numbers

In the preceding paragraphs of this section, we have demonstrated that the structure stiffness matrix **S** for plane frames can be formulated directly by algebraically adding the appropriate elements of the member global stiffness matrices **K** (see, for example, Eqs. (6.43) and (6.45), and Fig. 6.13). Furthermore, it has been shown that the structure fixed-joint force vector $\mathbf{P}_f$ for plane frames can also be established directly by algebraically adding the member global fixed-end forces $\mathbf{F}_f$ at the location, and in the direction, of each of the structure's degrees of freedom (see, for example, Eqs. (6.44) and (6.46), and Fig. 6.14).

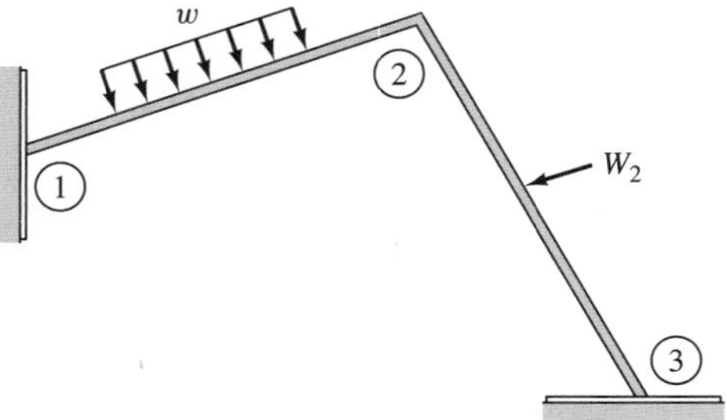

(a) Actual Frame Subjected to Member Loads

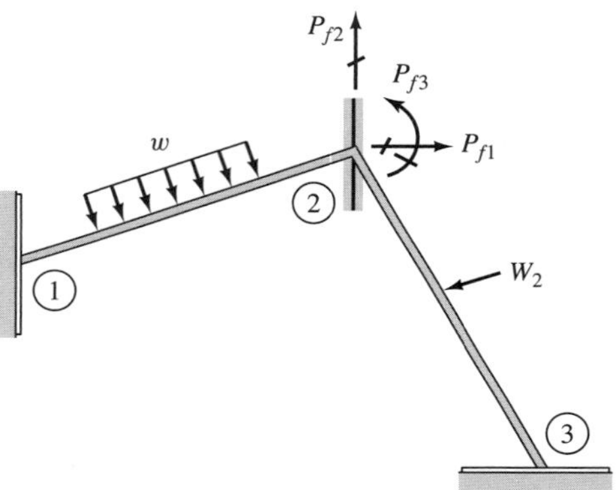

(b) Fixed Frame Subjected to Member Loads

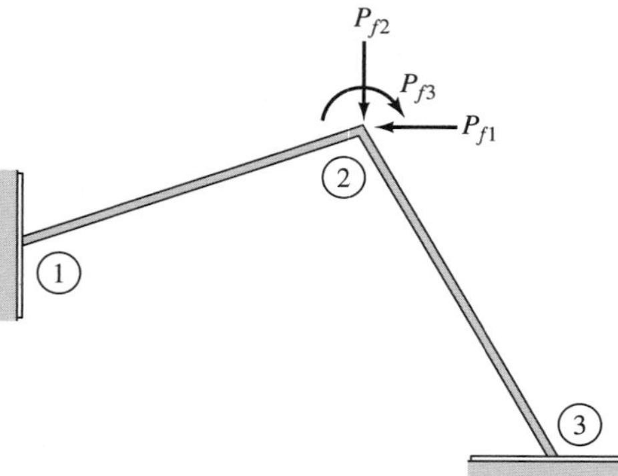

(c) Actual Frame Subjected to Equivalent Joint Loads

Fig. 6.15

The foregoing process of directly generating **S** and $\mathbf{P}_f$ can be conveniently implemented by employing the member code number technique described in detail in Chapters 3 and 5. The application of this technique for plane frames remains essentially the same as that for the case of beams, except that the member global (instead of local) stiffness matrices **K** and the member global fixed-end force vectors $\mathbf{F}_f$ must now be used to form **S** and $\mathbf{P}_f$, respectively. It should also be realized that each member of the plane frame has six code numbers, arranged in the sequential order of the member's global end displacements **v**. The application of the member code number technique for plane frames is illustrated in the following example.

EXAMPLE 6.5

Determine the structure stiffness matrix, the fixed-joint force vector, and the equivalent joint loads for the frame shown in Fig. 6.16(a).

SOLUTION

Analytical Model: See Fig. 6.16(b). The frame has four degrees of freedom and five restrained coordinates, as shown.

Structure Stiffness Matrix: The 4 × 4 structure stiffness matrix will be generated by evaluating each member's global stiffness matrix **K**, and storing its pertinent elements in **S** using the member code numbers.

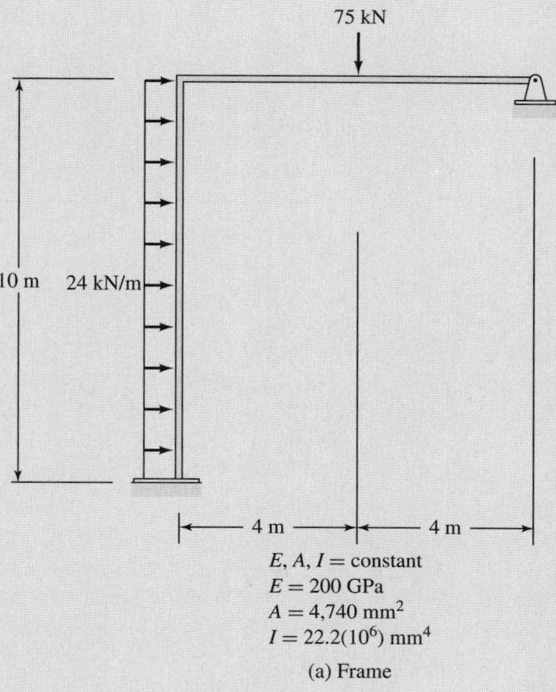

(a) Frame

Fig. 6.16

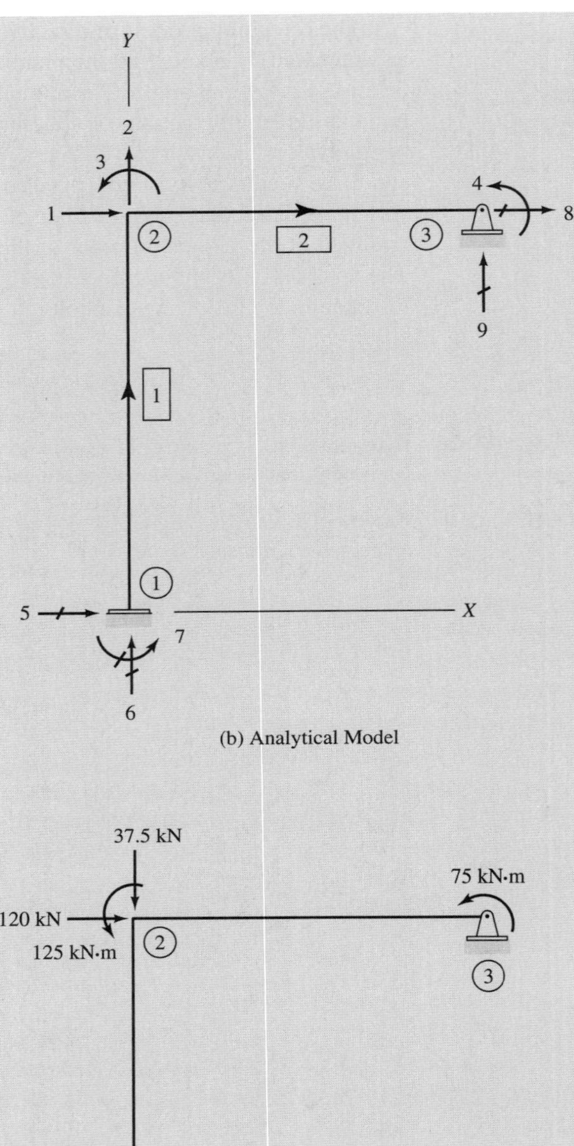

(b) Analytical Model

(c) Equivalent Joint Loads

Fig. 6.16 *(continued)*

Member 1 As shown in Fig. 6.16(b), joint 1 is the beginning joint and joint 2 is the end joint for this member. Thus,

$$\cos\theta = \frac{X_2 - X_1}{L} = \frac{0-0}{10} = 0$$

$$\sin\theta = \frac{Y_2 - Y_1}{L} = \frac{10-0}{10} = 1$$

By substituting $E = 200(10)^6$ kN/m^2, $A = 0.00474$ m^2, $I = 0.0000222$ m^4, $L = 10$ m, and the foregoing values of the direction cosines into the expression for **K** given in Eq. (6.31), we obtain

$$\mathbf{K}_1 = \begin{array}{c} \begin{matrix} 5 & 6 & 7 & 1 & 2 & 3 \end{matrix} \\ \left[\begin{array}{rrr|rrr} 53.28 & 0 & -266.4 & -53.28 & 0 & -266.4 \\ 0 & 94{,}800 & 0 & 0 & -94{,}800 & 0 \\ -266.4 & 0 & 1{,}776 & 266.4 & 0 & 888 \\ \hline -53.28 & 0 & 266.4 & 53.28 & 0 & 266.4 \\ 0 & -94{,}800 & 0 & 0 & 94{,}800 & 0 \\ -266.4 & 0 & 888 & 266.4 & 0 & 1{,}776 \end{array}\right] \end{array} \begin{matrix} 5 \\ 6 \\ 7 \\ 1 \\ 2 \\ 3 \end{matrix} \qquad \textbf{(1)}$$

From Fig. 6.16(b), we observe that the code numbers for member 1 are 5, 6, 7, 1, 2, 3. These numbers are written on the right side and at the top of $\mathbf{K}_1$ (Eq. (1)), to indicate the rows and columns of **S** in which the elements of $\mathbf{K}_1$ are to be stored. Thus,

$$\mathbf{S} = \begin{array}{c} \begin{matrix} 1 & 2 & 3 & 4 \end{matrix} \\ \left[\begin{array}{rrrr} 53.28 & 0 & 266.4 & 0 \\ 0 & 94{,}800 & 0 & 0 \\ 266.4 & 0 & 1{,}776 & 0 \\ 0 & 0 & 0 & 0 \end{array}\right] \end{array} \begin{matrix} 1 \\ 2 \\ 3 \\ 4 \end{matrix} \qquad \textbf{(2)}$$

Member 2 From Fig. 6.16(b), we can see that this member is horizontal, with its left-end joint 2 selected as the beginning joint, thereby orienting the positive directions of the member's local x and y axes in the positive directions of the global X and Y axes, respectively. Thus, no coordinate transformations are needed for this member (i.e., $\cos\theta = 1$, $\sin\theta = 0$, and $\mathbf{T} = \mathbf{I}$); and its stiffness relations, and fixed-end forces, are the same in the local and global coordinate systems.

By substituting the numerical values of E, A, and I, and $L = 8$ m into Eq. (6.6), we obtain

$$\mathbf{K}_2 = \mathbf{k}_2 = \begin{array}{c} \begin{matrix} 1 & 2 & 3 & 8 & 9 & 4 \end{matrix} \\ \left[\begin{array}{rrr|rr|r} 118{,}500 & 0 & 0 & -118{,}500 & 0 & 0 \\ 0 & 104.06 & 416.25 & 0 & -104.06 & 416.25 \\ 0 & 416.25 & 2{,}220 & 0 & -416.25 & 1{,}110 \\ \hline -118{,}500 & 0 & 0 & 118{,}500 & 0 & 0 \\ 0 & -104.06 & -416.25 & 0 & 104.06 & -416.25 \\ \hline 0 & 416.25 & 1{,}110 & 0 & -416.25 & 2{,}220 \end{array}\right] \end{array} \begin{matrix} 1 \\ 2 \\ 3 \\ 8 \\ 9 \\ 4 \end{matrix}$$

The code numbers for this member—1, 2, 3, 8, 9, 4 (see Fig. 6.16(b))—are now used to add the pertinent elements of $\mathbf{K}_2$ in their proper positions in the structure stiffness matrix **S** given in Eq. (2), which now becomes

$$\mathbf{S} = \begin{bmatrix} 53.28 + 118{,}500 & 0 & 266.4 & 0 \\ 0 & 94{,}800 + 104.06 & 416.25 & 416.25 \\ 266.4 & 416.25 & 1{,}776 + 2{,}220 & 1{,}110 \\ 0 & 416.25 & 1{,}110 & 2{,}220 \end{bmatrix} \begin{matrix} 1 \\ 2 \\ 3 \\ 4 \end{matrix}$$

(columns 1, 2, 3, 4)

Because the stiffnesses of both members of the frame have now been stored in **S**, the structure stiffness matrix for the given frame is

$$\mathbf{S} = \begin{bmatrix} 118{,}553 & 0 & 266.4 & 0 \\ 0 & 94{,}904 & 416.25 & 416.25 \\ 266.4 & 416.25 & 3{,}996 & 1{,}110 \\ 0 & 416.25 & 1{,}110 & 2{,}220 \end{bmatrix} \begin{matrix} 1 \\ 2 \\ 3 \\ 4 \end{matrix} \qquad \textbf{Ans}$$

(columns 1, 2, 3, 4)

Note that the structure stiffness matrix is symmetric.

Structure Fixed-Joint Force Vector: We will generate the 4 × 1 structure fixed-joint force vector by evaluating, for each member, the global fixed-end force vector $\mathbf{F}_f$, and storing its pertinent elements in $\mathbf{P}_f$ using the member code numbers.

Member 1 The 24 kN/m uniformly distributed load acting on this member is positive, because it acts in the negative direction of the member's local y axis. By substituting $w = 24$ kN/m, $L = 10$ m, and $l_1 = l_2 = 0$ into the fixed-end force expressions for loading type 3 listed inside the front cover, we evaluate

$$FS_b = FS_e = \frac{24(10)}{2} = 120 \text{ kN}$$

$$FM_b = -FM_e = \frac{24(10)^2}{12} = 200 \text{ kN} \cdot \text{m}$$

As the member is not subjected to any axial loads,

$$FA_b = FA_e = 0$$

By substituting the foregoing values of the member fixed-end forces, along with $\cos\theta = 0$ and $\sin\theta = 1$, into the explicit form of $\mathbf{F}_f$ given in Eq. (6.33), we obtain

$$\mathbf{F}_{f1} = \begin{bmatrix} -120 \\ 0 \\ 200 \\ \hdashline -120 \\ 0 \\ -200 \end{bmatrix} \begin{matrix} 5 \\ 6 \\ 7 \\ 1 \\ 2 \\ 3 \end{matrix} \qquad \textbf{(3)}$$

The code numbers of the member, 5, 6, 7, 1, 2, 3, are written on the right side of $\mathbf{F}_{f1}$ in Eq. (3) to indicate the rows of $\mathbf{P}_f$ in which the elements of $\mathbf{F}_{f1}$ are to be stored. Thus,

$$\mathbf{P}_f = \begin{bmatrix} -120 \\ 0 \\ -200 \\ 0 \end{bmatrix} \begin{matrix} 1 \\ 2 \\ 3 \\ 4 \end{matrix} \qquad \textbf{(4)}$$

Member 2 By substituting $P = 75$ kN, $L = 8$ m, and $l_1 = l_2 = 4$ m into the fixed-end force expressions for loading type 1, we determine the member fixed-end shears and moments to be

$$FS_b = FS_e = \frac{75}{2} = 37.5 \text{ kN}$$

$$FM_b = -FM_e = \frac{75(8)}{8} = 75 \text{ kN} \cdot \text{m}$$

As no axial loads are applied to this member,

$$FA_b = FA_e = 0$$

Thus,

$$\mathbf{F}_{f2} = \mathbf{Q}_{f2} = \begin{bmatrix} 0 \\ 37.5 \\ 75 \\ \hdashline 0 \\ 37.5 \\ \hdashline -75 \end{bmatrix} \begin{matrix} 1 \\ 2 \\ 3 \\ 8 \\ 9 \\ 4 \end{matrix}$$

Using the member code numbers 1, 2, 3, 8, 9, 4, we add the pertinent elements of $\mathbf{F}_{f2}$ in their proper positions in $\mathbf{P}_f$ (as given in Eq. (4)), which now becomes

$$\mathbf{P}_f = \begin{bmatrix} -120 \\ 37.5 \\ -200 + 75 \\ -75 \end{bmatrix} \begin{matrix} 1 \\ 2 \\ 3 \\ 4 \end{matrix}$$

Because the fixed-end forces for both members of the frame have now been stored in $\mathbf{P}_f$, the structure fixed-joint force vector for the given frame is

$$\mathbf{P}_f = \begin{bmatrix} -120 \\ 37.5 \\ -125 \\ -75 \end{bmatrix} \begin{matrix} 1 \\ 2 \\ 3 \\ 4 \end{matrix} \qquad \textbf{Ans}$$

Equivalent Joint Loads:

$$\mathbf{P}_e = -\mathbf{P}_f = \begin{bmatrix} 120 \\ -37.5 \\ 125 \\ 75 \end{bmatrix} \begin{matrix} 1 \\ 2 \\ 3 \\ 4 \end{matrix} \qquad \textbf{Ans}$$

The equivalent joint loads are depicted in Fig. 6.16(c). These equivalent joint loads cause the same joint displacements of the frame as the actual member loads of Fig. 6.16(a).

6.6 PROCEDURE FOR ANALYSIS

Using the concepts discussed in the previous sections, we can now develop the following step-by-step procedure for the analysis of plane frames by the matrix stiffness method.

1. Prepare an analytical model of the structure, identifying its degrees of freedom and restrained coordinates (as discussed in Section 6.1). Recall that for horizontal members, the coordinate transformations can be avoided by selecting the left-end joint of the member as the beginning joint.
2. Evaluate the structure stiffness matrix $\mathbf{S}(NDOF \times NDOF)$ and fixed-joint force vector $\mathbf{P}_f$ $(NDOF \times 1)$. For each member of the structure, perform the following operations:
   - **a.** Calculate the length and direction cosines (i.e., $\cos\theta$ and $\sin\theta$) of the member (Eqs. (3.62)).
   - **b.** Compute the member stiffness matrix in the global coordinate system, $\mathbf{K}$, using its explicit form given in Eq. (6.31). The member global stiffness matrix alternatively can be obtained by first forming the member local stiffness matrix $\mathbf{k}$ (Eq. (6.6)) and the transformation matrix $\mathbf{T}$ (Eq. (6.19)), and then evaluating the matrix triple product, $\mathbf{K} = \mathbf{T}^T\mathbf{k}\mathbf{T}$ (Eq. (6.29)). The matrix $\mathbf{K}$ must be symmetric.
   - **c.** If the member is subjected to external loads, then evaluate the member fixed-end force vector in the global coordinate system, $\mathbf{F}_f$, using the expressions for fixed-end forces given inside the front cover, and the explicit form of $\mathbf{F}_f$ given in Eq. (6.33). The member global fixed-end force vector can also be obtained by first forming the member local fixed-end force vector $\mathbf{Q}_f$ (Eq. (6.15)), and then using the relationship $\mathbf{F}_f = \mathbf{T}^T\mathbf{Q}_f$ (Eq. (6.30)).
   - **d.** Identify the member code numbers and store the pertinent elements of $\mathbf{K}$ and $\mathbf{F}_f$ in their proper positions in the structure stiffness matrix $\mathbf{S}$ and the fixed-joint force vector $\mathbf{P}_f$, respectively.

   The complete structure stiffness matrix $\mathbf{S}$, obtained by assembling the stiffness coefficients of all the members of the structure, must be symmetric.
3. If the structure is subjected to joint loads, then form the joint load vector $\mathbf{P}(NDOF \times 1)$.
4. Determine the joint displacements $\mathbf{d}$. Substitute $\mathbf{P}$, $\mathbf{P}_f$, and $\mathbf{S}$ into the structure stiffness relationship, $\mathbf{P} - \mathbf{P}_f = \mathbf{S}\mathbf{d}$ (Eq. (6.42)), and solve the resulting system of simultaneous equations for the unknown joint displacements $\mathbf{d}$. To check the solution for correctness, substitute the numerical values of the joint displacements $\mathbf{d}$ back into the stiffness relationship $\mathbf{P} - \mathbf{P}_f = \mathbf{S}\mathbf{d}$. If the solution is correct, then the stiffness relationship should be satisfied. Note that joint translations are considered positive when in the positive directions of the global X and Y axes, and joint rotations are considered positive when counterclockwise.
5. Compute member end displacements and end forces, and support reactions. For each member of the structure, carryout the following steps:
   - **a.** Obtain member end displacements in the global coordinate system, $\mathbf{v}$, from the joint displacements, $\mathbf{d}$, using the member code numbers.

  b. Form the member transformation matrix **T** (Eq. (6.19)), and determine the member end displacements in the local coordinate system, **u**, using the transformation relationship $\mathbf{u} = \mathbf{T}\mathbf{v}$ (Eq. (6.20)).
  c. Form the member local stiffness matrix **k** (Eq. (6.6)) and local fixed-end force vector $\mathbf{Q}_f$ (Eq. (6.15)); then calculate the member end forces in the local coordinate system, **Q**, using the stiffness relationship $\mathbf{Q} = \mathbf{k}\mathbf{u} + \mathbf{Q}_f$ (Eq. (6.4)).
  d. Determine the member end forces in the global coordinate system, **F**, using the transformation relationship $\mathbf{F} = \mathbf{T}^T\mathbf{Q}$ (Eq. (6.23)).
  e. If the member is attached to a support joint, then use the member code numbers to store the pertinent elements of **F** in their proper positions in the support reaction vector **R**.

6. Check the calculation of member end forces and support reactions by applying the equilibrium equations $\left(\sum F_X = 0, \sum F_Y = 0, \text{and} \sum M = 0\right)$ to the free body of the entire structure. If the calculations have been carried out correctly, then the equilibrium equations should be satisfied.

Instead of following steps 5(c) and (d) of this procedure, the member end forces alternatively can be obtained by first calculating the global forces **F** using the global stiffness relationship $\mathbf{F} = \mathbf{K}\mathbf{v} + \mathbf{F}_f$ (Eq. (6.28)), and then evaluating the local forces **Q** from the transformation relationship $\mathbf{Q} = \mathbf{T}\mathbf{F}$ (Eq. (6.18)). It should also be noted that it is usually not necessary to determine the global end forces for all the members of the structure, because such forces are not used for design purposes. However, **F** vectors for the members that are attached to supports are always evaluated, so that they can be used to form the support reaction vector **R**.

EXAMPLE 6.6

Determine the joint displacements, member end forces, and support reactions for the two-member frame shown in Fig. 6.17(a) on the next page, using the matrix stiffness method.

SOLUTION

Analytical Model: See Fig. 6.17(b). The frame has three degrees of freedom—the translations in the X and Y directions, and the rotation, of joint 2—which are numbered 1, 2, and 3, respectively. The six restrained coordinates of the frame are identified by numbers 4 through 9, as shown in Fig. 6.17(b).

Structure Stiffness Matrix and Fixed-Joint Force Vector:

Member 1 As shown in Fig. 6.17(b), we have selected joint 1 as the beginning joint, and joint 2 as the end joint for this member. By applying Eqs. (3.62), we determine

$$L = \sqrt{(X_2 - X_1)^2 + (Y_2 - Y_1)^2} = \sqrt{(10-0)^2 + (20-0)^2}$$

$$= 22.361 \text{ ft} = 268.33 \text{ in.} \tag{1a}$$

$$\cos\theta = \frac{X_2 - X_1}{L} = \frac{10-0}{22.361} = 0.44721 \tag{1b}$$

$$\sin\theta = \frac{Y_2 - Y_1}{L} = \frac{20-0}{22.361} = 0.89443 \tag{1c}$$

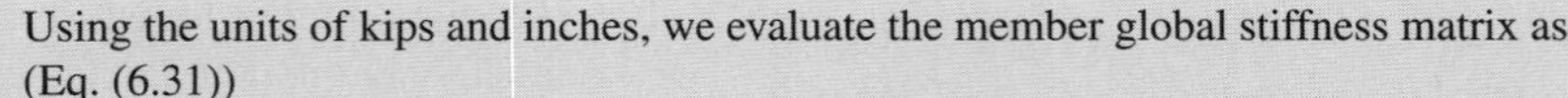

Using the units of kips and inches, we evaluate the member global stiffness matrix as (Eq. (6.31))

$$\mathbf{K}_1 = \begin{array}{c} \\ \end{array}\left[\begin{array}{rrr:rrr} 259.53 & 507.89 & -670.08 & -259.53 & -507.89 & -670.08 \\ 507.89 & 1{,}021.4 & 335.04 & -507.89 & -1{,}021.4 & 335.04 \\ -670.08 & 335.04 & 134{,}015 & 670.08 & -335.04 & 67{,}008 \\ \hdashline -259.53 & -507.89 & 670.08 & 259.53 & 507.89 & 670.08 \\ -507.89 & -1{,}021.4 & -335.04 & 507.89 & 1{,}021.4 & -335.04 \\ -670.08 & 335.04 & 67{,}008 & 670.08 & -335.04 & 134{,}015 \end{array}\right]\begin{array}{l} 4 \\ 5 \\ 6 \\ 1 \\ 2 \\ 3 \end{array}$$

(Column code numbers: 4, 5, 6, 1, 2, 3)

Note that $\mathbf{K}_1$ is symmetric.

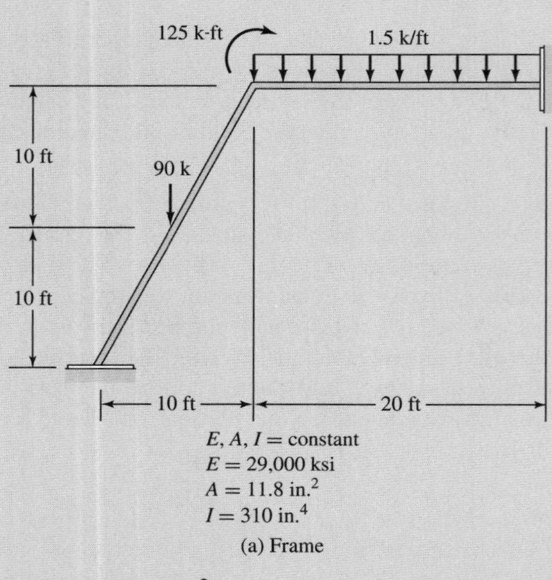

$E, A, I =$ constant
$E = 29{,}000$ ksi
$A = 11.8$ in.^{2}
$I = 310$ in.^{4}

(a) Frame

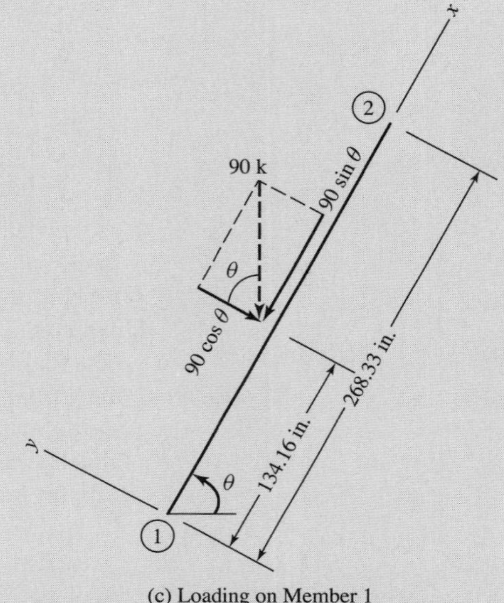

(c) Loading on Member 1

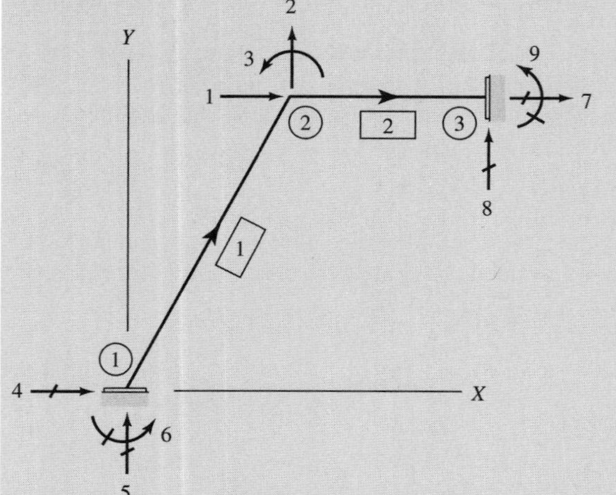

(b) Analytical Model

$$\mathbf{S} = \left[\begin{array}{ccc} 259.53 + 1{,}425.8 & 507.89 & 670.08 \\ 507.89 & 1{,}021.4 + 7.8038 & -335.04 + 936.46 \\ 670.08 & -335.04 + 936.46 & 134{,}015 + 149{,}833 \end{array}\right]\begin{array}{l} 1 \\ 2 \\ 3 \end{array}$$

$$= \left[\begin{array}{ccc} 1{,}685.3 & 507.89 & 670.08 \\ 507.89 & 1{,}029.2 & 601.42 \\ 670.08 & 601.42 & 283{,}848 \end{array}\right]\begin{array}{l} 1 \\ 2 \\ 3 \end{array}$$

$$\mathbf{P}_f = \left[\begin{array}{c} 0 \\ 45 + 15 \\ -1{,}350 + 600 \end{array}\right]\begin{array}{l} 1 \\ 2 \\ 3 \end{array} = \left[\begin{array}{r} 0 \\ 60 \\ -750 \end{array}\right]\begin{array}{l} 1 \\ 2 \\ 3 \end{array}$$

(d) Structure Stiffness Matrix and Fixed-Joint Force Vector

Fig. 6.17

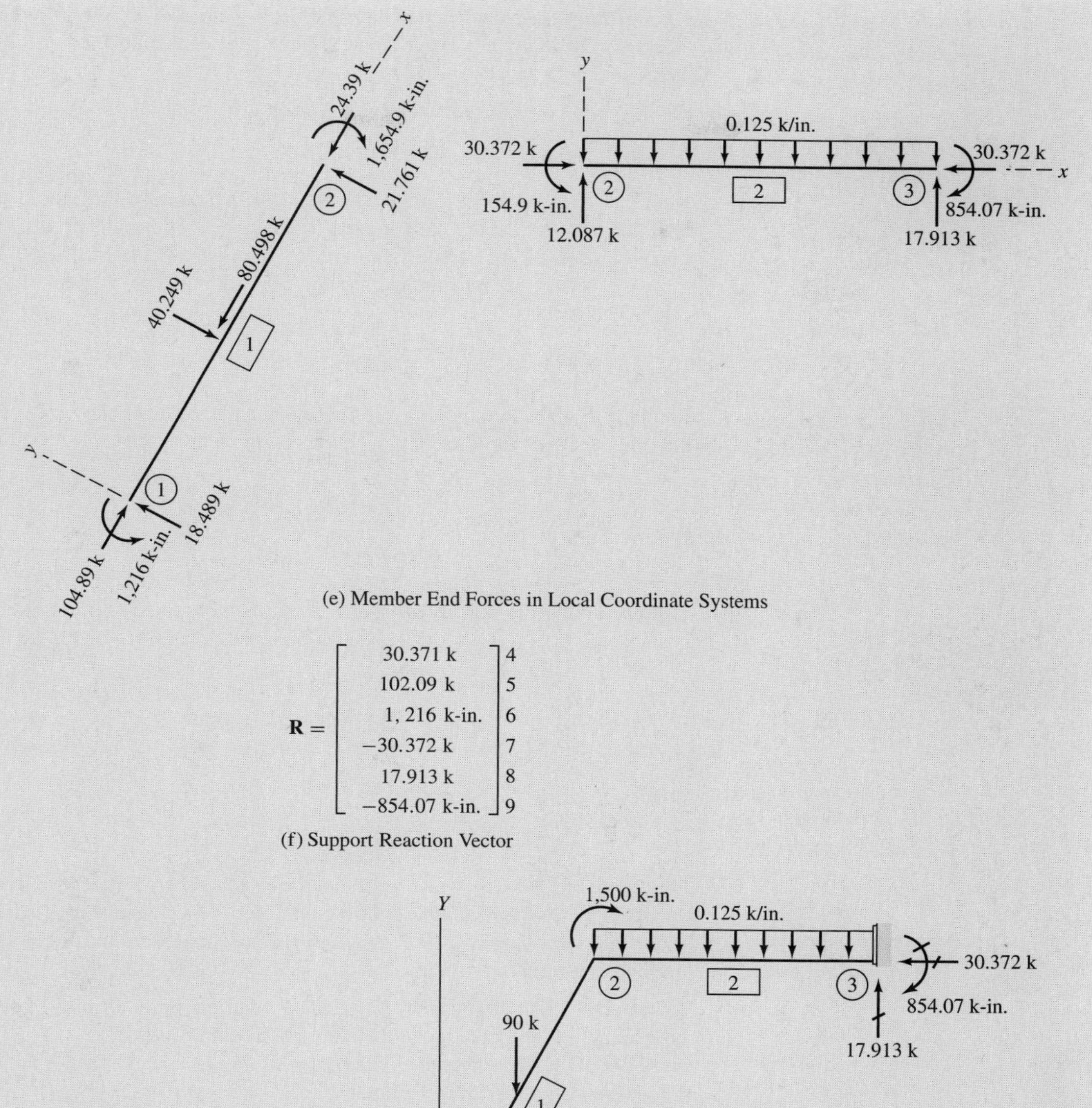

(e) Member End Forces in Local Coordinate Systems

$$\mathbf{R} = \begin{bmatrix} 30.371 \text{ k} \\ 102.09 \text{ k} \\ 1,216 \text{ k-in.} \\ -30.372 \text{ k} \\ 17.913 \text{ k} \\ -854.07 \text{ k-in.} \end{bmatrix} \begin{matrix} 4 \\ 5 \\ 6 \\ 7 \\ 8 \\ 9 \end{matrix}$$

(f) Support Reaction Vector

(g) Support Reactions

Fig. 6.17 *(continued)*

As the 90 k load applied to this member is inclined with respect to the member's local coordinate system, we evaluate the rectangular components of the load in the directions of the local x and y axes as (see Fig. 6.17(c))

$$W_x = 90 \sin\theta = 90(0.89443) = 80.498 \text{ k}$$

$$W_y = 90 \cos\theta = 90(0.44721) = 40.249 \text{ k}$$

Note that both W_x and W_y are considered positive because they act in the negative directions of the local x and y axes, respectively. The member's fixed-end axial forces can now be evaluated by substituting $W = W_x = 80.498$ k, $L = 268.33$ in., and $l_1 = l_2 = 134.16$ in. into the expressions for loading type 5 given inside the front cover. This yields

$$FA_b = FA_e = \frac{80.498}{2} = 40.249 \text{ k} \tag{2a}$$

Similarly, by substituting $W = W_y = 40.249$ k, and the numerical values of L, l_1, and l_2 into the equations for loading type 1, we obtain the fixed-end shears and moments as

$$FS_b = FS_e = \frac{40.249}{2} = 20.125 \text{ k} \tag{2b}$$

$$FM_b = -FM_e = \frac{40.249(268.33)}{8} = 1{,}350 \text{ k-in.} \tag{2c}$$

By substituting the numerical values of the member fixed-end forces and direction cosines into Eq. (6.33), we calculate the member global fixed-end force vector as

$$\mathbf{F}_{f1} = \begin{bmatrix} 0 \\ 45 \\ 1{,}350 \\ \hdashline 0 \\ 45 \\ -1{,}350 \end{bmatrix} \begin{matrix} 4 \\ 5 \\ 6 \\ 1 \\ 2 \\ 3 \end{matrix}$$

From Fig. 6.17(b), we observe that the code numbers for member 1 are 4, 5, 6, 1, 2, 3. Using these code numbers, we store the pertinent elements of $\mathbf{K}_1$ and $\mathbf{F}_{f1}$ in their proper positions in the 3×3 structure stiffness matrix $\mathbf{S}$ and the 3×1 structure fixed-joint force vector $\mathbf{P}_f$, respectively, as shown in Fig. 6.17(d).

Member 2 As this member is horizontal, with its left-end joint 2 selected as the beginning joint, no coordinate transformations are needed; that is, $\mathbf{T}_2 = \mathbf{I}$, $\mathbf{K}_2 = \mathbf{k}_2$, and $\mathbf{F}_{f2} = \mathbf{Q}_{f2}$. Thus, by substituting $L = 240$ in., $E = 29{,}000$ ksi, $A = 11.8$ in.^{2}, and $I = 310$ in.^{4} into Eq. (6.6), we obtain

$$\mathbf{K}_2 = \mathbf{k}_2 = \begin{matrix} 1 & 2 & 3 & 7 & 8 & 9 \\ \end{matrix}$$
$$\begin{bmatrix} 1{,}425.8 & 0 & 0 & -1{,}425.8 & 0 & 0 \\ 0 & 7.8038 & 936.46 & 0 & -7.8038 & 936.46 \\ 0 & 936.46 & 149{,}833 & 0 & -936.46 & 74{,}917 \\ -1{,}425.8 & 0 & 0 & 1{,}425.8 & 0 & 0 \\ 0 & -7.8038 & -936.46 & 0 & 7.8038 & -936.46 \\ 0 & 936.46 & 74{,}917 & 0 & -936.46 & 149{,}833 \end{bmatrix} \begin{matrix} 1 \\ 2 \\ 3 \\ 7 \\ 8 \\ 9 \end{matrix} \tag{3}$$

Using the equations given for loading type 3 inside the front cover, we obtain the fixed-end forces due to the uniformly distributed load of magnitude 0.125 k/in. (= 1.5 k/ft):

$$FA_b = FA_e = 0$$

$$FS_b = FS_e = \frac{0.125(240)}{2} = 15 \text{ k}$$

$$FM_b = -FM_e = \frac{0.125(240)^2}{12} = 600 \text{ k-in.}$$

Thus (Eq. (6.15)),

$$\mathbf{F}_{f2} = \mathbf{Q}_{f2} = \begin{bmatrix} 0 \\ 15 \\ 600 \\ \hdashline 0 \\ 15 \\ -600 \end{bmatrix} \begin{matrix} 1 \\ 2 \\ 3 \\ 7 \\ 8 \\ 9 \end{matrix} \qquad \textbf{(4)}$$

The relevant elements of $\mathbf{K}_2$ and $\mathbf{F}_{f2}$ are stored in $\mathbf{S}$ and $\mathbf{P}_f$, respectively, using the member code numbers 1, 2, 3, 7, 8, 9.

The completed structure stiffness matrix $\mathbf{S}$ and structure fixed-joint force vector $\mathbf{P}_f$ are given in Fig. 6.17(d). Note that $\mathbf{S}$ is symmetric.

Joint Load Vector: By comparing Figs. 6.17(a) and (b), we write the joint load vector, in kips and inches, as

$$\mathbf{P} = \begin{bmatrix} 0 \\ 0 \\ -1{,}500 \end{bmatrix} \begin{matrix} 1 \\ 2 \\ 3 \end{matrix}$$

Joint Displacements: By substituting the numerical values of $\mathbf{P}$, $\mathbf{P}_f$, and $\mathbf{S}$ into Eq. (6.42), we write the stiffness relations for the entire frame as

$$\mathbf{P} - \mathbf{P}_f = \mathbf{S}\mathbf{d}$$

$$\begin{bmatrix} 0 \\ 0 \\ -1{,}500 \end{bmatrix} - \begin{bmatrix} 0 \\ 60 \\ -750 \end{bmatrix} = \begin{bmatrix} 1{,}685.3 & 507.89 & 670.08 \\ 507.89 & 1{,}029.2 & 601.42 \\ 670.08 & 601.42 & 283{,}848 \end{bmatrix} \begin{bmatrix} d_1 \\ d_2 \\ d_3 \end{bmatrix}$$

or

$$\begin{bmatrix} 0 \\ -60 \\ -750 \end{bmatrix} = \begin{bmatrix} 1{,}685.3 & 507.89 & 670.08 \\ 507.89 & 1{,}029.2 & 601.42 \\ 670.08 & 601.42 & 283{,}848 \end{bmatrix} \begin{bmatrix} d_1 \\ d_2 \\ d_3 \end{bmatrix}$$

Solving these equations, we determine the joint displacements to be

$$\mathbf{d} = \begin{bmatrix} 0.021302 \text{ in.} \\ -0.06732 \text{ in.} \\ -0.0025499 \text{ rad} \end{bmatrix} \qquad \textbf{Ans}$$

To check the foregoing solution, we substitute the numerical values of **d** back into the structure stiffness relationship, as

$$\mathbf{P} - \mathbf{P}_f = \mathbf{Sd} = \begin{bmatrix} 1{,}685.3 & 507.89 & 670.08 \\ 507.89 & 1{,}029.2 & 601.42 \\ 670.08 & 601.42 & 283{,}848 \end{bmatrix} \begin{bmatrix} 0.021302 \\ -0.06732 \\ -0.0025499 \end{bmatrix}$$

$$= \begin{bmatrix} 0 \\ -60 \\ -750 \end{bmatrix} \qquad \text{Checks}$$

Member End Displacements and End Forces:

Member 1 As in the case of plane trusses, the global end displacements **v** for a plane frame member can be obtained by applying the member's compatibility equations, using its code numbers. Thus, for member 1 of the frame under consideration, the global end displacement vector can be established as

$$\mathbf{v}_1 = \begin{bmatrix} v_1 \\ v_2 \\ v_3 \\ v_4 \\ v_5 \\ v_6 \end{bmatrix} \begin{matrix} 4 \\ 5 \\ 6 \\ 1 \\ 2 \\ 3 \end{matrix} = \begin{bmatrix} 0 \\ 0 \\ 0 \\ d_1 \\ d_2 \\ d_3 \end{bmatrix} = \begin{bmatrix} 0 \\ 0 \\ 0 \\ 0.021302 \text{ in.} \\ -0.06732 \text{ in.} \\ -0.0025499 \text{ rad} \end{bmatrix} \tag{5}$$

As shown in Eq. (5), the code numbers for the member (4, 5, 6, 1, 2, 3) are first written on the right side of **v**. The fact that the code numbers corresponding to v_1, v_2, and v_3 are the restrained coordinate numbers 4, 5, and 6, respectively, indicates that $v_1 = v_2 = v_3 = 0$. Similarly, the code numbers 1, 2, and 3 corresponding to v_4, v_5, and v_6, respectively, indicate that $v_4 = d_1$, $v_5 = d_2$, and $v_6 = d_3$. Note that these compatibility equations can be verified easily by a visual inspection of the frame's line diagram given in Fig. 6.17(b).

To determine the member local end displacements, **u**, we first evaluate the transformation matrix **T** (Eq. (6.19)), using the direction cosines given in Eqs. (1):

$$\mathbf{T}_1 = \begin{bmatrix} 0.44721 & 0.89443 & 0 & 0 & 0 & 0 \\ -0.89443 & 0.44721 & 0 & 0 & 0 & 0 \\ 0 & 0 & 1 & 0 & 0 & 0 \\ 0 & 0 & 0 & 0.44721 & 0.89443 & 0 \\ 0 & 0 & 0 & -0.89443 & 0.44721 & 0 \\ 0 & 0 & 0 & 0 & 0 & 1 \end{bmatrix}$$

The member local end forces can now be calculated using the relationship $\mathbf{u} = \mathbf{Tv}$ (Eq. (6.20)), as

$$\mathbf{u}_1 = \mathbf{T}_1\mathbf{v}_1 = \begin{bmatrix} 0 \\ 0 \\ 0 \\ -0.050686 \text{ in.} \\ -0.04916 \text{ in.} \\ -0.0025499 \text{ rad} \end{bmatrix}$$

Before we can calculate the member's local end forces **Q**, we need to determine its local stiffness matrix **k** and fixed-end force vector $\mathbf{Q}_f$. Thus, using Eq. (6.6):

$$\mathbf{k}_1 = \begin{bmatrix} 1{,}275.3 & 0 & 0 & -1{,}275.3 & 0 & 0 \\ 0 & 5.584 & 749.17 & 0 & -5.584 & 749.17 \\ 0 & 749.17 & 134{,}015 & 0 & -749.17 & 67{,}008 \\ -1{,}275.3 & 0 & 0 & 1{,}275.3 & 0 & 0 \\ 0 & -5.584 & -749.17 & 0 & 5.584 & -749.17 \\ 0 & 749.17 & 67{,}008 & 0 & -749.17 & 134{,}015 \end{bmatrix}$$

and, by substituting Eqs. (2) into Eq. (6.15):

$$\mathbf{Q}_{f1} = \begin{bmatrix} 40.249 \\ 20.125 \\ 1{,}350 \\ 40.249 \\ 20.125 \\ -1{,}350 \end{bmatrix}$$

Now, using Eq. (6.4), we compute the member local end forces as

$$\mathbf{Q}_1 = \mathbf{k}_1\mathbf{u}_1 + \mathbf{Q}_{f1} = \begin{bmatrix} 104.89 \text{ k} \\ 18.489 \text{ k} \\ 1{,}216 \text{ k-in.} \\ -24.39 \text{ k} \\ 21.761 \text{ k} \\ -1{,}654.9 \text{ k-in.} \end{bmatrix}$$ **Ans**

The local end forces for member 1 are depicted in Fig. 6.17(e), and we can check our calculations for these forces by considering the equilibrium of the free body of the member, as follows.

$+\nearrow \sum F_x = 0 \quad 104.89 - 80.498 - 24.39 = 0.002 \cong 0$ **Checks**

$+\nwarrow \sum F_y = 0 \quad 18.489 - 40.249 + 21.761 = 0.001 \cong 0$ **Checks**

$+\circlearrowleft \sum M_{②} = 0 \quad 1{,}216 - 18.489(268.33) + 40.249(134.16) - 1{,}654.9$

$= -0.25 \cong 0$ **Checks**

The member global end forces **F** can now be determined by applying Eq. (6.23), as

$$\mathbf{F}_1 = \mathbf{T}_1^T\mathbf{Q}_1 = \begin{bmatrix} 30.371 \\ 102.09 \\ 1{,}216 \\ \hdashline -30.371 \\ -12.083 \\ -1{,}654.9 \end{bmatrix} \begin{matrix} 4 \\ 5 \\ 6 \\ 1 \\ 2 \\ 3 \end{matrix} \tag{6}$$

Next, to generate the support reaction vector **R**, we write the member code numbers (4, 5, 6, 1, 2, 3) on the right side of $\mathbf{F}_1$ as shown in Eq. (6), and store the pertinent elements of $\mathbf{F}_1$ in their proper positions in **R** by matching the code numbers on the side of $\mathbf{F}_1$ to the restrained coordinate numbers on the right side of **R** in Fig. 6.17(f).

Member 2 The global and local end displacements for this horizontal member are

$$\mathbf{u}_2 = \mathbf{v}_2 = \begin{bmatrix} v_1 \\ v_2 \\ v_3 \\ v_4 \\ v_5 \\ v_6 \end{bmatrix} \begin{matrix} 1 \\ 2 \\ 3 \\ 7 \\ 8 \\ 9 \end{matrix} = \begin{bmatrix} d_1 \\ d_2 \\ d_3 \\ 0 \\ 0 \\ 0 \end{bmatrix} = \begin{bmatrix} 0.021302 \text{ in.} \\ -0.06732 \text{ in.} \\ -0.0025499 \text{ rad} \\ 0 \\ 0 \\ 0 \end{bmatrix}$$

By using $\mathbf{k}_2$ from Eq. (3) and $\mathbf{Q}_{f2}$ from Eq. (4), we compute the member local and global end forces to be

$$\mathbf{F}_2 = \mathbf{Q}_2 = \mathbf{k}_2\mathbf{u}_2 + \mathbf{Q}_{f2} = \begin{bmatrix} 30.372 \text{ k} \\ 12.087 \text{ k} \\ 154.9 \text{ k-in.} \\ \hline -30.372 \text{ k} \\ 17.913 \text{ k} \\ -854.07 \text{ k-in.} \end{bmatrix} \begin{matrix} 1 \\ 2 \\ 3 \\ 7 \\ 8 \\ 9 \end{matrix}$$ **Ans**

These end forces for member 2 are depicted in Fig. 6.17(e). To check our calculations, we apply equilibrium equations to the free body of member 2, as follows.

$\stackrel{+}{\rightarrow} \sum F_x = 0$ $\quad 30.372 - 30.372 = 0$ **Checks**

$+\uparrow \sum F_y = 0$ $\quad 12.087 - 0.125(240) + 17.913 = 0$ **Checks**

$+\circlearrowleft \sum M_{②} = 0$ $\quad 154.9 - 0.125(240)(120) + 17.913(240) - 854.07 = -0.05 \cong 0$ **Checks**

Next, we store the pertinent elements of $\mathbf{F}_2$ in their proper positions in the reaction vector $\mathbf{R}$, using the member code numbers (1, 2, 3, 7, 8, 9), as shown in Fig. 6.17(f).

Support Reactions: The completed reaction vector $\mathbf{R}$ is given in Fig. 6.17(f), and the support reactions are depicted on a line diagram of the structure in Fig. 6.17(g). **Ans**

Equilibrium Check: Finally, we check our calculations by considering the equilibrium of the free body of the entire structure (Fig. 6.17(g)), as follows.

$\stackrel{+}{\rightarrow} \sum F_X = 0$ $\quad 30.371 - 30.372 = -0.001 \cong 0$ **Checks**

$+\uparrow \sum F_Y = 0$ $\quad 102.09 - 90 - 0.125(240) + 17.913 = 0.003 \cong 0$ **Checks**

$+\circlearrowleft \sum M_{①} = 0$ $\quad 1{,}216 - 90(60) - 1{,}500 - 0.125(240)(240) + 30.372(240) + 17.913(360) - 854.07 = -0.11 \cong 0$ **Checks**

EXAMPLE 6.7

Determine the joint displacements, member local end forces, and support reactions for the two-story frame, subjected to a wind loading, shown in Fig. 6.18(a).

SOLUTION *Analytical Model:* See Fig. 6.18(b). The frame has nine degrees of freedom, numbered 1 through 9; and six restrained coordinates, identified by the numbers 10 through 15.

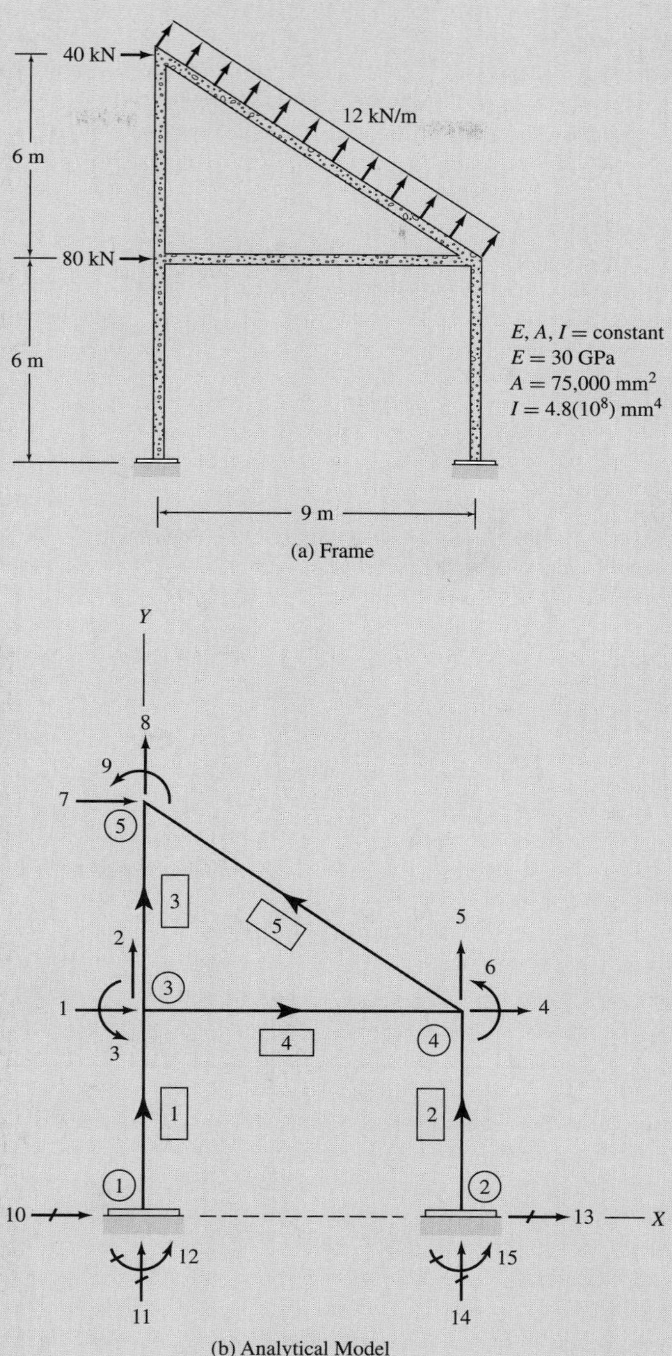
40 kN
12 kN/m
6 m
80 kN
6 m
E, A, I = constant
E = 30 GPa
A = 75,000 mm²
I = 4.8(10⁸) mm⁴
9 m
(a) Frame
Y
8
9
7
5
3
5
2
3
5
6
1
4
3
4
4
1
2
1
2
10
13
X
12
15
11
14
(b) Analytical Model

Fig. 6.18

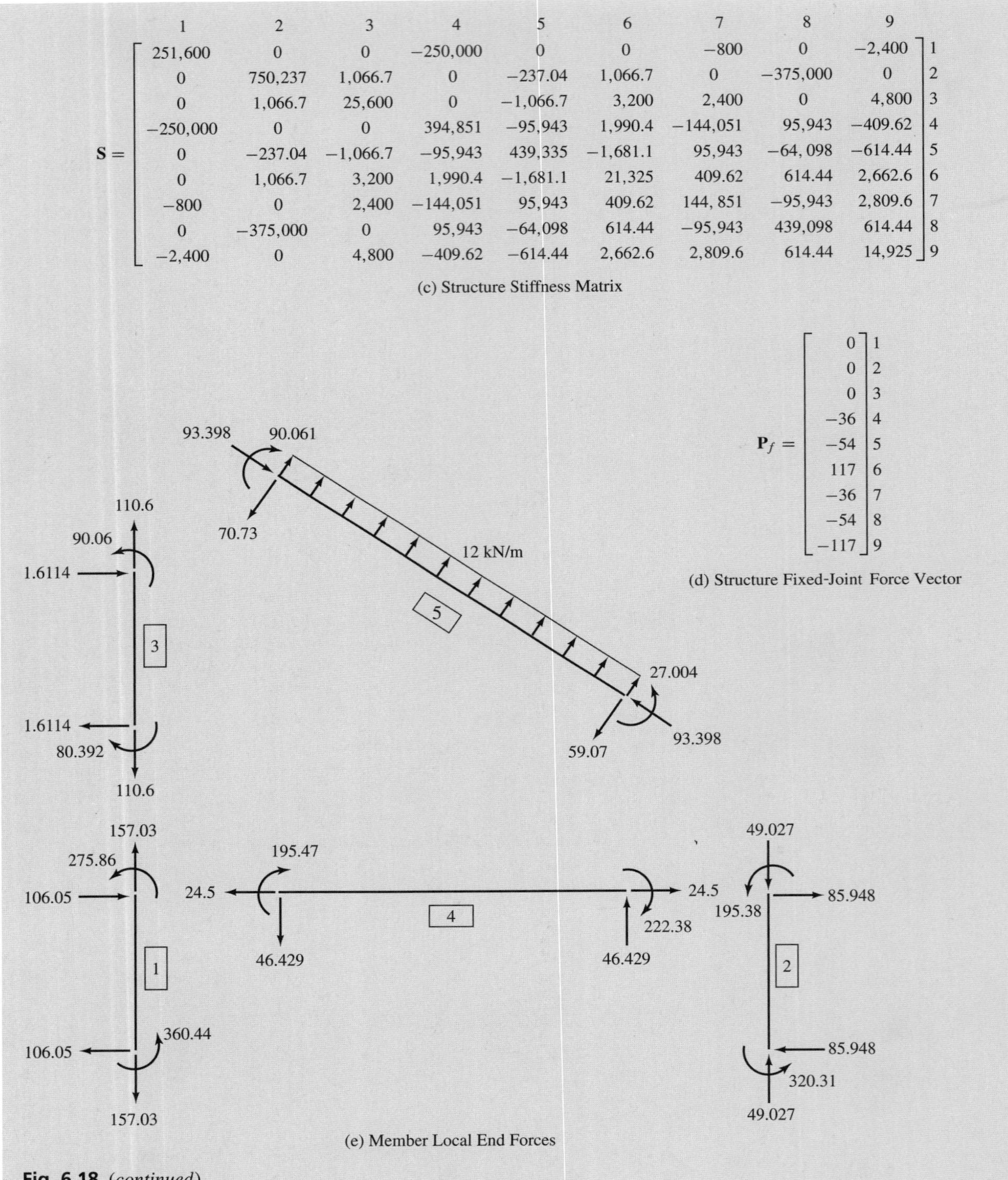

$$
\mathbf{S} = \begin{bmatrix}
251{,}600 & 0 & 0 & -250{,}000 & 0 & 0 & -800 & 0 & -2{,}400 \\
0 & 750{,}237 & 1{,}066.7 & 0 & -237.04 & 1{,}066.7 & 0 & -375{,}000 & 0 \\
0 & 1{,}066.7 & 25{,}600 & 0 & -1{,}066.7 & 3{,}200 & 2{,}400 & 0 & 4{,}800 \\
-250{,}000 & 0 & 0 & 394{,}851 & -95{,}943 & 1{,}990.4 & -144{,}051 & 95{,}943 & -409.62 \\
0 & -237.04 & -1{,}066.7 & -95{,}943 & 439{,}335 & -1{,}681.1 & 95{,}943 & -64{,}098 & -614.44 \\
0 & 1{,}066.7 & 3{,}200 & 1{,}990.4 & -1{,}681.1 & 21{,}325 & 409.62 & 614.44 & 2{,}662.6 \\
-800 & 0 & 2{,}400 & -144{,}051 & 95{,}943 & 409.62 & 144{,}851 & -95{,}943 & 2{,}809.6 \\
0 & -375{,}000 & 0 & 95{,}943 & -64{,}098 & 614.44 & -95{,}943 & 439{,}098 & 614.44 \\
-2{,}400 & 0 & 4{,}800 & -409.62 & -614.44 & 2{,}662.6 & 2{,}809.6 & 614.44 & 14{,}925
\end{bmatrix}
\begin{matrix} 1 \\ 2 \\ 3 \\ 4 \\ 5 \\ 6 \\ 7 \\ 8 \\ 9 \end{matrix}
$$

(c) Structure Stiffness Matrix

$$
\mathbf{P}_f = \begin{bmatrix} 0 \\ 0 \\ 0 \\ -36 \\ -54 \\ 117 \\ -36 \\ -54 \\ -117 \end{bmatrix}
\begin{matrix} 1 \\ 2 \\ 3 \\ 4 \\ 5 \\ 6 \\ 7 \\ 8 \\ 9 \end{matrix}
$$

(d) Structure Fixed-Joint Force Vector

(e) Member Local End Forces

Fig. 6.18 *(continued)*

$$
\mathbf{R} = \begin{bmatrix} -106.05 \text{ kN} \\ -157.03 \text{ kN} \\ 360.44 \text{ kN·m} \\ -85.948 \text{ kN} \\ 49.027 \text{ kN} \\ 320.31 \text{ kN·m} \end{bmatrix} \begin{matrix} 10 \\ 11 \\ 12 \\ 13 \\ 14 \\ 15 \end{matrix}
$$

(f) Support Reaction Vector

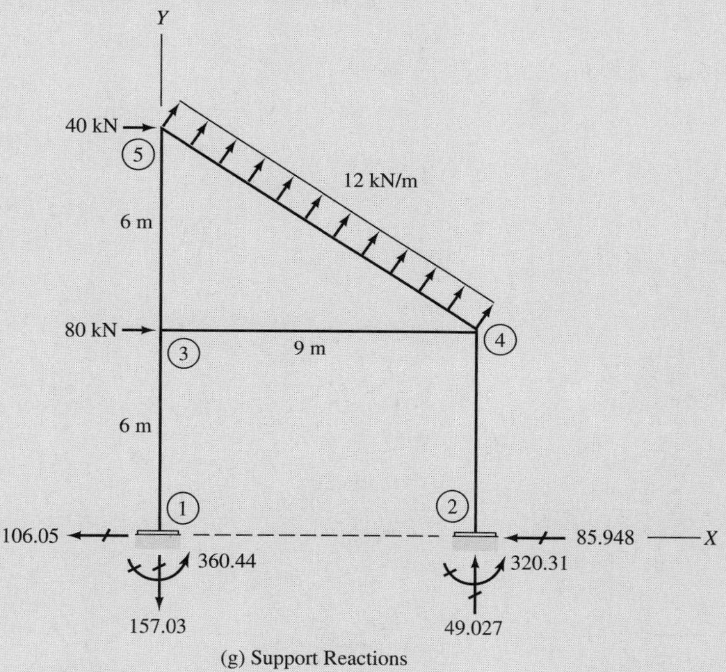

(g) Support Reactions

Fig. 6.18 (*continued*)

Structure Stiffness Matrix and Fixed-Joint Force Vector:

Members 1, 2, and 3 $E = 30(10^6)\,\text{kN/m}^2, A = 0.075\,\text{m}^2, I = 480(10^{-6})\,\text{m}^4, L = 6\,\text{m}$, $\cos\,\theta = 0$, and $\sin\,\theta = 1$. The member global stiffness matrix, in units of kN and meters, is given by the following (see Eq. (6.31)).

Member 3 →	1	2	3	7	8	9
Member 2 →	13	14	15	4	5	6
Member 1 →	10	11	12	1	2	3

$$
\mathbf{K}_1 = \mathbf{K}_2 = \mathbf{K}_3 = \begin{bmatrix} 800 & 0 & -2{,}400 & -800 & 0 & -2{,}400 \\ 0 & 375{,}000 & 0 & 0 & -375{,}000 & 0 \\ -2{,}400 & 0 & 9{,}600 & 2{,}400 & 0 & 4{,}800 \\ -800 & 0 & 2{,}400 & 800 & 0 & 2{,}400 \\ 0 & -375{,}000 & 0 & 0 & 375{,}000 & 0 \\ -2{,}400 & 0 & 4{,}800 & 2{,}400 & 0 & 9{,}600 \end{bmatrix} \begin{matrix} 10 \\ 11 \\ 12 \\ 1 \\ 2 \\ 3 \end{matrix} \; \begin{matrix} 13 \\ 14 \\ 15 \\ 4 \\ 5 \\ 6 \end{matrix} \; \begin{matrix} 1 \\ 2 \\ 3 \\ 7 \\ 8 \\ 9 \end{matrix}
$$

As these members are not subjected to any loads, their fixed-end forces are 0. Thus,

$$\mathbf{F}_{f1} = \mathbf{F}_{f2} = \mathbf{F}_{f3} = \mathbf{0}$$

Using the code numbers for member 1 (10, 11, 12, 1, 2, 3), member 2 (13, 14, 15, 4, 5, 6), and member 3 (1, 2, 3, 7, 8, 9), the relevant elements of $\mathbf{K}_1$, $\mathbf{K}_2$, and $\mathbf{K}_3$ are stored in their proper positions in the 9×9 structure stiffness matrix $\mathbf{S}$ (Fig. 6.18(c)).

Member 4 Substituting $L = 9$ m and the foregoing values of E, A, and I into Eq. (6.6), we obtain

$$\mathbf{K}_4 = \mathbf{k}_4 = \begin{array}{c} \\ \left[\begin{array}{cccccc} 1 & 2 & 3 & 4 & 5 & 6 \\ 250{,}000 & 0 & 0 & -250{,}000 & 0 & 0 \\ 0 & 237.04 & 1{,}066.7 & 0 & -237.04 & 1{,}066.7 \\ 0 & 1{,}066.7 & 6{,}400 & 0 & -1{,}066.7 & 3{,}200 \\ -250{,}000 & 0 & 0 & 250{,}000 & 0 & 0 \\ 0 & -237.04 & -1{,}066.7 & 0 & 237.04 & -1{,}066.7 \\ 0 & 1{,}066.7 & 3{,}200 & 0 & -1{,}066.7 & 6{,}400 \end{array}\right] \end{array} \begin{array}{c} \\ 1 \\ 2 \\ 3 \\ 4 \\ 5 \\ 6 \end{array} \quad \textbf{(1)}$$

As no loads are applied to this member,

$$\mathbf{F}_{f4} = \mathbf{Q}_{f4} = \mathbf{0}$$

The pertinent elements of $\mathbf{K}_4$ are stored in $\mathbf{S}$ using the member code numbers 1, 2, 3, 4, 5, 6.

Member 5

$$L = \sqrt{(X_5 - X_4)^2 + (Y_5 - Y_4)^2} = \sqrt{(0 - 9)^2 + (12 - 6)^2} = 10.817 \text{ m} \quad \textbf{(2a)}$$

$$\cos\theta = \frac{X_5 - X_4}{L} = \frac{0 - 9}{10.817} = -0.83205 \quad \textbf{(2b)}$$

$$\sin\theta = \frac{Y_5 - Y_4}{L} = \frac{12 - 6}{10.817} = 0.5547 \quad \textbf{(2c)}$$

$$\mathbf{K}_5 = \left[\begin{array}{cccccc} 4 & 5 & 6 & 7 & 8 & 9 \\ 144{,}051 & -95{,}943 & -409.62 & -144{,}051 & 95{,}943 & -409.62 \\ -95{,}943 & 64{,}098 & -614.44 & 95{,}943 & -64{,}098 & -614.44 \\ -409.62 & -614.44 & 5{,}325.1 & 409.62 & 614.44 & 2{,}662.6 \\ -144{,}051 & 95{,}943 & 409.62 & 144{,}051 & -95{,}943 & 409.62 \\ 95{,}943 & -64{,}098 & 614.44 & -95{,}943 & 64{,}098 & 614.44 \\ -409.62 & -614.44 & 2{,}662.6 & 409.62 & 614.44 & 5{,}325.1 \end{array}\right] \begin{array}{c} \\ 4 \\ 5 \\ 6 \\ 7 \\ 8 \\ 9 \end{array}$$

From Figs. 6.18(a) and (b), we observe that the 12 kN/m uniformly distributed load applied to member 5 acts in the negative direction of the member's local y axis; therefore, it is considered positive for the purpose of calculating fixed-end forces. Thus,

$$FA_b = FA_e = 0 \quad \textbf{(3a)}$$

$$FS_b = FS_e = \frac{wL}{2} = \frac{12(10.817)}{2} = 64.9 \text{ kN} \quad \textbf{(3b)}$$

$$FM_b = -FM_e = \frac{wL^2}{12} = \frac{12(10.817)^2}{12} = 117 \text{ kN} \cdot \text{m} \quad \textbf{(3c)}$$

Using Eq. (6.33), we determine the global fixed-end force vector for the member to be

$$\mathbf{F}_{f5} = \begin{bmatrix} -36 \\ -54 \\ 117 \\ -36 \\ -54 \\ -117 \end{bmatrix} \begin{matrix} 4 \\ 5 \\ 6 \\ 7 \\ 8 \\ 9 \end{matrix}$$

Using the code numbers 4, 5, 6, 7, 8, 9, we store the pertinent elements of $\mathbf{K}_5$ and $\mathbf{F}_{f5}$ in their proper positions in the $\mathbf{S}$ matrix and the $\mathbf{P}_f$ vector, respectively.

The complete structure stiffness matrix $\mathbf{S}$ and structure fixed-joint force vector $\mathbf{P}_f$ are shown in Figs. 6.18(c) and (d), respectively.

Joint Load Vector:

$$\mathbf{P} = \begin{bmatrix} 80 \\ 0 \\ 0 \\ 0 \\ 0 \\ 0 \\ 40 \\ 0 \\ 0 \end{bmatrix} \begin{matrix} 1 \\ 2 \\ 3 \\ 4 \\ 5 \\ 6 \\ 7 \\ 8 \\ 9 \end{matrix} \qquad \textbf{(4)}$$

Joint Displacements: By substituting the numerical values of $\mathbf{S}$ (Fig. 6.18(c)), $\mathbf{P}_f$ (Fig. 6.18(d)), and $\mathbf{P}$ (Eq. (4)) into the structural stiffness relationship $\mathbf{P} - \mathbf{P}_f = \mathbf{Sd}$ (Eq. (6.42)), and solving the resulting system of simultaneous equations, we obtain the following joint displacements:

$$\mathbf{d} = \begin{bmatrix} 0.185422 \text{ m} \\ 0.000418736 \text{ m} \\ -0.0176197 \text{ rad} \\ 0.18552 \text{ m} \\ -0.000130738 \text{ m} \\ -0.0260283 \text{ rad} \\ 0.186622 \text{ m} \\ 0.000713665 \text{ m} \\ 0.0178911 \text{ rad} \end{bmatrix} \begin{matrix} 1 \\ 2 \\ 3 \\ 4 \\ 5 \\ 6 \\ 7 \\ 8 \\ 9 \end{matrix} \qquad \textbf{Ans}$$

To check this solution, we evaluate the matrix product $\mathbf{Sd}$, using the foregoing values of the joint displacements $\mathbf{d}$, and substitute the results into the structure stiffness relationship, as

$$\mathbf{P} - \mathbf{P}_f = \mathbf{Sd}$$

$$\begin{bmatrix} 80 \\ 0 \\ 0 \\ 0 \\ 0 \\ 0 \\ 40 \\ 0 \\ 0 \end{bmatrix} - \begin{bmatrix} 0 \\ 0 \\ 0 \\ -36 \\ -54 \\ 117 \\ -36 \\ -54 \\ -117 \end{bmatrix} \cong \begin{bmatrix} 79.939 \\ -0.001466 \\ 0.0013239 \\ 36.051 \\ 54.006 \\ -116.992 \\ 76.007 \\ 53.994 \\ 116.994 \end{bmatrix} \qquad \textbf{Checks}$$

Member End Displacements and End Forces:

Member 1

$$\mathbf{v}_1 = \begin{bmatrix} v_1 \\ v_2 \\ v_3 \\ v_4 \\ v_5 \\ v_6 \end{bmatrix} \begin{matrix} 10 \\ 11 \\ 12 \\ 1 \\ 2 \\ 3 \end{matrix} = \begin{bmatrix} 0 \\ 0 \\ 0 \\ d_1 \\ d_2 \\ d_3 \end{bmatrix} = \begin{bmatrix} 0 \\ 0 \\ 0 \\ 0.185422 \\ 0.000418736 \\ -0.0176197 \end{bmatrix}$$

$\cos\theta = 0, \sin\theta = 1$

$$\mathbf{T}_1 = \mathbf{T}_2 = \mathbf{T}_3 = \begin{bmatrix} 0 & 1 & 0 & 0 & 0 & 0 \\ -1 & 0 & 0 & 0 & 0 & 0 \\ 0 & 0 & 1 & 0 & 0 & 0 \\ 0 & 0 & 0 & 0 & 1 & 0 \\ 0 & 0 & 0 & -1 & 0 & 0 \\ 0 & 0 & 0 & 0 & 0 & 1 \end{bmatrix} \quad \textbf{(5)}$$

$$\mathbf{u}_1 = \mathbf{T}_1\mathbf{v}_1 = \begin{bmatrix} 0 \\ 0 \\ 0 \\ 0.000418736 \\ -0.185422 \\ -0.0176197 \end{bmatrix}$$

$$\mathbf{k}_1 = \mathbf{k}_2 = \mathbf{k}_3 = \begin{bmatrix} 375{,}000 & 0 & 0 & -375{,}000 & 0 & 0 \\ 0 & 800 & 2{,}400 & 0 & -800 & 2{,}400 \\ 0 & 2{,}400 & 9{,}600 & 0 & -2{,}400 & 4{,}800 \\ -375{,}000 & 0 & 0 & 375{,}000 & 0 & 0 \\ 0 & -800 & -2{,}400 & 0 & 800 & -2{,}400 \\ 0 & 2{,}400 & 4{,}800 & 0 & -2{,}400 & 9{,}600 \end{bmatrix} \quad \textbf{(6)}$$

$$\mathbf{Q}_{f1} = \mathbf{0}$$

$$\mathbf{Q}_1 = \mathbf{k}_1\mathbf{u}_1 = \begin{bmatrix} -157.03 \text{ kN} \\ 106.05 \text{ kN} \\ 360.44 \text{ kN}\cdot\text{m} \\ 157.03 \text{ kN} \\ -106.05 \text{ kN} \\ 275.86 \text{ kN}\cdot\text{m} \end{bmatrix} \quad \textbf{Ans}$$

$$\mathbf{F}_1 = \mathbf{T}_1^T\mathbf{Q}_1 = \begin{bmatrix} -106.05 \\ -157.03 \\ 360.44 \\ 106.05 \\ 157.03 \\ 275.86 \end{bmatrix} \begin{matrix} 10 \\ 11 \\ 12 \\ 1 \\ 2 \\ 3 \end{matrix} \quad \textbf{(7)}$$

Member 2

$$\mathbf{v}_2 = \begin{bmatrix} 0 \\ 0 \\ 0 \\ 0.18552 \\ -0.000130738 \\ -0.0260283 \end{bmatrix} \begin{matrix} 13 \\ 14 \\ 15 \\ 4 \\ 5 \\ 6 \end{matrix}$$

Using $\mathbf{T}_2$ from Eq. (5), we obtain

$$\mathbf{u}_2 = \mathbf{T}_2\mathbf{v}_2 = \begin{bmatrix} 0 \\ 0 \\ 0 \\ -0.000130738 \\ -0.18552 \\ -0.0260283 \end{bmatrix}$$

Using $\mathbf{k}_2$ from Eq. (6), and realizing that $\mathbf{Q}_{f2} = \mathbf{0}$, we determine that

$$\mathbf{Q}_2 = \mathbf{k}_2\mathbf{u}_2 = \begin{bmatrix} 49.027 \text{ kN} \\ 85.948 \text{ kN} \\ 320.31 \text{ kN} \cdot \text{m} \\ -49.027 \text{ kN} \\ -85.948 \text{ kN} \\ 195.38 \text{ kN} \cdot \text{m} \end{bmatrix}$$ Ans

$$\mathbf{F}_2 = \mathbf{T}_2^T\mathbf{Q}_2 = \begin{bmatrix} -85.948 \\ 49.027 \\ 320.31 \\ 85.948 \\ -49.027 \\ 195.38 \end{bmatrix} \begin{matrix} 13 \\ 14 \\ 15 \\ 4 \\ 5 \\ 6 \end{matrix} \qquad (8)$$

Member 3

$$\mathbf{v}_3 = \begin{bmatrix} 0.185422 \\ 0.000418736 \\ -0.0176197 \\ 0.186622 \\ 0.000713665 \\ 0.0178911 \end{bmatrix} \begin{matrix} 1 \\ 2 \\ 3 \\ 7 \\ 8 \\ 9 \end{matrix}$$

Using $\mathbf{T}_3$ from Eq. (5), we compute

$$\mathbf{u}_3 = \mathbf{T}_3\mathbf{v}_3 = \begin{bmatrix} 0.000418736 \\ -0.185422 \\ -0.0176197 \\ 0.000713665 \\ -0.186622 \\ 0.0178911 \end{bmatrix}$$

Using $\mathbf{k}_3$ from Eq. (6), and realizing that $\mathbf{Q}_{f3} = \mathbf{0}$, we obtain

$$\mathbf{Q}_3 = \mathbf{k}_3\mathbf{u}_3 = \begin{bmatrix} -110.6 \text{ kN} \\ 1.6114 \text{ kN} \\ -80.392 \text{ kN}\cdot\text{m} \\ 110.6 \text{ kN} \\ -1.6114 \text{ kN} \\ 90.06 \text{ kN}\cdot\text{m} \end{bmatrix} \qquad \textbf{Ans}$$

Note that it is not necessary to compute the member global end force vector $\mathbf{F}_3$, because this member is not attached to any supports (and, therefore, none of the elements of $\mathbf{F}_3$ will appear in the support reaction vector $\mathbf{R}$).

Member 4 $\mathbf{T}_4 = \mathbf{I}$

$$\mathbf{u}_4 = \mathbf{v}_4 = \begin{bmatrix} 0.185422 \\ 0.000418736 \\ -0.0176197 \\ 0.18552 \\ -0.000130738 \\ -0.0260283 \end{bmatrix} \begin{matrix} 1 \\ 2 \\ 3 \\ 4 \\ 5 \\ 6 \end{matrix}$$

Using $\mathbf{k}_4$ from Eq. (1), and $\mathbf{Q}_{f4} = \mathbf{0}$, we obtain

$$\mathbf{Q}_4 = \mathbf{k}_4\mathbf{u}_4 = \begin{bmatrix} -24.5 \text{ kN} \\ -46.429 \text{ kN} \\ -195.47 \text{ kN}\cdot\text{m} \\ 24.5 \text{ kN} \\ 46.429 \text{ kN} \\ -222.38 \text{ kN}\cdot\text{m} \end{bmatrix} \qquad \textbf{Ans}$$

Member 5

$$\mathbf{v}_5 = \begin{bmatrix} 0.18552 \\ -0.000130738 \\ -0.0260283 \\ 0.186622 \\ 0.000713665 \\ 0.0178911 \end{bmatrix} \begin{matrix} 4 \\ 5 \\ 6 \\ 7 \\ 8 \\ 9 \end{matrix}$$

$\cos\theta = -0.83205$, $\sin\theta = 0.5547$ (Eqs. (2))

$$\mathbf{T}_5 = \begin{bmatrix} -0.83205 & 0.5547 & 0 & 0 & 0 & 0 \\ -0.5547 & -0.83205 & 0 & 0 & 0 & 0 \\ 0 & 0 & 1 & 0 & 0 & 0 \\ 0 & 0 & 0 & -0.83205 & 0.5547 & 0 \\ 0 & 0 & 0 & -0.5547 & -0.83205 & 0 \\ 0 & 0 & 0 & 0 & 0 & 1 \end{bmatrix}$$

$$\mathbf{u}_5 = \mathbf{T}_5\mathbf{v}_5 = \begin{bmatrix} -0.154434 \\ -0.102799 \\ -0.0260283 \\ -0.154883 \\ -0.104113 \\ 0.0178911 \end{bmatrix}$$

$$
\mathbf{k}_5 = \begin{bmatrix} 208{,}013 & 0 & 0 & -208{,}013 & 0 & 0 \\ 0 & 136.54 & 738.46 & 0 & -136.54 & 738.46 \\ 0 & 738.46 & 5{,}325.1 & 0 & -738.46 & 2{,}662.6 \\ -208{,}013 & 0 & 0 & 208{,}013 & 0 & 0 \\ 0 & -136.54 & -738.46 & 0 & 136.54 & -738.46 \\ 0 & 738.46 & 2{,}662.6 & 0 & -738.46 & 5{,}325.1 \end{bmatrix}
$$

From Eqs. (3), we obtain

$$
\mathbf{Q}_{f5} = \begin{bmatrix} 0 \\ 64.9 \\ 117 \\ 0 \\ 64.9 \\ -117 \end{bmatrix}
$$

$$
\mathbf{Q}_5 = \mathbf{k}_5\mathbf{u}_5 + \mathbf{Q}_{f5} = \begin{bmatrix} 93.398 \text{ kN} \\ 59.07 \text{ kN} \\ 27.004 \text{ kN} \cdot \text{m} \\ -93.398 \text{ kN} \\ 70.73 \text{ kN} \\ -90.061 \text{ kN} \cdot \text{m} \end{bmatrix}
$$ **Ans**

The member local end forces are shown in Fig. 6.18(e).

Support Reactions: The reaction vector **R**, as assembled from the appropriate elements of the member global end force vectors $\mathbf{F}_1$ and $\mathbf{F}_2$ (Eqs. (7) and (8), respectively), is given in Fig. 6.18(f). Also, Fig. 6.18(g) depicts these support reactions on a line diagram of the frame. **Ans**

Equilibrium Check: Considering the equilibrium of the entire frame, we write (Fig. 6.18(g))

$$
+ \rightarrow \sum F_x = 0 \quad 40 + 80 + (12\sqrt{117})\frac{6}{\sqrt{117}} - 106.05 - 85.948 = 0.002 \cong 0
$$ **Checks**

$$
+ \uparrow \sum F_Y = 0 \qquad (12\sqrt{117})\frac{9}{\sqrt{117}} - 157.03 + 49.027 = -0.003 \cong 0
$$ **Checks**

$$
\begin{aligned}
+ \circlearrowleft \sum M_{①} = 0 \qquad & 360.44 - 80(6) - 40(12) - (12\sqrt{117})\left(\frac{6}{\sqrt{117}}\right)9 \\
& + (12\sqrt{117})\left(\frac{9}{\sqrt{117}}\right)4.5 + 320.31 + 49.027(9) \\
& = -0.007 \cong 0
\end{aligned}
$$ **Checks**

6.7 COMPUTER PROGRAM

The overall organization and format of the computer program for the analysis of plane frames remains the same as the plane truss and beam analysis programs developed previously. All the parts, and many subroutines, of the new program can be replicated from the previous programs with no, or relatively minor, modifications. In this section, we discuss the development of this program for the analysis of plane frames, with emphasis on the programming aspects not considered in previous chapters.

Input Module

Joint Data The part of the computer program for reading and storing the joint data for plane frames (i.e., the number of joints, *NJ*, and *X* and *Y* coordinates of each joint) remains the same as Part I of the plane truss analysis program (see flowchart in Fig. 4.3(a)). As discussed in Section 4.1, the program stores the joint coordinates in a $NJ \times 2$ joint coordinate matrix **COORD** in computer memory. As an example, let us consider the gable frame of Fig. 6.19(a), with its analytical model shown in Fig. 6.19(b). Since the frame has five joints, its **COORD** matrix has five rows, with the *X* and *Y* coordinates of a joint *i* stored in the first and second columns, respectively, of the *i*th row, as shown in Fig. 6.19(c). An example of the input data file for the gable frame is given in Fig. 6.20 on page 320.

Support Data The computer code written for Part II of the plane truss analysis program (see flowchart in Fig. 4.3(b)) can be used to input the support data for plane frames, provided that the number of structure coordinates per joint is set equal to 3 (i.e., $NCJT = 3$) in the program. A three-digit code is now used to specify the restraints at a support joint, with the first two digits representing the translational restraint conditions in the global *X* and *Y* directions, respectively, and the third digit representing the rotational restraint condition at the joint. As in the case of plane trusses and beams, each digit of the restraint

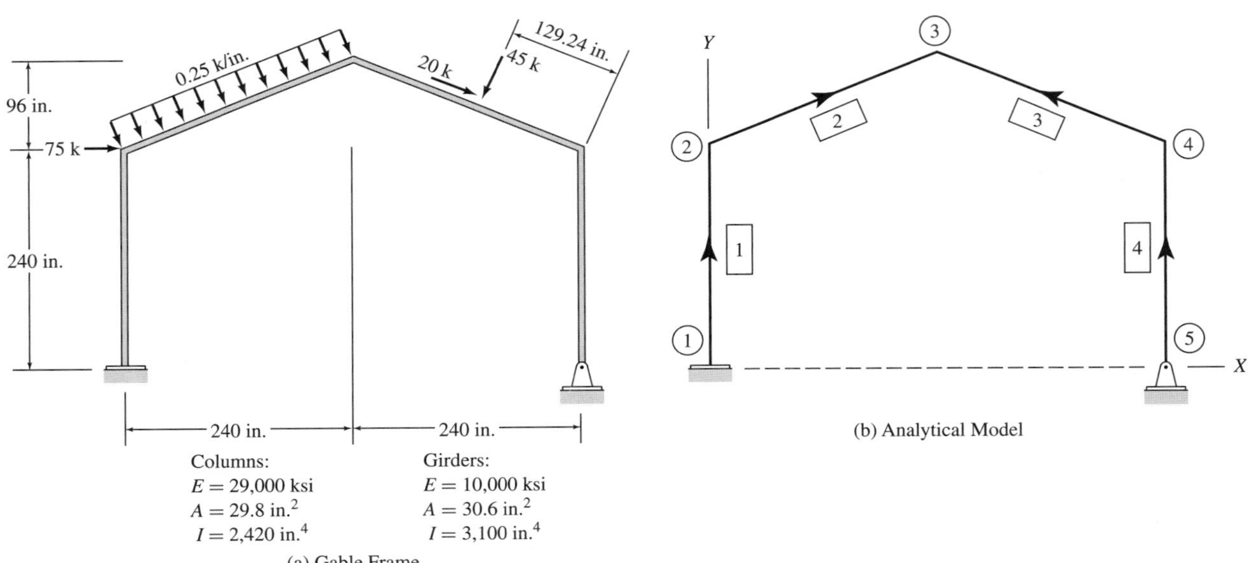

Fig. 6.19

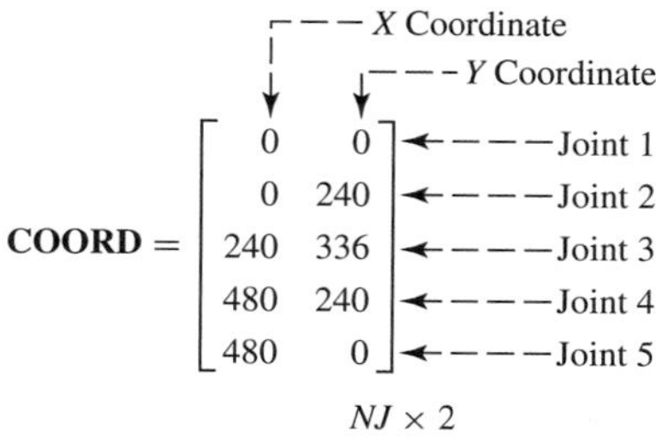

$$\begin{array}{rcl} & \begin{array}{cc} \text{--- } X \text{ Coordinate} & \\ \downarrow & \text{--- } Y \text{ Coordinate} \\ & \downarrow \end{array} & \\ \mathbf{COORD} = & \begin{bmatrix} 0 & 0 \\ 0 & 240 \\ 240 & 336 \\ 480 & 240 \\ 480 & 0 \end{bmatrix} & \begin{array}{l} \leftarrow\text{---Joint 1} \\ \leftarrow\text{---Joint 2} \\ \leftarrow\text{---Joint 3} \\ \leftarrow\text{---Joint 4} \\ \leftarrow\text{---Joint 5} \end{array} \\ & NJ \times 2 & \end{array}$$

(c) Joint Coordinate Matrix

$$\begin{array}{rc} & \begin{array}{cccc} \text{Joint Number} & \text{Restraint in } X \text{ Direction } (0 = \text{free, } 1 = \text{restrained}) & \text{Restraint in } Y \text{ Direction } (0 = \text{free, } 1 = \text{restrained}) & \text{Rotational Restraint } (0 = \text{free, } 1 = \text{restrained}) \\ \downarrow & \downarrow & \downarrow & \downarrow \end{array} \\ \mathbf{MSUP} = & \begin{bmatrix} 1 & 1 & 1 & 1 \\ 5 & 1 & 1 & 0 \end{bmatrix} \\ & NS \times (NCJT + 1) \end{array}$$

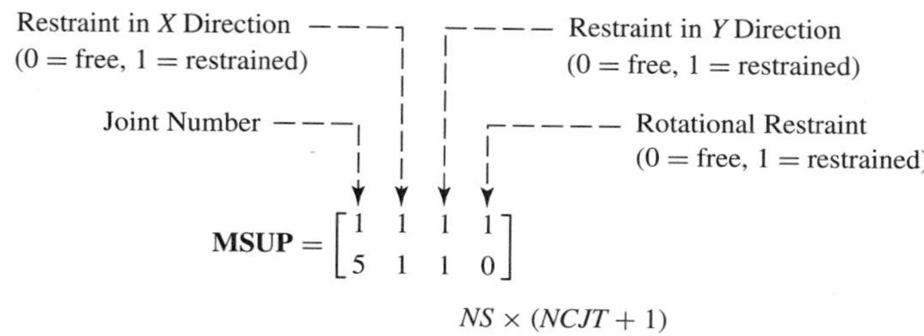

(d) Support Data Matrix

$$\mathbf{EM} = \underset{NMP \times 1}{\begin{bmatrix} 29000 \\ 10000 \end{bmatrix}} \begin{array}{l} \leftarrow\text{--- Material No. 1} \\ \leftarrow\text{--- Material No. 2} \end{array}$$

(e) Elastic Modulus Vector

$$\begin{array}{rcl} & \begin{array}{cc} \text{--- Area} & \\ \downarrow & \text{--- Moment of Inertia} \\ & \downarrow \end{array} & \\ \mathbf{CP} = & \begin{bmatrix} 29.8 & 2420 \\ 30.6 & 3100 \end{bmatrix} & \begin{array}{l} \leftarrow\text{--- Cross-Section Type No. 1} \\ \leftarrow\text{--- Cross-Section Type No. 2} \end{array} \\ & NCP \times 2 & \end{array}$$

(f) Cross-Sectional Property Matrix

$$\begin{array}{rcl} & \begin{array}{cccc} \text{--- Beginning Joint} & & & \\ \downarrow & \text{--- End Joint} & & \\ & \downarrow & \text{--- Material No.} & \\ & & \downarrow & \text{--- Cross-Section Type No.} \\ & & & \downarrow \end{array} & \\ \mathbf{MPRP} = & \begin{bmatrix} 1 & 2 & 1 & 1 \\ 2 & 3 & 2 & 2 \\ 4 & 3 & 2 & 2 \\ 5 & 4 & 1 & 1 \end{bmatrix} & \begin{array}{l} \leftarrow\text{---Member 1} \\ \leftarrow\text{---Member 2} \\ \leftarrow\text{---Member 3} \\ \leftarrow\text{---Member 4} \end{array} \\ & NM \times 4 & \end{array}$$

(g) Member Data Matrix

$$\begin{array}{rc} & \begin{array}{c} \text{--- Joint Number} \\ \downarrow \end{array} \\ \mathbf{JP} = & [2] \\ & NJL \times 1 \end{array} \qquad \begin{array}{rc} & \begin{array}{ccc} \text{--- Force in } X \text{ Direction} & & \\ \downarrow & \text{--- Force in } Y \text{ Direction} & \\ & \downarrow & \text{--- Moment} \\ & & \downarrow \end{array} \\ \mathbf{PJ} = & [75 \quad 0 \quad 0] \\ & NJL \times NCJT \end{array}$$

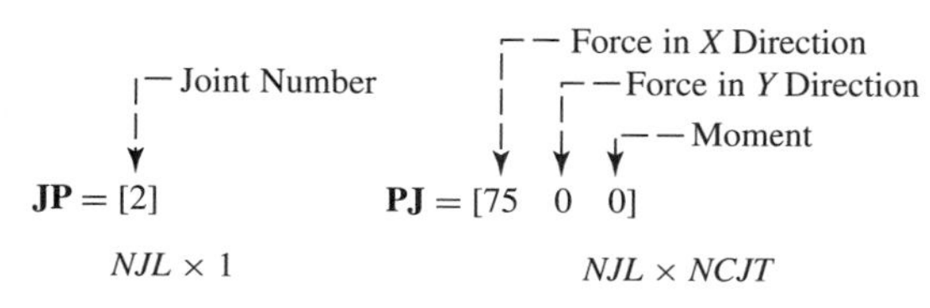

(h) Joint Load Data Matrices

$$\begin{array}{rc} & \begin{array}{cc} \text{--- Member Number} & \\ \downarrow & \text{--- Load Type Number} \\ & \downarrow \end{array} \\ \mathbf{MP} = & \begin{bmatrix} 2 & 3 \\ 3 & 1 \\ 3 & 5 \end{bmatrix} \\ & NML \times 2 \end{array} \qquad \begin{array}{rc} \mathbf{PM} = & \begin{bmatrix} 0.25 & 0 & 0 & 0 \\ -45 & 0 & 129.24 & 0 \\ 20 & 0 & 129.24 & 0 \end{bmatrix} \quad NML \times 4 \\ & \begin{array}{cccc} \uparrow & \uparrow & \uparrow & \uparrow \\ W, M, w \text{ or } w_1 & w_2 \text{ (if Load Type} = 4) \text{; } 0 \text{ (otherwise)} & l_1 & l_2 \text{ (if Load Type} = 3, 4 \text{ or } 6) \text{; } 0 \text{ (otherwise)} \end{array} \end{array}$$

(i) Member Load Data Matrices

Fig. 6.19 *(continued)*

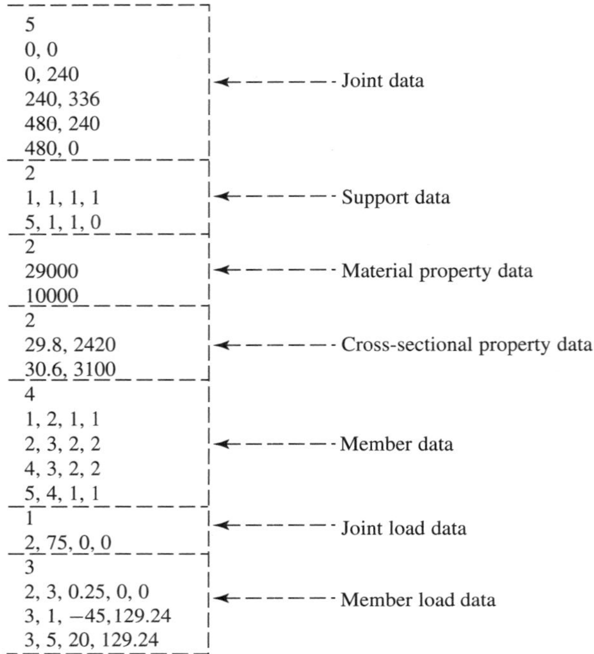

Fig. 6.20 *An Example of an Input Data File*

code can be either 0 (indicating no restraint) or 1 (indicating restraint). The restraint codes for some common types of supports for plane frames are given in Fig. 6.21. The program stores the support data in a $NS \times 4$ **MSUP** matrix, as shown in Fig. 6.19(d) for the gable frame, and an example of how this data may appear in an input file is given in Fig. 6.20.

Material Property Data This part of the program remains the same as Part III of the plane truss analysis program (see flowchart in Fig. 4.3(c)). The program stores the moduli of elasticity in a $NMP \times 1$ **EM** vector, as shown in Fig. 6.19(e) for the example gable frame; Fig. 6.20 illustrates how this data may appear in an input data file.

Cross-sectional Property Data As two cross-sectional properties (namely, area and moment of inertia) are needed in the analysis of plane frames, the code written previously for Part IV of the plane truss program should be modified to increase the number of columns of the *cross-sectional property matrix* **CP** from one to two, as indicated by the flowchart in Fig. 6.22(a). As before, the number of rows of **CP** equals the number of cross-section types (NCP), with the area and moment of inertia of the cross-section i now stored in the first and second columns, respectively, of the ith row of the **CP** matrix of order $NCP \times 2$. For example, the **CP** matrix for the gable frame of Fig. 6.19(a) is shown in Fig. 6.19(f), and Fig. 6.20 shows how this data may appear in an input data file.

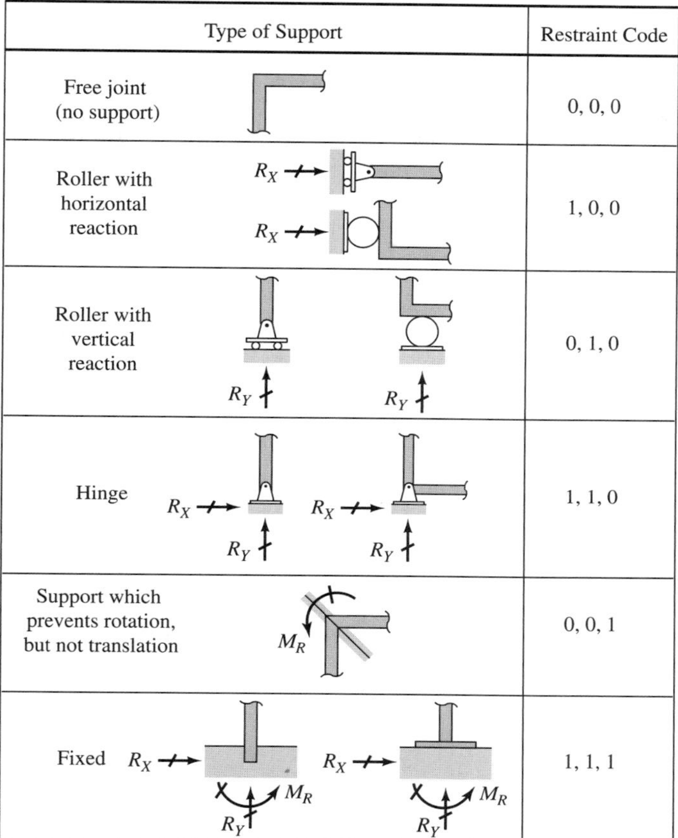

Type of Support	Restraint Code
Free joint (no support)	0, 0, 0
Roller with horizontal reaction	1, 0, 0
Roller with vertical reaction	0, 1, 0
Hinge	1, 1, 0
Support which prevents rotation, but not translation	0, 0, 1
Fixed	1, 1, 1

Fig. 6.21 *Restraint Codes for Plane Frames*

Member Data This part of the computer program remains the same as Part V of the plane truss analysis program (see flowchart in Fig. 4.3(e)). As discussed in Section 4.1, the program stores the member data in an integer matrix **MPRP** of order $NM \times 4$. The **MPRP** matrix for the example gable frame is shown in Fig. 6.19(g).

Joint Load Data The code written for Part VIa of the beam analysis program (see flowchart in Fig. 5.20(b)) can be used for inputting joint load data for plane frames, provided that $NCJT$ is set equal to 3. The program stores the numbers of the loaded joints in an integer vector **JP** of order $NJL \times 1$, with the corresponding loads in the X and Y directions and the couple being stored in the first, second, and third columns, respectively, of a real matrix **PJ** of order $NJL \times 3$. The joint load matrices for the example gable frame are shown in Fig. 6.19(h); Fig. 6.20 illustrates how this data may appear in an input data file.

Member Load Data As members of plane frames may be subjected to both axial and perpendicular loads, the code written for Part VIb of the beam

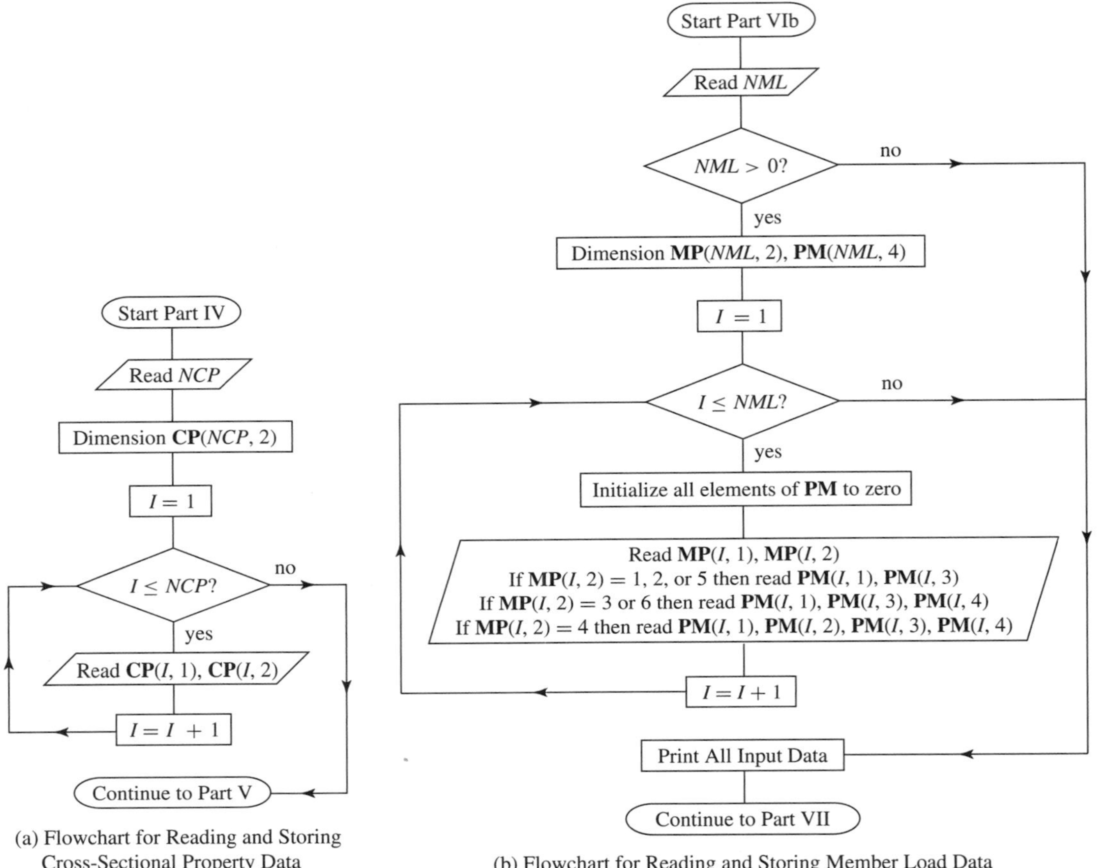

(a) Flowchart for Reading and Storing Cross-Sectional Property Data

(b) Flowchart for Reading and Storing Member Load Data

Fig. 6.22

analysis program should be modified to include inputting of the member axial loads. The flowchart shown in Fig. 6.22(b) can be used for programming the input of the four perpendicular, and two axial, member load types (i.e., load types 1 through 6) given inside the front cover. The format for reading and storing the member load data for plane frames remains the same as that for beams, as discussed in Section 5.8. The member load matrices, **MP** and **PM**, for the example gable frame are shown in Fig. 6.19(i); Fig. 6.20 shows this member load data in an input file that can be read by the program.

An example of a computer printout of the input data for the gable frame of Fig. 6.19 is given in Fig. 6.23.

Analysis Module

Assignment of Structure Coordinate Numbers The parts of the program for determining the number of degrees of freedom (*NDOF*) and forming the

```
*********************************
*       Computer Software       *
*              for              *
*  MATRIX ANALYSIS OF STRUCTURES *
*        Second Edition         *
*              by               *
*        Aslam Kassimali        *
*********************************
```

General Structural Data

Project Title: Figure 6-19
Structure Type: Plane Frame
Number of Joints: 5
Number of Members: 4
Number of Material Property Sets (E): 2
Number of Cross-Sectional Property Sets: 2

Joint Coordinates

Joint No.	X Coordinate	Y Coordinate
1	0.0000E+00	0.0000E+00
2	0.0000E+00	2.4000E+02
3	2.4000E+02	3.3600E+02
4	4.8000E+02	2.4000E+02
5	4.8000E+02	0.0000E+00

Supports

Joint No.	X Restraint	Y Restraint	Rotational Restraint
1	Yes	Yes	Yes
5	Yes	Yes	No

Material Properties

Material No.	Modulus of Elasticity (E)	Co-efficient of Thermal Expansion
1	2.9000E+04	0.0000E+00
2	1.0000E+04	0.0000E+00

Cross-Sectional Properties

Property No.	Area (A)	Moment of Inertia (I)
1	2.9800E+01	2.4200E+03
2	3.0600E+01	3.1000E+03

Fig. 6.23 *A Sample Printout of Input Data*

Member Data

Member No.	Beginning Joint	End Joint	Material No.	Cross-Sectional Property No.
1	1	2	1	1
2	2	3	2	2
3	4	3	2	2
4	5	4	1	1

Joint Loads

Joint No.	X Force	Y Force	Moment
2	7.5000E+01	0.0000E+00	0.0000E+00

Member Loads

Member No.	Load Type	Load Magnitude (W or M) or Intensity (w or w1)	Load Intensity w2	Distance l1	Distance l2
2	Uniform	2.500E-1	---	0.00E+0	0.00E+0
3	Axial-C	2.000E+1	---	1.29E+2	----
3	Conc.	-4.500E+1	---	1.29E+2	----

************** End of Input Data **************

Fig. 6.23 *(continued)*

structure coordinate number vector (**NSC**), for plane frames, remain the same as Parts VII and VIII of the plane truss analysis program (see flowcharts in Figs. 4.8(a) and (b)), provided that *NCJT* is set equal to 3 in these programs.

Generation of the Structure Stiffness Matrix and Equivalent Joint Load Vector A flowchart for writing this part of the plane frame analysis program is given in Fig. 6.24. Comparing this flowchart with that for Part IX of the beam analysis program in Fig. 5.24, we can see that the two programs are similar; the present program, however, transforms the stiffness matrix and fixed-end force vector of each member from its local to the global coordinate system before storing their elements in the structure stiffness matrix and equivalent joint load vector, respectively (see Fig. 6.24). Recall from Chapter 5 that such coordinate transformations are not necessary for beams, because the local and global coordinate systems of such structures are oriented in the same direction. From the flowchart in Fig. 6.24, we can see that for each member of the plane frame, the program first reads the member's material and cross-sectional

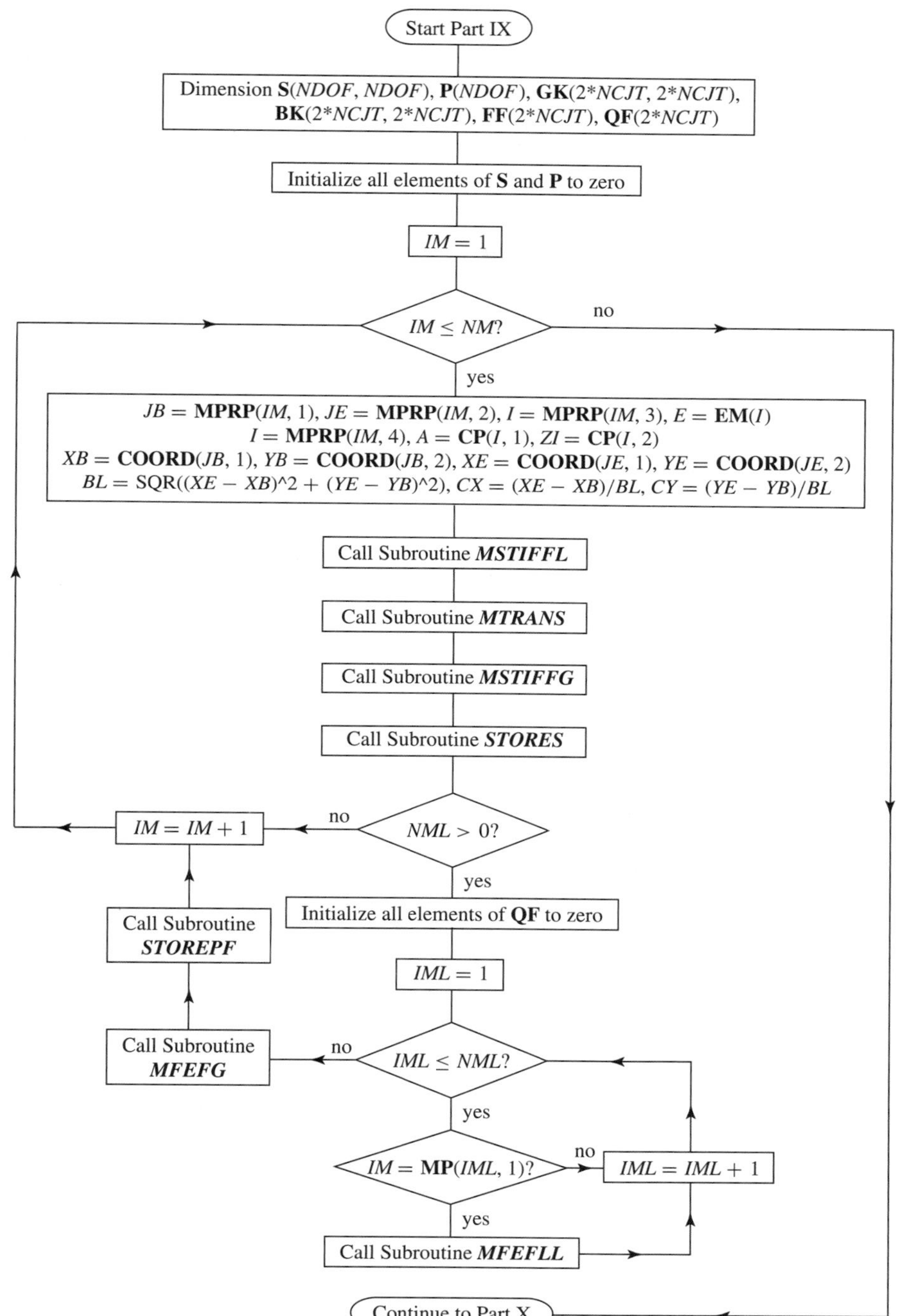

Fig. 6.24 *Flowchart for Generating Structure Stiffness Matrix and Equivalent Joint Load Vector*

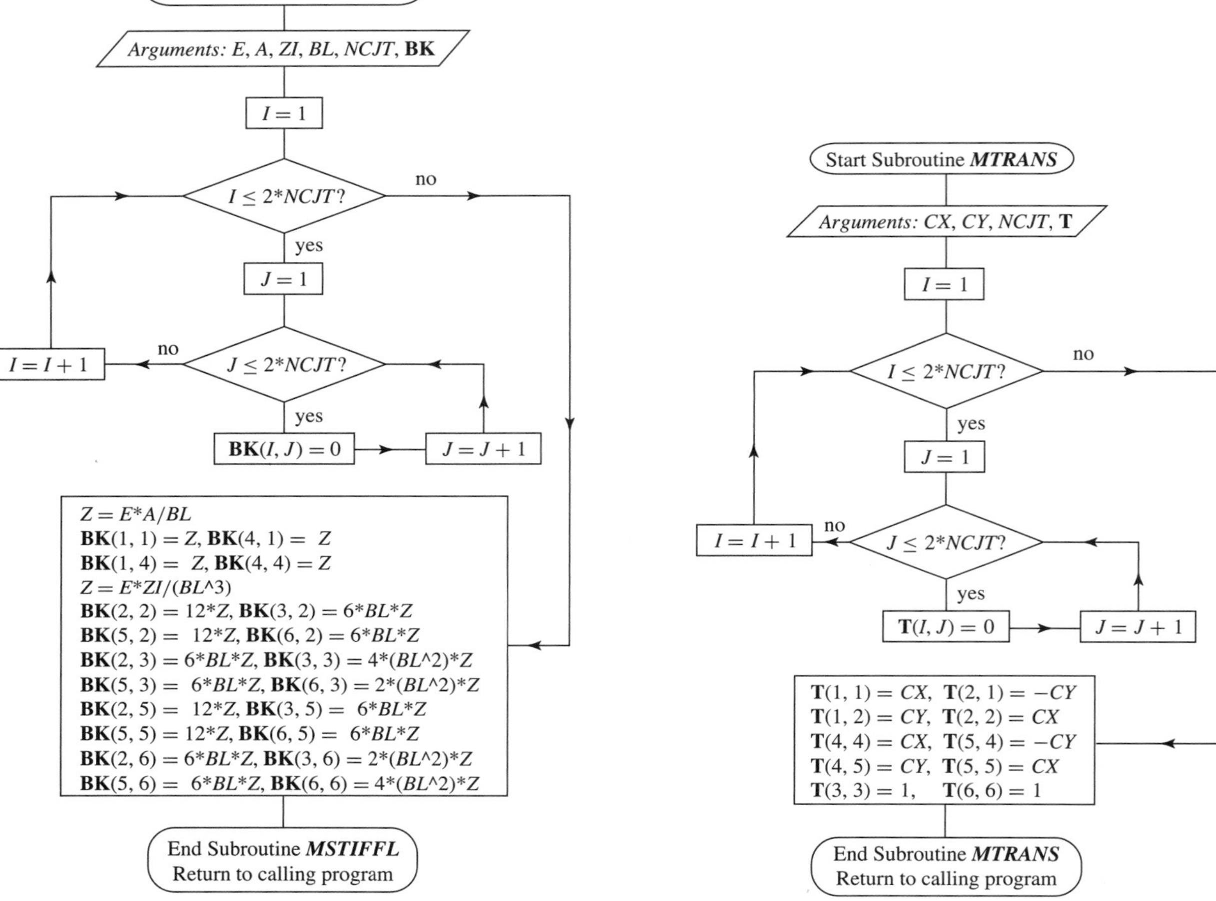

Fig. 6.25 *Flowchart of Subroutine* ***MSTIFFL*** *for Determining Member Local Stiffness Matrix for Plane Frames*

Fig. 6.26 *Flowchart of Subroutine* ***MTRANS*** *for Determining Member Transformation Matrix for Plane Frames*

properties, and calculates its length and direction cosines. Next, the program calls the subroutines ***MSTIFFL*** and ***MTRANS*** to form the member local stiffness matrix **BK** and transformation matrix **T**, respectively. As the flowcharts given in Figs. 6.25 and 6.26 indicate, these subroutines calculate the matrices **BK** and **T** in accordance with Eqs. (6.6) and (6.19), respectively. The program then calls the subroutine ***MSTIFFG*** to obtain the member global stiffness matrix **GK**. A flowchart of this subroutine, which evaluates the member global stiffness matrix using the matrix triple product $\mathbf{K} = \mathbf{T}^T\mathbf{k}\mathbf{T}$ (Eq. (6.29)), is given in Fig. 6.27. As this flowchart indicates, the subroutine ***MSTIFFG*** uses two nested *Do Loops* to calculate the member global stiffness matrix **GK** (= **K**). In the first loop, the member local stiffness matrix **BK** (= **k**) is post-multiplied by its transformation matrix **T** to obtain an intermediate matrix **TS**;

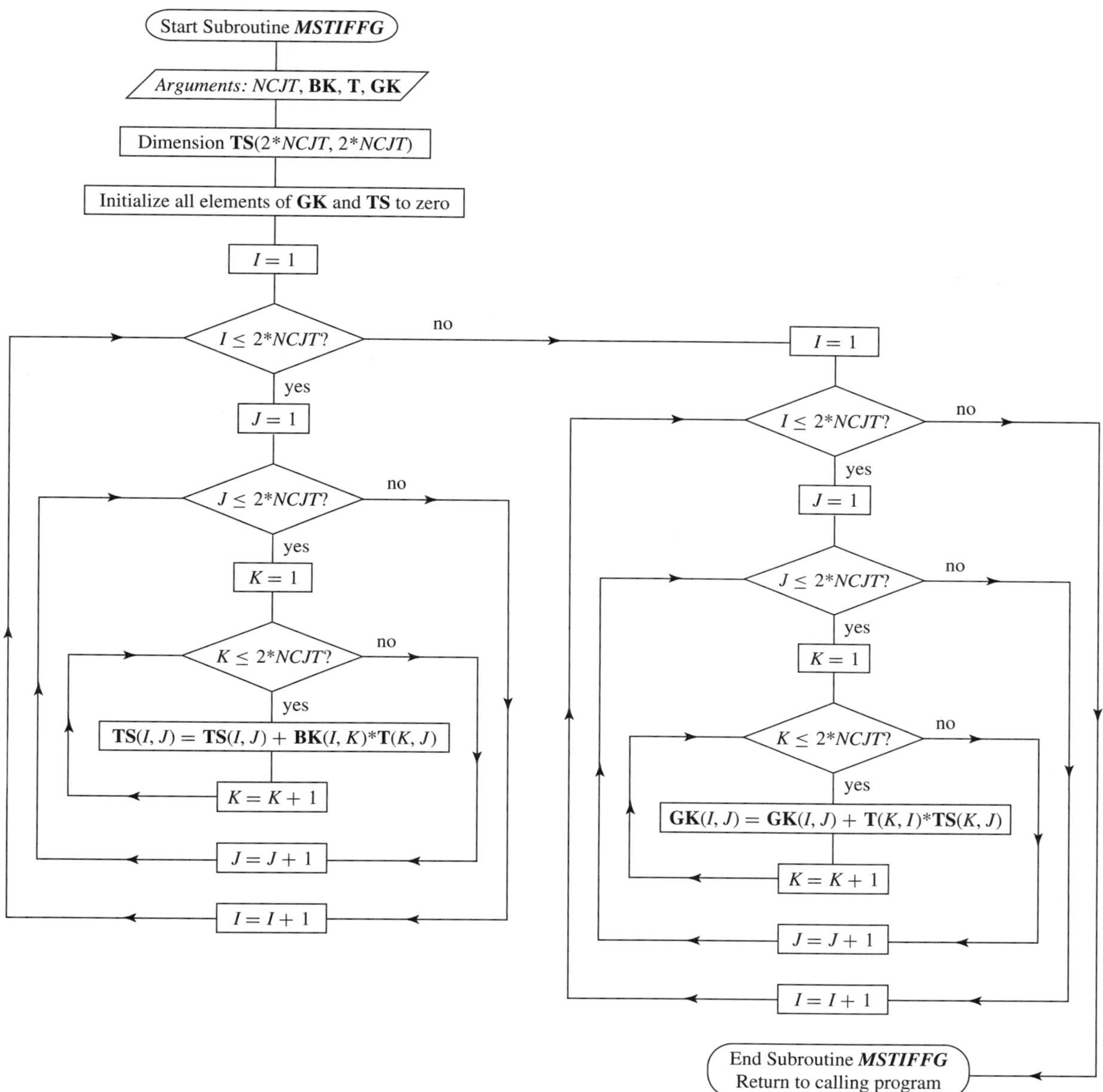

Fig. 6.27 *Flowchart of Subroutine* ***MSTIFFG*** *for Determining Member Global Stiffness Matrix*

in the second, the matrix **TS** is premultiplied by the transpose of the transformation matrix (i.e., $\mathbf{T}^T$) to obtain the desired member global stiffness matrix **GK** (= **K**). Returning our attention to Fig. 6.24, we can see that the program then calls the subroutine ***STORES*** to store the pertinent elements of **GK** in the structure stiffness matrix **S**. This subroutine remains the same as the ***STORES*** subroutine of the plane truss analysis program (see flowchart in Fig. 4.11). After ***STORES*** has been executed, the program (Fig. 6.24) forms the member

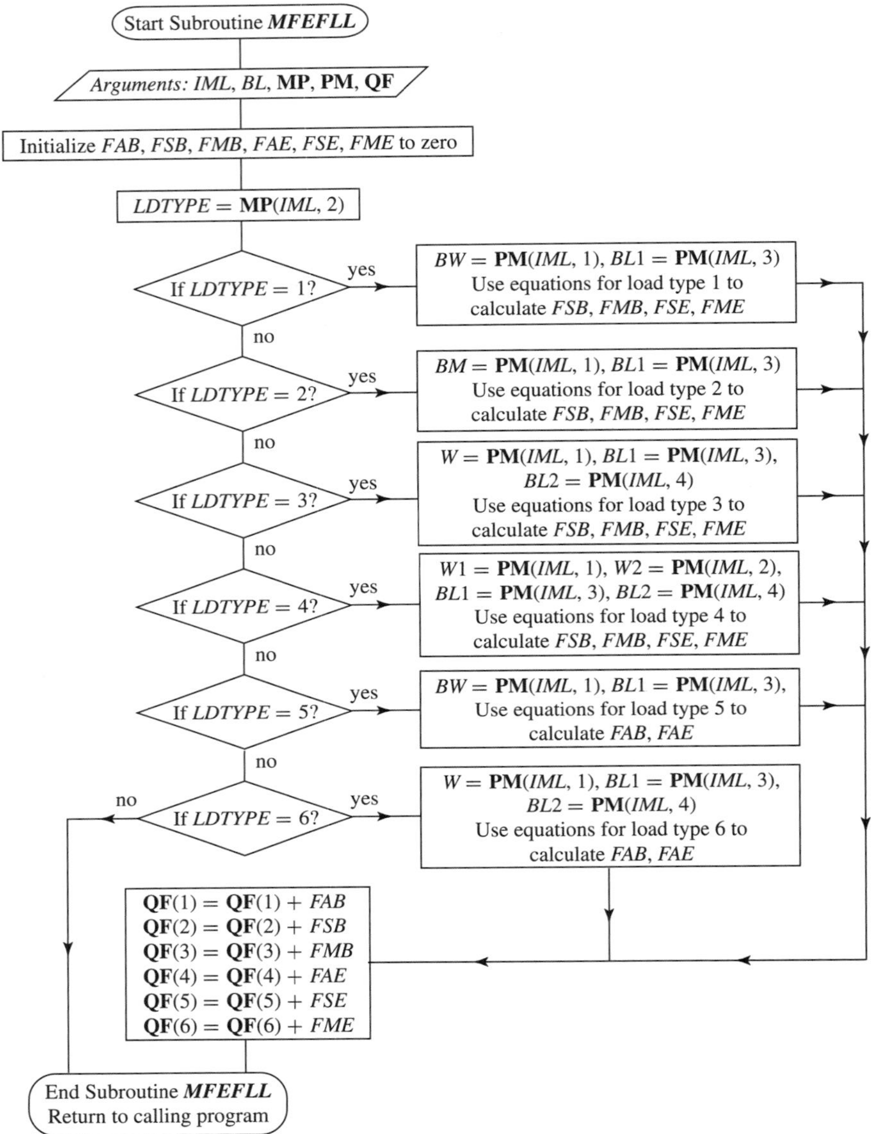

Fig. 6.28 *Flowchart of Subroutine **MFEFLL** for Determining Member Local Fixed-End Force Vector for Plane Frames*

local fixed-end force vector **QF** (= $\mathbf{Q}_f$) by calling the subroutine ***MFEFLL*** (Fig. 6.28), which calculates the values of the member fixed-end forces, for load types 1 through 6, using the equations given inside the front cover. Next, the program calls the subroutine ***MFEFG*** (Fig. 6.29), which evaluates the member global fixed-end force vector **FF** (= $\mathbf{F}_f$), using the relationship $\mathbf{F}_f = \mathbf{T}^T\mathbf{Q}_f$ (Eq. (6.30)). Finally, the program calls the subroutine ***STOREPF*** (Fig. 6.30) to store the negative values of the pertinent elements of **FF** in their

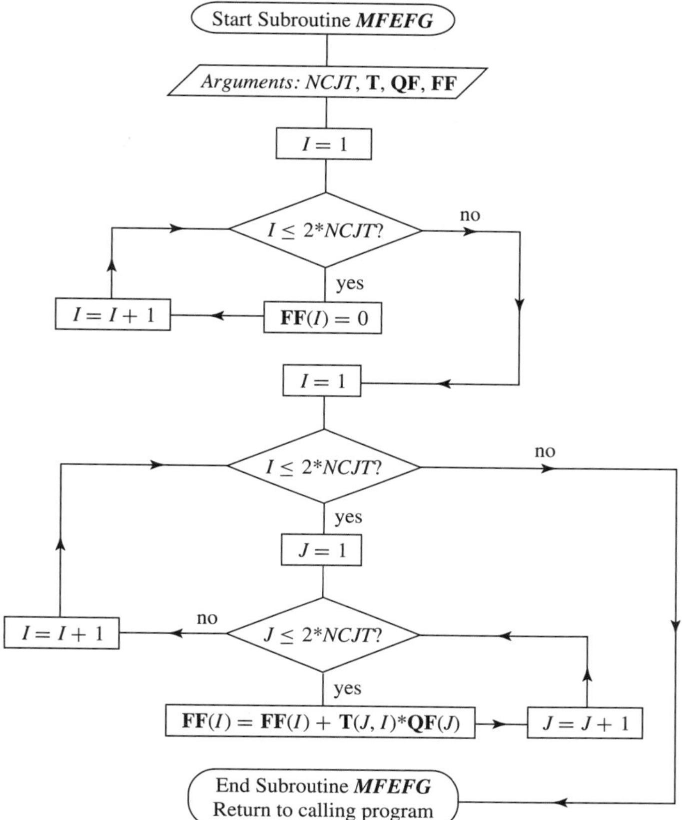

Fig. 6.29 *Flowchart of Subroutine **MFEFG** for Determining Member Global Fixed-End Force Vector*

proper positions in the structure load vector **P**. When all the operations shown in Fig. 6.24 have been performed for each member of the frame, the structure stiffness matrix **S** is complete, and the structure load vector **P** equals the negative of the structure fixed-joint force vector (i.e., $\mathbf{P} = -\mathbf{P}_f = \mathbf{P}_e$).

Storage of Joint Loads into the Structure Load Vector This is the same as Part X of the beam analysis program (see flowchart in Fig. 5.29).

Solution for Joint Displacements This part of the program remains the same as Part XI of the plane truss analysis program (see flowchart in Fig. 4.13). Recall that upon completion of this part, the vector **P** contains the values of the joint displacements **d**.

Calculation of Member Forces and Support Reactions A flowchart for writing this last part of the program is presented in Fig. 6.31. Note that this part of

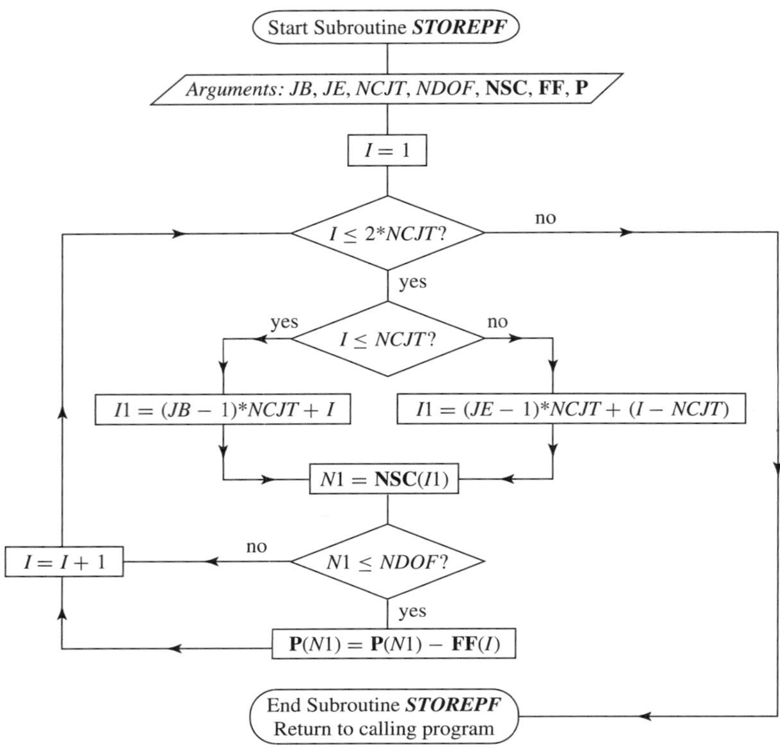

Fig. 6.30 *Flowchart of Subroutine **STOREPF** for Storing Member Global Fixed-End Force Vector in Structure Load Vector*

the plane frame analysis program is essentially a combination of the corresponding parts (XII) of the plane truss and beam analysis programs developed previously. From the flowchart in Fig. 6.31, we can see that, for each member of the frame, the program first reads the member's material and cross-sectional properties, and calculates its length and direction cosines. Next, the program calls the subroutine ***MDISPG*** to form the member global end displacement vector **V** (= **v**). This subroutine is the same as the ***MDISPG*** subroutine of the plane truss analysis program (see flowchart in Fig. 4.15). The program then calls the subroutine ***MTRANS*** (Fig. 6.26) to form the member transformation matrix **T**, and the subroutine ***MDISPL***, which calculates the member local end displacement vector **U** (= **u**), using the relationship $\mathbf{u} = \mathbf{T}\mathbf{v}$ (Eq. (6.20)). The subroutine ***MDISPL*** remains the same as the corresponding subroutine of the plane truss program (see flowchart in Fig. 4.17). Next, the subroutine ***MSTIFFL*** (Fig. 6.25) is called by the program to form the member local stiffness matrix **BK** (= **k**); if the member under consideration is subjected to loads, then its local fixed-end force vector **QF** (= $\mathbf{Q}_f$) is generated using the subroutine ***MFEFLL*** (Fig. 6.28). The program then calls the subroutines ***MFORCEL*** and ***MFORCEG***, respectively, to calculate the member's local and global end

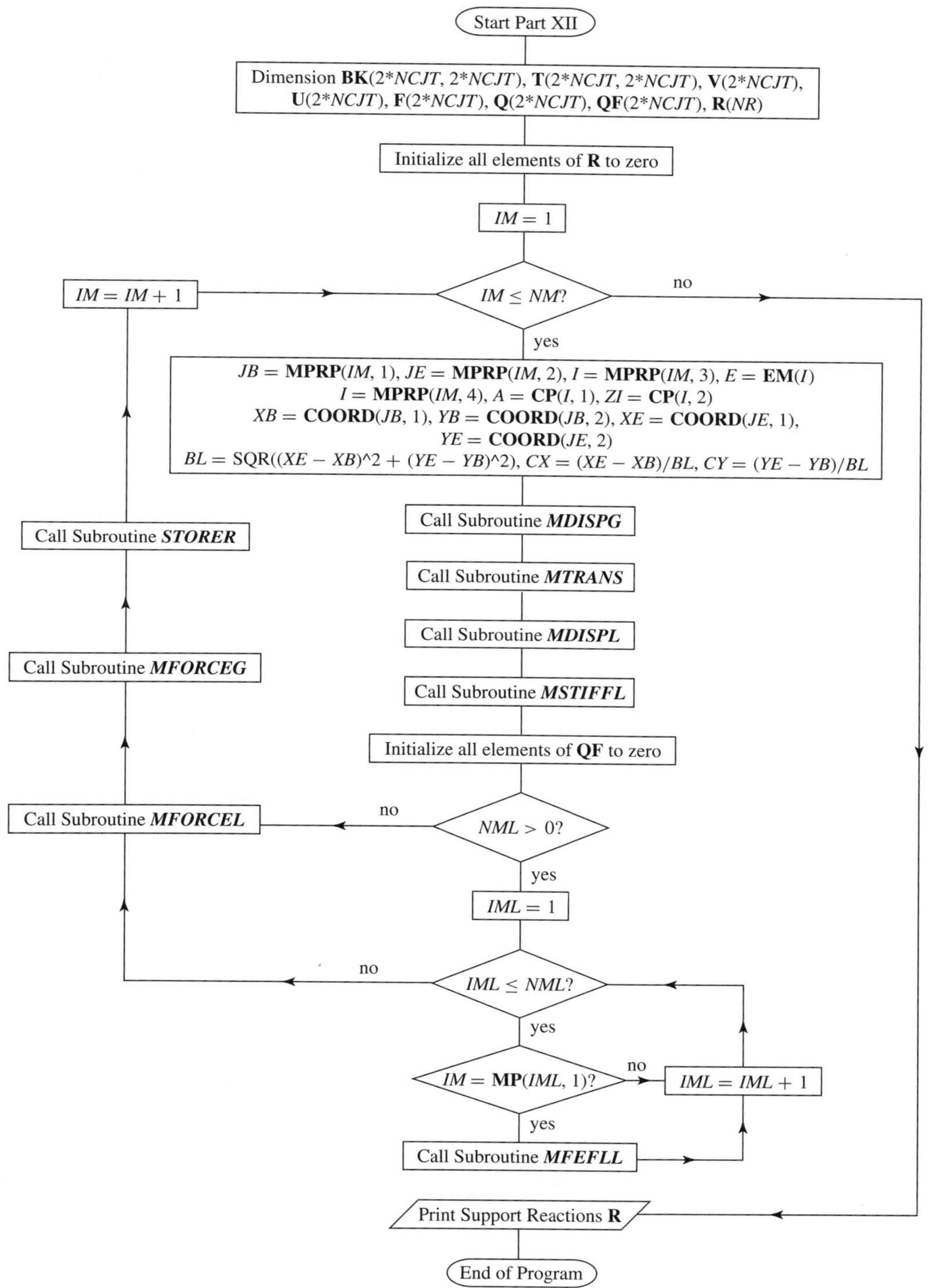

Fig. 6.31 *Flowchart for Determination of Member Forces and Support Reactions for Plane Frames*

force vectors **Q** and **F**. The subroutine ***MFORCEL***, which evaluates **Q** using the relationship $\mathbf{Q} = \mathbf{ku} + \mathbf{Q}_f$ (Eq. (6.4)), is the same as the corresponding subroutine of the beam analysis program (see flowchart in Fig. 5.32); the subroutine ***MFORCEG***, which computes **F** using the relationship $\mathbf{F} = \mathbf{T}^T\mathbf{Q}$ (Eq. (6.23)), remains the same as the corresponding subroutine of the plane truss program (see flowchart in Fig. 4.20). Finally, the program stores the pertinent elements of **F** in the support reaction vector **R** by calling the subroutine ***STORER***, which remains the same as the corresponding subroutine of the plane truss program (see flowchart in Fig. 4.21). The computational process depicted in Fig. 6.31 can be somewhat expedited by calling the subroutines

```
*****************************************************
*                 Results of Analysis               *
*****************************************************

                  ======================
                  Joint Displacements
                  ======================

Joint No.  X Translation  Y Translation   Rotation (Rad)
    1       0.0000E+00     0.0000E+00      0.0000E+00
    2       3.4472E+00    -9.1684E-03     -1.9513E-02
    3       3.9520E+00    -1.3152E+00      7.0646E-03
    4       4.4247E+00    -2.1160E-02     -9.2709E-03
    5       0.0000E+00     0.0000E+00     -2.3019E-02

          ========================================
          Member End Forces in Local Coordinates
          ========================================

Member  Joint    Axial Force   Shear Force    Moment
  1       1      3.3014E+01    6.7356E+01    1.3789E+04
          2     -3.3014E+01   -6.7356E+01    2.3767E+03

  2       2      1.9358E+01    2.7814E+01   -2.3767E+03
          3     -1.9358E+01    3.6808E+01    1.2142E+03

  3       4      5.9404E+01   -5.8303E+01   -8.0403E+03
          3     -3.9404E+01    1.3303E+01   -1.2142E+03

  4       5      7.6195E+01    3.3501E+01    1.5378E-03
          4     -7.6195E+01   -3.3501E+01    8.0403E+03

                  =================
                  Support Reactions
                  =================

  Joint No.     X Force        Y Force        Moment
     1       -6.7356E+01    3.3014E+01    1.3789E+04
     5       -3.3501E+01    7.6195E+01    0.0000E+00

**************** End of Analysis ****************
```

Fig. 6.32 *A Sample Printout of Analysis Results*

MFORCEG and ***STORER*** for only those members of the frame that are attached to supports. To check whether or not a member is attached to a support, its beginning and end joint numbers (i.e., *JB* and *JE*) can be compared with the support joint numbers stored in the first column of the support data matrix **MSUP**.

A sample printout, showing the results of analysis for the example gable frame of Fig. 6.19, is presented in Fig. 6.32, and the entire computer program for the analysis of plane frames is summarized in Table 6.1. As indicated in this table, the computer program consists of a main program (which is divided into twelve parts) and twelve subroutines. Of these, eight parts of the main program and six subroutines can be replicated from the previously developed plane truss and beam analysis programs without any modifications. Finally, it should be realized that the computer program, developed herein for the analysis of plane frames, can also be used to analyze beams, although it is not as efficient for beam analysis as the program developed specifically for that purpose in Chapter 5.

Table 6.1 *Computer Program for Analysis of Plane Frames*

Main program part	Description
I	Reads and stores joint data (Fig. 4.3(a))
II	Reads and stores support data (Fig. 4.3(b))
III	Reads and stores material properties (Fig. 4.3(c))
IV	Reads and stores cross-sectional properties (Fig. 6.22(a))
V	Reads and stores member data (Fig. 4.3(e))
VIa	Reads and stores joint loads (Fig. 5.20(b))
VIb	Reads and stores member loads (Fig. 6.22(b))
VII	Determines the number of degrees of freedom *NDOF* of the structure (Fig. 4.8(a))
VIII	Forms the structure coordinate number vector **NSC** (Fig. 4.8(b))
IX	Generates the structure stiffness matrix **S** and the structure load vector $\mathbf{P} = \mathbf{P}_e = -\mathbf{P}_f$ due to member loads (Fig. 6.24) Subroutines called: ***MSTIFFL***, ***MTRANS***, ***MSTIFFG***, ***STORES***, ***MFEFLL***, ***MFEFG***, and ***STOREPF***
X	Stores joint loads in the structure load vector **P** (Fig. 5.29)
XI	Calculates the structure's joint displacements by solving the stiffness relationship, $\mathbf{Sd} = \mathbf{P}$, using Gauss–Jordan elimination. The vector **P** now contains joint displacements (Fig. 4.13).
XII	Determines the member end force vectors **Q** and **F**, and the support reaction vector **R** (Fig. 6.31) Subroutines called: ***MDISPG***, ***MTRANS***, ***MDISPL***, ***MSTIFFL***, ***MFEFLL***, ***MFORCEL***, ***MFORCEG***, and ***STORER***

(*continued*)

Table 6.1 *(continued)*

Subroutine	Description
MDISPG	Forms the member global displacement vector **V** from the joint displacement vector **P** (Fig. 4.15)
MDISPL	Evaluates the member local displacement vector **U** = **TV** (Fig. 4.17)
MFEFG	Determines the member global fixed-end force vector **FF** = $\mathbf{T}^T$**QF** (Fig. 6.29)
MFEFLL	Calculates the member local fixed-end force vector **QF** (Fig. 6.28)
MFORCEG	Evaluates the member global force vector **F** = $\mathbf{T}^T$ **Q** (Fig. 4.20)
MFORCEL	Calculates the member local force vector **Q** = **BK U** + **QF** (Fig. 5.32)
MSTIFFG	Determines the member global stiffness matrix **GK** = $\mathbf{T}^T$ **BK T** (Fig. 6.27)
MSTIFFL	Forms the member local stiffness matrix **BK** (Fig. 6.25)
MTRANS	Forms the member transformation matrix **T** (Fig. 6.26)
STOREPF	Stores the negative values of the pertinent elements of the member global fixed-end force vector **FF** in the structure load vector **P** (Fig. 6.30)
STORER	Stores the pertinent elements of the member global force vector **F** in the reaction vector **R** (Fig. 4.21)
STORES	Stores the pertinent elements of the member global stiffness matrix **GK** in the structure stiffness matrix **S** (Fig. 4.11)

SUMMARY

In this chapter, we have developed the matrix stiffness method for the analysis of rigidly connected plane frames subjected to external loads. A block diagram summarizing the various steps of the analysis is shown in Fig. 6.33.

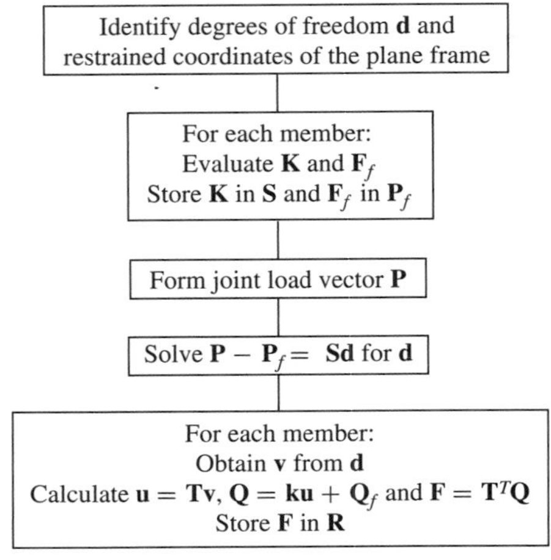
Identify degrees of freedom **d** and restrained coordinates of the plane frame
For each member: Evaluate **K** and $\mathbf{F}_f$ Store **K** in **S** and $\mathbf{F}_f$ in $\mathbf{P}_f$
Form joint load vector **P**
Solve $\mathbf{P} - \mathbf{P}_f = \mathbf{Sd}$ for **d**
For each member: Obtain **v** from **d** Calculate $\mathbf{u} = \mathbf{Tv}$, $\mathbf{Q} = \mathbf{ku} + \mathbf{Q}_f$ and $\mathbf{F} = \mathbf{T}^T\mathbf{Q}$ Store **F** in **R**

Fig. 6.33

PROBLEMS

Section 6.1

6.1 through 6.6 Identify by numbers the degrees of freedom and restrained coordinates of the frames shown in Figs. P6.1 through P6.6. Also, form the joint load vector **P** for these frames.

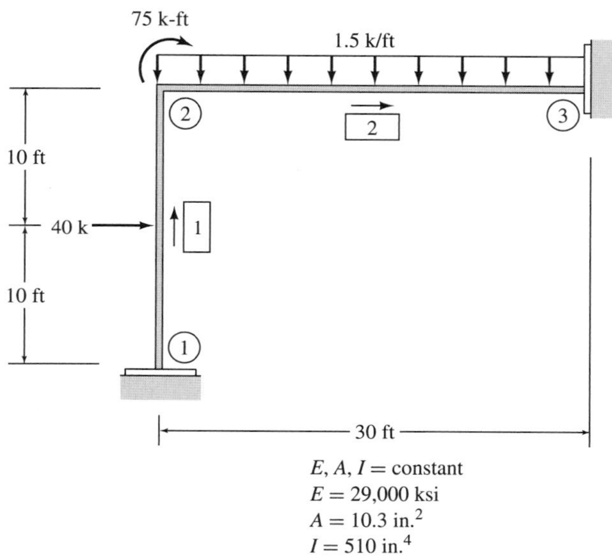

$E, A, I =$ constant
$E = 29{,}000$ ksi
$A = 10.3$ in.^{2}
$I = 510$ in.^{4}

Fig. P6.1, P6.7, P6.16, P6.24, P6.32, P6.42

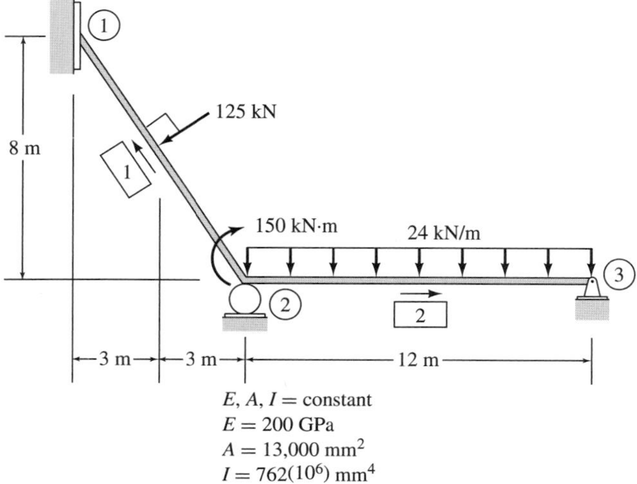

$E, A, I =$ constant
$E = 200$ GPa
$A = 13{,}000$ mm^2
$I = 762(10^6)$ mm^4

Fig. P6.2, P6.8, P6.17, P6.25, P6.33, P6.43

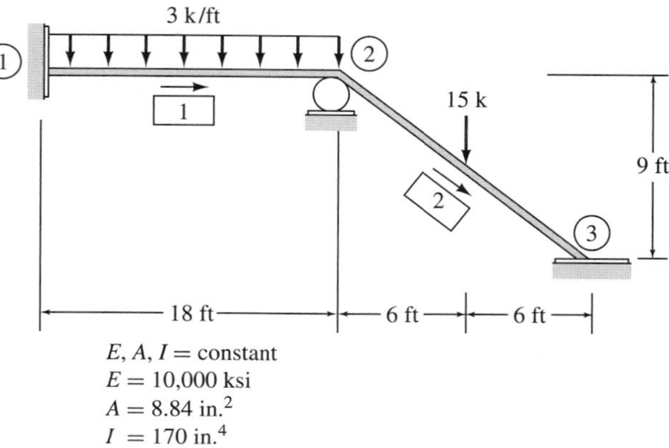

$E, A, I =$ constant
$E = 10{,}000$ ksi
$A = 8.84$ in.^{2}
$I = 170$ in.^{4}

Fig. P6.3, P6.9, P6.18, P6.26, P6.34, P6.44

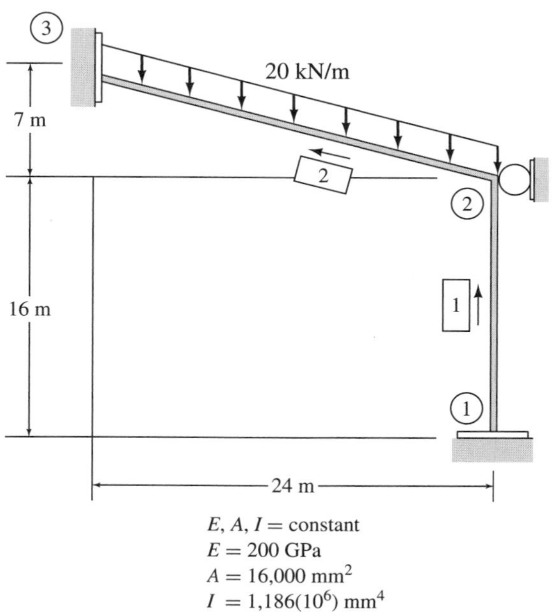

$E, A, I =$ constant
$E = 200$ GPa
$A = 16{,}000$ mm^2
$I = 1{,}186(10^6)$ mm^4

Fig. P6.4, P6.10, P6.19, P6.27, P6.35, P6.45

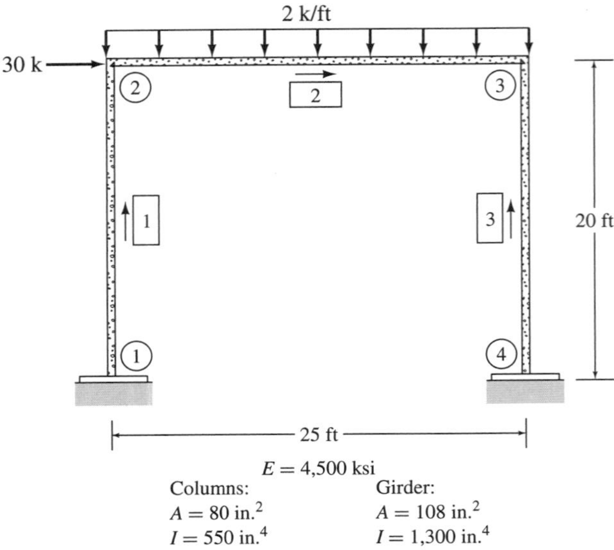

Fig. P6.5, P6.11, P6.20, P6.28, P6.36, P6.48

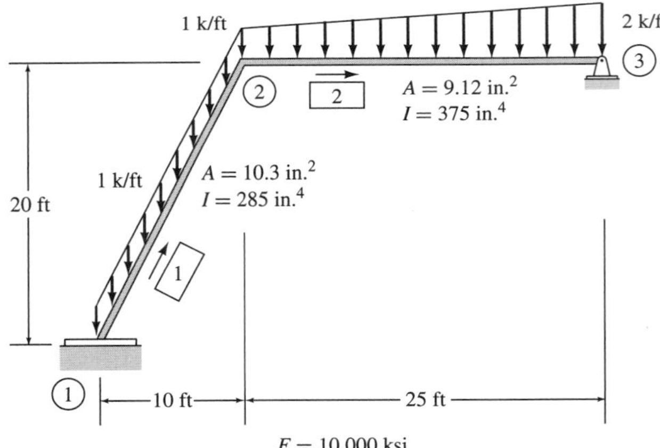

Fig. P6.6, P6.12, P6.21, P6.29, P6.37, P6.47

Section 6.2

6.7 through 6.12 Determine the local stiffness matrix **k**, and the fixed-end force vector $\mathbf{Q}_f$, for each member of the frames shown in Figs. P6.7 through P6.12. Use the fixed-end force equations given inside the front cover.

6.13 Using the integration approach, derive the equations of fixed-end forces due to the concentrated axial member load shown in Fig. P6.13. Check the results using the fixed-end force expressions given inside the front cover.

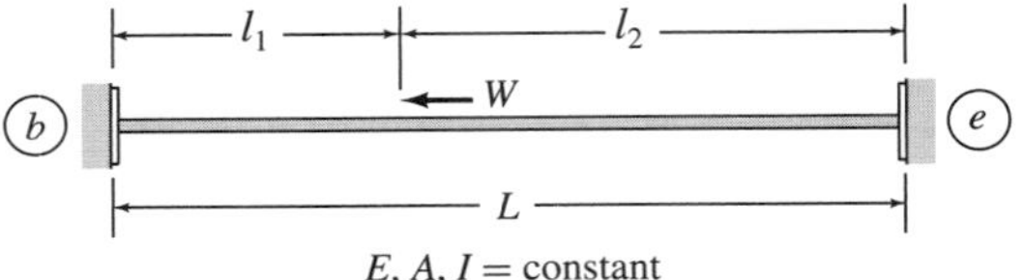

Fig. P6.13

6.14 Assume that the local end displacements for the members of the frame shown in Fig. P6.14 are

$$\mathbf{u}_1 = \begin{bmatrix} 0 \\ 0 \\ 0 \\ -5.2507 \text{ mm} \\ -12.251 \text{ mm} \\ -0.12416 \text{ rad} \end{bmatrix}; \qquad \mathbf{u}_2 = \begin{bmatrix} 9.2888 \text{ mm} \\ -9.5586 \text{ mm} \\ -0.12416 \text{ rad} \\ 0 \\ 0 \\ 0 \end{bmatrix}$$

Calculate the member local end force vectors. Are the members in equilibrium under these forces?

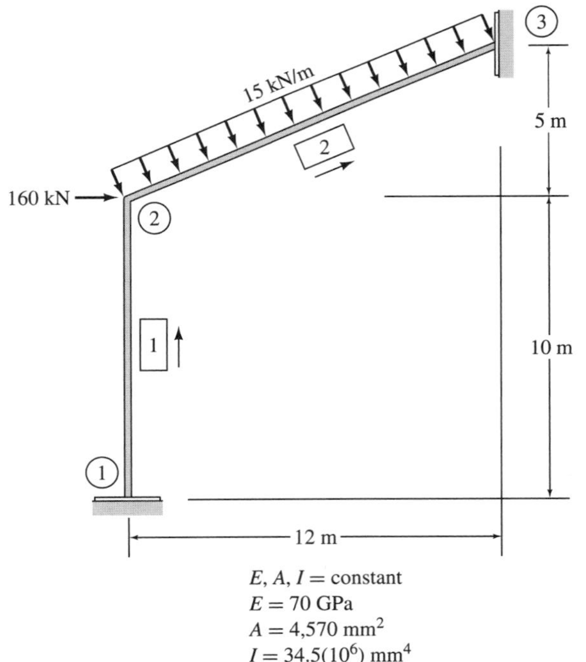

Fig. P6.14, P6.22, P6.30

6.15 Assume that the local end displacements for the members of the frame shown in Fig. P6.15 are

$$\mathbf{u}_1 = \begin{bmatrix} 0 \\ 0 \\ 0 \\ -0.05147 \text{ in.} \\ 2.0939 \text{ in.} \\ 0.0079542 \text{ rad} \end{bmatrix}; \quad \mathbf{u}_2 = \begin{bmatrix} 2.0184 \text{ in.} \\ 0.14398 \text{ in.} \\ 0.0028882 \text{ rad} \\ 1.9526 \text{ in.} \\ -0.75782 \text{ in.} \\ 0.0079542 \text{ rad} \end{bmatrix};$$

$$\mathbf{u}_3 = \begin{bmatrix} 0 \\ 0 \\ 0 \\ 0.04847 \text{ in.} \\ 2.023 \text{ in.} \\ 0.0028882 \text{ rad} \end{bmatrix}$$

Calculate the member local end force vectors. Are the members in equilibrium under these forces?

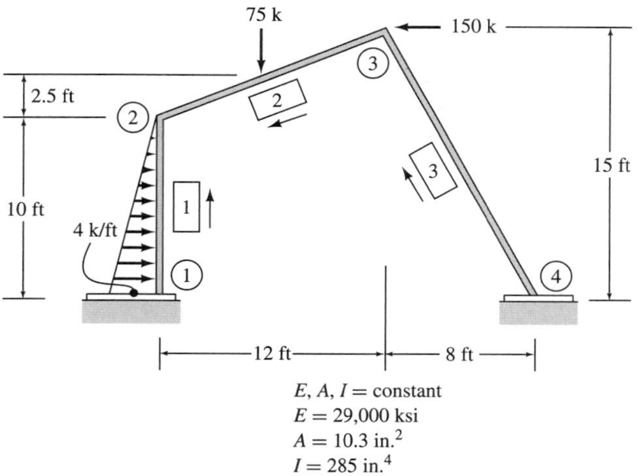

Fig. P6.15, P6.23, P6.31

Section 6.3

6.16 through 6.21 Determine the transformation matrix **T** for each member of the frames shown in Figs. P6.16 through P6.21.

6.22 Using the local end displacements given in Problem 6.14 for the members of the frame of Fig. P6.22, calculate the global end displacement vector and the global end force vector for each member of the frame. Are the members in equilibrium under the global end forces?

6.23 Using the local end displacements given in Problem 6.15 for the members of the frame of Fig. P6.23, calculate the global end displacement vector and the global end force vector for each member of the frame. Are the members in equilibrium under the global end forces?

Section 6.4

6.24 through 6.29 Determine the global stiffness matrix **K**, and fixed-end force vector $\mathbf{F}_f$, for each member of the frames shown in Figs. P6.24 through P6.29.

6.30 Calculate the member global end force vectors required in Problem 6.22 using the member global stiffness relationship $\mathbf{F} = \mathbf{K}\mathbf{v} + \mathbf{F}_f$.

6.31 Calculate the member global end force vectors required in Problem 6.23 using the member global stiffness relationship $\mathbf{F} = \mathbf{K}\mathbf{v} + \mathbf{F}_f$.

Section 6.5

6.32 through 6.37 Determine the structure stiffness matrix, the fixed-joint force vector, and the equivalent joint loads for the frames shown in Figs. P6.32 through P6.37.

6.38 Assume that the joint displacements for the frame of Fig. P6.38 are

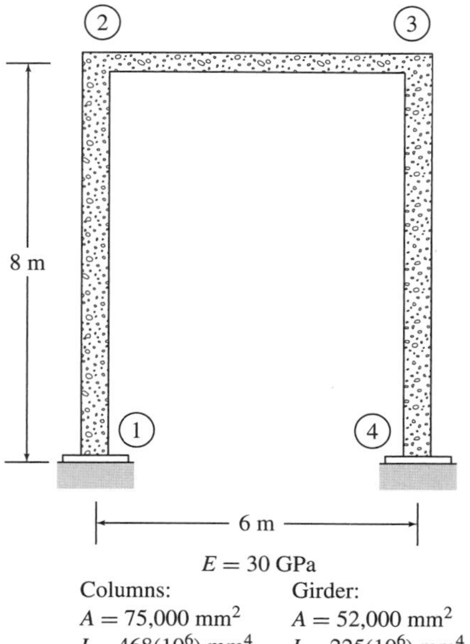

Fig. P6.38

$$\mathbf{d} = \begin{bmatrix} 0.1965 \text{ m} \\ 0.00016452 \text{ m} \\ -0.017932 \text{ rad} \\ 0.19637 \text{ m} \\ -0.00016452 \text{ m} \\ -0.023307 \text{ rad} \end{bmatrix}$$

Calculate the joint loads causing these displacements. (No loads are applied to the members of the frame.)

6.39 Assume that the joint displacements for the frame of Fig. P6.39 are

$$\mathbf{d} = \begin{bmatrix} -0.059209 \text{ rad} \\ 4.1192 \text{ in.} \\ -2.7371 \text{ in.} \\ 0.0041099 \text{ rad} \\ 3.3212 \text{ in.} \\ 1.7637 \text{ in.} \\ 0.02198 \text{ rad} \\ -0.048502 \text{ rad} \end{bmatrix}$$

Calculate the joint loads causing these displacements. (No loads are applied to the members of the frame.)

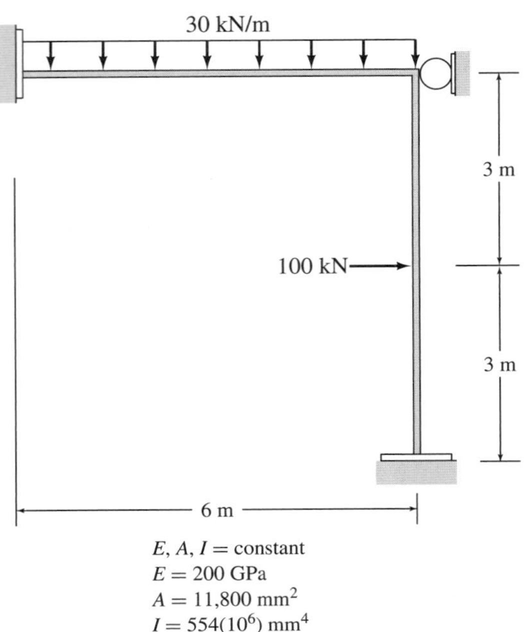

Fig. P6.40

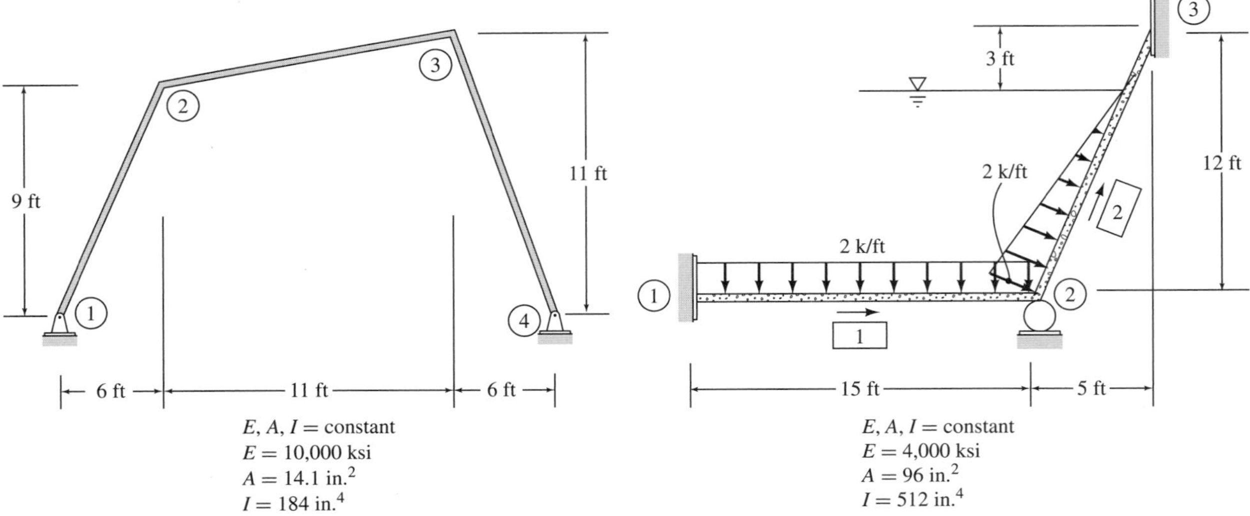

Fig. P6.39

Fig. P6.41

Section 6.6

6.40 through 6.50 Determine the joint displacements, member local end forces, and support reactions for the frames shown in Figs. P6.40 through P6.50, using the matrix stiffness method. Check the hand-calculated results by using the computer program which can be downloaded from the publisher's website for this book, or by using any other general purpose structural analysis program available.

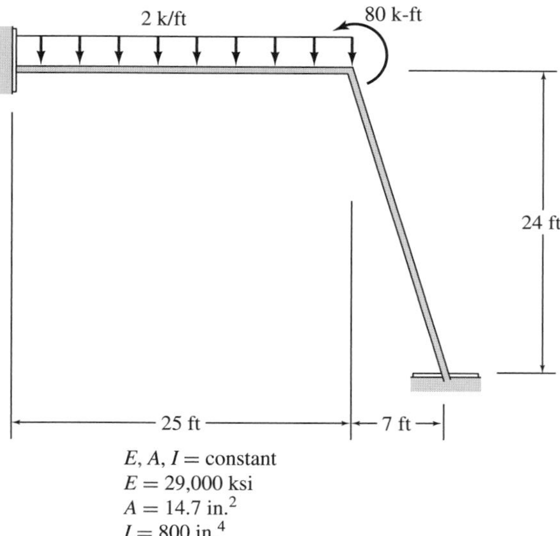

Fig. P6.46

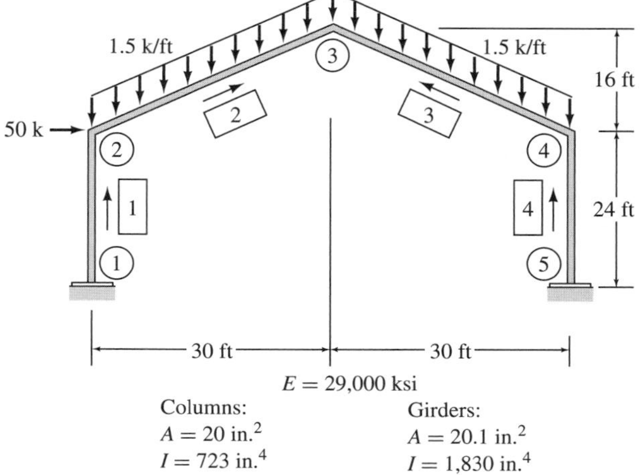

Fig. P6.49

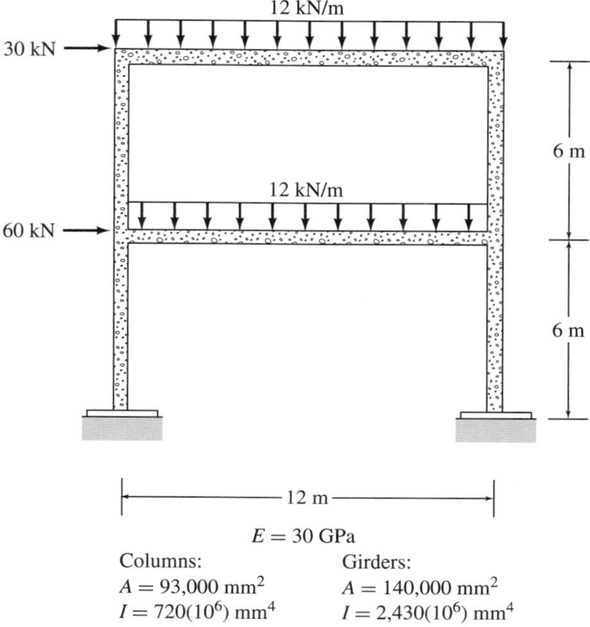

Fig. P6.50

Section 6.7

6.51 Develop a general computer program for the analysis of rigidly connected plane frames by the matrix stiffness method. Use the program to analyze the frames of Problems 6.40 through 6.50, and compare the computer-generated results to those obtained by hand calculations.

7 MEMBER RELEASES AND SECONDARY EFFECTS

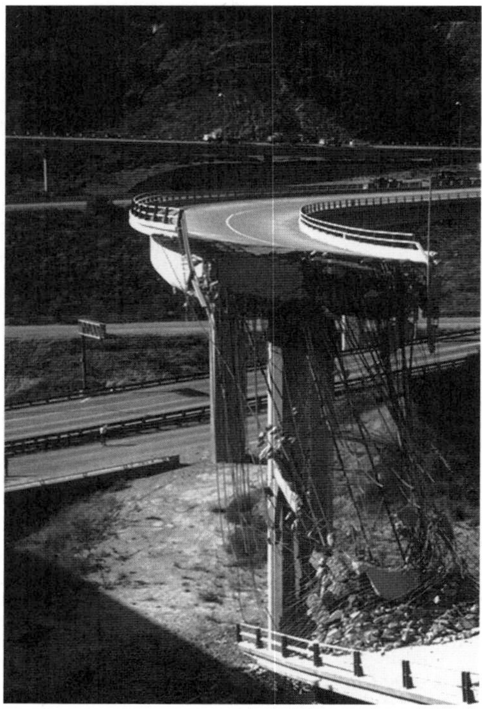

Earthquake-Damaged Bridge
(Courtesy of the USGS)

The matrix stiffness analysis of beams and plane frames, as developed in Chapters 5 and 6, is based on the assumption that each member of a structure is rigidly connected to joints at both ends, so that the member end rotations are equal to the rotations of the adjacent joints. Whereas these methods of analysis, as presented in preceding chapters, cannot be used to analyze beams and plane frames containing members connected by hinged connections, they can be modified relatively easily to include the effects of member hinges in the analysis. When the end of a member in a plane frame or beam is connected to the adjacent joint by a hinged connection, the moment at the hinged end must be zero. Because of this moment-releasing characteristic, member hinges are often referred to as *member releases*. In this chapter, we discuss modifications of the matrix stiffness methods that allow them to be used to analyze plane frames and beams containing members connected to joints by rigid (i.e., moment-resisting) and/or hinged (i.e., simple or shear) connections.

In this chapter, we also consider the procedures for including in matrix stiffness methods of analysis the effects of support displacements (due to weak foundations), temperature changes, and fabrication errors. Such secondary effects can induce significant stresses in statically indeterminate structures, and must be considered in their designs.

We begin the chapter by deriving the stiffness relationships for members of plane frames and beams with hinges. A procedure for the analysis of structures containing member releases is also developed in Section 7.1; the computer implementation of this procedure is presented in Section 7.2. We develop the analysis for the effects of support displacements in Section 7.3, and discuss the extension of the previously developed computer programs to include the effects of support displacements in Section 7.4. Finally, the procedure for including in the analysis the effects of temperature changes and fabrication errors is presented in Section 7.5.

7.1 MEMBER RELEASES IN PLANE FRAMES AND BEAMS

The effects of member releases can be conveniently incorporated in our stiffness methods by modifying the member local stiffness relationships to account for such releases. Only moment releases, in the form of hinges located at one or both ends of a member (see Fig. 7.1), are considered herein, because such releases are by far the most commonly encountered in civil engineering practice. However, the concepts presented can be readily used to introduce the effects of other types of member releases (e.g., shear and axial force releases) into the analysis.

Figure 7.1 depicts the types of member releases considered herein. From a computer programming viewpoint, it is usually convenient to classify each member of a beam or a plane frame into one of the four *member types* (MT) shown in the figure. Thus, as indicated in Fig. 7.1(a), a member that is rigidly connected to joints at both ends (i.e., has no hinges), is considered to be of type 0 (i.e., $MT = 0$). If end b of a member is connected to the adjacent joint by a hinged connection, while its opposite end e is rigidly connected to the

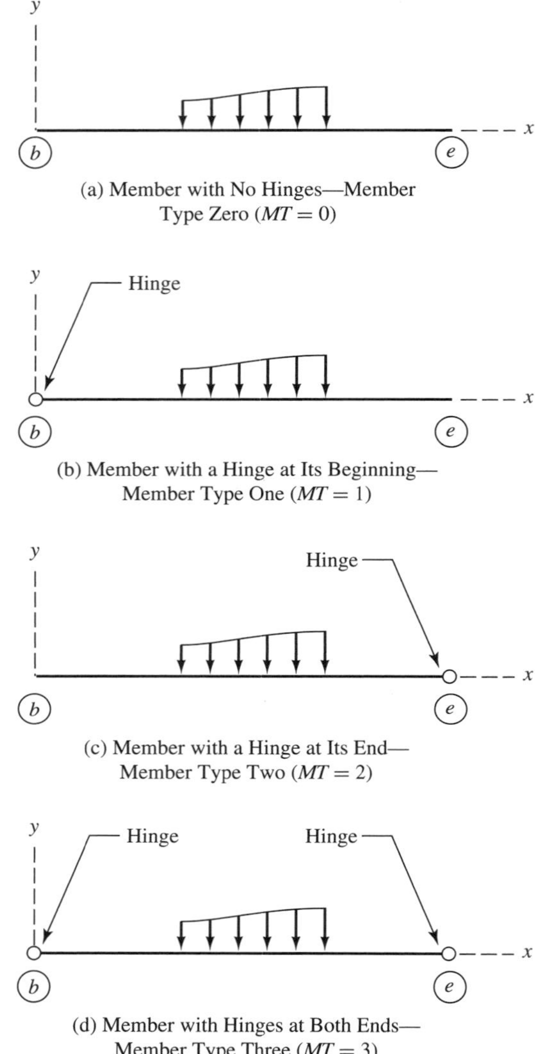

Fig. 7.1 *Member Releases*

adjacent joint (Fig. 7.1(b)), then the member is classified as type 1 (i.e., $MT = 1$). Conversely, if end b of a member is rigidly attached to the adjacent joint, but its end e is connected by a hinged connection to the adjacent joint (Fig. 7.1(c)), then the member is considered to be of type 2 (i.e., $MT = 2$). Finally, if a member is attached to joints at both ends by hinged connections (Fig. 7.1(d)), then it is classified as type 3 (i.e., $MT = 3$).

The expressions for the member local stiffness matrices $\mathbf{k}$ (Eqs. (5.53) and (6.6)) and the member local fixed-end force vectors $\mathbf{Q}_f$ (Eqs. (5.99) and (6.15)) derived for beams and plane frames can be used only for members of type 0

($MT = 0$), because they are based on the condition that the member is rigidly connected to joints at both ends, so that the member end rotations are equal to the rotations of the adjacent joints. When an end of a member is connected to the adjacent joint by a hinged connection, the moment at the hinged end must be zero. The previous expressions for **k** and $\mathbf{Q}_f$ can be easily modified to reflect the conditions of zero moments at the hinged member ends, as explained in the following paragraphs.

Local Stiffness Relations for Plane Frame Members with Hinges

We begin the development of the modified expressions by first writing the previously derived stiffness relations for a plane frame member with no hinges, in explicit form. By substituting the expressions for **k** and $\mathbf{Q}_f$ from Eqs. (6.6) and (6.15), respectively, into the member local stiffness relation $\mathbf{Q} = \mathbf{ku} + \mathbf{Q}_f$ (Eq. (6.4)), and carrying out the necessary matrix multiplication and addition, we obtain

$$Q_1 = \frac{EA}{L}(u_1 - u_4) + FA_b \tag{7.1a}$$

$$Q_2 = \frac{EI}{L^3}(12u_2 + 6Lu_3 - 12u_5 + 6Lu_6) + FS_b \tag{7.1b}$$

$$Q_3 = \frac{EI}{L^3}(6Lu_2 + 4L^2u_3 - 6Lu_5 + 2L^2u_6) + FM_b \tag{7.1c}$$

$$Q_4 = \frac{EA}{L}(-u_1 + u_4) + FA_e \tag{7.1d}$$

$$Q_5 = \frac{EI}{L^3}(-12u_2 - 6Lu_3 + 12u_5 - 6Lu_6) + FS_e \tag{7.1e}$$

$$Q_6 = \frac{EI}{L^3}(6Lu_2 + 2L^2u_3 - 6Lu_5 + 4L^2u_6) + FM_e \tag{7.1f}$$

***Members with a Hinge at the Beginning* ($MT = 1$)** When end b of a member is connected to the adjacent joint by a hinged connection, then from Fig. 6.3(b) we can see that its end moment Q_3 must be 0. By substituting $Q_3 = 0$ into Eq. (7.1c), and solving the resulting equation for the end rotation u_3, we obtain

$$u_3 = \frac{3}{2L}(-u_2 + u_5) - \frac{1}{2}u_6 - \frac{L}{4EI}FM_b \tag{7.2}$$

This equation indicates that the rotation u_3 (of the hinged end b of the member) is no longer an independent member coordinate (or degree of freedom), but is now a function of the end displacements u_2, u_5, and u_6. Thus, the number of independent member coordinates—that is, the independent end displacements required to define the displaced member configuration—is now reduced to five (i.e., u_1, u_2, u_4, u_5, and u_6). To eliminate the released coordinate u_3 from the member stiffness relations, we substitute Eq. (7.2) into Eqs. (7.1). This yields the following member stiffness equations:

$$Q_1 = \frac{EA}{L}(u_1 - u_4) + FA_b \tag{7.3a}$$

$$Q_2 = \frac{EI}{L^3}(3u_2 - 3u_5 + 3Lu_6) + \left(FS_b - \frac{3}{2L}FM_b\right) \tag{7.3b}$$

$$Q_3 = 0 \tag{7.3c}$$

$$Q_4 = \frac{EA}{L}(-u_1 + u_4) + FA_e \tag{7.3d}$$

$$Q_5 = \frac{EI}{L^3}(-3u_2 + 3u_5 - 3Lu_6) + \left(FS_e + \frac{3}{2L}FM_b\right) \tag{7.3e}$$

$$Q_6 = \frac{EI}{L^3}(3Lu_2 - 3Lu_5 + 3L^2u_6) + \left(FM_e - \frac{1}{2}FM_b\right) \tag{7.3f}$$

Equations (7.3), which represent the modified local stiffness relations for member type 1 ($MT = 1$), can be expressed in matrix form as

$$\begin{bmatrix} Q_1 \\ Q_2 \\ Q_3 \\ Q_4 \\ Q_5 \\ Q_6 \end{bmatrix} = \frac{EI}{L^3}\begin{bmatrix} \frac{AL^2}{I} & 0 & 0 & -\frac{AL^2}{I} & 0 & 0 \\ 0 & 3 & 0 & 0 & -3 & 3L \\ 0 & 0 & 0 & 0 & 0 & 0 \\ -\frac{AL^2}{I} & 0 & 0 & \frac{AL^2}{I} & 0 & 0 \\ 0 & -3 & 0 & 0 & 3 & -3L \\ 0 & 3L & 0 & 0 & -3L & 3L^2 \end{bmatrix}\begin{bmatrix} u_1 \\ u_2 \\ u_3 \\ u_4 \\ u_5 \\ u_6 \end{bmatrix} + \begin{bmatrix} FA_b \\ FS_b - \frac{3}{2L}FM_b \\ 0 \\ FA_e \\ FS_e + \frac{3}{2L}FM_b \\ FM_e - \frac{1}{2}FM_b \end{bmatrix} \tag{7.4}$$

or, symbolically, as

$$\mathbf{Q} = \mathbf{ku} + \mathbf{Q}_f$$

with

$$\mathbf{k} = \frac{EI}{L^3}\begin{bmatrix} \frac{AL^2}{I} & 0 & 0 & -\frac{AL^2}{I} & 0 & 0 \\ 0 & 3 & 0 & 0 & -3 & 3L \\ 0 & 0 & 0 & 0 & 0 & 0 \\ -\frac{AL^2}{I} & 0 & 0 & \frac{AL^2}{I} & 0 & 0 \\ 0 & -3 & 0 & 0 & 3 & -3L \\ 0 & 3L & 0 & 0 & -3L & 3L^2 \end{bmatrix} \tag{7.5}$$

and

$$\mathbf{Q}_f = \begin{bmatrix} FA_b \\ FS_b - \dfrac{3}{2L}FM_b \\ 0 \\ FA_e \\ FS_e + \dfrac{3}{2L}FM_b \\ FM_e - \dfrac{1}{2}FM_b \end{bmatrix} \qquad \textbf{(7.6)}$$

The **k** matrix in Eq. (7.5) and the $\mathbf{Q}_f$ vector in Eq. (7.6) now represent the modified local stiffness matrix and the modified local fixed-end force vector, respectively, for plane frame members of type 1 ($MT = 1$).

Members with a Hinge at the End (MT = 2) When end e of a member is hinged, then its end moment Q_6 (Fig. 6.3(b)) must be 0. By substituting $Q_6 = 0$ into Eq. 7.1(f), and solving the resulting equation for the end rotation u_6, we obtain

$$u_6 = \frac{3}{2L}(-u_2 + u_5) - \frac{1}{2}u_3 - \frac{L}{4EI}FM_e \qquad \textbf{(7.7)}$$

Next, we substitute Eq. (7.7) into Eqs. (7.1) to eliminate u_6 from the member stiffness relations. This yields

$$Q_1 = \frac{EA}{L}(u_1 - u_4) + FA_b \qquad \textbf{(7.8a)}$$

$$Q_2 = \frac{EI}{L^3}(3u_2 + 3Lu_3 - 3u_5) + \left(FS_b - \frac{3}{2L}FM_e\right) \qquad \textbf{(7.8b)}$$

$$Q_3 = \frac{EI}{L^3}(3Lu_2 + 3L^2u_3 - 3Lu_5) + \left(FM_b - \frac{1}{2}FM_e\right) \qquad \textbf{(7.8c)}$$

$$Q_4 = \frac{EA}{L}(-u_1 + u_4) + FA_e \qquad \textbf{(7.8d)}$$

$$Q_5 = \frac{EI}{L^3}(-3u_2 - 3Lu_3 + 3u_5) + \left(FS_e + \frac{3}{2L}FM_e\right) \qquad \textbf{(7.8e)}$$

$$Q_6 = 0 \qquad \textbf{(7.8f)}$$

The foregoing equations can be expressed in matrix form as

$$\mathbf{Q} = \mathbf{ku} + \mathbf{Q}_f$$

with

$$\mathbf{k} = \frac{EI}{L^3}\begin{bmatrix} \frac{AL^2}{I} & 0 & 0 & -\frac{AL^2}{I} & 0 & 0 \\ 0 & 3 & 3L & 0 & -3 & 0 \\ 0 & 3L & 3L^2 & 0 & -3L & 0 \\ -\frac{AL^2}{I} & 0 & 0 & \frac{AL^2}{I} & 0 & 0 \\ 0 & -3 & -3L & 0 & 3 & 0 \\ 0 & 0 & 0 & 0 & 0 & 0 \end{bmatrix} \tag{7.9}$$

and

$$\mathbf{Q}_f = \begin{bmatrix} FA_b \\ FS_b - \frac{3}{2L}FM_e \\ FM_b - \frac{1}{2}FM_e \\ FA_e \\ FS_e + \frac{3}{2L}FM_e \\ 0 \end{bmatrix} \tag{7.10}$$

The **k** matrix as given in Eq. (7.9) and the $\mathbf{Q}_f$ vector in Eq. (7.10) represent the modified local stiffness matrix and fixed-end force vector, respectively, for plane frame members of type 2 ($MT = 2$).

Members with Hinges at Both Ends (MT = 3) If both ends of a member are hinged, then both of its end moments, Q_3 and Q_6, must be 0. Thus, by substituting $Q_3 = 0$ and $Q_6 = 0$ into Eqs. 7.1(c) and (f), respectively, and solving the resulting simultaneous equations for the end rotations u_3 and u_6, we obtain

$$u_3 = \frac{1}{L}(-u_2 + u_5) - \frac{L}{6EI}(2FM_b - FM_e) \tag{7.11a}$$

$$u_6 = \frac{1}{L}(-u_2 + u_5) - \frac{L}{6EI}(2FM_e - FM_b) \tag{7.11b}$$

Next, we substitute the foregoing equations into Eqs. (7.1) to obtain the local stiffness relations for the member type 3:

$$Q_1 = \frac{EA}{L}(u_1 - u_4) + FA_b \tag{7.12a}$$

$$Q_2 = FS_b - \frac{1}{L}(FM_b + FM_e) \tag{7.12b}$$

$$Q_3 = 0 \tag{7.12c}$$

$$Q_4 = \frac{EA}{L}(-u_1 + u_4) + FA_e \tag{7.12d}$$

$$Q_5 = FS_e + \frac{1}{L}(FM_b + FM_e) \tag{7.12e}$$

$$Q_6 = 0 \tag{7.12f}$$

Equations (7.12) can be expressed in matrix form as

$$\mathbf{Q} = \mathbf{ku} + \mathbf{Q}_f$$

with

$$\mathbf{k} = \frac{EA}{L}\begin{bmatrix} 1 & 0 & 0 & -1 & 0 & 0 \\ 0 & 0 & 0 & 0 & 0 & 0 \\ 0 & 0 & 0 & 0 & 0 & 0 \\ -1 & 0 & 0 & 1 & 0 & 0 \\ 0 & 0 & 0 & 0 & 0 & 0 \\ 0 & 0 & 0 & 0 & 0 & 0 \end{bmatrix} \tag{7.13}$$

and

$$\mathbf{Q}_f = \begin{bmatrix} FA_b \\ FS_b - \frac{1}{L}(FM_b + FM_e) \\ 0 \\ FA_e \\ FS_e + \frac{1}{L}(FM_b + FM_e) \\ 0 \end{bmatrix} \tag{7.14}$$

The foregoing **k** matrix (Eq. (7.13)) and $\mathbf{Q}_f$ vector (Eq. (7.14)) represent the modified local stiffness matrix and fixed-end force vector, respectively, for plane frame members of type 3 ($MT = 3$). Interestingly, from Eq. (7.13), we observe that the deletion of the third and sixth rows and columns (which correspond to the rotational coordinates of the plane frame members) from the **k** matrix for $MT = 3$ reduces it to the **k** matrix for members of plane trusses (Eq. (3.27)).

It should be realized that, although the number of independent member coordinates is reduced due to member releases, the orders of the modified stiffness matrices **k** (Eqs. (7.5), (7.9), and (7.13)) and the fixed-end force vector $\mathbf{Q}_f$ (Eqs. (7.6), (7.10), and (7.14)) are maintained as 6×6 and 6×1, respectively, with 0 elements in the rows and columns that correspond to the released coordinates. This form of **k** and $\mathbf{Q}_f$ eliminates the need to modify the expression for the member transformation matrix, **T**, derived in Chapter 6 (Eq. (6.19)), and provides an efficient means of incorporating the effect of member releases in the computer program developed in Section 6.7.

Local Stiffness Relations for Beam Members with Hinges

As discussed in Chapter 5, in beams subjected to lateral loads, the axial displacements of members are 0. Thus, a member of a beam can have up to four degrees of freedom: namely, a translation perpendicular to the member's centroidal axis and a rotation, at each end. The modified stiffness relations for beam members with releases can be derived by applying the same procedure just used for members of plane frames. However, it is more convenient to obtain the modified member stiffness matrices **k** for beams by simply deleting the first and fourth rows and columns of the corresponding **k** matrices for plane-frame members. Similarly, the modified fixed-end force vectors $\mathbf{Q}_f$ for beam members can be obtained by deleting the first and fourth rows of the corresponding $\mathbf{Q}_f$ vectors for plane-frame members.

***Members with a Hinge at the Beginning* (*MT* = 1)** To obtain the modified stiffness matrix **k** for beam members of type 1, we delete rows 1 and 4 and columns 1 and 4 from the **k** matrix given in Eq. (7.5) for plane-frame members of the same type. This yields

$$\mathbf{k} = \frac{EI}{L^3}\begin{bmatrix} 3 & 0 & -3 & 3L \\ 0 & 0 & 0 & 0 \\ -3 & 0 & 3 & -3L \\ 3L & 0 & -3L & 3L^2 \end{bmatrix} \tag{7.15}$$

Similarly, the modified fixed-end force vector $\mathbf{Q}_f$ for beam members of type 1 can be obtained by deleting rows 1 and 4 from the $\mathbf{Q}_f$ vector given in Eq. (7.6) for plane-frame members of type 1. Thus,

$$\mathbf{Q}_f = \begin{bmatrix} FS_b - \dfrac{3}{2L}FM_b \\ 0 \\ FS_e + \dfrac{3}{2L}FM_b \\ FM_e - \dfrac{1}{2}FM_b \end{bmatrix} \tag{7.16}$$

The rotation u_2 of the hinged end b of the member, if desired, can be evaluated by using the following relationship:

$$u_2 = \frac{3}{2L}(-u_1 + u_3) - \frac{1}{2}u_4 - \frac{L}{4EI}FM_b \tag{7.17}$$

Equation (7.17) is obtained simply by replacing u_2, u_3, u_5, and u_6 in Eq. (7.2) with u_1, u_2, u_3, and u_4, respectively.

***Members with a Hinge at the End* (*MT* = 2)** By deleting the first and fourth rows and columns from the **k** matrix given in Eq. (7.9), we obtain the modified stiffness matrix for the beam members of type 2 (i.e., $MT = 2$):

$$\mathbf{k} = \frac{EI}{L^3}\begin{bmatrix} 3 & 3L & -3 & 0 \\ 3L & 3L^2 & -3L & 0 \\ -3 & -3L & 3 & 0 \\ 0 & 0 & 0 & 0 \end{bmatrix} \tag{7.18}$$

and by deleting the first and fourth rows from the $\mathbf{Q}_f$ vector given in Eq. (7.10), we determine the modified fixed-end force vector for beam members of type 2 ($MT = 2$):

$$\mathbf{Q}_f = \begin{bmatrix} FS_b - \dfrac{3}{2L}FM_e \\ FM_b - \dfrac{1}{2}FM_e \\ FS_e + \dfrac{3}{2L}FM_e \\ 0 \end{bmatrix} \tag{7.19}$$

The expression for the rotation u_4, of the hinged end e of the member, can be obtained by substituting the subscripts 1, 2, 3, and 4 for the subscripts 2, 3, 5, and 6, respectively, in Eq. (7.7). This yields

$$u_4 = \frac{3}{2L}(-u_1 + u_3) - \frac{1}{2}u_2 - \frac{L}{4EI}FM_e \tag{7.20}$$

***Members with Hinges at Both Ends (MT* = 3)** By deleting the first and fourth rows and columns from the $\mathbf{k}$ matrix given in Eq. (7.13), we realize that the modified stiffness matrix for beam members of type 3 (i.e., $MT = 3$) is a null matrix; that is,

$$\mathbf{k} = \mathbf{0} \tag{7.21}$$

which indicates that a beam member hinged at both ends offers no resistance against small end displacements in the direction perpendicular to its centroidal axis. (Recall from Section 3.3 that the members of trusses behave in a similar manner when subjected to lateral end displacements—see Figs. 3.3(d) and (f).) By deleting the first and fourth rows from the $\mathbf{Q}_f$ vector given in Eq. (7.14), we obtain the modified fixed-end force vector for beam members of type 3 (i.e., $MT = 3$):

$$\mathbf{Q}_f = \begin{bmatrix} FS_b - \dfrac{1}{L}(FM_b + FM_e) \\ 0 \\ FS_e + \dfrac{1}{L}(FM_b + FM_e) \\ 0 \end{bmatrix} \tag{7.22}$$

and from Eqs. (7.11) we obtain the following expressions for the rotations u_2 and u_4 of the hinged ends b and e, respectively, of the member.

$$u_2 = \frac{1}{L}(-u_1 + u_3) - \frac{L}{6EI}(2FM_b - FM_e) \tag{7.23a}$$

$$u_4 = \frac{1}{L}(-u_1 + u_3) - \frac{L}{6EI}(2FM_e - FM_b) \tag{7.23b}$$

Procedure for Analysis

The analysis procedures developed in Chapters 5 and 6 can be applied to beams and plane frames, respectively, containing member releases, provided that the modified expressions for the stiffness matrices **k** and fixed-end force vectors $\mathbf{Q}_f$, developed in this section are used for the members with releases (i.e., $MT = 1, 2,$ or 3). Furthermore, in the analysis of plane frames, the global stiffness matrix **K** for members with releases is now evaluated using the matrix triple product $\mathbf{K} = \mathbf{T}^T\mathbf{k}\mathbf{T}$ (Eq. (6.29)), instead of the explicit form of **K** given in Eq. (6.31), which is valid only for members with no releases (i.e., $MT = 0$). Similarly, the global fixed-end force vector $\mathbf{F}_f$ for plane frame members with releases is evaluated using the relationship $\mathbf{F}_f = \mathbf{T}^T\mathbf{Q}_f$ (Eq. (6.30)), instead of the explicit form given in Eq. (6.33). The rotations of the hinged member ends, if desired, can be evaluated using Eqs. (7.2), (7.7), and (7.11) when analyzing plane frames, and Eqs. (7.17), (7.20), and (7.23) in the case of beams.

Hinged Joints in Beams and Plane Frames

If all the members meeting at a joint are connected to it by hinged connections, then the joint is considered to be a hinged joint. For example, joint 4 of the two-story plane frame shown in Fig. 7.2(a) is considered to be a hinged joint,

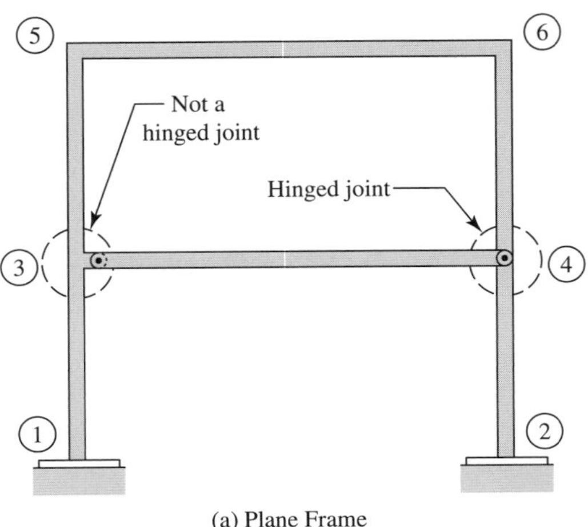

(a) Plane Frame

Fig. 7.2

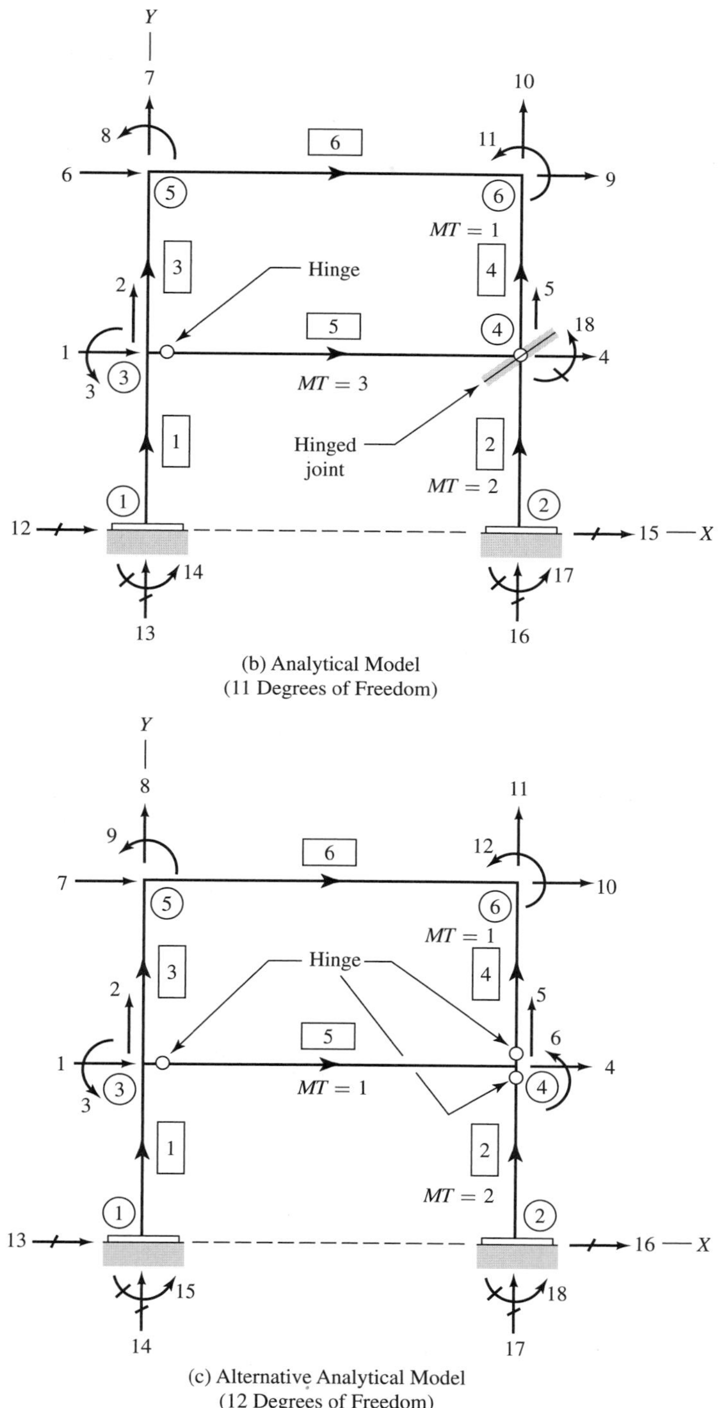

(b) Analytical Model
(11 Degrees of Freedom)

(c) Alternative Analytical Model
(12 Degrees of Freedom)

Fig. 7.2 *(continued)*

because all three members meeting at this joint are attached to it by hinged connections. However, joint 3 of this frame is not considered to be hinged, because only one of the three members meeting at this joint is attached by a hinged connection; the remaining two members are rigidly connected to the joint.

As hinged joints cannot transmit any moments and are free to rotate, their rotational stiffnesses are 0. Thus, inclusion of the rotational degrees of freedom of such joints in the analysis causes the structure stiffness matrix **S** to become *singular,* with 0 elements in the rows and columns that correspond to the rotational degrees of freedom of the hinged joints. (Recall from your previous course in *college algebra* that the coefficient matrix of a system of linear equations is considered to be singular if its determinant is 0; and that such a system of equations does not yield a unique solution.) Perhaps the most straightforward and efficient way to remedy this difficulty is to eliminate the rotational degrees of freedom of hinged joints from the analysis by modeling such joints as restrained (or fixed) against rotations. This approach is based on the realization that because hinged joints are not subjected to any moments, their rotations are 0; even though the released ends of the members connected to such a joint can, and do, rotate. In Fig. 7.2(b), the hinged joint 4 of the example frame is modeled using this approach. As indicated in this figure, an imaginary clamp is applied to hinged joint 4 to restrain (or fix) it against rotation, while allowing it to freely translate in any direction. Joint 4, therefore, has two degrees of freedom—the translations in the X and Y directions—which are identified as d_4 and d_5, respectively; and one restrained coordinate, R_{18}, which represents the reaction moment that develops at the imaginary clamp. It should be realized that because hinged joints are not subjected to any external moments (or couples), the magnitudes of the imaginary reaction moments at such joints are always 0. However, the assignment of restrained coordinate numbers to these imaginary reactions, in accordance with the previously established scheme for numbering structure coordinates, enables us to include the effect of hinged joints in the computer programs developed in Chapters 5 and 6 without any reprogramming.

An alternative approach that can be used to overcome the problem of singularity (due to the lack of rotational stiffnesses of a hinged joint) is to model such a joint as rigidly connected to one (and only one) of the members meeting at the joint. This approach is based on the following concept: as no external moment is applied to the hinged joint, and because the moments at the ends of all but one of the members meeting at the joint are 0, the moment at the end of the one member that is rigidly connected to the joint must also be 0, to satisfy the moment equilibrium equation ($\sum M = 0$) for the joint. This alternative approach is used in Fig. 7.2(c) to model hinged joint 4 of the example frame. As shown in this figure, whereas members 2 and 4 are still attached by hinged connections to joint 4, the third member 5 is now rigidly connected to this joint. Note that because the end of member 5 is now rigidly connected, its member type, which was 3 (i.e., $MT = 3$) in the previous analytical model (Fig. 7.2(b)), is now 1 (i.e., $MT = 1$), as shown in Fig. 7.2(c). Joint 4 can now be treated as any other rigid joint of the plane frame, and is assigned three degrees of freedom, d_4, d_5, and d_6, as shown in the figure—with d_6 representing the rotation of joint 4, which in turn equals the rotation of the end of member 5.

Needless to state, the two approaches we have discussed for modeling hinged joints yield identical analysis results. However, the first approach generally provides a more efficient analytical model in terms of the number of degrees of freedom of the structure. From Figs. 7.2(b) and (c), we can see that the analytical models of the example frame, based on the first and the alternative approaches, involve 11 and 12 degrees of freedom, respectively.

EXAMPLE 7.1

Determine the joint displacements, member end forces, and support reactions for the plane frame shown in Fig. 7.3(a), using the matrix stiffness method.

SOLUTION

Analytical Model: The analytical model of the frame is depicted in Fig. 7.3(b). Since both members 1 and 2, meeting at joint 2, are attached to it by hinged connections, joint 2 is modeled as a hinged joint with its rotation restrained by an imaginary clamp. Thus, joint 2 has two degrees of freedom—the translations in the X and Y directions—which are identified as d_1 and d_2, respectively. Also, for member 1, $MT = 2$, because the end of this member is hinged, whereas $MT = 1$ for member 2, which is hinged at its beginning.

As far as the modeling of joint 4 is concerned, recall that in Chapters 5 and 6 (e.g., see Examples 5.7 and 6.5) we modeled such a joint as a rigid joint, free to rotate, with its rotation treated as a degree of freedom of the structure. However, in light of the discussion of member releases and hinged joints presented in this section, we can now eliminate the rotational degree of freedom of joint 4 from the analysis by modeling member 3 as hinged at its beginning (i.e., $MT = 1$), which allows us to model joint 4 as a hinged joint with its rotation restrained by an imaginary clamp. Note that the end

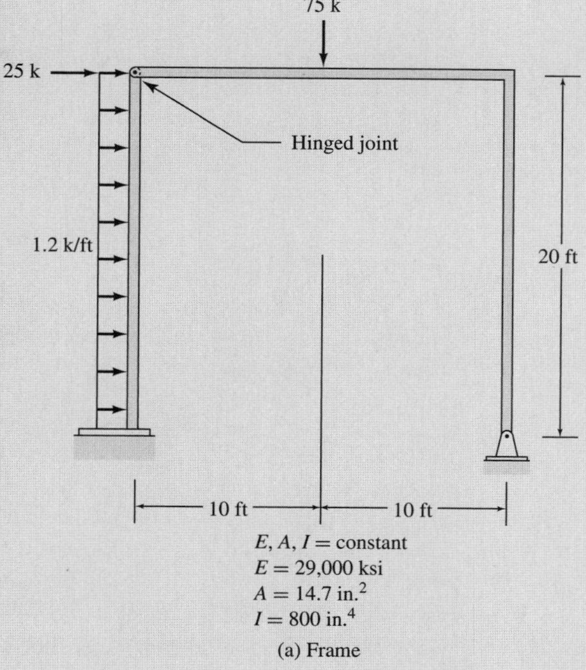

(a) Frame

Fig. 7.3

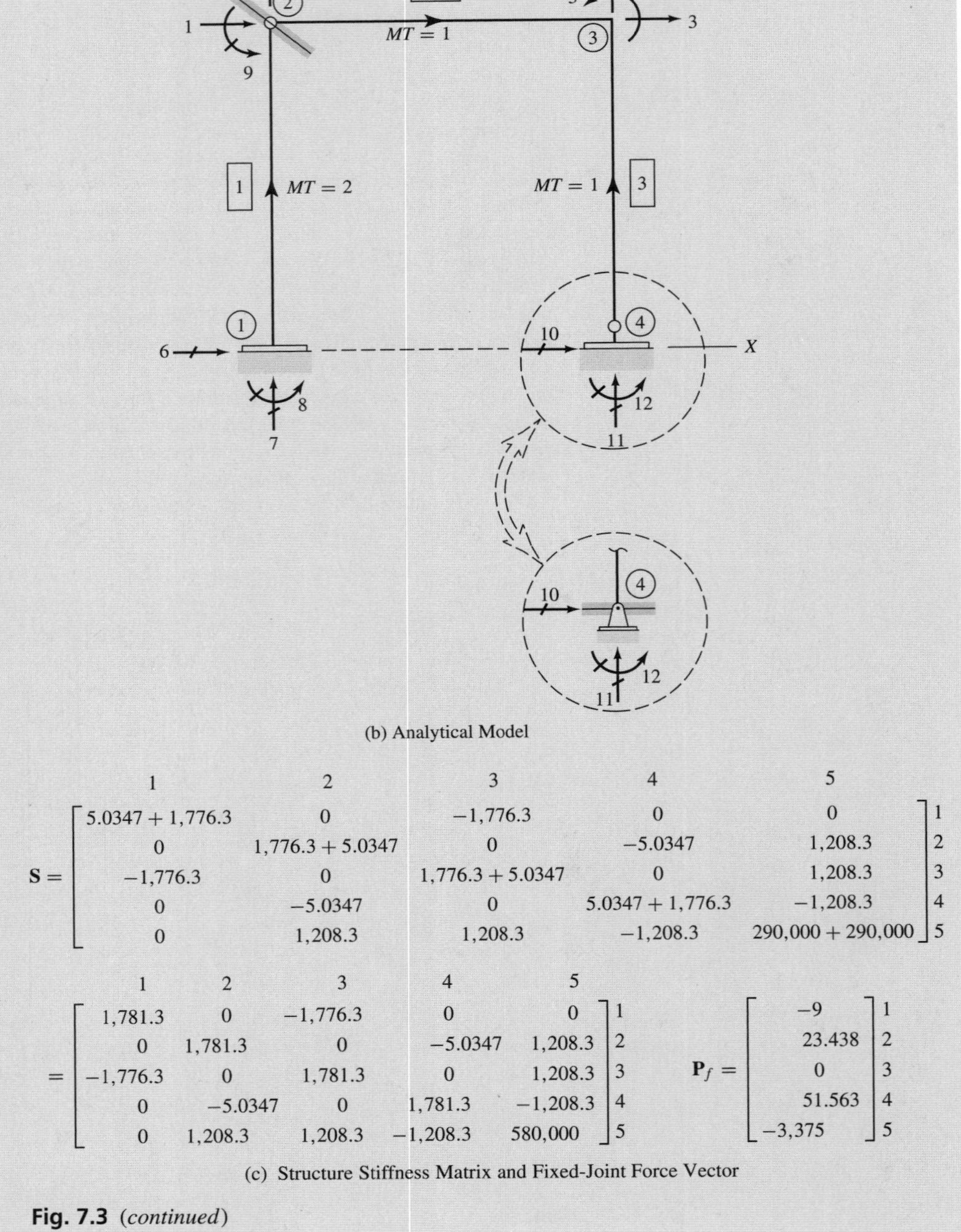

(b) Analytical Model

$$
\mathbf{S} = \begin{bmatrix} 5.0347 + 1{,}776.3 & 0 & -1{,}776.3 & 0 & 0 \\ 0 & 1{,}776.3 + 5.0347 & 0 & -5.0347 & 1{,}208.3 \\ -1{,}776.3 & 0 & 1{,}776.3 + 5.0347 & 0 & 1{,}208.3 \\ 0 & -5.0347 & 0 & 5.0347 + 1{,}776.3 & -1{,}208.3 \\ 0 & 1{,}208.3 & 1{,}208.3 & -1{,}208.3 & 290{,}000 + 290{,}000 \end{bmatrix}
$$

$$
= \begin{bmatrix} 1{,}781.3 & 0 & -1{,}776.3 & 0 & 0 \\ 0 & 1{,}781.3 & 0 & -5.0347 & 1{,}208.3 \\ -1{,}776.3 & 0 & 1{,}781.3 & 0 & 1{,}208.3 \\ 0 & -5.0347 & 0 & 1{,}781.3 & -1{,}208.3 \\ 0 & 1{,}208.3 & 1{,}208.3 & -1{,}208.3 & 580{,}000 \end{bmatrix} \qquad \mathbf{P}_f = \begin{bmatrix} -9 \\ 23.438 \\ 0 \\ 51.563 \\ -3{,}375 \end{bmatrix}
$$

(Rows and columns of both matrices, and the rows of $\mathbf{P}_f$, are numbered 1–5.)

(c) Structure Stiffness Matrix and Fixed-Joint Force Vector

Fig. 7.3 *(continued)*

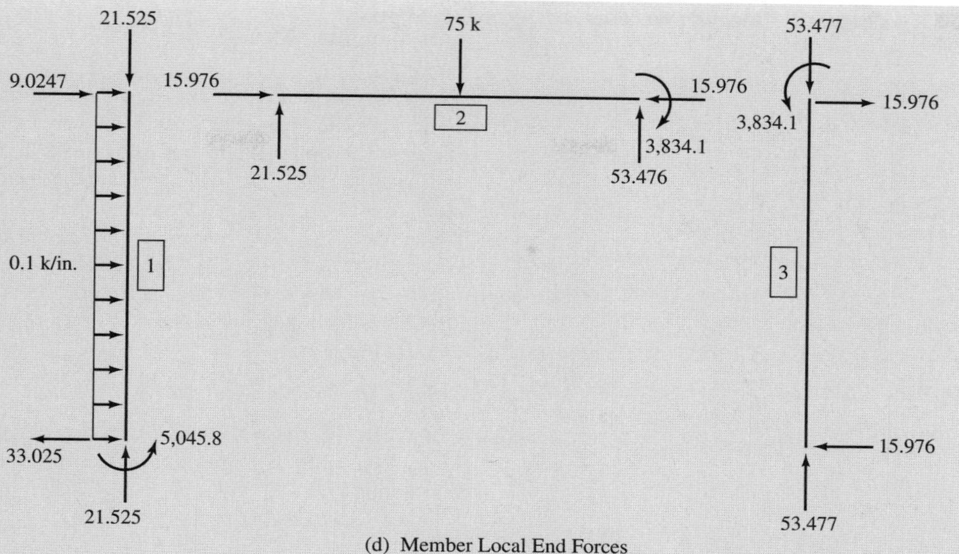

(d) Member Local End Forces

$$\mathbf{R} = \begin{bmatrix} -33.025 \text{ k} \\ 21.525 \text{ k} \\ 5{,}045.8 \text{ k-in.} \\ 0 \\ -15.976 \text{ k} \\ 53.477 \text{ k} \\ 0 \end{bmatrix} \begin{matrix} 6 \\ 7 \\ 8 \\ 9 \\ 10 \\ 11 \\ 12 \end{matrix}$$

(e) Support Reaction Vector

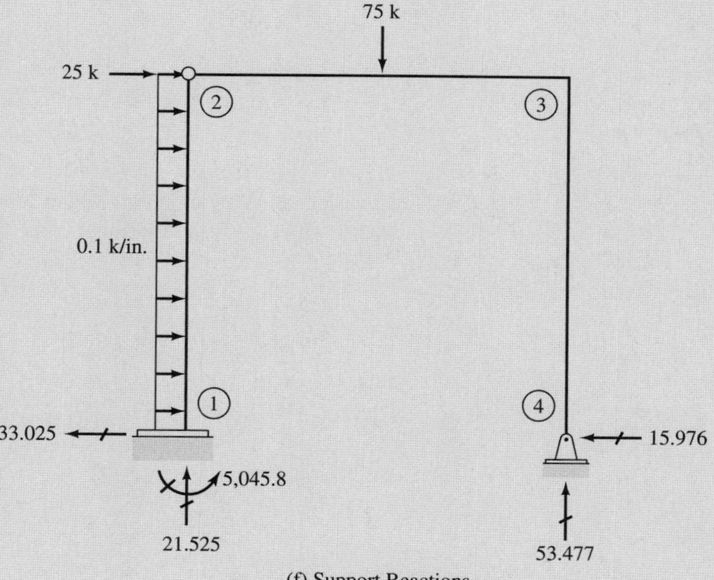

(f) Support Reactions

Fig. 7.3 (*continued*)

of member 3, which is connected to joint 4, can be considered to be hinged, because there is only one member connected to the joint that is not subjected to any external couple. Furthermore, the joint is supported by a hinged support which cannot exert any reaction moment at the joint. Thus, the moment at the end of member 3, which is connected to joint 4, must be 0; therefore, the member end can be treated as a hinged end. With its rotation restrained by the imaginary clamp, and its translations in the X and Y directions restrained by the actual hinged support, joint 4 is modeled as if it is attached to a fixed support, with no degrees of freedom, as depicted in Fig. 7.3(b).

Thus, the entire frame has five degrees of freedom and seven restrained coordinates, as shown in Fig. 7.3(b).

Structure Stiffness Matrix and Fixed-Joint Force Vector:

Member 1 ($MT = 2$) Because $MT = 2$ for this member, we use Eqs. (7.9) and (7.10) to determine its local stiffness matrix $\mathbf{k}$ and fixed-end force vector $\mathbf{Q}_f$, respectively. Thus, by substituting $E = 29{,}000$ ksi, $A = 14.7$ in.^{2}, $I = 800$ in.^{4}, and $L = 20$ ft $=$ 240 in. into Eq. (7.9), we obtain

$$\mathbf{k}_1 = \begin{bmatrix} 1{,}776.3 & 0 & 0 & -1{,}776.3 & 0 & 0 \\ 0 & 5.0347 & 1{,}208.3 & 0 & -5.0347 & 0 \\ 0 & 1{,}208.3 & 290{,}000 & 0 & -1{,}208.3 & 0 \\ -1{,}776.3 & 0 & 0 & 1{,}776.3 & 0 & 0 \\ 0 & -5.0347 & -1{,}208.3 & 0 & 5.0347 & 0 \\ 0 & 0 & 0 & 0 & 0 & 0 \end{bmatrix} \tag{1}$$

To determine the local fixed-end force vector due to the member load $w = 1.2$ k/ft $=$ 0.1 k/in., we first evaluate the fixed-end axial forces, shears, and moments in a corresponding rigidly connected member by using the expressions given inside the front cover:

$$FA_b = FA_e = 0$$

$$FS_b = FS_e = \frac{0.1(240)}{2} = 12 \text{ k}$$

$$FM_b = -FM_e = \frac{0.1(240)^2}{12} = 480 \text{ k-in.}$$

Next, we substitute the foregoing values into Eq. (7.10) to obtain the local fixed-end force vector for the released member under consideration:

$$\mathbf{Q}_{f1} = \begin{bmatrix} 0 \\ 12 - \dfrac{3(-480)}{2(240)} \\ 480 - \dfrac{1}{2}(-480) \\ 0 \\ 12 + \dfrac{3(-480)}{2(240)} \\ 0 \end{bmatrix} = \begin{bmatrix} 0 \\ 15 \text{ k} \\ 720 \text{ k-in.} \\ 0 \\ 9 \text{ k} \\ 0 \end{bmatrix} \tag{2}$$

To obtain the member's stiffness matrix $\mathbf{K}$ and the fixed-end force vector $\mathbf{F}_f$ in the global coordinate system, we first substitute its direction cosines, $\cos\theta = 0$ and $\sin\theta = 1$, into Eq. (6.19), to obtain the transformation matrix.

$$\mathbf{T}_1 = \mathbf{T}_3 = \begin{bmatrix} 0 & 1 & 0 & 0 & 0 & 0 \\ -1 & 0 & 0 & 0 & 0 & 0 \\ 0 & 0 & 1 & 0 & 0 & 0 \\ 0 & 0 & 0 & 0 & 1 & 0 \\ 0 & 0 & 0 & -1 & 0 & 0 \\ 0 & 0 & 0 & 0 & 0 & 1 \end{bmatrix} \quad (3)$$

Next, by substituting $\mathbf{k}_1$ (Eq. (1)) and $\mathbf{T}_1$ (Eq. (3)) into the relationship $\mathbf{K} = \mathbf{T}^T\mathbf{k}\mathbf{T}$ (Eq. (6.29)), and performing the necessary matrix multiplications, we obtain

$$\mathbf{K}_1 = \begin{array}{c} \\ \end{array}\begin{bmatrix} 5.0347 & 0 & -1{,}208.3 & -5.0347 & 0 & 0 \\ 0 & 1{,}776.3 & 0 & 0 & -1{,}776.3 & 0 \\ -1{,}208.3 & 0 & 290{,}000 & 1{,}208.3 & 0 & 0 \\ -5.0347 & 0 & 1{,}208.3 & 5.0347 & 0 & 0 \\ 0 & -1{,}776.3 & 0 & 0 & 1{,}776.3 & 0 \\ 0 & 0 & 0 & 0 & 0 & 0 \end{bmatrix}\begin{array}{c} 6 \\ 7 \\ 8 \\ 1 \\ 2 \\ 9 \end{array}$$

(column code numbers: 6, 7, 8, 1, 2, 9)

Note that $\mathbf{K}_1$ is symmetric.

Similarly, by substituting $\mathbf{Q}_{f1}$ (Eq. (2)) and $\mathbf{T}_1$ (Eq. (3)) into the relationship $\mathbf{F}_f = \mathbf{T}^T\mathbf{Q}_f$ (Eq. (6.30)), we obtain

$$\mathbf{F}_{f1} = \begin{bmatrix} -15 \\ 0 \\ 720 \\ -9 \\ 0 \\ 0 \end{bmatrix}\begin{array}{c} 6 \\ 7 \\ 8 \\ 1 \\ 2 \\ 9 \end{array}$$

From Fig. 7.3(b), we observe that the code numbers for member 1 are 6, 7, 8, 1, 2, 9. Using these code numbers, we store the pertinent elements of $\mathbf{K}_1$ and $\mathbf{F}_{f1}$ in their proper positions in the 5×5 structure stiffness matrix $\mathbf{S}$ and the 5×1 structure fixed-joint force vector $\mathbf{P}_f$, respectively, as shown in Fig. 7.3(c).

Member 2 ($MT = 1$) No coordinate transformations are needed for this horizontal member; that is, $\mathbf{T}_2 = \mathbf{I}$, $\mathbf{K}_2 = \mathbf{k}_2$, and $\mathbf{F}_{f2} = \mathbf{Q}_{f2}$. As $MT = 1$, we use Eq. (7.5) to obtain

$$\mathbf{K}_2 = \mathbf{k}_2 = \mathbf{k}_3 = \begin{bmatrix} 1{,}776.3 & 0 & 0 & -1{,}776.3 & 0 & 0 \\ 0 & 5.0347 & 0 & 0 & -5.0347 & 1{,}208.3 \\ 0 & 0 & 0 & 0 & 0 & 0 \\ -1{,}776.3 & 0 & 0 & 1{,}776.3 & 0 & 0 \\ 0 & -5.0347 & 0 & 0 & 5.0347 & -1{,}208.3 \\ 0 & 1{,}208.3 & 0 & 0 & -1{,}208.3 & 290{,}000 \end{bmatrix}\begin{array}{c} 1 \\ 2 \\ 9 \\ 3 \\ 4 \\ 5 \end{array} \quad (4)$$

(column code numbers: 1, 2, 9, 3, 4, 5)

Using the fixed-end force expressions given for loading type 1 for the 75 k member load, we obtain

$$FA_b = FA_e = 0$$

$$FS_b = FS_e = 37.5 \text{ k}$$

$$FM_b = -FM_e = 2{,}250 \text{ k-in.}$$

Substitution of the foregoing values into Eq. (7.6) yields

$$\mathbf{F}_{f2} = \mathbf{Q}_{f2} = \begin{bmatrix} 0 \\ 37.5 - \dfrac{3(2{,}250)}{2(240)} \\ 0 \\ 0 \\ 37.5 + \dfrac{3(2{,}250)}{2(240)} \\ -2{,}250 - \dfrac{1}{2}(2{,}250) \end{bmatrix} = \begin{bmatrix} 0 \\ 23.438\text{ k} \\ \hdashline 0 \\ \hdashline 0 \\ 51.563\text{ k} \\ -3{,}375\text{ k-in.} \end{bmatrix} \begin{matrix} 1 \\ 2 \\ 9 \\ 3 \\ 4 \\ 5 \end{matrix} \tag{5}$$

The relevant elements of $\mathbf{K}_2$ and $\mathbf{F}_{f2}$ are stored in $\mathbf{S}$ and $\mathbf{P}_f$, respectively, using the member code numbers 1, 2, 9, 3, 4, 5.

Member 3 ($MT = 1$) As E, A, I, L, and MT for member 3 are the same as for member 2, $\mathbf{k}_3 = \mathbf{k}_2$ as given in Eq. (4). Also, since the member is not subjected to any loads,

$$\mathbf{F}_{f3} = \mathbf{Q}_{f3} = \mathbf{0}$$

Furthermore, since the direction cosines of member 3 are identical to those of member 1, $\mathbf{T}_3 = \mathbf{T}_1$ as given in Eq. (3).

To determine the member global stiffness matrix, we substitute $\mathbf{k}_3$ from Eq. (4), and $\mathbf{T}_3$ from Eq. (3), into the relationship $\mathbf{K} = \mathbf{T}^T\mathbf{kT}$ (Eq. (6.29)), and perform the necessary matrix multiplications. This yields

$$\mathbf{K}_3 = \begin{matrix} \begin{matrix} 10 & 11 & 12 & 3 & 4 & 5 \end{matrix} \\ \left[\begin{array}{ccc:ccc} 5.0347 & 0 & 0 & -5.0347 & 0 & -1{,}208.3 \\ 0 & 1{,}776.3 & 0 & 0 & -1{,}776.3 & 0 \\ 0 & 0 & 0 & 0 & 0 & 0 \\ \hdashline -5.0347 & 0 & 0 & 5.0347 & 0 & 1{,}208.3 \\ 0 & -1{,}776.3 & 0 & 0 & 1{,}776.3 & 0 \\ -1{,}208.3 & 0 & 0 & 1{,}208.3 & 0 & 290{,}000 \end{array}\right] \end{matrix} \begin{matrix} 10 \\ 11 \\ 12 \\ 3 \\ 4 \\ 5 \end{matrix}$$

The pertinent elements of $\mathbf{K}_3$ are stored in $\mathbf{S}$ using the member code numbers 10, 11, 12, 3, 4, 5. The completed structure stiffness matrix $\mathbf{S}$ and the structure fixed-joint force vector $\mathbf{P}_f$ are given in Fig. 7.3(c).

Joint Load Vector: By comparing Figs. 7.3(a) and (b), we write

$$\mathbf{P} = \begin{bmatrix} 25 \\ 0 \\ 0 \\ 0 \\ 0 \end{bmatrix} \begin{matrix} 1 \\ 2 \\ 3 \\ 4 \\ 5 \end{matrix}$$

Joint Displacements: By solving the system of simultaneous equations representing the structure stiffness relationship $\mathbf{P} - \mathbf{P}_f = \mathbf{Sd}$ (Eq. (6.42)), we obtain the following joint displacements:

$$\mathbf{d} = \begin{bmatrix} 3.5801\text{ in.} \\ -0.012118\text{ in.} \\ 3.5711\text{ in.} \\ -0.030106\text{ in.} \\ -0.0016582\text{ rad} \end{bmatrix} \begin{matrix} 1 \\ 2 \\ 3 \\ 4 \\ 5 \end{matrix}$$ **Ans**

Member End Displacements and End Forces:

Member 1 (***MT*** **= 2)** Using the member code numbers 6, 7, 8, 1, 2, 9, we write the global end displacement vector as

$$\mathbf{v}_1 = \begin{bmatrix} v_1 \\ v_2 \\ v_3 \\ v_4 \\ v_5 \\ v_6 \end{bmatrix} \begin{matrix} 6 \\ 7 \\ 8 \\ 1 \\ 2 \\ 9 \end{matrix} = \begin{bmatrix} 0 \\ 0 \\ 0 \\ d_1 \\ d_2 \\ 0 \end{bmatrix} = \begin{bmatrix} 0 \\ 0 \\ 0 \\ 3.5801 \\ -0.012118 \\ 0 \end{bmatrix} \tag{6}$$

Next, we obtain the local end displacement vector $\mathbf{u}$ by substituting the foregoing $\mathbf{v}_1$ and $\mathbf{T}_1$ (Eq. (3)) into the relationship $\mathbf{u} = \mathbf{Tv}$ (Eq. (6.20)). This yields

$$\mathbf{u}_1 = \mathbf{T}_1\mathbf{v}_1 = \begin{bmatrix} 0 \\ 0 \\ 0 \\ -0.012118 \\ -3.5801 \\ 0 \end{bmatrix} \tag{7}$$

We can now determine the member local end forces $\mathbf{Q}$ by substituting $\mathbf{u}_1$, $\mathbf{k}_1$ (Eq. (1)), and $\mathbf{Q}_{f1}$ (Eq. (2)) in the member stiffness relationship $\mathbf{Q} = \mathbf{ku} + \mathbf{Q}_f$ (Eq. (6.4)). Thus,

$$\mathbf{Q}_1 = \mathbf{k}_1\mathbf{u}_1 + \mathbf{Q}_{f1} = \begin{bmatrix} 21.525 \text{ k} \\ 33.025 \text{ k} \\ 5{,}045.8 \text{ k-in.} \\ -21.525 \text{ k} \\ -9.0247 \text{ k} \\ 0 \end{bmatrix}$$ **Ans**

These end forces for member 1 are depicted in Fig. 7.3(d).

To generate the support reaction vector $\mathbf{R}$ for the frame, we evaluate the global end forces $\mathbf{F}$ for the member by applying Eq. (6.23) as

$$\mathbf{F}_1 = \mathbf{T}_1^T\mathbf{Q}_1 = \begin{bmatrix} -33.025 \\ 21.525 \\ 5{,}045.8 \\ \hdashline 9.0247 \\ -21.525 \\ \hdashline 0 \end{bmatrix} \begin{matrix} 6 \\ 7 \\ 8 \\ 1 \\ 2 \\ 9 \end{matrix}$$

The pertinent elements of $\mathbf{F}_1$ are stored in $\mathbf{R}$, as shown in Fig. 7.3(e).

It should be realized that because the member end displacement vectors $\mathbf{v}$ and $\mathbf{u}$ are based on the compatibility of the joint and the member end displacements, such vectors (in the case of members with releases) contain 0 elements in the rows that correspond to the rotations of the released (or hinged) member ends. Thus, we can see from Eqs. (6) and (7) that the vectors $\mathbf{v}_1$ and $\mathbf{u}_1$ for member 1 (with $MT = 2$) contain 0 elements in their sixth rows. We can evaluate the rotation u_6 of the released end of this member by using Eq. (7.7), as

$$u_6 = \frac{3}{2(240)}(0 - 3.5801) - \frac{1}{2}(0) - \frac{240(-480)}{4(29{,}000)(800)}$$
$$= -0.021134 \text{ rad} = 0.021134 \text{ rad} \circlearrowright$$

Because the member end rotations are the same in the local and global coordinate systems,

$$v_6 = u_6 = 0.021134 \text{ rad} \circlearrowright$$

Member 2 ($MT = 1$)

$$\mathbf{u}_2 = \mathbf{v}_2 = \begin{bmatrix} v_1 \\ v_2 \\ v_3 \\ v_4 \\ v_5 \\ v_6 \end{bmatrix} \begin{matrix} 1 \\ 2 \\ 9 \\ 3 \\ 4 \\ 5 \end{matrix} = \begin{bmatrix} d_1 \\ d_2 \\ 0 \\ d_3 \\ d_4 \\ d_5 \end{bmatrix} = \begin{bmatrix} 3.5801 \\ -0.012118 \\ 0 \\ 3.5711 \\ -0.030106 \\ -0.0016582 \end{bmatrix}$$

By using $\mathbf{k}_2$ from Eq. (4) and $\mathbf{Q}_{f2}$ from Eq. (5), we compute the member end forces to be

$$\mathbf{F}_2 = \mathbf{Q}_2 = \mathbf{k}_2\mathbf{u}_2 + \mathbf{Q}_{f2} = \begin{bmatrix} 15.976 \text{ k} \\ 21.525 \text{ k} \\ 0 \\ -15.976 \text{ k} \\ 53.476 \text{ k} \\ -3{,}834.1 \text{ k-in.} \end{bmatrix} \begin{matrix} 1 \\ 2 \\ 9 \\ 3 \\ 4 \\ 5 \end{matrix} \quad \textbf{Ans}$$

The rotation, u_3, of the released end of this member, if desired, can be calculated by using Eq. (7.2).

Member 3 ($MT = 1$)

$$\mathbf{v}_3 = \begin{bmatrix} v_1 \\ v_2 \\ v_3 \\ v_4 \\ v_5 \\ v_6 \end{bmatrix} \begin{matrix} 10 \\ 11 \\ 12 \\ 3 \\ 4 \\ 5 \end{matrix} = \begin{bmatrix} 0 \\ 0 \\ 0 \\ d_3 \\ d_4 \\ d_5 \end{bmatrix} = \begin{bmatrix} 0 \\ 0 \\ 0 \\ 3.5711 \\ -0.030106 \\ -0.0016582 \end{bmatrix}$$

By using $\mathbf{T}_3$ from Eq. (3), we obtain

$$\mathbf{u}_3 = \mathbf{T}_3\mathbf{v}_3 = \begin{bmatrix} 0 \\ 0 \\ 0 \\ -0.030106 \\ -3.5711 \\ -0.0016582 \end{bmatrix}$$

Using $\mathbf{k}_3$ from Eq. (4) and $\mathbf{Q}_{f3} = \mathbf{0}$, we obtain the member local end forces as

$$\mathbf{Q}_3 = \mathbf{k}_3\mathbf{u}_3 = \begin{bmatrix} 53.477 \text{ k} \\ 15.976 \text{ k} \\ 0 \\ -53.477 \text{ k} \\ -15.976 \text{ k} \\ 3{,}834.1 \text{ k-in.} \end{bmatrix} \quad \textbf{Ans}$$

The member local end forces are shown in Fig. 7.3(d).

$$\mathbf{F}_3 = \mathbf{T}_3^T \mathbf{Q}_3 = \begin{bmatrix} -15.976 \\ 53.477 \\ 0 \\ \hdashline 15.976 \\ -53.477 \\ 3{,}834.1 \end{bmatrix} \begin{matrix} 10 \\ 11 \\ 12 \\ 3 \\ 4 \\ 5 \end{matrix}$$

Support Reactions: See Figs. 7.3(e) and (f). **Ans**

7.2 COMPUTER IMPLEMENTATION OF ANALYSIS FOR MEMBER RELEASES

The computer programs developed in Chapters 5 and 6 for the analysis of rigidly connected beams and plane frames can be extended, with only minor modifications, to include the effects of member releases. In this section, we discuss the modifications in the program for the analysis of plane frames (Section 6.7) that are necessary to consider member releases. While the beam analysis program (Section 5.8) can be modified in a similar manner, the implementation of these modifications is left as an exercise for the reader.

The overall organization and format of the plane frame analysis program, as summarized in Table 6.1, remains the same when considering member releases. However, parts V, IX, and XII, and the subroutines ***MSTIFFL*** and ***MFEFLL***, must be revised as follows:

Member Data (Part V) This part of the program (see flowchart in Fig. 4.3(e)) should be modified to include the reading and storing of the member type, *MT*, for each member of the frame. The number of columns of the *member data matrix* **MPRP** should be increased from four to five, with the value of *MT* (= 0, 1, 2, or 3) for a member *i* stored in the fifth column of the *i*th row of **MPRP**.

Generation of the Structure Stiffness Matrix and Equivalent Joint Load Vector (Part IX), and Calculation of Member Forces and Support Reactions (Part XII) In parts IX and XII of the computer program (see flowcharts in Figs. 6.24 and 6.31, respectively) a statement should be added to read, for each member, the value of *MT* from the fifth column of the **MPRP** matrix (i.e., *MT* = **MPRP** (*IM*, 5)), before the subroutines ***MSTIFFL*** and ***MFEFLL*** are called to form the member local stiffness matrix **BK**, and the local fixed-end force vector **QF**, respectively.

Subroutine MSTIFFL A flowchart for programming the modified version of this subroutine is given in Fig. 7.4 on the next page. As this flowchart indicates, the subroutine calculates the **BK** matrix using Eq. (6.6) if *MT* equals 0, Eq. (7.5) if *MT* equals 1, Eq. (7.9) if *MT* equals 2, or Eq. (7.13) if *MT* equals 3.

Subroutine MFEFLL A flowchart of the modified version of this subroutine is shown in Fig. 7.5 on page 363. The subroutine first calculates the fixed-end forces

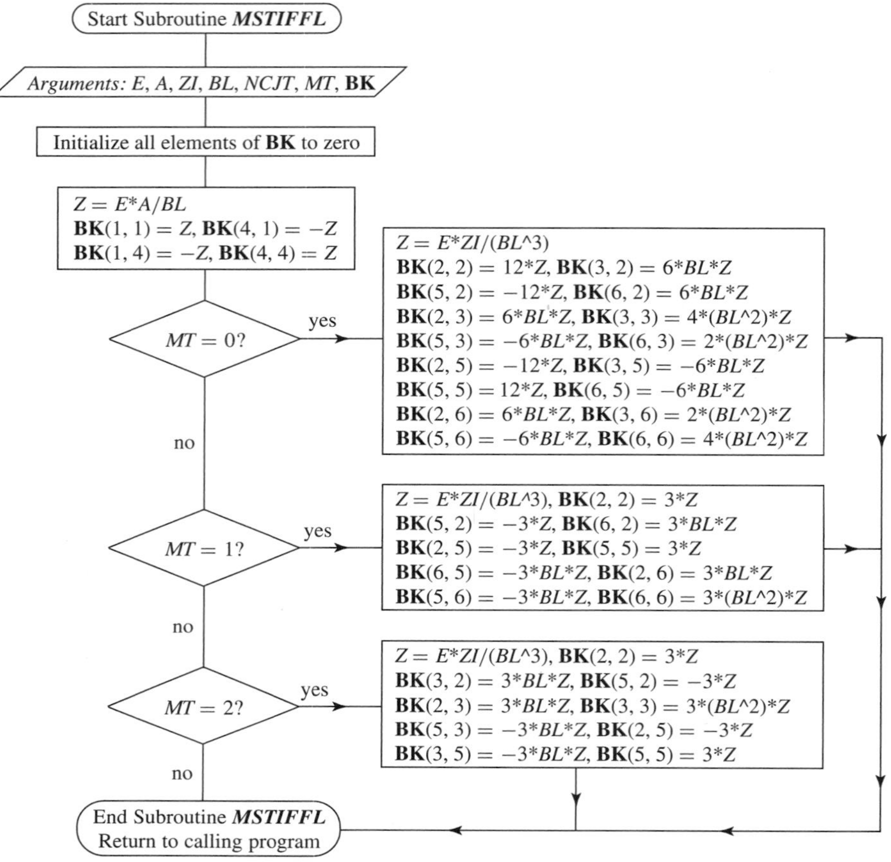

Fig. 7.4 *Flowchart of Subroutine* ***MSTIFFL*** *for Determining Member Local Stiffness Matrix for Plane Frames with Member Releases*

(*FAB*, *FSB*, *FMB*, *FAE*, *FSE*, and *FME*) in a corresponding rigidly connected member using the equations given inside the front cover. The **QF** vector is then formed in accordance with Eq. (6.15) if $MT = 0$, Eq. (7.6) if $MT = 1$, Eq. (7.10) if $MT = 2$, or Eq. (7.14) if $MT = 3$.

7.3 SUPPORT DISPLACEMENTS

The effect of small support displacements, due to weak foundations or other causes, can be conveniently included in the matrix stiffness method of analysis using the concept of equivalent joint loads [14]. This approach, which was discussed in Sections 5.6 and 6.5 for the case of member loads, essentially involves applying the prescribed external action (such as a system of member loads, support settlements, etc.) to the structure, with all of its joint displacements restrained by imaginary restraints. The structure fixed-joint forces that develop in the hypothetical fixed structure, as reactions at the imaginary

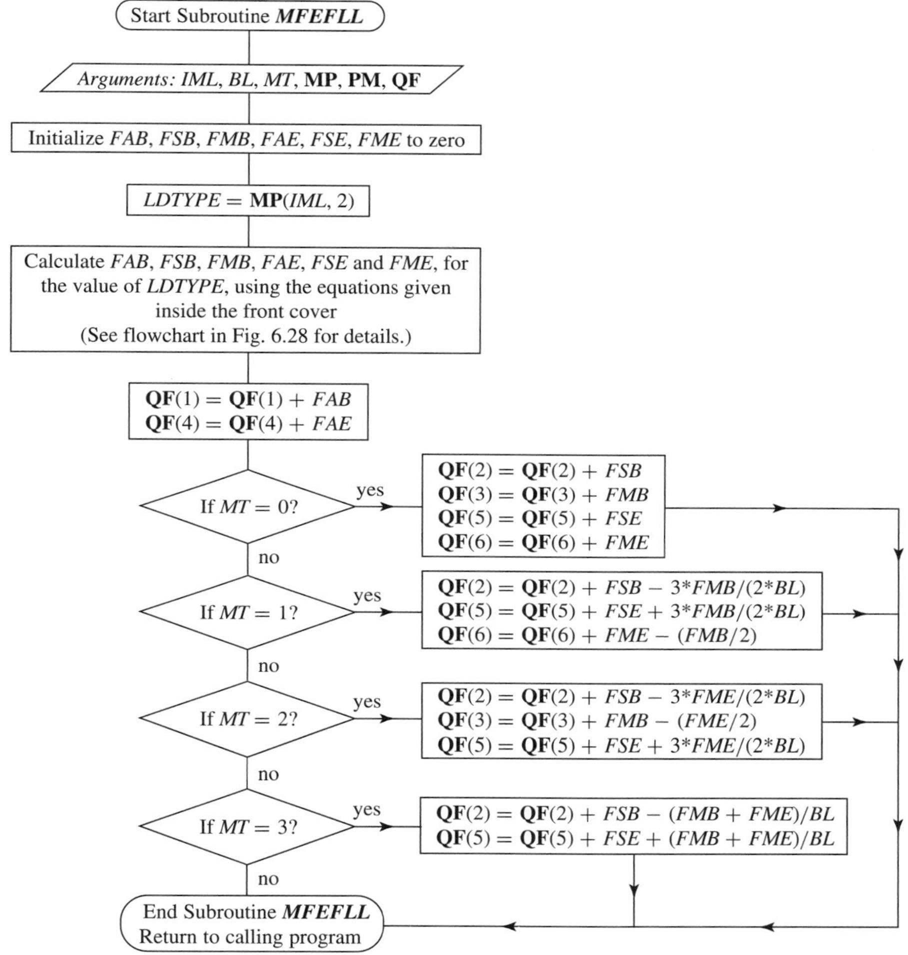

Fig. 7.5 *Flowchart of Subroutine **MFEFLL** for Determining Member Local Fixed-End Force Vector for Plane Frames with Member Releases*

restraints (i.e., at the location and in the direction of each degree of freedom of the actual structure), are then evaluated. The structure fixed-joint forces, with their directions reversed, now represent the equivalent joint loads, in the sense that when applied to the actual structure, they cause the same joint displacements as the original action (i.e., member loads, support settlements, etc.). Once the response of the structure to the equivalent joint loads has been determined, the actual structural response due to the original action is obtained by superposition of the responses of the fixed structure to the original action and the actual structure to the equivalent joint loads.

The foregoing approach is illustrated in Fig. 7.6 on the next page, for the case of support displacements, using an arbitrary three-degree-of-freedom frame as an example. Figure 7.6(a) shows the actual frame, whose supports 3 and 4 undergo small

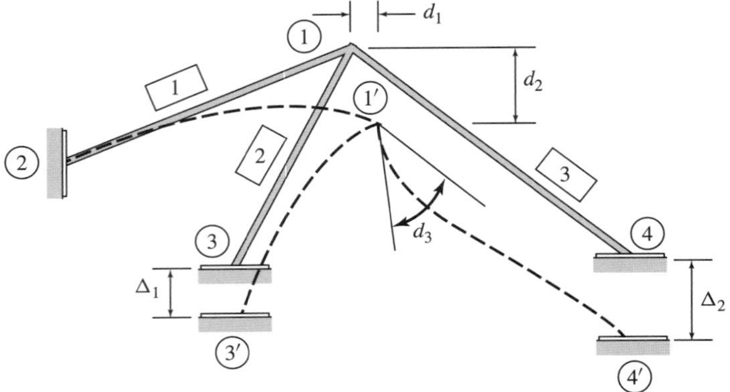

(a) Actual Frame Subjected to Support Settlements

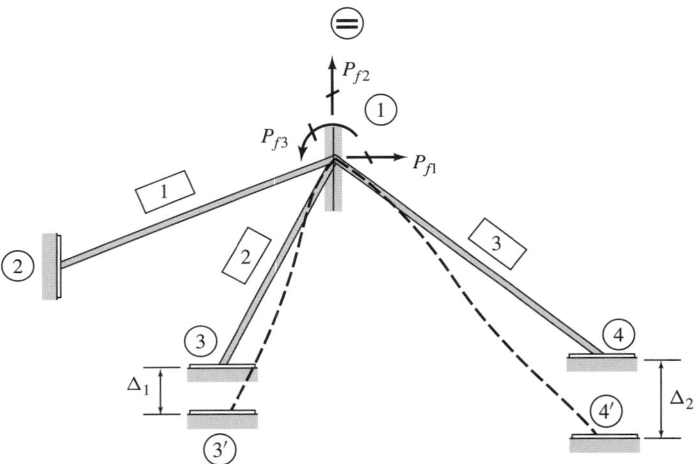

(b) Fixed Frame Subjected to Support Settlements

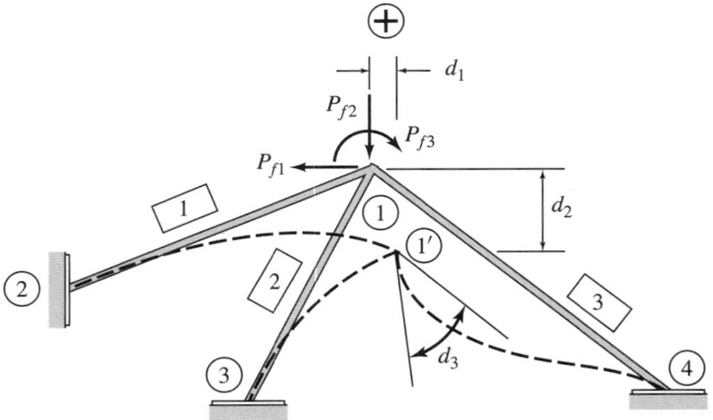

(c) Actual Frame Subjected to Equivalent Joint Loads

Fig. 7.6

settlements Δ_1 and Δ_2, respectively, causing the displacements d_1, d_2, and d_3 of free joint 1. To determine the response (i.e., joint displacements, member forces, and support reactions) of the frame to the support settlements, we first restrain all the joint displacements of the frame by applying an imaginary restraint at joint 1, and subject this completely fixed frame to the prescribed support settlements Δ_1 and Δ_2, as shown in Fig. 7.6(b). As joint 1 of the frame, initially free, is now restrained from translating and rotating by the imaginary restraint, the structure fixed-joint forces P_{f1}, P_{f2}, and P_{f3} develop at the imaginary restraint at this joint. (A procedure for evaluating structure fixed-joint forces due to support settlements is developed in a subsequent part of this section.) Next, as shown in Fig. 7.6(c), we apply the foregoing structure fixed-joint forces P_{f1}, P_{f2}, and P_{f3}, with their directions reversed, as external loads at joint 1 of the actual frame.

A comparison of Figs. 7.6(a), (b), and (c) indicates that the superposition of the support settlements and joint loads applied to the frame in Figs. 7.6(b) and (c) yields only the support settlements the frame is subjected to in Fig. 7.6(a), because each of the fixed-joint forces in Fig. 7.6(b) is canceled by its negative counterpart applied as a load in Fig. 7.6(c). Thus, according to the principle of superposition, the joint displacements d_1, d_2, and d_3 of the frame due to the support settlements Δ_1 and Δ_2 (Fig. 7.6(a)) must equal the algebraic sums of the corresponding joint displacements of the fixed frame subjected to the support settlements (Fig. 7.6(b)), and the actual frame, subjected to no settlements, but to the negatives of the fixed-joint forces (Fig. 7.6(c)). However, since the displacements of joint 1 of the fixed frame (Fig. 7.6(b)) are 0, the joint displacements of the frame subjected to the negatives of fixed-joint forces (Fig. 7.6(c)) must equal the actual joint displacements d_1, d_2, and d_3 of the frame due to the support settlements Δ_1 and Δ_2 (Fig. 7.6(a)). In other words, the negatives of the structure fixed-joint forces cause the same displacements at the locations and in the directions of the frame's degrees of freedom as the prescribed support settlements; and, in that sense, such forces can be considered as equivalent joint loads.

It should be realized that the foregoing equivalency is valid only for joint displacements. From Fig. 7.6(b), we can see that the end displacements of the members of the fixed frame are not 0. Therefore, the member end forces and support reactions of the actual frame due to settlements (Fig. 7.6(a)) must be obtained by superposition of the corresponding responses of the fixed frame (Fig. 7.6(b)) and the actual frame subjected to the equivalent joint loads (Fig. 7.6(c)).

It may be recalled from Chapters 5 and 6 that, in the case of member loads, the response of the fixed structure was evaluated using the fixed-end force expressions for various types of member loads, as given inside the front cover; and that the fixed-joint force vector $\mathbf{P}_f$ was obtained by algebraically adding the fixed-end forces of members meeting at the joints (via the member code numbers). A procedure for evaluating the member fixed-end forces, and the structure fixed-joint forces, due to support settlements is presented in the following paragraphs. With the fixed-joint forces known, the response of the structure to the equivalent joint loads can be determined, using the standard matrix stiffness methods described in Chapters 3 through 6.

Evaluation of Structure Fixed-Joint Forces Due to Support Displacements

We begin by establishing a systematic way of identifying the support displacements of a structure. For that purpose, let us reconsider the three-degree-of-freedom frame of Fig. 7.6(a), subjected to the support settlements Δ_1 and Δ_2. The frame is redrawn in Fig. 7.7(a), with its analytical model depicted in Fig. 7.7(b). From Fig. 7.7(b), we observe that the frame has nine support reactions, which are identified by the restrained coordinate numbers 4 through 12. Thus, the frame can be subjected to a maximum of nine support displacements. The numbers assigned to the restrained coordinates are also used to identify the support displacements, with a support displacement at the location and in the direction of a support reaction R_i denoted by the symbol d_{si}. Thus, a comparison of Figs. 7.7(a) and (b) shows that for the frame under consideration,

$$d_{s8} = -\Delta_1 \qquad \text{and} \qquad d_{s11} = -\Delta_2 \tag{7.24}$$

with the remaining seven support displacements being 0. The negative signs

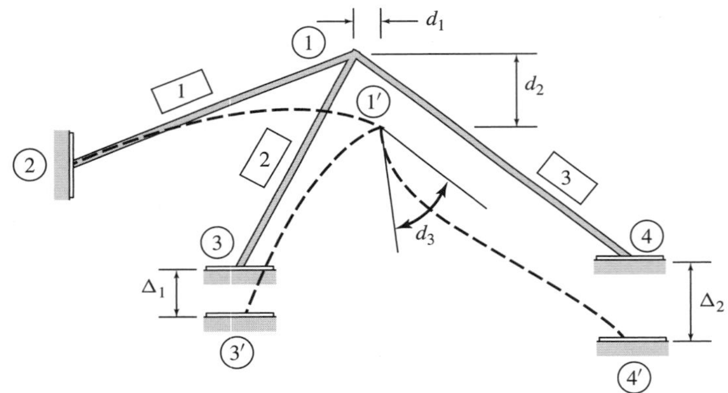

(a) Frame Subjected to Support Settlements

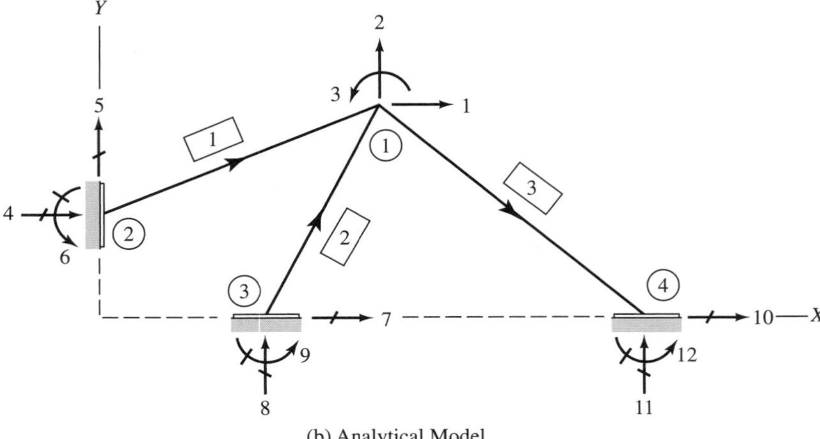

(b) Analytical Model

Fig. 7.7

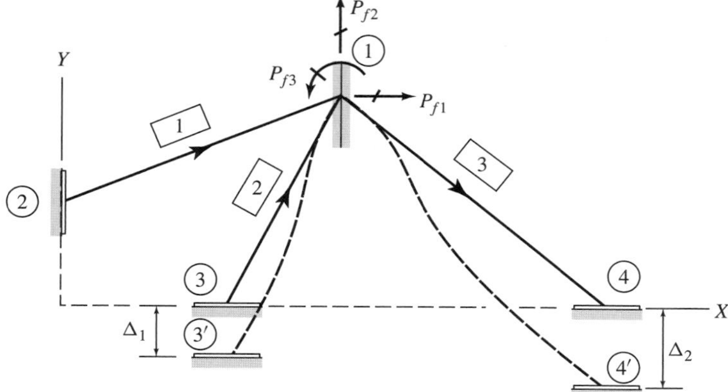

(c) Fixed Frame Subjected to Support Settlements

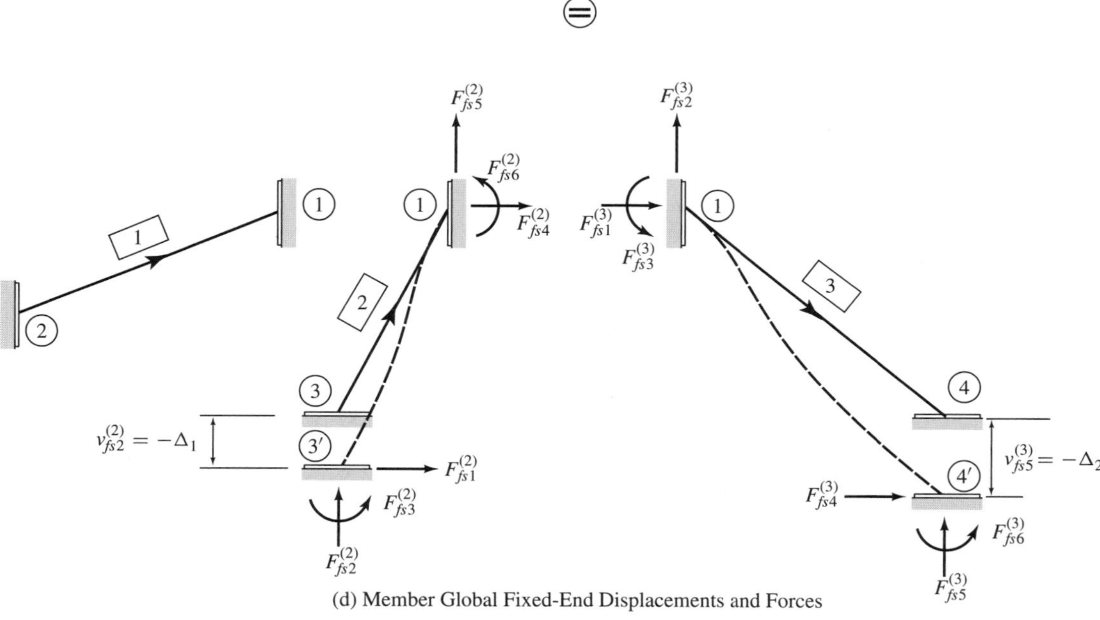

(d) Member Global Fixed-End Displacements and Forces

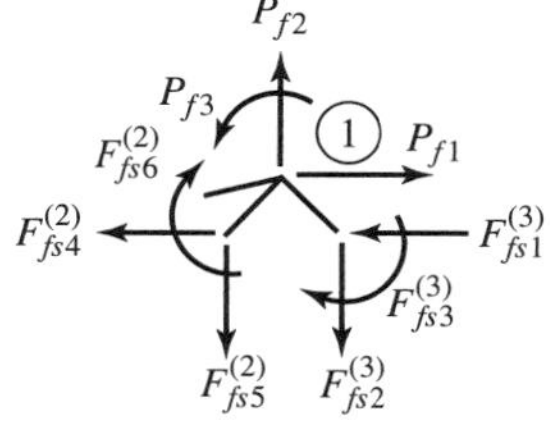

(e) Free Body of Joint 1 —Fixed Frame

Fig. 7.7 *(continued)*

assigned to the magnitudes Δ_1 and Δ_2 of d_{s8} and d_{s11} indicate that these support displacements occur in the negative Y (i.e., downward) direction.

To illustrate the process of evaluating a structure's fixed-joint forces due to a prescribed set of support settlements, we restrain the joint displacements of the example frame by applying an imaginary restraint at joint 1, and subject this hypothetical completely fixed frame to the given support settlements Δ_1 and Δ_2, as shown in Fig. 7.7(c). The structure fixed-joint forces that develop at the imaginary restraint at joint 1 are denoted by P_{f1}, P_{f2}, and P_{f3} in the figure, with the fixed-joint force corresponding to an ith degree of freedom denoted by P_{fi}. To evaluate the fixed-joint forces, we first determine the displacements that the support settlements Δ_1 and Δ_2 cause at the ends of the members of the fixed frame. The free-body diagrams of the three members of the hypothetical fixed frame are depicted in Fig. 7.7(d). From Figs. 7.7(c) and (d), we observe that, while the settlements of supports 3 and 4 do not cause any displacement in member 1, they induce downward displacements of magnitudes Δ_1 and Δ_2, respectively, at the lower ends of members 2 and 3. Note that all other member end displacements are 0, because all the joint displacements of the fixed frame are 0, with the exception of the known support settlements. Thus, the end displacements of members 2 and 3, respectively, can be expressed in vector form, as

$$\mathbf{v}_{fs2} = \begin{bmatrix} 0 \\ -\Delta_1 \\ 0 \\ 0 \\ 0 \\ 0 \end{bmatrix} \begin{matrix} 7 \\ 8 \\ 9 \\ 1 \\ 2 \\ 3 \end{matrix} \quad \text{and} \quad \mathbf{v}_{fs3} = \begin{bmatrix} 0 \\ 0 \\ 0 \\ 0 \\ -\Delta_2 \\ 0 \end{bmatrix} \begin{matrix} 1 \\ 2 \\ 3 \\ 10 \\ 11 \\ 12 \end{matrix} \tag{7.25}$$

in which $\mathbf{v}_{fs}$ represents the *member global fixed-end displacement vector due to support displacements.* The foregoing member global fixed-end displacement vectors can be directly generated using the member code numbers, which define the member compatibility equations. For example, from Fig. 7.7(b), we can see that the code numbers for member 2 are 7, 8, 9, 1, 2, 3. By comparing these member code numbers with the support displacements of the frame, $d_{s8} = -\Delta_1$ and $d_{s11} = -\Delta_2$ (see Eq. (7.24)), we conclude that all the elements of $\mathbf{v}_{fs2}$ are zero, with the exception of the element in the second row which equals $-\Delta_1$ (i.e., $v_{fs2}^{(2)} = d_{s8} = -\Delta_1$). Similarly, by examining the code numbers 1, 2, 3, 10, 11, 12 of member 3, we realize that the only nonzero element of $\mathbf{v}_{fs3}$ is in the fifth row and it equals $-\Delta_2$ (i.e., $v_{fs5}^{(3)} = d_{s11} = -\Delta_2$).

Once the member fixed-end displacement vectors $\mathbf{v}_{fs}$ have been determined, they are used to calculate the corresponding *member global fixed-end force vectors due to support displacements,* $\mathbf{F}_{fs}$, through the member global stiffness relationship (Eq. (6.28)) derived in Chapter 6. By substituting $\mathbf{F} = \mathbf{F}_{fs}$, $\mathbf{v} = \mathbf{v}_{fs}$, and $\mathbf{F}_f = \mathbf{0}$ into Eq. (6.28), we obtain the following relationship between $\mathbf{F}_{fs}$ and $\mathbf{v}_{fs}$:

$$\boxed{\mathbf{F}_{fs} = \mathbf{K}\mathbf{v}_{fs}} \tag{7.26}$$

With the member global fixed-end forces known, the structure fixed-joint forces due to the support displacements can be evaluated using joint equilibrium equations. Thus, for the example frame, we apply the three equations of equilibrium, $\sum F_X = 0$, $\sum F_Y = 0$, and $\sum M = 0$, to the free body of joint 1 (see Fig. 7.7(e)) to obtain the following expressions for the fixed-joint forces in terms of the member fixed-end forces:

$$P_{f1} = F_{fs4}^{(2)} + F_{fs1}^{(3)} \tag{7.27a}$$

$$P_{f2} = F_{fs5}^{(2)} + F_{fs2}^{(3)} \tag{7.27b}$$

$$P_{f3} = F_{fs6}^{(2)} + F_{fs3}^{(3)} \tag{7.27c}$$

The structure fixed-joint force vector for the support settlements of the example frame can, therefore, be expressed as

$$\mathbf{P}_f = \begin{bmatrix} F_{fs4}^{(2)} + F_{fs1}^{(3)} \\ F_{fs5}^{(2)} + F_{fs2}^{(3)} \\ F_{fs6}^{(2)} + F_{fs3}^{(3)} \end{bmatrix} \tag{7.28}$$

As demonstrated in Chapters 5 and 6 for the case of member loads, the structure fixed-joint force vectors $\mathbf{P}_f$ can be conveniently generated by employing the member code number technique. The application of the technique remains the same in the case of support displacements, except that the elements of the member global fixed-end force vectors due to support displacements, $\mathbf{F}_{fs}$, must now be added into $\mathbf{P}_f$. When a structure is subjected to more than one type of action requiring evaluation of fixed-joint forces (e.g., member loads and support settlements), then the fixed-joint forces representing different types of actions can be conveniently combined into a single $\mathbf{P}_f$ vector. For example, in the case of a frame subjected to member loads and support settlements, the elements of the two types of member fixed-end force vectors—that is, due to member loads ($\mathbf{F}_f$) and support displacements ($\mathbf{F}_{fs}$)—can be stored in a single $\mathbf{P}_f$ vector using the member code number technique.

Once the structure fixed-joint forces due to support displacements have been evaluated, the structure stiffness relations $\mathbf{P} - \mathbf{P}_f = \mathbf{Sd}$ (Eq. (6.42)) can be solved for the unknown joint displacements $\mathbf{d}$. With $\mathbf{d}$ known, the member global end displacement vector $\mathbf{v}$ for each member is determined by applying the compatibility equations defined by its code numbers. For members that are attached to the supports that undergo displacements, the displacements of the supported ends, due to the corresponding support displacements, must be included in the member global end displacement vectors $\mathbf{v}$. The inclusion of support displacements in the $\mathbf{v}$ vectors automatically adds the response of the fixed structure to support settlements (see, for example, Fig. 7.6(b)) into the analysis, thereby enabling us to evaluate the member local and global end forces, and support reactions, using the procedures developed in previous chapters.

Procedure for Analysis

Based on the discussion presented in this section, we can develop the following step-by-step procedure for the matrix stiffness analysis of framed structures due to support displacements.

1. Prepare an analytical model of the structure, and determine its structure stiffness matrix **S**. If the structure is subjected to member loads, then evaluate its fixed-joint force vector $\mathbf{P}_f$ due to the member loads. If the structure is subjected to joint loads, then form its joint load vector **P**.
2. Calculate the structure fixed-joint force vector $\mathbf{P}_f$ $(NDOF \times 1)$ due to the given support displacements. If a $\mathbf{P}_f$ vector was formed in step 1 for member loads, then store the member fixed-end forces due to support displacements in the previously formed $\mathbf{P}_f$ vector. For each member that is attached to the supports that undergo displacements, perform the following operations:
   a. Identify the member code numbers, and form the member global fixed-end displacement vector, $\mathbf{v}_{fs}$, from the specified support displacements, d_{si}. Note that the support translations are considered positive when in the positive directions of the global X and Y axes, and support rotations are considered positive when counterclockwise. For beams, form the member local fixed-end displacement vector due to support displacements, $\mathbf{u}_{fs}$, using the same process.
   b. Evaluate the member global fixed-end force vector due to support displacements, $\mathbf{F}_{fs}$, using the relationship $\mathbf{F}_{fs} = \mathbf{K}\mathbf{v}_{fs}$ (Eq. (7.26)). For beams, evaluate the member local fixed-end force vector due to support displacements, $\mathbf{Q}_{fs}$, using the relationship $\mathbf{Q}_{fs} = \mathbf{k}\mathbf{u}_{fs}$.
   c. Using member code numbers, store the pertinent elements of $\mathbf{F}_{fs}$, or $\mathbf{Q}_{fs}$ for beams, in their proper positions in the structure fixed-joint force vector $\mathbf{P}_f$.
3. Determine the unknown joint displacements **d** by solving the structure stiffness relationship, $\mathbf{P} - \mathbf{P}_f = \mathbf{S}\mathbf{d}$.
4. Compute member end displacements and end forces, and support reactions. For each member of the structure, carry out the following steps.
   a. Obtain member end displacements in the global coordinate system, **v**, from the joint displacements **d** and the specified support displacements d_{si}, by using the member code numbers. For beams, obtain the member local end displacements, **u**, using the same process, and then go to step 4c.
   b. Determine the member end displacements in the local coordinate system, **u**, by using the transformation relationship $\mathbf{u} = \mathbf{T}\mathbf{v}$.
   c. Calculate the member end forces in the local coordinate system, **Q**, by using the stiffness relationship $\mathbf{Q} = \mathbf{k}\mathbf{u} + \mathbf{Q}_f$. If the member is not subjected to any member loads, then $\mathbf{Q}_f = \mathbf{0}$. For beams, go to step 4e.
   d. Compute the member end forces in the global coordinate system, **F**, using the transformation relationship $\mathbf{F} = \mathbf{T}^T\mathbf{Q}$.

e. If the member is attached to a support joint, then use the member code numbers to store the pertinent elements of **F**, or **Q** for beams, in their proper positions in the support reaction vector **R**.

EXAMPLE 7.2

Determine the joint displacements, member axial forces, and support reactions for the plane truss shown in Fig. 7.8(a) due to a settlement of $\frac{1}{2}$ in. of support 4. Use the matrix stiffness method.

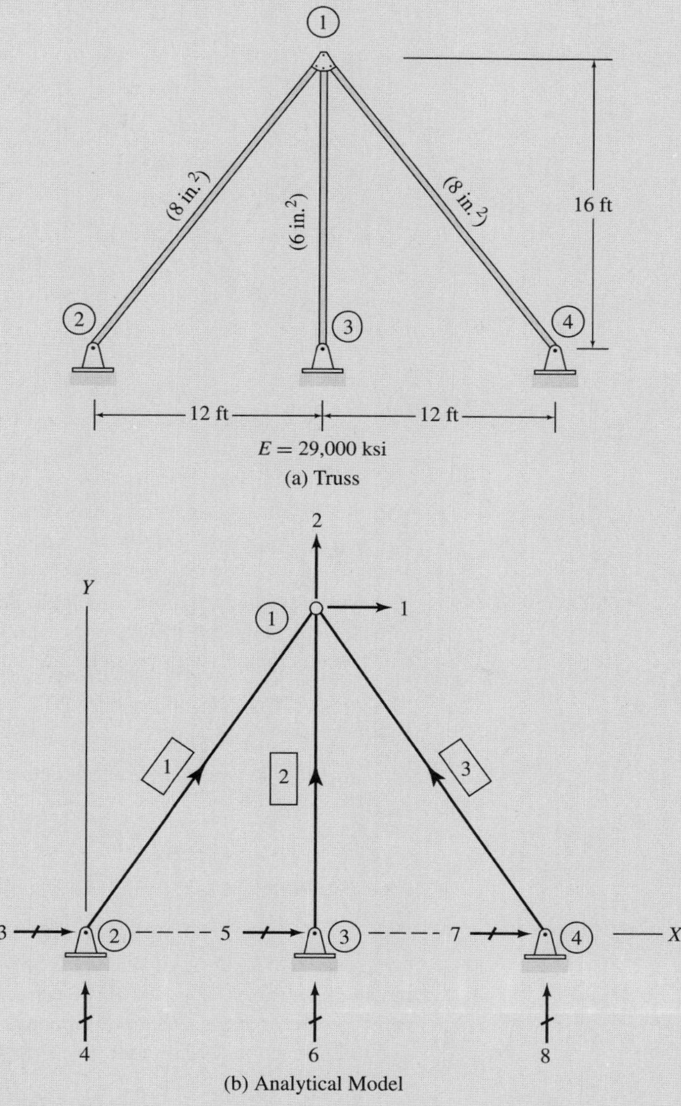

Fig. 7.8

$$\mathbf{R} = \begin{bmatrix} -49.039 \\ -65.386 \\ 0 \\ 130.78 \\ 49.039 \\ -65.386 \end{bmatrix} \begin{matrix} 3 \\ 4 \\ 5 \\ 6 \\ 7 \\ 8 \end{matrix} \text{ k}$$

(c) Support Reaction Vector

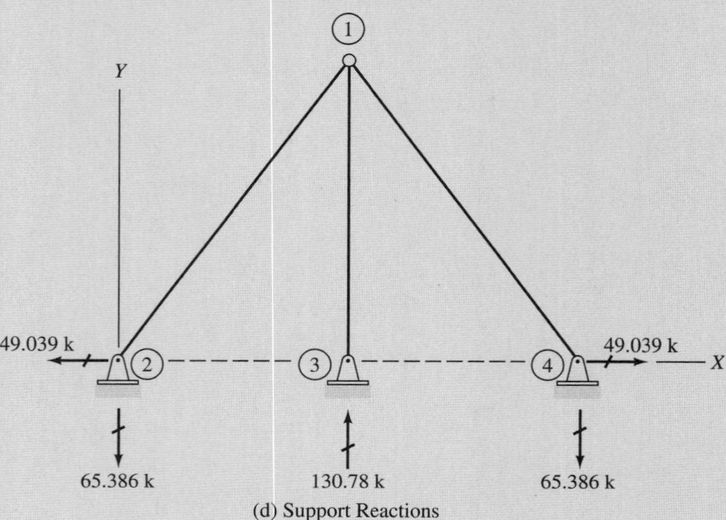

(d) Support Reactions

Fig. 7.8 *(continued)*

SOLUTION This truss was analyzed in Example 3.8 for joint loads. In this example, we use the same analytical model of the truss, so that the various member and structure matrices calculated in the previous example can be reused herein.

Analytical Model: See Fig. 7.8(b). The truss has two degrees of freedom and six restrained coordinates.

Structure Stiffness Matrix: From Example 3.8,

$$\mathbf{S} = \begin{bmatrix} 696 & 0 \\ 0 & 2,143.6 \end{bmatrix} \text{ k/in.} \tag{1}$$

Joint Load Vector: As the truss is not subjected to any loads,

$$\mathbf{P} = \mathbf{0} \tag{2}$$

Structure Fixed-Joint Force Vector Due to Support Displacements: From the analytical model of the truss in Fig. 7.8(b), we can see that the given .5 in. settlement (i.e., vertically downward displacement) of support joint 4 occurs at the location and in the direction of the reaction R_8. Thus, the given support displacement can be expressed as

$$d_{s8} = -0.5 \text{ in.}$$

From Fig. 7.8(b), we observe that member 3 is the only member attached to support 4 that undergoes displacement. Thus, using the member's code numbers 7, 8, 1, 2, we form its global fixed-end displacement vector due to support displacement as

$$\mathbf{v}_{fs3} = \begin{bmatrix} v_{fs1} \\ v_{fs2} \\ v_{fs3} \\ v_{fs4} \end{bmatrix} \begin{matrix} 7 \\ 8 \\ 1 \\ 2 \end{matrix} = \begin{bmatrix} 0 \\ d_{s8} \\ 0 \\ 0 \end{bmatrix} = \begin{bmatrix} 0 \\ -0.5 \\ 0 \\ 0 \end{bmatrix} \text{ in.}$$

Next, we evaluate the global fixed-end force vector $\mathbf{F}_{fs3}$ due to the support settlement, for member 3, using the member global stiffness matrix $\mathbf{K}_3$ calculated in Example 3.8, and Eq. (7.26). Thus,

$$\mathbf{F}_{fs3} = \mathbf{K}_3\mathbf{v}_{fs3} = \begin{bmatrix} 348 & -464 & -348 & 464 \\ -464 & 618.67 & 464 & -618.67 \\ -348 & 464 & 348 & -464 \\ 464 & -618.67 & -464 & 618.67 \end{bmatrix} \begin{bmatrix} 0 \\ -0.5 \\ 0 \\ 0 \end{bmatrix} = \begin{bmatrix} 232 \\ -309.33 \\ \hdashline -232 \\ 309.33 \end{bmatrix} \begin{matrix} 7 \\ 8 \\ 1 \\ 2 \end{matrix} \text{ k}$$

From the member code numbers, which are written on the right side of $\mathbf{F}_{fs3}$, we realize that the elements in the third and fourth rows of $\mathbf{F}_{fs3}$ should be stored in rows 1 and 2, respectively, of the $\mathbf{P}_f$ vector. Thus, the structure fixed-joint force vector, due to the support settlement, is given by

$$\mathbf{P}_f = \begin{bmatrix} -232 \\ 309.33 \end{bmatrix} \begin{matrix} 1 \\ 2 \end{matrix} \text{ k} \tag{3}$$

Joint Displacements: By substituting **P** (Eq. (2)), $\mathbf{P}_f$ (Eq. (3)), and **S** (Eq. (1)) into the structure stiffness relationship, we write

$$\mathbf{P} - \mathbf{P}_f = \mathbf{S}\mathbf{d}$$

$$\begin{bmatrix} 0 \\ 0 \end{bmatrix} - \begin{bmatrix} -232 \\ 309.33 \end{bmatrix} = \begin{bmatrix} 232 \\ -309.33 \end{bmatrix} = \begin{bmatrix} 696 & 0 \\ 0 & 2{,}143.6 \end{bmatrix} \begin{bmatrix} d_1 \\ d_2 \end{bmatrix}$$

By solving these equations, we determine the joint displacements to be

$$\mathbf{d} = \begin{bmatrix} 0.33333 \\ -0.14431 \end{bmatrix} \begin{matrix} 1 \\ 2 \end{matrix} \text{ in.}$$ **Ans**

Member End Displacements and End Forces:

Member 1 Using the member code numbers 3, 4, 1, 2, we write the global end displacement vector as

$$\mathbf{v}_1 = \begin{bmatrix} v_1 \\ v_2 \\ v_3 \\ v_4 \end{bmatrix} \begin{matrix} 3 \\ 4 \\ 1 \\ 2 \end{matrix} = \begin{bmatrix} 0 \\ 0 \\ d_1 \\ d_2 \end{bmatrix} = \begin{bmatrix} 0 \\ 0 \\ 0.33333 \\ -0.14431 \end{bmatrix} \text{ in.}$$

Next, we determine the member local end displacement vector $\mathbf{u}_1$, using the transformation matrix $\mathbf{T}_1$ from Example 3.8, and Eq. (3.63), as

$$\mathbf{u}_1 = \mathbf{T}_1\mathbf{v}_1 = \begin{bmatrix} 0.6 & 0.8 & 0 & 0 \\ -0.8 & 0.6 & 0 & 0 \\ 0 & 0 & 0.6 & 0.8 \\ 0 & 0 & -0.8 & 0.6 \end{bmatrix} \begin{bmatrix} 0 \\ 0 \\ 0.33333 \\ -0.14431 \end{bmatrix} = \begin{bmatrix} 0 \\ 0 \\ 0.08455 \\ -0.35325 \end{bmatrix} \text{ in.}$$

We can now calculate the member local end forces $\mathbf{Q}_1$ by applying the member stiffness relationship, $\mathbf{Q} = \mathbf{ku}$ (Eq. (3.7)). Thus, using $\mathbf{k}_1$ from Example 3.8, we obtain

$$\mathbf{Q}_1 = \mathbf{k}_1\mathbf{u}_1 = \begin{bmatrix} 966.67 & 0 & -966.67 & 0 \\ 0 & 0 & 0 & 0 \\ -966.67 & 0 & 966.67 & 0 \\ 0 & 0 & 0 & 0 \end{bmatrix} \begin{bmatrix} 0 \\ 0 \\ 0.08455 \\ -0.35325 \end{bmatrix} = \begin{bmatrix} -81.732 \\ 0 \\ 81.732 \\ 0 \end{bmatrix} \text{k}$$

Recall from Chapter 3 that the member axial force equals the first element of the $\mathbf{Q}_1$ vector; that is,

$$Q_{a1} = -81.732 \text{ k}$$

in which the negative sign indicates that the axial force is tensile, or

$$Q_{a1} = 81.732 \text{ k (T)}$$ **Ans**

By applying Eq. (3.66), we determine the member global end forces as

$$\mathbf{F}_1 = \mathbf{T}_1^T\mathbf{Q}_1 = \begin{bmatrix} 0.6 & -0.8 & 0 & 0 \\ 0.8 & 0.6 & 0 & 0 \\ 0 & 0 & 0.6 & -0.8 \\ 0 & 0 & 0.8 & 0.6 \end{bmatrix} \begin{bmatrix} -81.732 \\ 0 \\ 81.732 \\ 0 \end{bmatrix} = \begin{bmatrix} -49.039 \\ -65.386 \\ 49.039 \\ 65.386 \end{bmatrix} \begin{matrix} 3 \\ 4 \\ 1 \\ 2 \end{matrix} \text{ k}$$

Using the member code numbers 3, 4, 1, 2, we store the pertinent elements of $\mathbf{F}_1$ in the support reaction vector $\mathbf{R}$ (see Fig. 7.8(c)).

Member 2

$$\mathbf{v}_2 = \begin{bmatrix} v_1 \\ v_2 \\ v_3 \\ v_4 \end{bmatrix} \begin{matrix} 5 \\ 6 \\ 1 \\ 2 \end{matrix} = \begin{bmatrix} 0 \\ 0 \\ d_1 \\ d_2 \end{bmatrix} = \begin{bmatrix} 0 \\ 0 \\ 0.33333 \\ -0.14431 \end{bmatrix} \text{ in.}$$

Using $\mathbf{T}_2$ from Example 3.8, we calculate

$$\mathbf{u}_2 = \mathbf{T}_2\mathbf{v}_2 = \begin{bmatrix} 0 & 1 & 0 & 0 \\ -1 & 0 & 0 & 0 \\ 0 & 0 & 0 & 1 \\ 0 & 0 & -1 & 0 \end{bmatrix} \begin{bmatrix} 0 \\ 0 \\ 0.33333 \\ -0.14431 \end{bmatrix} = \begin{bmatrix} 0 \\ 0 \\ -0.14431 \\ -0.33333 \end{bmatrix} \text{ in.}$$

Next, using $\mathbf{k}_2$ from Example 3.8, we determine the member local end forces to be

$$\mathbf{Q}_2 = \mathbf{k}_2\mathbf{u}_2 = \begin{bmatrix} 906.25 & 0 & -906.25 & 0 \\ 0 & 0 & 0 & 0 \\ -906.25 & 0 & 906.25 & 0 \\ 0 & 0 & 0 & 0 \end{bmatrix} \begin{bmatrix} 0 \\ 0 \\ -0.14431 \\ -0.33333 \end{bmatrix} = \begin{bmatrix} 130.78 \\ 0 \\ -130.78 \\ 0 \end{bmatrix} \text{k}$$

$$Q_{a2} = 130.78 \text{ k (C)}$$ **Ans**

$$\mathbf{F}_2 = \mathbf{T}_2^T\mathbf{Q}_2 = \begin{bmatrix} 0 & -1 & 0 & 0 \\ 1 & 0 & 0 & 0 \\ 0 & 0 & 0 & -1 \\ 0 & 0 & 1 & 0 \end{bmatrix} \begin{bmatrix} 130.78 \\ 0 \\ -130.78 \\ 0 \end{bmatrix} = \begin{bmatrix} 0 \\ 130.78 \\ 0 \\ -130.78 \end{bmatrix} \begin{matrix} 5 \\ 6 \\ 1 \\ 2 \end{matrix} \text{ k}$$

Member 3

$$\mathbf{v}_3 = \begin{bmatrix} v_1 \\ v_2 \\ v_3 \\ v_4 \end{bmatrix} \begin{matrix} 7 \\ 8 \\ 1 \\ 2 \end{matrix} = \begin{bmatrix} 0 \\ d_{s8} \\ d_1 \\ d_2 \end{bmatrix} = \begin{bmatrix} 0 \\ -0.5 \\ 0.33333 \\ -0.14431 \end{bmatrix} \text{in.}$$

Note that the support settlement $d_{s8} = -0.5$ in. is included in the foregoing global end displacement vector $\mathbf{v}_3$ for member 3. Next, using the member's direction cosines, $\cos\theta = -0.6$ and $\sin\theta = 0.8$, and Eq. (3.61), we evaluate its transformation matrix:

$$\mathbf{T}_3 = \begin{bmatrix} -0.6 & 0.8 & 0 & 0 \\ -0.8 & -0.6 & 0 & 0 \\ 0 & 0 & -0.6 & 0.8 \\ 0 & 0 & -0.8 & -0.6 \end{bmatrix}$$

and determine the member local end displacements as

$$\mathbf{u}_3 = \mathbf{T}_3\mathbf{v}_3 = \begin{bmatrix} -0.6 & 0.8 & 0 & 0 \\ -0.8 & -0.6 & 0 & 0 \\ 0 & 0 & -0.6 & 0.8 \\ 0 & 0 & -0.8 & -0.6 \end{bmatrix} \begin{bmatrix} 0 \\ -0.5 \\ 0.33333 \\ -0.14431 \end{bmatrix} = \begin{bmatrix} -0.4 \\ 0.3 \\ -0.31545 \\ -0.18008 \end{bmatrix} \text{in.}$$

To obtain the member local stiffness matrix, we substitute $E = 29{,}000$ ksi, $A = 8$ in.^{2}, and $L = 240$ in. into Eq. (3.27):

$$\mathbf{k}_3 = \begin{bmatrix} 966.67 & 0 & -966.67 & 0 \\ 0 & 0 & 0 & 0 \\ -966.67 & 0 & 966.67 & 0 \\ 0 & 0 & 0 & 0 \end{bmatrix} \text{k/in.}$$

The member local end forces can now be computed as

$$\mathbf{Q}_3 = \mathbf{k}_3\mathbf{u}_3 = \begin{bmatrix} 966.67 & 0 & -966.67 & 0 \\ 0 & 0 & 0 & 0 \\ -966.67 & 0 & 966.67 & 0 \\ 0 & 0 & 0 & 0 \end{bmatrix} \begin{bmatrix} -0.4 \\ 0.3 \\ -0.31545 \\ -0.18008 \end{bmatrix} = \begin{bmatrix} -81.732 \\ 0 \\ 81.732 \\ 0 \end{bmatrix} \text{k}$$

Thus,

$Q_{a3} = -81.732 \text{ k} = 81.732 \text{ k (T)}$ **Ans**

Finally, we calculate the member global end forces as

$$\mathbf{F}_3 = \mathbf{T}_3^T\mathbf{Q}_3 = \begin{bmatrix} -0.6 & -0.8 & 0 & 0 \\ 0.8 & -0.6 & 0 & 0 \\ 0 & 0 & -0.6 & -0.8 \\ 0 & 0 & 0.8 & -0.6 \end{bmatrix} \begin{bmatrix} -81.732 \\ 0 \\ 81.732 \\ 0 \end{bmatrix} = \begin{bmatrix} 49.039 \\ -65.386 \\ -49.039 \\ 65.386 \end{bmatrix} \begin{matrix} 7 \\ 8 \\ 1 \\ 2 \end{matrix}$$

and store the pertinent elements of $\mathbf{F}_3$ in the reaction vector $\mathbf{R}$, as shown in Fig. 7.8(c).

Support Reactions: The completed reaction vector $\mathbf{R}$ is shown in Fig. 7.8(c), and the support reactions are depicted on a line diagram of the truss in Fig. 7.8(d). **Ans**

Equilibrium Check: Applying the equilibrium equations to the free body of the entire structure (Fig. 7.8(d)), we write

$$+\rightarrow \sum F_X = 0 \qquad -49.039 + 49.039 = 0 \qquad \textbf{Checks}$$

$$+\uparrow \sum F_Y = 0 \qquad -65.386 + 130.78 - 65.386 = 0.008 \approx 0 \qquad \textbf{Checks}$$

$$+\circlearrowleft \sum M_{②} = 0 \qquad 130.78(12) - 65.386(24) = 0.096 \text{ k-ft} \approx 0 \qquad \textbf{Checks}$$

EXAMPLE 7.3

Determine the joint displacements, member end forces, and support reactions for the continuous beam shown in Fig. 7.9(a), due to the combined effect of the uniformly distributed load shown and the settlements of 45 mm and 15 mm, respectively, of supports 3 and 4. Use the matrix stiffness method.

SOLUTION *Analytical Model:* See Fig. 7.9(b). The structure has two degrees of freedom and six restrained coordinates. Note that member 3 is modeled as being hinged at its right end

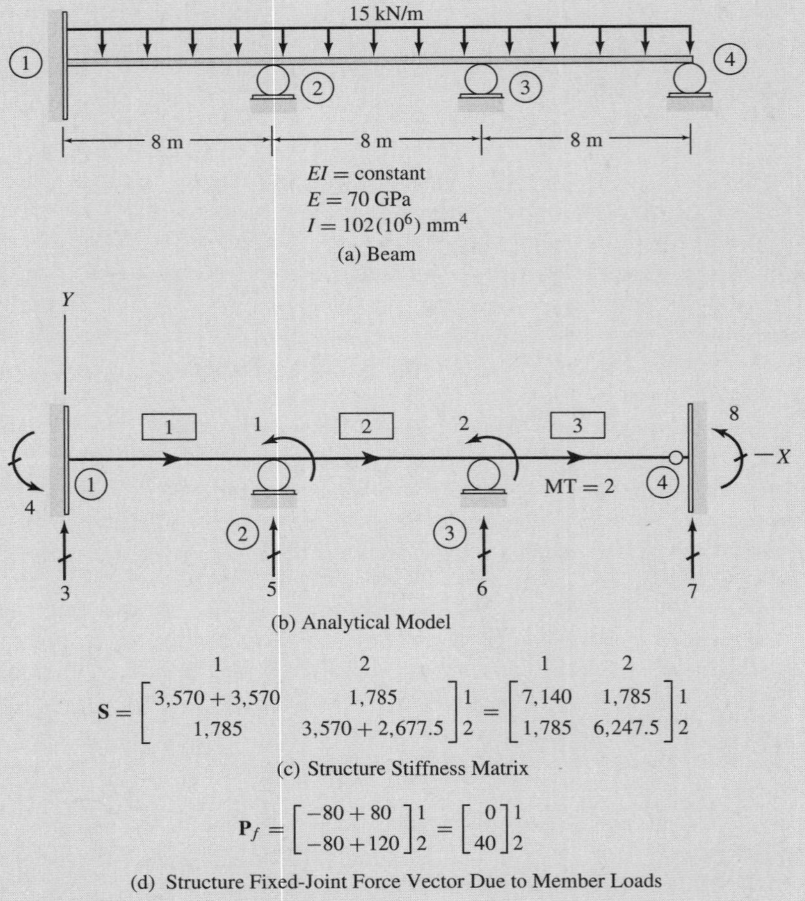

(a) Beam

(b) Analytical Model

$$\mathbf{S} = \begin{bmatrix} 3{,}570 + 3{,}570 & 1{,}785 \\ 1{,}785 & 3{,}570 + 2{,}677.5 \end{bmatrix} = \begin{bmatrix} 7{,}140 & 1{,}785 \\ 1{,}785 & 6{,}247.5 \end{bmatrix}$$

(c) Structure Stiffness Matrix

$$\mathbf{P}_f = \begin{bmatrix} -80 + 80 \\ -80 + 120 \end{bmatrix} = \begin{bmatrix} 0 \\ 40 \end{bmatrix}$$

(d) Structure Fixed-Joint Force Vector Due to Member Loads

Fig. 7.9

$$\mathbf{P}_f = \begin{bmatrix} 0 + 30.122 \\ 40 + 30.122 - 10.041 \end{bmatrix} \begin{matrix} 1 \\ 2 \end{matrix} = \begin{bmatrix} 30.122 \\ 60.081 \end{bmatrix} \begin{matrix} 1 \\ 2 \end{matrix}$$

(e) Structure Fixed-Joint Force Vector Due to Member Loads and Support Displacements

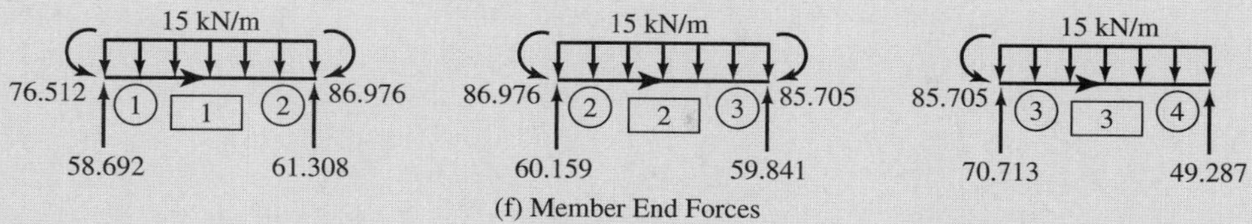

(f) Member End Forces

$$\mathbf{R} = \begin{bmatrix} 58.692 \\ 76.512 \\ 61.308 + 60.159 \\ 59.841 + 70.713 \\ 49.287 \\ 0 \end{bmatrix} \begin{matrix} 3 \\ 4 \\ 5 \\ 6 \\ 7 \\ 8 \end{matrix} = \begin{bmatrix} 58.692 \text{ kN} \\ 76.512 \text{ kN·m} \\ 121.47 \text{ kN} \\ 130.55 \text{ kN} \\ 49.287 \text{ kN} \\ 0 \end{bmatrix}$$

(g) Support Reaction Vector

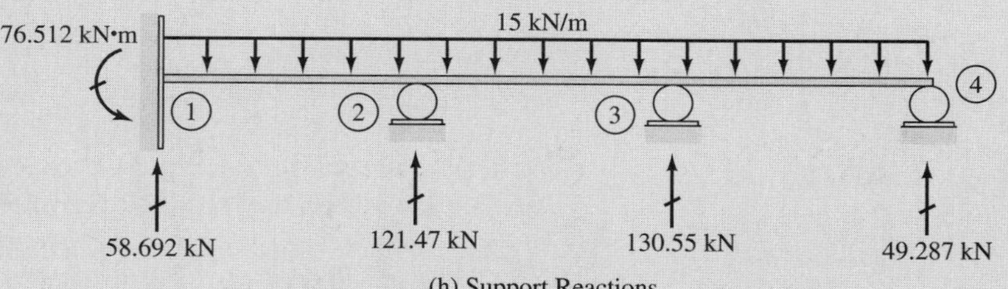

(h) Support Reactions

Fig. 7.9 *(continued)*

(i.e., $MT = 2$), because the moment at that end of the member must be 0. This approach enables us to eliminate the rotational degrees of freedom of joint 4 from the analysis, by modeling it as a hinged joint with its rotation restrained by an imaginary clamp.

Structure Stiffness Matrix and Fixed-Joint Forces Due to Member Loads:

Members 1 and 2 ($MT = 0$) By substituting $E = 70(10^6)$ kN/m², $I = 102(10^{-6})$ m⁴, and $L = 8$ m into Eq. (5.53), we evaluate the member stiffness matrices **k** as

$$\begin{matrix} \text{Member 2} \longrightarrow & 5 & 1 & 6 & 2 \\ \text{Member 1} \longrightarrow & 3 & 4 & 5 & 1 \end{matrix}$$

$$\mathbf{k}_1 = \mathbf{k}_2 = \begin{bmatrix} 167.34 & 669.38 & -167.34 & 669.38 \\ 669.38 & 3{,}570 & -669.38 & 1{,}785 \\ -167.34 & -669.38 & 167.34 & -669.38 \\ 669.38 & 1{,}785 & -669.38 & 3{,}570 \end{bmatrix} \begin{matrix} 3 \\ 4 \\ 5 \\ 1 \end{matrix} \begin{matrix} 5 \\ 1 \\ 6 \\ 2 \end{matrix} \qquad (1)$$

Using the equations given inside the front cover, we evaluate the fixed-end shears and moments due to the 15 kN/m uniformly distributed load as

$$FS_b = FS_e = 60 \text{ kN} \qquad FM_b = -FM_e = 80 \text{ kN} \cdot \text{m} \tag{2}$$

Thus, using Eq. (5.99), we obtain the member fixed-end force vectors:

$$\mathbf{Q}_{f1} = \mathbf{Q}_{f2} = \begin{bmatrix} 60 \\ 80 \\ 60 \\ -80 \end{bmatrix} \begin{matrix} 3 & 5 \\ 4 & 1 \\ 5 & 6 \\ 1 & 2 \end{matrix} \tag{3}$$

Member 1 (left column of code numbers), Member 2 (right column of code numbers)

Next, using the code numbers for member 1 (3, 4, 5, 1) and member 2 (5, 1, 6, 2), we store the pertinent elements of $\mathbf{k}_1$ and $\mathbf{k}_2$ into the structure stiffness matrix $\mathbf{S}$, as shown in Fig. 7.9(c). Similarly, the pertinent elements of $\mathbf{Q}_{f1}$ and $\mathbf{Q}_{f2}$ are stored in the structure fixed-joint force vector $\mathbf{P}_f$, as shown in Fig. 7.9(d).

Member 3 (MT = 2) Because $MT = 2$ for this member, we use Eqs. (7.18) and (7.19) to determine its stiffness matrix $\mathbf{k}$ and fixed-end force vector $\mathbf{Q}_f$, respectively. Thus, by applying Eq. (7.18), we obtain

$$\mathbf{k}_3 = \begin{matrix} 6 & 2 & 7 & 8 \end{matrix} \\ \begin{bmatrix} 41.836 & 334.69 & -41.836 & 0 \\ 334.69 & 2{,}677.5 & -334.69 & 0 \\ -41.836 & -334.69 & 41.836 & 0 \\ 0 & 0 & 0 & 0 \end{bmatrix} \begin{matrix} 6 \\ 2 \\ 7 \\ 8 \end{matrix} \tag{4}$$

Next, by substituting the values of the fixed-end shears and moments from Eq. (2) into Eq. (7.19), we obtain the fixed-end force vector for the released member 3 as

$$\mathbf{Q}_{f3} = \begin{bmatrix} 60 - \dfrac{3(-80)}{2(8)} \\ 80 - \dfrac{1}{2}(-80) \\ 60 + \dfrac{3(-80)}{2(8)} \\ 0 \end{bmatrix} = \begin{bmatrix} 75 \\ 120 \\ 45 \\ 0 \end{bmatrix} \begin{matrix} 6 \\ 2 \\ 7 \\ 8 \end{matrix} \tag{5}$$

The relevant elements of $\mathbf{k}_3$ and $\mathbf{Q}_{f3}$ are stored in $\mathbf{S}$ and $\mathbf{P}_f$, respectively, using the member code numbers 6, 2, 7, 8. The completed structure stiffness matrix $\mathbf{S}$, and the $\mathbf{P}_f$ vector containing the structure fixed-joint forces due to member loads, are shown in Figs. 7.9(c) and (d), respectively.

Structure Fixed-Joint Forces Due to Support Displacements: From the analytical model given in Fig. 7.9(b), we observe that the given support displacements can be expressed as

$$d_{s6} = -0.045 \text{ m} \qquad d_{s7} = -0.015 \text{ m}$$

As members 2 and 3 are attached to the supports that undergo displacements, we compute, for these members, the fixed-end forces due to support displacements, and add them to the previously formed $\mathbf{P}_f$ vector due to member loads.

Member 2 Using the member code numbers 5, 1, 6, 2, we form its fixed-end displacement vector due to support displacements, as

$$\mathbf{u}_{fs2} = \begin{bmatrix} u_{fs1} \\ u_{fs2} \\ u_{fs3} \\ u_{fs4} \end{bmatrix} \begin{matrix} 5 \\ 1 \\ 6 \\ 2 \end{matrix} = \begin{bmatrix} 0 \\ 0 \\ d_{s6} \\ 0 \end{bmatrix} = \begin{bmatrix} 0 \\ 0 \\ -0.045 \\ 0 \end{bmatrix} \text{m}$$

Next, using the member stiffness matrix from Eq. (1) and the member stiffness relationship $\mathbf{Q}_{fs} = \mathbf{k}\mathbf{u}_{fs}$, we evaluate the fixed-end force vector due to support displacements, as

$$\mathbf{Q}_{fs2} = \mathbf{k}_2\mathbf{u}_{fs2} = \begin{bmatrix} 167.34 & 669.38 & -167.34 & 669.38 \\ 669.38 & 3{,}570 & -669.38 & 1{,}785 \\ -167.34 & -669.38 & 167.34 & -669.38 \\ 669.38 & 1{,}785 & -669.38 & 3{,}570 \end{bmatrix} \begin{bmatrix} 0 \\ 0 \\ -0.045 \\ 0 \end{bmatrix}$$

$$= \begin{bmatrix} 7.53 \\ 30.122 \\ -7.53 \\ 30.122 \end{bmatrix} \begin{matrix} 5 \\ 1 \\ 6 \\ 2 \end{matrix}$$

The relevant elements of $\mathbf{Q}_{fs2}$ are now added into the previously formed $\mathbf{P}_f$, using the member code numbers, as indicated in Fig. 7.9(e).

Member 3 Based on the member code numbers 6, 2, 7, 8, its fixed-end displacement vector, due to support displacements, is written as

$$\mathbf{u}_{fs3} = \begin{bmatrix} u_{fs1} \\ u_{fs2} \\ u_{fs3} \\ u_{fs4} \end{bmatrix} \begin{matrix} 6 \\ 2 \\ 7 \\ 8 \end{matrix} = \begin{bmatrix} d_{s6} \\ 0 \\ d_{s7} \\ 0 \end{bmatrix} = \begin{bmatrix} -0.045 \\ 0 \\ -0.015 \\ 0 \end{bmatrix} \text{m}$$

Using $\mathbf{k}_3$ from Eq. (4), we calculate

$$\mathbf{Q}_{fs3} = \mathbf{k}_3\mathbf{u}_{fs3} = \begin{bmatrix} 41.836 & 334.69 & -41.836 & 0 \\ 334.69 & 2{,}677.5 & -334.69 & 0 \\ -41.836 & -334.69 & 41.836 & 0 \\ 0 & 0 & 0 & 0 \end{bmatrix} \begin{bmatrix} -0.045 \\ 0 \\ -0.015 \\ 0 \end{bmatrix}$$

$$= \begin{bmatrix} -1.2551 \\ -10.041 \\ 1.2551 \\ 0 \end{bmatrix} \begin{matrix} 6 \\ 2 \\ 7 \\ 8 \end{matrix}$$

The pertinent elements of $\mathbf{Q}_{fs3}$ are stored in $\mathbf{P}_f$ using the member code numbers. The completed structure fixed-joint force vector $\mathbf{P}_f$, due to member loads and support displacements, is given in Fig. 7.9(e).

Joint Load Vector: Since no external loads are applied to the joints of the beam, its joint load vector is 0; that is

$$\mathbf{P} = \mathbf{0}$$

Joint Displacements: By substituting the numerical values of **P**, $\mathbf{P}_f$, and **S** into Eq. (5.109), we write the stiffness relations for the entire beam as

$$\mathbf{P} - \mathbf{P}_f = \mathbf{S}\mathbf{d}$$

$$\begin{bmatrix} 0 \\ 0 \end{bmatrix} - \begin{bmatrix} 30.122 \\ 60.081 \end{bmatrix} = \begin{bmatrix} -30.122 \\ -60.081 \end{bmatrix} = \begin{bmatrix} 7{,}140 & 1{,}785 \\ 1{,}785 & 6{,}247.5 \end{bmatrix} \begin{bmatrix} d_1 \\ d_2 \end{bmatrix}$$

By solving these equations, we determine the joint displacements to be

$$\mathbf{d} = \begin{bmatrix} -1.9541 \\ -9.0585 \end{bmatrix} \begin{matrix} 1 \\ 2 \end{matrix} \times 10^{-3} \text{ rad}$$ **Ans**

Member End Displacements and End Forces:

Member 1 Using the member code numbers 3, 4, 5, 1, we write its end displacement vector as

$$\mathbf{u}_1 = \begin{bmatrix} u_1 \\ u_2 \\ u_3 \\ u_4 \end{bmatrix} \begin{matrix} 3 \\ 4 \\ 5 \\ 1 \end{matrix} = \begin{bmatrix} 0 \\ 0 \\ 0 \\ d_1 \end{bmatrix} = \begin{bmatrix} 0 \\ 0 \\ 0 \\ -1.9541 \end{bmatrix} \times 10^{-3} \tag{6}$$

The member end forces can now be calculated using the member stiffness relationship $\mathbf{Q} = \mathbf{ku} + \mathbf{Q}_f$ (Eq. (5.4)). Thus, using $\mathbf{k}_1$ from Eq. (1), $\mathbf{Q}_{f1}$ from Eq. (3), and $\mathbf{u}_1$ from Eq. (6), we calculate

$$\mathbf{Q}_1 = \mathbf{k}_1\mathbf{u}_1 + \mathbf{Q}_{f1} = \begin{bmatrix} 58.692 \text{ kN} \\ 76.512 \text{ kN} \cdot \text{m} \\ 61.308 \text{ kN} \\ -86.976 \text{ kN} \cdot \text{m} \end{bmatrix} \begin{matrix} 3 \\ 4 \\ 5 \\ 1 \end{matrix}$$ **Ans**

The end forces for member 1 are depicted in Fig. 7.9(f). To generate the support reaction vector **R**, we store the pertinent elements of $\mathbf{Q}_1$ in **R**, using the member code numbers, as shown in Fig. 7.9(g).

Member 2

$$\mathbf{u}_2 = \begin{bmatrix} u_1 \\ u_2 \\ u_3 \\ u_4 \end{bmatrix} \begin{matrix} 5 \\ 1 \\ 6 \\ 2 \end{matrix} = \begin{bmatrix} 0 \\ d_1 \\ d_{s6} \\ d_2 \end{bmatrix} = \begin{bmatrix} 0 \\ -1.9541 \\ -45 \\ -9.0585 \end{bmatrix} \times 10^{-3} \tag{7}$$

Note that the support displacement d_{s6} is included in the foregoing end displacement vector for member 2. Using $\mathbf{k}_2$ from Eq. (1), $\mathbf{Q}_{f2}$ from Eq. (3), and $\mathbf{u}_2$ from Eq. (7), we determine

$$\mathbf{Q}_2 = \mathbf{k}_2\mathbf{u}_2 + \mathbf{Q}_{f2} = \begin{bmatrix} 60.159 \text{ kN} \\ 86.976 \text{ kN} \cdot \text{m} \\ 59.841 \text{ kN} \\ -85.705 \text{ kN} \cdot \text{m} \end{bmatrix} \begin{matrix} 5 \\ 1 \\ 6 \\ 2 \end{matrix}$$ **Ans**

Member 3

$$\mathbf{u}_3 = \begin{bmatrix} u_1 \\ u_2 \\ u_3 \\ u_4 \end{bmatrix} \begin{matrix} 6 \\ 2 \\ 7 \\ 8 \end{matrix} = \begin{bmatrix} d_{s6} \\ d_2 \\ d_{s7} \\ 0 \end{bmatrix} = \begin{bmatrix} -45 \\ -9.0585 \\ -15 \\ 0 \end{bmatrix} \times 10^{-3} \tag{8}$$

The rotation, u_4, of the released end of this member, if desired, can be evaluated using Eq. (7.20). Finally, using $\mathbf{k}_3$ from Eq. (4), $\mathbf{Q}_{f3}$ from Eq. (5), and $\mathbf{u}_3$ from Eq. (8), we calculate the member end forces as

$$\mathbf{Q}_3 = \mathbf{k}_3\mathbf{u}_3 + \mathbf{Q}_{f3} = \begin{bmatrix} 70.713 \text{ kN} \\ 85.705 \text{ kN} \cdot \text{m} \\ 49.287 \text{ kN} \\ 0 \end{bmatrix} \begin{matrix} 6 \\ 2 \\ 7 \\ 8 \end{matrix}$$ **Ans**

The member end forces are shown in Fig. 7.9(f).

Support Reactions: The completed reaction vector **R** is shown in Fig. 7.9(g), and the support reactions are depicted on a line diagram of the beam in Fig. 7.9(h). **Ans**

EXAMPLE 7.4

Determine the joint displacements, member local end forces, and support reactions for the plane frame of Fig. 7.10(a), due to the combined effect of the loading shown and a settlement of 1 in. of the left support. Use the matrix stiffness method.

SOLUTION

This frame was analyzed in Example 6.6 for external loading. In this example, we use the same analytical model of the frame, so that the various member and structure matrices calculated previously can be reused in the present example.

Analytical Model: See Fig. 7.10(b). The frame has three degrees of freedom and six restrained coordinates.

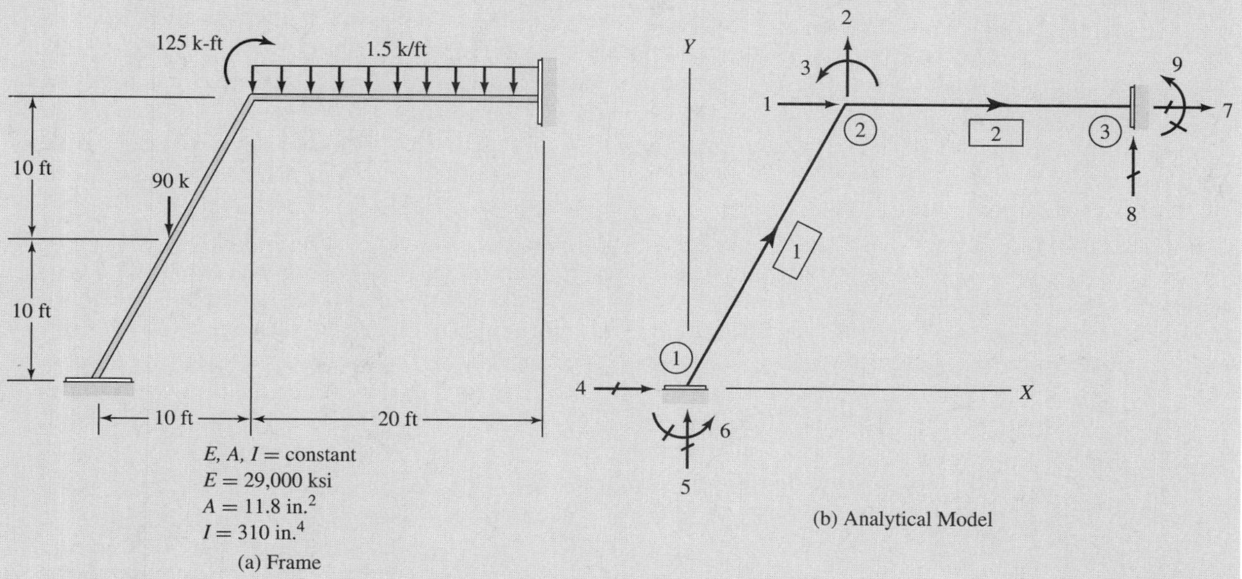

(a) Frame

(b) Analytical Model

Fig. 7.10

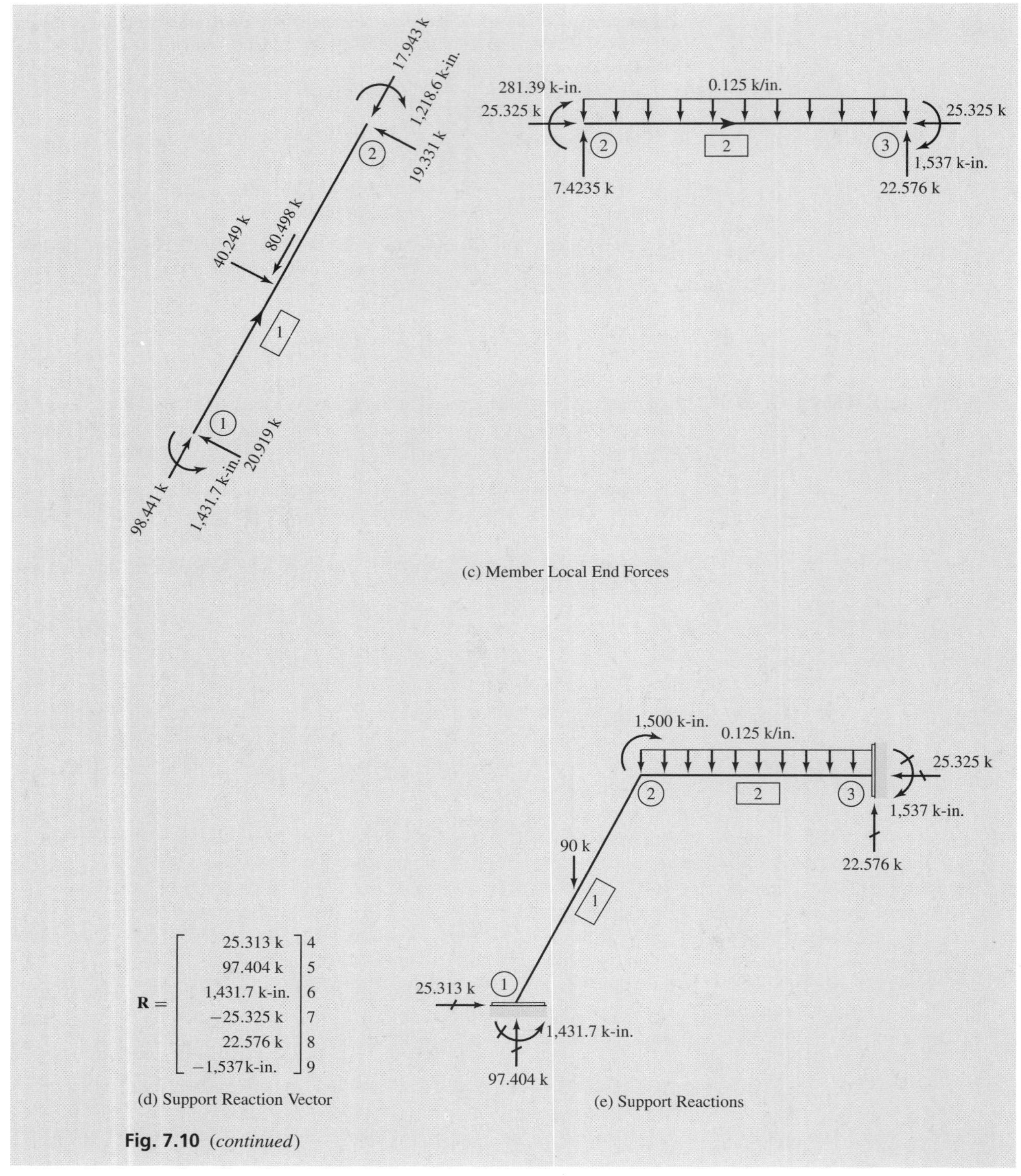

(c) Member Local End Forces

$$\mathbf{R} = \begin{bmatrix} 25.313 \text{ k} \\ 97.404 \text{ k} \\ 1{,}431.7 \text{ k-in.} \\ -25.325 \text{ k} \\ 22.576 \text{ k} \\ -1{,}537 \text{k-in.} \end{bmatrix} \begin{matrix} 4 \\ 5 \\ 6 \\ 7 \\ 8 \\ 9 \end{matrix}$$

(d) Support Reaction Vector

(e) Support Reactions

Fig. 7.10 *(continued)*

Structure Stiffness Matrix: As determined in Example 6.6, the structure stiffness matrix for the frame, in units of kips and inches, is given by

$$\mathbf{S} = \begin{bmatrix} 1{,}685.3 & 507.89 & 670.08 \\ 507.89 & 1{,}029.2 & 601.42 \\ 670.08 & 601.42 & 2{,}838.48 \end{bmatrix} \tag{1}$$

Structure Fixed-Joint Forces Due to Member Loads: From Example 6.6,

$$\mathbf{P}_f = \begin{bmatrix} 0 \\ 60 \\ -750 \end{bmatrix} \begin{matrix} 1 \\ 2 \\ 3 \end{matrix} \tag{2}$$

Joint Load Vector: From Example 6.6,

$$\mathbf{P} = \begin{bmatrix} 0 \\ 0 \\ -1{,}500 \end{bmatrix} \begin{matrix} 1 \\ 2 \\ 3 \end{matrix} \tag{3}$$

Structure Fixed-Joint Forces Due to Support Displacement: From Fig. 7.10(b), we observe that the given 1 in. downward displacement of support 1 can be expressed as

$$d_{s5} = -1 \text{ in.}$$

As member 1 is the only member attached to support 1, we form its global fixed-end displacement vector due to support displacement, using the member code numbers 4, 5, 6, 1, 2, 3, as

$$\mathbf{v}_{fs1} = \begin{bmatrix} v_{fs1} \\ v_{fs2} \\ v_{fs3} \\ v_{fs4} \\ v_{fs5} \\ v_{fs6} \end{bmatrix} \begin{matrix} 4 \\ 5 \\ 6 \\ 1 \\ 2 \\ 3 \end{matrix} = \begin{bmatrix} 0 \\ d_{s5} \\ 0 \\ 0 \\ 0 \\ 0 \end{bmatrix} = \begin{bmatrix} 0 \\ -1 \\ 0 \\ 0 \\ 0 \\ 0 \end{bmatrix} \text{ in.}$$

Next, we substitute the member global stiffness matrix $\mathbf{K}_1$ (given in Example 6.6) and the foregoing $\mathbf{v}_{fs1}$ vector into Eq. (7.26), to evaluate the member global fixed-end force vector, $\mathbf{F}_{fs1}$, due to support settlement:

$$\mathbf{F}_{fs1} = \mathbf{K}_1 \mathbf{v}_{fs1} = \begin{bmatrix} -507.89 \\ -1{,}021.4 \\ -335.04 \\ \hdashline 507.89 \\ 1{,}021.4 \\ -335.04 \end{bmatrix} \begin{matrix} 4 \\ 5 \\ 6 \\ 1 \\ 2 \\ 3 \end{matrix}$$

Based on the member code numbers, we add the elements in the fourth, fifth, and sixth rows of $\mathbf{F}_{fs1}$ into rows 1, 2, and 3, respectively, of the previously formed $\mathbf{P}_f$ vector (Eq. (2)), to obtain the structure fixed-joint force vector due to the combined effect of the member loads and support displacement, as

$$\mathbf{P}_f = \begin{bmatrix} 0 + 507.89 \\ 60 + 1{,}021.4 \\ -750 - 335.04 \end{bmatrix} \begin{matrix} 1 \\ 2 \\ 3 \end{matrix} = \begin{bmatrix} 507.89 \\ 1{,}081.4 \\ -1{,}085.04 \end{bmatrix} \tag{4}$$

Joint Displacements: By substituting **P** (Eq. (3)), $\mathbf{P}_f$ (Eq. (4)), and **S** (Eq. (1)) into Eq. (6.42), we write the stiffness relations for the entire frame as

$$\mathbf{P} - \mathbf{P}_f = \mathbf{S}\mathbf{d}$$

$$\begin{bmatrix} 0 \\ 0 \\ -1{,}500 \end{bmatrix} - \begin{bmatrix} 507.89 \\ 1{,}081.4 \\ -1{,}085.04 \end{bmatrix} = \begin{bmatrix} -507.89 \\ -1{,}081.4 \\ -414.96 \end{bmatrix} = \begin{bmatrix} 1{,}685.3 & 507.89 & 670.08 \\ 507.89 & 1{,}029.2 & 601.42 \\ 670.08 & 601.42 & 283{,}848 \end{bmatrix} \begin{bmatrix} d_1 \\ d_2 \\ d_3 \end{bmatrix}$$

Solving these equations, we determine the joint displacements to be

$$\mathbf{d} = \begin{bmatrix} 0.017762 \text{ in.} \\ -1.0599 \text{ in.} \\ 0.00074192 \text{ rad} \end{bmatrix} \begin{matrix} 1 \\ 2 \\ 3 \end{matrix}$$ **Ans**

Member End Displacements and End Forces:

Member 1

$$\mathbf{v}_1 = \begin{bmatrix} v_1 \\ v_2 \\ v_3 \\ v_4 \\ v_5 \\ v_6 \end{bmatrix} \begin{matrix} 4 \\ 5 \\ 6 \\ 1 \\ 2 \\ 3 \end{matrix} = \begin{bmatrix} 0 \\ d_{s5} \\ 0 \\ d_1 \\ d_2 \\ d_3 \end{bmatrix} = \begin{bmatrix} 0 \\ -1 \\ 0 \\ 0.017762 \\ -1.0599 \\ 0.00074192 \end{bmatrix}$$

Using the member transformation matrix $\mathbf{T}_1$ from Example 6.6, and Eq. (6.20), we calculate

$$\mathbf{u}_1 = \mathbf{T}_1 \mathbf{v}_1 = \begin{bmatrix} -0.89443 \\ -0.44721 \\ 0 \\ -0.94006 \\ -0.48988 \\ 0.00074192 \end{bmatrix}$$

Next, we use the member local stiffness matrix $\mathbf{k}_1$ and fixed-end force vector $\mathbf{Q}_{f1}$ from Example 6.6, and Eq. (6.4), to compute the local end forces as

$$\mathbf{Q}_1 = \mathbf{k}_1 \mathbf{u}_1 + \mathbf{Q}_{f1} = \begin{bmatrix} 98.441 \text{ k} \\ 20.919 \text{ k} \\ 1{,}431.7 \text{ k-in.} \\ -17.943 \text{ k} \\ 19.331 \text{ k} \\ -1{,}218.6 \text{ k-in.} \end{bmatrix}$$ **Ans**

The local member end forces are depicted in Fig. 7.10(c).

The member global end forces **F** can now be determined by applying Eq. (6.23), as

$$\mathbf{F}_1 = \mathbf{T}^T \mathbf{Q}_1 = \begin{bmatrix} 25.313 \\ 97.404 \\ 1{,}431.7 \\ \hdashline -25.314 \\ -7.404 \\ -1{,}218.6 \end{bmatrix} \begin{matrix} 4 \\ 5 \\ 6 \\ 1 \\ 2 \\ 3 \end{matrix}$$

Using the member code numbers, the pertinent elements of $\mathbf{F}_1$ are stored in the reaction vector $\mathbf{R}$ (see Fig. 7.10(d)).

Member 2 The global and local end displacements for this horizontal member are

$$\mathbf{u}_2 = \mathbf{v}_2 = \begin{bmatrix} v_1 \\ v_2 \\ v_3 \\ v_4 \\ v_5 \\ v_6 \end{bmatrix} \begin{matrix} 1 \\ 2 \\ 3 \\ 7 \\ 8 \\ 9 \end{matrix} = \begin{bmatrix} d_1 \\ d_2 \\ d_3 \\ 0 \\ 0 \\ 0 \end{bmatrix} = \begin{bmatrix} 0.017762 \\ -1.0599 \\ 0.00074192 \\ 0 \\ 0 \\ 0 \end{bmatrix}$$

Using $\mathbf{k}_2$ and $\mathbf{Q}_{f2}$ from Example 6.6, we compute the member local and global end forces to be

$$\mathbf{F}_2 = \mathbf{Q}_2 = \mathbf{k}_2\mathbf{u}_2 + \mathbf{Q}_{f2} = \begin{bmatrix} 25.325 \text{ k} \\ 7.4235 \text{ k} \\ -281.39 \text{ k-in.} \\ \hline -25.325 \text{ k} \\ 22.576 \text{ k} \\ -1{,}537 \text{ k-in.} \end{bmatrix} \begin{matrix} 1 \\ 2 \\ 3 \\ 7 \\ 8 \\ 9 \end{matrix}$$

Ans

The pertinent elements of $\mathbf{F}_2$ are stored in $\mathbf{R}$.

Support Reactions: See Figs. 7.10(d) and (e). **Ans**

7.4 COMPUTER IMPLEMENTATION OF SUPPORT DISPLACEMENT EFFECTS

The computer programs developed previously can be extended with relative ease, and without changing their overall organization, to include the effects of support displacements in the analysis. From the analysis procedure developed in Section 7.3, we realize that inclusion of support displacement effects essentially involves extension of the existing programs to perform three additional tasks: (a) reading and storing of the support displacement data, (b) evaluation of the structure fixed-joint forces due to support displacements, and (c) inclusion of support displacements in the member end displacement vectors, before calculation of the final member end forces and support reactions.

In this section, we consider the programming of these tasks, with particular reference to the program for the analysis of plane frames (Section 6.7). The modifications necessary in the plane truss and beam analysis programs are also described.

Input of Support Displacement Data The process of reading and storing the support displacements is similar to that for inputting the joint load data (e.g., see flowcharts in Figs. 4.3(f) and 5.20(b)). This process can be conveniently programmed using the flowchart given in Fig. 7.11 on the next page. The support displacement data consists of (a) the number of supports that undergo displacements (*NSD*), and (b) the joint number, and the magnitudes of the displacements, for each such support. As indicated in Fig. 7.11, the joint numbers of the

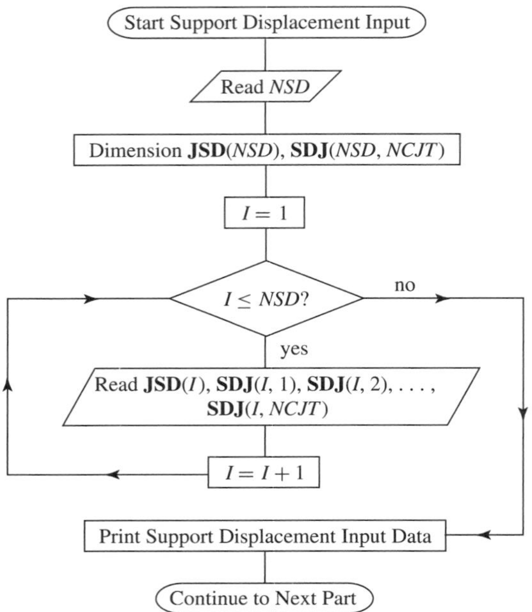

Fig. 7.11 *Flowchart for Reading and Storing Support Displacement Data*

supports that undergo displacements are stored in an integer vector **JSD** of order $NSD \times 1$, with their displacements stored in the corresponding rows of a real matrix **SDJ** of order $NSD \times NCJT$. For example, in the case of plane frames, the support displacement matrix **SDJ** would be of order $NSD \times 3$, with the support translations in the X and Y directions and the rotations being stored in the first, second, and third columns, respectively, of the matrix **SDJ**.

This subprogram for inputting support displacement data can be conveniently added as Part VIc in the computer programs for the analysis of plane frames (see Table 6.1) and beams (see Table 5.1); and it can be inserted between Parts VI and VII of the plane truss computer program (see Table 4.1).

Evaluation of Structure Equivalent Joint Loads Due to Support Displacements In this part of the program, the equivalent joint loads, or the negatives of the structure fixed-joint forces (i.e., $-\mathbf{P}_f$) due to support displacements, are added to the structure load vector **P**. A flowchart for constructing this part of the plane frame analysis program is presented in Fig. 7.12. As this flowchart indicates, the program essentially performs the following operations for each member of the structure.

1. First, the program determines whether the member under consideration, IM, is attached to a support that undergoes displacement, by comparing the member beginning and end joint numbers to those stored in the support displacement vector **JSD**. If the member is not attached to such a support, then no further action is taken for that member.

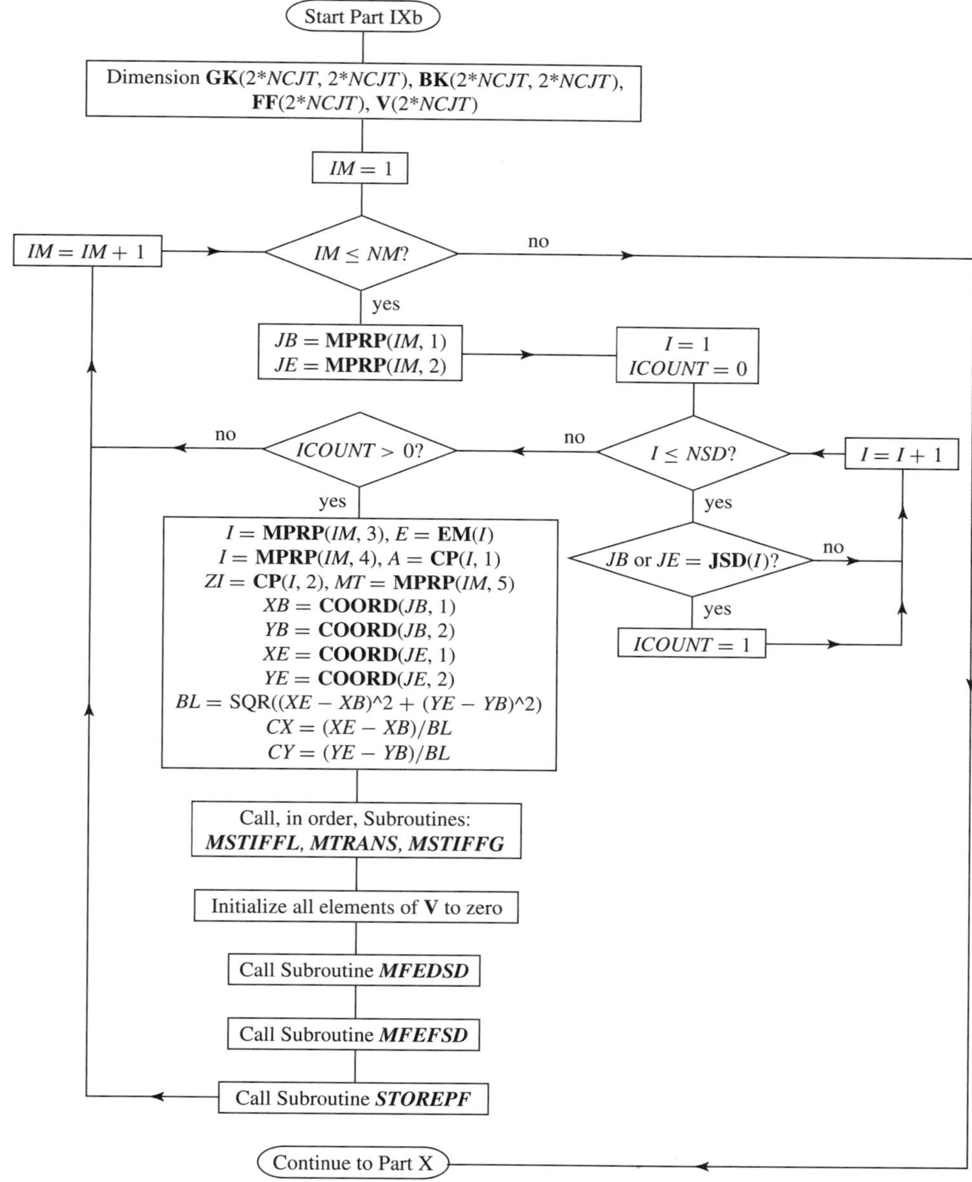

Fig. 7.12 *Flowchart for Generating Structure Equivalent Joint Load Vector Due to Support Displacements*

2. If the member is attached to a support that undergoes displacement(s), then its global stiffness matrix **GK** (= **K**) is obtained by calling, in order, the subroutines ***MSTIFFL*** (Fig. 7.4), ***MTRANS*** (Fig. 6.26), and ***MSTIFFG*** (Fig. 6.27).

3. Next, the program calls the subroutine ***MFEDSD*** to form the member global fixed-end displacement vector, **V** (= $\mathbf{v}_{fs}$), due to support displacements.

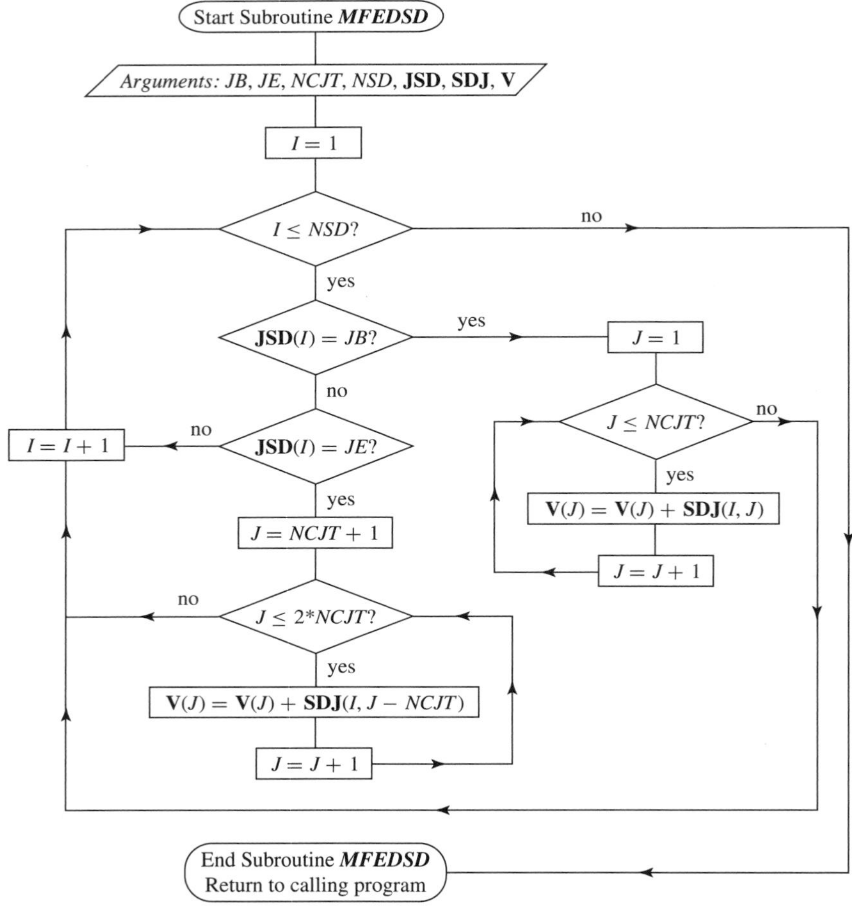

Fig. 7.13 *Flowchart of Subroutine **MFEDSD** for Determining Member Global Fixed-End Displacement Vector Due to Support Displacements*

As the flowchart in Fig. 7.13 indicates, this subroutine first checks the support displacement vector **JSD** to determine if the beginning joint of the member, *JB*, is a support joint subjected to displacements. If *JB* is such a joint, then the values of its displacements are read from the corresponding row of the support displacement matrix **SDJ**, and stored in the appropriate elements of the upper half of the member fixed-end displacement vector **V**. The process is then repeated for the end joint of the member, *JE*, with any corresponding support displacements being stored in the lower half of **V**.

4. Returning our attention to Fig. 7.12, we can see that the program then calls the subroutine ***MFEFSD*** (Fig. 7.14), which evaluates the member global fixed-end force vector due to support settlements **FF** ($= \mathbf{F}_{fs}$), using the relationship $\mathbf{F}_{fs} = \mathbf{K}\mathbf{v}_{fs}$ (Eq. (7.26)).

5. Finally, the negative values of the pertinent element of **FF** are added in their proper positions in the structural load vector **P**, using the subroutine

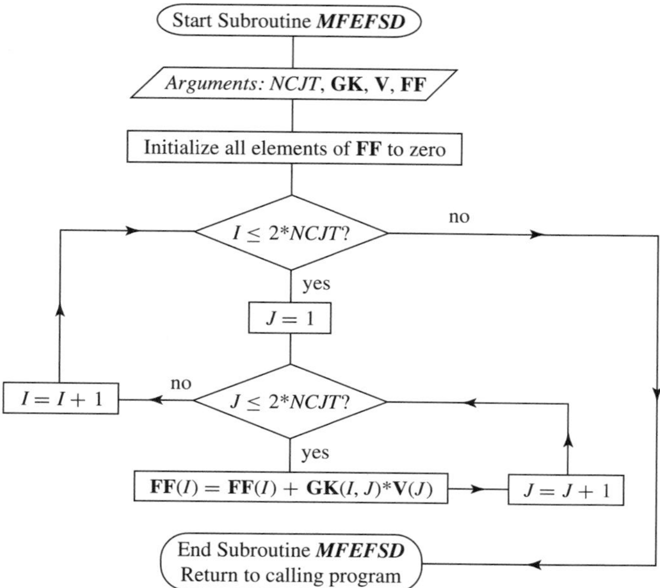

Fig. 7.14 *Flowchart of Subroutine **MFEFSD** for Determining Member Global Fixed-End Force Vector Due to Support Displacements*

STOREPF, which was developed in Chapter 6 (Fig. 6.30). When these operations have been completed for each member of the frame, the structure load vector **P** contains the equivalent joint loads (or the negatives of the structure fixed-joint forces) due to support displacements.

This subprogram, designated Part IXb in Fig. 7.12, can be conveniently inserted between Parts IX and X of the program for the analysis of plane frames (see Table 6.1). The flowcharts given in Figs. 7.12 through 7.14 can be used to develop the corresponding part of the beam analysis program (see Table 5.1), provided that: (a) the member global vectors **V** ($= \mathbf{v}_{fs}$) and **FF** ($= \mathbf{F}_{fs}$) are replaced by the local vectors **U** ($= \mathbf{u}_{fs}$) and **QF** ($= \mathbf{Q}_{fs}$), respectively; (b) the member local stiffness matrix **BK** ($= \mathbf{k}$) is used, instead of the global matrix **GK** ($= \mathbf{K}$), in subroutine ***MFEFSD***; and (c) the subroutine ***STOREPF*** developed in Chapter 5 (Fig. 5.28) is employed to store the negative elements of **QF** in the structure load vector **P**. The process of programming the corresponding part of the plane truss analysis program is essentially the same as discussed herein for the case of plane frames, except that the subroutine ***STOREPF*** (Fig. 6.30) should be copied from the plane frame analysis program and added to the plane truss program.

Calculation of Member Forces and Support Reactions (Part XII) Parts XII of the programs developed previously (see flowcharts in Figs. 6.31, 5.30, and 4.14) should be modified to include support displacements in the end displacement vectors of members attached to supports, before the end forces for such members are calculated. This can be achieved by simply calling the subroutine ***MFEDSD*** (Fig. 7.13) in these programs, to add the compatible support displacements to the member end displacement vectors. In the plane frame and truss analysis programs (Figs. 6.31 and 4.14, respectively), the subroutine ***MFEDSD*** should be called *after* the subroutine ***MDISPG*** has been used to form the member global end displacement vector **V** (= **v**) due to the joint displacements **d**, but *before* the subroutine ***MDISPL*** is called to evaluate the member local end displacement vector **U** (= **u**). In the program for the analysis of beams (Fig. 5.30), however, the subroutine ***MFEDSD*** should be called *after* the subroutine ***MDISPL*** has been used to form the member end displacement vector **U** (= **u**) from the joint displacements **d**, but *before* the member end forces are calculated using the subroutine ***MFORCEL***. Furthermore, as discussed previously, before it can be used in the beam analysis program, the subroutine ***MFEDSD*** (as given in Fig. 7.13) must be modified to replace **V** with **U**.

It may be of interest to note that the program for the analysis of plane frames, which was initially developed in Chapter 6 and has been extended in this chapter, is quite general, in the sense that it can also be used to analyze beams and plane trusses. When analyzing a truss using the frame analysis program, all of the truss members are modeled as hinged at both ends with $MT = 3$, and all the joints of the truss are modeled as hinged joints restrained against rotations by imaginary clamps.

7.5 TEMPERATURE CHANGES AND FABRICATION ERRORS

Like support displacements, changes in temperature and small fabrication errors can cause considerable stresses in statically indeterminate structures, which must be taken into account in their designs. However, unlike support displacements, which are generally specified with reference to the global coordinate systems of structures, temperature changes and fabrication errors, like member loads, are usually defined relative to the local coordinate systems of members. Therefore, the stiffness methods developed previously for the analysis of structures subjected to member loads, can be used without modifications to determine the structural responses to temperature changes and fabrication errors. The only difference is that the fixed-end forces, which develop in members due to temperature changes and fabrication errors, must now be included in the member local fixed-end force vectors **QF**.

In this section, we derive the expressions for the fixed-end forces that develop in the members of framed structures due to temperature changes and two common types of fabrication errors. The application of these fixed-end force expressions in analysis is then illustrated by some examples.

Member Fixed-End Forces Due to Temperature Changes

We can develop the desired relationships by first determining the displacements caused by temperature changes at the ends of members that are free to

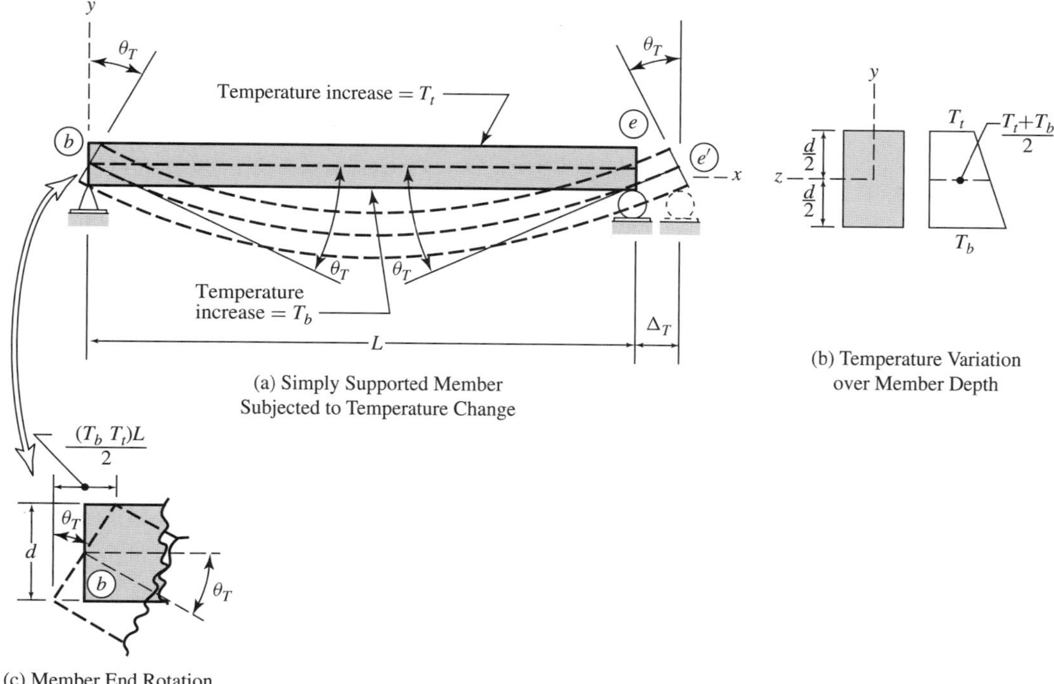

(a) Simply Supported Member Subjected to Temperature Change

(b) Temperature Variation over Member Depth

(c) Member End Rotation

Fig. 7.15

deform. The fixed-end forces required to suppress these member end displacements can then be obtained, using the member stiffness matrices.

To examine member end displacements due to temperature changes, let us consider an arbitrary simply supported member of a plane frame, as shown in Fig. 7.15(a). Now, assume that the member is heated so that the temperature increase of its top surface is T_t and that of its bottom surface is T_b, with the temperature increase varying linearly between T_t and T_b over the depth d of the member cross-section, as shown in Fig. 7.15(b). Note that the temperature does not vary along the length of the member. Because the member is simply supported (so that it is statically determinate), it is free to expand in the longitudinal direction. If we assume that the member cross-section is symmetric about the xz plane (Fig. 7.15(b)) containing its centroidal axis, then the temperature increase at the level of the centroidal axis (i.e., at the distance $d/2$ from the top or bottom of the member) would be $(T_b + T_t)/2$. This temperature increase causes the member's centroidal axis to elongate by an amount Δ_T:

$$\Delta_T = \alpha\left(\frac{T_b + T_t}{2}\right)L \tag{7.29}$$

in which α denotes the coefficient of thermal expansion.

In addition to the axial deformation Δ_T, the member also undergoes bending as its top and bottom surfaces elongate by different amounts (because they are subjected to different temperature increases). For example, as depicted in Fig. 7.15, if $T_b > T_t$, then the member bends concave upward, causing the cross-sections at its ends b and e to rotate inward, as shown. Since the temperature increase is uniform along the member's length, the rotations of its two end cross-sections must be equal in magnitude. From Fig. 7.15, we can see that these member end rotations can be related to the temperature change by dividing one-half of the difference between the elongations of the bottom and top fibers of the member, by its depth. Thus,

$$\theta_T = \frac{\alpha(T_b - T_t)L}{2d} \tag{7.30}$$

in which θ_T represents the magnitude of the rotations of the member end cross-sections which, in turn, equal the slopes of the elastic curve of the member at its ends, as shown in Figs. 7.15(a) and (c).

Using the sign convention for member local end displacements established in Chapter 6, we can express the local end displacement vector $\mathbf{u}_T$ for the simply supported member, due to the temperature change, as

$$\mathbf{u}_T = \begin{bmatrix} 0 \\ 0 \\ -\theta_T \\ \Delta_T \\ 0 \\ \theta_T \end{bmatrix} \begin{matrix} 1 \\ 2 \\ 3 \\ 4 \\ 5 \\ 6 \end{matrix} = \frac{\alpha L}{2} \begin{bmatrix} 0 \\ 0 \\ -(T_b - T_t)/d \\ T_b + T_t \\ 0 \\ (T_b - T_t)/d \end{bmatrix} \tag{7.31}$$

in which the rotation of the beginning, b, of the member is negative, because it is clockwise, whereas the rotation of the member end, e, which has a counterclockwise sense, is positive.

The member fixed-end forces necessary to suppress its end displacements $\mathbf{u}_T$ can now be established by applying the principle of superposition, as illustrated in Fig. 7.16. Figure 7.16(a) shows a fixed member of a plane frame subjected to a temperature increase, causing fixed-end forces to develop at its ends. In Fig. 7.16(b), the corresponding simply supported member is subjected to the same temperature change, causing the displacements $\mathbf{u}_T$ (Eq. (7.31)) at its ends, but no end forces; in Fig. 7.16(c), the simply supported member is subjected to the same fixed-end forces that develop in the fixed member of Fig. 7.16(a), but no temperature change. By comparing Figs. 7.16(a) through (c), we realize that the response of the fixed member of Fig. 7.16(a) must equal the superposition of the responses of the two simply supported members of Figs. 7.16(b) and (c). Therefore, since the end displacements of the fixed member due to the temperature change are 0 (Fig. 7.16(a)), its fixed-end forces, when applied to the simply supported beam (Fig. 7.16(c)), must cause the end displacements, $-\mathbf{u}_T$, that are equal in magnitude but opposite in direction to those due to the temperature change, $\mathbf{u}_T$ (Fig. 7.16(b)). The forces that can cause the end displacements $-\mathbf{u}_T$ in the simply supported member can be conveniently obtained by premultiplying the negative of the $\mathbf{u}_T$ vector given in

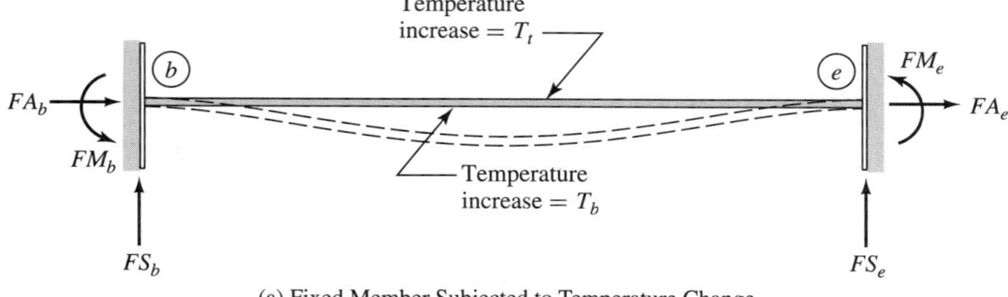

(a) Fixed Member Subjected to Temperature Change and Fixed-End Forces (No End Displacements)

=

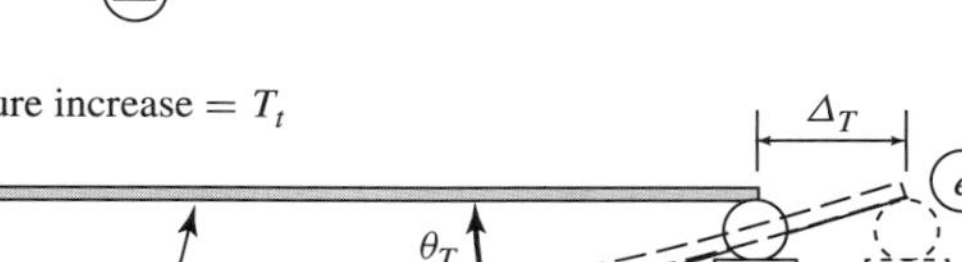

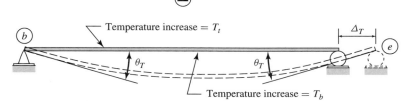

(b) Simply Supported Member Subjected to Temperature Change Only (End Displacements = $\mathbf{u}_T$)

+

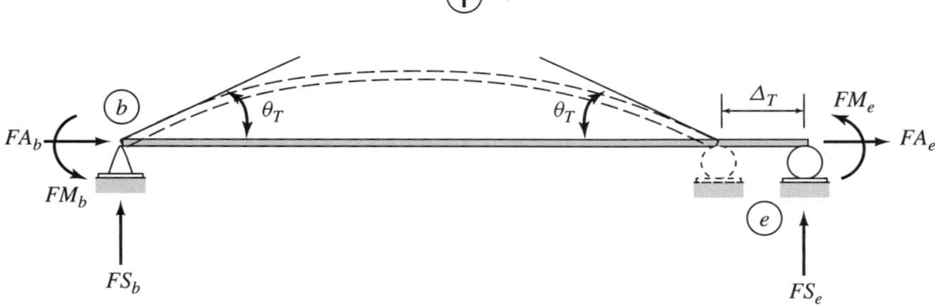

(c) Simply Supported Member Subjected to Fixed-End Forces Only (End Displacements = $-\mathbf{u}_T$)

Fig. 7.16

Eq. (7.31), by the member local stiffness matrix $\mathbf{k}$ (Eq. (6.6)). Thus,

$$\begin{bmatrix} FA_b \\ FS_b \\ FM_b \\ FA_e \\ FS_e \\ FM_e \end{bmatrix} = \frac{EI}{L^3} \begin{bmatrix} \frac{AL^2}{I} & 0 & 0 & -\frac{AL^2}{I} & 0 & 0 \\ 0 & 12 & 6L & 0 & -12 & 6L \\ 0 & 6L & 4L^2 & 0 & -6L & 2L^2 \\ -\frac{AL^2}{I} & 0 & 0 & \frac{AL^2}{I} & 0 & 0 \\ 0 & -12 & -6L & 0 & 12 & -6L \\ 0 & 6L & 2L^2 & 0 & -6L & 4L^2 \end{bmatrix} \left(-\frac{\alpha L}{2}\right) \begin{bmatrix} 0 \\ 0 \\ -(T_b - T_t)/d \\ T_b + T_t \\ 0 \\ (T_b - T_t)/d \end{bmatrix}$$

From which we obtain

$$\begin{bmatrix} FA_b \\ FS_b \\ FM_b \\ FA_e \\ FS_e \\ FM_e \end{bmatrix} = E\alpha \begin{bmatrix} A(T_b + T_t)/2 \\ 0 \\ I(T_b - T_t)/d \\ -A(T_b + T_t)/2 \\ 0 \\ -I(T_b - T_t)/d \end{bmatrix} \tag{7.32}$$

Thus, the fixed-end forces for the members of plane frames can be expressed as

$$\boxed{\begin{aligned} FA_b &= -FA_e = EA\alpha\left(\frac{T_b + T_t}{2}\right) \\ FM_b &= -FM_e = EI\alpha\left(\frac{T_b - T_t}{d}\right) \end{aligned}} \tag{7.33}$$

The expressions for the fixed-end moments, given in Eqs. (7.33), can also be used for the members of beams. However, as the beam members are free to expand axially, their fixed-end axial forces are 0 (i.e., $FA_b = FA_e = 0$). Similarly, the expressions for the fixed-end axial forces, given in Eqs. (7.33), can be used for the members of trusses; however, the fixed-end moments must now be set equal to 0 (i.e., $FM_b = FM_e = 0$) in Eqs. (7.33), because the ends of truss members are free to rotate.

The fixed-end force expressions given in Eqs. (7.33) are based on a linearly varying temperature change over the depth of the member cross-section. If the member is subjected to a uniform temperature increase, T_u, over its depth, then the corresponding expressions for fixed-end forces can be obtained by simply substituting $T_b = T_t = T_u$ into Eqs. (7.33). This yields

$$\boxed{FA_b = -FA_e = EA\alpha T_u} \tag{7.34}$$

As Eqs. (7.34) indicate, the member fixed-end moments would be 0 in the case of a uniform temperature change, because such a temperature change has no tendency to bend the member, but only to cause axial deformation. Equations (7.34) can be used to determine the fixed-end forces for the members of plane frames and trusses subjected to uniform temperature changes. As stated previously, the members of beams are free to expand in their axial directions; therefore, a uniform temperature change does not cause any fixed-end forces in such members.

Member Fixed-End Forces Due to Fabrication Errors

In structural analysis terminology, *fabrication error* is used to refer to a small initial deformation of a member in its unstressed state. The expressions for

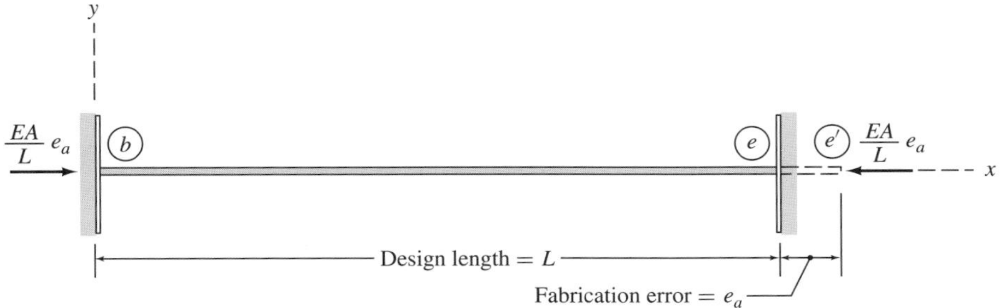

Fig. 7.17

member fixed-end forces due to fabrication errors can be derived in a manner similar to that for the case of temperature changes. In the following paragraphs, we develop the fixed-end force expressions for two common types of fabrication errors.

Errors in Initial Member Length Consider a member of a plane frame with a specified design length L. Now, suppose that the member is fabricated so that its initial unstressed length is longer than the specified length L by an amount e_a, as shown in Fig. 7.17. As the distance between the fixed supports is L, the supports must exert a compressive axial force of magnitude EAe_a/L on the member to reduce its length from $L + e_a$ to L, so that it can fit between the supports. Thus, the fixed-end forces that develop in the member due to its fabricated length being too long by an amount e_a are

$$FA_b = -FA_e = \frac{EA}{L} e_a \tag{7.35}$$

Equations (7.35) can also be used to obtain fixed-end forces for the members of trusses due to fabrication errors in their lengths.

Errors in Initial Member Straightness Another type of fabrication error commonly encountered in structural design involves a lack of initial straightness of the members of beams and plane frames. Figure 7.18(a) on the next page shows such a member of a beam, which somehow has been fabricated with an initial bend, causing a small deflection e_b at a distance l_1 from the member's left end. To determine the fixed-end forces for this member, we first express the member end rotations θ_b and θ_e in terms of the fabrication error e_b, as (see Fig. 7.18(a))

$$\theta_b = \frac{e_b}{l_1} \quad \text{and} \quad \theta_e = \frac{e_b}{l_2} \tag{7.36}$$

Using the sign convention established for beam members in Chapter 5, we write the local end displacement vector $\mathbf{u}_e$ for the member, due to the fabrication

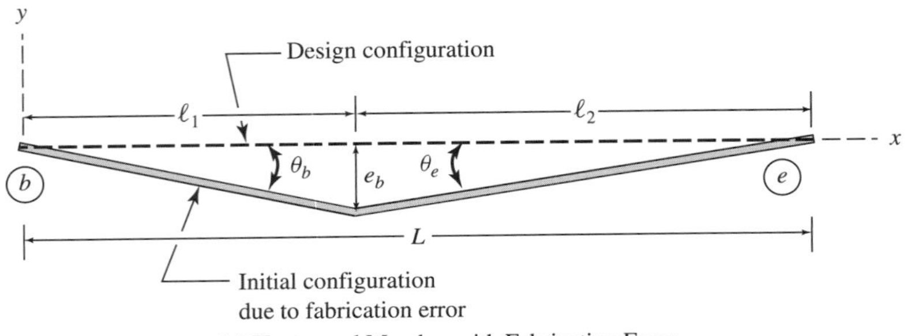

(a) Unstressed Member with Fabrication Error

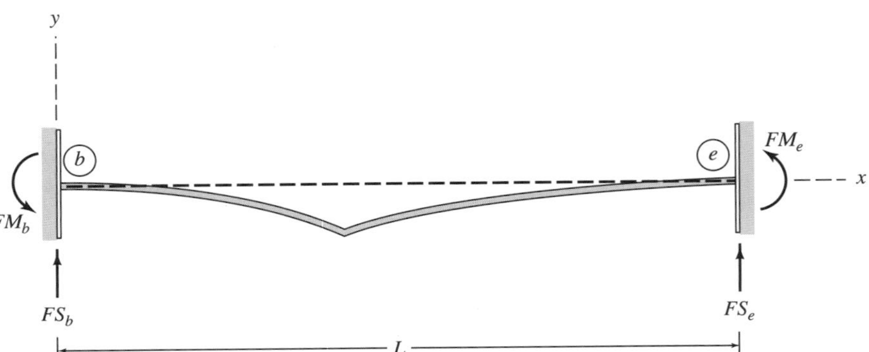

(b) Fixed Member with Fabrication Error

Fig. 7.18

error, as

$$\mathbf{u}_e = \begin{bmatrix} 0 \\ -\theta_b \\ 0 \\ \theta_e \end{bmatrix} \begin{matrix} 1 \\ 2 \\ 3 \\ 4 \end{matrix} = e_b \begin{bmatrix} 0 \\ -1/l_1 \\ 0 \\ 1/l_2 \end{bmatrix} \tag{7.37}$$

in which the rotation of the beginning, b, of the member is considered to be negative, because it has a clockwise sense. The member fixed-end forces (Fig. 7.18(b)) necessary to suppress the end displacements $\mathbf{u}_e$ can now be determined by premultiplying the negative of the $\mathbf{u}_e$ vector by the member local stiffness matrix $\mathbf{k}$ (Eq. (5.53)). Thus,

$$\begin{bmatrix} FS_b \\ FM_b \\ FS_e \\ FM_e \end{bmatrix} = \frac{EI}{L^3} \begin{bmatrix} 12 & 6L & -12 & 6L \\ 6L & 4L^2 & -6L & 2L^2 \\ -12 & -6L & 12 & -6L \\ 6L & 2L^2 & -6L & 4L^2 \end{bmatrix} (-e_b) \begin{bmatrix} 0 \\ -1/l_1 \\ 0 \\ 1/l_2 \end{bmatrix}$$

$$= \frac{2EI\, e_b}{L^2 l_1 l_2} \begin{bmatrix} 3(l_2 - l_1) \\ L(2l_2 - l_1) \\ 3(l_1 - l_2) \\ L(l_2 - 2l_1) \end{bmatrix} \tag{7.38}$$

Therefore, the fixed-end forces for the members of beams are:

$$\boxed{\begin{aligned} FS_b &= -FS_e = \frac{6EI\,e_b}{L^2 l_1 l_2}(l_2 - l_1) \\ FM_b &= \frac{2EI\,e_b}{L l_1 l_2}(2l_2 - l_1) \\ FM_e &= \frac{2EI\,e_b}{L l_1 l_2}(l_2 - 2l_1) \end{aligned}} \tag{7.39}$$

As the fabrication error e_b is assumed to be small, it does not cause any axial deformation of the member; therefore, no axial force develops in the fixed member (i.e., $FA_b = FA_e = 0$). Thus, the expressions for the fixed-end forces given in Eq. (7.39) can also be used for the members of plane frames.

Procedure for Analysis

As stated at the beginning of this section, the procedures for the analysis of beams and plane frames, including the effects of temperature changes and fabrication errors, remain the same as developed in Chapters 5 and 6, respectively—provided that the member fixed-end forces caused by the temperature changes and fabrication errors are now included in the member local fixed-end force vectors $\mathbf{Q}_f$. In the case of plane trusses, however, the member and structure stiffness relationships must now be modified, to include the effects of temperature changes and fabrication errors, as follows: (a) the member local stiffness relationship given in Eq. (3.7) should be modified to $\mathbf{Q} = \mathbf{ku} + \mathbf{Q}_f$, (b) the member global stiffness relationship (Eq. (3.71)) now becomes $\mathbf{F} = \mathbf{Kv} + \mathbf{F}_f$, and (c) the structure stiffness relationship (Eq. (3.89)) should be updated to include the structure fixed-joint forces as $\mathbf{P} - \mathbf{P}_f = \mathbf{Sd}$. The structure fixed-joint force vector $\mathbf{P}_f$ can be generated using the member code number technique as discussed in Chapter 6 for the case of plane frames.

EXAMPLE 7.5 Determine the joint displacements, member axial forces, and support reactions for the plane truss shown in Fig. 7.19(a) on the next page, due to the combined effect of the following: (a) the joint loads shown in the figure, (b) a temperature drop of 30° F in member 1, and (c) the fabricated length of member 3 being $\frac{1}{8}$ in. too short. Use the matrix stiffness method.

SOLUTION This truss was analyzed in Example 3.8 for joint loads only, and in Example 7.2 for a support displacement.

Analytical Model: See Fig. 7.19(b). The analytical model used herein is the same as used in Examples 3.8 and 7.2.

Structure Stiffness Matrix: From Example 3.8,

$$\mathbf{S} = \begin{bmatrix} 696 & 0 \\ 0 & 2{,}143.6 \end{bmatrix} \text{ k/in.} \tag{1}$$

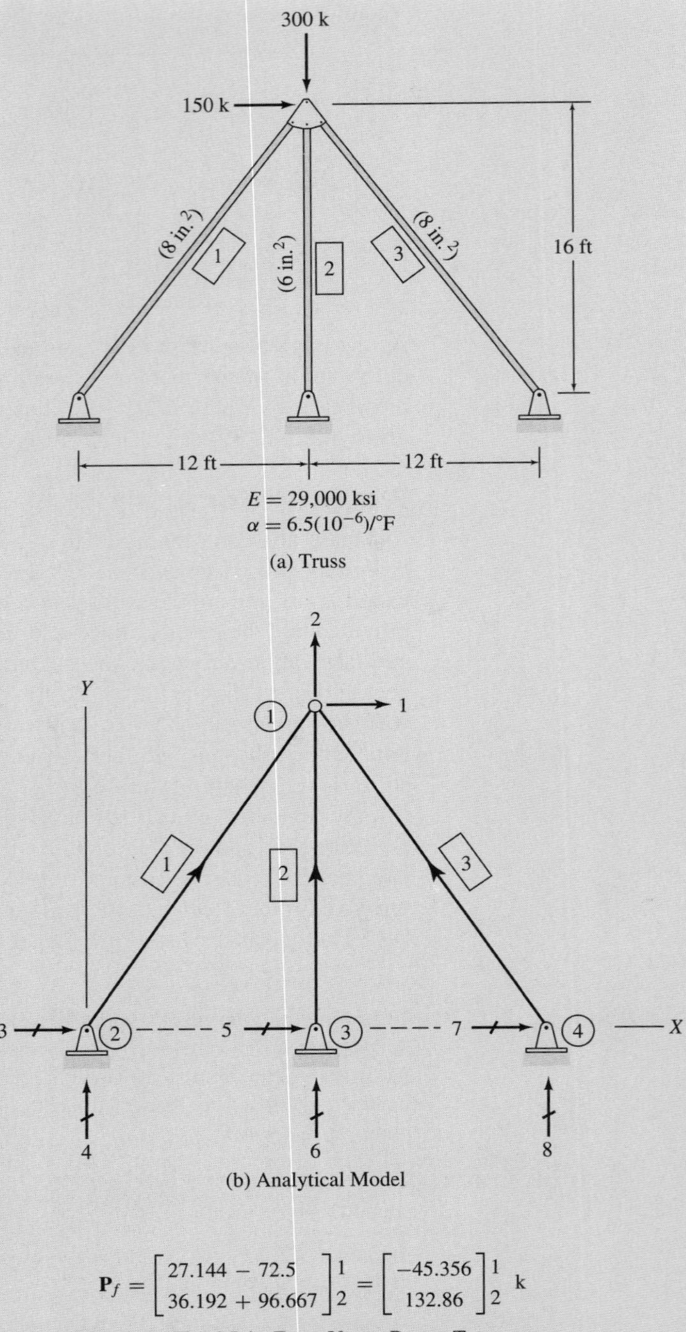

$$\mathbf{P}_f = \begin{bmatrix} 27.144 - 72.5 \\ 36.192 + 96.667 \end{bmatrix} \begin{matrix} 1 \\ 2 \end{matrix} = \begin{bmatrix} -45.356 \\ 132.86 \end{bmatrix} \begin{matrix} 1 \\ 2 \end{matrix} \text{ k}$$

(c) Structure Fixed-Joint Force Vector Due to Temperature Changes and Fabrication Errors

Fig. 7.19

$$\mathbf{R} = \begin{bmatrix} -31.125 \\ -41.5 \\ 0 \\ 183 \\ -118.87 \\ 158.5 \end{bmatrix} \begin{matrix} 3 \\ 4 \\ 5 \\ 6 \\ 7 \\ 8 \end{matrix} \text{ k}$$

(d) Support Reaction Vector

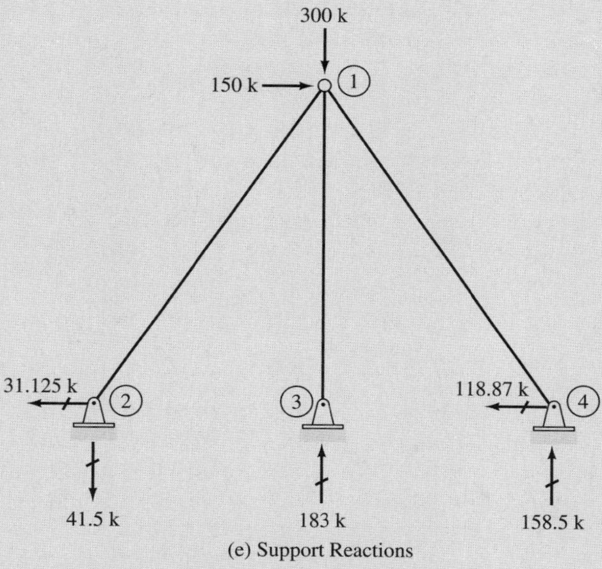

(e) Support Reactions

Fig. 7.19 *(continued)*

Joint Load Vector: From Example 3.8,

$$\mathbf{P} = \begin{bmatrix} 150 \\ -300 \end{bmatrix} \text{ k} \tag{2}$$

Structure Fixed-Joint Force Vector Due to Temperature Changes and Fabrication Errors:

Member 1 By substituting $E = 29{,}000$ ksi, $A = 8$ in.^{2}, $\alpha = 6.5(10^{-6})/^\circ$ F, and $T_u = -30^\circ$ F into Eqs. (7.34), we evaluate the member fixed-end forces, due to the specified temperature change, as

$$FA_b = -FA_e = -45.24 \text{ k}$$

Thus, the local fixed-end force vector for member 1 can be expressed as

$$\mathbf{Q}_{f1} = \begin{bmatrix} FA_b \\ FS_b \\ FA_e \\ FS_e \end{bmatrix} = \begin{bmatrix} -45.24 \\ 0 \\ 45.24 \\ 0 \end{bmatrix} \text{ k} \tag{3}$$

Next, we obtain the global fixed-end force vector for this member by applying the transformation relationship $\mathbf{F}_f = \mathbf{T}^T\mathbf{Q}_f$, while using the transformation matrix $\mathbf{T}_1$ from

Example 3.8. Thus,

$$\mathbf{F}_{f1} = \mathbf{T}_1^T \mathbf{Q}_{f1} = \begin{bmatrix} 0.6 & -0.8 & 0 & 0 \\ 0.8 & 0.6 & 0 & 0 \\ 0 & 0 & 0.6 & -0.8 \\ 0 & 0 & 0.8 & 0.6 \end{bmatrix} \begin{bmatrix} -45.24 \\ 0 \\ 45.24 \\ 0 \end{bmatrix} = \begin{bmatrix} -27.144 \\ -36.192 \\ \hdashline 27.144 \\ 36.192 \end{bmatrix} \begin{matrix} 3 \\ 4 \\ 1 \\ 2 \end{matrix} \text{ k}$$

From the member code numbers 3, 4, 1, 2, which are written on the right side of $\mathbf{F}_{f1}$, we realize that the elements in the third and fourth rows of $\mathbf{F}_{f1}$ should be stored in rows 1 and 2, respectively, of the 2×1 structure fixed-joint force vector $\mathbf{P}_f$, as shown in Fig. 7.19(c).

Member 3 By substituting $e_a = -\frac{1}{8}$ in. into Eq. (7.35), we obtain

$$FA_b = -FA_e = \frac{29{,}000(8)}{20(12)}\left(-\frac{1}{8}\right) = -120.83 \text{ k}$$

Thus,

$$\mathbf{Q}_{f3} = \begin{bmatrix} FA_b \\ FS_b \\ FA_e \\ FS_e \end{bmatrix} = \begin{bmatrix} -120.83 \\ 0 \\ 120.83 \\ 0 \end{bmatrix} \text{ k} \quad \textbf{(4)}$$

Using $\mathbf{T}_3$ from Example 7.2, we calculate

$$\mathbf{F}_{f3} = \mathbf{T}_3^T \mathbf{Q}_{f3} = \begin{bmatrix} -0.6 & -0.8 & 0 & 0 \\ 0.8 & -0.6 & 0 & 0 \\ 0 & 0 & -0.6 & -0.8 \\ 0 & 0 & 0.8 & -0.6 \end{bmatrix} \begin{bmatrix} -120.83 \\ 0 \\ 120.83 \\ 0 \end{bmatrix} = \begin{bmatrix} 72.5 \\ -96.667 \\ \hdashline -72.5 \\ 96.667 \end{bmatrix} \begin{matrix} 7 \\ 8 \\ 1 \\ 2 \end{matrix} \text{ k}$$

The relevant elements of $\mathbf{F}_{f3}$ are stored in $\mathbf{P}_f$ using the member code numbers. The completed structure fixed-joint force vector $\mathbf{P}_f$, due to temperature change and fabrication error, is given in Fig. 7.19(c).

Joint Displacements: By substituting $\mathbf{P}$ (Eq. (2)), $\mathbf{P}_f$ (Fig. 7.19(c)), and $\mathbf{S}$ (Eq. (1)) into the structure stiffness relationship, we write

$$\mathbf{P} - \mathbf{P}_f = \mathbf{S}\mathbf{d}$$

$$\begin{bmatrix} 150 \\ -300 \end{bmatrix} - \begin{bmatrix} -45.356 \\ 132.86 \end{bmatrix} = \begin{bmatrix} 195.36 \\ -432.86 \end{bmatrix} = \begin{bmatrix} 696 & 0 \\ 0 & 2{,}143.6 \end{bmatrix} \begin{bmatrix} d_1 \\ d_2 \end{bmatrix}$$

Solving the foregoing equations,

$$\mathbf{d} = \begin{bmatrix} 0.28068 \\ -0.20193 \end{bmatrix} \begin{matrix} 1 \\ 2 \end{matrix} \text{ in.} \quad \textbf{Ans}$$

Member End Displacements and End Forces:

Member 1

$$\mathbf{v}_1 = \begin{bmatrix} 0 \\ 0 \\ d_1 \\ d_2 \end{bmatrix} \begin{matrix} 3 \\ 4 \\ 1 \\ 2 \end{matrix} = \begin{bmatrix} 0 \\ 0 \\ 0.28068 \\ -0.20193 \end{bmatrix} \text{ in.}$$

$$\mathbf{u}_1 = \mathbf{T}_1\mathbf{v}_1 = \begin{bmatrix} 0 \\ 0 \\ 0.006864 \\ -0.3457 \end{bmatrix} \text{ in.}$$

Next, we calculate the member local end forces by applying the member stiffness relationship $\mathbf{Q} = \mathbf{ku} + \mathbf{Q}_f$. Thus, using $\mathbf{k}_1$ from Example 3.8, and $\mathbf{Q}_{f1}$ from Eq. (3), we obtain

$$\mathbf{Q}_1 = \mathbf{k}_1\mathbf{u}_1 + \mathbf{Q}_{f1} = \begin{bmatrix} 966.67 & 0 & -966.67 & 0 \\ 0 & 0 & 0 & 0 \\ -966.67 & 0 & 966.67 & 0 \\ 0 & 0 & 0 & 0 \end{bmatrix} \begin{bmatrix} 0 \\ 0 \\ 0.006864 \\ -0.3457 \end{bmatrix} + \begin{bmatrix} -45.24 \\ 0 \\ 45.24 \\ 0 \end{bmatrix}$$

from which,

$$\mathbf{Q}_1 = \begin{bmatrix} -51.875 \\ 0 \\ 51.875 \\ 0 \end{bmatrix} \text{ k}$$

Thus,

$$Q_{a1} = -51.875 \text{ k} = 51.875 \text{ k (T)}$$ **Ans**

$$\mathbf{F}_1 = \mathbf{T}_1^T\mathbf{Q}_1 = \begin{bmatrix} -31.125 \\ -41.5 \\ \hdashline 31.125 \\ 41.5 \end{bmatrix} \begin{matrix} 3 \\ 4 \\ 1 \\ 2 \end{matrix} \text{ k}$$

The pertinent elements of $\mathbf{F}_1$ are stored in the support reaction vector $\mathbf{R}$ in Fig. 7.19(d).

Member 2

$$\mathbf{v}_2 = \begin{bmatrix} 0 \\ 0 \\ d_1 \\ d_2 \end{bmatrix} \begin{matrix} 5 \\ 6 \\ 1 \\ 2 \end{matrix} = \begin{bmatrix} 0 \\ 0 \\ 0.28068 \\ -0.20193 \end{bmatrix} \text{ in.}$$

Using $\mathbf{T}_2$ from Example 3.8, we obtain

$$\mathbf{u}_2 = \mathbf{T}_2\mathbf{v}_2 = \begin{bmatrix} 0 & 1 & 0 & 0 \\ -1 & 0 & 0 & 0 \\ 0 & 0 & 0 & 1 \\ 0 & 0 & -1 & 0 \end{bmatrix} \begin{bmatrix} 0 \\ 0 \\ 0.28068 \\ -0.20193 \end{bmatrix} = \begin{bmatrix} 0 \\ 0 \\ -0.20193 \\ -0.28068 \end{bmatrix} \text{ in.}$$

With $\mathbf{Q}_{f2} = \mathbf{0}$ and $\mathbf{k}_2$ obtained from Example 3.8, we determine the member local end forces to be

$$\mathbf{Q}_2 = \mathbf{k}_2\mathbf{u}_2 = \begin{bmatrix} 906.25 & 0 & -906.25 & 0 \\ 0 & 0 & 0 & 0 \\ -906.25 & 0 & 906.25 & 0 \\ 0 & 0 & 0 & 0 \end{bmatrix} \begin{bmatrix} 0 \\ 0 \\ -0.20193 \\ -0.28068 \end{bmatrix} = \begin{bmatrix} 183 \\ 0 \\ -183 \\ 0 \end{bmatrix} \text{ k}$$

$$Q_{a2} = 183 \text{ k (C)}$$ **Ans**

$$\mathbf{F}_2 = \mathbf{T}_2^T \mathbf{Q}_2 = \begin{bmatrix} 0 \\ 183 \\ \hdashline 0 \\ -183 \end{bmatrix} \begin{matrix} 5 \\ 6 \\ 1 \\ 2 \end{matrix}$$

Member 3

$$\mathbf{v}_3 = \begin{bmatrix} 0 \\ 0 \\ d_1 \\ d_2 \end{bmatrix} \begin{matrix} 7 \\ 8 \\ 1 \\ 2 \end{matrix} = \begin{bmatrix} 0 \\ 0 \\ 0.28068 \\ -0.20193 \end{bmatrix} \text{in.}$$

$$\mathbf{u}_3 = \mathbf{T}_3 \mathbf{v}_3 = \begin{bmatrix} 0 \\ 0 \\ -0.32995 \\ -0.10339 \end{bmatrix} \text{in.}$$

Using $\mathbf{k}_3$ from Example 7.2 and $\mathbf{Q}_{f3}$ from Eq. (4), we calculate

$$\mathbf{Q}_3 = \mathbf{k}_3 \mathbf{u}_3 + \mathbf{Q}_{f3} = \begin{bmatrix} 966.67 & 0 & -966.67 & 0 \\ 0 & 0 & 0 & 0 \\ -966.67 & 0 & 966.67 & 0 \\ 0 & 0 & 0 & 0 \end{bmatrix} \begin{bmatrix} 0 \\ 0 \\ -0.32995 \\ -0.10339 \end{bmatrix}$$

$$+ \begin{bmatrix} -120.83 \\ 0 \\ 120.83 \\ 0 \end{bmatrix} = \begin{bmatrix} 198.12 \\ 0 \\ -198.12 \\ 0 \end{bmatrix} \text{k}$$

$Q_{a3} = 198.12$ k (C) **Ans**

$$\mathbf{F}_3 = \mathbf{T}_3^T \mathbf{Q}_3 = \begin{bmatrix} -118.87 \\ 158.5 \\ \hdashline 118.87 \\ -158.5 \end{bmatrix} \begin{matrix} 7 \\ 8 \\ 1 \\ 2 \end{matrix}$$

Support Reactions: See Figs. 7.19(d) and (e). **Ans**

EXAMPLE 7.6 Determine the joint displacements, member end forces, and support reactions for the three-span continuous beam shown in Fig. 7.20(a), due to a temperature increase of 10° C at the top surface and 70° C at the bottom surface, of all spans. The temperature increase varies linearly over the depth $d = 600$ mm of the beam cross-section. Use the matrix stiffness method.

SOLUTION This beam was analyzed in Example 7.3 for member loads and support settlements.

Analytical Model: See Fig. 7.20(b). The analytical model used herein is the same as used in Example 7.3.

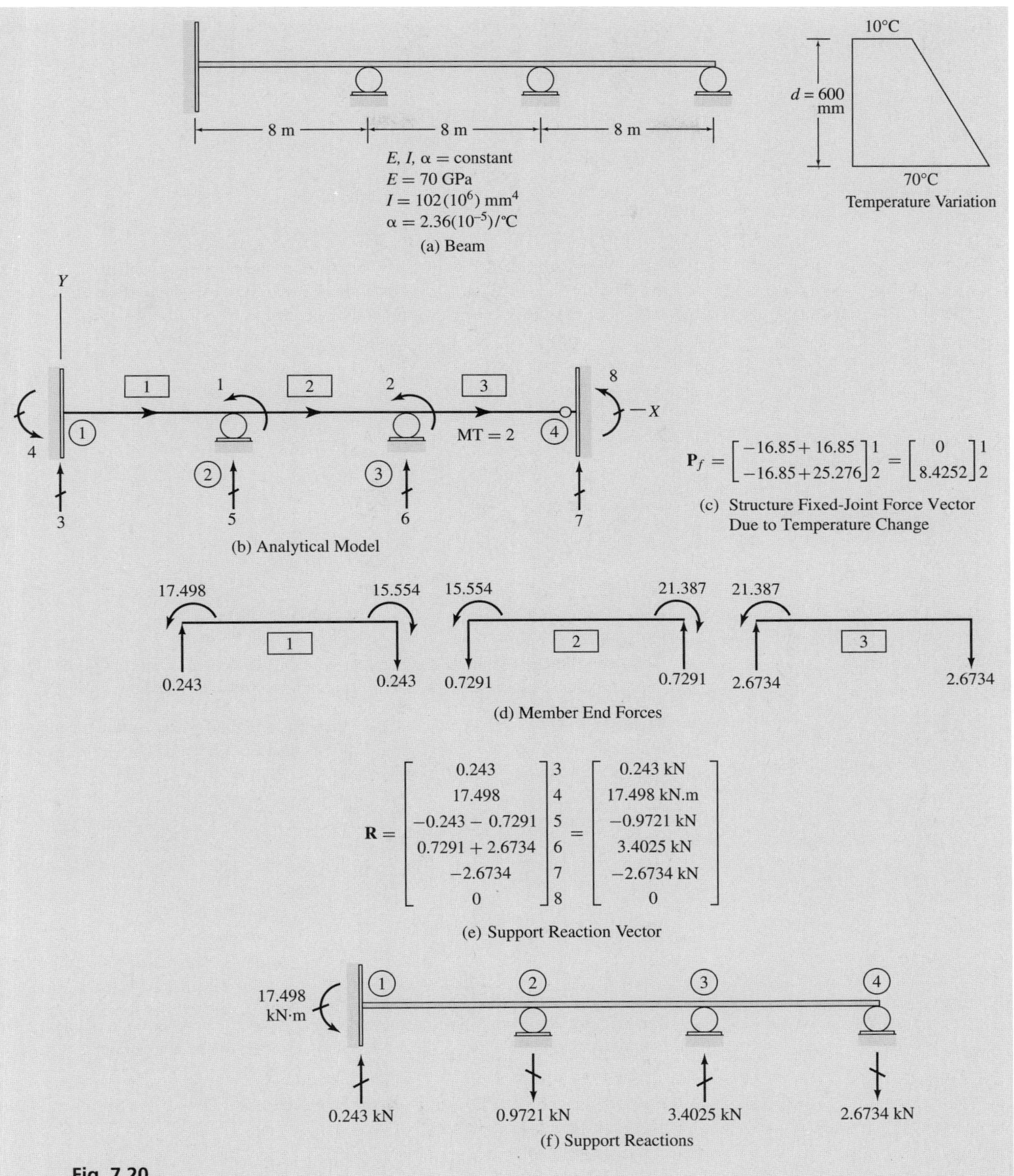

10°C
d = 600 mm
70°C
Temperature Variation
8 m
8 m
8 m
E, I, α = constant
E = 70 GPa
I = 102(10^6) mm^4
α = 2.36(10^-5)/°C
(a) Beam
Y
X
MT = 2
(b) Analytical Model
$\mathbf{P}_f = \begin{bmatrix} -16.85 + 16.85 \\ -16.85 + 25.276 \end{bmatrix} \begin{matrix} 1 \\ 2 \end{matrix} = \begin{bmatrix} 0 \\ 8.4252 \end{bmatrix} \begin{matrix} 1 \\ 2 \end{matrix}$
(c) Structure Fixed-Joint Force Vector Due to Temperature Change
17.498
15.554
15.554
21.387
21.387
0.243
0.243
0.7291
0.7291
2.6734
2.6734
(d) Member End Forces
$\mathbf{R} = \begin{bmatrix} 0.243 \\ 17.498 \\ -0.243 - 0.7291 \\ 0.7291 + 2.6734 \\ -2.6734 \\ 0 \end{bmatrix} \begin{matrix} 3 \\ 4 \\ 5 \\ 6 \\ 7 \\ 8 \end{matrix} = \begin{bmatrix} 0.243 \text{ kN} \\ 17.498 \text{ kN.m} \\ -0.9721 \text{ kN} \\ 3.4025 \text{ kN} \\ -2.6734 \text{ kN} \\ 0 \end{bmatrix}$
(e) Support Reaction Vector
17.498 kN·m
0.243 kN
0.9721 kN
3.4025 kN
2.6734 kN
(f) Support Reactions

Fig. 7.20

Structure Stiffness Matrix: As determined in Example 7.3, the structure stiffness matrix for the beam, in units of kN and meters, is

$$\mathbf{S} = \begin{bmatrix} 7{,}140 & 1{,}785 \\ 1{,}785 & 6{,}247.5 \end{bmatrix} \quad \textbf{(1)}$$

Joint Load Vector:

$$\mathbf{P} = \mathbf{0} \quad \textbf{(2)}$$

Structure Fixed-Joint Force Vector Due to Temperature Change:

Members 1 and 2 ($MT = 0$) By substituting the numerical values of E, I, L, α, $d = 0.6$ m, $T_t = 10°$C and $T_b = 70°$C into Eq. (7.33), we evaluate the member fixed-end moments due to the given temperature change as

$$FM_b = -FM_e = 70(10^6)(102)(10^{-6})(2.36)(10^{-5})\left(\frac{70-10}{0.6}\right) = 16.85\ \text{kN·m} \quad \textbf{(3a)}$$

$$FS_b = FS_e = 0 \quad \textbf{(3b)}$$

Thus,

$$\mathbf{Q}_{f1} = \mathbf{Q}_{f2} = \begin{bmatrix} 0 \\ 16.85 \\ 0 \\ -16.85 \end{bmatrix} \begin{matrix} 3 & 5 \\ 4 & 1 \\ 5 & 6 \\ 1 & 2 \end{matrix} \quad \textbf{(4)}$$

Member 1 (code numbers 3, 4, 5, 1) Member 2 (code numbers 5, 1, 6, 2)

Next, using the member code numbers, we store the pertinent elements of $\mathbf{Q}_{f1}$ and $\mathbf{Q}_{f2}$ in their proper positions in the structure fixed-joint force vector $\mathbf{P}_f$, as shown in Fig. 7.20(c).

Member 3 ($MT = 2$) Because $MT = 2$ for this member, we substitute the values of fixed-end moments and shears from Eqs. (3) into Eq. (7.19) to obtain the fixed-end force vector for the released member 3 as

$$\mathbf{Q}_{f3} = \begin{bmatrix} 0 - \dfrac{3(-16.85)}{2(8)} \\ 16.85 - \dfrac{1}{2}(-16.85) \\ 0 + \dfrac{3(-16.85)}{2(8)} \\ 0 \end{bmatrix} = \begin{bmatrix} 3.1595 \\ 25.276 \\ -3.1595 \\ 0 \end{bmatrix} \begin{matrix} 6 \\ 2 \\ 7 \\ 8 \end{matrix} \quad \textbf{(5)}$$

The relevant elements of $\mathbf{Q}_{f3}$ are stored in $\mathbf{P}_f$ using the member code numbers. The completed structure fixed-joint force vector $\mathbf{P}_f$ due to the temperature change, is shown in Fig. 7.20(c).

Joint Displacements: The structure stiffness relationship $\mathbf{P} - \mathbf{P}_f = \mathbf{Sd}$ for the entire beam can be written as

$$\begin{bmatrix} 0 \\ -8.4252 \end{bmatrix} = \begin{bmatrix} 7{,}140 & 1{,}785 \\ 1{,}785 & 6{,}247.5 \end{bmatrix} \begin{bmatrix} d_1 \\ d_2 \end{bmatrix}$$

By solving these equations, we determine the joint displacements to be

$$\mathbf{d} = \begin{bmatrix} 3.6308 \\ -14.523 \end{bmatrix} \begin{matrix} 1 \\ 2 \end{matrix} \times 10^{-4} \text{ rad}$$ **Ans**

Member End Displacements and End Forces:

Member 1

$$\mathbf{u}_1 = \begin{bmatrix} 0 \\ 0 \\ 0 \\ d_1 \end{bmatrix} \begin{matrix} 3 \\ 4 \\ 5 \\ 1 \end{matrix} = \begin{bmatrix} 0 \\ 0 \\ 0 \\ 3.6308 \end{bmatrix} \times 10^{-4}$$

Using $\mathbf{k}_1$ from Example 7.3 and $\mathbf{Q}_{f1}$ from Eq. (4),

$$\mathbf{Q}_1 = \mathbf{k}_1\mathbf{u}_1 + \mathbf{Q}_{f1} = \begin{bmatrix} 0.243 \text{ kN} \\ 17.498 \text{ kN·m} \\ -0.243 \text{ kN} \\ -15.554 \text{ kN·m} \end{bmatrix} \begin{matrix} 3 \\ 4 \\ 5 \\ 1 \end{matrix}$$ **Ans**

The end forces for member 1 are depicted in Fig. 7.20(d). To generate the support reaction vector **R**, the pertinent elements of $\mathbf{Q}_1$ are stored in **R**, as shown in Fig. 7.20(e).

Member 2

$$\mathbf{u}_2 = \begin{bmatrix} 0 \\ d_1 \\ 0 \\ d_2 \end{bmatrix} \begin{matrix} 5 \\ 1 \\ 6 \\ 2 \end{matrix} = \begin{bmatrix} 0 \\ 3.6308 \\ 0 \\ -14.523 \end{bmatrix} \times 10^{-4}$$

Using $\mathbf{k}_2$ from Example 7.3 and $\mathbf{Q}_{f2}$ from Eq. (4),

$$\mathbf{Q}_2 = \mathbf{k}_2\mathbf{u}_2 + \mathbf{Q}_{f2} = \begin{bmatrix} -0.7291 \text{ kN} \\ 15.554 \text{ kN·m} \\ 0.7291 \text{ kN} \\ -21.387 \text{ kN·m} \end{bmatrix} \begin{matrix} 5 \\ 1 \\ 6 \\ 2 \end{matrix}$$ **Ans**

Member 3

$$\mathbf{u}_3 = \begin{bmatrix} 0 \\ d_2 \\ 0 \\ 0 \end{bmatrix} \begin{matrix} 6 \\ 2 \\ 7 \\ 8 \end{matrix} = \begin{bmatrix} 0 \\ -14.523 \\ 0 \\ 0 \end{bmatrix} \times 10^{-4}$$

Using $\mathbf{k}_3$ from Example 7.3 and $\mathbf{Q}_{f3}$ from Eq. (5),

$$\mathbf{Q}_3 = \mathbf{k}_3\mathbf{u}_3 + \mathbf{Q}_{f3} = \begin{bmatrix} 2.6734 \text{ kN} \\ 21.387 \text{ kN·m} \\ -2.6734 \text{ kN} \\ 0 \end{bmatrix} \begin{matrix} 6 \\ 2 \\ 7 \\ 8 \end{matrix}$$ **Ans**

The member end forces are shown in Fig. 7.20(d).

Support Reactions: See Figs. 7.20(e) and (f). **Ans**

EXAMPLE 7.7 Determine the joint displacements, member end forces, and support reactions for the plane frame shown in Fig. 7.21(a) due to the combined effect of the following: (a) a temperature increase of 75° F in the girder, and (b) the fabricated length of the left column being $\frac{1}{4}$ in. too short. Use the matrix stiffness method.

SOLUTION This frame, subjected to joint and member loads, was analyzed in Example 7.1.

Analytical Model: See Fig. 7.21(b). The analytical model used herein is the same as used in Example 7.1.

Structure Stiffness Matrix: As determined in Example 7.1, the structure stiffness matrix for the frame, in units of kips and inches, is

$$\mathbf{S} = \begin{bmatrix} 1,781.3 & 0 & -1,776.3 & 0 & 0 \\ 0 & 1,781.3 & 0 & -5.0347 & 1,208.3 \\ -1,776.3 & 0 & 1,781.3 & 0 & 1,208.3 \\ 0 & -5.0347 & 0 & 1,781.3 & -1,208.3 \\ 0 & 1,208.3 & 1,208.3 & -1,208.3 & 580,000 \end{bmatrix} \tag{1}$$

Joint Load Vector:

$$\mathbf{P} = \mathbf{0} \tag{2}$$

Structure Fixed-Joint Force Vector Due to Temperature Changes and Fabrication Errors:

Member 1 ($MT = 2$) By substituting the numerical values of $E, A, L,$ and $e_a = -\frac{1}{4}$ in. into Eq. (7.35), we obtain the member fixed-end forces, due to the specified fabrication error, as

$$FA_b = -FA_e = -444.06 \text{ k}$$
$$FS_b = FS_e = FM_b = FM_e = 0$$

As $MT = 2$, we use Eq. (7.10) to form the member local fixed-end force vector. Thus,

$$\mathbf{Q}_{f1} = \begin{bmatrix} -444.06 \\ 0 \\ 0 \\ 444.06 \\ 0 \\ 0 \end{bmatrix} \tag{3}$$

Using the member transformation matrix, $\mathbf{T}_1$, from Example 7.1, we evaluate the global fixed-end force vector as

$$\mathbf{F}_{f1} = \mathbf{T}_1^T \mathbf{Q}_{f1} = \begin{bmatrix} 0 \\ -444.06 \\ 0 \\ \hline 0 \\ 444.06 \\ \hline 0 \end{bmatrix} \begin{matrix} 6 \\ 7 \\ 8 \\ 1 \\ 2 \\ 9 \end{matrix}$$

Next, using the member code numbers, we store the pertinent elements of $\mathbf{F}_{f1}$ in their proper positions in the 5×1 structure fixed-joint force vector $\mathbf{P}_f$, as shown in Fig. 7.21(c).

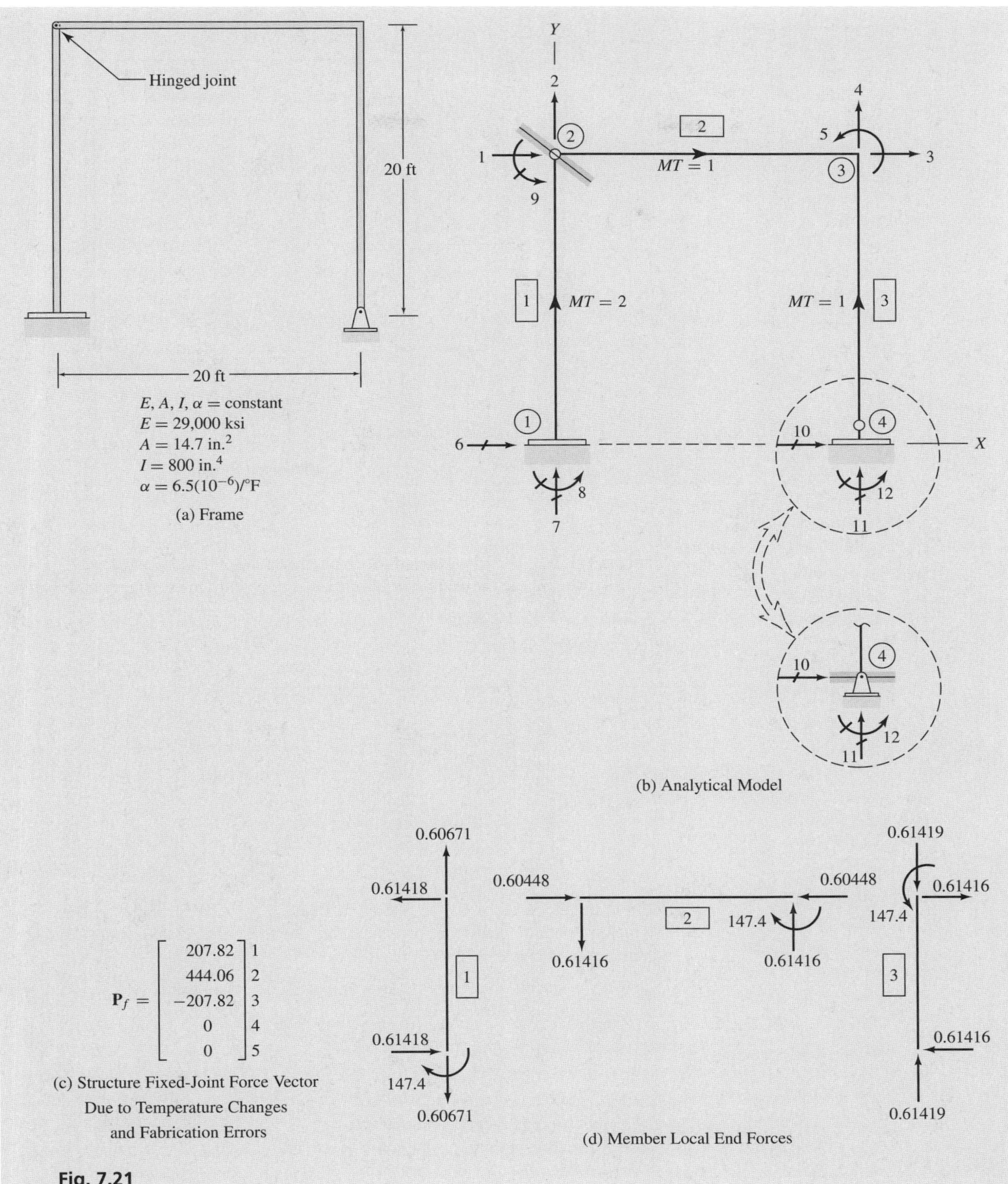

(a) Frame

(b) Analytical Model

$$\mathbf{P}_f = \begin{bmatrix} 207.82 \\ 444.06 \\ -207.82 \\ 0 \\ 0 \end{bmatrix} \begin{matrix} 1 \\ 2 \\ 3 \\ 4 \\ 5 \end{matrix}$$

(c) Structure Fixed-Joint Force Vector Due to Temperature Changes and Fabrication Errors

(d) Member Local End Forces

Fig. 7.21

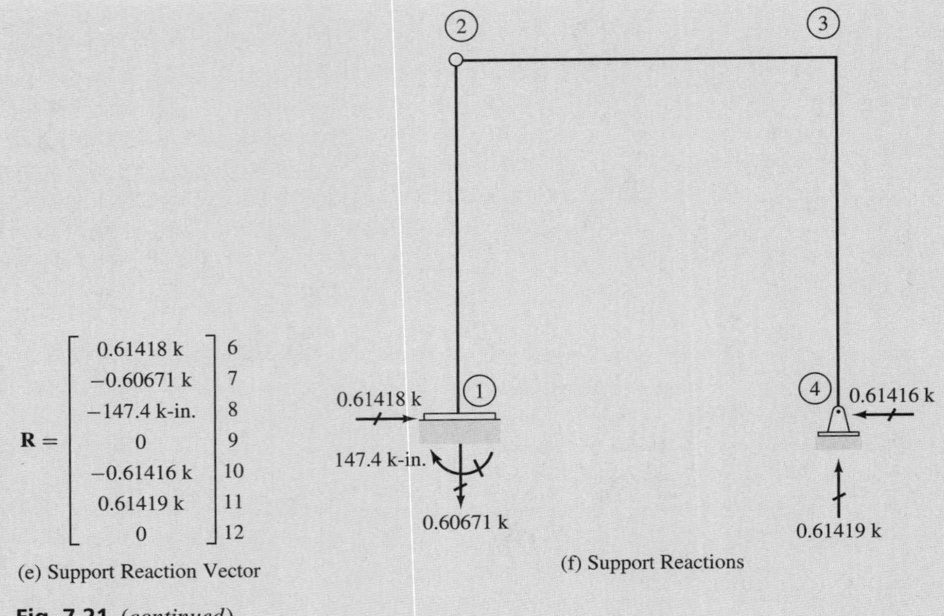

(e) Support Reaction Vector

(f) Support Reactions

Fig. 7.21 *(continued)*

Member 2 ($MT = 1$) By substituting the numerical values of E, A, α, and $T_u = 75°$ F into Eq. (7.34), we evaluate the member fixed-end forces, due to the given temperature change:

$$FA_b = -FA_e = 207.82\,\text{k}$$

$$FS_b = FS_e = FM_b = FM_e = 0$$

As $MT = 1$, we use Eq. (7.6) to form this horizontal member's local and global fixed-end force vectors:

$$\mathbf{F}_{f2} = \mathbf{Q}_{f2} = \begin{bmatrix} 207.82 \\ 0 \\ \hdashline 0 \\ \hdashline -207.82 \\ 0 \\ 0 \end{bmatrix} \begin{matrix} 1 \\ 2 \\ 9 \\ 3 \\ 4 \\ 5 \end{matrix} \tag{4}$$

The relevant elements of $\mathbf{F}_{f2}$ are stored in $\mathbf{P}_f$ using the member code numbers. The completed structure fixed-joint force vector $\mathbf{P}_f$, due to the temperature change and fabrication error, is given in Fig. 7.21(c).

Joint Displacements: Solving the structure stiffness relationship $\mathbf{P} - \mathbf{P}_f = \mathbf{S}\mathbf{d}$, we obtain the following joint displacements.

$$\mathbf{d} = \begin{bmatrix} -0.12199\text{ in.} \\ -0.24965\text{ in.} \\ -0.0053343\text{ in.} \\ -0.00034577\text{ in.} \\ 0.00053051\text{ rad} \end{bmatrix} \begin{matrix} 1 \\ 2 \\ 3 \\ 4 \\ 5 \end{matrix} \qquad \textbf{Ans}$$

Member End Displacements and End Forces:

Member 1 ($MT = 2$)

$$\mathbf{v}_1 = \begin{bmatrix} 0 \\ 0 \\ 0 \\ d_1 \\ d_2 \\ 0 \end{bmatrix} \begin{matrix} 6 \\ 7 \\ 8 \\ 1 \\ 2 \\ 9 \end{matrix} = \begin{bmatrix} 0 \\ 0 \\ 0 \\ -0.12199 \\ -0.24965 \\ 0 \end{bmatrix}$$

$$\mathbf{u}_1 = \mathbf{T}_1 \mathbf{v}_1 = \begin{bmatrix} 0 \\ 0 \\ 0 \\ -0.24965 \\ 0.12199 \\ 0 \end{bmatrix}$$

Using $\mathbf{k}_1$ from Example 7.1 and $\mathbf{Q}_{f1}$ from Eq. (3),

$$\mathbf{Q}_1 = \mathbf{k}_1 \mathbf{u}_1 + \mathbf{Q}_{f1} = \begin{bmatrix} -0.60671 \text{ k} \\ -0.61418 \text{ k} \\ -147.4 \text{ k-in.} \\ 0.60671 \text{ k} \\ 0.61418 \text{ k} \\ 0 \end{bmatrix}$$ **Ans**

These end forces are depicted in Fig. 7.21(d).

$$\mathbf{F}_1 = \mathbf{T}_1^T \mathbf{Q}_1 = \begin{bmatrix} 0.61418 \\ -0.60671 \\ -147.4 \\ \hdashline -0.61418 \\ 0.60671 \\ \hdashline 0 \end{bmatrix} \begin{matrix} 6 \\ 7 \\ 8 \\ 1 \\ 2 \\ 9 \end{matrix}$$

The pertinent elements of $\mathbf{F}_1$ are stored in $\mathbf{R}$, as shown in Fig. 7.21(e).

Member 2 ($MT = 1$)

$$\mathbf{u}_2 = \mathbf{v}_2 = \begin{bmatrix} d_1 \\ d_2 \\ 0 \\ d_3 \\ d_4 \\ d_5 \end{bmatrix} \begin{matrix} 1 \\ 2 \\ 9 \\ 3 \\ 4 \\ 5 \end{matrix} = \begin{bmatrix} -0.12199 \\ -0.24965 \\ 0 \\ -0.0053343 \\ -0.00034577 \\ 0.00053051 \end{bmatrix}$$

Using $\mathbf{k}_2$ from Example 7.1 and $\mathbf{Q}_{f2}$ from Eq. (4),

$$\mathbf{F}_2 = \mathbf{Q}_2 = \mathbf{k}_2\mathbf{u}_2 + \mathbf{Q}_{f2} = \begin{bmatrix} 0.60448 \text{ k} \\ -0.61416 \text{ k} \\ \hdashline 0 \\ \hdashline -0.60448 \text{ k} \\ 0.61416 \text{ k} \\ -147.4 \text{ k-in.} \end{bmatrix} \begin{matrix} 1 \\ 2 \\ 9 \\ 3 \\ 4 \\ 5 \end{matrix}$$ **Ans**

Member 3 ($MT = 1$)

$$\mathbf{v}_3 = \begin{bmatrix} 0 \\ 0 \\ 0 \\ d_3 \\ d_4 \\ d_5 \end{bmatrix} \begin{matrix} 10 \\ 11 \\ 12 \\ 3 \\ 4 \\ 5 \end{matrix} = \begin{bmatrix} 0 \\ 0 \\ 0 \\ -0.0053343 \\ -0.00034577 \\ 0.00053051 \end{bmatrix}$$

Using $\mathbf{T}_3$ from Example 7.1,

$$\mathbf{u}_3 = \mathbf{k}_3\mathbf{v}_3 = \begin{bmatrix} 0 \\ 0 \\ 0 \\ -0.00034577 \\ 0.0053343 \\ 0.00053051 \end{bmatrix}$$

Using $\mathbf{k}_3$ from Example 7.1 and $\mathbf{Q}_{f3} = \mathbf{0}$, we calculate,

$$\mathbf{Q}_3 = \mathbf{k}_3\mathbf{u}_3 = \begin{bmatrix} 0.61419 \text{ k} \\ 0.61416 \text{ k} \\ 0 \\ -0.61419 \text{ k} \\ -0.61416 \text{ k} \\ 147.4 \text{ k-in.} \end{bmatrix}$$ **Ans**

The member local end forces are shown in Fig. 7.21(d).

$$\mathbf{F}_3 = \mathbf{T}_3^T\mathbf{Q}_3 = \begin{bmatrix} -0.61416 \\ 0.61419 \\ 0 \\ \hdashline 0.61416 \\ -0.61419 \\ 147.4 \end{bmatrix} \begin{matrix} 10 \\ 11 \\ 12 \\ 3 \\ 4 \\ 5 \end{matrix}$$

Support Reactions: See Figs. 7.21(e) and (f). **Ans**

SUMMARY

In this chapter, we have extended the matrix stiffness formulation so that it can be used to analyze plane-framed structures containing member releases. Furthermore, the formulation has been extended to include in the analysis, the secondary effects of support displacements, temperature changes, and fabrication errors.

In the presence of member releases, the overall analysis procedure remains the same as before, except that the modified expressions for the member local stiffness matrices **k** and the fixed-end force vectors $\mathbf{Q}_f$, developed in Section 7.1, must be used for members with releases. If all the members meeting at a joint are connected to it by hinged connections, then such a joint can be modeled as a hinged joint with its rotation restrained by an imaginary clamp.

The effects of support displacements are included in the analysis using the concept of equivalent joint loads. The structure fixed-joint forces, due to the support displacements, are added to the $\mathbf{P}_f$ vector by performing the following operations for each member that is attached to a support that undergoes displacements: (a) forming the fixed-end displacement vector $\mathbf{v}_{fs}$ from the support displacements, (b) evaluating the fixed-end force vector $\mathbf{F}_{fs} = \mathbf{K}\mathbf{v}_{fs}$, and (c) storing the relevant elements of $\mathbf{F}_{fs}$ in $\mathbf{P}_f$ using the member code numbers. Once the structure's joint displacements have been determined by solving its stiffness relationship $\mathbf{P} - \mathbf{P}_f = \mathbf{S}\mathbf{d}$, the member end displacement vectors **v** are formed using both the joint displacements **d** and the specified support displacements. The rest of the procedure for evaluating member forces and support reactions remains the same as for the case of external loads.

The effects of temperature changes and fabrication errors can be included in the analysis methods developed previously, simply by including the member fixed-end forces due to these actions in the local fixed-end force vectors $\mathbf{Q}_f$. The expressions for member fixed-end forces, due to temperature changes and fabrication errors, are given in Section 7.5.

PROBLEMS

Section 7.1

7.1 and 7.2 Determine the joint displacements, member end forces, and support reactions for the beams shown in Figs. P7.1 and P7.2, using the matrix stiffness method.

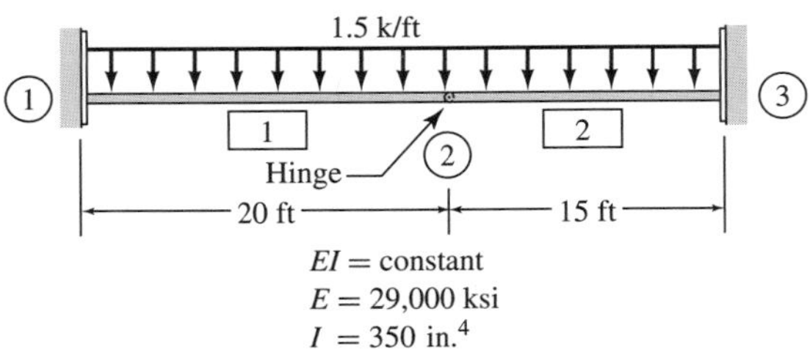

Fig. P7.1

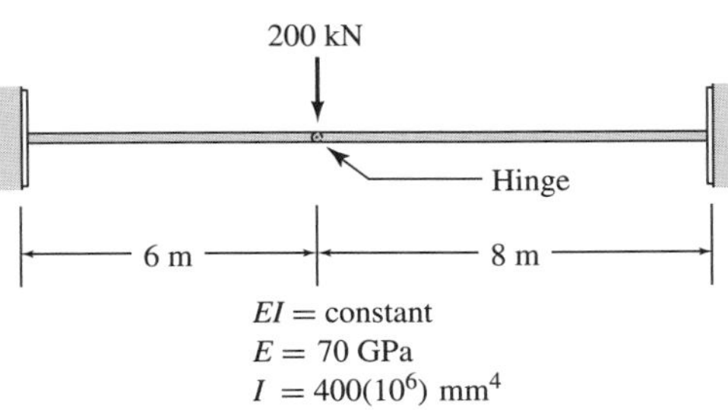

Fig. P7.2

7.3 Determine the joint displacements, member end forces, and support reactions for the beam shown in Fig. P7.3, by modeling member 3 as being hinged at its right end.

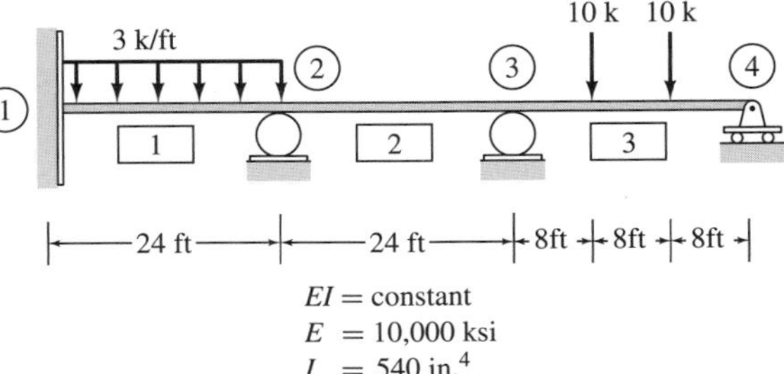

Fig. P7.3

7.4 Determine the joint displacements, member end forces, and support reactions for the beam shown in Fig. P7.4, by modeling member 1 as being hinged at its left end and member 3 as being hinged at its right end.

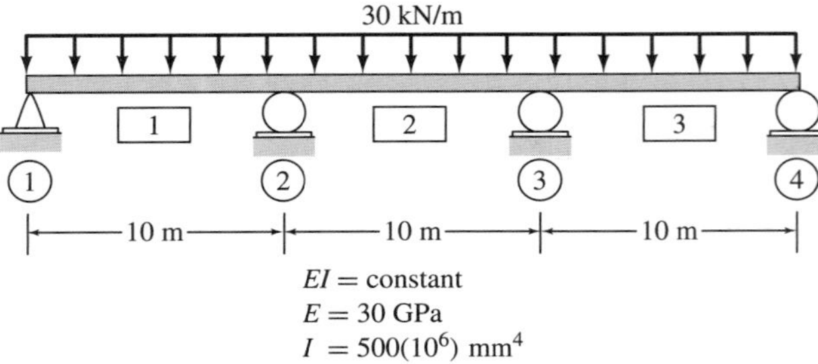

Fig. P7.4, P7.20

7.5 Determine the joint displacements, member end forces, and support reactions for the beam shown in Fig. P7.5 by modeling member 1 as being hinged at its left end.

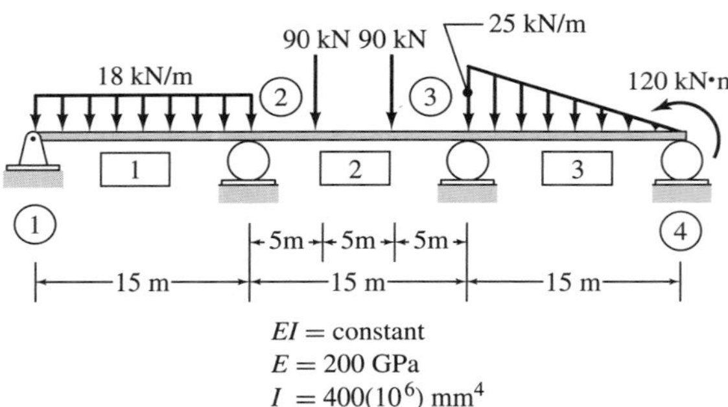

Fig. P7.5

7.6 Determine the joint displacements, member local end forces, and support reactions for the plane frame shown in Fig. P7.6, using the matrix stiffness method.

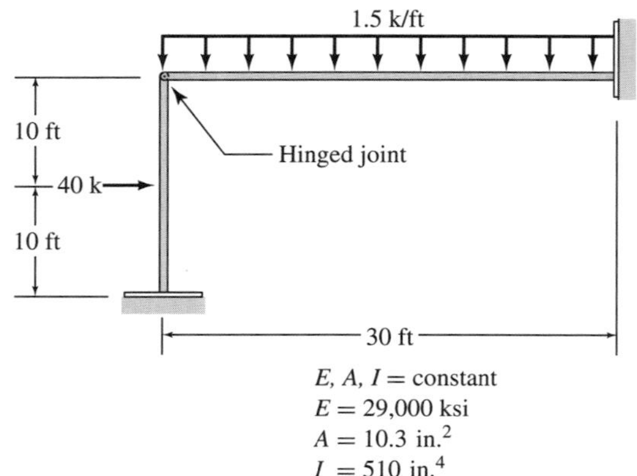

Fig. P7.6

7.7 and 7.8 Determine the joint displacements, member local end forces, and support reactions for the plane frames shown in Figs. P7.7 and P7.8, by modeling the horizontal member as being hinged at its right end.

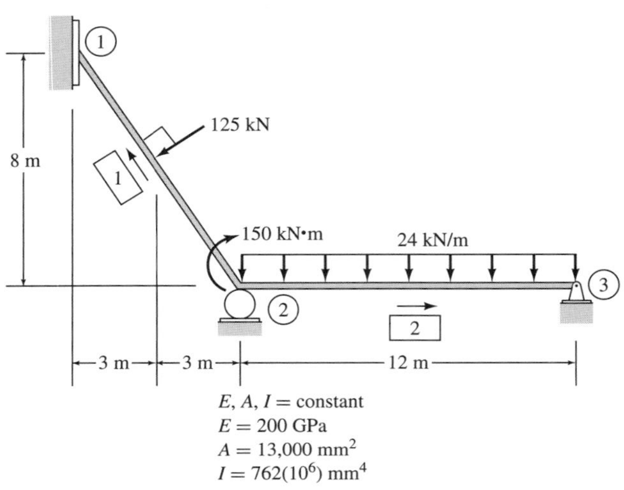

Fig. P7.7

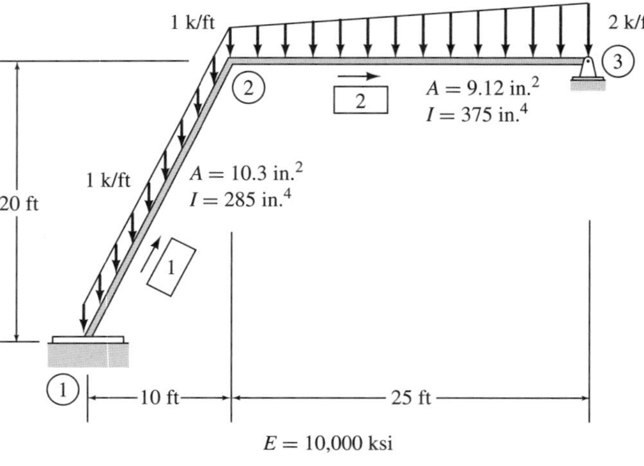

Fig. P7.8

7.9 Determine the joint displacements, member local end forces, and support reactions for the plane frame shown in Fig. P7.9, by modeling the horizontal member as being hinged at its left end and the inclined member as being hinged at its lower end.

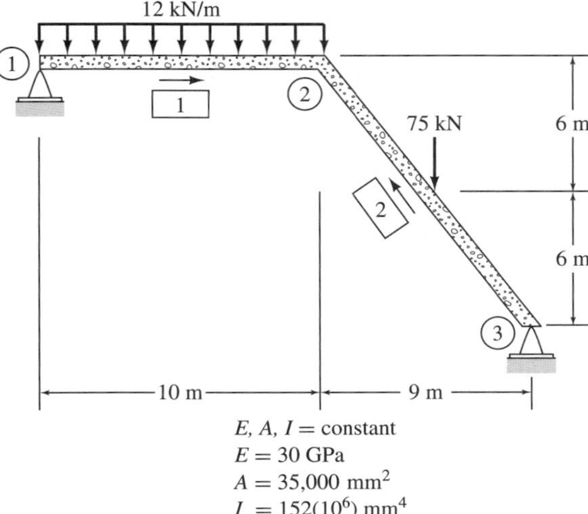

Fig. P7.9, P7.23, P7.33

7.10 Determine the joint displacements, member local end forces, and support reactions for the plane frame shown in Fig. P7.10, using the matrix stiffness method.

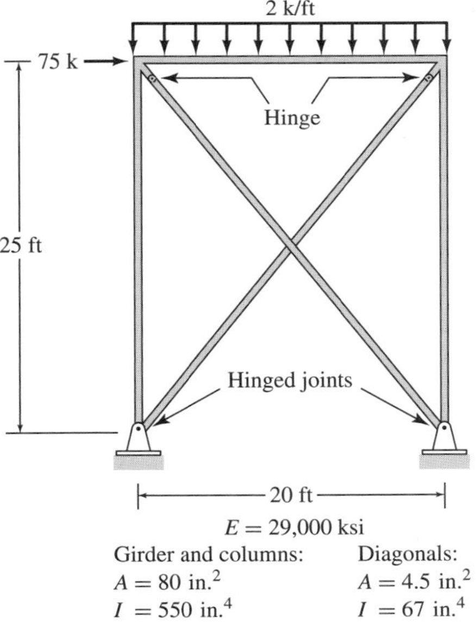

Fig. P7.10

7.11 Using a structural analysis computer program, determine the joint displacements, member local end forces, and reactions for the frame shown in Fig. P7.11 for the value of the load parameter $P = 300$ kN. What is the largest value of P that can be applied to the frame without exceeding the drift (maximum horizontal deflection) limitation of one percent of the frame height? Assume that the braces (inclined members) are connected by hinged connections at both ends.

7.12 Solve Problem 7.11 by assuming that the braces are connected by rigid (moment-resisting) connections at both ends.

7.13 Solve Problem 7.11 by assuming that the frame is unbraced. Note that, instead of developing a new analytical model for the unbraced frame, the previously developed models of the corresponding braced frames can be modified to eliminate the effect of bracing by simply using a very small value for the modulus of elasticity, E, of the bracing members (e.g., $E = 0.000001$ kN/m²).

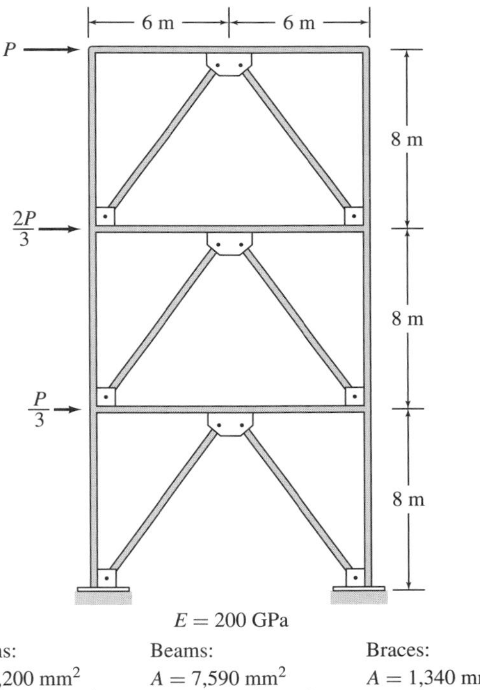

Fig. P7.11, P7.12, P7.13

Section 7.2

7.14 Modify the computer program developed in Chapter 5 for the analysis of rigidly connected beams, to include the effect of member releases. Use the modified program to analyze the beams of Problems 7.1 through 7.5, and compare the computer-generated results to those obtained by hand calculations.

7.15 Modify the program developed in Chapter 6 for the analysis of rigidly connected plane frames, to include the effect of member releases. Use the modified program to analyze the frames of Problems 7.6 through 7.10, and compare the computer-generated results to those obtained by hand calculations.

Section 7.3

7.16 Determine the joint displacements, member axial forces, and support reactions for the plane truss shown in Fig. P7.16, due to the combined effect of the loading shown and a settlement of $\frac{1}{2}$ in. of support 2. Use the matrix stiffness method.

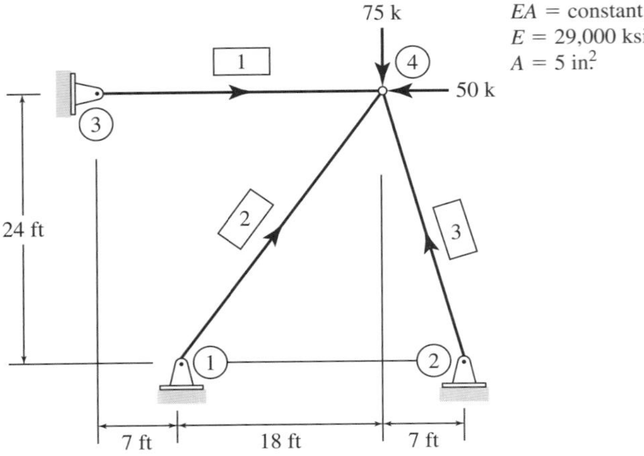

Fig. P7.16, P7.28

7.17 Determine the joint displacements, member axial forces, and support reactions for the plane truss shown in Fig. P7.17, due to the combined effect of the loading shown and a settlement of $\frac{1}{4}$ in. of support 3. Use the matrix stiffness method.

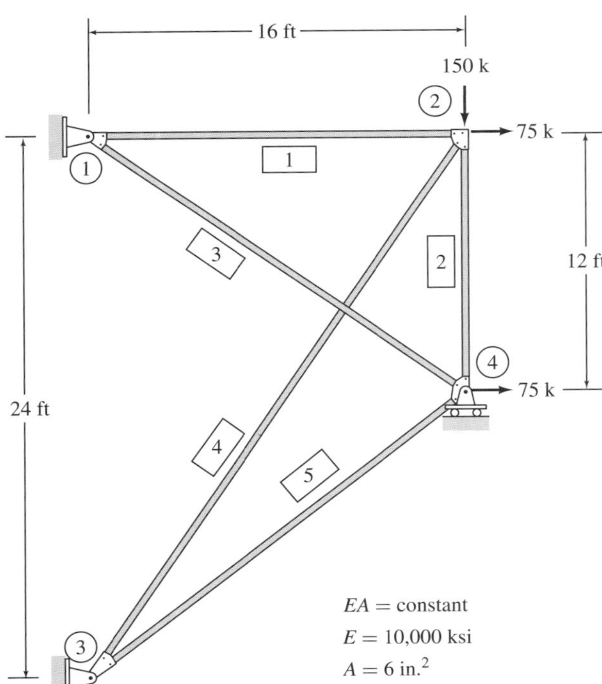

Fig. P7.17, P7.29, P7.30

7.18 Determine the joint displacements, member axial forces, and support reactions for the plane truss shown in Fig. P7.18, due to the combined effect of the loading shown and settlements of $\frac{1}{2}$ and $\frac{1}{4}$ in., respectively, of supports 3 and 4. Use the matrix stiffness method.

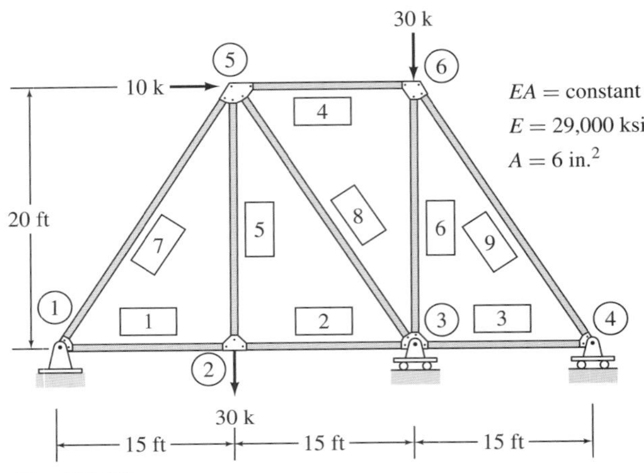

Fig. P7.18

7.19 Determine the joint displacements, member end forces, and support reactions for the three-span continuous beam shown in Fig. P7.19, due to settlements of 8 and 30 mm, respectively, of supports 2 and 3. Use the matrix stiffness method.

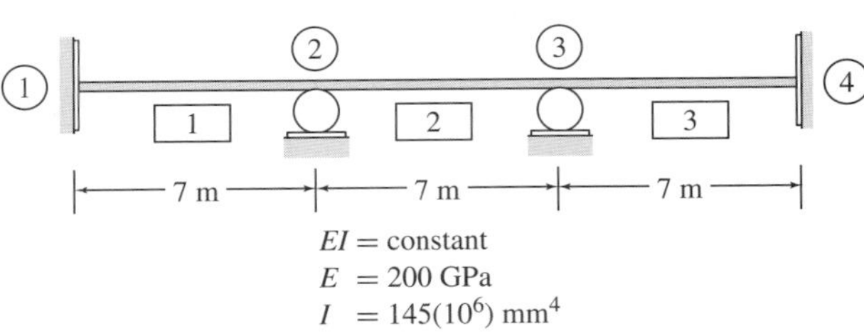

Fig. P7.19, P7.34

7.20 Solve Problem 7.4 for the loading shown in Fig. P7.4 and settlements of 12,75,60 and 25 mm, respectively, of supports 1, 2, 3, and 4.

7.21 Determine the joint displacements, member end forces, and support reactions for the beam shown in Fig. P7.21, due to the combined effect of the loading shown and a settlement of $1\frac{1}{4}$ in. of the middle support. Use the matrix stiffness method.

7.22 Determine the joint displacements, member local end forces, and support reactions for the plane frame shown in

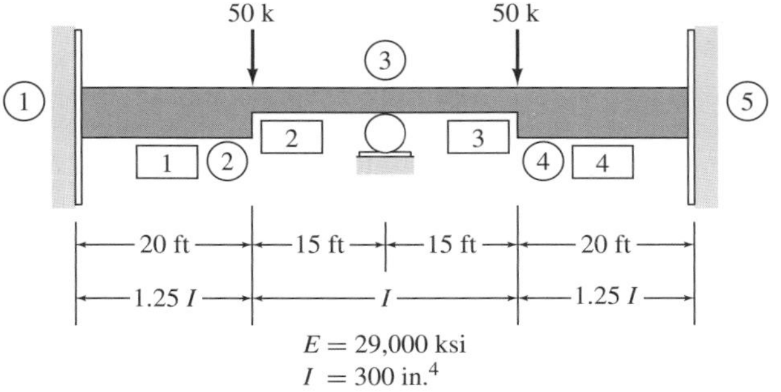

Fig. P7.21

Fig. P7.22, due to a settlement of 25 mm of the right support. Use the matrix stiffness method.

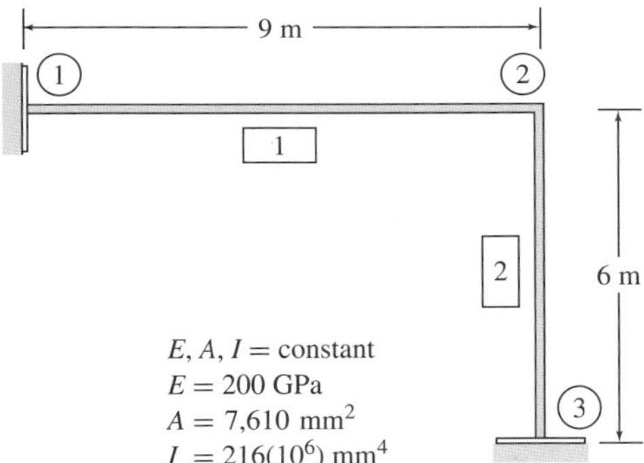

Fig. P7.22, P7.31, P7.32

7.23 Solve Problem 7.9 for the loading shown in Fig. P7.9 and a settlement of 50 mm of the right support.

7.24 Determine the joint displacements, member local end forces, and support reactions for the plane frame shown in Fig. P7.24, due to the combined effect of the following: (a) the loading shown in the figure, (b) a clockwise rotation of 0.017 radians of the left support, and (c) a settlement of $\frac{3}{4}$ in. of the right support. Use the matrix stiffness method.

Section 7.4

7.25 Extend the program developed in Chapter 4 for the analysis of plane trusses subjected to joint loads, to include the effect of support displacements. Use the modified program to analyze the trusses of Problems 7.16 through 7.18, and compare the computer-generated results to those obtained by hand calculations.

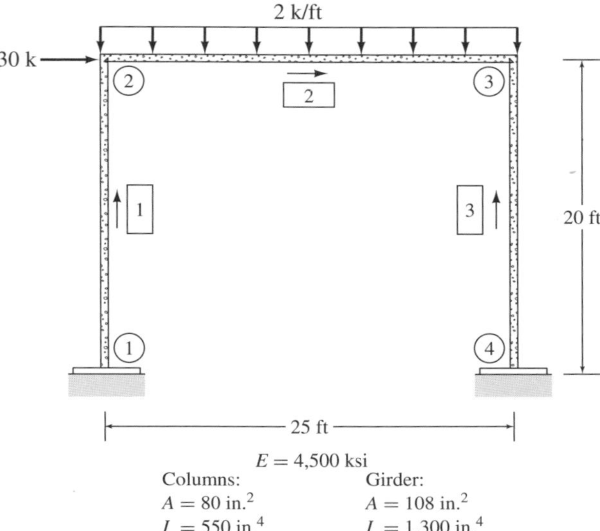

Fig. P7.24

7.26 Extend the program developed in Problem 7.14 for the analysis of beams subjected to external loads, to include the effect of support displacements. Use the modified program to analyze the beams of Problems 7.19 through 7.21, and compare the computer-generated results to those obtained by hand calculations.

7.27 Extend the program developed in Problem 7.15 for the analysis of plane frames subjected to external loads, to include the effect of support displacements. Use the modified program to analyze the frames of Problems 7.22 through 7.24, and compare the computer-generated results to those obtained by hand calculations.

Section 7.5

7.28 Determine the joint displacements, member axial forces, and support reactions for the plane truss shown in Fig. P7.28, due to a temperature drop of 100° F in member 2. Neglect the joint loads shown in the figure. Use the matrix stiffness method; $\alpha = 6.5(10^{-6})/°$ F.

7.29 Determine the joint displacements, member axial forces, and support reactions for the plane truss shown in Fig. P7.29, due to the combined effect of the following: (a) the joint loads shown in the figure, (b) a temperature increase of 70° F in member 2, (c) a temperature drop of 30° F in member 5, and (d) the fabricated length of member 4 being $\frac{1}{4}$ in. too long. Use the matrix stiffness method; $\alpha = 1.3(10^{-5})/°$ F.

7.30 Determine the joint displacements, member axial forces, and support reactions for the plane truss shown in Fig. P7.30, due to the fabricated lengths of members 3 and 4 being $\frac{1}{4}$ in. too short. Neglect the joint loads shown in the figure, and use the matrix stiffness method.

7.31 Determine the joint displacements, member local end forces, and support reactions for the plane frame of Fig. P7.31, due to a temperature increase of 50° C in the two members. Use the matrix stiffness method; $\alpha = 1.2(10^{-5})/^\circ$ C.

7.32 Determine the joint displacements, member local end forces, and support reactions for the plane frame of Fig. P7.32, due to the fabricated lengths of the two members being 15 mm too short. Use the matrix stiffness method.

7.33 Determine the joint displacements, member local end forces, and support reactions for the plane frame of Fig. P7.33, due to the combined effect of the following: (a) the external loads shown in the figure, and (b) a temperature drop of 60° C in the two members. Use the matrix stiffness method; $\alpha = 10^{-5}/^\circ$ C.

7.34 Determine the joint displacements, member end forces, and support reactions for the beam of Fig. P7.34, due to a linearly varying temperature increase of 55° C at the top surface and 5° C at the bottom surface, of all the members. Use the matrix stiffness method; $\alpha = 1.2(10^{-5})/^\circ$ C and $d = 300$ mm.

8 THREE-DIMENSIONAL FRAMED STRUCTURES

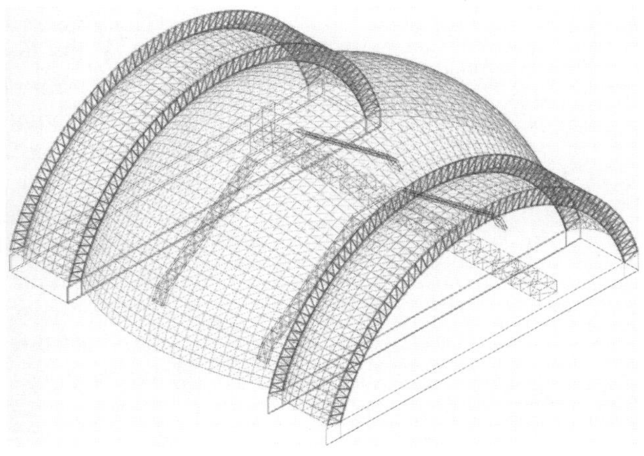

Space Truss and its Analytical Model
(Courtesy of Triodetic)

Up to this point, we have focused our attention on the analysis of plane-framed structures. While many actual three-dimensional structures can be divided into planar parts for the purpose of analysis, there are others (e.g., lattice domes and transmission towers) that, because of the arrangement of their members or applied loading, cannot be divided into plane structures. Such structures are analyzed as space structures subjected to three-dimensional loadings. The matrix stiffness analysis of space structures is similar to that of plane structures—except, of course, that member stiffness and transformation matrices appropriate for the particular type of space structure under consideration are now used in the analysis.

In this chapter, we extend the matrix stiffness formulation, developed for plane structures, to the analysis of three-dimensional or space structures. Three types of space-framed structures are considered: space trusses, grids, and space frames, with methods for their analysis presented in Sections 8.1, 8.2, and 8.3, respectively.

The computer programs for the analysis of space-framed structures can be conveniently adapted from those for plane structures, via relatively straightforward modifications that should become apparent as the analysis of space structures is developed in this chapter. Therefore, the details of programming the analysis of space structures are not covered herein; they are, instead, left as exercises for the reader.

8.1 SPACE TRUSSES

A *space truss* is defined as *a three-dimensional assemblage of straight prismatic members connected at their ends by frictionless ball-and-socket joints, and subjected to loads and reactions that act only at the joints.* Like plane trusses, the members of space trusses develop only axial forces. The matrix stiffness analysis of space trusses is similar to that of plane trusses developed in Chapter 3 (and modified in Chapter 7).

The process of developing the analytical models of space trusses (and numbering the degrees of freedom and restrained coordinates) is essentially the same as that for plane trusses (Chapter 3). The overall geometry of the space truss, and its joint loads and displacements, are described with reference to a global Cartesian or rectangular right-handed XYZ coordinate system, with three global (X, Y, and Z) coordinates now used to specify the location of each joint. Furthermore, since an unsupported joint of a space truss can translate in any direction in the three-dimensional space, three displacements—the translations in the X, Y, and Z directions—are needed to completely establish its deformed position. Thus, a free joint of a space truss has three degrees of freedom, and three structure coordinates (i.e., free and/or restrained coordinates) need to be defined at each joint, for the purpose of analysis. Thus,

$$\left.\begin{array}{l} NCJT = 3 \\ NDOF = 3(NJ) - NR \end{array}\right\} \quad \text{for space trusses} \tag{8.1}$$

The procedure for assigning numbers to the structure coordinates of a space truss is analogous to that for plane trusses. The degrees of freedom of the space truss are numbered first by beginning at the lowest-numbered joint with a degree of freedom, and proceeding sequentially to the highest-numbered joint. If a joint has more than one degree of freedom, then the translation in the X direction is numbered first, followed by the translation in the Y direction, and then the translation in the Z direction. After all the degrees of freedom have been numbered, the restrained coordinates of the space truss are numbered in the same manner as the degrees of freedom.

Consider, for example, the three-member space truss shown in Fig. 8.1(a). As the analytical model of the truss depicted in Fig. 8.1(b) indicates, the structure has three degrees of freedom ($NDOF = 3$), which are the translations d_1, d_2, and d_3 of joint 2 in the X, Y, and Z directions, respectively; and nine restrained coordinates ($NR = 9$), which are identified as R_4 through R_{12} at the support joints 1, 3, and 4.

As in the case of plane trusses, a local right-handed xyz coordinate system is established for each member of the space truss. The origin of the local coordinate system is located at one of the ends (which is referred to as the *beginning* of the member), with the x axis directed along the member's centroidal axis in its undeformed state. Since the space truss members can only develop axial forces, the positive directions of the y and z axes can be chosen

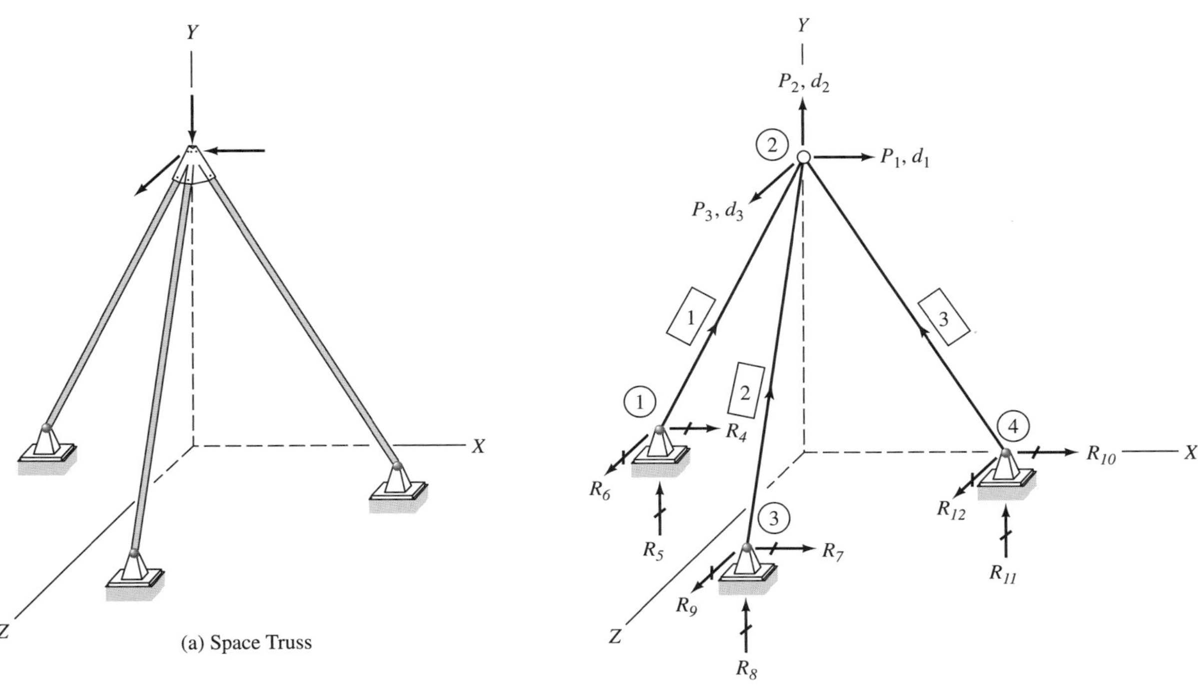

Fig. 8.1

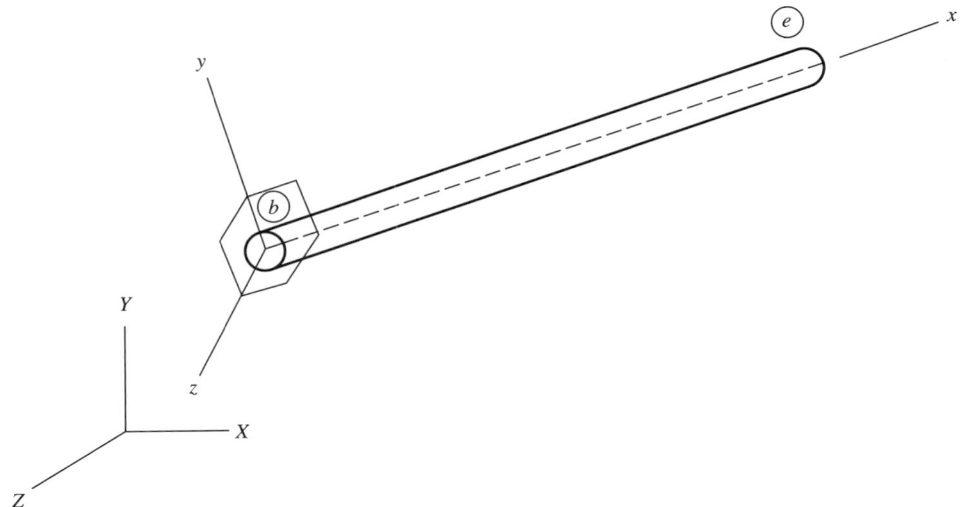

Fig. 8.2 *Local Coordinate System for Members of Space Trusses*

arbitrarily, provided that the x, y, and z axes are mutually perpendicular and form a right-handed coordinate system (Fig. 8.2).

Member Stiffness Relations in the Local Coordinate System

To establish the member local stiffness relations, let us focus our attention on an arbitrary prismatic member m of a space truss. When the truss is subjected to external loads, member m deforms and axial forces are induced at its ends. The initial and displaced positions of the member are shown in Fig. 8.3(a). As this figure indicates, three displacements—translations in the x, y, and z directions—are needed to completely specify the displaced position of each end of the member. Thus, the member has a total of six degrees of freedom or end displacements. However, as discussed in Section 3.3 (see Figs. 3.3(d) and (f)), small end displacements in the directions perpendicular to a truss member's centroidal axis do not cause any forces in the member. Thus, the end displacements u_{by}, u_{bz}, u_{ey}, and u_{ez} in the directions of the local y and z axes of the member, as shown in Fig. 8.3(a), are usually not evaluated in the analysis; and for analytical purposes, the member is considered to have only two degrees of freedom, u_1 and u_2, in its local coordinate system. Thus, the local end displacement vector $\mathbf{u}$ for a member of a space truss is expressed as

$$\mathbf{u} = \begin{bmatrix} u_1 \\ u_2 \end{bmatrix}$$

in which u_1 and u_2 represent the displacements of the member ends b and e, respectively, in the direction of the member's local x axis, as shown in Fig. 8.3(a). As this figure also indicates, the member end forces corresponding to the end displacements u_1 and u_2 are denoted by Q_1 and Q_2, respectively.

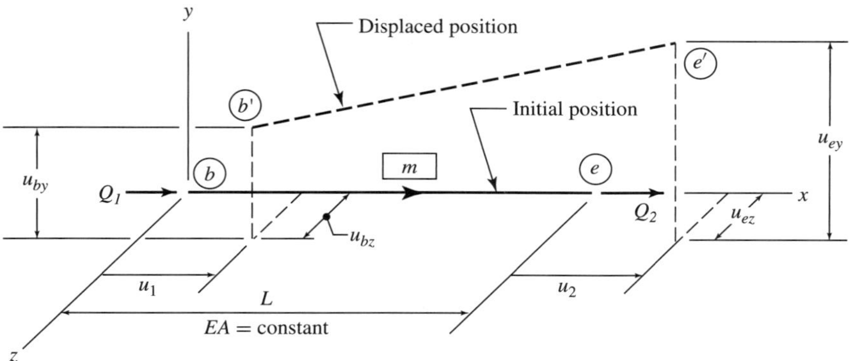

(a) Member Forces and Displacements in the Local Coordinate System

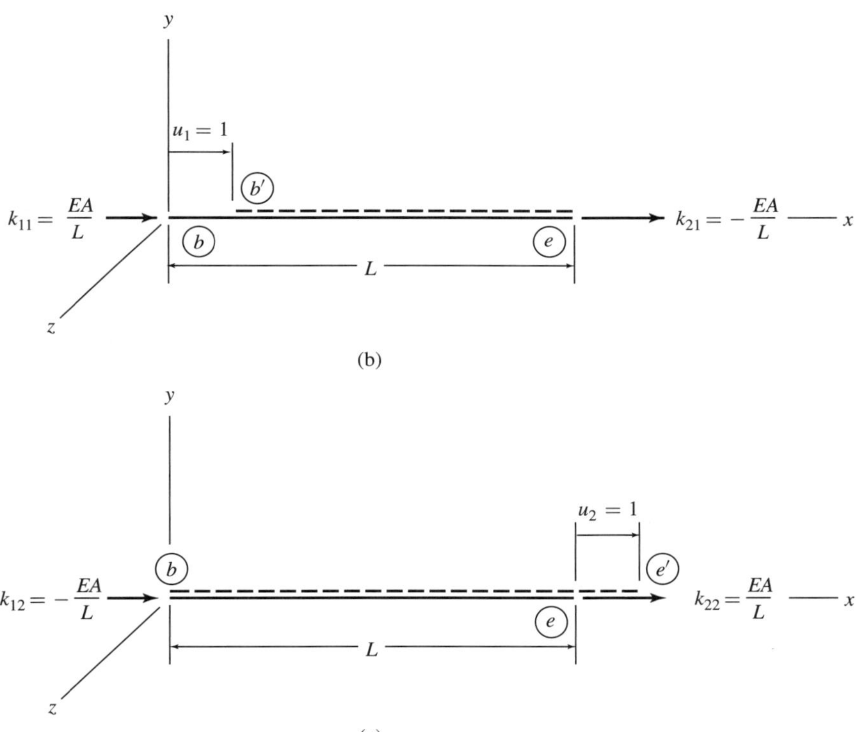

Fig. 8.3

The relationship between the local end forces $\mathbf{Q}$ and the end displacements $\mathbf{u}$, for the members of space trusses, is written as

$$\boxed{\mathbf{Q} = \mathbf{k}\mathbf{u}} \tag{8.2}$$

in which **k** represents the 2×2 member stiffness matrix in the local coordinate system. The explicit form of **k** can be obtained by subjecting the member to the unit end displacements, $u_1 = 1$ and $u_2 = 1$, as shown in Figs. 8.3(b) and (c), respectively, and evaluating the corresponding member end forces. Thus, the local stiffness matrix for the members of space trusses can be explicitly expressed as

$$\mathbf{k} = \frac{EA}{L}\begin{bmatrix} 1 & -1 \\ -1 & 1 \end{bmatrix} \tag{8.3}$$

Coordinate Transformations

Consider an arbitrary member m of a space truss, as shown in Fig. 8.4(a), and let X_b, Y_b, Z_b, and X_e, Y_e, Z_e be the global coordinates of the joints to which the member ends b and e, respectively, are attached. The length and the direction

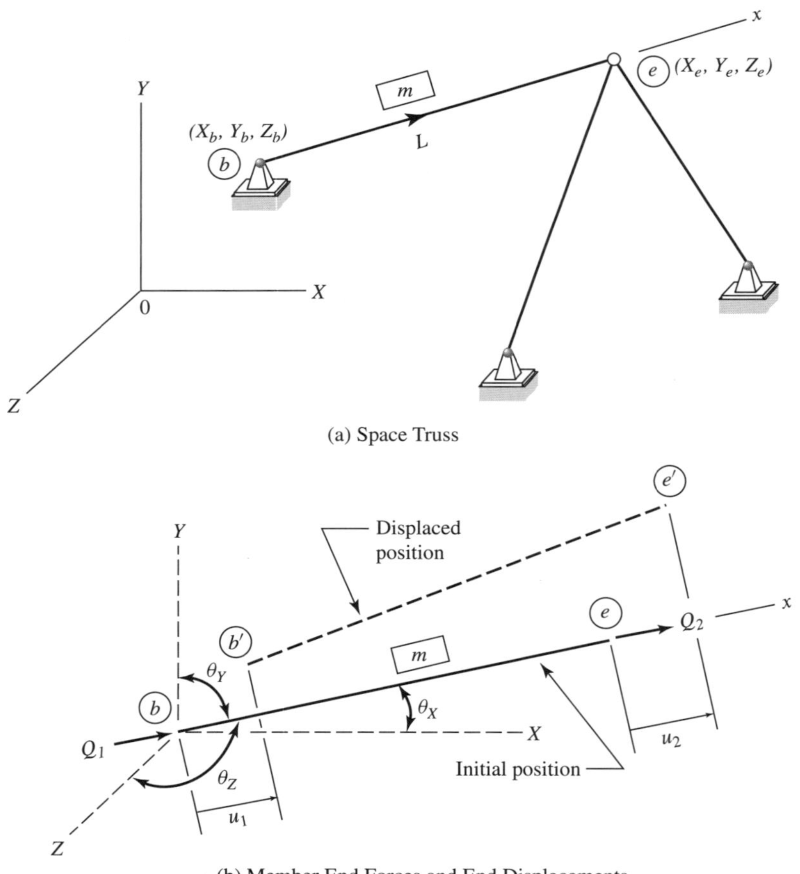

Fig. 8.4

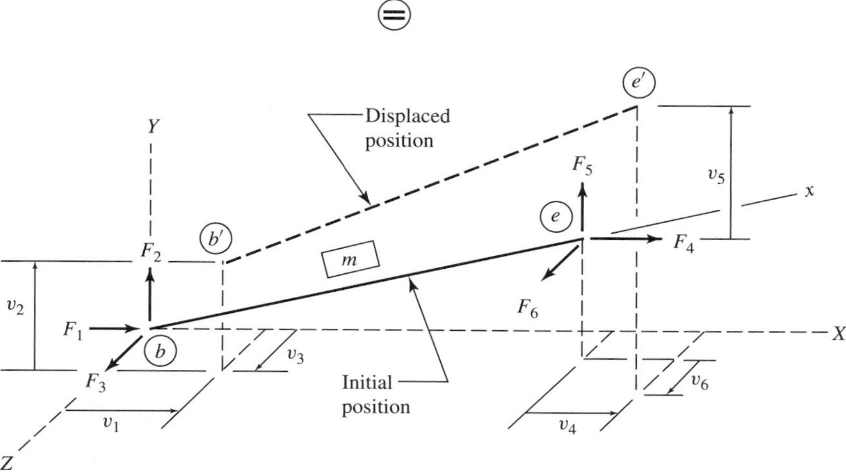

(c) Member End Forces and End Displacements in the Global Coordinate System

Fig. 8.4 *(continued)*

cosines of the member can be expressed in terms of the global coordinates of its ends by the following relationships:

$$L = \sqrt{(X_e - X_b)^2 + (Y_e - Y_b)^2 + (Z_e - Z_b)^2} \tag{8.4a}$$

$$\cos\theta_X = \frac{X_e - X_b}{L} \tag{8.4b}$$

$$\cos\theta_Y = \frac{Y_e - Y_b}{L} \tag{8.4c}$$

$$\cos\theta_Z = \frac{Z_e - Z_b}{L} \tag{8.4d}$$

in which θ_X, θ_Y, and θ_Z represent the angles between the positive directions of the global X, Y, and Z axes, respectively, and the positive direction of the member's local x axis, as shown in Fig. 8.4(b). Note that the origin of the global coordinate system is shown to coincide with that of the local coordinate system in this figure. With no loss in generality of the formulation, this convenient arrangement allows the angles between the local and global axes to be clearly visualized. It is important to realize that the member transformation matrix depends only on the angles between the local and global axes, regardless of whether or not the origins of the local and global coordinate systems coincide. Also shown in Fig. 8.4(b) are the member end displacements **u** and end forces **Q** in the local coordinate system; the equivalent systems of end displacements **v** and end forces **F**, in the global coordinate system, are depicted in Fig. 8.4(c). As indicated in Fig. 8.4(c), the global member end displacements **v** and end forces **F** are numbered by beginning at member end b, with the translation and force in the X direction numbered first, followed by the translation and force in the Y direction, and then the translation and force in the Z direction. The displacements and forces at the member's opposite end e are then numbered in the same sequential order.

Let us consider the transformation of member end forces and end displacements from a global to a local coordinate system. By comparing Figs. 8.4(b) and (c), we observe that at end b of the member, the local force Q_1 must be equal to the algebraic sum of the components of the global forces F_1, F_2, and F_3 in the direction of the local x axis; that is,

$$Q_1 = F_1 \cos\theta_X + F_2 \cos\theta_Y + F_3 \cos\theta_Z \tag{8.5a}$$

Similarly, at end e of the member, we can express Q_2 in terms of F_4, F_5, and F_6 as

$$Q_2 = F_4 \cos\theta_X + F_5 \cos\theta_Y + F_6 \cos\theta_Z \tag{8.5b}$$

Equations 8.5(a) and (b) can be written in matrix form as

$$\begin{bmatrix} Q_1 \\ Q_2 \end{bmatrix} = \begin{bmatrix} \cos\theta_X & \cos\theta_Y & \cos\theta_Z & 0 & 0 & 0 \\ 0 & 0 & 0 & \cos\theta_X & \cos\theta_Y & \cos\theta_Z \end{bmatrix} \begin{bmatrix} F_1 \\ F_2 \\ F_3 \\ F_4 \\ F_5 \\ F_6 \end{bmatrix} \tag{8.6}$$

Equation (8.6) can be symbolically expressed as $\mathbf{Q} = \mathbf{TF}$, with the 2×6 transformation matrix $\mathbf{T}$ given by

$$\mathbf{T} = \begin{bmatrix} \cos\theta_X & \cos\theta_Y & \cos\theta_Z & 0 & 0 & 0 \\ 0 & 0 & 0 & \cos\theta_X & \cos\theta_Y & \cos\theta_Z \end{bmatrix} \tag{8.7}$$

Since member end displacements, like end forces, are vectors, which are defined in the same directions as the corresponding forces, the foregoing transformation matrix $\mathbf{T}$ can also be used to transform member end displacements from the global to the local coordinate system; that is, $\mathbf{u} = \mathbf{Tv}$.

Next, we examine the transformation of member end forces from the local to the global coordinate system. A comparison of Figs. 8.4(b) and (c) indicates that at end b of the member, the global forces F_1, F_2, and F_3 must be the components of the local force Q_1 in the directions of the global X, Y, and Z axes, respectively; that is,

$$F_1 = Q_1 \cos\theta_X \qquad F_2 = Q_1 \cos\theta_Y \qquad F_3 = Q_1 \cos\theta_Z \tag{8.8a}$$

Similarly, at end e of the member, the global forces F_4, F_5, and F_6 can be expressed as the components of the local force Q_2, as

$$F_4 = Q_2 \cos\theta_X \qquad F_5 = Q_2 \cos\theta_Y \qquad F_6 = Q_2 \cos\theta_Z \tag{8.8b}$$

We can write Eqs. 8.8(a) and (b) in matrix form as

$$\begin{bmatrix} F_1 \\ F_2 \\ F_3 \\ F_4 \\ F_5 \\ F_6 \end{bmatrix} = \begin{bmatrix} \cos\theta_X & 0 \\ \cos\theta_Y & 0 \\ \cos\theta_Z & 0 \\ 0 & \cos\theta_X \\ 0 & \cos\theta_Y \\ 0 & \cos\theta_Z \end{bmatrix} \begin{bmatrix} Q_1 \\ Q_2 \end{bmatrix} \tag{8.9}$$

As the first matrix on the right side of Eq. (8.9) is the transpose of the transformation matrix **T** (Eq. (8.7)), the equation can be symbolically expressed as

$$\underset{6\times 1}{\mathbf{F}} = \underset{6\times 2}{\mathbf{T}^T}\;\underset{2\times 1}{\mathbf{Q}} \tag{8.10}$$

It may be of interest to note that the transformation relationship analogous to Eq. (8.10) for member end displacements (i.e., $\mathbf{v} = \mathbf{T}^T\mathbf{u}$) is not defined for space truss members, with two degrees of freedom, as used herein. This is because the local end displacement vectors **u** for such members do not contain the displacements of the member ends in the local y and z directions. As discussed previously, while the end forces in the local y and z directions of the members of space trusses are always 0, the displacements of the member ends in the local y and z directions are generally nonzero (see Fig. 8.3(a)). However, the foregoing limitation of the two-degree-of-freedom member model has no practical consequences, because the transformation relation $\mathbf{v} = \mathbf{T}^T\mathbf{u}$ is needed neither in the formulation of the matrix stiffness method of analysis, nor in its application.

Member Stiffness Relations in the Global Coordinate System

As in the case of plane trusses, the relationship between the global end forces **F** and the end displacements **v** for the members of space trusses is expressed as $\mathbf{F} = \mathbf{K}\mathbf{v}$, with the member global stiffness matrix **K** given by the equation

$$\underset{6\times 6}{\mathbf{K}} = \underset{6\times 2}{\mathbf{T}^T}\;\underset{2\times 2}{\mathbf{k}}\;\underset{2\times 6}{\mathbf{T}} \tag{8.11}$$

The explicit form of the 6 × 6 **K** matrix can be determined by substituting Eqs. (8.3) and (8.7) into Eq. (8.11) and performing the required matrix multiplications. The explicit form of the member global stiffness matrix **K**, thus obtained, is given in Eq. (8.12).

$$\mathbf{K} = \frac{EA}{L}\begin{bmatrix} \cos^2\theta_X & \cos\theta_X\cos\theta_Y & \cos\theta_X\cos\theta_Z & -\cos^2\theta_X & -\cos\theta_X\cos\theta_Y & -\cos\theta_X\cos\theta_Z \\ \cos\theta_X\cos\theta_Y & \cos^2\theta_Y & \cos\theta_Y\cos\theta_Z & -\cos\theta_X\cos\theta_Y & -\cos^2\theta_Y & -\cos\theta_Y\cos\theta_Z \\ \cos\theta_X\cos\theta_Z & \cos\theta_Y\cos\theta_Z & \cos^2\theta_Z & -\cos\theta_X\cos\theta_Z & -\cos\theta_Y\cos\theta_Z & -\cos^2\theta_Z \\ -\cos^2\theta_X & -\cos\theta_X\cos\theta_Y & -\cos\theta_X\cos\theta_Z & \cos^2\theta_X & \cos\theta_X\cos\theta_Y & \cos\theta_X\cos\theta_Z \\ -\cos\theta_X\cos\theta_Y & -\cos^2\theta_Y & -\cos\theta_Y\cos\theta_Z & \cos\theta_X\cos\theta_Y & \cos^2\theta_Y & \cos\theta_Y\cos\theta_Z \\ -\cos\theta_X\cos\theta_Z & -\cos\theta_Y\cos\theta_Z & -\cos^2\theta_Z & \cos\theta_X\cos\theta_Z & \cos\theta_Y\cos\theta_Z & \cos^2\theta_Z \end{bmatrix} \tag{8.12}$$

Procedure for Analysis

The procedure for the analysis of plane trusses developed in Chapter 3 (see block diagram in Fig. 3.20), and modified in Chapter 7, can be used to analyze space trusses provided that: (a) three structure coordinates (i.e., degrees of freedom and/or restrained coordinates), in the global X, Y, and Z directions, are defined at each joint; and (b) the member stiffness and transformation matrices developed in this section (Eqs. (8.3), (8.7), and (8.12)) are used in the analysis. The procedure is illustrated by the following example.

EXAMPLE 8.1 Determine the joint displacements, member axial forces, and support reactions for the space truss shown in Fig. 8.5(a) by the matrix stiffness method.

SOLUTION *Analytical Model:* See Fig. 8.5(b). The truss has three degrees of freedom, which are the translations of joint 5 in the X, Y, and Z directions. These are numbered 1, 2, and 3, respectively. The twelve restrained coordinates of the truss are identified by numbers 4 through 15 in the figure.

Structure Stiffness Matrix:

Member 1 From Fig. 8.5(b), we can see that joint 1 is the beginning joint, and joint 5 is the end joint, for this member. By applying Eqs. (8.4), we determine

$$L = \sqrt{(X_5 - X_1)^2 + (Y_5 - Y_1)^2 + (Z_5 - Z_1)^2}$$

$$= \sqrt{(0+6)^2 + (24-0)^2 + (0-8)^2} = 26 \text{ ft} = 312 \text{ in.}$$

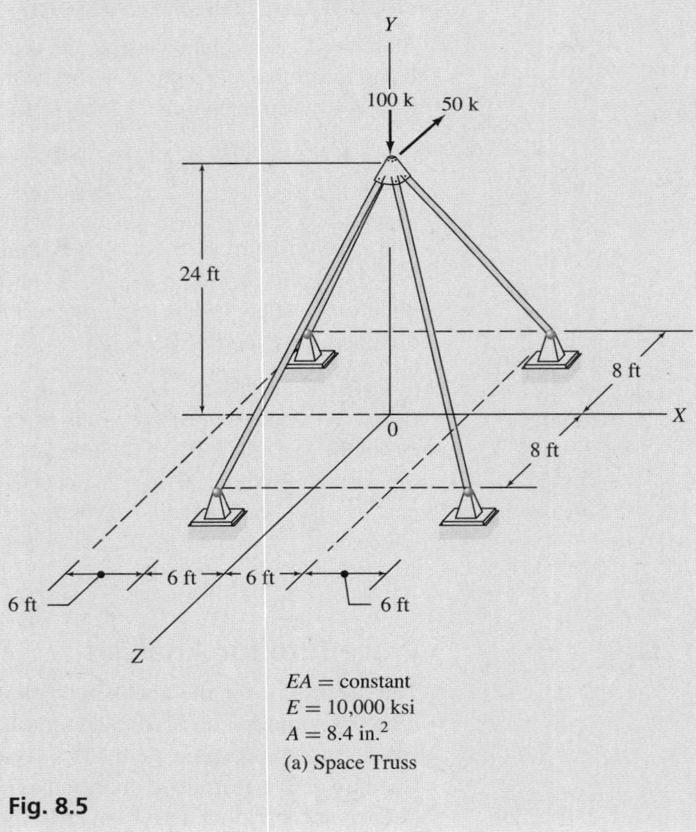

(a) Space Truss

Fig. 8.5

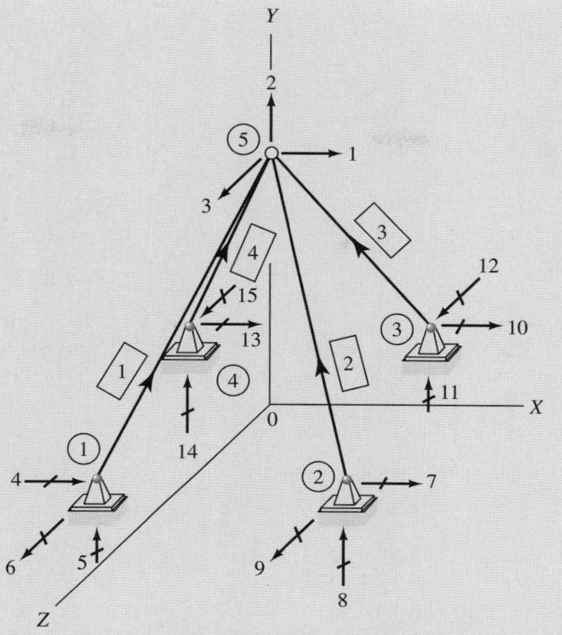

(b) Analytical Model

$$\mathbf{S} = \begin{array}{c} \begin{array}{ccc} \quad 1 \quad & \quad 2 \quad & \quad 3 \quad \end{array} \\ \left[\begin{array}{ccc} \begin{array}{c}(14.338 + 45.918\\ +14.338 + 45.918)\end{array} & \begin{array}{c}(57.351 - 91.837\\ -57.351 + 91.837)\end{array} & \begin{array}{c}(-19.117 + 30.612\\ -19.117 + 30.612)\end{array} \\ \begin{array}{c}(57.351 - 91.837\\ -57.351 + 91.837)\end{array} & \begin{array}{c}(229.4 + 183.67\\ +229.4 + 183.67)\end{array} & \begin{array}{c}(-76.468 - 61.224\\ +76.468 + 61.224)\end{array} \\ \begin{array}{c}(-19.117 + 30.612\\ -19.117 + 30.612)\end{array} & \begin{array}{c}(-76.468 - 61.224\\ +76.468 + 61.224)\end{array} & \begin{array}{c}(25.489 + 20.408\\ +25.489 + 20.408)\end{array} \end{array}\right] \end{array} \begin{array}{c} \\ 1 \\ \\ 2 \\ \\ 3 \end{array} = \begin{array}{c} \begin{array}{ccc} 1 & 2 & 3 \end{array} \\ \left[\begin{array}{ccc} 120.51 & 0 & 22.99 \\ 0 & 826.14 & 0 \\ 22.99 & 0 & 91.794 \end{array}\right] \end{array} \begin{array}{c} \\ 1 \\ 2 \\ 3 \end{array} \text{ k/in.}$$

(c) Structure Stiffness Matrix

$$\mathbf{R} = \left[\begin{array}{r} -5.5581 \\ -22.232 \\ 7.4108 \\ 1.3838 \\ -2.7677 \\ 0.92255 \\ -19.442 \\ 77.768 \\ 25.923 \\ 23.616 \\ 47.232 \\ 15.744 \end{array}\right] \begin{array}{l} 4 \\ 5 \\ 6 \\ 7 \\ 8 \\ 9 \\ 10 \\ 11 \\ 12 \\ 13 \\ 14 \\ 15 \end{array} \text{ k}$$

(d) Support Reaction Vector

Fig. 8.5 *(continued)*

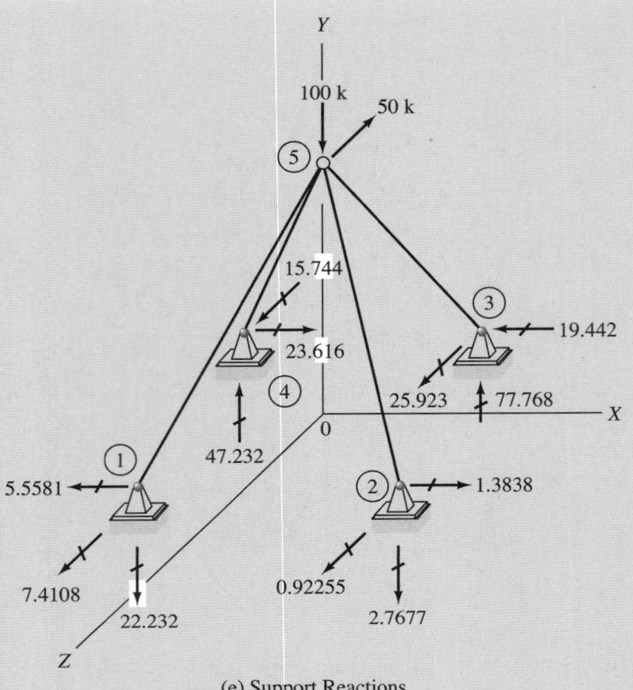

(e) Support Reactions

Fig. 8.5 *(continued)*

$$\cos\theta_X = \frac{X_5 - X_1}{L} = \frac{0+6}{26} = 0.23077$$

$$\cos\theta_Y = \frac{Y_5 - Y_1}{L} = \frac{24-0}{26} = 0.92308$$

$$\cos\theta_Z = \frac{Z_5 - Z_1}{L} = \frac{0-8}{26} = -0.30769$$

By substituting $E = 10{,}000$ ksi, $A = 8.4$ in.^{2}, $L = 312$ in., and the foregoing direction cosines, into Eq. (8.12), we calculate the member's global stiffness matrix to be

$$\mathbf{K}_1 = \begin{array}{c} \begin{matrix} 4 & 5 & 6 & 1 & 2 & 3 \end{matrix} \\ \left[\begin{array}{ccc|ccc} 14.338 & 57.351 & -19.117 & -14.338 & -57.351 & 19.117 \\ 57.351 & 229.4 & -76.468 & -57.351 & -229.4 & 76.468 \\ -19.117 & -76.468 & 25.489 & 19.117 & 76.468 & -25.489 \\ \hline -14.338 & -57.351 & 19.117 & 14.338 & 57.351 & -19.117 \\ -57.351 & -229.4 & 76.468 & 57.351 & 229.4 & -76.468 \\ 19.117 & 76.468 & -25.489 & -19.117 & -76.468 & 25.489 \end{array}\right] \begin{matrix} 4 \\ 5 \\ 6 \\ 1 \\ 2 \\ 3 \end{matrix} \end{array} \text{ k/in.}$$

Next, by using the member code numbers 4, 5, 6, 1, 2, 3, we store the pertinent elements of $\mathbf{K}_1$ in the 3×3 structure stiffness matrix $\mathbf{S}$ in Fig. 8.5(c).

Member 2

$$L = \sqrt{(X_5 - X_2)^2 + (Y_5 - Y_2)^2 + (Z_5 - Z_2)^2}$$

$$= \sqrt{(0 - 12)^2 + (24 - 0)^2 + (0 - 8)^2} = 28 \text{ ft} = 336 \text{ in.}$$

$$\cos\theta_X = \frac{X_5 - X_2}{L} = \frac{0 - 12}{28} = -0.42857$$

$$\cos\theta_Y = \frac{Y_5 - Y_2}{L} = \frac{24 - 0}{28} = 0.85714$$

$$\cos\theta_Z = \frac{Z_5 - Z_2}{L} = \frac{0 - 8}{28} = -0.28571$$

$$\mathbf{K}_2 = \begin{array}{c} \begin{matrix} 7 & \quad 8 & \quad 9 & \quad 1 & \quad 2 & \quad 3 \end{matrix} \\ \left[\begin{array}{rrrrrr} 45.918 & -91.837 & 30.612 & -45.918 & 91.837 & -30.612 \\ -91.837 & 183.67 & -61.224 & 91.837 & -183.67 & 61.224 \\ 30.612 & -61.224 & 20.408 & -30.612 & 61.224 & -20.408 \\ -45.918 & 91.837 & -30.612 & 45.918 & -91.837 & 30.612 \\ 91.837 & -183.67 & 61.224 & -91.837 & 183.67 & -61.224 \\ -30.612 & 61.224 & -20.408 & 30.612 & -61.224 & 20.408 \end{array}\right] \end{array} \begin{array}{c} \\ 7 \\ 8 \\ 9 \\ 1 \\ 2 \\ 3 \end{array} \text{ k/in.}$$

Member 3

$$L = \sqrt{(X_5 - X_3)^2 + (Y_5 - Y_3)^2 + (Z_5 - Z_3)^2}$$

$$= \sqrt{(0 - 6)^2 + (24 - 0)^2 + (0 + 8)^2} = 26 \text{ ft} = 312 \text{ in.}$$

$$\cos\theta_X = \frac{X_5 - X_3}{L} = \frac{0 - 6}{26} = -0.23077$$

$$\cos\theta_Y = \frac{Y_5 - Y_3}{L} = \frac{24 - 0}{26} = 0.92308$$

$$\cos\theta_Z = \frac{Z_5 - Z_3}{L} = \frac{0 + 8}{26} = 0.30769$$

$$\mathbf{K}_3 = \begin{array}{c} \begin{matrix} 10 & \quad 11 & \quad 12 & \quad 1 & \quad 2 & \quad 3 \end{matrix} \\ \left[\begin{array}{rrrrrr} 14.338 & -57.351 & -19.117 & -14.338 & 57.351 & 19.117 \\ -57.351 & 229.4 & 76.468 & 57.351 & -229.4 & -76.468 \\ -19.117 & 76.468 & 25.489 & 19.117 & -76.468 & -25.489 \\ -14.338 & 57.351 & 19.117 & 14.338 & -57.351 & -19.117 \\ 57.351 & -229.4 & -76.468 & -57.351 & 229.4 & 76.468 \\ 19.117 & -76.468 & -25.489 & -19.117 & 76.468 & 25.489 \end{array}\right] \end{array} \begin{array}{c} \\ 10 \\ 11 \\ 12 \\ 1 \\ 2 \\ 3 \end{array} \text{ k/in.}$$

Member 4

$$L = \sqrt{(X_5 - X_4)^2 + (Y_5 - Y_4)^2 + (Z_5 - Z_4)^2}$$

$$= \sqrt{(0 + 12)^2 + (24 - 0)^2 + (0 + 8)^2} = 28 \text{ ft} = 336 \text{ in.}$$

$$\cos\theta_X = \frac{X_5 - X_4}{L} = \frac{0+12}{28} = 0.42857$$

$$\cos\theta_Y = \frac{Y_5 - Y_4}{L} = \frac{24-0}{28} = 0.85714$$

$$\cos\theta_Z = \frac{Z_5 - Z_4}{L} = \frac{0+8}{28} = 0.28571$$

$$\mathbf{K}_4 = \begin{array}{c} \begin{array}{cccccc} 13 & 14 & 15 & 1 & 2 & 3 \end{array} \\ \left[\begin{array}{ccc|ccc} 45.918 & 91.837 & 30.612 & -45.918 & -91.837 & -30.612 \\ 91.837 & 183.67 & 61.224 & -91.837 & -183.67 & -61.224 \\ 30.612 & 61.224 & 20.408 & -30.612 & -61.224 & -20.408 \\ \hline -45.918 & -91.837 & -30.612 & 45.918 & 91.837 & 30.612 \\ -91.837 & -183.67 & -61.224 & 91.837 & 183.67 & 61.224 \\ -30.612 & -61.224 & -20.408 & 30.612 & 61.224 & 20.408 \end{array}\right] \end{array} \begin{array}{c} 13 \\ 14 \\ 15 \\ 1 \\ 2 \\ 3 \end{array} \text{ k/in.}$$

The complete structure stiffness matrix **S**, obtained by assembling the pertinent stiffness coefficients of the four members of the truss, is given in Fig. 8.5(c).

Joint Load Vector: By comparing Fig. 8.5(a) and (b), we obtain

$$\mathbf{P} = \begin{bmatrix} 0 \\ -100 \\ -50 \end{bmatrix} \text{ k}$$

Joint Displacements: By substituting **P** and **S** into the structure stiffness relationship, **P** = **Sd**, we write

$$\begin{bmatrix} 0 \\ -100 \\ -50 \end{bmatrix} = \begin{bmatrix} 120.51 & 0 & 22.99 \\ 0 & 826.14 & 0 \\ 22.99 & 0 & 91.794 \end{bmatrix} \begin{bmatrix} d_1 \\ d_2 \\ d_3 \end{bmatrix}$$

By solving the foregoing equations, we determine the joint displacements to be

$$\mathbf{d} = \begin{bmatrix} 0.10913 \\ -0.12104 \\ -0.57202 \end{bmatrix} \begin{array}{c} 1 \\ 2 \\ 3 \end{array} \text{ in.}$$ **Ans**

Member End Displacements and End Forces:

Member 1 Using its code numbers, we determine the member's global end displacements to be

$$\mathbf{v}_1 = \begin{bmatrix} 0 \\ 0 \\ 0 \\ 0.10913 \\ -0.12104 \\ -0.57202 \end{bmatrix} \begin{array}{c} 4 \\ 5 \\ 6 \\ 1 \\ 2 \\ 3 \end{array} \text{ in.}$$

To determine the member's end displacements in the local coordinate system, we first evaluate its transformation matrix as defined in Eq. (8.7):

$$\mathbf{T}_1 = \begin{bmatrix} 0.23077 & 0.92308 & -0.30769 & 0 & 0 & 0 \\ 0 & 0 & 0 & 0.23077 & 0.92308 & -0.30769 \end{bmatrix}$$

The member local end displacements can now be calculated by using the relationship $\mathbf{u} = \mathbf{Tv}$, as

$$\mathbf{u}_1 = \mathbf{T}_1\mathbf{v}_1 = \begin{bmatrix} 0 \\ 0.089459 \end{bmatrix} \text{ in.}$$

Before we can evaluate the member local end forces, we need to determine the local stiffness matrix $\mathbf{k}$, using Eq. (8.3):

$$\mathbf{k}_1 = \begin{bmatrix} 269.23 & -269.23 \\ -269.23 & 269.23 \end{bmatrix} \text{ k/in.}$$

Now, we can compute the member local end forces by using the relationship $\mathbf{Q} = \mathbf{ku}$, as

$$\mathbf{Q}_1 = \mathbf{k}_1\mathbf{u}_1 = \begin{bmatrix} -24.085 \\ 24.085 \end{bmatrix} \text{ k}$$

in which the negative sign of the first element of $\mathbf{Q}_1$ indicates that the member axial force is tensile; that is,

$Q_{a1} = 24.085$ k (T) **Ans**

By applying the relationship $\mathbf{F} = \mathbf{T}^T\mathbf{Q}$, we determine the member end forces in the global coordinate system to be

$$\mathbf{F}_1 = \mathbf{T}_1^T\mathbf{Q}_1 = \begin{bmatrix} -5.5581 \\ -22.232 \\ 7.4108 \\ \hdashline 5.5581 \\ 22.232 \\ -7.4108 \end{bmatrix} \begin{matrix} 4 \\ 5 \\ 6 \\ 1 \\ 2 \\ 3 \end{matrix} \text{ k}$$

Using the member code numbers 4, 5, 6, 1, 2, 3, the pertinent elements of $\mathbf{F}_1$ are stored in their proper positions in the support reaction vector $\mathbf{R}$, as shown in Fig. 8.5(d).

Member 2

$$\mathbf{v}_2 = \begin{bmatrix} 0 \\ 0 \\ 0 \\ 0.10913 \\ -0.12104 \\ -0.57202 \end{bmatrix} \begin{matrix} 7 \\ 8 \\ 9 \\ 1 \\ 2 \\ 3 \end{matrix} \text{ in.}$$

$$\mathbf{T}_2 = \begin{bmatrix} -0.42857 & 0.85714 & -0.28571 & 0 & 0 & 0 \\ 0 & 0 & 0 & -0.42857 & 0.85714 & -0.28571 \end{bmatrix}$$

$$\mathbf{u}_2 = \mathbf{T}_2\mathbf{v}_2 = \begin{bmatrix} 0 \\ 0.012916 \end{bmatrix} \text{ in.}$$

$$\mathbf{k}_2 = \begin{bmatrix} 250 & -250 \\ -250 & 250 \end{bmatrix} \text{ k/in.}$$

$$\mathbf{Q}_2 = \mathbf{k}_2\mathbf{u}_2 = \begin{bmatrix} -3.2289 \\ 3.2289 \end{bmatrix} \text{ k}$$

$Q_{a2} = 3.2289$ k (T) **Ans**

$$\mathbf{F}_2 = \mathbf{T}_2^T\mathbf{Q}_2 = \begin{bmatrix} 1.3838 \\ -2.7677 \\ 0.92255 \\ \hdashline -1.3838 \\ 2.7677 \\ -0.92255 \end{bmatrix} \begin{matrix} 7 \\ 8 \\ 9 \\ 1 \\ 2 \\ 3 \end{matrix} \text{ k}$$

Member 3

$$\mathbf{v}_3 = \begin{bmatrix} 0 \\ 0 \\ 0 \\ 0.10913 \\ -0.12104 \\ -0.57202 \end{bmatrix} \begin{matrix} 10 \\ 11 \\ 12 \\ 1 \\ 2 \\ 3 \end{matrix} \text{ in.}$$

$$\mathbf{T}_3 = \begin{bmatrix} -0.23077 & 0.92308 & 0.30769 & 0 & 0 & 0 \\ 0 & 0 & 0 & -0.23077 & 0.92308 & 0.30769 \end{bmatrix}$$

$$\mathbf{u}_3 = \mathbf{T}_3\mathbf{v}_3 = \begin{bmatrix} 0 \\ -0.31292 \end{bmatrix} \text{ in.}$$

$$\mathbf{k}_3 = \mathbf{k}_1$$

$$\mathbf{Q}_3 = \mathbf{k}_3\mathbf{u}_3 = \begin{bmatrix} 84.248 \\ -84.248 \end{bmatrix} \text{ k}$$

$Q_{a3} = 84.248$ k (C) **Ans**

$$\mathbf{F}_3 = \mathbf{T}_3^T\mathbf{Q}_3 = \begin{bmatrix} -19.442 \\ 77.768 \\ 25.923 \\ \hdashline 19.442 \\ -77.768 \\ -25.923 \end{bmatrix} \begin{matrix} 10 \\ 11 \\ 12 \\ 1 \\ 2 \\ 3 \end{matrix} \text{ k}$$

Member 4

$$\mathbf{v}_4 = \begin{bmatrix} 0 \\ 0 \\ 0 \\ 0.10913 \\ -0.12104 \\ -0.57202 \end{bmatrix} \begin{matrix} 13 \\ 14 \\ 15 \\ 1 \\ 2 \\ 3 \end{matrix} \text{ in.}$$

$$\mathbf{T}_4 = \begin{bmatrix} 0.42857 & 0.85714 & 0.28571 & 0 & 0 & 0 \\ 0 & 0 & 0 & 0.42857 & 0.85714 & 0.28571 \end{bmatrix}$$

$$\mathbf{u}_4 = \mathbf{T}_4\mathbf{v}_4 = \begin{bmatrix} 0 \\ -0.22042 \end{bmatrix} \text{ in.}$$

$$\mathbf{k}_4 = \mathbf{k}_2$$

$$\mathbf{Q}_4 = \mathbf{k}_4\mathbf{u}_4 = \begin{bmatrix} 55.104 \\ -55.104 \end{bmatrix} \text{k}$$

$Q_{a4} = 55.104$ k (C) **Ans**

$$\mathbf{F}_4 = \mathbf{T}_4^T\mathbf{Q}_4 = \begin{bmatrix} 23.616 \\ 47.232 \\ 15.744 \\ \hdashline -23.616 \\ -47.232 \\ -15.744 \end{bmatrix} \begin{matrix} 13 \\ 14 \\ 15 \\ 1 \\ 2 \\ 3 \end{matrix} \text{k}$$

Support Reactions: The completed reaction vector **R** is shown in Fig. 8.5(d), and the support reactions are depicted on a line diagram of the truss in Fig. 8.5(e). **Ans**

Equilibrium Check: Applying the equations of equilibrium to the free body of the entire space truss (Fig. 8.5(e)), we obtain

$+\rightarrow \sum F_X = 0 \qquad -5.5581 + 1.3838 - 19.442 + 23.616 \approx 0$ **Checks**

$+\uparrow \sum F_Y = 0 \qquad -22.232 - 2.7677 + 77.768 + 47.232 - 100 \approx 0$ **Checks**

$+\swarrow \sum F_Z = 0 \qquad 7.4108 + 0.92255 + 25.923 + 15.744 - 50 \approx 0$ **Checks**

$+\circlearrowleft \sum M_X = 0 \qquad 22.232(8) + 2.7677(8) + 77.768(8) + 47.232(8) - 50(24) \approx 0$ **Checks**

$+\circlearrowleft \sum M_Y = 0 \qquad -5.5581(8) + 7.4108(6) + 1.3838(8) - 0.92255(12) + 19.442(8) - 25.923(6) - 23.616(8) + 15.744(12) \approx 0$ **Checks**

$+\circlearrowleft \sum M_Z = 0 \qquad 22.232(6) - 2.7677(12) + 77.768(6) - 47.232(12) \approx 0$ **Checks**

8.2 GRIDS

A *grid* is defined as *a two-dimensional framework of straight members connected together by rigid and/or flexible connections, and subjected to loads and reactions perpendicular to the plane of the structure.* Because of their widespread use as supporting structures for long-span roofs and floors, the analysis of grids is usually formulated with the structural framework lying in a horizontal plane (unlike plane frames, which are oriented in a vertical plane), and subjected to external loads acting in the vertical direction, as shown in Fig. 8.6(a) on the next page.

Grids are composed of members that have doubly symmetric cross-sections, with each member oriented so that one of the planes of symmetry of its cross-section is in the vertical direction; that is, perpendicular to the plane of the structure, and in (or parallel to) the direction of the external loads (Fig. 8.6(a)). Under the action of vertical external loads, the joints of a grid can translate in the vertical direction and can rotate about axes in the

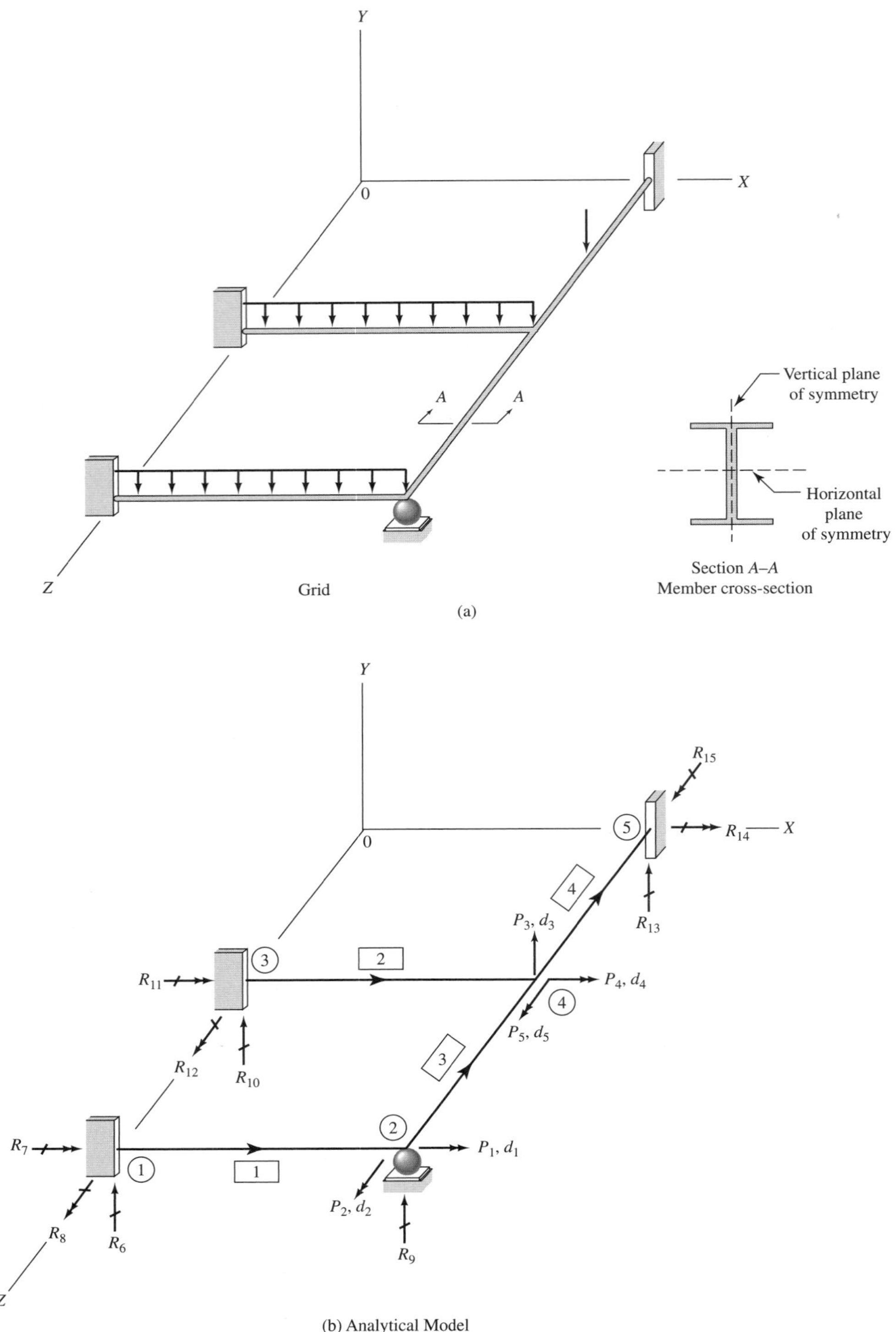
Y
0
X
A
A
Z
Grid
(a)
Vertical plane
of symmetry
Horizontal
plane
of symmetry
Section A–A
Member cross-section
Y
0
X
R_{15}
R_{14}
R_{13}
5
4
P_3, d_3
3
2
P_4, d_4
4
R_{11}
P_5, d_5
R_{12}
R_{10}
3
2
R_7
P_1, d_1
1
1
P_2, d_2
R_8
R_6
R_9
Z
(b) Analytical Model

Fig. 8.6

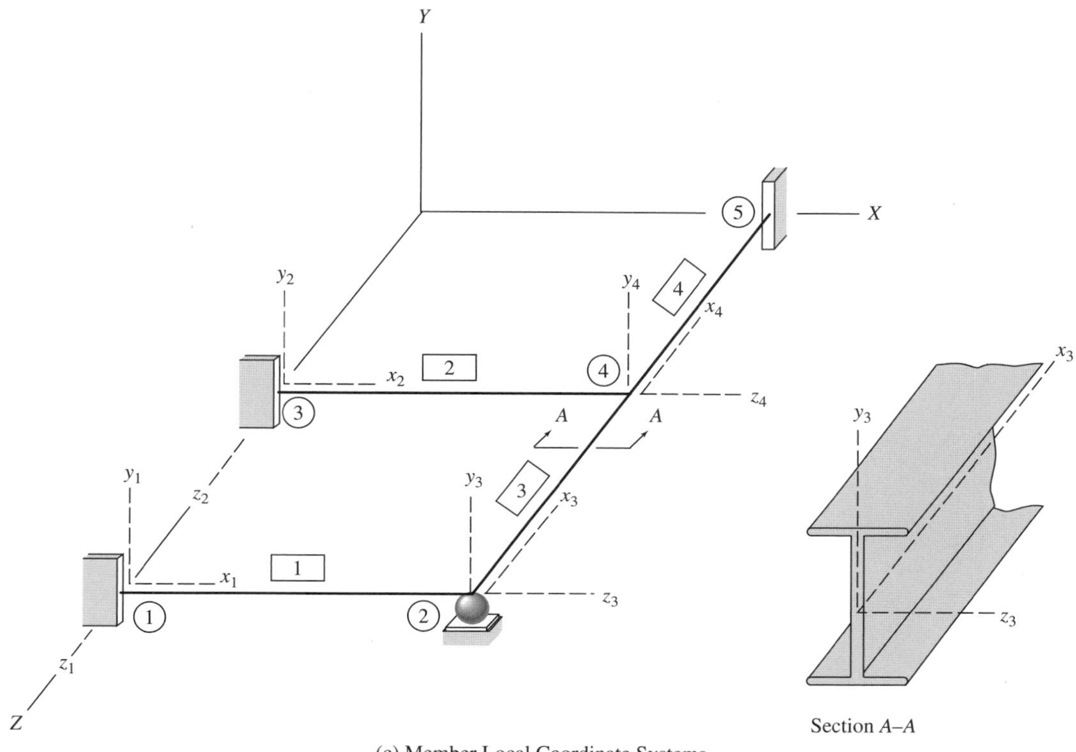

(c) Member Local Coordinate Systems

Fig. 8.6 *(continued)*

(horizontal) plane of the structure, while the grid members may be subjected to torsion, and uniaxial bending out of the plane of the structure.

Analytical Model and Degrees of Freedom

The process of dividing grids into members and joints, for the purpose of analysis, is the same as that for beams and plane frames—that is, a grid is divided into members and joints so that all of the members are straight and prismatic, and all the external reactions act only at the joints. Consider, for example, the grid of Fig. 8.6(a). The analytical model of the grid, as depicted in Fig. 8.6(b), shows that, for analysis, the grid is considered to be composed of four members and five joints. The overall geometry of the grid, and its joint loads and displacements, are described with reference to a global right-handed *XYZ* coordinate system, with the structure lying in the horizontal *XZ* plane, as shown in Fig. 8.6(b). Two global (*X* and *Z*) coordinates are needed to specify the location of each joint.

For each member of the grid, a local *xyz* coordinate system is established, with its origin at an end of the member and the *x* axis directed along the member's centroidal axis in the undeformed state. The local *y* and *z* axes are oriented, respectively, parallel to the vertical and horizontal axes of symmetry (or the principal axes of inertia) of the member cross-section. The positive direction of the local x axis is defined from the *beginning* toward the *end* of the member; the local y axis is considered positive upward (i.e., in the positive direction of the global Y axis); and the positive direction of the local z axis is defined so that the local xyz coordinate system is right-handed. The local coordinate systems selected for the four members of the example grid are depicted in Fig. 8.6(c).

As discussed previously, an unsupported joint of a grid can translate in the global Y direction and rotate about any axis in the XZ plane. Since small rotations can be treated as vector quantities, the foregoing joint rotation can be conveniently represented by its component rotations about the X and Z axes. Thus, a free joint of a grid has three degrees of freedom—the translation in the Y direction and the rotations about the X and Z axes. Therefore, three structure coordinates (i.e., free and/or restrained coordinates) need to be defined at each joint of the grid for the purpose of analysis; that is,

$$\left.\begin{aligned} NCJT &= 3 \\ NDOF &= 3(NJ) - NR \end{aligned}\right\} \quad \text{for grids} \tag{8.13}$$

The procedure for numbering the structure coordinates of grids is analogous to that for other types of framed structures. The degrees of freedom are numbered before the restrained coordinates. In the case of a joint with multiple degrees of freedom, the translation in the Y direction is numbered first, followed by the rotation about the X axis, and then the rotation about the Z axis. After all the degrees of freedom have been numbered, the grid's restrained coordinates are numbered in the same manner as the degrees of freedom. In Fig. 8.6(b), the degrees of freedom and restrained coordinates of the example grid are numbered using this procedure. It should be noted from this figure that the rotations and moments are now represented by *double-headed arrows* ($\twoheadrightarrow$), instead of the curved arrows ($\circlearrowleft$) used previously for plane structures. The double-headed arrows provide a convenient and unambiguous means of representing rotations and moments in three-dimensional space. To represent a rotation (or a moment/couple), an arrow is drawn pointing in the positive direction of the axis about which the rotation occurs (or the moment/couple acts). The positive sense (i.e., clockwise or counterclockwise) of the rotation (or moment/couple) is indicated by the curved fingers of the right hand with the extended thumb pointing in the direction of the arrowheads, as shown in Fig. 8.7.

Member Stiffness Relations in the Local Coordinate System

When a member with a noncircular (e.g., rectangular or I-shaped) cross-section is subjected to torsion, its initially plane cross-sections become warped surfaces; restraint of this *warping,* or out-of-plane deformation, of cross-sections can induce bending stresses in the member. Thus, in the analysis of grids and space frames (to be developed in the next section), it is commonly

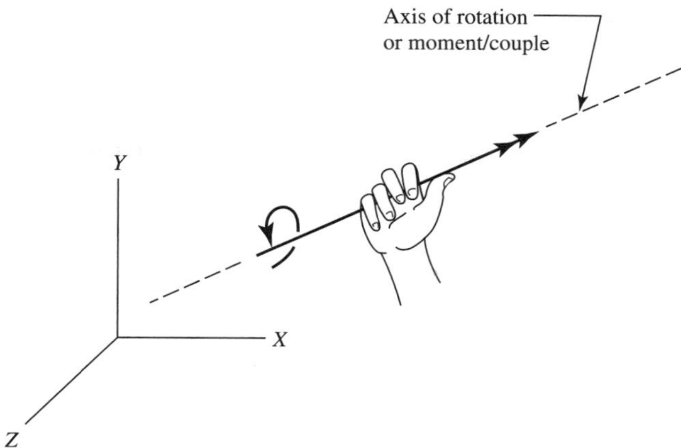

Fig. 8.7 *Representation of Rotation or Moment/Couple in Three-Dimensional Space*

assumed that *the cross sections of all the members are free to warp out of their planes under the action of torsional moments.* This assumption, together with the previously stated condition about the cross-sections of grid members being doubly symmetric with one of the planes of symmetry oriented parallel to the direction of applied loads, has the effect of uncoupling the member's torsional and bending stiffnesses so that a twisting (or torsional deformation) of the member induces only torsional moments but no bending moments, and vice versa. With the torsional and bending effects uncoupled, the local stiffness relations for the members of grids can be obtained by simply extending the stiffness relations for beams (Chapter 5) to include the familiar torsional stiffness relations found in textbooks on *mechanics of materials.*

Consider an arbitrary member m of a grid, as shown in Fig. 8.8(a) on the next page. Like a joint of a grid, three displacements are needed to completely specify the displaced position of each end of the grid member. Thus, the member has a total of six degrees of freedom. In the local coordinate system of the member, the six member end displacements are denoted by u_1 through u_6, and the associated member end forces are denoted by Q_1 through Q_6, as shown in Fig. 8.8(a). As indicated in this figure, *a member's local end displacements and end forces are numbered by beginning at its end b, with the translation and the force in the y direction numbered first, followed by the rotation and moment about the x axis, and then the rotation and moment about the z axis. The displacements and forces at the member's opposite end e are then numbered in the same sequential order.*

The relationship between the end forces $\mathbf{Q}$ and the end displacements $\mathbf{u}$, for the members of grids, can be expressed as

$$\boxed{\mathbf{Q} = \mathbf{k}\mathbf{u} + \mathbf{Q}_f} \tag{8.14}$$

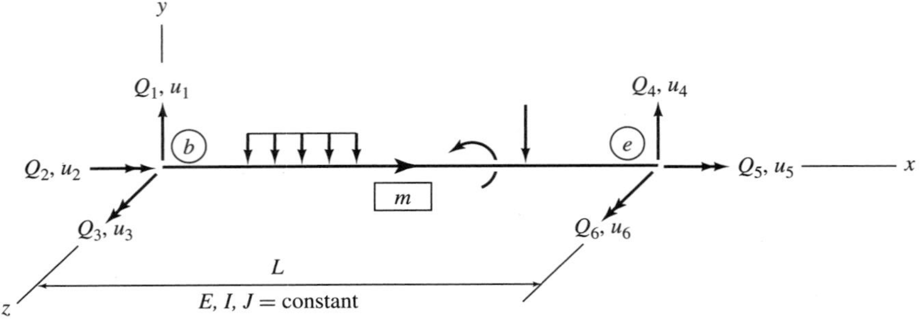

(a) Member Forces and Displacements in the Local Coordinate System

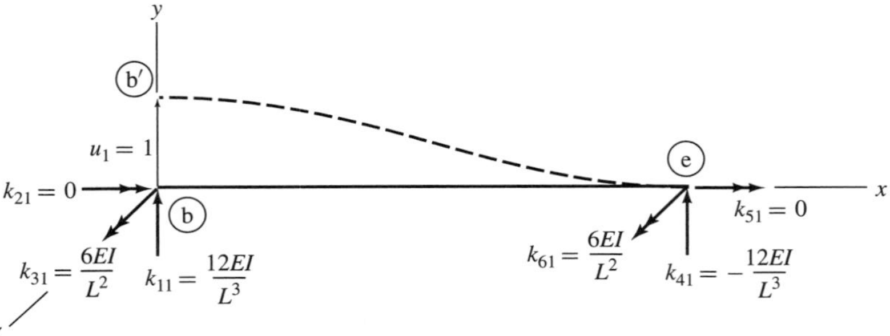

(b) $u_1 = 1,\ u_2 = u_3 = u_4 = u_5 = u_6 = 0$

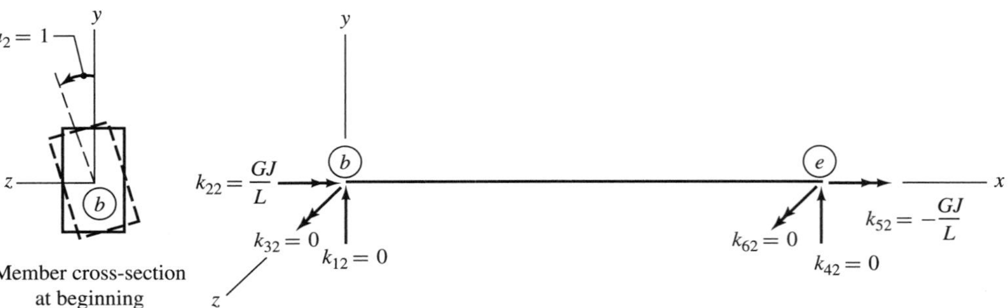

(c) $u_2 = 1,\ u_1 = u_3 = u_4 = u_5 = u_6 = 0$

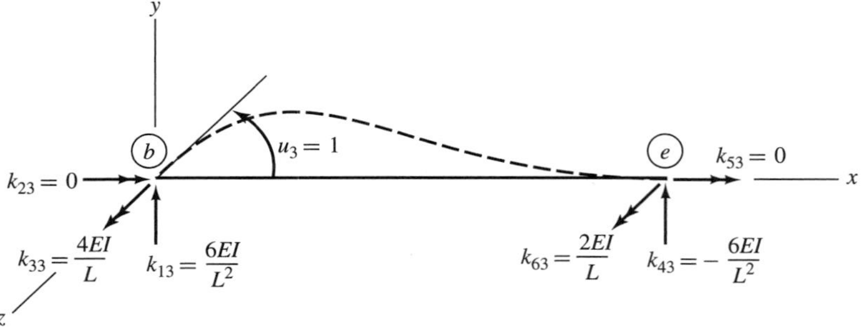

(d) $u_3 = 1,\ u_1 = u_2 = u_4 = u_5 = u_6 = 0$

Fig. 8.8

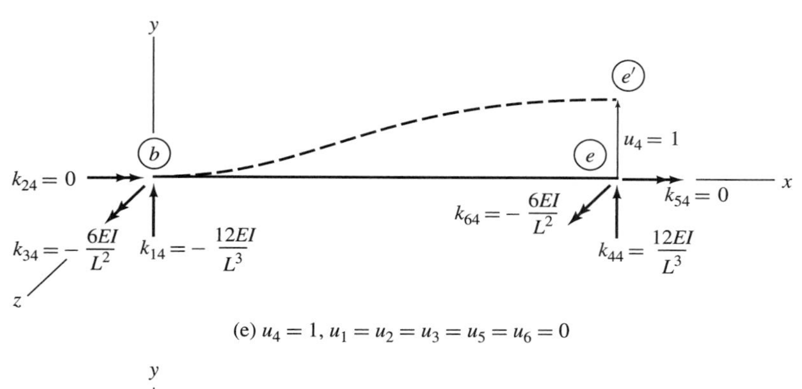

(e) $u_4 = 1, u_1 = u_2 = u_3 = u_5 = u_6 = 0$

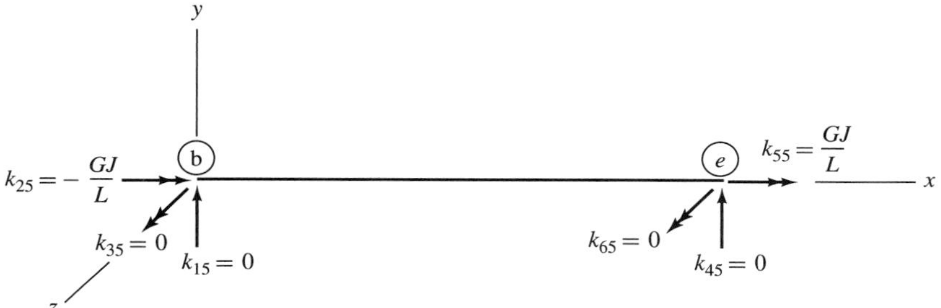

(f) $u_5 = 1, u_1 = u_2 = u_3 = u_4 = u_6 = 0$

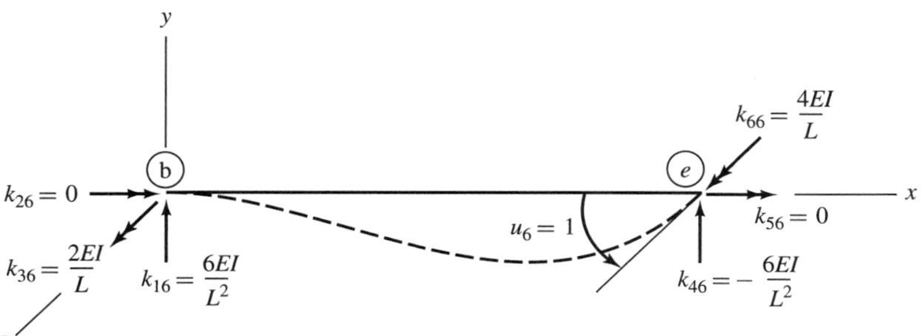

(g) $u_6 = 1, u_1 = u_2 = u_3 = u_4 = u_5 = 0$

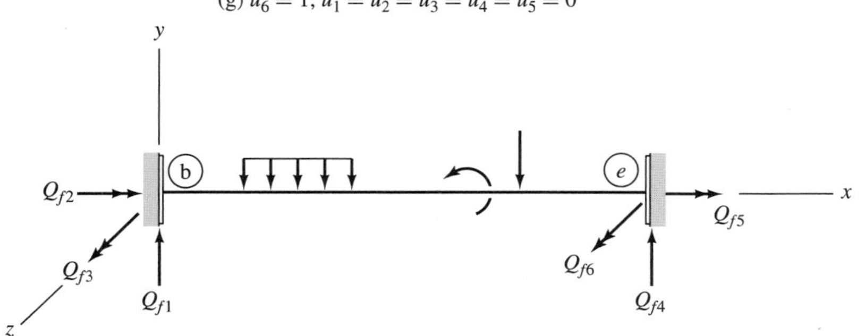

(h) Member Fixed-End Forces in the Local Coordinate System ($u_1 = u_2 = u_3 = u_4 = u_5 = u_6 = 0$)

Fig. 8.8 *(continued)*

in which $\mathbf{k}$ represents the 6×6 member stiffness matrix in the local coordinate system, and $\mathbf{Q}_f$ denotes the 6×1 member local fixed-end force vector.

Like the other types of framed structures, the explicit form of $\mathbf{k}$ for grid members can be obtained by subjecting a member, separately, to unit values of each of the six end displacements, as shown in Figs. 8.8(b) through (g), and evaluating the corresponding member end forces.

The stiffness coefficients required to cause the unit values of the member end displacements u_1, u_3, u_4, and u_6, are shown in Figs. 8.8(b), (d), (e), and (g), respectively. The expressions for these stiffness coefficients were derived in Section 5.2. To derive the expressions for the torsional stiffness coefficients, recall from a previous course on mechanics of materials that the relationship between a torsional moment (or torque) M_T applied at the free end of a cantilever circular shaft, and the resulting angle of twist ϕ (see Fig. 8.9), can be expressed as

$$\phi = \frac{M_T L}{GJ} \tag{8.15}$$

in which G denotes the shear modulus of the material, and J denotes the polar moment of inertia of the shaft.

For members with noncircular cross sections, the relationship between the torsional moment M_T and the angle of twist ϕ can be quite complicated because of warping [40]. However, if warping is not restrained, then Eq. (8.15) can be used to approximate the torsional behavior of members with noncircular cross-sections—provided that J is now considered to be the *Saint-Venant's torsion constant,* or simply the *torsion constant,* of the member's cross-section, instead of its polar moment of inertia. Although the derivation of the expressions for torsion constant J for various cross-sectional shapes is beyond the scope of this text, such derivations can be found in textbooks on the *theory of elasticity* and *advanced mechanics of materials* [40]. The expressions for J for some common cross-sectional shapes are listed in Table 8.1. Furthermore, the torsion constant for any thin-walled, open cross-section can be approximated

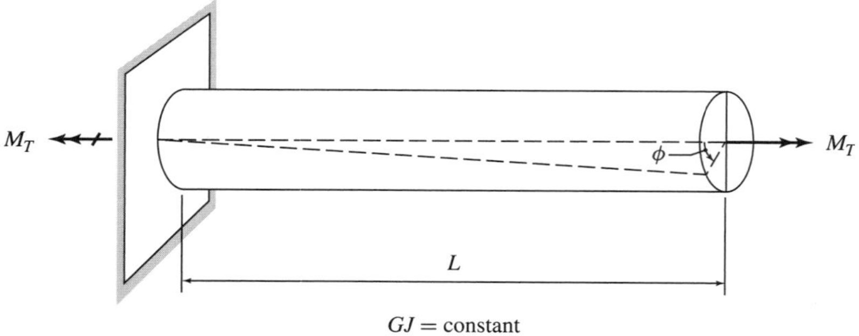

Fig. 8.9 *Circular Shaft Subjected to Torsional Moment*

Table 8.1 *Torsion Constants for Common Member Cross-Sections* [40, 52]

Cross-Section	Torsion Constant
Solid circle, radius r	$J = \frac{1}{2}\pi r^4$
Thin-walled circular tube, radius r, thickness t	$J = 2\pi r^3 t$
Solid rectangle, width b, depth d	$J = \beta b^3 d$ for $b \leq d$ $\beta = \frac{1}{3} - 0.21\frac{b}{d}\left[1 - \frac{1}{12}\left(\frac{b}{d}\right)^4\right]$
I-section, flange width b_f, flange thickness t_f, web thickness t_w, height h	$J = \frac{1}{3}\left(2b_f t_f^3 + h t_w^3\right)$
Rectangular box section, width b, height h, flange thickness t_f, web thickness t_w	$J = \dfrac{2b^2h^2}{b/t_f + h/t_w}$

by the relationship

$$J = \frac{1}{3}\sum bt^3 \tag{8.16}$$

in which b and t denote, respectively, the width and thickness of each rectangular segment of the cross section.

Returning our attention to Fig. 8.8(c), we realize that the expression for the stiffness coefficient k_{22} can be obtained by substituting $\phi = u_2 = 1$ and

$M_T = k_{22}$ into Eq. (8.15), and solving the resulting equation for k_{22}. This yields

$$\boxed{k_{22} = \frac{GJ}{L}} \tag{8.17}$$

The other torsional stiffness coefficient k_{52} can now be determined by applying the following equilibrium equation.

$$\overset{+}{\twoheadrightarrow} \sum M_X = 0 \qquad \frac{GJ}{L} + k_{52} = 0$$

$$\boxed{k_{52} = -\frac{GJ}{L}} \tag{8.18}$$

The expressions for coefficients k_{25} and k_{55} (Fig. 8.8(f)) can be obtained in a similar manner. Substitution of $\phi = u_5 = 1$ and $M_T = k_{55}$ into Eq. (8.15) yields

$$\boxed{k_{55} = \frac{GJ}{L}} \tag{8.19}$$

and by considering the equilibrium of the free body of the member, we obtain k_{25} as

$$\overset{+}{\twoheadrightarrow} \sum M_x = 0 \qquad k_{25} + \frac{GJ}{L} = 0$$

$$\boxed{k_{25} = -\frac{GJ}{L}} \tag{8.20}$$

Thus, by arranging all the stiffness coefficients shown in Figs. 8.8(b) through (g) into a matrix, we obtain the following expression for the local stiffness matrix for the members of grids.

$$\mathbf{k} = \frac{EI}{L^3}\begin{bmatrix} 12 & 0 & 6L & -12 & 0 & 6L \\ 0 & \dfrac{GJL^2}{EI} & 0 & 0 & -\dfrac{GJL^2}{EI} & 0 \\ 6L & 0 & 4L^2 & -6L & 0 & 2L^2 \\ -12 & 0 & -6L & 12 & 0 & -6L \\ 0 & -\dfrac{GJL^2}{EI} & 0 & 0 & \dfrac{GJL^2}{EI} & 0 \\ 6L & 0 & 2L^2 & -6L & 0 & 4L^2 \end{bmatrix} \tag{8.21}$$

The local fixed-end force vector for the members of grids is expressed as (Fig. 8.8(h))

$$\mathbf{Q}_f = \begin{bmatrix} FS_b \\ FT_b \\ FM_b \\ FS_e \\ FT_e \\ FM_e \end{bmatrix} \tag{8.22}$$

in which the fixed-end shears (FS_b and FS_e) and bending moments (FM_b and FM_e) can be calculated by using the fixed-end force equations given for loading types 1 through 4 inside the front cover. (The procedure for deriving those fixed-end shear and bending moment equations was discussed in Section 5.4.)

The expressions for the fixed-end torsional moments (FT_b and FT_e), due to an external torque M_T applied to the member, are also given inside the front cover (see loading type 7). To derive these expressions, let us consider a fixed member of a grid, subjected to a torque M_T, as shown in Fig. 8.10(a). If the end e of the member were free to rotate, then its cross-section would twist clockwise as shown in Fig. 8.10(b). Let ϕ be the angle of twist at end e of the released member. As portion Ae of the released member (Fig. 8.10(b)) is not subjected to any torsional moments, the angle of twist, ϕ, at end e equals that at point A, and its magnitude can be obtained by substituting $L = l_1$ into Eq. (8.15); that is,

$$\phi = \frac{M_T l_1}{GJ} \tag{8.23}$$

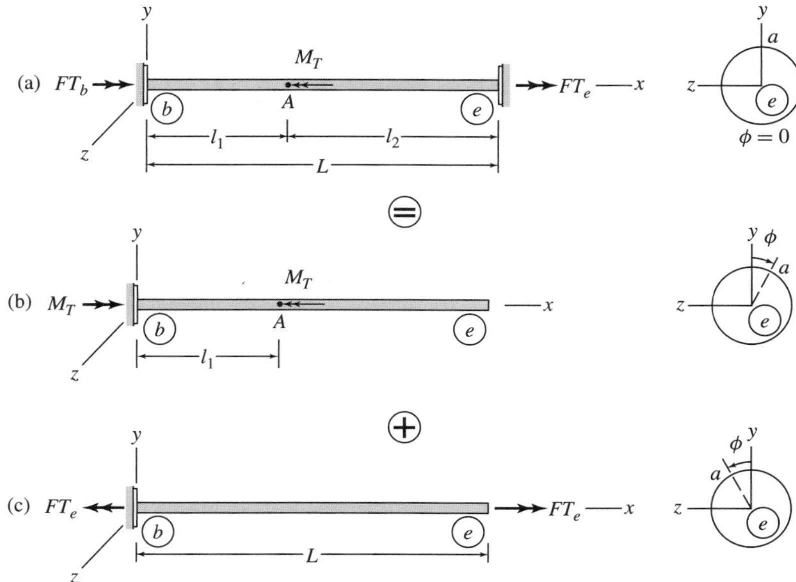

Fig. 8.10

Since the angle of twist at end e of the actual fixed member (Fig. 8.10(a)) is 0, the fixed-end torsional moment FT_e must be of such magnitude that, when applied to the released member as shown in Fig. 8.10(c), it should twist the cross-section at end e by an angle equal in magnitude to the angle ϕ due to the torque M_T, but in the opposite (i.e., counterclockwise) direction. The angle of twist due to FT_e can be obtained by substituting $M_T = FT_e$ into Eq. (8.15); that is,

$$\phi = \frac{FT_e L}{GJ} \tag{8.24}$$

and the relationship between FT_e and the external torque M_T can be established by equating Eqs. (8.23) and (8.24), as

$$\phi = \frac{FT_e L}{GJ} = \frac{M_T l_1}{GJ}$$

from which we obtain the expression for the fixed-end torsional moment FT_e:

$$FT_e = \frac{M_T l_1}{L} \tag{8.25}$$

The expression for the other fixed-end torsional moment, FT_b, can now be determined by applying the equilibrium condition that the algebraic sum of the three torsional moments acting on the fixed member (Fig. 8.10(a)) must be 0; that is,

$$\xrightarrow{+}\!\!\!\!\!\!\rightarrow \sum M_x = 0 \qquad FT_b - M_T + FT_e = 0$$

By substituting Eq. (8.25) into the foregoing equation and rearranging terms, we obtain the expression for FT_b:

$$FT_b = M_T\left(\frac{L - l_1}{L}\right) = \frac{M_T l_2}{L} \tag{8.26}$$

Member Releases The expressions for the local stiffness matrix $\mathbf{k}$ and the fixed-end force vector $\mathbf{Q}_f$, as given in Eqs. (8.21) and (8.22), respectively, are valid only for members of type 0 (i.e., $MT = 0$), which are rigidly connected to joints at both ends. For grid members with moment releases, the foregoing expressions for $\mathbf{k}$ and $\mathbf{Q}_f$ need to be modified using the procedure described in Section 7.1. If the member releases are assumed to be in the form of spherical hinges (or ball-and-socket type of connections), so that both bending and torsional moments are 0 at the released member ends, then the modified local stiffness matrices $\mathbf{k}$ and fixed-end force vectors $\mathbf{Q}_f$ for the grid members with releases can be expressed as follows.

For a member with a hinge at the beginning ($MT = 1$):

$$\mathbf{k} = \frac{EI}{L^3}\begin{bmatrix} 3 & 0 & 0 & -3 & 0 & 3L \\ 0 & 0 & 0 & 0 & 0 & 0 \\ 0 & 0 & 0 & 0 & 0 & 0 \\ -3 & 0 & 0 & 3 & 0 & -3L \\ 0 & 0 & 0 & 0 & 0 & 0 \\ 3L & 0 & 0 & -3L & 0 & 3L^2 \end{bmatrix} \tag{8.27}$$

$$\mathbf{Q}_f = \begin{bmatrix} FS_b - \dfrac{3}{2L}FM_b \\ 0 \\ 0 \\ FS_e + \dfrac{3}{2L}FM_b \\ FT_e + FT_b \\ FM_e - \dfrac{1}{2}FM_b \end{bmatrix} \tag{8.28}$$

For a member with a hinge at the end ($MT = 2$):

$$\mathbf{k} = \frac{EI}{L^3}\begin{bmatrix} 3 & 0 & 3L & -3 & 0 & 0 \\ 0 & 0 & 0 & 0 & 0 & 0 \\ 3L & 0 & 3L^2 & -3L & 0 & 0 \\ -3 & 0 & -3L & 3 & 0 & 0 \\ 0 & 0 & 0 & 0 & 0 & 0 \\ 0 & 0 & 0 & 0 & 0 & 0 \end{bmatrix} \tag{8.29}$$

$$\mathbf{Q}_f = \begin{bmatrix} FS_b - \dfrac{3}{2L}FM_e \\ FT_b + FT_e \\ FM_b - \dfrac{1}{2}FM_e \\ FS_e + \dfrac{3}{2L}FM_e \\ 0 \\ 0 \end{bmatrix} \tag{8.30}$$

For a member with hinges at both ends ($MT = 3$):

$$\mathbf{k} = \mathbf{0} \tag{8.31}$$

$$\mathbf{Q}_f = \begin{bmatrix} FS_b - \dfrac{1}{L}(FM_b + FM_e) \\ 0 \\ 0 \\ FS_e + \dfrac{1}{L}(FM_b + FM_e) \\ 0 \\ 0 \end{bmatrix} \tag{8.32}$$

Note that the members of type 3 offer no resistance against twisting and, therefore, cannot be subjected to any torques or torsional member loading.

Coordinate Transformations

Consider an arbitrary member m of a grid, as shown in Fig. 8.11(a). The orientation of the member in the horizontal (XZ) plane is defined by an angle θ between the positive directions of the global X axis and the member's local x axis, as shown in the figure. The member's length, and its direction cosines, can be expressed in terms of the global coordinates of the member end joints, b and e, by the following relationships.

$$L = \sqrt{(X_e - X_b)^2 + (Z_e - Z_b)^2} \tag{8.33a}$$

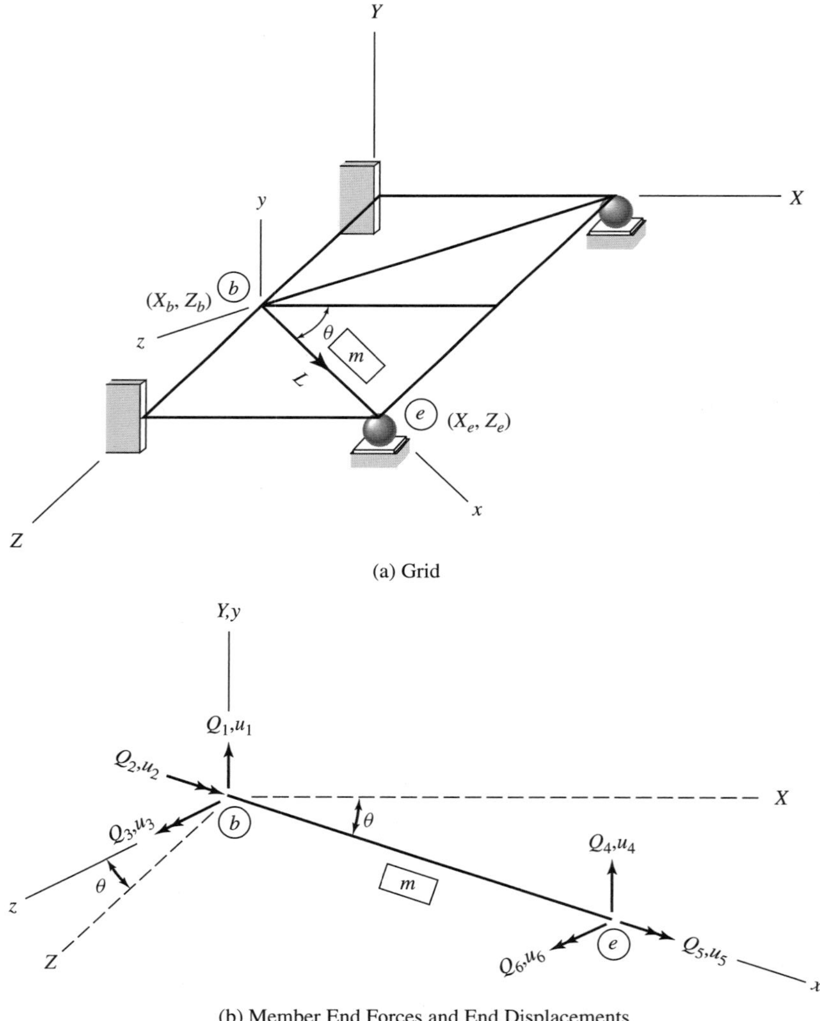

Fig. 8.11

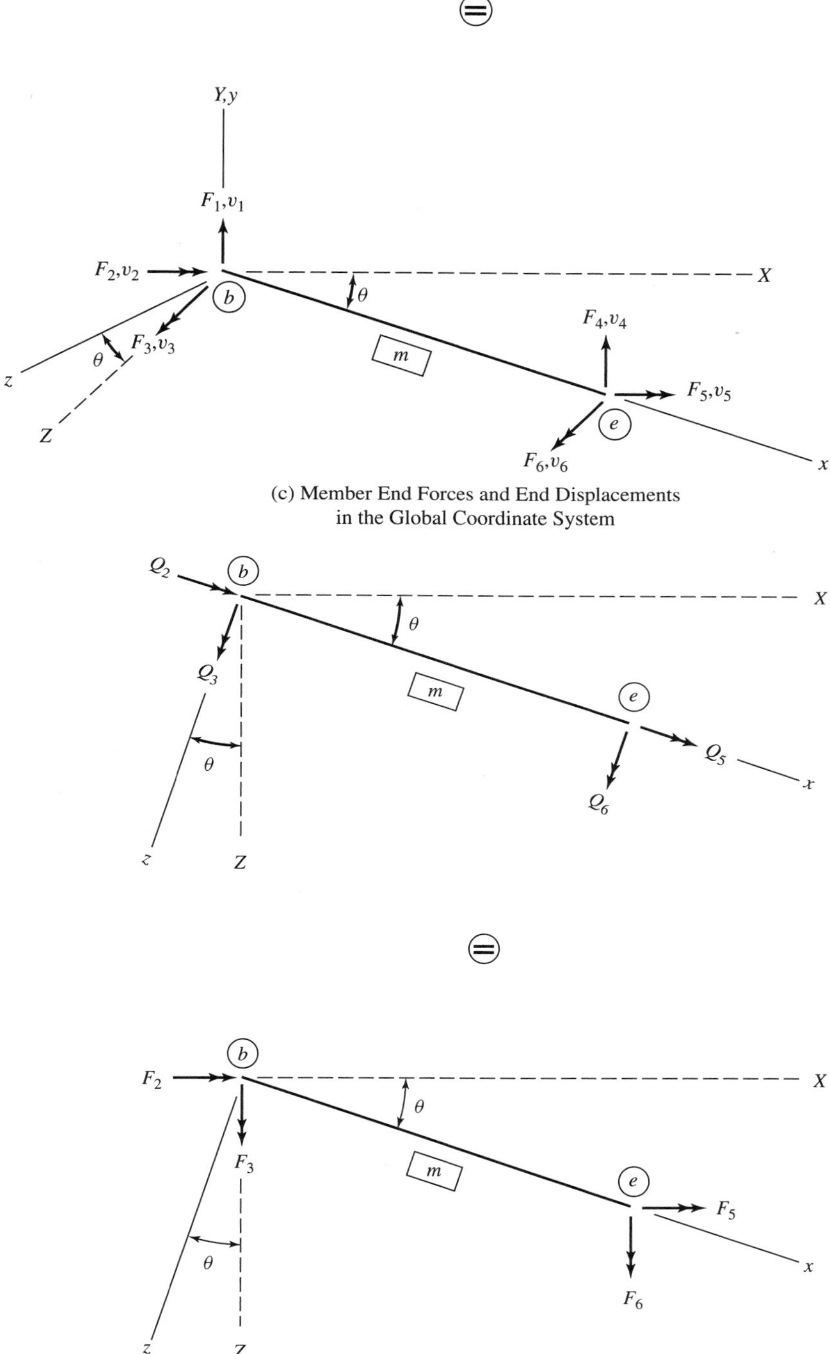

(c) Member End Forces and End Displacements in the Global Coordinate System

(d) Member Local and Global End Forces in the Horizontal (*XZ*) or (*xz*) Plane

Fig. 8.11 (*continued*)

$$\cos\theta = \frac{X_e - X_b}{L} \tag{8.33b}$$

$$\sin\theta = \frac{Z_e - Z_b}{L} \tag{8.33c}$$

The member local end forces **Q** and end displacements **u** are shown in Fig. 8.11(b); Fig. 8.11(c) depicts the equivalent system of end forces **F** and end displacements **v**, in the global coordinate system. As indicated in Fig. 8.11(c), the global member end forces and end displacements are numbered by beginning at member end b, with the force and translation in the Y direction numbered first, followed by the moment and rotation about the X axis, and then the moment and rotation about the Z axis. The forces and displacements at the member's opposite end e are then numbered in the same sequential order.

By comparing Figs. 8.11(b) and (c), we realize that at member end b, the local forces Q_1, Q_2, and Q_3 must be equal to the algebraic sums of the components of the global forces F_1, F_2, and F_3 in the directions of the local y, x, and z axes, respectively; that is (also, see Fig. 8.11(d)),

$$Q_1 = F_1 \tag{8.34a}$$

$$Q_2 = F_2\cos\theta + F_3\sin\theta \tag{8.34b}$$

$$Q_3 = -F_2\sin\theta + F_3\cos\theta \tag{8.34c}$$

Similarly, the local forces at member end e can be expressed in terms of the global forces as

$$Q_4 = F_4 \tag{8.34d}$$

$$Q_5 = F_5\cos\theta + F_6\sin\theta \tag{8.34e}$$

$$Q_6 = -F_5\sin\theta + F_6\cos\theta \tag{8.34f}$$

Equations 8.34(a) through (f) can be expressed in matrix form as

$$\mathbf{Q} = \mathbf{TF} \tag{8.35}$$

with the transformation matrix **T** given by

$$\mathbf{T} = \begin{bmatrix} 1 & 0 & 0 & 0 & 0 & 0 \\ 0 & \cos\theta & \sin\theta & 0 & 0 & 0 \\ 0 & -\sin\theta & \cos\theta & 0 & 0 & 0 \\ 0 & 0 & 0 & 1 & 0 & 0 \\ 0 & 0 & 0 & 0 & \cos\theta & \sin\theta \\ 0 & 0 & 0 & 0 & -\sin\theta & \cos\theta \end{bmatrix} \tag{8.36}$$

Because the translations and small rotations of the member ends can be treated as vector quantities, the foregoing transformation matrix also defines the transformation of member end displacements from the global to the local coordinate system; that is, $\mathbf{u} = \mathbf{Tv}$. Furthermore, the transformation matrix **T**, as given in Eq. (8.36), can be used to transform member end forces and displacements from the local to the global coordinate system via the relationships $\mathbf{F} = \mathbf{T}^T\mathbf{Q}$ and $\mathbf{v} = \mathbf{T}^T\mathbf{u}$, respectively.

Procedure for Analysis

The procedure for analysis of grids remains the same as that for plane frames developed in Chapter 6 (and modified in Chapter 7); provided, of course, that the member local stiffness and transformation matrices, and local fixed-end force vectors, developed in this section are used in the analysis. The procedure is illustrated by the following example.

EXAMPLE 8.2 Determine the joint displacements, member end forces, and support reactions for the three-member grid shown in Fig. 8.12(a), using the matrix stiffness method.

SOLUTION *Analytical Model:* The grid has three degrees of freedom and nine restrained coordinates, as shown in Fig. 8.12(b).

Structure Stiffness Matrix:

Member 1 From Fig. 8.12(b), we can see that joint 1 is the beginning joint, and joint 4 the end joint, for this member. By applying Eqs. (8.33), we determine the length, and the direction cosines, for the member to be

$$L = \sqrt{(X_4 - X_1)^2 + (Z_4 - Z_1)^2} = \sqrt{(8-0)^2 + (6-0)^2} = 10 \text{ m}$$

$$\cos\theta = \frac{X_4 - X_1}{L} = \frac{8-0}{10} = 0.8$$

$$\sin\theta = \frac{Z_4 - Z_1}{L} = \frac{6-0}{10} = 0.6$$

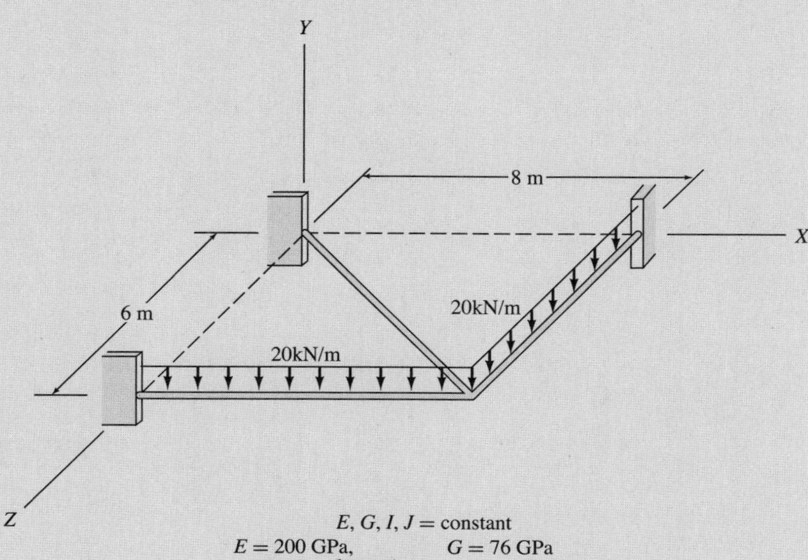

(a) Grid

Fig. 8.12

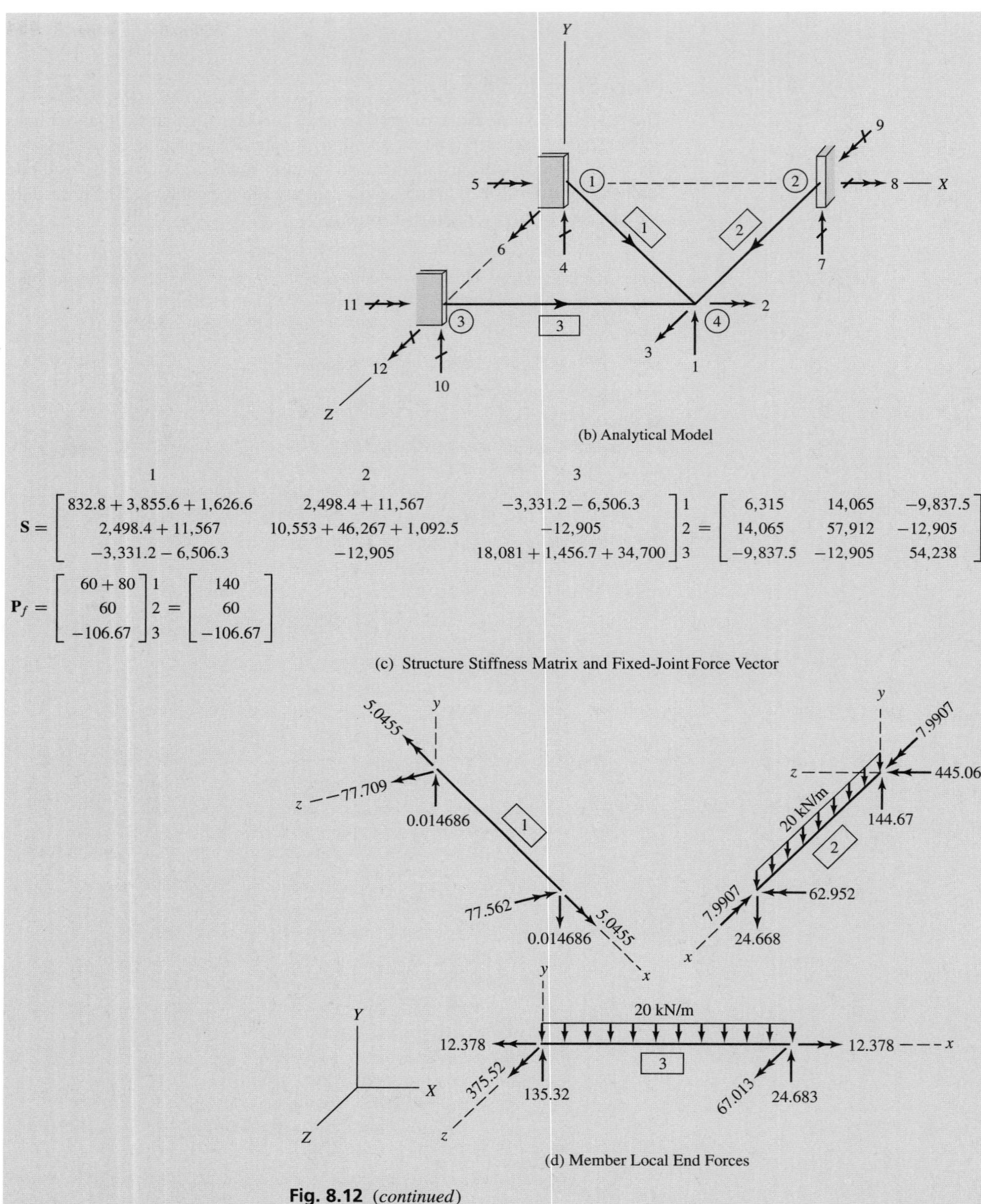

(b) Analytical Model

$$\mathbf{S} = \begin{bmatrix} 832.8 + 3{,}855.6 + 1{,}626.6 & 2{,}498.4 + 11{,}567 & -3{,}331.2 - 6{,}506.3 \\ 2{,}498.4 + 11{,}567 & 10{,}553 + 46{,}267 + 1{,}092.5 & -12{,}905 \\ -3{,}331.2 - 6{,}506.3 & -12{,}905 & 18{,}081 + 1{,}456.7 + 34{,}700 \end{bmatrix} \begin{matrix} 1 \\ 2 \\ 3 \end{matrix} = \begin{bmatrix} 6{,}315 & 14{,}065 & -9{,}837.5 \\ 14{,}065 & 57{,}912 & -12{,}905 \\ -9{,}837.5 & -12{,}905 & 54{,}238 \end{bmatrix}$$

$$\mathbf{P}_f = \begin{bmatrix} 60 + 80 \\ 60 \\ -106.67 \end{bmatrix} \begin{matrix} 1 \\ 2 \\ 3 \end{matrix} = \begin{bmatrix} 140 \\ 60 \\ -106.67 \end{bmatrix}$$

(c) Structure Stiffness Matrix and Fixed-Joint Force Vector

(d) Member Local End Forces

Fig. 8.12 *(continued)*

$$
\mathbf{R} = \begin{bmatrix} 0.014686 \text{ kN} \\ -50.662 \text{ kN}\cdot\text{m} \\ 59.14 \text{ kN}\cdot\text{m} \\ 144.67 \text{ kN} \\ -445.06 \text{ kN}\cdot\text{m} \\ 7.9907 \text{ kN}\cdot\text{m} \\ 135.32 \text{ kN} \\ -12.378 \text{ kN}\cdot\text{m} \\ 375.52 \text{ kN}\cdot\text{m} \end{bmatrix} \begin{matrix} 4 \\ 5 \\ 6 \\ 7 \\ 8 \\ 9 \\ 10 \\ 11 \\ 12 \end{matrix}
$$

(e) Support Reaction Vector

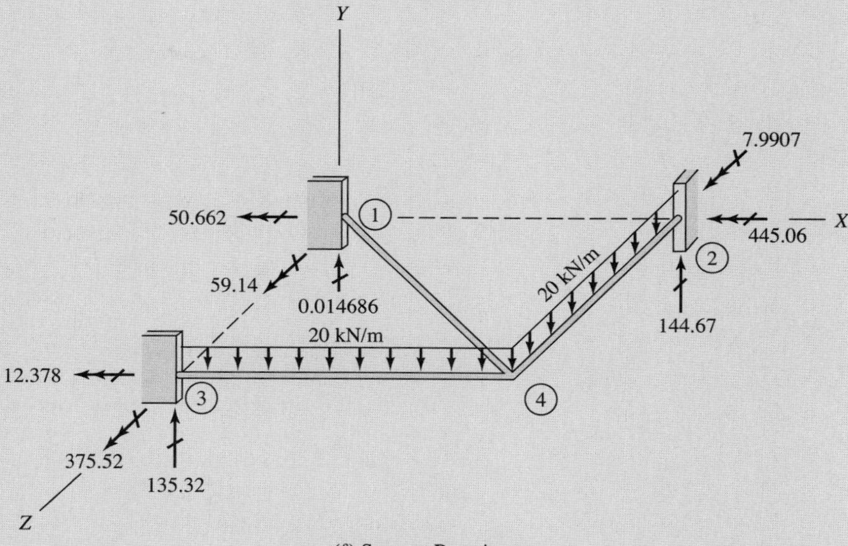

(f) Support Reactions

Fig. 8.12 *(continued)*

Since $MT = 0$ for this member, we use Eq. (8.21) to determine its local stiffness matrix $\mathbf{k}$. Thus, by substituting $E = 200(10^6)$ kN/m^2, $G = 76(10^6)$ kN/m^2, $L = 10$ m, $I = 347(10^{-6})$ m^4, and $J = 115(10^{-6})$ m^4 into Eq. (8.21), we obtain

$$
\mathbf{k}_1 = \begin{bmatrix} 832.8 & 0 & 4,164 & -832.8 & 0 & 4,164 \\ 0 & 874 & 0 & 0 & -874 & 0 \\ 4,164 & 0 & 27,760 & -4,164 & 0 & 13,880 \\ -832.8 & 0 & -4,164 & 832.8 & 0 & -4,164 \\ 0 & -874 & 0 & 0 & 874 & 0 \\ 4,164 & 0 & 13,880 & -4,164 & 0 & 27,760 \end{bmatrix} \quad \textbf{(1)}
$$

As the member is not subjected to any loads, its global and local fixed-end force vectors are 0; that is,

$$\mathbf{F}_{f1} = \mathbf{Q}_{f1} = \mathbf{0}$$

Before we can calculate the member global stiffness matrix $\mathbf{K}$, we need to evaluate its transformation matrix $\mathbf{T}$. Thus, by substituting $\cos\theta = 0.8$ and $\sin\theta = 0.6$ into

Eq. (8.36), we obtain

$$\mathbf{T}_1 = \begin{bmatrix} 1 & 0 & 0 & 0 & 0 & 0 \\ 0 & 0.8 & 0.6 & 0 & 0 & 0 \\ 0 & -0.6 & 0.8 & 0 & 0 & 0 \\ 0 & 0 & 0 & 1 & 0 & 0 \\ 0 & 0 & 0 & 0 & 0.8 & 0.6 \\ 0 & 0 & 0 & 0 & -0.6 & 0.8 \end{bmatrix} \tag{2}$$

Next, by substituting $\mathbf{k}_1$ (Eq. (1)) and $\mathbf{T}_1$ (Eq. (2)) into the relationship $\mathbf{K} = \mathbf{T}^T\mathbf{k}\mathbf{T}$, and performing the necessary matrix multiplications, we obtain the following global stiffness matrix for member 1:

$$\mathbf{K}_1 = \begin{array}{c} \begin{matrix} 4 & 5 & 6 & 1 & 2 & 3 \end{matrix} \\ \left[\begin{array}{rrr|rrr} 832.8 & -2{,}498.4 & 3{,}331.2 & -832.8 & -2{,}498.4 & 3{,}331.2 \\ -2{,}498.4 & 10{,}553 & -12{,}905 & 2{,}498.4 & 4{,}437.4 & -7{,}081.9 \\ 3{,}331.2 & -12{,}905 & 18{,}081 & -3{,}331.2 & -7{,}081.9 & 8{,}568.6 \\ \hline -832.8 & 2{,}498.4 & -3{,}331.2 & 832.8 & 2{,}498.4 & -3{,}331.2 \\ -2{,}498.4 & 4{,}437.4 & -7{,}081.9 & 2{,}498.4 & 10{,}553 & -12{,}905 \\ 3{,}331.2 & -7{,}081.9 & 8{,}568.6 & -3{,}331.2 & -12{,}905 & 18{,}081 \end{array}\right] \end{array} \begin{matrix} 4 \\ 5 \\ 6 \\ 1 \\ 2 \\ 3 \end{matrix}$$

From Fig. 8.12(b), we observe that the code numbers for member 1 are 4, 5, 6, 1, 2, 3. By using these code numbers, we store the pertinent elements of $\mathbf{K}_1$ in the 3×3 structure stiffness matrix $\mathbf{S}$, as shown in Fig. 8.12(c).

Member 2 $L = 6$ m, $\cos\theta = 0$, $\sin\theta = 1$

$$\mathbf{k}_2 = \begin{bmatrix} 3{,}855.6 & 0 & 11{,}567 & -3{,}855.6 & 0 & 11{,}567 \\ 0 & 1{,}456.7 & 0 & 0 & -1{,}456.7 & 0 \\ 11{,}567 & 0 & 46{,}267 & -11{,}567 & 0 & 23{,}133 \\ -3{,}855.6 & 0 & -11{,}567 & 3{,}855.6 & 0 & -11{,}567 \\ 0 & -1{,}456.7 & 0 & 0 & 1{,}456.7 & 0 \\ 11{,}567 & 0 & 23{,}133 & -11{,}567 & 0 & 46{,}267 \end{bmatrix} \tag{3}$$

$$\mathbf{T}_2 = \begin{bmatrix} 1 & 0 & 0 & 0 & 0 & 0 \\ 0 & 0 & 1 & 0 & 0 & 0 \\ 0 & -1 & 0 & 0 & 0 & 0 \\ 0 & 0 & 0 & 1 & 0 & 0 \\ 0 & 0 & 0 & 0 & 0 & 1 \\ 0 & 0 & 0 & 0 & -1 & 0 \end{bmatrix} \tag{4}$$

$$\mathbf{K}_2 = \mathbf{T}_2^T\mathbf{k}_2\mathbf{T}_2 = \begin{array}{c} \begin{matrix} 7 & 8 & 9 & 1 & 2 & 3 \end{matrix} \\ \left[\begin{array}{rrr|rrr} 3{,}855.6 & -11{,}567 & 0 & -3{,}855.6 & -11{,}567 & 0 \\ -11{,}567 & 46{,}267 & 0 & 11{,}567 & 23{,}133 & 0 \\ 0 & 0 & 1{,}456.7 & 0 & 0 & -1{,}456.7 \\ \hline -3{,}855.6 & 11{,}567 & 0 & 3{,}855.6 & 11{,}567 & 0 \\ -11{,}567 & 23{,}133 & 0 & 11{,}567 & 46{,}267 & 0 \\ 0 & 0 & -1{,}456.7 & 0 & 0 & 1{,}456.7 \end{array}\right] \end{array} \begin{matrix} 7 \\ 8 \\ 9 \\ 1 \\ 2 \\ 3 \end{matrix}$$

To determine the local fixed-end force vector due to the 20 kN/m member load, we first evaluate the fixed-end shears and moments by using the expressions for loading type 3 given inside the front cover. This yields

$$FS_b = FS_e = 60 \text{ kN}$$
$$FM_b = -FM_e = 60 \text{ kN·m}$$
$$FT_b = FT_e = 0$$

Since $MT = 0$ for this member, we use Eq. (8.22) to obtain its local fixed-end force vector:

$$\mathbf{Q}_{f2} = \begin{bmatrix} 60 \\ 0 \\ 60 \\ 60 \\ 0 \\ -60 \end{bmatrix} \tag{5}$$

Next, by substituting $\mathbf{T}_2$ (Eq. (4)) and $\mathbf{Q}_{f2}$ (Eq. (5)) into the transformation relationship $\mathbf{F}_f = \mathbf{T}^T\mathbf{Q}_f$, we obtain the global fixed-end force vector for member 2:

$$\mathbf{F}_{f2} = \begin{bmatrix} 60 \\ -60 \\ 0 \\ \hdashline 60 \\ 60 \\ 0 \end{bmatrix} \begin{matrix} 7 \\ 8 \\ 9 \\ 1 \\ 2 \\ 3 \end{matrix}$$

The relevant elements of $\mathbf{K}_2$ and $\mathbf{F}_{f2}$ are stored in $\mathbf{S}$ and the 3×1 structure fixed-joint force vector $\mathbf{P}_f$, respectively, as shown in Fig. 8.12(c).

Member 3 As the local x axis of this member is oriented in the positive direction of the global X axis, no coordinate transformations are needed; that is, $\mathbf{T}_3 = \mathbf{I}$. By using Eq. (8.21) with $L = 8$ m, we obtain

$$\mathbf{K}_3 = \mathbf{k}_3 = \begin{matrix} \begin{matrix} 10 & 11 & 12 & 1 & 2 & 3 \end{matrix} \\ \begin{bmatrix} 1{,}626.6 & 0 & 6{,}506.3 & -1{,}626.6 & 0 & 6{,}506.3 \\ 0 & 1{,}092.5 & 0 & 0 & -1{,}092.5 & 0 \\ 6{,}506.3 & 0 & 34{,}700 & -6{,}506.3 & 0 & 17{,}350 \\ -1{,}626.6 & 0 & -6{,}506.3 & 1{,}626.6 & 0 & -6{,}506.3 \\ 0 & -1{,}092.5 & 0 & 0 & 1{,}092.5 & 0 \\ 6{,}506.3 & 0 & 17{,}350 & -6{,}506.3 & 0 & 34{,}700 \end{bmatrix} \end{matrix} \begin{matrix} 10 \\ 11 \\ 12 \\ 1 \\ 2 \\ 3 \end{matrix} \tag{6}$$

$$FS_b = FS_e = 80 \text{ kN}$$
$$FM_b = -FM_e = 106.67 \text{ kN} \cdot \text{m}$$
$$FT_b = FT_e = 0$$

$$\mathbf{F}_{f3} = \mathbf{Q}_{f3} = \begin{bmatrix} 80 \\ 0 \\ 106.67 \\ \hdashline 80 \\ 0 \\ -106.67 \end{bmatrix} \begin{matrix} 10 \\ 11 \\ 12 \\ 1 \\ 2 \\ 3 \end{matrix} \tag{7}$$

The complete structure stiffness matrix **S** and the structure fixed-joint force vector $\mathbf{P}_f$ are given in Fig. 8.12(c).

Joint Load Vector: Because the grid is not subjected to any external loads at its joints, the joint load vector is 0; that is,

$$\mathbf{P} = \mathbf{0}$$

Joint Displacements: By substituting **P**, $\mathbf{P}_f$, and **S** into the structure stiffness relationship, $\mathbf{P} - \mathbf{P}_f = \mathbf{Sd}$, we write

$$\begin{bmatrix} 0 \\ 0 \\ 0 \end{bmatrix} - \begin{bmatrix} 140 \\ 60 \\ -106.67 \end{bmatrix} = \begin{bmatrix} 6{,}315 & 14{,}065 & -9{,}837.5 \\ 14{,}065 & 57{,}912 & -12{,}905 \\ -9{,}837.5 & -12{,}905 & 54{,}238 \end{bmatrix} \begin{bmatrix} d_1 \\ d_2 \\ d_3 \end{bmatrix}$$

or

$$\begin{bmatrix} -140 \\ -60 \\ 106.67 \end{bmatrix} = \begin{bmatrix} 6{,}315 & 14{,}065 & -9{,}837.5 \\ 14{,}065 & 57{,}912 & -12{,}905 \\ -9{,}837.5 & -12{,}905 & 54{,}238 \end{bmatrix} \begin{bmatrix} d_1 \\ d_2 \\ d_3 \end{bmatrix}$$

By solving the foregoing simultaneous equations, we determine the joint displacements to be

$$\mathbf{d} = \begin{bmatrix} -55.951 \text{ m} \\ 11.33 \text{ rad} \\ -5.4856 \text{ rad} \end{bmatrix} \times 10^{-3} \qquad \textbf{Ans}$$

Member End Displacements and End Forces:

Member 1

$$\mathbf{v}_1 = \begin{bmatrix} v_1 \\ v_2 \\ v_3 \\ v_4 \\ v_5 \\ v_6 \end{bmatrix} \begin{matrix} 4 \\ 5 \\ 6 \\ 1 \\ 2 \\ 3 \end{matrix} = \begin{bmatrix} 0 \\ 0 \\ 0 \\ d_1 \\ d_2 \\ d_3 \end{bmatrix} = \begin{bmatrix} 0 \\ 0 \\ 0 \\ -55.951 \\ 11.33 \\ -5.4856 \end{bmatrix} \times 10^{-3}$$

$$\mathbf{u}_1 = \mathbf{T}_1 \mathbf{v}_1 = \begin{bmatrix} 0 \\ 0 \\ 0 \\ -55.951 \\ 5.7728 \\ -11.187 \end{bmatrix} \times 10^{-3}$$

$$\mathbf{Q}_1 = \mathbf{k}_1 \mathbf{u}_1 = \begin{bmatrix} 0.014686 \text{ kN} \\ -5.0455 \text{ kN·m} \\ 77.709 \text{ kN·m} \\ -0.014686 \text{ kN} \\ 5.0455 \text{ kN·m} \\ -77.562 \text{ kN·m} \end{bmatrix} \qquad \textbf{Ans}$$

The member local end forces are depicted in Fig. 8.12(d).

$$\mathbf{F}_1 = \mathbf{T}_1^T \mathbf{Q}_1 = \begin{bmatrix} 0.014686 \\ -50.662 \\ 59.14 \\ \hdashline -0.014686 \\ 50.574 \\ -59.022 \end{bmatrix} \begin{matrix} 4 \\ 5 \\ 6 \\ 1 \\ 2 \\ 3 \end{matrix}$$

The pertinent elements of $\mathbf{F}_1$ are stored in the reaction vector $\mathbf{R}$, as shown in Fig. 8.12(e).

Member 2

$$\mathbf{v}_2 = \begin{bmatrix} 0 \\ 0 \\ 0 \\ -55.951 \\ 11.33 \\ -5.4856 \end{bmatrix} \begin{matrix} 7 \\ 8 \\ 9 \\ 1 \\ 2 \\ 3 \end{matrix} \times 10^{-3}, \quad \mathbf{u}_2 = \mathbf{T}_2 \mathbf{v}_2 = \begin{bmatrix} 0 \\ 0 \\ 0 \\ -55.951 \\ -5.4856 \\ -11.33 \end{bmatrix} \times 10^{-3}$$

$$\mathbf{Q}_2 = \mathbf{k}_2 \mathbf{u}_2 + \mathbf{Q}_{f2} = \begin{bmatrix} 144.67 \text{ kN} \\ 7.9907 \text{ kN·m} \\ 445.06 \text{ kN·m} \\ -24.668 \text{ kN} \\ -7.9907 \text{ kN·m} \\ 62.952 \text{ kN·m} \end{bmatrix}$$ **Ans**

$$\mathbf{F}_2 = \mathbf{T}_2^T \mathbf{Q}_2 = \begin{bmatrix} 144.67 \\ -445.06 \\ 7.9907 \\ \hdashline -24.668 \\ -62.952 \\ -7.9907 \end{bmatrix} \begin{matrix} 7 \\ 8 \\ 9 \\ 1 \\ 2 \\ 3 \end{matrix}$$

Member 3

$$\mathbf{u}_3 = \mathbf{v}_3 = \begin{bmatrix} 0 \\ 0 \\ 0 \\ -55.951 \\ 11.33 \\ -5.4856 \end{bmatrix} \begin{matrix} 10 \\ 11 \\ 12 \\ 1 \\ 2 \\ 3 \end{matrix} \times 10^{-3}$$

$$\mathbf{F}_3 = \mathbf{Q}_3 = \mathbf{k}_3 \mathbf{u}_3 + \mathbf{Q}_{f3} = \begin{bmatrix} 135.32 \text{ kN} \\ -12.378 \text{ kN·m} \\ 375.52 \text{ kN·m} \\ \hdashline 24.683 \text{ kN} \\ 12.378 \text{ kN·m} \\ 67.013 \text{ kN·m} \end{bmatrix} \begin{matrix} 10 \\ 11 \\ 12 \\ 1 \\ 2 \\ 3 \end{matrix}$$ **Ans**

Support Reactions: The completed reaction vector $\mathbf{R}$ is shown in Fig. 8.12(e), and the support reactions are depicted on a line diagram of the grid in Fig. 8.12(f). **Ans**

Equilibrium Check: The three equilibrium equations ($\sum F_Y = 0$, $\sum M_X = 0$, and $\sum M_Z = 0$) are satisfied.

8.3 SPACE FRAMES

Space frames constitute the most general type of framed structures. The members of such frames may be oriented in any directions in three-dimensional space, and may be connected by rigid and/or flexible connections. Furthermore, external loads oriented in any arbitrary directions can be applied to the joints, as well as members, of space frames (Fig. 8.13(a)). Under the action of external loads, the members of a space frame are generally subjected to bending moments about both principal axes, shears in both principal directions, torsional moments, and axial forces.

As with grids, the analysis of space frames is commonly based on the assumption that the cross-sections of all the members are symmetric about at least two mutually perpendicular axes, and are free to warp out of their planes under the action of torsional moments. As discussed previously in the case of grids, the bending and torsional stiffnesses of a member are uncoupled if it satisfies the foregoing assumption.

The process of developing the analytical models, and numbering the degrees of freedom and restrained coordinates, of space frames is analogous to that for

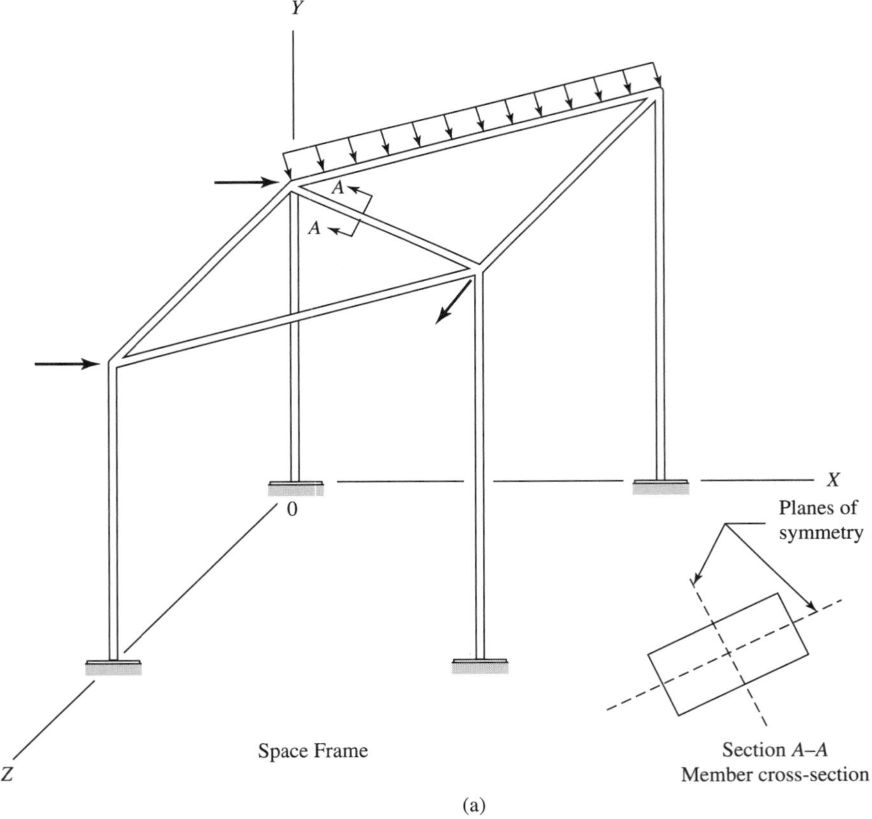

Fig. 8.13

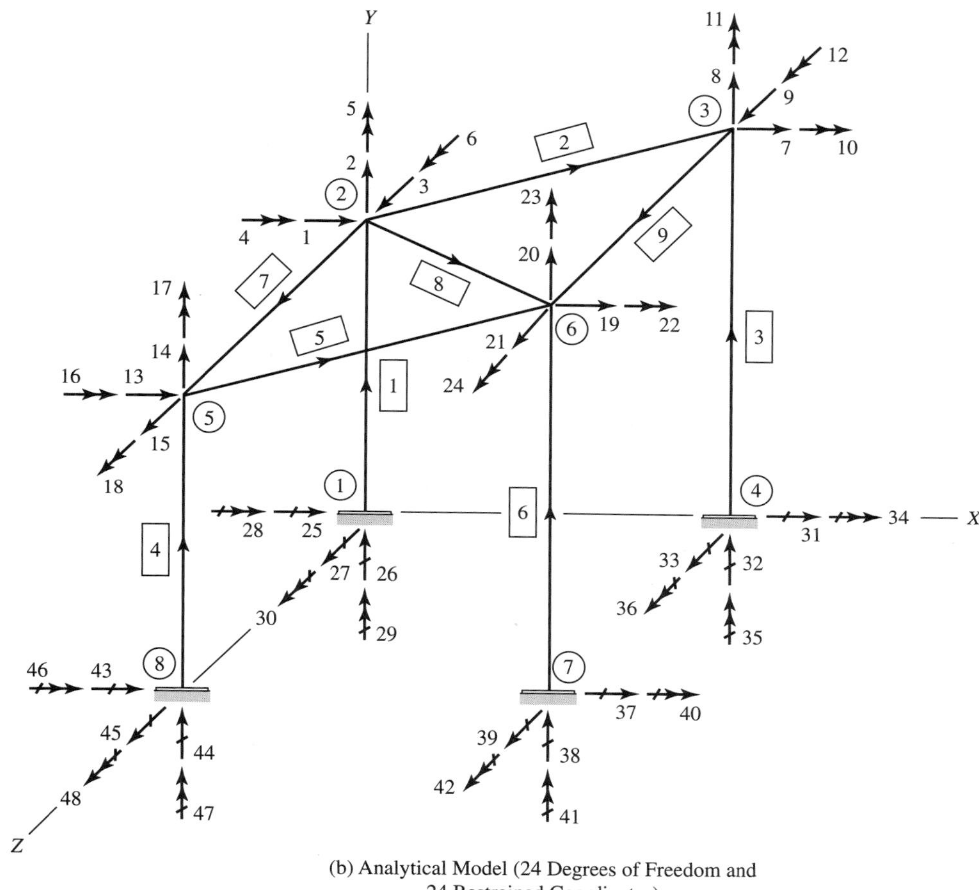

(b) Analytical Model (24 Degrees of Freedom and 24 Restrained Coordinates)

Fig. 8.13 (*continued*)

other types of framed structures. The overall geometry of the space frame, and its joint loads and displacements, are described with reference to a global right-handed *XYZ* coordinate system, with three global (*X*, *Y*, and *Z*) coordinates used to specify the location of each joint. An unsupported joint of a space frame can translate in any direction, and rotate about any axis, in three-dimensional space. Since small rotations can be treated as vector quantities, the rotation of a joint can be conveniently represented by its component rotations about the *X*, *Y*, and *Z* axes. Thus, a free joint of a space frame has six degrees of freedom—the translations in the *X*, *Y*, and *Z* directions and the rotations about the *X*, *Y*, and *Z* axes. Therefore, six structure coordinates (i.e., free and/or restrained coordinates) need to be defined at each joint of the space frame for the purpose of analysis; that is,

$$\left.\begin{aligned} NCJT &= 6 \\ NDOF &= 6(NJ) - NR \end{aligned}\right\} \quad \text{for space frames} \tag{8.37}$$

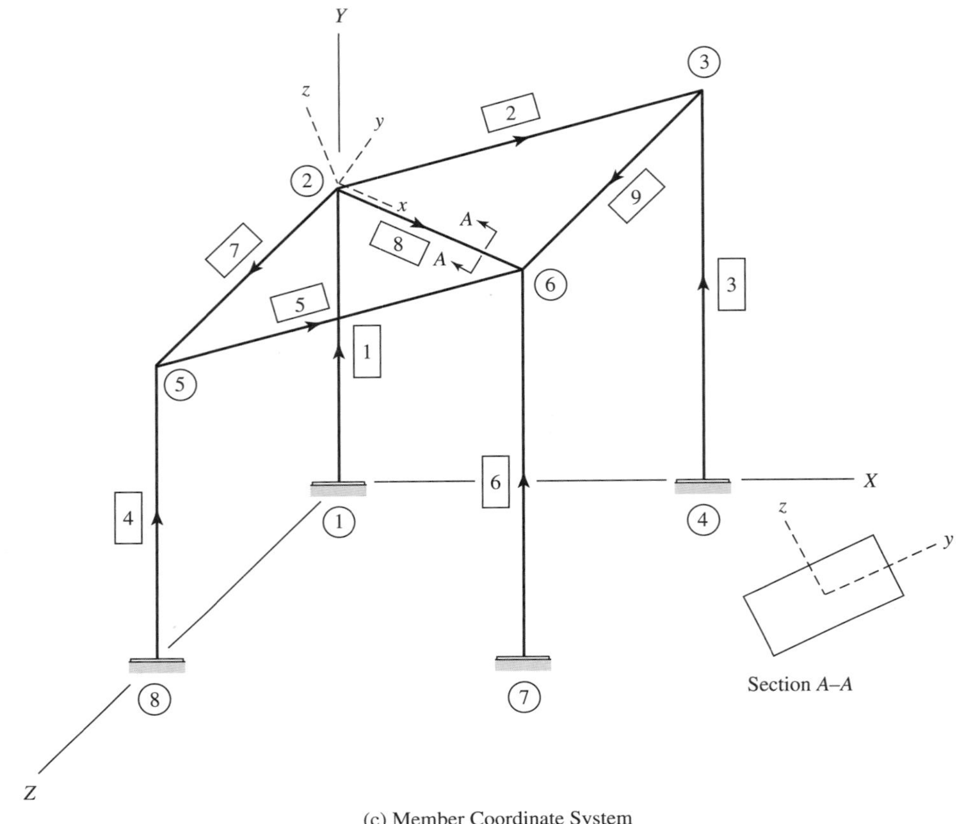

(c) Member Coordinate System

Fig. 8.13 *(continued)*

The procedure for assigning numbers to the structure coordinates of a space frame is similar to that for other types of framed structures, with the degrees of freedom numbered before the restrained coordinates. In the case of a joint with multiple degrees of freedom (or restrained coordinates), the translations (or forces) in the X, Y, and Z directions are numbered first in sequential order, followed by the rotations (or moments) about the X, Y, and Z axes, respectively, as shown in Fig. 8.13(b).

For each member of a space frame, a local xyz coordinate system is established, with its origin at an end of the member and the x axis directed along the member's centroidal axis in the undeformed state. The local y and z axes are oriented, respectively, parallel to the two axes of symmetry (or the principal axes of inertia) of the member cross-section, with their positive directions defined so that the local xyz coordinate system is right-handed (Fig. 8.13(c)).

Member Stiffness Relations in the Local Coordinate System

To establish the local stiffness relations, let us consider an arbitrary member m of a space frame, as shown in Fig. 8.14(a). Like a joint of a space frame, six

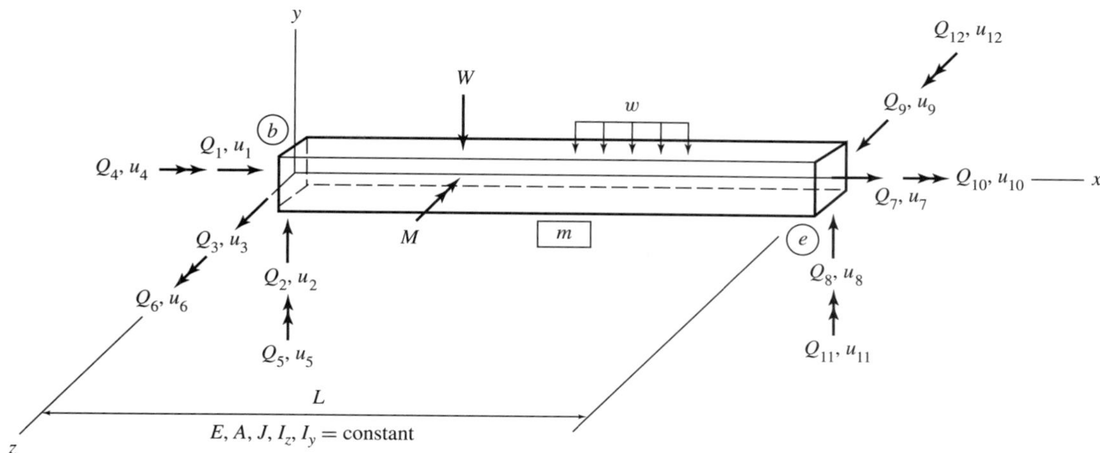

(a) Member Forces and Displacements in the Local Coordinate System

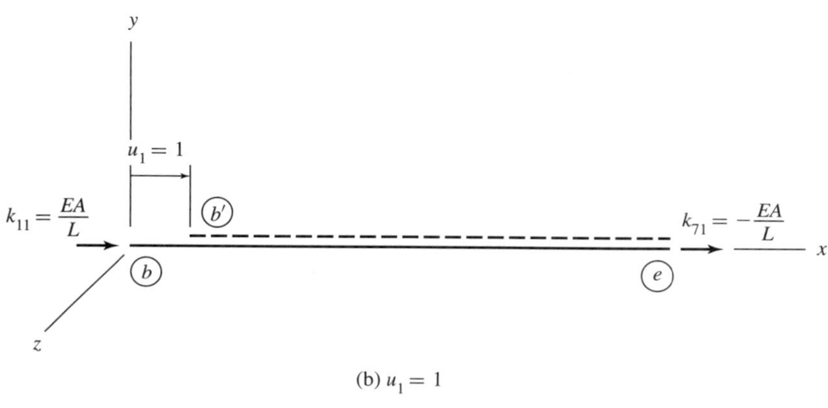

(b) $u_1 = 1$

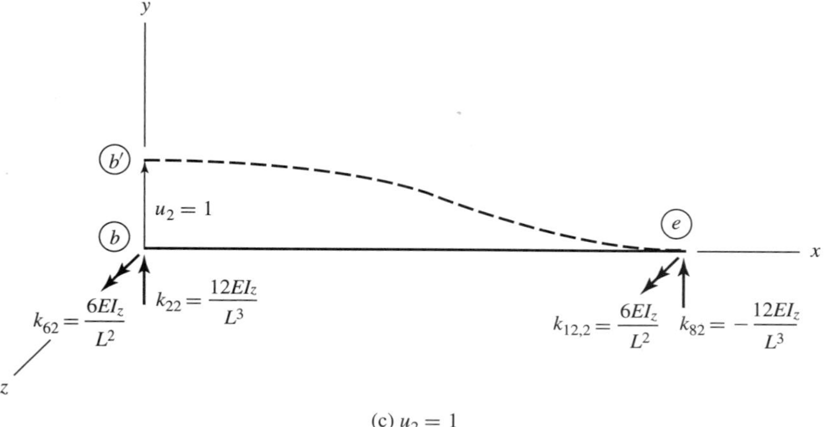

(c) $u_2 = 1$

Fig. 8.14

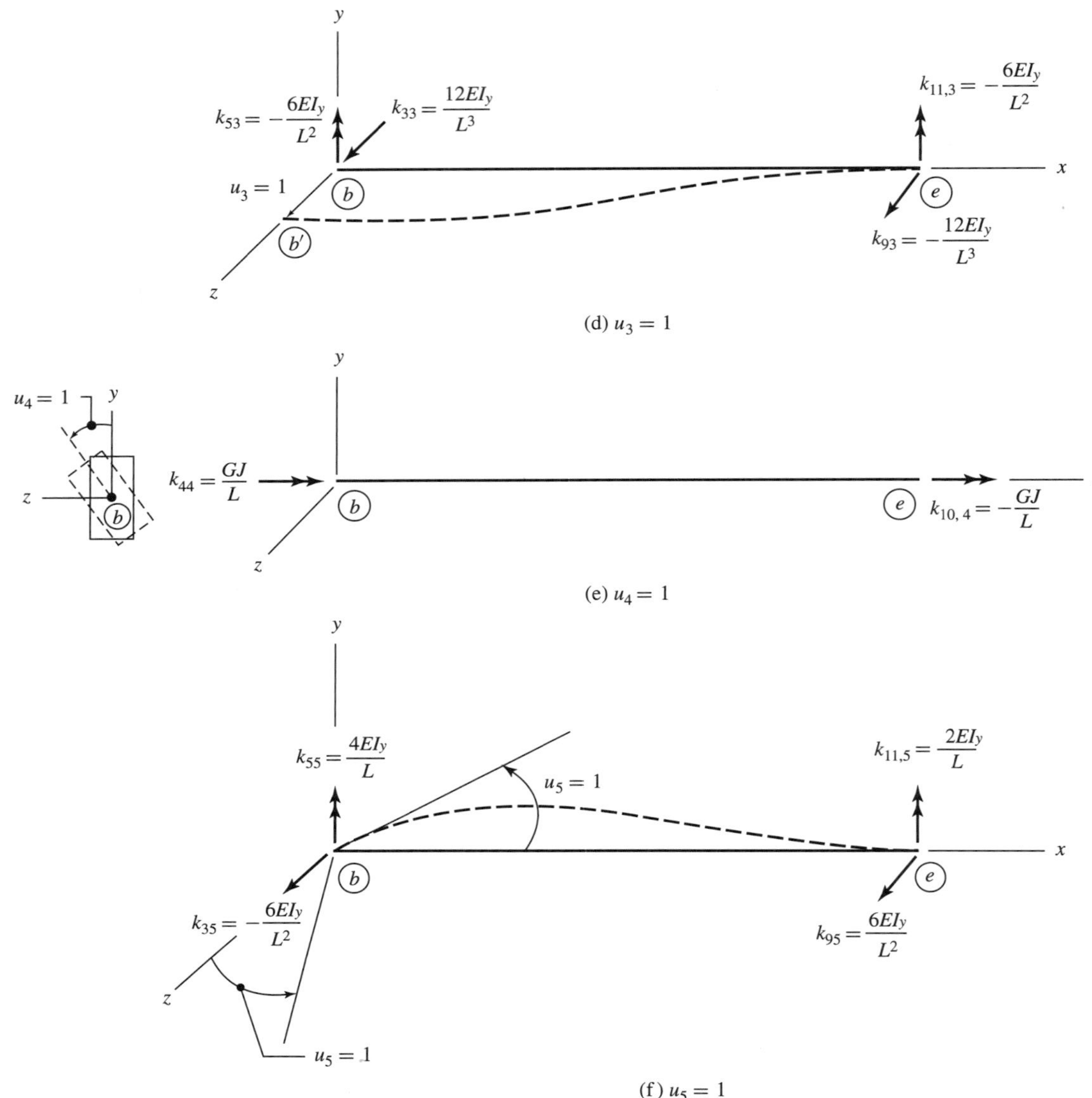

Fig. 8.14 *(continued)*

displacements are needed to completely specify the displaced position of each end of the space frame member. Thus, a member of a space frame has 12 degrees of freedom. In the member local coordinate system, the 12 end displacements are denoted by u_1 through u_{12}, and the corresponding member end forces are denoted by Q_1 through Q_{12}, as shown in Fig. 8.14(a). As indicated in this figure, *a member's local end displacements (or end forces) are numbered by beginning at its end b, with the translations (or forces) in the x, y, and z directions numbered first in sequential order, followed by the rotations (or moments) about the x, y, and z axes, respectively. The displacements (or forces) at the member's opposite end e are then numbered in the same sequential order.*

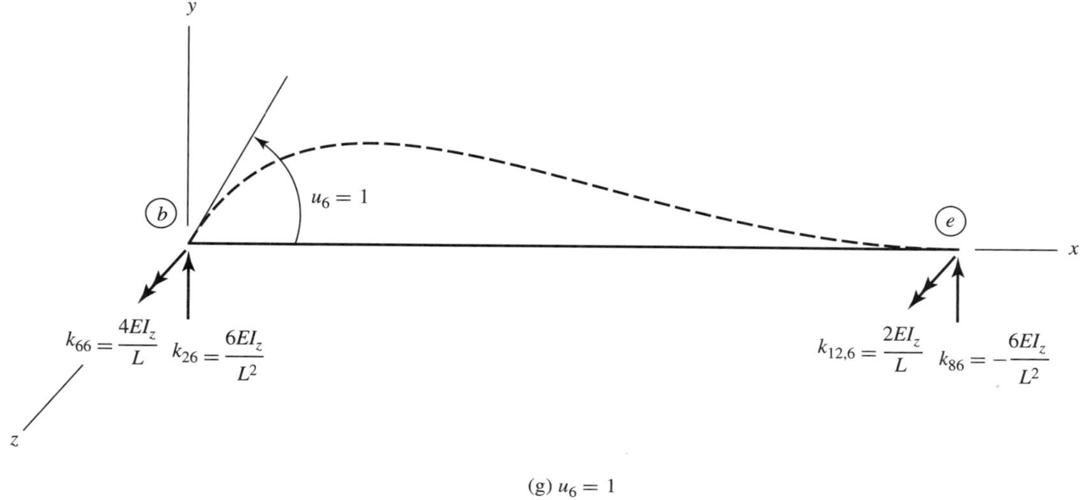

(g) $u_6 = 1$

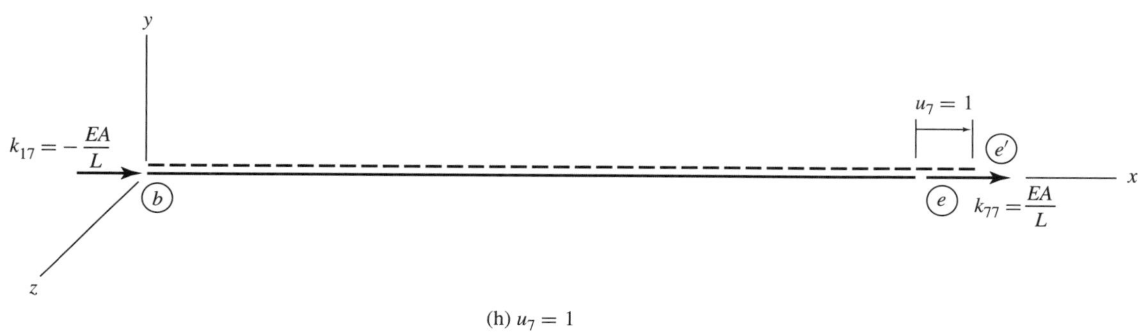

(h) $u_7 = 1$

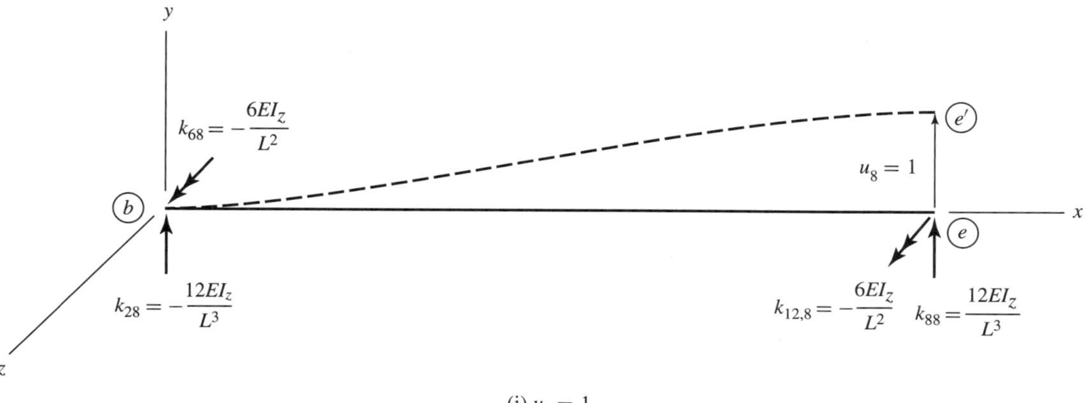

(i) $u_8 = 1$

Fig. 8.14 *(continued)*

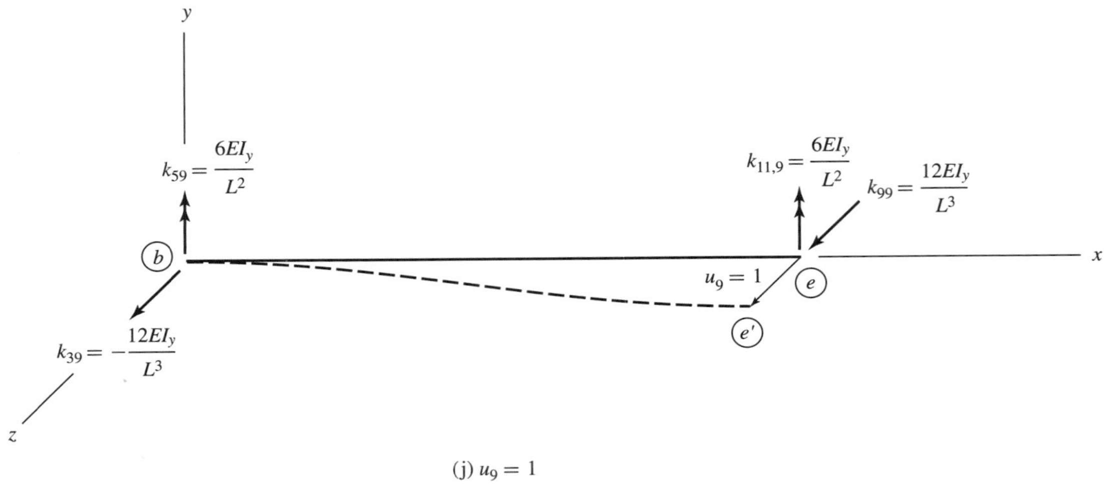

(j) $u_9 = 1$

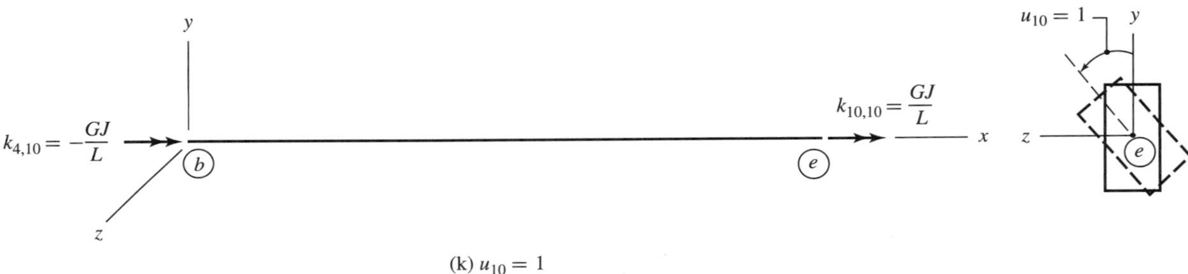

(k) $u_{10} = 1$

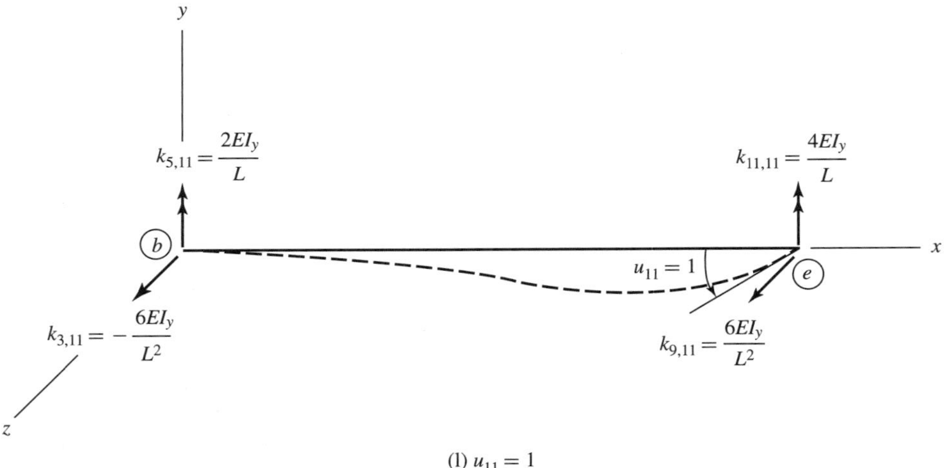

(l) $u_{11} = 1$

Fig. 8.14 *(continued)*

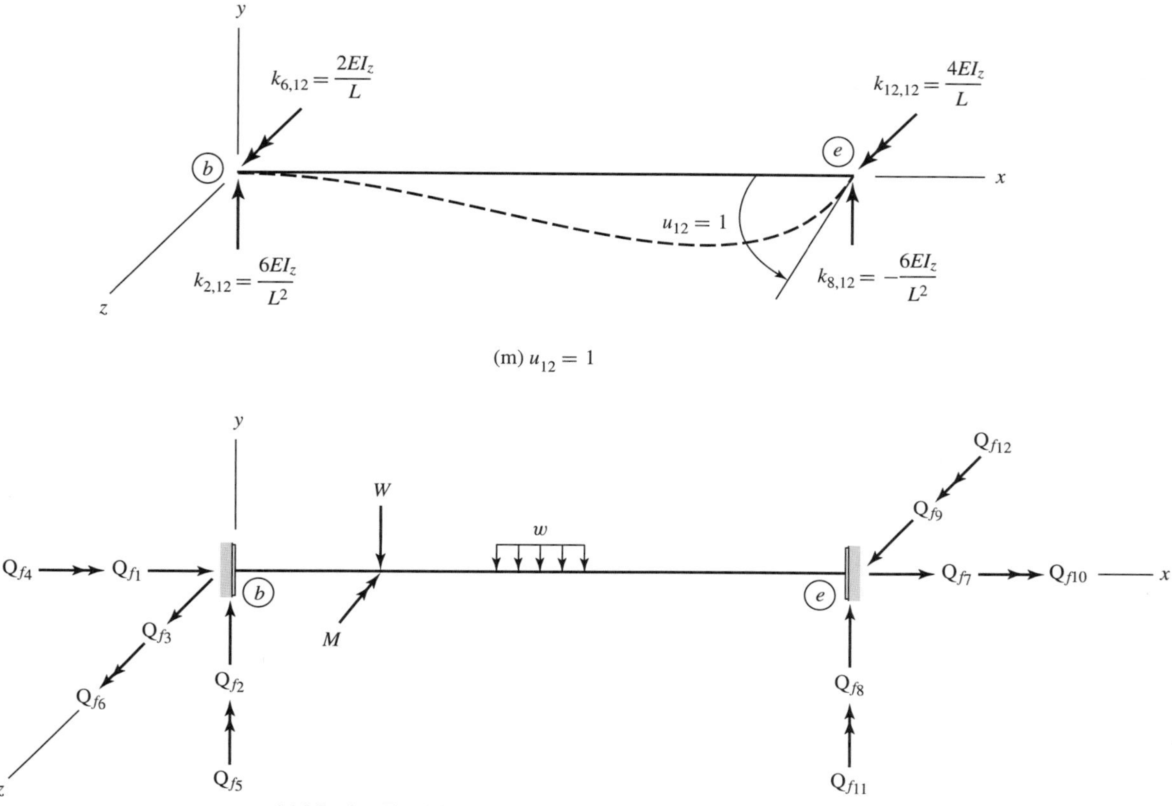

Fig. 8.14 (*continued*)

The relationship between the end forces **Q** and the end displacements **u**, for space frame members, can be expressed in the following, now familiar, form:

$$\mathbf{Q} = \mathbf{k}\mathbf{u} + \mathbf{Q}_f \tag{8.38}$$

with **k** now representing the 12 × 12 member local stiffness matrix, and $\mathbf{Q}_f$ denoting the 12 × 1 member local fixed-end force vector.

The explicit form of **k** for members of space frames can be conveniently obtained using the expressions of the stiffness coefficients derived previously for prismatic members subjected to axial deformations (Section 3.3), bending deformations (Section 5.2), and torsional deformations (Section 8.2). The stiffness coefficients for a space frame member thus obtained, due to the unit values of the 12 end displacements (u_1 through u_{12}, respectively), are given in Figs. 8.14(b) through (m). Note that in Figs. 8.14(c), (g), (i), and (m), the moment of inertia of the member cross-section about its local z axis, I_z, is used in the expressions for the stiffness coefficients, because the end displacements u_2,

u_6, u_8, and u_{12} cause the member to bend about the z axis. However, in Figs. 8.14(d), (f), (j), and (l), because the end displacements u_3, u_5, u_9, and u_{11} cause the member to bend about its local y axis, the moment of inertia about the y axis, I_y, is used in the expressions for the corresponding stiffness coefficients. The explicit form of the local stiffness matrix **k** for members of space frames, obtained by arranging all the stiffness coefficients shown in Figs. 8.14(b) through (m) in a 12 × 12 matrix, is given in Eq. (8.39).

$$\mathbf{k} = \frac{E}{L^3}\begin{bmatrix} AL^2 & 0 & 0 & 0 & 0 & 0 & -AL^2 & 0 & 0 & 0 & 0 & 0 \\ 0 & 12I_z & 0 & 0 & 0 & 6LI_z & 0 & -12I_z & 0 & 0 & 0 & 6LI_z \\ 0 & 0 & 12I_y & 0 & -6LI_y & 0 & 0 & 0 & -12I_y & 0 & -6LI_y & 0 \\ 0 & 0 & 0 & \frac{GJL^2}{E} & 0 & 0 & 0 & 0 & 0 & -\frac{GJL^2}{E} & 0 & 0 \\ 0 & 0 & -6LI_y & 0 & 4L^2I_y & 0 & 0 & 0 & 6LI_y & 0 & 2L^2I_y & 0 \\ 0 & 6LI_z & 0 & 0 & 0 & 4L^2I_z & 0 & -6LI_z & 0 & 0 & 0 & 2L^2I_z \\ -AL^2 & 0 & 0 & 0 & 0 & 0 & AL^2 & 0 & 0 & 0 & 0 & 0 \\ 0 & -12I_z & 0 & 0 & 0 & -6LI_z & 0 & 12I_z & 0 & 0 & 0 & -6LI_z \\ 0 & 0 & -12I_y & 0 & 6LI_y & 0 & 0 & 0 & 12I_y & 0 & 6LI_y & 0 \\ 0 & 0 & 0 & -\frac{GJL^2}{E} & 0 & 0 & 0 & 0 & 0 & \frac{GJL^2}{E} & 0 & 0 \\ 0 & 0 & -6LI_y & 0 & 2L^2I_y & 0 & 0 & 0 & 6LI_y & 0 & 4L^2I_y & 0 \\ 0 & 6LI_z & 0 & 0 & 0 & 2L^2I_z & 0 & -6LI_z & 0 & 0 & 0 & 4L^2I_z \end{bmatrix} \quad \textbf{(8.39)}$$

The local fixed-end force vector for the members of space frames is expressed as follows (Fig. 8.14(n)).

$$\mathbf{Q}_f = \begin{bmatrix} FA_b \\ FS_{by} \\ FS_{bz} \\ FT_b \\ FM_{by} \\ FM_{bz} \\ FA_e \\ FS_{ey} \\ FS_{ez} \\ FT_e \\ FM_{ey} \\ FM_{ez} \end{bmatrix} \quad \textbf{(8.40)}$$

in which FS_{by} and FS_{bz} denote the fixed-end shears at member end b in the local y and z directions, respectively; and FM_{by} and FM_{bz} represent the fixed-end moments at the same member end about the y and z axes, respectively. The fixed-end shears and moments at the opposite end e of the member are defined in a similar manner. The fixed-end forces due to a prescribed member loading can be conveniently evaluated, using the fixed-end force expressions given inside the front cover. Any inclined member loads must be resolved into their components in the directions of the member's local x, y, and z axes before proceeding with the calculation of the fixed-end forces.

Member Releases The expressions for **k** and $\mathbf{Q}_f$, as given in Eqs. (8.39) and (8.40), respectively, are valid only for members rigidly connected to joints at both ends (i.e., members of type 0, or $MT = 0$). For members of space frames with moment releases, the foregoing expressions need to be modified, using the procedure described in Section 7.1. If the member releases are assumed to be in the form of spherical hinges (or ball-and-socket type of connections), so that all three moments (i.e., the bending moments about the y and z axes, and the torsional moment) are 0 at the released member ends, then the modified local stiffness matrices **k** and fixed-end force vectors $\mathbf{Q}_f$ for the members with releases can be expressed as follows.

For members with a hinge at the beginning ($MT = 1$), the modified **k** is given in Eq. (8.41):

$$\mathbf{k} = \frac{E}{L^3}\begin{bmatrix} AL^2 & 0 & 0 & 0 & 0 & 0 & -AL^2 & 0 & 0 & 0 & 0 & 0 \\ 0 & 3I_z & 0 & 0 & 0 & 0 & 0 & -3I_z & 0 & 0 & 0 & 3LI_z \\ 0 & 0 & 3I_y & 0 & 0 & 0 & 0 & 0 & -3I_y & 0 & -3LI_y & 0 \\ 0 & 0 & 0 & 0 & 0 & 0 & 0 & 0 & 0 & 0 & 0 & 0 \\ 0 & 0 & 0 & 0 & 0 & 0 & 0 & 0 & 0 & 0 & 0 & 0 \\ 0 & 0 & 0 & 0 & 0 & 0 & 0 & 0 & 0 & 0 & 0 & 0 \\ -AL^2 & 0 & 0 & 0 & 0 & 0 & AL^2 & 0 & 0 & 0 & 0 & 0 \\ 0 & -3I_z & 0 & 0 & 0 & 0 & 0 & 3I_z & 0 & 0 & 0 & -3LI_z \\ 0 & 0 & -3I_y & 0 & 0 & 0 & 0 & 0 & 3I_y & 0 & 3LI_y & 0 \\ 0 & 0 & 0 & 0 & 0 & 0 & 0 & 0 & 0 & 0 & 0 & 0 \\ 0 & 0 & -3LI_y & 0 & 0 & 0 & 0 & 0 & 3LI_y & 0 & 3L^2I_y & 0 \\ 0 & 3LI_z & 0 & 0 & 0 & 0 & 0 & -3LI_z & 0 & 0 & 0 & 3L^2I_z \end{bmatrix} \quad \textbf{(8.41)}$$

and

$$\mathbf{Q}_f = \begin{bmatrix} FA_b \\ FS_{by} - \dfrac{3}{2L}FM_{bz} \\ FS_{bz} + \dfrac{3}{2L}FM_{by} \\ 0 \\ 0 \\ 0 \\ FA_e \\ FS_{ey} + \dfrac{3}{2L}FM_{bz} \\ FS_{ez} - \dfrac{3}{2L}FM_{by} \\ FT_b + FT_e \\ FM_{ey} - \dfrac{1}{2}FM_{by} \\ FM_{ez} - \dfrac{1}{2}FM_{bz} \end{bmatrix} \quad \textbf{(8.42)}$$

For members with a hinge at the end ($MT = 2$), the modified **k** is given in Eq. (8.43):

$$\mathbf{k} = \frac{E}{L^3}\begin{bmatrix} AL^2 & 0 & 0 & 0 & 0 & 0 & -AL^2 & 0 & 0 & 0 & 0 & 0 \\ 0 & 3I_z & 0 & 0 & 0 & 3LI_z & 0 & -3I_z & 0 & 0 & 0 & 0 \\ 0 & 0 & 3I_y & 0 & -3LI_y & 0 & 0 & 0 & -3I_y & 0 & 0 & 0 \\ 0 & 0 & 0 & 0 & 0 & 0 & 0 & 0 & 0 & 0 & 0 & 0 \\ 0 & 0 & -3LI_y & 0 & 3L^2I_y & 0 & 0 & 0 & 3LI_y & 0 & 0 & 0 \\ 0 & 3LI_z & 0 & 0 & 0 & 3L^2I_z & 0 & -3LI_z & 0 & 0 & 0 & 0 \\ -AL^2 & 0 & 0 & 0 & 0 & 0 & AL^2 & 0 & 0 & 0 & 0 & 0 \\ 0 & -3I_z & 0 & 0 & 0 & -3LI_z & 0 & 3I_z & 0 & 0 & 0 & 0 \\ 0 & 0 & -3I_y & 0 & 3LI_y & 0 & 0 & 0 & 3I_y & 0 & 0 & 0 \\ 0 & 0 & 0 & 0 & 0 & 0 & 0 & 0 & 0 & 0 & 0 & 0 \\ 0 & 0 & 0 & 0 & 0 & 0 & 0 & 0 & 0 & 0 & 0 & 0 \\ 0 & 0 & 0 & 0 & 0 & 0 & 0 & 0 & 0 & 0 & 0 & 0 \end{bmatrix} \quad \textbf{(8.43)}$$

and

$$\mathbf{Q}_f = \begin{bmatrix} FA_b \\ FS_{by} - \dfrac{3}{2L}FM_{ez} \\ FS_{bz} + \dfrac{3}{2L}FM_{ey} \\ FT_b + FT_e \\ FM_{by} - \dfrac{1}{2}FM_{ey} \\ FM_{bz} - \dfrac{1}{2}FM_{ez} \\ FA_e \\ FS_{ey} + \dfrac{3}{2L}FM_{ez} \\ FS_{ez} - \dfrac{3}{2L}FM_{ey} \\ 0 \\ 0 \\ 0 \end{bmatrix} \quad \textbf{(8.44)}$$

For members with hinges at both ends ($MT = 3$), the modified **k** is given in Eq. (8.45):

$$\mathbf{k} = \frac{EA}{L}\begin{bmatrix} 1 & 0 & 0 & 0 & 0 & 0 & -1 & 0 & 0 & 0 & 0 & 0 \\ 0 & 0 & 0 & 0 & 0 & 0 & 0 & 0 & 0 & 0 & 0 & 0 \\ 0 & 0 & 0 & 0 & 0 & 0 & 0 & 0 & 0 & 0 & 0 & 0 \\ 0 & 0 & 0 & 0 & 0 & 0 & 0 & 0 & 0 & 0 & 0 & 0 \\ 0 & 0 & 0 & 0 & 0 & 0 & 0 & 0 & 0 & 0 & 0 & 0 \\ 0 & 0 & 0 & 0 & 0 & 0 & 0 & 0 & 0 & 0 & 0 & 0 \\ -1 & 0 & 0 & 0 & 0 & 0 & 1 & 0 & 0 & 0 & 0 & 0 \\ 0 & 0 & 0 & 0 & 0 & 0 & 0 & 0 & 0 & 0 & 0 & 0 \\ 0 & 0 & 0 & 0 & 0 & 0 & 0 & 0 & 0 & 0 & 0 & 0 \\ 0 & 0 & 0 & 0 & 0 & 0 & 0 & 0 & 0 & 0 & 0 & 0 \\ 0 & 0 & 0 & 0 & 0 & 0 & 0 & 0 & 0 & 0 & 0 & 0 \\ 0 & 0 & 0 & 0 & 0 & 0 & 0 & 0 & 0 & 0 & 0 & 0 \end{bmatrix} \quad \textbf{(8.45)}$$

and

$$\mathbf{Q}_f = \begin{bmatrix} FA_b \\ FS_{by} - \frac{1}{L}(FM_{bz} + FM_{ez}) \\ FS_{bz} + \frac{1}{L}(FM_{by} + FM_{ey}) \\ 0 \\ 0 \\ 0 \\ FA_e \\ FS_{ey} + \frac{1}{L}(FM_{bz} + FM_{ez}) \\ FS_{ez} - \frac{1}{L}(FM_{by} + FM_{ey}) \\ 0 \\ 0 \\ 0 \end{bmatrix} \tag{8.46}$$

Coordinate Transformations

The expression of the transformation matrix **T** for members of space frames can be derived using a procedure essentially similar to that used previously for other types of framed structures. However, unlike the transformation matrices for trusses, plane frames, and grids, which contain direction cosines of only the member's longitudinal (or x) axis, the transformation matrix for members of space frames involves direction cosines of all three (x, y, and z) axes of the member local coordinate system with respect to the structure's global (XYZ) coordinate system.

Consider an arbitrary member m of a space frame, as shown in Fig. 8.15(a) on the next page. The member end forces **Q** and end displacements **u**, in the local coordinate system, are shown in Fig. 8.15(b), and Fig. 8.15(c) depicts the equivalent system of member end forces **F** and end displacements **v**, in the global coordinate system. As indicated in Fig. 8.15(c), the global member end forces and displacements are numbered in a manner analogous to the local forces and displacements, except that they act in the directions of the global X, Y, and Z axes.

The orientation of a member of a space frame is defined by the angles between its local x, y, and z axes and the global X, Y, and Z axes. As shown in Fig. 8.16(a) on page 469, the angles between the local x axis and the global X, Y, and Z axes are denoted by θ_{xX}, θ_{xY}, and θ_{xZ}, respectively. Similarly, the angles between the local y axis and the global X, Y, and Z axes are denoted by θ_{yX}, θ_{yY}, and θ_{yZ}, respectively (Fig. 8.16(b)); and the angles between the local z axis and the global X, Y, and Z axes are denoted by θ_{zX}, θ_{zY}, and θ_{zZ}, respectively (Fig. 8.16(c)).

Now, let us consider the transformation of member end forces from the global to a local coordinate system. By comparing Figs. 8.15(b) and (c), we realize that, at member end b, the local forces Q_1, Q_2, and Q_3 must be equal to the

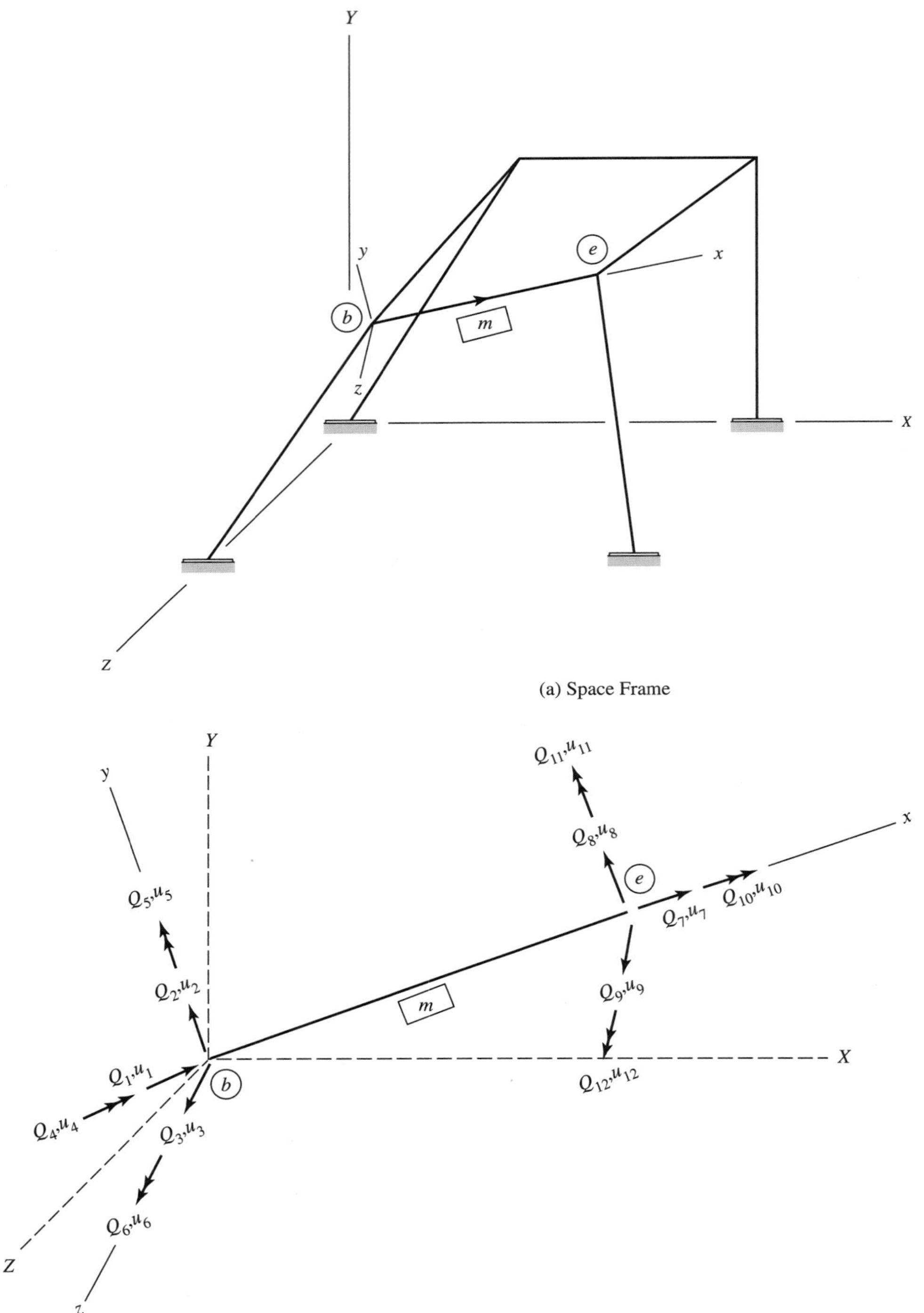

(a) Space Frame

(b) Member End Forces and End Displacements in the Local Coordinate System

Fig. 8.15

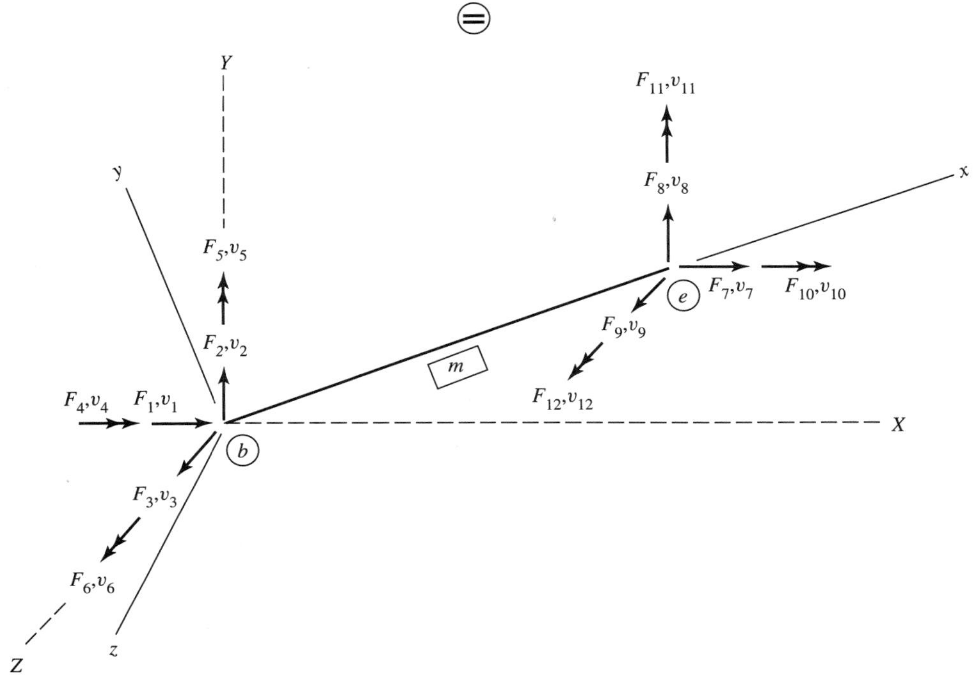

(c) Member End Forces and End Displacements in the Global Coordinate System

Fig. 8.15 *(continued)*

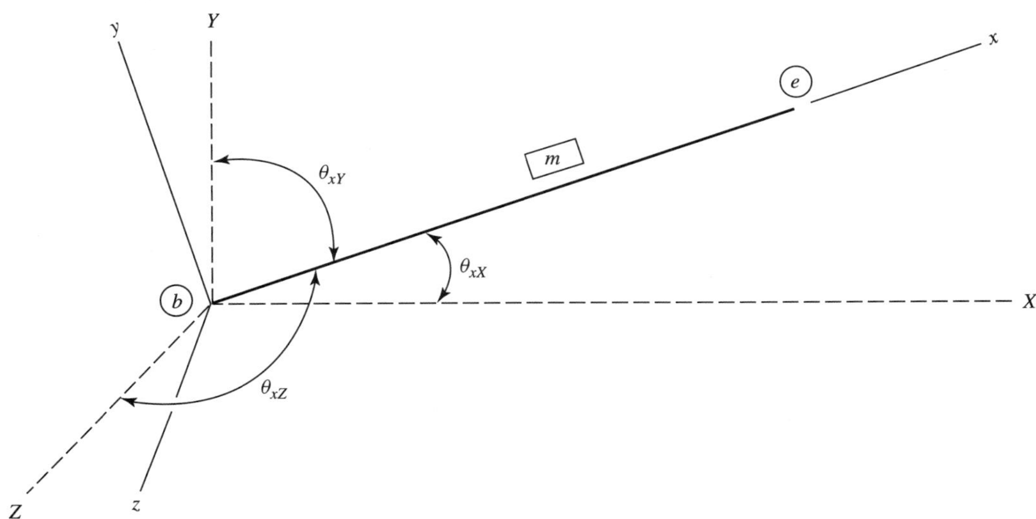

(a) Orientation of Member Local x Axis

Fig. 8.16

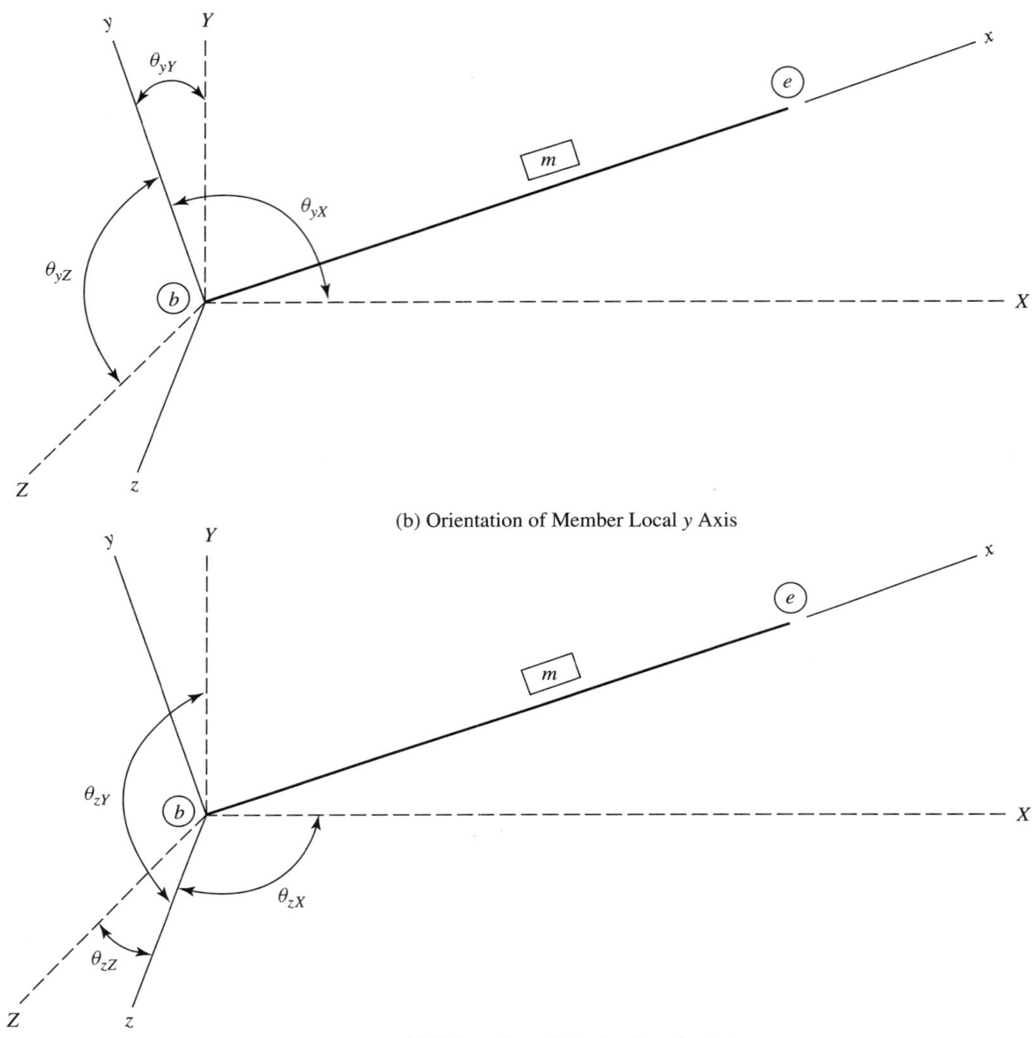

(b) Orientation of Member Local y Axis

(c) Orientation of Member Local z Axis

Fig. 8.16 *(continued)*

algebraic sums of the components of the global forces F_1, F_2, and F_3 in the directions of the local x, y, and z axes, respectively; that is (also, see Fig. 8.16),

$$Q_1 = F_1 \cos\theta_{xX} + F_2 \cos\theta_{xY} + F_3 \cos\theta_{xZ} \tag{8.47a}$$

$$Q_2 = F_1 \cos\theta_{yX} + F_2 \cos\theta_{yY} + F_3 \cos\theta_{yZ} \tag{8.47b}$$

$$Q_3 = F_1 \cos\theta_{zX} + F_2 \cos\theta_{zY} + F_3 \cos\theta_{zZ} \tag{8.47c}$$

Equations (8.47) can be written in matrix form as

$$\begin{bmatrix} Q_1 \\ Q_2 \\ Q_3 \end{bmatrix} = \begin{bmatrix} r_{xX} & r_{xY} & r_{xZ} \\ r_{yX} & r_{yY} & r_{yZ} \\ r_{zX} & r_{zY} & r_{zZ} \end{bmatrix} \begin{bmatrix} F_1 \\ F_2 \\ F_3 \end{bmatrix} \tag{8.48}$$

in which

$$r_{iJ} = \cos\theta_{iJ} \qquad i = x, y, \text{or } z \quad \text{and} \quad J = X, Y, \text{or } Z \tag{8.49}$$

The local moments Q_4, Q_5, and Q_6, at member end b, can be similarly expressed in terms of their global counterparts F_4, F_5, and F_6, as

$$\begin{bmatrix} Q_4 \\ Q_5 \\ Q_6 \end{bmatrix} = \begin{bmatrix} r_{xX} & r_{xY} & r_{xZ} \\ r_{yX} & r_{yY} & r_{yZ} \\ r_{zX} & r_{zY} & r_{zZ} \end{bmatrix} \begin{bmatrix} F_4 \\ F_5 \\ F_6 \end{bmatrix} \tag{8.50}$$

Similarly, the local forces and moments at member end e can be expressed in terms of the global forces and moments by the following relationships.

$$\begin{bmatrix} Q_7 \\ Q_8 \\ Q_9 \end{bmatrix} = \begin{bmatrix} r_{xX} & r_{xY} & r_{xZ} \\ r_{yX} & r_{yY} & r_{yZ} \\ r_{zX} & r_{zY} & r_{zZ} \end{bmatrix} \begin{bmatrix} F_7 \\ F_8 \\ F_9 \end{bmatrix} \tag{8.51}$$

and

$$\begin{bmatrix} Q_{10} \\ Q_{11} \\ Q_{12} \end{bmatrix} = \begin{bmatrix} r_{xX} & r_{xY} & r_{xZ} \\ r_{yX} & r_{yY} & r_{yZ} \\ r_{zX} & r_{zY} & r_{zZ} \end{bmatrix} \begin{bmatrix} F_{10} \\ F_{11} \\ F_{12} \end{bmatrix} \tag{8.52}$$

By combining Eqs. (8.48) and Eqs. (8.50) through (8.52), we can now express the transformation relationship between the 12 × 1 member local end force vector **Q** and the 12 × 1 member global end force vector **F**, in the standard form of

$$\mathbf{Q} = \mathbf{TF} \tag{8.53}$$

in which **T** represents the 12 × 12 transformation matrix for the members of space frames. The explicit form of **T** is given in Eq. (8.54).

$$\mathbf{T} = \left[\begin{array}{ccc:ccc:ccc:ccc} r_{xX} & r_{xY} & r_{xZ} & 0 & 0 & 0 & 0 & 0 & 0 & 0 & 0 & 0 \\ r_{yX} & r_{yY} & r_{yZ} & 0 & 0 & 0 & 0 & 0 & 0 & 0 & 0 & 0 \\ r_{zX} & r_{zY} & r_{zZ} & 0 & 0 & 0 & 0 & 0 & 0 & 0 & 0 & 0 \\ \hdashline 0 & 0 & 0 & r_{xX} & r_{xY} & r_{xZ} & 0 & 0 & 0 & 0 & 0 & 0 \\ 0 & 0 & 0 & r_{yX} & r_{yY} & r_{yZ} & 0 & 0 & 0 & 0 & 0 & 0 \\ 0 & 0 & 0 & r_{zX} & r_{zY} & r_{zZ} & 0 & 0 & 0 & 0 & 0 & 0 \\ \hdashline 0 & 0 & 0 & 0 & 0 & 0 & r_{xX} & r_{xY} & r_{xZ} & 0 & 0 & 0 \\ 0 & 0 & 0 & 0 & 0 & 0 & r_{yX} & r_{yY} & r_{yZ} & 0 & 0 & 0 \\ 0 & 0 & 0 & 0 & 0 & 0 & r_{zX} & r_{zY} & r_{zZ} & 0 & 0 & 0 \\ \hdashline 0 & 0 & 0 & 0 & 0 & 0 & 0 & 0 & 0 & r_{xX} & r_{xY} & r_{xZ} \\ 0 & 0 & 0 & 0 & 0 & 0 & 0 & 0 & 0 & r_{yX} & r_{yY} & r_{yZ} \\ 0 & 0 & 0 & 0 & 0 & 0 & 0 & 0 & 0 & r_{zX} & r_{zY} & r_{zZ} \end{array}\right] \tag{8.54}$$

The transformation matrix **T** is usually expressed in a compact form in terms of its submatrices as

$$\mathbf{T} = \begin{bmatrix} \mathbf{r} & \mathbf{O} & \mathbf{O} & \mathbf{O} \\ \mathbf{O} & \mathbf{r} & \mathbf{O} & \mathbf{O} \\ \mathbf{O} & \mathbf{O} & \mathbf{r} & \mathbf{O} \\ \mathbf{O} & \mathbf{O} & \mathbf{O} & \mathbf{r} \end{bmatrix} \tag{8.55}$$

in which **O** represents a 3 × 3 null matrix; and the 3 × 3 matrix **r**, which is commonly referred to as the *member rotation matrix,* is given by

$$\mathbf{r} = \begin{bmatrix} r_{xX} & r_{xY} & r_{xZ} \\ r_{yX} & r_{yY} & r_{yZ} \\ r_{zX} & r_{zY} & r_{zZ} \end{bmatrix} \tag{8.56}$$

The rotation matrix **r** plays a key role in the analysis of space frames, and an alternate form of this matrix, which enables us to specify the member orientations more conveniently, is developed subsequently.

Since the member local and global end displacements, **u** and **v**, are also vector quantities, which are defined in the same directions as the corresponding forces, the foregoing transformation matrix **T** can also be used to transform member end displacements from the global to the local coordinate system; that is, $\mathbf{u} = \mathbf{T}\mathbf{v}$. Furthermore, by employing a procedure similar to that used in the preceding paragraphs, it can be shown that the inverse transformations of the member end forces and end displacements, from the local to the global coordinate system, are defined by the transpose of the transformation matrix given in Eq. (8.54) (or Eqs. (8.55) and (8.56)); that is, $\mathbf{F} = \mathbf{T}^T\mathbf{Q}$ and $\mathbf{v} = \mathbf{T}^T\mathbf{u}$. Once the transformation matrix **T** has been established for a member of a space frame, its global stiffness matrix and fixed-end force vector can be obtained via the standard relationships $\mathbf{K} = \mathbf{T}^T\mathbf{k}\mathbf{T}$ and $\mathbf{F}_f = \mathbf{T}^T\mathbf{Q}_f$, respectively.

Member Rotation Matrix in Terms of the Angle of Roll From Eq. (8.56), we can see that the rotation matrix **r** consists of nine elements, with each element representing the direction cosine of a local axis with respect to a global axis, in accordance with Eq. (8.49). Of these nine direction cosines, the three in the first row of **r**, which represent the direction cosines of the local x axis, can be directly evaluated using the global coordinates of the two joints to which the member ends are attached. Thus, if X_b, Y_b, and Z_b and X_e, Y_e, and Z_e denote the global coordinates of the joints to which member ends b and e, respectively, are attached, then the direction cosines of the local x axis, with respect to the global X, Y, and Z axes, respectively, can be expressed as

$$r_{xX} = \cos\theta_{xX} = \frac{X_e - X_b}{L} \tag{8.57a}$$

$$r_{xY} = \cos\theta_{xY} = \frac{Y_e - Y_b}{L} \tag{8.57b}$$

$$r_{xZ} = \cos\theta_{xZ} = \frac{Z_e - Z_b}{L} \tag{8.57c}$$

in which the member length L is given by

$$L = \sqrt{(X_e - X_b)^2 + (Y_e - Y_b)^2 + (Z_e - Z_b)^2} \tag{8.57d}$$

With the direction cosines of the member x axis now established, we focus our attention on the question of how to determine the direction cosines of the local y and z axes using the information about the member orientation that can be conveniently input by the user of the computer program. Since the x, y, and

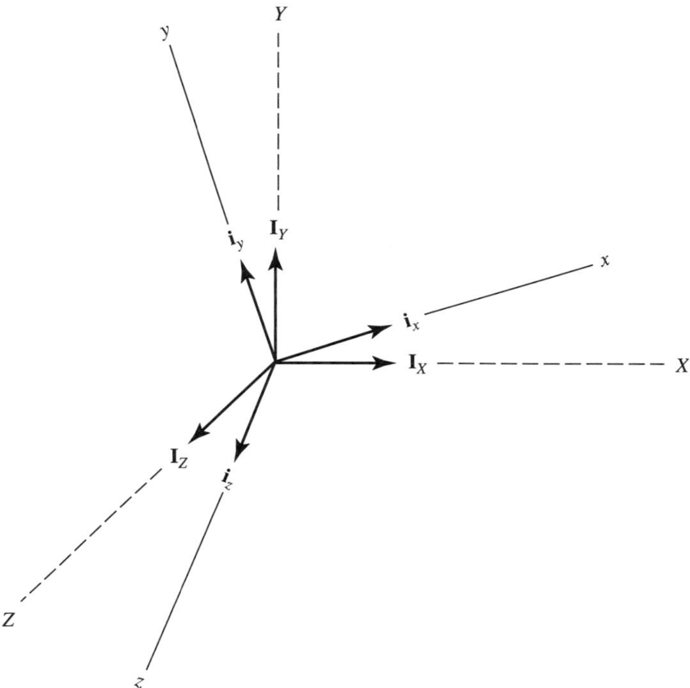

Fig. 8.17 *Unit Vectors in the Directions of the Local and Global Axes*

z axes form a mutually perpendicular right-handed coordinate system, it usually is convenient to define their directions by those of the unit vectors directed along these axes. Thus, if $\mathbf{i}_x$, $\mathbf{i}_y$, and $\mathbf{i}_z$ denote, respectively, the unit vectors in the directions of the local x, y, and z axes, and $\mathbf{I}_X$, $\mathbf{I}_Y$, and $\mathbf{I}_Z$ denote, respectively, the unit vectors directed along the global X, Y, and Z axes (see Fig. 8.17), then the relationship between the local and global unit vectors is defined by the member rotation matrix $\mathbf{r}$, as

$$\begin{bmatrix} \mathbf{i}_x \\ \mathbf{i}_y \\ \mathbf{i}_z \end{bmatrix} = \begin{bmatrix} r_{xX} & r_{xY} & r_{xZ} \\ r_{yX} & r_{yY} & r_{yZ} \\ r_{zX} & r_{zY} & r_{zZ} \end{bmatrix} \begin{bmatrix} \mathbf{I}_X \\ \mathbf{I}_Y \\ \mathbf{I}_Z \end{bmatrix} \tag{8.58}$$

The reader may recall from a previous course in *statics* that if the direction cosines of two of the three unit vectors, directed along the axes of an orthogonal coordinate system, are known, then those of the third unit vector can be obtained by using a cross (or vector) product of the two known vectors. In the case under consideration, as discussed previously, the direction cosines of one of the unit vectors, $\mathbf{i}_x$, are defined by the global coordinates of the member ends (Eqs. (8.57)). Thus, if the user of the computer program can provide, as input, the direction cosines of either $\mathbf{i}_y$ or $\mathbf{i}_z$ (i.e., either the y or the z axis), then the direction cosines of the remaining third vector can be conveniently established via the cross product of the two known vectors. However, as the hand calculation of direction cosines of the y or z axis for each member of a structure can be a tedious and time-consuming chore, this approach is not considered user-friendly and is seldom used by practitioners.

Instead, most computer programs allow the users to specify the orientation of the member y and z axes by means of the so-called *angle of roll* [3]. To define the angle of roll and to express the direction cosines of the member y and z axes in terms of this angle, we imagine that the member's desired (or actual design) orientation is reached in two steps, as shown in Figs. 8.18(a) and (b). In the first step, while the member's x axis is oriented in the desired direction, its y and z axes are oriented so that the xy plane is vertical and the z axis lies in a horizontal plane. The foregoing (imaginary) orientation of the member is depicted in Fig. 8.18(a), in which the member's principal axes are designated as $\bar{y}$ and $\bar{z}$ (instead of y and z, respectively), to indicate that they have not yet been positioned in their desired (or actual design) directions. As discussed previously, the direction of the local x axis is known from the global coordinates of the member ends. Since the $\bar{z}$ axis is perpendicular to the vertical $x\bar{y}$ plane, a vector $\bar{\mathbf{z}}$ directed along the $\bar{z}$ axis can be determined by the cross product of the vector $\mathbf{i}_x$ and a vertical unit vector $\mathbf{I}_Y$; that is,

$$\bar{\mathbf{z}} = \mathbf{i}_x \times \mathbf{I}_Y = \det \begin{vmatrix} \mathbf{I}_X & \mathbf{I}_Y & \mathbf{I}_Z \\ r_{xX} & r_{xY} & r_{xZ} \\ 0 & 1 & 0 \end{vmatrix} = -r_{xZ}\mathbf{I}_X + r_{xX}\mathbf{I}_Z \tag{8.59}$$

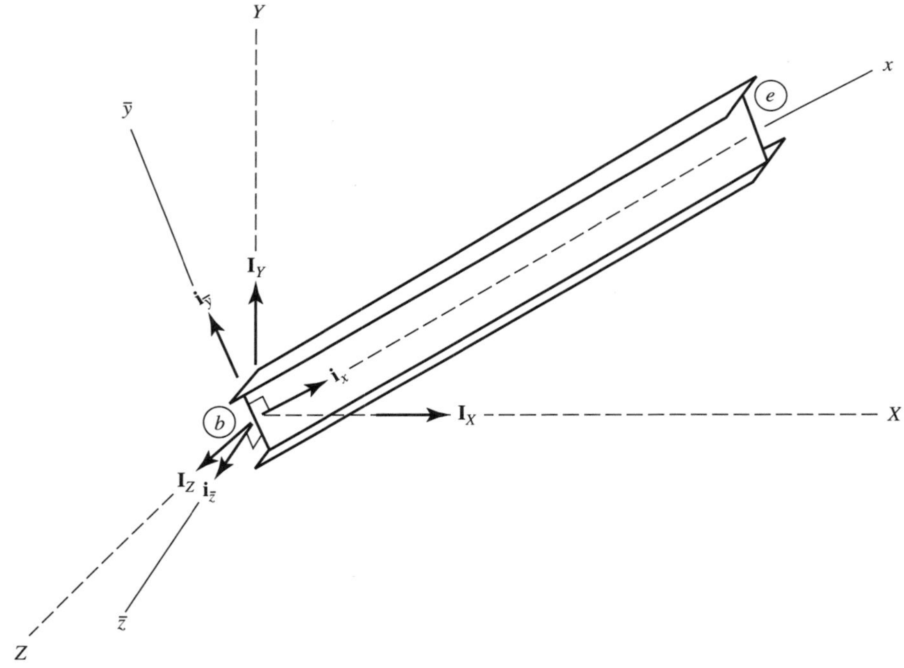

(a) Member Orientation with $x\bar{y}$ Plane Vertical

Fig. 8.18

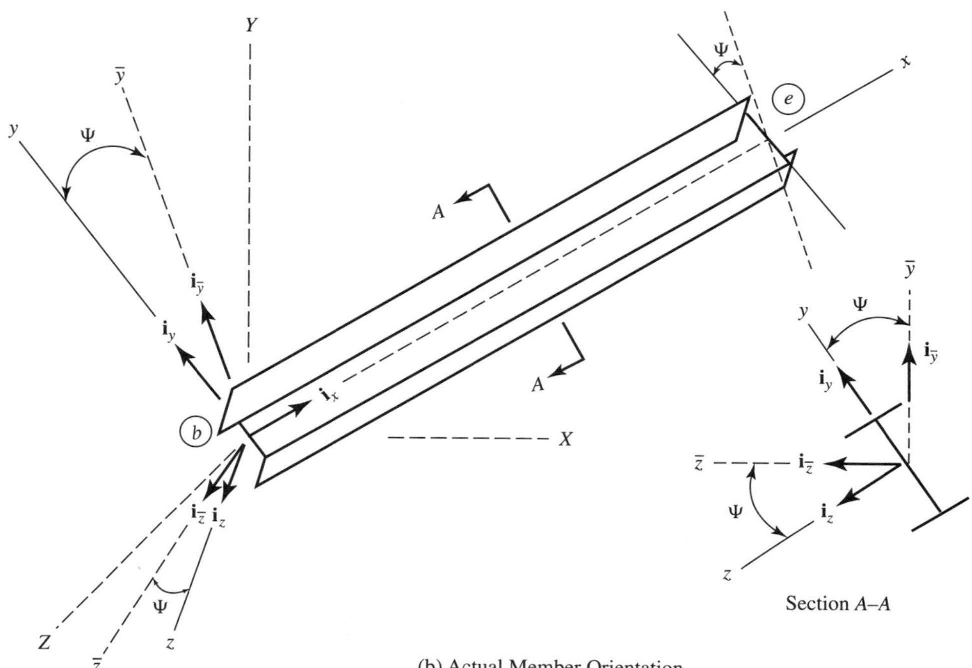

(b) Actual Member Orientation

Fig. 8.18 *(continued)*

To obtain the unit vector $\mathbf{i}_{\bar{z}}$ along the local $\bar{z}$ axis, we divide the vector $\bar{\mathbf{z}}$ by its magnitude $\sqrt{r_{xX}^2 + r_{xZ}^2}$. This yields

$$\mathbf{i}_{\bar{z}} = -\frac{r_{xZ}}{\sqrt{r_{xX}^2 + r_{xZ}^2}}\mathbf{I}_X + \frac{r_{xX}}{\sqrt{r_{xX}^2 + r_{xZ}^2}}\mathbf{I}_Z \tag{8.60}$$

The unit vector $\mathbf{i}_{\bar{y}}$ can now be established by using the cross product $\mathbf{i}_{\bar{z}} \times \mathbf{i}_x$, as

$$\mathbf{i}_{\bar{y}} = \mathbf{i}_{\bar{z}} \times \mathbf{i}_x = \det\begin{vmatrix} \mathbf{I}_X & \mathbf{I}_Y & \mathbf{I}_Z \\ -\dfrac{r_{xZ}}{\sqrt{r_{xX}^2 + r_{xZ}^2}} & 0 & \dfrac{r_{xX}}{\sqrt{r_{xX}^2 + r_{xZ}^2}} \\ r_{xX} & r_{xY} & r_{xZ} \end{vmatrix}$$

from which we obtain

$$\mathbf{i}_{\bar{y}} = \left(-\frac{r_{xX}r_{xY}}{\sqrt{r_{xX}^2 + r_{xZ}^2}}\right)\mathbf{I}_X + \left(\sqrt{r_{xX}^2 + r_{xZ}^2}\right)\mathbf{I}_Y - \left(\frac{r_{xY}r_{xZ}}{\sqrt{r_{xX}^2 + r_{xZ}^2}}\right)\mathbf{I}_Z \tag{8.61}$$

From Eqs. (8.60) and (8.61), we can see that the transformation relationship between the global *XYZ* and the auxiliary local $x\bar{y}\bar{z}$ coordinate systems can be expressed as

$$\begin{bmatrix} \mathbf{i}_x \\ \mathbf{i}_{\bar{y}} \\ \mathbf{i}_{\bar{z}} \end{bmatrix} = \begin{bmatrix} r_{xX} & r_{xY} & r_{xZ} \\ -\dfrac{r_{xX}r_{xY}}{\sqrt{r_{xX}^2 + r_{xZ}^2}} & \sqrt{r_{xX}^2 + r_{xZ}^2} & -\dfrac{r_{xY}r_{xZ}}{\sqrt{r_{xX}^2 + r_{xZ}^2}} \\ -\dfrac{r_{xZ}}{\sqrt{r_{xX}^2 + r_{xZ}^2}} & 0 & \dfrac{r_{xX}}{\sqrt{r_{xX}^2 + r_{xZ}^2}} \end{bmatrix} \begin{bmatrix} \mathbf{I}_X \\ \mathbf{I}_Y \\ \mathbf{I}_Z \end{bmatrix} \tag{8.62}$$

In the next step, we rotate the auxiliary $x\bar{y}\bar{z}$ coordinate system about its x axis, in a counterclockwise sense, by the angle of roll Ψ, until the member's principal axes are in their desired orientations. The final orientation of the member thus obtained is depicted in Fig. 8.18(b), in which the member's principal axes are now designated as y and z axes. From this figure, we can see that the unit vectors along the y and z axes can be expressed in terms of those directed along the $\bar{y}$ and $\bar{z}$ axes, as

$$\mathbf{i}_y = \cos\Psi \mathbf{i}_{\bar{y}} + \sin\Psi \mathbf{i}_{\bar{z}} \tag{8.63a}$$

$$\mathbf{i}_z = -\sin\Psi \mathbf{i}_{\bar{y}} + \cos\Psi \mathbf{i}_{\bar{z}} \tag{8.63b}$$

Thus, the transformation relationship between the auxiliary $x\bar{y}\bar{z}$ and the actual xyz coordinate systems is given by

$$\begin{bmatrix} \mathbf{i}_x \\ \mathbf{i}_y \\ \mathbf{i}_z \end{bmatrix} = \begin{bmatrix} 1 & 0 & 0 \\ 0 & \cos\Psi & \sin\Psi \\ 0 & -\sin\Psi & \cos\Psi \end{bmatrix} \begin{bmatrix} \mathbf{i}_x \\ \mathbf{i}_{\bar{y}} \\ \mathbf{i}_{\bar{z}} \end{bmatrix} \tag{8.64}$$

Finally, to obtain the transformation relationship between the global XYZ and the actual local xyz coordinate systems, we substitute Eq. (8.62) into Eq. (8.64) and carry out the required matrix multiplication. This yields

$$\begin{bmatrix} \mathbf{i}_x \\ \mathbf{i}_y \\ \mathbf{i}_z \end{bmatrix} = \begin{bmatrix} r_{xX} & r_{xY} & r_{xZ} \\ \dfrac{-r_{xX}r_{xY}\cos\Psi - r_{xZ}\sin\Psi}{\sqrt{r_{xX}^2 + r_{xZ}^2}} & \sqrt{r_{xX}^2 + r_{xZ}^2}\cos\Psi & \dfrac{-r_{xY}r_{xZ}\cos\Psi + r_{xX}\sin\Psi}{\sqrt{r_{xX}^2 + r_{xZ}^2}} \\ \dfrac{r_{xX}r_{xY}\sin\Psi - r_{xZ}\cos\Psi}{\sqrt{r_{xX}^2 + r_{xZ}^2}} & -\sqrt{r_{xX}^2 + r_{xZ}^2}\sin\Psi & \dfrac{r_{xY}r_{xZ}\sin\Psi + r_{xX}\cos\Psi}{\sqrt{r_{xX}^2 + r_{xZ}^2}} \end{bmatrix} \begin{bmatrix} \mathbf{I}_X \\ \mathbf{I}_Y \\ \mathbf{I}_Z \end{bmatrix} \tag{8.65}$$

By comparing Eqs. (8.58) and (8.65), we can see that the member rotation matrix $\mathbf{r}$ can be expressed as

$$\mathbf{r} = \begin{bmatrix} r_{xX} & r_{xY} & r_{xZ} \\ \dfrac{-r_{xX}r_{xY}\cos\Psi - r_{xZ}\sin\Psi}{\sqrt{r_{xX}^2 + r_{xZ}^2}} & \sqrt{r_{xX}^2 + r_{xZ}^2}\cos\Psi & \dfrac{-r_{xY}r_{xZ}\cos\Psi + r_{xX}\sin\Psi}{\sqrt{r_{xX}^2 + r_{xZ}^2}} \\ \dfrac{r_{xX}r_{xY}\sin\Psi - r_{xZ}\cos\Psi}{\sqrt{r_{xX}^2 + r_{xZ}^2}} & -\sqrt{r_{xX}^2 + r_{xZ}^2}\sin\Psi & \dfrac{r_{xY}r_{xZ}\sin\Psi + r_{xX}\cos\Psi}{\sqrt{r_{xX}^2 + r_{xZ}^2}} \end{bmatrix} \tag{8.66}$$

Note that the rotation matrix depends only on the global coordinates of the member ends and its angle of roll Ψ. Based on the foregoing derivation, the *angle of roll* Ψ is defined as *the angle, measured clockwise positive when looking in the negative x direction, through which the local xyz coordinate system must be rotated around its x axis, so that the xy plane becomes vertical with the y axis pointing upward (i.e., in the positive direction of the global Y axis).*

The expression of the rotation matrix **r**, as given by Eq. (8.66), can be used to determine the transformation matrices **T** for the members of space frames oriented in any arbitrary directions, except for vertical members. This is because for such members r_{xX} and r_{xZ} are zero, causing some elements of **r** in Eq. (8.66) to become undefined. This situation can be remedied by defining the angle of roll differently for vertical members, as follows. For the special case of vertical members (i.e., members with centroidal or local x axis parallel to the global Y axis), the angle of roll Ψ is defined as *the angle, measured clockwise positive when looking in the negative x direction, through which the local xyz coordinate system must be rotated around its x axis, so that the local z axis becomes parallel to, and points in the positive direction of, the global Z axis* (Fig. 8.19(b)).

The expression of the rotation matrix **r** for vertical members can be derived using a procedure similar to that used previously for members with other orientations. We imagine that the vertical member's desired (or actual design) orientation is reached in two steps, as shown in Figs. 8.19(a) and (b) on the next page. In the first step, while the member's x axis is oriented in the desired (vertical) direction, its y and z axes are oriented so that the local z axis is parallel to the global Z axis, as shown in Fig. 8.19(a). As indicated there, the member's principal axes in this (imaginary) orientation are designated as $\bar{y}$ and $\bar{z}$ (instead of y and z, respectively). The direction of the local x axis (known from the global coordinates of the member ends) is represented by the unit vector $\mathbf{i}_x = r_{xY}\mathbf{I}_Y$, while the direction of the $\bar{z}$ axis is given by the unit vector $\mathbf{i}_{\bar{z}} = \mathbf{I}_Z$. The unit vector $\mathbf{i}_{\bar{y}}$, directed along the $\bar{y}$ axis, can therefore be conveniently established using the cross product $\mathbf{i}_{\bar{z}} \times \mathbf{i}_x$, as

$$\mathbf{i}_{\bar{y}} = \mathbf{i}_{\bar{z}} \times \mathbf{i}_x = \det \begin{vmatrix} \mathbf{I}_X & \mathbf{I}_Y & \mathbf{I}_Z \\ 0 & 0 & 1 \\ 0 & r_{xY} & 0 \end{vmatrix} = -r_{xY}\mathbf{I}_X \tag{8.67}$$

Thus, the transformation relationship between the global XYZ and the auxiliary local $x\bar{y}\bar{z}$ coordinate system is given by

$$\begin{bmatrix} \mathbf{i}_x \\ \mathbf{i}_{\bar{y}} \\ \mathbf{i}_{\bar{z}} \end{bmatrix} = \begin{bmatrix} 0 & r_{xY} & 0 \\ -r_{xY} & 0 & 0 \\ 0 & 0 & 1 \end{bmatrix} \begin{bmatrix} \mathbf{I}_X \\ \mathbf{I}_Y \\ \mathbf{I}_Z \end{bmatrix} \tag{8.68}$$

In the next step, we rotate the auxiliary $x\bar{y}\bar{z}$ coordinate system about its x axis, in a counterclockwise sense, by the angle of roll Ψ, until the member's principal axes are in their desired orientations. This final orientation of the member is depicted in Fig. 8.19(b), in which the member principal axes are

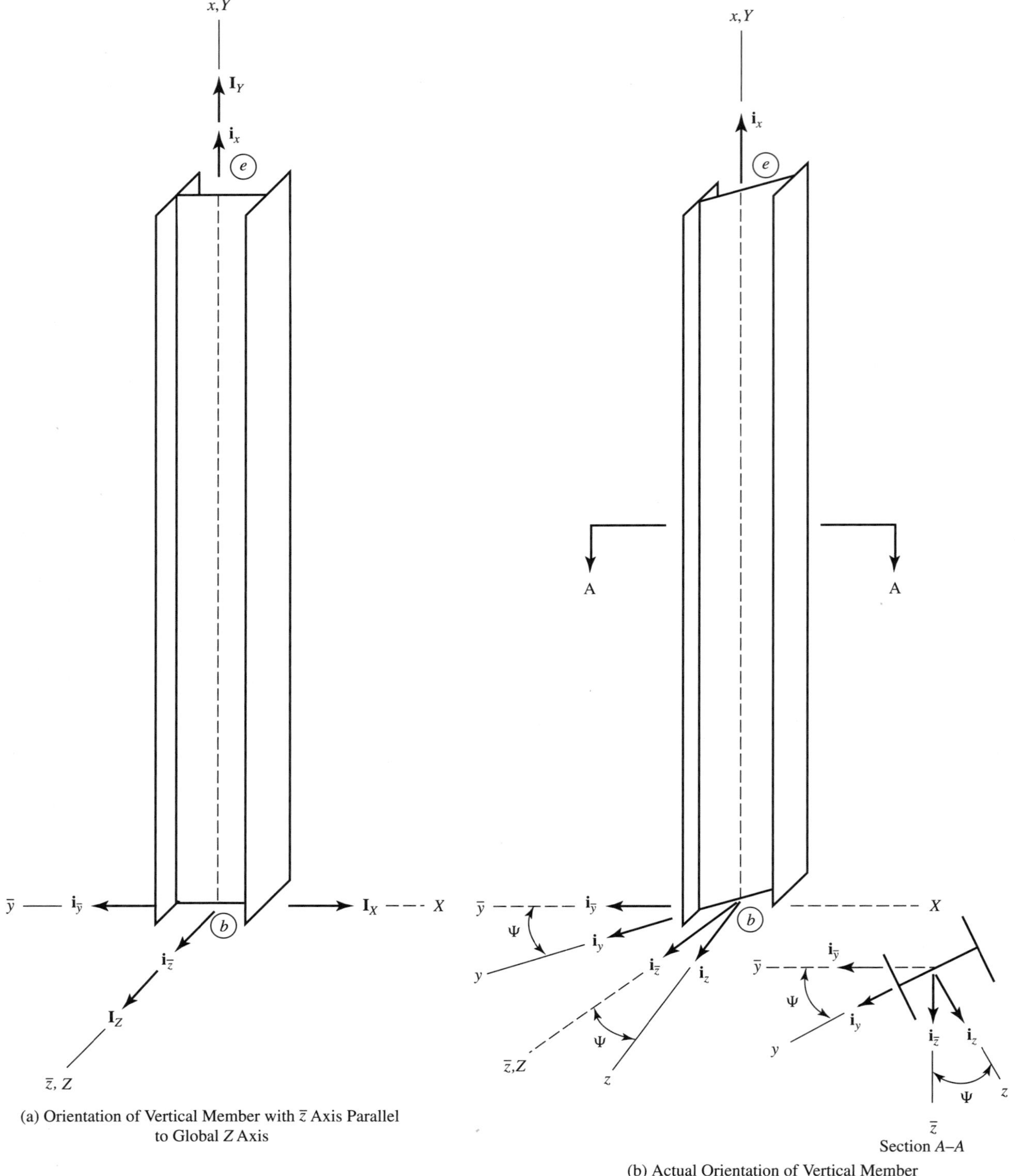

(a) Orientation of Vertical Member with $\bar{z}$ Axis Parallel to Global Z Axis

(b) Actual Orientation of Vertical Member

Fig. 8.19

now designated as the y and z axes. From this figure, we can see that the transformation relationship between the auxiliary $x\bar{y}\bar{z}$ and the actual xyz coordinate systems is the same as given previously in Eq. (8.64). Thus, the desired transformation from the global XYZ coordinate system to the local xyz coordinate system can be obtained by substituting Eq. (8.68) into Eq. (8.64) and performing the required matrix multiplication. This yields

$$\begin{bmatrix} \mathbf{i}_x \\ \mathbf{i}_y \\ \mathbf{i}_z \end{bmatrix} = \begin{bmatrix} 0 & r_{xY} & 0 \\ -r_{xY}\cos\Psi & 0 & \sin\Psi \\ r_{xY}\sin\Psi & 0 & \cos\Psi \end{bmatrix} \begin{bmatrix} \mathbf{I}_X \\ \mathbf{I}_Y \\ \mathbf{I}_Z \end{bmatrix} \tag{8.69}$$

from which we obtain the rotation matrix $\mathbf{r}$ for vertical members:

$$\mathbf{r} = \begin{bmatrix} 0 & r_{xY} & 0 \\ -r_{xY}\cos\Psi & 0 & \sin\Psi \\ r_{xY}\sin\Psi & 0 & \cos\Psi \end{bmatrix} \tag{8.70}$$

Member Rotation Matrix in Terms of a Reference Point In most space frames, members are usually oriented so that their angles of roll can be found by inspection. There are structures, however, in which the orientations of some members may be such that their angles of roll cannot be conveniently determined. The orientation of such a member can alternatively be specified by means of the global coordinates of a reference point that lies in one of the principal (xy or xz) planes of the member, but not on its centroidal (x) axis.

To discuss the process of determining the member rotation matrix $\mathbf{r}$ using such a reference point, consider the space-frame member shown in Fig. 8.20 on the next page, and let X_P, Y_P, and Z_P denote the global coordinates of an arbitrarily chosen reference point P, which is located in the member's local xy plane, but not on its x axis. Since the global coordinates of the member end b are X_b, Y_b, and Z_b, the position vector $\mathbf{p}$, directed from member end b to reference point P, can be written as

$$\mathbf{p} = (X_P - X_b)\mathbf{I}_X + (Y_P - Y_b)\mathbf{I}_Y + (Z_P - Z_b)\mathbf{I}_z \tag{8.71}$$

Note that both points b and P are located in the local xy plane and, therefore, vector $\mathbf{p}$ also lies in that plane. Since the direction cosines of the local x axis are already known from the global coordinates of the member ends, the direction cosines of the local z axis can be conveniently established using the following relationship.

$$\mathbf{i}_z = \frac{\mathbf{i}_x \times \mathbf{p}}{|\mathbf{i}_x \times \mathbf{p}|} \tag{8.72}$$

in which $|\mathbf{i}_x \times \mathbf{p}|$ represents the magnitude of the vector that results from the cross product of the vectors $\mathbf{i}_x$ and $\mathbf{p}$. With both $\mathbf{i}_x$ and $\mathbf{i}_z$ now known, the direction cosines of the local y axis can be obtained via the cross product,

$$\mathbf{i}_y = \mathbf{i}_z \times \mathbf{i}_x \tag{8.73}$$

In the case that the reference point P is specified in the local xz plane of the member, the direction cosines of the local y axis need to be determined first

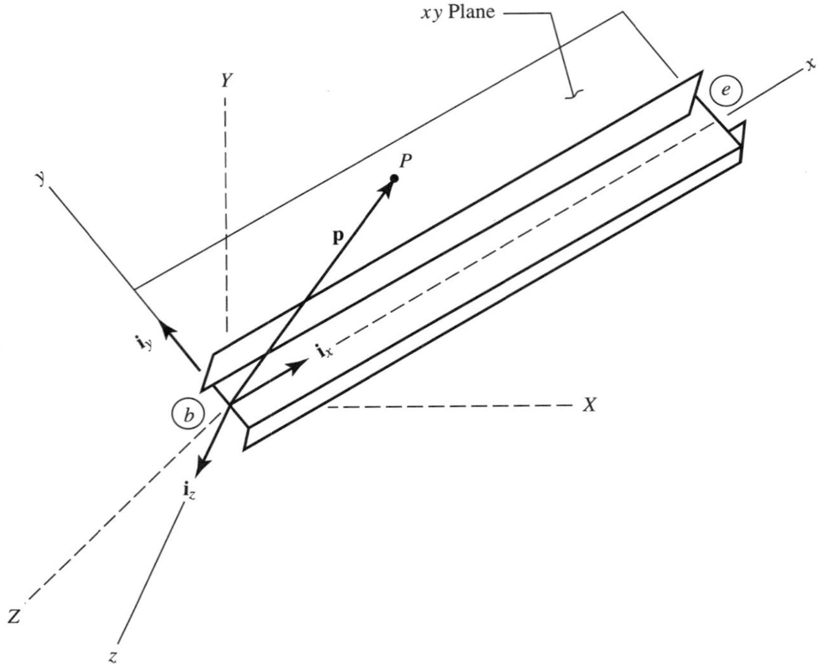

Fig. 8.20

using the relationship

$$\mathbf{i}_y = \frac{\mathbf{p} \times \mathbf{i}_x}{|\mathbf{p} \times \mathbf{i}_x|} \tag{8.74}$$

and then the direction cosines of the local z axis are obtained via the cross product

$$\mathbf{i}_z = \mathbf{i}_x \times \mathbf{i}_y \tag{8.75}$$

It should be realized that the procedure described by Eqs. (8.71) through (8.75) enables us to obtain the member rotation matrix $\mathbf{r}$ directly by means of a reference point, without involving the angle of roll of the member. However, if desired, the angle of roll can also be obtained from the global coordinates of a reference point. To establish the relationship between the angle of roll Ψ and a reference point P of a member, we first determine the components of the position vector $\mathbf{p}$ in the auxiliary $x\bar{y}\bar{z}$ coordinate system, by applying the transformation relationship given in Eq. (8.62), as

$$\begin{bmatrix} p_x \\ p_{\bar{y}} \\ p_{\bar{z}} \end{bmatrix} = \begin{bmatrix} r_{xX} & r_{xY} & r_{xZ} \\ -\dfrac{r_{xX}r_{xY}}{\sqrt{r_{xX}^2 + r_{xZ}^2}} & \sqrt{r_{xX}^2 + r_{xZ}^2} & -\dfrac{r_{xY}r_{xZ}}{\sqrt{r_{xX}^2 + r_{xZ}^2}} \\ -\dfrac{r_{xZ}}{\sqrt{r_{xX}^2 + r_{xZ}^2}} & 0 & \dfrac{r_{xX}}{\sqrt{r_{xX}^2 + r_{xZ}^2}} \end{bmatrix} \begin{bmatrix} (X_P - X_b) \\ (Y_P - Y_b) \\ (Z_P - Z_b) \end{bmatrix}$$

from which we obtain

$$p_x = r_{xX}(X_P - X_b) + r_{xY}(Y_P - Y_b) + r_{xZ}(Z_P - Z_b) \tag{8.76a}$$

$$p_{\bar{y}} = -\frac{r_{xX} r_{xY}}{\sqrt{r_{xX}^2 + r_{xZ}^2}}(X_P - X_b) + \sqrt{r_{xX}^2 + r_{xZ}^2}(Y_P - Y_b) - \frac{r_{xY} r_{xZ}}{\sqrt{r_{xX}^2 + r_{xZ}^2}}(Z_P - Z_b) \tag{8.76b}$$

$$p_{\bar{z}} = -\frac{r_{xZ}}{\sqrt{r_{xX}^2 + r_{xZ}^2}}(X_P - X_b) + \frac{r_{xX}}{\sqrt{r_{xX}^2 + r_{xZ}^2}}(Z_P - Z_b) \tag{8.76c}$$

in which p_x, $p_{\bar{y}}$, and $p_{\bar{z}}$ represent, respectively, the components of the position vector **p** in the directions of the local x axis and the $\bar{y}$ and $\bar{z}$ axes of the auxiliary $x\bar{y}\bar{z}$ coordinate system. Now, if the reference point P lies in the xy plane of the member as shown in Fig. 8.21, then we can see from this figure that the angle of roll Ψ and the components of **p** are related by the following equations:

$$\sin\Psi = \frac{p_{\bar{z}}}{\sqrt{p_{\bar{y}}^2 + p_{\bar{z}}^2}} \qquad \text{and} \qquad \cos\Psi = \frac{p_{\bar{y}}}{\sqrt{p_{\bar{y}}^2 + p_{\bar{z}}^2}} \tag{8.77}$$

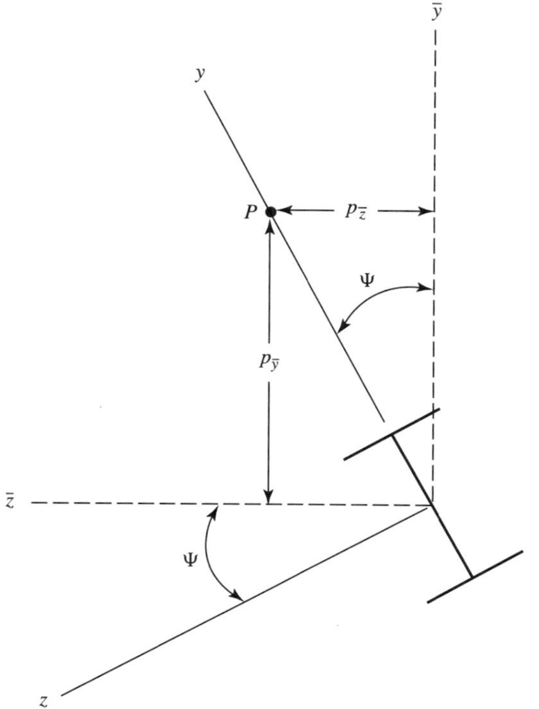

Fig. 8.21

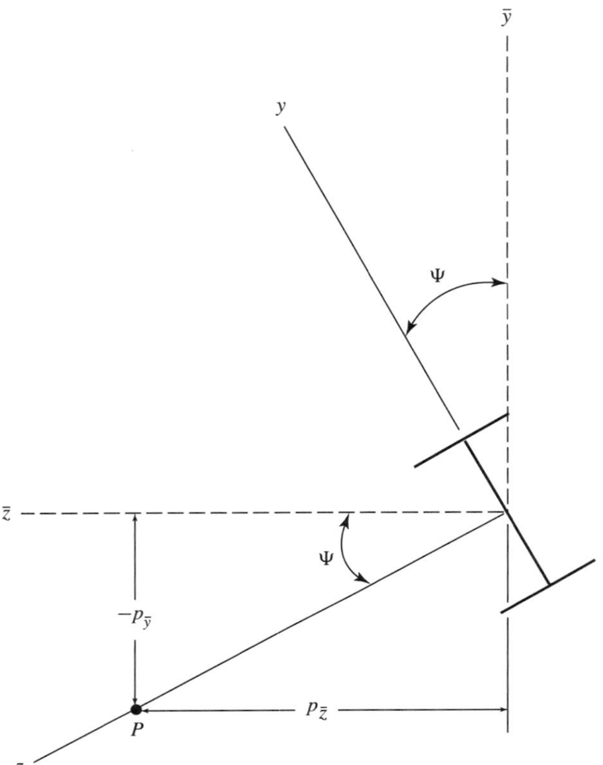

Fig. 8.22

In the case that the reference point P is specified in the local xz plane, then from Fig. 8.22 we can see that the relationships between the sine and cosine of Ψ and the components $\mathbf{p}$ can be expressed as

$$\sin\Psi = -\frac{p_{\bar{y}}}{\sqrt{p_{\bar{y}}^2 + p_{\bar{z}}^2}} \qquad \text{and} \qquad \cos\Psi = \frac{p_{\bar{z}}}{\sqrt{p_{\bar{y}}^2 + p_{\bar{z}}^2}} \tag{8.78}$$

Equations (8.77) and (8.78) are valid for space-frame members oriented in any arbitrary directions, including vertical members. However, since $r_{xX} = r_{xZ} = 0$ for vertical members, the expressions for $p_{\bar{y}}$ and $p_{\bar{z}}$, as given in Eqs. 8.76(b) and (c), cannot be used; appropriate expressions for the components of the position vector $\mathbf{p}$ in the auxiliary $x\bar{y}\bar{z}$ coordinate system must be derived by applying Eq. (8.68), as

$$\begin{bmatrix} p_x \\ p_{\bar{y}} \\ p_{\bar{z}} \end{bmatrix} = \begin{bmatrix} 0 & r_{xY} & 0 \\ -r_{xY} & 0 & 0 \\ 0 & 0 & 1 \end{bmatrix} \begin{bmatrix} (X_P - X_b) \\ (Y_P - Y_b) \\ (Z_P - Z_b) \end{bmatrix}$$

which yields

$$p_x = r_{xY}(Y_P - Y_b) \tag{8.79a}$$

$$p_{\bar{y}} = -r_{xY}(X_P - X_b) \quad \textbf{(8.79b)}$$
$$p_{\bar{z}} = Z_P - Z_b \quad \textbf{(8.79c)}$$

It is important to realize that, for vertical members, Eqs. 8.79(b) and (c) should be used to evaluate $p_{\bar{y}}$ and $p_{\bar{z}}$, whereas for members with other orientations, these components are obtained from Eqs. 8.76(b) and (c). After $p_{\bar{y}}$ and $p_{\bar{z}}$ have been evaluated, the sine and cosine of the member's angle of roll Ψ can be determined either by Eq. (8.77) if the reference point lies in the xy plane, or via Eq. (8.78) if the reference point is located in the xz plane. Once $\sin\Psi$ and $\cos\Psi$ are known, the member rotation matrix **r** can be determined by Eq. (8.66) if the member is not oriented in the vertical direction, or via Eq. (8.70) if the member is vertical.

EXAMPLE 8.3 The global coordinates of the joints to which the beginning and end of a space-frame member are attached are (4, 7, 6) ft and (20, 15, 17) ft, respectively. If the global coordinates of a reference point located in the local xy plane of the member are (10.75, 13.6, 13.85) ft, determine the rotation matrix of the member.

SOLUTION We determine the member rotation matrix **r** using the direct approach involving cross products of vectors, and then check our results using the angle-of-roll approach.

Using the given coordinates of the two ends of the member, we evaluate its length and the direction cosines of the local x axis, as (see Eqs. (8.57))

$$L = \sqrt{(X_e - X_b)^2 + (Y_e - Y_b)^2 + (Z_b - Z_e)^2}$$
$$= \sqrt{(20-4)^2 + (15-7)^2 + (17-6)^2} = 21 \text{ ft}$$

$$r_{xX} = \frac{X_e - X_b}{L} = \frac{20-4}{21} = 0.7619 \quad \textbf{(1a)}$$

$$r_{xY} = \frac{Y_e - Y_b}{L} = \frac{15-7}{21} = 0.38095 \quad \textbf{(1b)}$$

$$r_{xZ} = \frac{Z_e - Z_b}{L} = \frac{17-6}{21} = 0.52381 \quad \textbf{(1c)}$$

Thus, the unit vector directed along the member local x (or centroidal) axis is

$$\mathbf{i}_x = 0.7619\,\mathbf{I}_X + 0.38095\,\mathbf{I}_Y + 0.52381\,\mathbf{I}_Z \quad \textbf{(2)}$$

Next, we form the position vector **p**, directed from member end b to reference point P, as (Eq. (8.71))

$$\mathbf{p} = (X_P - X_b)\mathbf{I}_X + (Y_P - Y_b)\mathbf{I}_Y + (Z_P - Z_b)\mathbf{I}_Z$$
$$= (10.75-4)\mathbf{I}_X + (13.6-7)\mathbf{I}_Y + (13.85-6)\mathbf{I}_Z$$

or

$$\mathbf{p} = 6.75\,\mathbf{I}_X + 6.6\,\mathbf{I}_Y + 7.85\,\mathbf{I}_Z$$

With $\mathbf{i}_x$ and **p** known, we can now apply Eq. (8.72) to determine the unit vector in the local z direction. For that purpose, we first obtain a vector **z** along the local z axis by evaluating the cross product of $\mathbf{i}_x$ and $\mathbf{p}$. Thus,

$$\mathbf{z} = \mathbf{i}_x \times \mathbf{p} = \det \begin{vmatrix} \mathbf{I}_X & \mathbf{I}_Y & \mathbf{I}_Z \\ 0.7619 & 0.38095 & 0.52381 \\ 6.75 & 6.6 & 7.85 \end{vmatrix}$$

$$= [(0.38095)(7.85) - (6.6)(0.52381)]\,\mathbf{I}_X$$
$$- [(0.7619)(7.85) - (6.75)(0.52381)]\,\mathbf{I}_Y$$
$$+ [(0.7619)(6.6) - (6.75)(0.38095)]\,\mathbf{I}_Z$$

or

$$\mathbf{z} = -0.46669\,\mathbf{I}_X - 2.4452\,\mathbf{I}_Y + 2.4571\,\mathbf{I}_Z$$

Note that $\mathbf{z}$ is not a unit vector. To obtain the unit vector $\mathbf{i}_z$, we need to divide $\mathbf{z}$ by its magnitude $|\mathbf{z}|$, which equals

$$|\mathbf{z}| = |\mathbf{i}_x \times \mathbf{p}| = \sqrt{(-0.46669)^2 + (-2.4452)^2 + (2.4571)^2} = 3.4977 \text{ ft}$$

Thus, the unit vector $\mathbf{i}_z$ is given by

$$\mathbf{i}_z = \frac{\mathbf{z}}{|\mathbf{z}|} = -0.13343\,\mathbf{I}_X - 0.69909\,\mathbf{I}_Y + 0.70249\,\mathbf{I}_Z \qquad \textbf{(3)}$$

The third unit vector, $\mathbf{i}_y$, can now be evaluated using the cross product of $\mathbf{i}_z$ (Eq. (3)) and $\mathbf{i}_x$ (Eq. (2)). Thus,

$$\mathbf{i}_y = \mathbf{i}_z \times \mathbf{i}_x = \det \begin{vmatrix} \mathbf{I}_X & \mathbf{I}_Y & \mathbf{I}_Z \\ -0.13343 & -0.69909 & 0.70249 \\ 0.7619 & 0.38095 & 0.52381 \end{vmatrix}$$

$$= [(-0.69909)(0.52381) - (0.38095)(0.70249)]\,\mathbf{I}_X$$
$$- [(-0.13343)(0.52381) - (0.7619)(0.70249)]\,\mathbf{I}_Y$$
$$+ [(-0.13343)(0.38095) - (0.7619)(-0.69909)]\,\mathbf{I}_Z$$

or

$$\mathbf{i}_y = -0.6338\,\mathbf{I}_X + 0.60512\,\mathbf{I}_Y + 0.48181\,\mathbf{I}_Z \qquad \textbf{(4)}$$

The member rotation matrix $\mathbf{r}$ can now be obtained by arranging the components of $\mathbf{i}_x$ (Eq. (2)), $\mathbf{i}_y$ (Eq. (4)) and $\mathbf{i}_z$ (Eq. (3)) in the first, second, and third rows, respectively, of a 3 × 3 matrix. The member rotation matrix thus obtained is

$$\mathbf{r} = \begin{bmatrix} 0.7619 & 0.38095 & 0.52381 \\ -0.6338 & 0.60512 & 0.48181 \\ -0.13343 & -0.69909 & 0.70249 \end{bmatrix} \qquad \textbf{Ans}$$

Alternative Method: The member rotation matrix $\mathbf{r}$ can alternatively be determined by applying Eq. (8.66), which contains the sine and cosine of the angle of roll Ψ. We first evaluate the components $p_{\bar{y}}$ and $p_{\bar{z}}$ of the position vector $\mathbf{p}$ using Eqs. 8.76(b) and (c), respectively. By substituting the numerical values of r_{xX}, r_{xY}, and r_{xZ} (from Eqs. (1)) and the given coordinates of member end b and reference point P into these equations, we obtain

$$p_{\bar{y}} = 2.2892 \text{ ft}$$
$$p_{\bar{z}} = 2.6446 \text{ ft}$$

By substituting these values of $p_{\bar{y}}$ and $p_{\bar{z}}$ into Eqs. (8.77), we obtain the sine and cosine of the angle of roll:

$$\sin\Psi = 0.75608 \qquad \cos\Psi = 0.65448 \tag{5}$$

Finally, by substituting the numerical values from Eqs. (1) and (5) into Eq. (8.66), we obtain the following rotation matrix for the member under consideration:

$$\mathbf{r} = \begin{bmatrix} 0.7619 & 0.38095 & 0.52381 \\ -0.6338 & 0.60512 & 0.48179 \\ -0.13343 & -0.69907 & 0.70249 \end{bmatrix}$$

Checks

Procedure for Analysis

The general procedure for analysis of space frames remains the same as that for plane frames developed in Chapter 6 (and modified in Chapter 7)—provided that the member local stiffness and transformation matrices, and local fixed-end force vectors, developed in this section, are used in the analysis.

EXAMPLE 8.4 Determine the joint displacements, member end forces, and support reactions for the three-member space frame shown in Fig. 8.23(a) on the next page, using the matrix stiffness method.

SOLUTION *Analytical Model:* The space frame has six degrees of freedom and 18 restrained coordinates, as shown in Fig. 8.23(b).

Structure Stiffness Matrix:

Member 1 By substituting $L = 240$ in., and the material and cross-sectional properties given in Fig. 8.23(a), into Eq. (8.39), we obtain the local stiffness matrix $\mathbf{k}$ for member 1:

$$\mathbf{K}_1 = \mathbf{k}_1 = \begin{array}{c} \begin{array}{cccccccccccc} 7 & 8 & 9 & 10 & 11 & 12 & 1 & 2 & 3 & 4 & 5 & 6 \end{array} \\ \left[\begin{array}{cccccc|cccccc} 3,975.4 & 0 & 0 & 0 & 0 & 0 & -3,975.4 & 0 & 0 & 0 & 0 & 0 \\ 0 & 18.024 & 0 & 0 & 0 & 2,162.9 & 0 & -18.024 & 0 & 0 & 0 & 2,162.9 \\ 0 & 0 & 5.941 & 0 & -712.92 & 0 & 0 & 0 & -5.941 & 0 & -712.92 & 0 \\ 0 & 0 & 0 & 723.54 & 0 & 0 & 0 & 0 & 0 & -723.54 & 0 & 0 \\ 0 & 0 & -712.92 & 0 & 114,067 & 0 & 0 & 0 & 712.92 & 0 & 57,033 & 0 \\ 0 & 2,162.9 & 0 & 0 & 0 & 346,067 & 0 & -2,162.9 & 0 & 0 & 0 & 173,033 \\ \hline -3,975.4 & 0 & 0 & 0 & 0 & 0 & 3,975.4 & 0 & 0 & 0 & 0 & 0 \\ 0 & -18.024 & 0 & 0 & 0 & -2,162.9 & 0 & 18.024 & 0 & 0 & 0 & -2,162.9 \\ 0 & 0 & -5.941 & 0 & 712.92 & 0 & 0 & 0 & 5.941 & 0 & 712.92 & 0 \\ 0 & 0 & 0 & -723.54 & 0 & 0 & 0 & 0 & 0 & 723.54 & 0 & 0 \\ 0 & 0 & -712.92 & 0 & 57,033 & 0 & 0 & 0 & 712.92 & 0 & 114,067 & 0 \\ 0 & 2,162.9 & 0 & 0 & 0 & 173,033 & 0 & -2,162.9 & 0 & 0 & 0 & 346,067 \end{array}\right] \end{array} \begin{array}{c} \\ 7 \\ 8 \\ 9 \\ 10 \\ 11 \\ 12 \\ 1 \\ 2 \\ 3 \\ 4 \\ 5 \\ 6 \end{array} \tag{1}$$

Since the member's local x, y, and z axes are oriented in the directions of the global X, Y, and Z axes, respectively (see Fig. 8.23(a)), no coordinate transformations are necessary (i.e., $\mathbf{T}_1 = \mathbf{I}$); thus, $\mathbf{K}_1 = \mathbf{k}_1$.

To determine the fixed-end force vector due to the 0.25 k/in. (= 3 k/ft) member load, we apply the fixed-end force expressions for loading type 3 given inside the front cover. This yields

$$FS_{by} = FS_{ey} = 30 \text{ k}$$
$$FM_{bz} = -FM_{ez} = -1{,}200 \text{ k-in.}$$

with the remaining fixed-end forces being 0. Thus, using Eq. (8.40), we obtain

$$\mathbf{F}_{f1} = \mathbf{Q}_{f1} = \left[\begin{array}{c} 0 \\ 30 \\ 0 \\ 0 \\ 0 \\ 1{,}200 \\ \hdashline 0 \\ 30 \\ 0 \\ 0 \\ 0 \\ -1{,}200 \end{array}\right] \begin{array}{c} 7 \\ 8 \\ 9 \\ 10 \\ 11 \\ 12 \\ 1 \\ 2 \\ 3 \\ 4 \\ 5 \\ 6 \end{array} \tag{2}$$

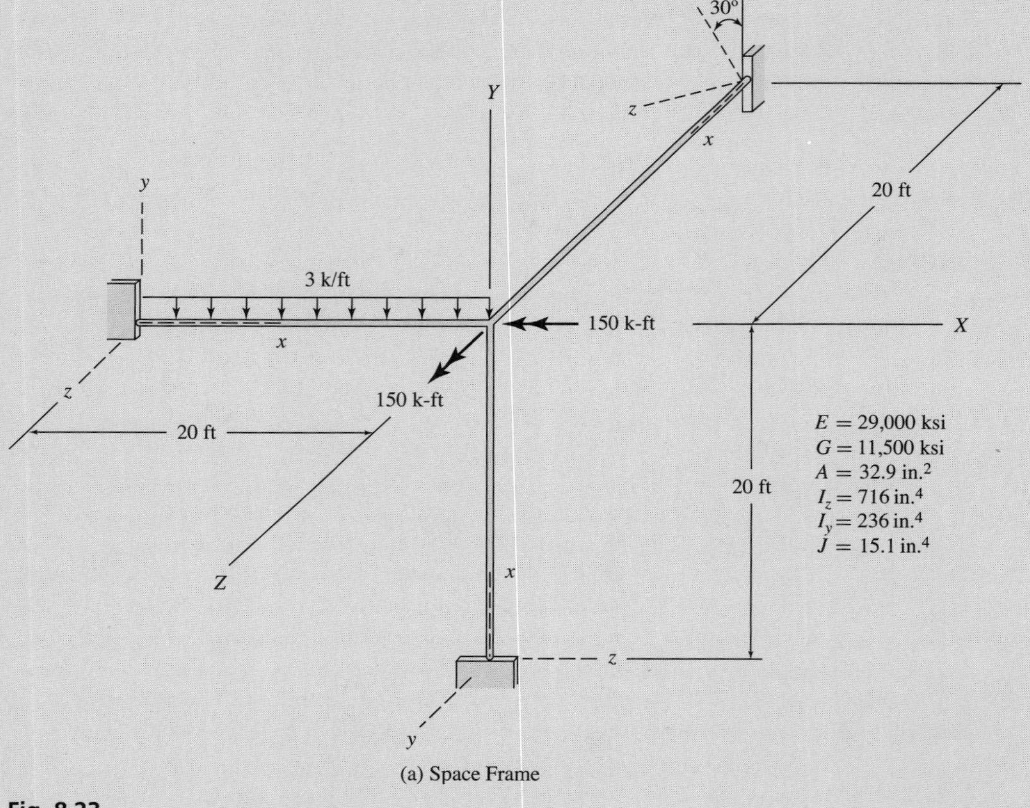

(a) Space Frame

Fig. 8.23

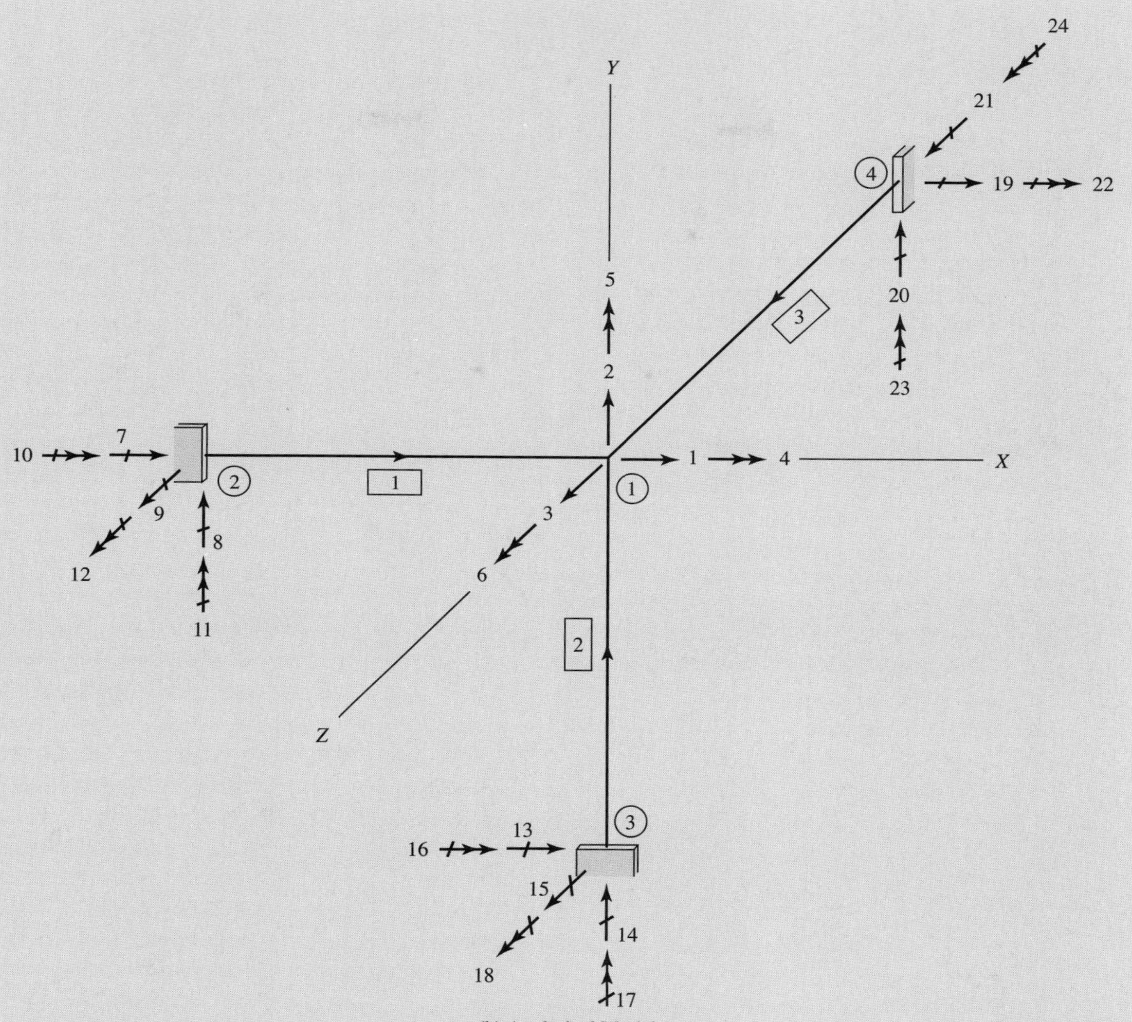

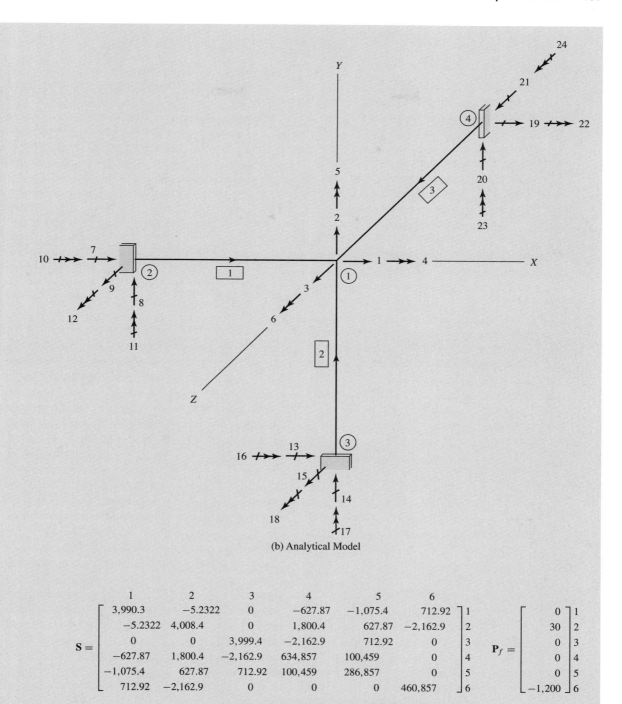

(b) Analytical Model

$$
\begin{array}{c}
\begin{array}{cccccc} 1 & 2 & 3 & 4 & 5 & 6 \end{array} \\
\mathbf{S} = \left[\begin{array}{cccccc}
3{,}990.3 & -5.2322 & 0 & -627.87 & -1{,}075.4 & 712.92 \\
-5.2322 & 4{,}008.4 & 0 & 1{,}800.4 & 627.87 & -2{,}162.9 \\
0 & 0 & 3{,}999.4 & -2{,}162.9 & 712.92 & 0 \\
-627.87 & 1{,}800.4 & -2{,}162.9 & 634{,}857 & 100{,}459 & 0 \\
-1{,}075.4 & 627.87 & 712.92 & 100{,}459 & 286{,}857 & 0 \\
712.92 & -2{,}162.9 & 0 & 0 & 0 & 460{,}857
\end{array}\right]
\begin{array}{c} 1 \\ 2 \\ 3 \\ 4 \\ 5 \\ 6 \end{array}
\end{array}
\qquad
\mathbf{P}_f = \left[\begin{array}{c} 0 \\ 30 \\ 0 \\ 0 \\ 0 \\ -1{,}200 \end{array}\right]
\begin{array}{c} 1 \\ 2 \\ 3 \\ 4 \\ 5 \\ 6 \end{array}
$$

(c) Structure Stiffness Matrix and Fixed-Joint Force Vector

Fig. 8.23 *(continued)*

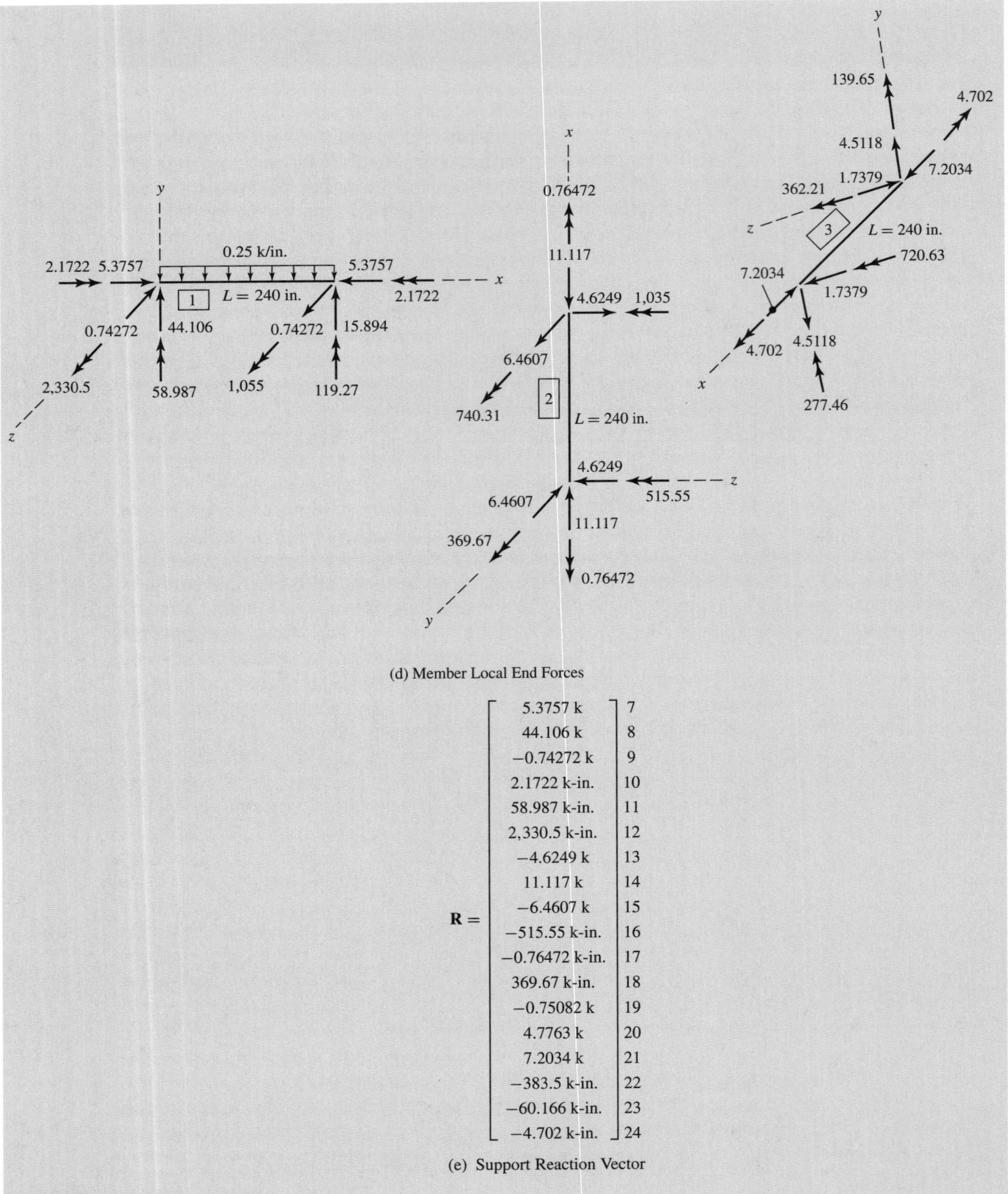

(d) Member Local End Forces

(e) Support Reaction Vector

Fig. 8.23 *(continued)*

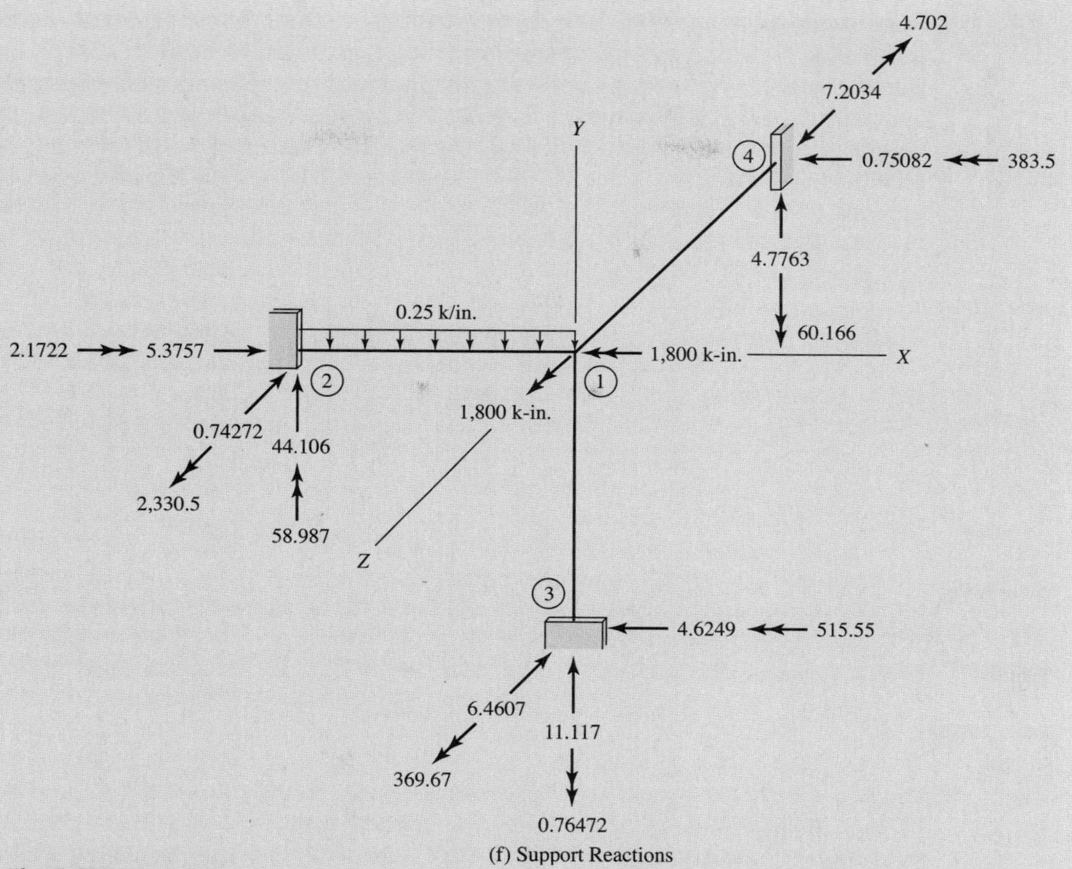

(f) Support Reactions

Fig. 8.23 *(continued)*

From Fig. 8.23(b), we observe that the code numbers for member 1 are 7, 8, 9, 10, 11, 12, 1, 2, 3, 4, 5, 6. Using these code numbers, we store the pertinent elements of $\mathbf{K}_1$ (Eq. (1)) and $\mathbf{F}_{f1}$ (Eq. (2)) in the 6×6 structure stiffness matrix $\mathbf{S}$ and the 6×1 structure fixed-joint force vector $\mathbf{P}_f$, respectively (Fig. 8.23(c)).

Member 2 Because the length, as well as the material and cross-sectional properties, of member 2 are identical to those of member 1, $\mathbf{k}_2 = \mathbf{k}_1$ (Eq. (1)).

To obtain the transformation matrix $\mathbf{T}$ for member 2, we first determine the direction cosines of its local x axis using Eqs. (8.57), as

$$r_{xX} = \frac{X_e - X_b}{L} = 0$$

$$r_{xY} = \frac{Y_e - Y_b}{L} = \frac{0 - (-20)}{20} = 1$$

$$r_{xZ} = \frac{Z_e - Z_b}{L} = 0$$

From Fig. 8.23(a), we can see that the angle of roll Ψ for this vertical member is 90°. Thus,

$$\cos\Psi = 0 \qquad \text{and} \qquad \sin\Psi = 1$$

By substituting the foregoing numerical values of r_{xX}, r_{xY}, r_{xZ}, $\cos\Psi$, and $\sin\Psi$ into Eq. (8.70), we determine the rotation matrix $\mathbf{r}$ for member 2 to be

$$\mathbf{r}_2 = \begin{bmatrix} 0 & 1 & 0 \\ 0 & 0 & 1 \\ 1 & 0 & 0 \end{bmatrix}$$

By substituting this rotation matrix into Eq. (8.55), we obtain the following 12×12 transformation matrix for member 2.

$$\mathbf{T}_2 = \left[\begin{array}{ccc|ccc|ccc|ccc}
0 & 1 & 0 & 0 & 0 & 0 & 0 & 0 & 0 & 0 & 0 & 0 \\
0 & 0 & 1 & 0 & 0 & 0 & 0 & 0 & 0 & 0 & 0 & 0 \\
1 & 0 & 0 & 0 & 0 & 0 & 0 & 0 & 0 & 0 & 0 & 0 \\ \hline
0 & 0 & 0 & 0 & 1 & 0 & 0 & 0 & 0 & 0 & 0 & 0 \\
0 & 0 & 0 & 0 & 0 & 1 & 0 & 0 & 0 & 0 & 0 & 0 \\
0 & 0 & 0 & 1 & 0 & 0 & 0 & 0 & 0 & 0 & 0 & 0 \\ \hline
0 & 0 & 0 & 0 & 0 & 0 & 0 & 1 & 0 & 0 & 0 & 0 \\
0 & 0 & 0 & 0 & 0 & 0 & 0 & 0 & 1 & 0 & 0 & 0 \\
0 & 0 & 0 & 0 & 0 & 0 & 1 & 0 & 0 & 0 & 0 & 0 \\ \hline
0 & 0 & 0 & 0 & 0 & 0 & 0 & 0 & 0 & 0 & 1 & 0 \\
0 & 0 & 0 & 0 & 0 & 0 & 0 & 0 & 0 & 0 & 0 & 1 \\
0 & 0 & 0 & 0 & 0 & 0 & 0 & 0 & 0 & 1 & 0 & 0
\end{array}\right] \tag{3}$$

The global stiffness matrix for member 2 can now be evaluated by substituting $\mathbf{k}_2$ (from Eq. (1)) and $\mathbf{T}_2$ (Eq. (3)) into the relationship $\mathbf{K} = \mathbf{T}^T\mathbf{k}\mathbf{T}$, and performing the necessary matrix multiplications. This yields

$$\mathbf{K}_2 = \begin{array}{c} \begin{array}{cccccccccccc} 13 & 14 & 15 & 16 & 17 & 18 & 1 & 2 & 3 & 4 & 5 & 6 \end{array} \\
\left[\begin{array}{cccccc|cccccc}
5.941 & 0 & 0 & 0 & 0 & -712.92 & -5.941 & 0 & 0 & 0 & 0 & -712.92 \\
0 & 3{,}975.4 & 0 & 0 & 0 & 0 & 0 & -3{,}975.4 & 0 & 0 & 0 & 0 \\
0 & 0 & 18.024 & 2{,}162.9 & 0 & 0 & 0 & 0 & -18.024 & 2{,}162.9 & 0 & 0 \\
0 & 0 & 2{,}162.9 & 346{,}067 & 0 & 0 & 0 & 0 & -2{,}162.9 & 173{,}033 & 0 & 0 \\
0 & 0 & 0 & 0 & 723.54 & 0 & 0 & 0 & 0 & 0 & -723.54 & 0 \\
-712.92 & 0 & 0 & 0 & 0 & 114{,}067 & 712.92 & 0 & 0 & 0 & 0 & 57{,}033 \\ \hline
-5.941 & 0 & 0 & 0 & 0 & 712.92 & 5.941 & 0 & 0 & 0 & 0 & 712.92 \\
0 & -3{,}975.4 & 0 & 0 & 0 & 0 & 0 & 3{,}975.4 & 0 & 0 & 0 & 0 \\
0 & 0 & -18.024 & -2{,}162.9 & 0 & 0 & 0 & 0 & 18.024 & -2{,}162.9 & 0 & 0 \\
0 & 0 & 2{,}162.9 & 173{,}033 & 0 & 0 & 0 & 0 & -2{,}162.9 & 346{,}067 & 0 & 0 \\
0 & 0 & 0 & 0 & -723.54 & 0 & 0 & 0 & 0 & 0 & 723.54 & 0 \\
-712.92 & 0 & 0 & 0 & 0 & 57{,}033 & 712.92 & 0 & 0 & 0 & 0 & 114{,}067
\end{array}\right] \end{array}
\begin{array}{c} \\ 13 \\ 14 \\ 15 \\ 16 \\ 17 \\ 18 \\ 1 \\ 2 \\ 3 \\ 4 \\ 5 \\ 6 \end{array}$$

The relevant elements of $\mathbf{K}_2$ are stored in $\mathbf{S}$ (Fig. 8.23(c)).

Member 3 $\mathbf{k}_3 = \mathbf{k}_1$ (given in Eq. (1)).

$$r_{xX} = \frac{X_e - X_b}{L} = 0$$

$$r_{xY} = \frac{Y_e - Y_b}{L} = 0$$

$$r_{xZ} = \frac{Z_e - Z_b}{L} = \frac{0 - (-20)}{20} = 1$$

From Fig. 8.23(a), we can see that $\Psi = 30°$. Thus,

$$\cos\Psi = 0.86603 \qquad \text{and} \qquad \sin\Psi = 0.5$$

By applying Eq. (8.66), we determine the rotation matrix for member 3 to be

$$\mathbf{r}_3 = \begin{bmatrix} 0 & 0 & 1 \\ -0.5 & 0.86603 & 0 \\ -0.86603 & -0.5 & 0 \end{bmatrix}$$

Thus, the transformation matrix for this member is given by

$$\mathbf{T}_3 = \left[\begin{array}{ccc|ccc|ccc|ccc}
0 & 0 & 1 & 0 & 0 & 0 & 0 & 0 & 0 & 0 & 0 & 0 \\
-0.5 & 0.86603 & 0 & 0 & 0 & 0 & 0 & 0 & 0 & 0 & 0 & 0 \\
-0.86603 & -0.5 & 0 & 0 & 0 & 0 & 0 & 0 & 0 & 0 & 0 & 0 \\
\hline
0 & 0 & 0 & 0 & 0 & 1 & 0 & 0 & 0 & 0 & 0 & 0 \\
0 & 0 & 0 & -0.5 & 0.86603 & 0 & 0 & 0 & 0 & 0 & 0 & 0 \\
0 & 0 & 0 & -0.86603 & -0.5 & 0 & 0 & 0 & 0 & 0 & 0 & 0 \\
\hline
0 & 0 & 0 & 0 & 0 & 0 & 0 & 0 & 1 & 0 & 0 & 0 \\
0 & 0 & 0 & 0 & 0 & 0 & -0.5 & 0.86603 & 0 & 0 & 0 & 0 \\
0 & 0 & 0 & 0 & 0 & 0 & -0.86603 & -0.5 & 0 & 0 & 0 & 0 \\
\hline
0 & 0 & 0 & 0 & 0 & 0 & 0 & 0 & 0 & 0 & 0 & 1 \\
0 & 0 & 0 & 0 & 0 & 0 & 0 & 0 & 0 & -0.5 & 0.86603 & 0 \\
0 & 0 & 0 & 0 & 0 & 0 & 0 & 0 & 0 & -0.86603 & -0.5 & 0
\end{array}\right] \qquad \textbf{(4)}$$

and the member global stiffness matrix $\mathbf{K}_3 = \mathbf{T}_3^T \mathbf{k}_3 \mathbf{T}_3$ is

$$\mathbf{K}_3 = \begin{array}{c} \begin{array}{cccccccccccc} 19 & 20 & 21 & 22 & 23 & 24 & 1 & 2 & 3 & 4 & 5 & 6 \end{array} \\ \left[\begin{array}{cccccc|cccccc}
8.9618 & -5.2322 & 0 & 627.87 & 1{,}075.4 & 0 & -8.9618 & 5.2322 & 0 & 627.87 & 1{,}075.4 & 0 \\
-5.2322 & 15.003 & 0 & -1{,}800.4 & -627.87 & 0 & 5.2322 & -15.003 & 0 & -1{,}800.4 & -627.87 & 0 \\
0 & 0 & 3{,}975.4 & 0 & 0 & 0 & 0 & 0 & -3{,}975.4 & 0 & 0 & 0 \\
627.87 & -1{,}800.4 & 0 & 288{,}067 & 100{,}459 & 0 & -627.87 & 1{,}800.4 & 0 & 144{,}033 & 50{,}229 & 0 \\
1{,}075.4 & -627.87 & 0 & 100{,}459 & 172{,}067 & 0 & -1{,}075.4 & 627.87 & 0 & 50{,}229 & 86{,}033 & 0 \\
0 & 0 & 0 & 0 & 0 & 723.54 & 0 & 0 & 0 & 0 & 0 & -723.54 \\
\hline
-8.9618 & 5.2322 & 0 & -627.87 & -1{,}075.4 & 0 & 8.9618 & -5.2322 & 0 & -627.87 & -1{,}075.4 & 0 \\
5.2322 & -15.003 & 0 & 1{,}800.4 & 627.87 & 0 & -5.2322 & 15.003 & 0 & 1{,}800.4 & 627.87 & 0 \\
0 & 0 & -3{,}975.4 & 0 & 0 & 0 & 0 & 0 & 3{,}975.4 & 0 & 0 & 0 \\
627.87 & -1{,}800.4 & 0 & 144{,}033 & 50{,}229 & 0 & -627.87 & 1{,}800.4 & 0 & 288{,}067 & 100{,}459 & 0 \\
1{,}075.4 & -627.87 & 0 & 50{,}229 & 86{,}033 & 0 & -1{,}075.4 & 627.87 & 0 & 100{,}459 & 172{,}067 & 0 \\
0 & 0 & 0 & 0 & 0 & -723.54 & 0 & 0 & 0 & 0 & 0 & 723.54
\end{array}\right] \end{array} \begin{array}{c} \\ 19 \\ 20 \\ 21 \\ 22 \\ 23 \\ 24 \\ 1 \\ 2 \\ 3 \\ 4 \\ 5 \\ 6 \end{array}$$

The complete structure stiffness matrix $\mathbf{S}$ and the structure fixed-joint force vector $\mathbf{P}_f$ are given in Fig. 8.23(c).

Joint Load Vector: By comparing Figs. 8.23(a) and (b), we obtain

$$\mathbf{P} = \begin{bmatrix} 0 \\ 0 \\ 0 \\ -1{,}800 \text{ k-in.} \\ 0 \\ 1{,}800 \text{ k-in.} \end{bmatrix} \begin{array}{c} 1 \\ 2 \\ 3 \\ 4 \\ 5 \\ 6 \end{array}$$

Joint Displacements: By substituting $\mathbf{P}$, $\mathbf{P}_f$, and $\mathbf{S}$ into the structure stiffness relationship, $\mathbf{P} - \mathbf{P}_f = \mathbf{Sd}$, and solving the resulting simultaneous equations, we determine the joint displacements to be

$$\mathbf{d} = \begin{bmatrix} -1.3522 \text{ in.} \\ -2.7965 \text{ in.} \\ -1.812 \text{ in.} \\ -3.0021 \text{ rad} \\ 1.0569 \text{ rad} \\ 6.4986 \text{ rad} \end{bmatrix} \times 10^{-3} \qquad \textbf{Ans}$$

Member End Displacements and End Forces:

Member 1

$$\mathbf{u}_1 = \mathbf{v}_1 = \begin{bmatrix} 0 \\ 0 \\ 0 \\ 0 \\ 0 \\ 0 \\ -1.3522 \\ -2.7965 \\ -1.812 \\ -3.0021 \\ 1.0569 \\ 6.4986 \end{bmatrix} \times 10^{-3} \qquad \textbf{(5)}$$

$$\mathbf{F}_1 = \mathbf{Q}_1 = \mathbf{k}_1\mathbf{u}_1 + \mathbf{Q}_{f1} = \begin{bmatrix} 5.3757 \text{ k} \\ 44.106 \text{ k} \\ -0.74272 \text{ k} \\ 2.1722 \text{ k-in.} \\ 58.987 \text{ k-in.} \\ 2{,}330.5 \text{ k-in.} \\ \hdashline -5.3757 \text{ k} \\ 15.894 \text{ k} \\ 0.74272 \text{ k} \\ -2.1722 \text{ k-in.} \\ 119.27 \text{ k-in.} \\ 1{,}055 \text{ k-in.} \end{bmatrix} \begin{matrix} 7 \\ 8 \\ 9 \\ 10 \\ 11 \\ 12 \\ 1 \\ 2 \\ 3 \\ 4 \\ 5 \\ 6 \end{matrix} \qquad \textbf{Ans}$$

The member local end forces are depicted in Fig. 8.23(d), and the pertinent elements of $\mathbf{F}_1$ are stored in the support reaction vector $\mathbf{R}$ (Fig. 8.23(e)).

Member 2 $\mathbf{v}_2 = \mathbf{v}_1$ (see Eq. (5)).

$$\mathbf{u}_2 = \mathbf{T}_2\mathbf{v}_2 = \begin{bmatrix} 0 \\ 0 \\ 0 \\ 0 \\ 0 \\ 0 \\ -2.7965 \\ -1.812 \\ -1.3522 \\ 1.0569 \\ 6.4986 \\ -3.0021 \end{bmatrix} \times 10^{-3} \qquad \mathbf{Q}_2 = \mathbf{k}_2\mathbf{u}_2 = \begin{bmatrix} 11.117 \text{ k} \\ -6.4607 \text{ k} \\ -4.6249 \text{ k} \\ -0.76472 \text{ k-in.} \\ 369.67 \text{ k-in.} \\ -515.55 \text{ k-in.} \\ -11.117 \text{ k} \\ 6.4607 \text{ k} \\ 4.6249 \text{ k} \\ 0.76472 \text{ k-in.} \\ 740.31 \text{ k-in.} \\ -1{,}035 \text{ k-in.} \end{bmatrix} \qquad \textbf{Ans}$$

$$\mathbf{F}_2 = \mathbf{T}_2^T \mathbf{Q}_2 = \left[\begin{array}{c} -4.6249 \\ 11.117 \\ -6.4607 \\ -515.55 \\ -0.76472 \\ 369.67 \\ \hdashline 4.6249 \\ -11.117 \\ 6.4607 \\ -1{,}035 \\ 0.76472 \\ 740.31 \end{array}\right] \begin{array}{c} 13 \\ 14 \\ 15 \\ 16 \\ 17 \\ 18 \\ 1 \\ 2 \\ 3 \\ 4 \\ 5 \\ 6 \end{array}$$

Member 3 $\mathbf{v}_3 = \mathbf{v}_1$ (see Eq. (5)).

$$\mathbf{u}_3 = \mathbf{T}_3 \mathbf{v}_3 = \left[\begin{array}{c} 0 \\ 0 \\ 0 \\ 0 \\ 0 \\ 0 \\ -1.812 \\ -1.7457 \\ 2.5693 \\ 6.4986 \\ 2.4164 \\ 2.0714 \end{array}\right] \times 10^{-3} \qquad \mathbf{Q}_3 = \mathbf{k}_3 \mathbf{u}_3 = \left[\begin{array}{c} 7.2034 \text{ k} \\ 4.5118 \text{ k} \\ -1.7379 \text{ k} \\ -4.702 \text{ k-in.} \\ 139.65 \text{ k-in.} \\ 362.21 \text{ k-in.} \\ -7.2034 \text{ k} \\ -4.5118 \text{ k} \\ 1.7379 \text{ k} \\ 4.702 \text{ k-in.} \\ 277.46 \text{ k-in.} \\ 720.63 \text{ k-in.} \end{array}\right]$$

Ans

$$\mathbf{F}_3 = \mathbf{T}_3^T \mathbf{Q}_3 = \left[\begin{array}{c} -0.75082 \\ 4.7763 \\ 7.2034 \\ -383.5 \\ -60.166 \\ -4.702 \\ \hdashline 0.75082 \\ -4.7763 \\ -7.2034 \\ -762.82 \\ -120.03 \\ 4.702 \end{array}\right] \begin{array}{c} 19 \\ 20 \\ 21 \\ 22 \\ 23 \\ 24 \\ 1 \\ 2 \\ 3 \\ 4 \\ 5 \\ 6 \end{array}$$

Support Reactions: The completed reaction vector **R** is shown in Fig. 8.23(e), and the support reactions are depicted on a line diagram of the space frame in Fig. 8.23(f).

Ans

Equilibrium checks: The six equations of equilibrium ($\sum F_x = 0$, $\sum F_y = 0$, $\sum F_z = 0$, $\sum M_x = 0$, $\sum M_y = 0$, and $\sum M_z = 0$) are satisfied for each member of the space frame shown in Fig. 8.23(d). Furthermore, the six equilibrium equations in the directions of the global coordinate axes ($\sum F_X = 0$, $\sum F_Y = 0$, $\sum F_Z = 0$, $\sum M_X = 0$, $\sum M_Y = 0$, and $\sum M_Z = 0$) are satisfied for the entire structure shown in Fig. 8.23(f).

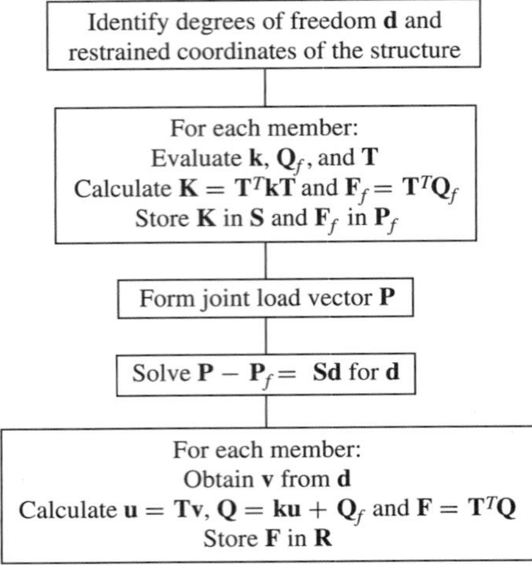
Identify degrees of freedom **d** and restrained coordinates of the structure
For each member:
Evaluate **k**, $\mathbf{Q}_f$, and **T**
Calculate $\mathbf{K} = \mathbf{T}^T\mathbf{k}\mathbf{T}$ and $\mathbf{F}_f = \mathbf{T}^T\mathbf{Q}_f$
Store **K** in **S** and $\mathbf{F}_f$ in $\mathbf{P}_f$
Form joint load vector **P**
Solve $\mathbf{P} - \mathbf{P}_f = \mathbf{S}\mathbf{d}$ for **d**
For each member:
Obtain **v** from **d**
Calculate $\mathbf{u} = \mathbf{T}\mathbf{v}$, $\mathbf{Q} = \mathbf{k}\mathbf{u} + \mathbf{Q}_f$ and $\mathbf{F} = \mathbf{T}^T\mathbf{Q}$
Store **F** in **R**

Fig. 8.24 *Stiffness Method of Analysis*

SUMMARY

In this chapter, we have extended the matrix stiffness method to the analysis of three-dimensional framed structures. The stiffness and transformation relationships for the members of space trusses, grids, and space frames are developed in Sections 8.1, 8.2, and 8.3, respectively. It should be noted that the overall format of the stiffness method of analysis remains the same for all types of (two- and three-dimensional) framed structures—provided that the member stiffness and transformation relations, appropriate for the particular type of structure being analyzed, are used in the analysis. A block diagram summarizing the overall format of the stiffness method is shown in Fig. 8.24.

PROBLEMS

Section 8.1

8.1 through 8.5 Determine the joint displacements, member axial forces, and support reactions for the space trusses shown in Figs. P8.1 through P8.5, using the matrix stiffness method. Check the hand-calculated results by using the computer program which can be downloaded from the publisher's website for this book, or by using any other general purpose structural analysis program available.

8.6 Develop a computer program for the analysis of space trusses by the matrix stiffness method. Use the program to analyze the trusses of Problems 8.1 through 8.5, and compare the computer-generated results to those obtained by hand calculations.

Section 8.2

8.7 through 8.12 Determine the joint displacements, member local end forces, and support reactions for the grids shown in Figs. P8.7 through P8.12, using the matrix stiffness method. Check the hand-calculated results by using the computer program which can be downloaded from the publisher's website for this book, or by using any other general purpose structural analysis program available.

8.13 Develop a program for the analysis of grids by the matrix stiffness method. Use the program to analyze the grids of Problems 8.7 through 8.12, and compare the computer-generated results to those obtained by hand calculations.

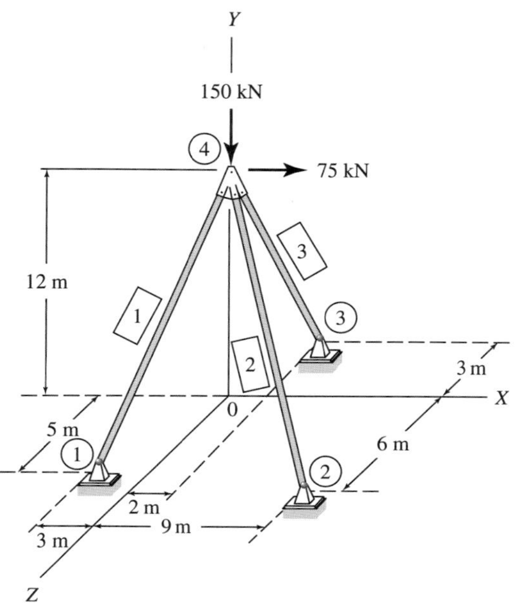

EA = constant
E = 70 GPa
A = 2,000 mm^2

Fig. P8.1

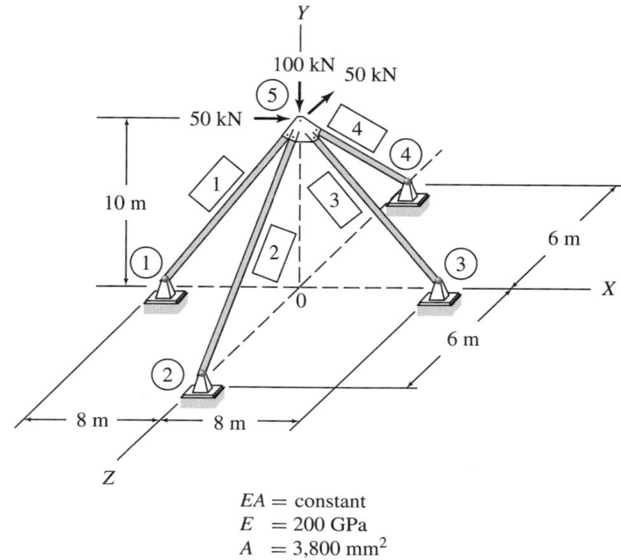

EA = constant
E = 200 GPa
A = 3,800 mm^2

Fig. P8.3

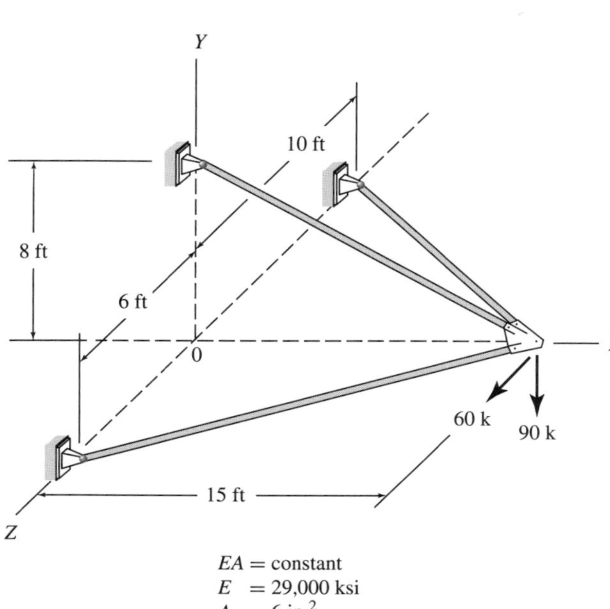

EA = constant
E = 29,000 ksi
A = 6 in.^{2}

Fig. P8.2

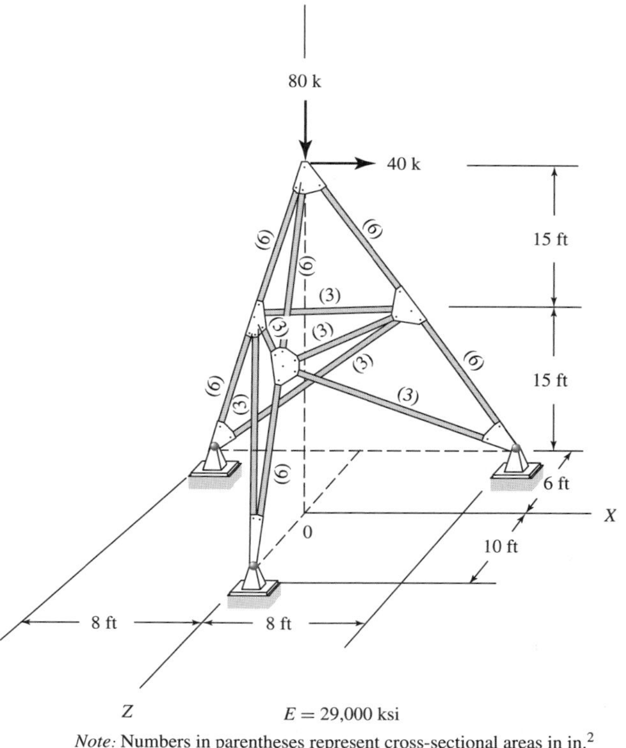

E = 29,000 ksi

Note: Numbers in parentheses represent cross-sectional areas in in.^{2}

Fig. P8.4

Section 8.3

8.14 through 8.17 Determine the joint displacements, member local end forces, and support reactions for the space frames shown in Figs. 8.14 through 8.17, using the matrix stiffness method. Check the hand-calculated results by using the computer program which can be downloaded from the publisher's website for this book, or by using any other general purpose structural analysis program available.

8.18 Develop a program for the analysis of space frames by the matrix stiffness method. Use the program to analyze the frames of Problems 8.14 through 8.17, and compare the computer-generated results to those obtained by hand calculations.

8.19 Develop a general computer program that can be used to analyze any type of framed structure.

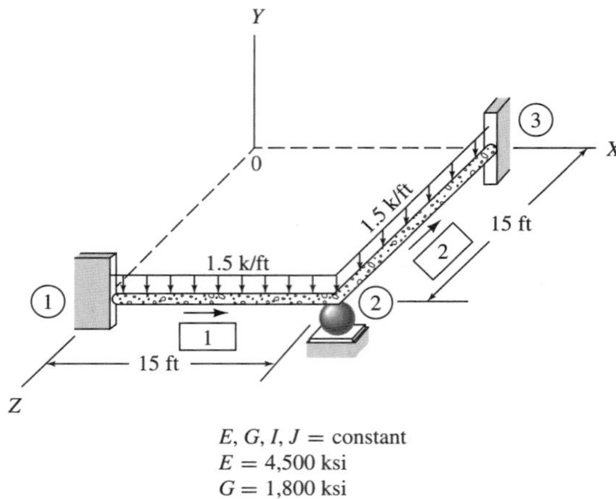

Fig. P8.7

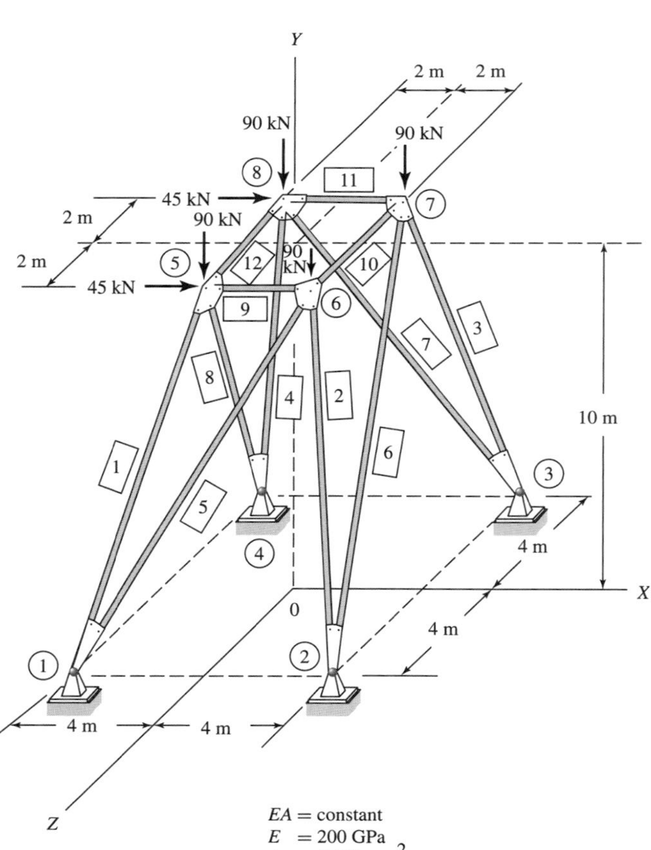

Fig. P8.5

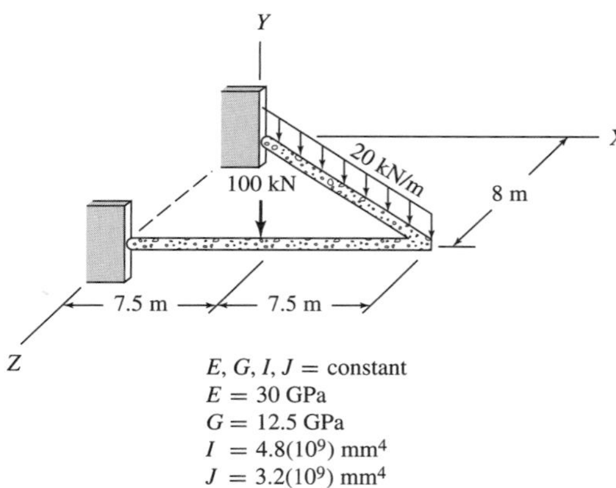

Fig. P8.8

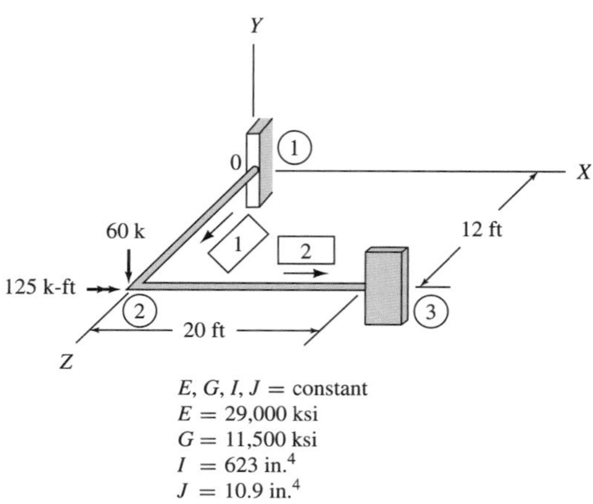

Fig. P8.9

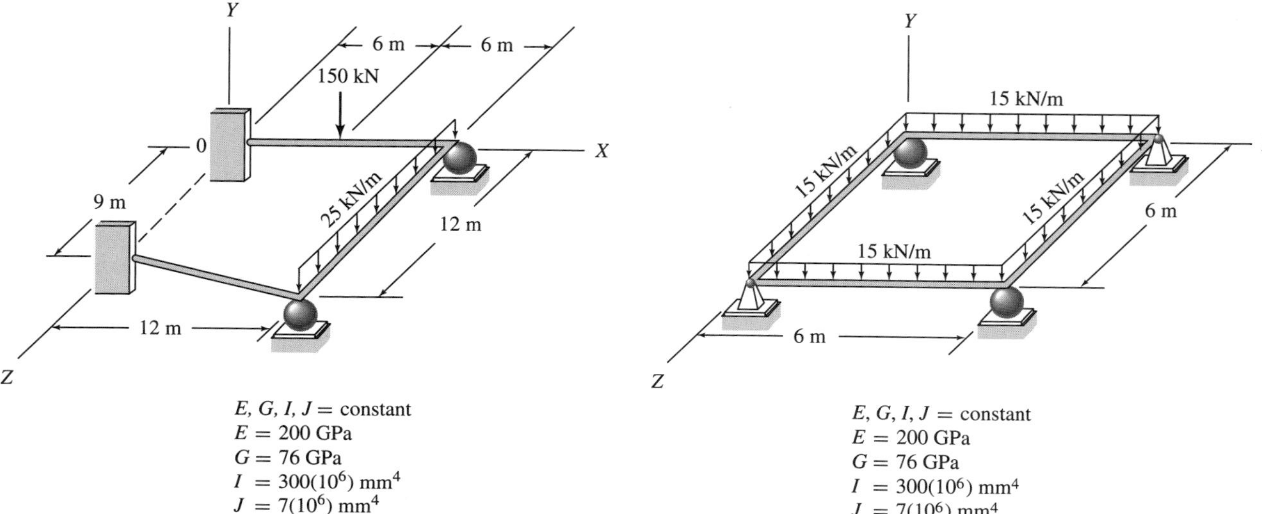

Fig. P8.10

Fig. P8.12

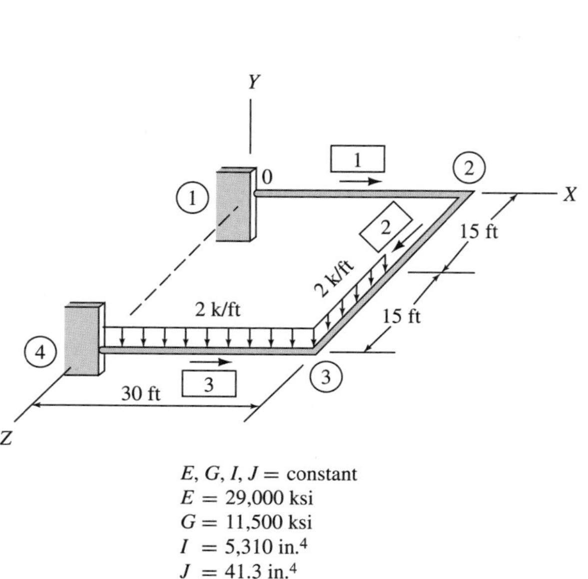

Fig. P8.11

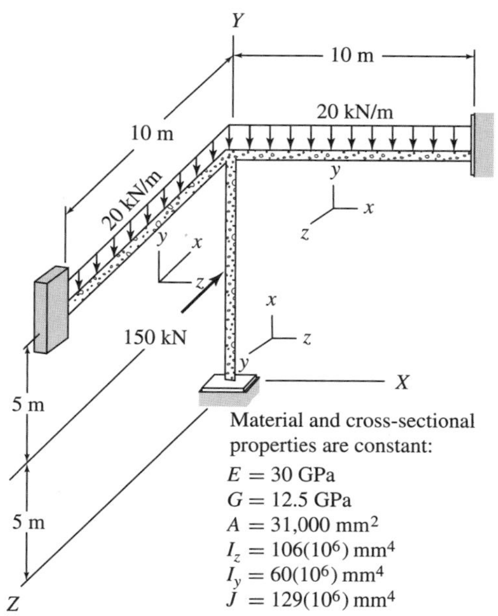

Fig. P8.14

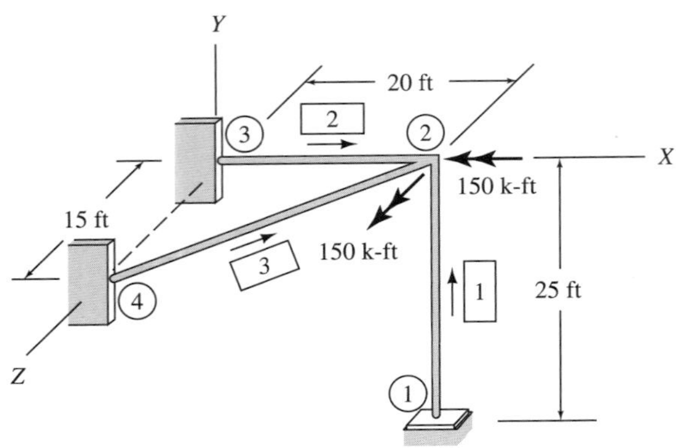

Material properties are constant:
$E = 10{,}000$ ksi
$G = 4{,}000$ ksi
All members have circular cross-sections with
$A = 4.52$ in.^{2}
$I_z = I_y = 18.7$ in.^{4}
$J = 37.4$ in.^{4}

Fig. P8.15

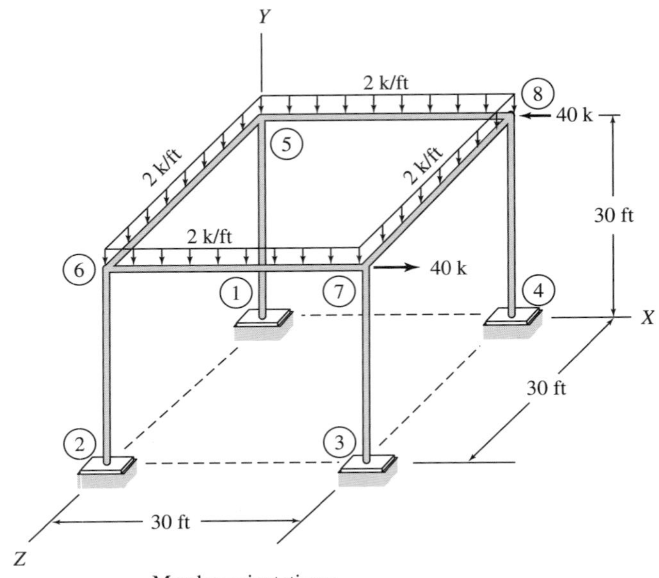

Member orientations:
Girders: local y axis is vertical
Columns:

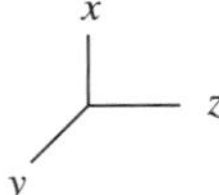

Material and cross-sectional properties are constant:
$E = 29{,}000$ ksi
$G = 11{,}500$ ksi
$A = 47.7$ in.^{2}
$I_z = 5{,}170$ in.^{4}
$I_y = 443$ in.^{4}
$J = 18.5$ in.^{4}

Fig. P8.17

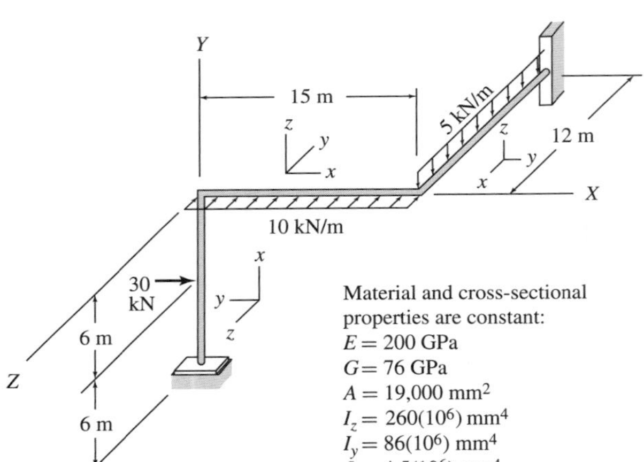

Material and cross-sectional properties are constant:
$E = 200$ GPa
$G = 76$ GPa
$A = 19{,}000$ mm^2
$I_z = 260(10^6)$ mm^4
$I_y = 86(10^6)$ mm^4
$J = 4.5(10^6)$ mm^4

Fig. P8.16

9 SPECIAL TOPICS AND MODELING TECHNIQUES

The Eiffel Tower, Paris
(Filip Fuxa/Shutterstock)

In this chapter, we consider some modifications and extensions of the matrix stiffness method developed in the preceding chapters. Also considered herein are techniques for modeling certain special features (or details) of structures, so that more realistic analytical models can be created and more accurate structural responses predicted from the analysis.

We begin by discussing an alternative formulation of the stiffness method in Section 9.1. As this alternative formulation involves the structure stiffness matrix for all the coordinates (including the restrained coordinates) of the structure, it is less efficient for computer implementation than the formulation used in the preceding chapters. Nonetheless, an understanding of this alternative formulation provides some important insights into the stiffness method of analysis. In Sections 9.2 and 9.3, we consider some techniques for reducing a structure's degrees of freedom, and/or the number of structure stiffness equations to be processed simultaneously. These techniques are useful in handling the analysis of large structures. Section 9.4 is devoted to the modeling of inclined roller supports; in the following two sections, we develop techniques for modeling the effects of offset connections (Section 9.5), and semirigid connections (Section 9.6), in the analysis. The inclusion of shear deformation effects in the analysis of beams, grids and frames is considered in Section 9.7; and in Section 9.8, we cover the analysis of structures composed of nonprismatic members. Finally, we conclude the chapter by discussing a procedure for efficiently storing and solving the systems of linear equations that arise in the analysis of large structures.

9.1 THE STRUCTURE STIFFNESS MATRIX INCLUDING RESTRAINED COORDINATES—AN ALTERNATIVE FORMULATION OF THE STIFFNESS METHOD

In the formulation of the matrix stiffness method, as developed in the preceding chapters, the conditions of zero (or known) joint displacements corresponding to restrained coordinates are applied to the member force-displacement relationships before the stiffness relations for the entire structure are assembled. This approach yields structure stiffness relations that contain only the degrees of freedom as unknowns. Alternatively, the stiffness method can be formulated by first establishing the stiffness relations for all the coordinates (free and restrained) of the structure, and then applying the restraint conditions to the structure stiffness relations, which now contain both the degrees of freedom, and the support reactions, as unknowns.

In the alternative formulation, the structure's restrained coordinates are initially treated as free coordinates, and the stiffness relations for all the coordinates of the structure are expressed as

$$\underset{NC \times 1}{\mathbf{P}^*} - \underset{NC \times 1}{\mathbf{P}_f^*} = \underset{NC \times NC}{\mathbf{S}^*} \;\; \underset{NC \times 1}{\mathbf{d}^*} \tag{9.1}$$

with

$$NC = NCJT(NJ) = NDOF + NR \tag{9.2}$$

In Eqs. (9.1) and (9.2), NC denotes the number of structure coordinates; $\mathbf{P}^*$ represents the joint forces (i.e., the known external loads and the unknown support reactions); $\mathbf{P}_f^*$ denotes the fixed-joint forces, due to member loads, temperature changes, and fabrication errors, at the locations, and in the directions, of the structure coordinates; $\mathbf{S}^*$ represents the stiffness matrix for the structure coordinates (free and restrained); and $\mathbf{d}^*$ denotes the joint displacements (i.e., the unknown degrees of freedom and the known displacements corresponding to the restrained coordinates). The structure stiffness matrix $\mathbf{S}^*$ and fixed-joint force vector $\mathbf{P}_f^*$ can be determined by assembling the member global stiffness matrices $\mathbf{K}$ and fixed-end force vectors $\mathbf{F}_f$, respectively, using the member code number technique described in the preceding chapters. The application of this technique remains essentially the same, except that now those elements of $\mathbf{K}$ and $\mathbf{F}_f$ that correspond to the restrained coordinates are no longer discarded, but are added (stored) in their proper positions in $\mathbf{S}^*$ and $\mathbf{P}_f^*$.

As indicated in the preceding paragraph, the structure stiffness relations (Eq. (9.1)) contain two types of unknown quantities; namely, the unknown joint displacements and the unknown support reactions. To separate the two types of unknowns, we rewrite Eq. (9.1) in partitioned-matrix form:

$$\begin{bmatrix} \underset{NDOF \times 1}{\mathbf{P}} \\ \hline \underset{NR \times 1}{\mathbf{R}} \end{bmatrix} - \begin{bmatrix} \underset{NDOF \times 1}{\mathbf{P}_f} \\ \hline \underset{NR \times 1}{\mathbf{R}_f} \end{bmatrix} = \begin{bmatrix} \underset{NDOF \times NDOF}{\mathbf{S}} & \underset{NDOF \times NR}{\mathbf{S}_{FR}} \\ \hline \underset{NR \times NDOF}{\mathbf{S}_{RF}} & \underset{NR \times NR}{\mathbf{S}_{RR}} \end{bmatrix} \begin{bmatrix} \underset{NDOF \times 1}{\mathbf{d}} \\ \hline \underset{NR \times 1}{\mathbf{d}_R} \end{bmatrix} \tag{9.3}$$

in which, $\mathbf{P}$, $\mathbf{R}$, $\mathbf{P}_f$, $\mathbf{S}$, and $\mathbf{d}$ denote the same quantities as in the preceding chapters; $\mathbf{R}_f$ denotes the structure fixed-joint forces corresponding to the restrained coordinates; and $\mathbf{d}_R$ denotes the *support displacement vector*. Note that the $\mathbf{P}_f$ and $\mathbf{R}_f$ vectors contain structure fixed-joint forces due to member loads, temperature changes, and fabrication errors. The effects of support displacements are not included in $\mathbf{P}_f$ and $\mathbf{R}_f$, but are directly incorporated into the analysis through the support displacement vector $\mathbf{d}_R$. Each element of the submatrix $\mathbf{S}_{FR}$ in Eq. (9.3) represents the force at a free coordinate caused by a unit displacement of a restrained coordinate. The other two submatrices, $\mathbf{S}_{RF}$ and $\mathbf{S}_{RR}$, can be interpreted in an analogous manner. By multiplying the two partitioned matrices on the right-hand side of Eq. (9.3), we obtain two matrix equations,

$$\mathbf{P} - \mathbf{P}_f = \mathbf{S}\mathbf{d} + \mathbf{S}_{FR}\mathbf{d}_R \tag{9.4a}$$

$$\mathbf{R} - \mathbf{R}_f = \mathbf{S}_{RF}\mathbf{d} + \mathbf{S}_{RR}\mathbf{d}_R \tag{9.4b}$$

which can be rearranged as

$$\mathbf{P} - \mathbf{P}_f - \mathbf{S}_{FR}\mathbf{d}_R = \mathbf{S}\mathbf{d} \tag{9.5a}$$

$$\mathbf{R} = \mathbf{R}_f + \mathbf{S}_{RF}\mathbf{d} + \mathbf{S}_{RR}\mathbf{d}_R \tag{9.5b}$$

The procedure for analysis essentially consists of first solving Eq. (9.5a) for the unknown joint displacements **d**, and then substituting **d** into Eq. (9.5b) to evaluate the support reactions **R**. With **d** known, the member end displacements and end forces can be obtained using the procedures described in the preceding chapters. In the case of structures with no support displacements, $\mathbf{d}_R = \mathbf{0}$, and Eqs. (9.5) reduce to

$$\mathbf{P} - \mathbf{P}_f = \mathbf{S}\mathbf{d} \tag{9.6a}$$

$$\mathbf{R} = \mathbf{R}_f + \mathbf{S}_{RF}\mathbf{d} \tag{9.6b}$$

The main advantages of the alternative formulation are that support displacements can be incorporated into the analysis in a direct and straightforward manner, and the reactions can be more conveniently calculated by using the structure stiffness relations. However, since the alternative formulation uses the stiffness matrix for all of the structure's coordinates, it requires significantly more computer memory space than the standard formulation developed in the preceding chapters, which uses the stiffness matrix for only the free coordinates of the structure. For this reason, the alternative formulation is not considered to be as efficient for computer implementation as the formulation developed in the preceding chapters [14].

The application of the alternative formulation is illustrated by the following example.

EXAMPLE 9.1 Determine the joint displacements, member local end forces, and support reactions for the plane frame of Fig. 9.1(a), due to the combined effect of the loading shown and a

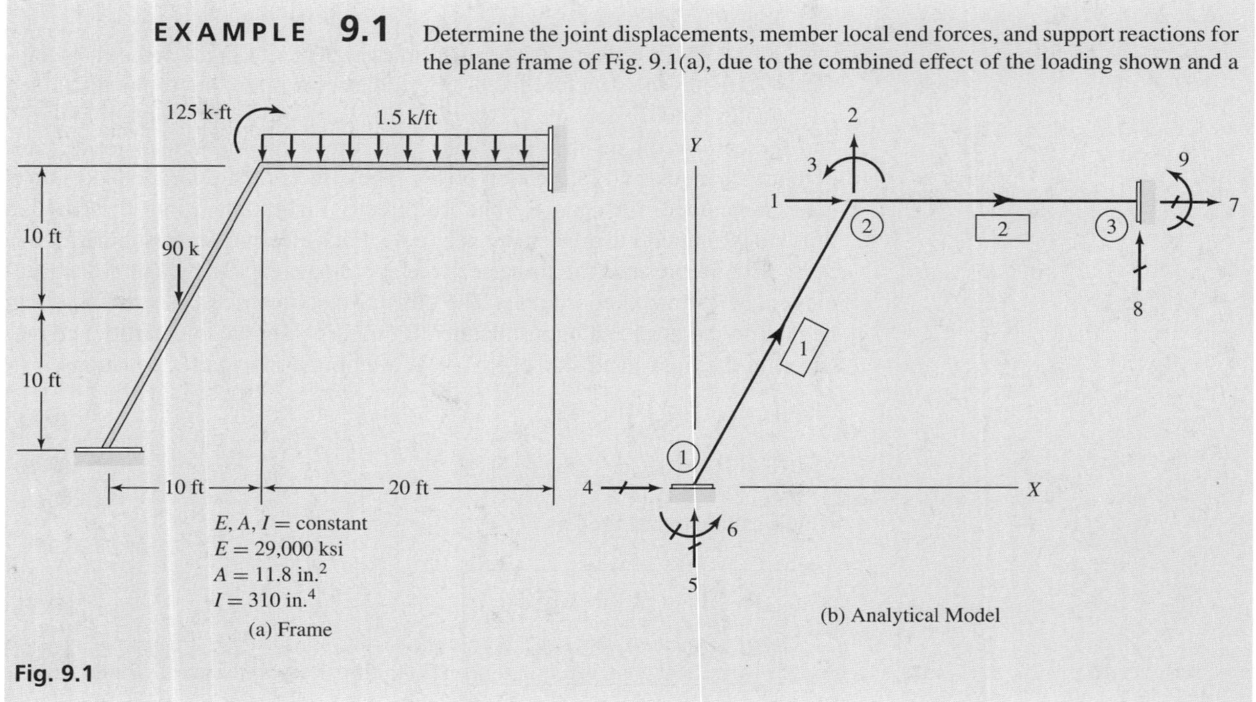

Fig. 9.1

settlement of 1 in. of the left support. Use the alternative formulation of the matrix stiffness method.

SOLUTION This frame was analyzed in Example 6.6 for external loading, and in Example 7.4 for the combined effect of the loading and the support settlement, using the standard formulation.

Analytical Model: See Fig. 9.1(b). In this example, we use the same analytical model of the frame as used previously, so that the various member matrices calculated in Example 6.6 can be reused. The frame has three degrees of freedom and six restrained coordinates. Thus, the total number of structure coordinates is nine.

Structure Stiffness Matrix: By storing the element of the member global stiffness matrices $\mathbf{K}_1$ and $\mathbf{K}_2$, calculated in Example 6.6, in their proper positions in the 9×9 structure stiffness matrix $\mathbf{S}^*$, we obtain, in units of kips and inches, the following stiffness matrix for all the structure coordinates.

$$\mathbf{S}^* = \left[\begin{array}{c|c} \mathbf{S} & \mathbf{S}_{FR} \\ \hline \mathbf{S}_{RF} & \mathbf{S}_{RR} \end{array}\right] = \begin{array}{c} \begin{array}{ccccccccc} 1 & 2 & 3 & 4 & 5 & 6 & 7 & 8 & 9 \end{array} \\ \left[\begin{array}{ccc|cccccc} 1{,}685.3 & 507.89 & 670.08 & -259.53 & -507.89 & 670.08 & -1{,}425.8 & 0 & 0 \\ 507.89 & 1{,}029.2 & 601.42 & -507.89 & -1{,}021.4 & -335.04 & 0 & -7.8038 & 936.46 \\ 670.08 & 601.42 & 283{,}848 & -670.08 & 335.04 & 67{,}008 & 0 & -936.46 & 74{,}917 \\ \hline -259.53 & -507.89 & -670.08 & 259.53 & 507.89 & -670.08 & 0 & 0 & 0 \\ -507.89 & -1{,}021.4 & 335.04 & 507.89 & 1{,}021.4 & 335.04 & 0 & 0 & 0 \\ 670.08 & -335.04 & 67{,}008 & -670.08 & 335.04 & 13{,}401.5 & 0 & 0 & 0 \\ -1{,}425.8 & 0 & 0 & 0 & 0 & 0 & 1{,}425.8 & 0 & 0 \\ 0 & -7.8038 & -936.46 & 0 & 0 & 0 & 0 & 7.8038 & -936.46 \\ 0 & 936.46 & 74{,}917 & 0 & 0 & 0 & 0 & -936.46 & 149{,}833 \end{array}\right] \end{array} \begin{array}{c} 1 \\ 2 \\ 3 \\ 4 \\ 5 \\ 6 \\ 7 \\ 8 \\ 9 \end{array} \tag{1}$$

Structure Fixed-Joint Force Vector Due to Member Loads: Similarly, by storing the elements of the member global fixed-end force vectors $\mathbf{F}_{f1}$ and $\mathbf{F}_{f2}$, calculated in Example 6.6, in the 9×1 structure fixed-joint force vector $\mathbf{P}_f^*$, we obtain

$$\mathbf{P}_f^* = \left[\begin{array}{c} \mathbf{P}_f \\ \hline \mathbf{R}_f \end{array}\right] = \left[\begin{array}{r} 0 \\ 60 \\ -750 \\ \hline 0 \\ 45 \\ 1{,}350 \\ 0 \\ 15 \\ -600 \end{array}\right] \begin{array}{c} 1 \\ 2 \\ 3 \\ 4 \\ 5 \\ 6 \\ 7 \\ 8 \\ 9 \end{array} \tag{2}$$

Joint Load Vector: From Example 6.6,

$$\mathbf{P} = \left[\begin{array}{r} 0 \\ 0 \\ -1{,}500 \end{array}\right] \begin{array}{c} 1 \\ 2 \\ 3 \end{array} \tag{3}$$

Support Displacement Vector: From the analytical model of the structure in Fig. 9.1(b), we observe that the given 1 in. settlement of the left support occurs at the location and in the direction of restraint coordinate 5. Thus, the support displacement vector can be expressed as

$$\mathbf{d}_R = \begin{bmatrix} 0 \\ -1 \\ 0 \\ 0 \\ 0 \\ 0 \end{bmatrix} \begin{matrix} 4 \\ 5 \\ 6 \\ 7 \\ 8 \\ 9 \end{matrix} \tag{4}$$

Joint Displacements: By substituting $\mathbf{S}$ and $\mathbf{S}_{FR}$ from Eq. (1), $\mathbf{P}_f$ from Eq. (2), $\mathbf{P}$ from Eq. (3), and $\mathbf{d}_R$ from Eq. (4) into Eq. (9.5a), we write the stiffness relations for the free coordinates of the frame as

$$\mathbf{P} - \mathbf{P}_f - \mathbf{S}_{FR}\mathbf{d}_R = \mathbf{S}\mathbf{d}$$

$$\begin{bmatrix} 0 \\ 0 \\ -1{,}500 \end{bmatrix} - \begin{bmatrix} 0 \\ 60 \\ -750 \end{bmatrix} - \begin{bmatrix} -259.53 & -507.89 & 670.08 & -1{,}425.8 & 0 & 0 \\ -507.89 & -1{,}021.4 & -335.04 & 0 & -7.8038 & 936.46 \\ -670.08 & 335.04 & 67{,}008 & 0 & -936.46 & 74{,}917 \end{bmatrix} \begin{bmatrix} 0 \\ -1 \\ 0 \\ 0 \\ 0 \\ 0 \end{bmatrix}$$

$$= \begin{bmatrix} 1{,}685.3 & 507.89 & 670.08 \\ 507.89 & 1{,}029.2 & 601.42 \\ 670.08 & 601.42 & 283{,}848 \end{bmatrix} \begin{bmatrix} d_1 \\ d_2 \\ d_3 \end{bmatrix}$$

or

$$\begin{bmatrix} -507.89 \\ -1{,}081.4 \\ -414.96 \end{bmatrix} = \begin{bmatrix} 1{,}685.3 & 507.89 & 670.08 \\ 507.89 & 1{,}029.2 & 601.42 \\ 670.08 & 601.42 & 283{,}848 \end{bmatrix} \begin{bmatrix} d_1 \\ d_2 \\ d_3 \end{bmatrix}$$

By solving these equations, we determine the joint displacements to be

$$\mathbf{d} = \begin{bmatrix} 0.017762 \text{ in.} \\ -1.0599 \text{ in.} \\ 0.00074192 \text{ rad} \end{bmatrix} \begin{matrix} 1 \\ 2 \\ 3 \end{matrix} \tag{5}$$

Ans

Note that these joint displacements are identical to those calculated in Example 7.4. The joint displacement vector for all the coordinates (free and restrained) of the structure can be expressed as

$$\mathbf{d}^* = \begin{bmatrix} \mathbf{d} \\ \hline \mathbf{d}_R \end{bmatrix} = \begin{bmatrix} 0.017762 \text{ in.} \\ -1.0599 \text{ in.} \\ 0.00074192 \text{ rad} \\ \hline 0 \\ -1 \text{ in.} \\ 0 \\ 0 \\ 0 \\ 0 \end{bmatrix} \begin{matrix} 1 \\ 2 \\ 3 \\ 4 \\ 5 \\ 6 \\ 7 \\ 8 \\ 9 \end{matrix} \tag{6}$$

Support Reactions: To evaluate the support reaction vector $\mathbf{R}$, we substitute $\mathbf{S}_{RF}$ and $\mathbf{S}_{RR}$ from Eq. (1), $\mathbf{R}_f$ from Eq. (2), $\mathbf{d}_R$ from Eq. (4), and $\mathbf{d}$ from Eq. (5) into Eq. 9.5(b): $\mathbf{R} = \mathbf{R}_f + \mathbf{S}_{RF}\mathbf{d} + \mathbf{S}_{RR}\mathbf{d}_R$. This yields

$$\mathbf{R} = \begin{bmatrix} 25.316 \text{ k} \\ 97.409 \text{ k} \\ 1{,}431.7 \text{ k-in.} \\ -25.325 \text{ k} \\ 22.576 \text{ k} \\ -1{,}537 \text{ k-in.} \end{bmatrix} \begin{matrix} 4 \\ 5 \\ 6 \\ 7 \\ 8 \\ 9 \end{matrix}$$ **Ans**

Note that these support reactions are the same as those calculated in Example 7.4.

Member End Displacements and End Forces:

Member 1 Using member code numbers and Eq. (6), we obtain

$$\mathbf{v}_1 = \begin{bmatrix} v_1 \\ v_2 \\ v_3 \\ v_4 \\ v_5 \\ v_6 \end{bmatrix} \begin{matrix} 4 \\ 5 \\ 6 \\ 1 \\ 2 \\ 3 \end{matrix} = \begin{bmatrix} d_4^* \\ d_5^* \\ d_6^* \\ d_1^* \\ d_2^* \\ d_3^* \end{bmatrix} = \begin{bmatrix} 0 \\ -1 \\ 0 \\ 0.017762 \\ -1.0599 \\ 0.00074192 \end{bmatrix}$$

Next, we use the member transformation matrix $\mathbf{T}_1$ from Example 6.6, to calculate

$$\mathbf{u}_1 = \mathbf{T}_1\mathbf{v}_1 = \begin{bmatrix} -0.89443 \\ -0.44721 \\ 0 \\ -0.94006 \\ -0.48988 \\ 0.00074192 \end{bmatrix}$$

The member local end forces can now be obtained by using the member local stiffness matrix $\mathbf{k}_1$ and fixed-end force vector $\mathbf{Q}_{f1}$, from Example 6.6, as

$$\mathbf{Q}_1 = \mathbf{k}_1\mathbf{u}_1 + \mathbf{Q}_{f1} = \begin{bmatrix} 98.441 \text{ k} \\ 20.919 \text{ k} \\ 1{,}431.7 \text{ k-in.} \\ -17.943 \text{ k} \\ 19.331 \text{ k} \\ -1{,}218.6 \text{ k-in.} \end{bmatrix}$$ **Ans**

Member 2 The global and local end displacements for this horizontal member are

$$\mathbf{u}_1 = \mathbf{v}_1 = \begin{bmatrix} v_1 \\ v_2 \\ v_3 \\ v_4 \\ v_5 \\ v_6 \end{bmatrix} \begin{matrix} 1 \\ 2 \\ 3 \\ 7 \\ 8 \\ 9 \end{matrix} = \begin{bmatrix} d_1^* \\ d_2^* \\ d_3^* \\ d_7^* \\ d_8^* \\ d_9^* \end{bmatrix} = \begin{bmatrix} 0.017762 \\ -1.0599 \\ 0.00074192 \\ 0 \\ 0 \\ 0 \end{bmatrix}$$

By using $\mathbf{k}_2$ and $\mathbf{Q}_{f2}$ from Example 6.6, we compute the member local end forces to be

$$\mathbf{Q}_2 = \mathbf{k}_2\mathbf{u}_2 + \mathbf{Q}_{f2} = \begin{bmatrix} 25.325 \text{ k} \\ 7.4235 \text{ k} \\ -281.39 \text{ k-in.} \\ -25.325 \text{ k} \\ 22.576 \text{ k} \\ -1{,}537 \text{ k-in.} \end{bmatrix} \qquad \textbf{Ans}$$

As expected, the foregoing member local end force vectors $\mathbf{Q}_1$ and $\mathbf{Q}_2$ are identical to those calculated in Example 7.4.

9.2 APPROXIMATE MATRIX ANALYSIS OF RECTANGULAR BUILDING FRAMES

In building frames of low to medium height, the axial deformations of members are generally much smaller than the bending deformations. Therefore, the number of degrees of freedom of such frames can be reduced, without significantly compromising the accuracy of the analysis results, by neglecting the axial deformations of members, or by assuming that the members are *inextensible*. In this section, we consider the analysis of rectangular plane frames composed of horizontal and vertical members which are assumed to be inextensible (i.e., they cannot undergo any axial elongation or shortening).

Consider, for example, the portal frame shown in Fig. 9.2. Recall from Chapter 6 that the frame actually has six degrees of freedom, when both axial

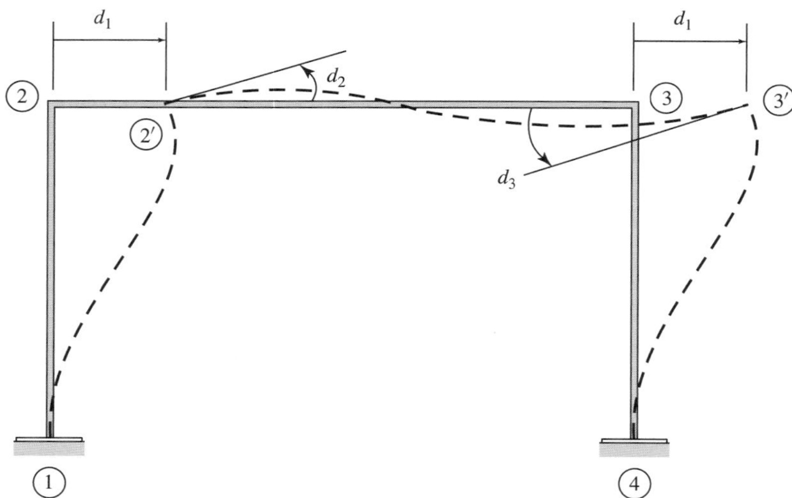

Fig. 9.2 *Portal Frame with Inextensible Members (Three Degrees of Freedom)*

and bending deformations of members are taken into account in the analysis. However, if the members of the frame are assumed to be inextensible, then the number of degrees of freedom is reduced to only three. From the deformed shape of the arbitrarily loaded frame given in Fig. 9.2, we can see that fixed joints 1 and 4 can neither rotate nor translate, whereas joints 2 and 3 can rotate and translate in the horizontal direction, but not in the vertical direction because their vertical translations are prevented by the left and right columns, respectively, which are assumed to be inextensible. Furthermore, since the girder (i.e., the horizontal member) of the frame is assumed to be inextensible, the horizontal translations of joints 2 and 3 must be equal. Thus, the portal frame has three degrees of freedom, namely d_1, d_2, and d_3, as shown in the figure.

As another example, consider the two-story three-bay building frame shown in Fig. 9.3. The frame actually has 24 degrees of freedom when both axial and bending deformations are included in the analysis. However, if the members are assumed to be inextensible, then the number of degrees of freedom is reduced to 10, as shown in the figure. As this example indicates, the assumption of member inextensibility provides a means for a significant reduction in the number of degrees of freedom of large structures. Needless to say, this approximate approach is appropriate only for frames in which the member axial deformations are small enough to have a negligible effect on their response. As the axial deformations in the columns of tall building frames can have a significant effect on the structural response, the approximate method under consideration is usually not considered suitable for the analysis of such structures.

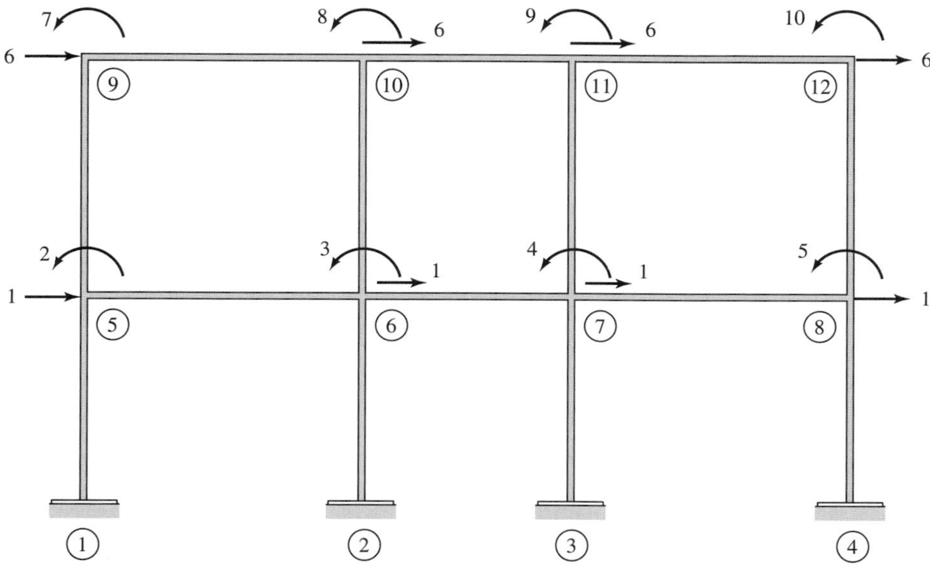

Fig. 9.3 *Inextensible Building Frame (Ten Degrees of Freedom)*

The overall procedure for the approximate analysis of rectangular plane frames remains the same as that for general plane frames, developed in Chapter 6—provided that the member stiffness relations are modified to exclude the axial effects. As the frame is composed of only horizontal and vertical members, each member now has four degrees of freedom in both the local and global coordinate systems. The local and global end forces and end displacements for the girders (i.e., horizontal members), and the columns (i.e., vertical members), of the frame, are given in Fig. 9.4. To simplify the analysis, the member local x axis is oriented positive to the right for girders (Fig. 9.4(a)) and positive upward for columns (Fig. 9.4(b)). With the axial effects neglected, the relationship between the member local end forces, $\mathbf{Q}$, and end displacements, $\mathbf{u}$, is expressed by the local stiffness matrix $\mathbf{k}$ and fixed-end force vector $\mathbf{Q}_f$ for

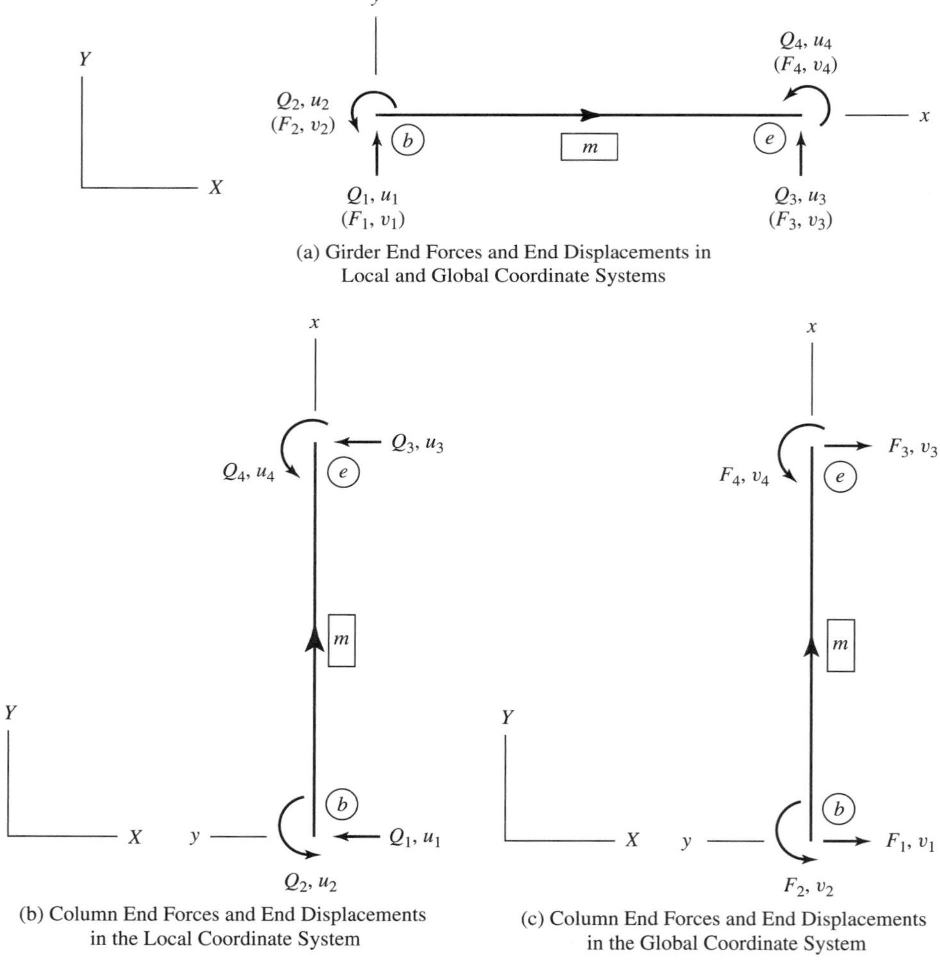

(a) Girder End Forces and End Displacements in Local and Global Coordinate Systems

(b) Column End Forces and End Displacements in the Local Coordinate System

(c) Column End Forces and End Displacements in the Global Coordinate System

Fig. 9.4

beam members, derived in Chapter 5 (Eqs. (5.53) and (5.99)). Thus, $\mathbf{Q} = \mathbf{ku} + \mathbf{Q}_f$, with

$$\mathbf{k} = \frac{EI}{L^3}\begin{bmatrix} 12 & 6L & -12 & 6L \\ 6L & 4L^2 & -6L & 2L^2 \\ -12 & -6L & 12 & -6L \\ 6L & 2L^2 & -6L & 4L^2 \end{bmatrix} \tag{9.7}$$

and

$$\mathbf{Q}_f = \begin{bmatrix} FS_b \\ FM_b \\ FS_e \\ FM_e \end{bmatrix} \tag{9.8}$$

As for the member stiffness relations in the global coordinate system, for girders (Fig. 9.4(a)) no coordinate transformations are needed; that is, $\mathbf{K}$ (girder) $= \mathbf{k}$ and $\mathbf{F}_f$ (girder) $= \mathbf{Q}_f$. For columns, the transformation matrix, $\mathbf{T}$ (column), can be established via the following relationships between the local end forces $\mathbf{Q}$ and the global end forces $\mathbf{F}$ (see Figs. 9.4(b) and (c)):

$$Q_1 = -F_1 \qquad Q_2 = F_2 \qquad Q_3 = -F_3 \qquad Q_4 = F_4$$

or

$$\begin{bmatrix} Q_1 \\ Q_2 \\ Q_3 \\ Q_4 \end{bmatrix} = \begin{bmatrix} -1 & 0 & 0 & 0 \\ 0 & 1 & 0 & 0 \\ 0 & 0 & -1 & 0 \\ 0 & 0 & 0 & 1 \end{bmatrix} \begin{bmatrix} F_1 \\ F_2 \\ F_3 \\ F_4 \end{bmatrix}$$

from which,

$$\mathbf{T}\text{ (column)} = \begin{bmatrix} -1 & 0 & 0 & 0 \\ 0 & 1 & 0 & 0 \\ 0 & 0 & -1 & 0 \\ 0 & 0 & 0 & 1 \end{bmatrix} \tag{9.9}$$

The expression of the global stiffness matrix for columns, $\mathbf{K}$ (column), can now be obtained by applying the relationship $\mathbf{K} = \mathbf{T}^T\mathbf{kT}$, which yields

$$\mathbf{K}\text{ (column)} = \frac{EI}{L^3}\begin{bmatrix} 12 & -6L & -12 & -6L \\ -6L & 4L^2 & 6L & 2L^2 \\ -12 & 6L & 12 & 6L \\ -6L & 2L^2 & 6L & 4L^2 \end{bmatrix} \tag{9.10}$$

It is important to realize that the assumption of negligibly small axial deformations, as used herein, does not imply that the member axial forces are also negligibly small. As the axial forces do not appear in the member stiffness relations, the application of the matrix stiffness method yields only member end shears and end moments. Once the member end shears are known, the member axial forces can be evaluated by considering the equilibrium of the free bodies of the joints and members of the structure.

EXAMPLE 9.2 Determine the approximate joint displacements, member local end forces, and support reactions for the portal frame shown in Fig. 9.5(a), assuming the members to be inextensible.

SOLUTION *Analytical Model:* See Fig. 9.5(b). The frame has three degrees of freedom—the translation of the girder in the X direction, and the rotations of joints 2 and 3. The six restrained coordinates of the frame are identified by numbers 4 through 9 as usual, as shown in Fig. 9.5(b).

Structure Stiffness Matrix and Fixed-Joint Force Vector: By applying Eq. (9.10) for members 1 and 3, and Eq. (9.7) for member 2, we obtain the following member global stiffness matrices (in units of kips and inches):

$$
\begin{array}{l}
\text{Member 3} \longrightarrow \quad 7 \qquad 9 \qquad 1 \qquad 3 \\
\text{Member 1} \longrightarrow \quad 4 \qquad 6 \qquad 1 \qquad 2
\end{array}
$$

$$
\mathbf{K}_1 = \mathbf{K}_3 = \left[\begin{array}{cccc}
5.3107 & -955.93 & -5.3107 & -955.93 \\
-955.93 & 229{,}422 & 955.93 & 114{,}711 \\
-5.3107 & 955.93 & 5.3107 & 955.93 \\
-955.93 & 114{,}711 & 955.93 & 229{,}422
\end{array}\right]
\begin{array}{cc}
4 & 7 \\
6 & 9 \\
1 & 1 \\
2 & 3
\end{array}
$$

$$
\mathbf{K}_2 = \mathbf{k}_2 = \mathbf{k}_1 = \mathbf{k}_3 =
\begin{array}{c}
\begin{array}{cccc} 0 & 2 & 0 & 3 \end{array} \\
\left[\begin{array}{cccc}
5.3107 & 955.93 & -5.3107 & 955.93 \\
955.93 & \boxed{229{,}422} & -955.93 & \boxed{114{,}711} \\
-5.3107 & -955.93 & 5.3107 & -955.93 \\
955.93 & \boxed{114{,}711} & -955.93 & \boxed{229{,}422}
\end{array}\right]
\end{array}
\begin{array}{c}
0 \\ 2 \\ 0 \\ 3
\end{array}
\qquad \textbf{(1)}
$$

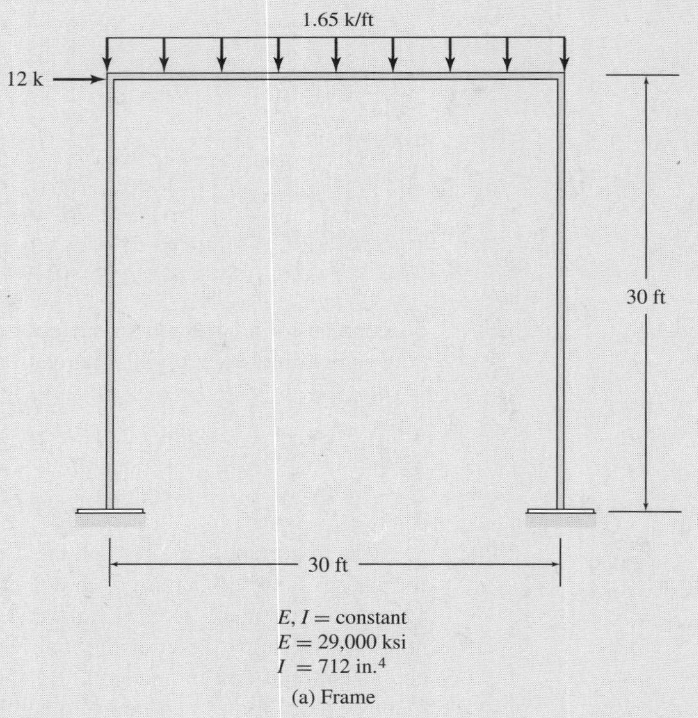

(a) Frame

Fig. 9.5

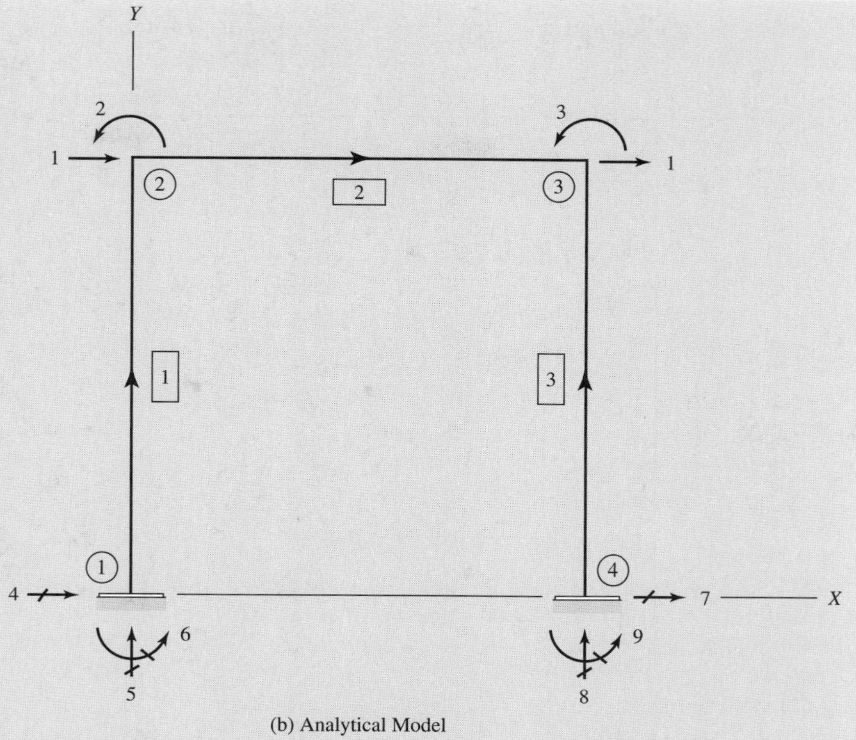

(b) Analytical Model

Fig. 9.5 *(continued)*

From Fig. 9.5(b), we can see that for member 1, the structure coordinates in the directions of the member end shears and end moments are numbered 4, 6, 1, and 2. Thus, the code numbers for this member are 4, 6, 1, 2. Similarly, the code numbers for member 3 are 7, 9, 1, 3. Since the structure coordinates corresponding to the end shears of member 2 are not defined (because the corresponding joint displacements are 0), we use 0s for the corresponding member code numbers. Thus, the code numbers for member 1 are 0, 2, 0, 3. By using the foregoing member code numbers, the relevant elements of $\mathbf{K}_1$, $\mathbf{K}_2$, and $\mathbf{K}_3$ are stored in the 3×3 structure stiffness matrix $\mathbf{S}$. Note that the elements of $\mathbf{K}_2$ that correspond to 0 code numbers are simply disregarded. The structure stiffness matrix thus obtained is

$$\mathbf{S} = \begin{matrix} 1 & 2 & 3 & \\ \left[\begin{matrix} 10.621 \\ 955.93 \\ 955.93 \end{matrix}\right. & \begin{matrix} 955.93 \\ 458{,}844 \\ 114{,}711 \end{matrix} & \left.\begin{matrix} 955.93 \\ 114{,}711 \\ 458{,}844 \end{matrix}\right] & \begin{matrix} 1 \\ 2 \\ 3 \end{matrix} \end{matrix} \tag{2}$$

The fixed-end shears and moments due to the 0.1375 k/in. (= 1.65 k/ft) uniformly distributed load applied to member 2 are calculated as

$$FS_b = FS_e = \frac{wL}{2} = \frac{0.1375(360)}{2} = 24.75 \text{ k}$$

$$FM_b = -FM_e = \frac{wL^2}{12} = \frac{0.1375(360)^2}{12} = 1{,}485 \text{ k-in.}$$

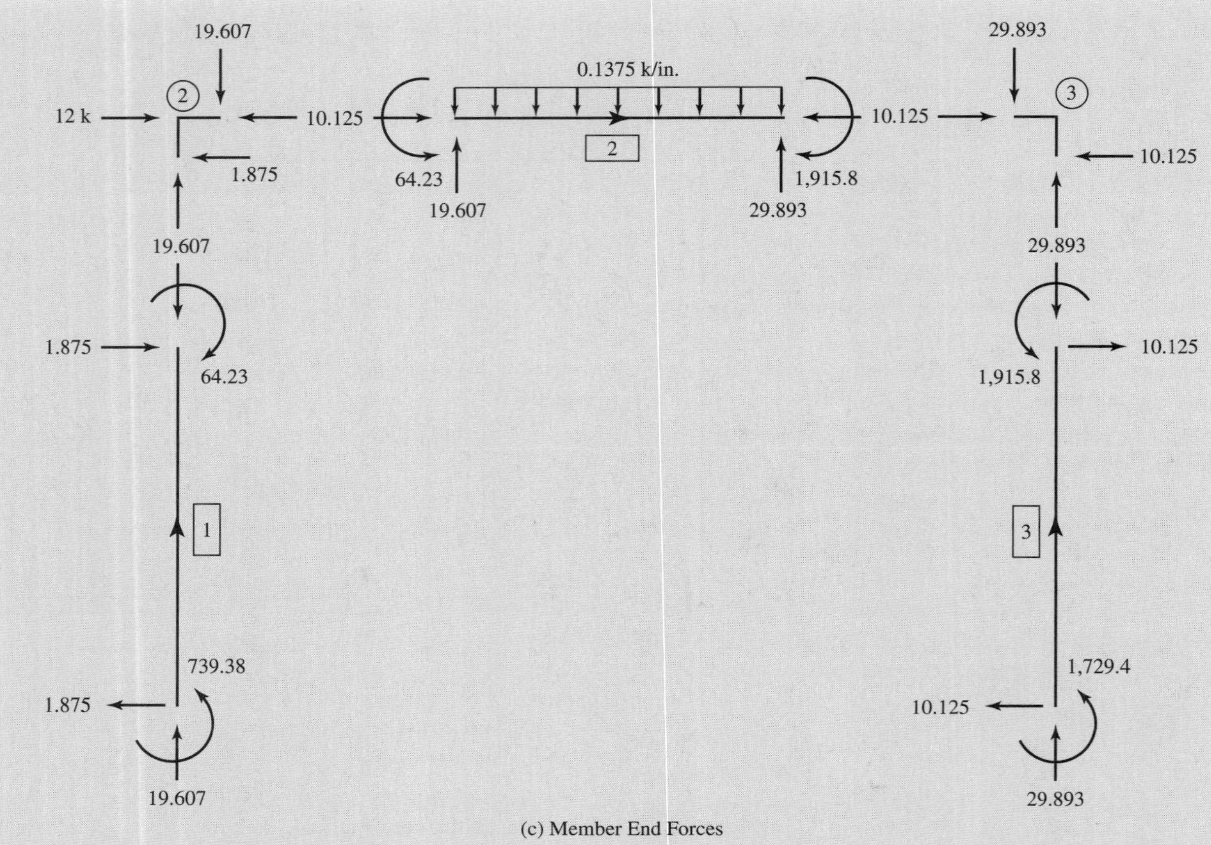

(c) Member End Forces

Fig. 9.5 *(continued)*

Using Eq. (9.8), we obtain

$$\mathbf{F}_{f2} = \mathbf{Q}_{f2} = \begin{bmatrix} 24.75 \\ 1{,}485 \\ 24.75 \\ -1{,}485 \end{bmatrix} \begin{matrix} 0 \\ 2 \\ 0 \\ 3 \end{matrix} \tag{3}$$

Thus, the structure fixed-joint force vector $\mathbf{P}_f$ is given by

$$\mathbf{P}_f = \begin{bmatrix} 0 \\ 1{,}485 \\ -1{,}485 \end{bmatrix} \begin{matrix} 1 \\ 2 \\ 3 \end{matrix} \tag{4}$$

Joint Load Vector:

$$\mathbf{P} = \begin{bmatrix} 12 \\ 0 \\ 0 \end{bmatrix} \begin{matrix} 1 \\ 2 \\ 3 \end{matrix} \tag{5}$$

Joint Displacements: By substituting the numerical values of $\mathbf{S}$ (Eq. (2)), $\mathbf{P}_f$ (Eq. (4)), and $\mathbf{P}$ (Eq. (5)) into the structure stiffness relationship $\mathbf{P} - \mathbf{P}_f = \mathbf{Sd}$, and solving the resulting system of simultaneous equations, we obtain the following joint displacements.

$$\mathbf{d} = \begin{bmatrix} 1.6141 \text{ in.} \\ -0.0070053 \text{ rad} \\ 0.001625 \text{ rad} \end{bmatrix} \begin{matrix} 1 \\ 2 \\ 3 \end{matrix}$$ **Ans**

Member End Shears and End Moments:

Member 1

$$\mathbf{v}_1 = \begin{bmatrix} v_1 \\ v_2 \\ v_3 \\ v_4 \end{bmatrix} \begin{matrix} 4 \\ 6 \\ 1 \\ 2 \end{matrix} = \begin{bmatrix} 0 \\ 0 \\ d_1 \\ d_2 \end{bmatrix} = \begin{bmatrix} 0 \\ 0 \\ 1.6141 \\ -0.0070053 \end{bmatrix}$$

From Eq. (9.9):

$$\mathbf{T}_1 = \mathbf{T}_3 = \begin{bmatrix} -1 & 0 & 0 & 0 \\ 0 & 1 & 0 & 0 \\ 0 & 0 & -1 & 0 \\ 0 & 0 & 0 & 1 \end{bmatrix} \quad (6)$$

$$\mathbf{u}_1 = \mathbf{T}_1 \mathbf{v}_1 = \begin{bmatrix} 0 \\ 0 \\ -1.6141 \\ -0.0070053 \end{bmatrix}$$

By using $\mathbf{k}_1$ from Eq. (1) and $\mathbf{Q}_{f1} = \mathbf{0}$, we obtain

$$\mathbf{Q}_1 = \mathbf{k}_1 \mathbf{u}_1 = \begin{bmatrix} 1.875 \text{ k} \\ 739.38 \text{ k-in.} \\ -1.875 \text{ k} \\ -64.23 \text{ k-in.} \end{bmatrix}$$ **Ans**

Member 2

$$\mathbf{u}_2 = \mathbf{v}_2 = \begin{bmatrix} 0 \\ -0.0070053 \\ 0 \\ 0.001625 \end{bmatrix} \begin{matrix} 0 \\ 2 \\ 0 \\ 3 \end{matrix}$$

By using $\mathbf{k}_2$ from Eq. (1) and $\mathbf{Q}_{f2}$ from Eq. (3), we calculate

$$\mathbf{Q}_2 = \mathbf{k}_2 \mathbf{u}_2 + \mathbf{Q}_{f2} = \begin{bmatrix} 19.607 \text{ k} \\ 64.23 \text{ k-in.} \\ 29.893 \text{ k} \\ -1{,}915.8 \text{ k-in.} \end{bmatrix}$$ **Ans**

Member 3

$$\mathbf{v}_3 = \begin{bmatrix} 0 \\ 0 \\ 1.6141 \\ 0.001625 \end{bmatrix} \begin{matrix} 7 \\ 9 \\ 1 \\ 3 \end{matrix}$$

Using $\mathbf{T}_3$ from Eq. (6), we obtain

$$\mathbf{u}_3 = \mathbf{T}_3\mathbf{v}_3 = \begin{bmatrix} 0 \\ 0 \\ -1.6141 \\ 0.001625 \end{bmatrix}$$

Using $\mathbf{k}_3$ from Eq. (1) and $\mathbf{Q}_{f3} = \mathbf{0}$, we calculate

$$\mathbf{Q}_3 = \mathbf{k}_3\mathbf{u}_3 + \mathbf{Q}_{f3} = \begin{bmatrix} 10.125 \text{ k} \\ 1{,}729.4 \text{ k-in.} \\ -10.125 \text{ k} \\ 1{,}915.8 \text{ k-in.} \end{bmatrix}$$ **Ans**

The member end shears and end moments, as given by the foregoing local end force vectors $\mathbf{Q}_1$, $\mathbf{Q}_2$, and $\mathbf{Q}_3$, are depicted in Fig. 9.5(c).

Member Axial Forces: With the member end shears now known, we can calculate the axial forces for the three members of the frame by applying the equations of equilibrium, $\sum F_X = 0$ and $\sum F_Y = 0$, to the free bodies of joints 2 and 3. The member axial forces thus obtained are shown in Fig. 9.5(c). **Ans**

Support Reactions: By comparing Figs. 9.5(b) and (c), we realize that the forces at the lower ends of the columns of the frame represent its support reactions; that is,

$$\mathbf{R} = \begin{bmatrix} -1.875 \text{ k} \\ 19.607 \text{ k} \\ 739.38 \text{ k-in.} \\ -10.125 \text{ k} \\ 29.893 \text{ k} \\ 1{,}729.4 \text{ k-in.} \end{bmatrix} \begin{matrix} 4 \\ 5 \\ 6 \\ 7 \\ 8 \\ 9 \end{matrix}$$ **Ans**

9.3 CONDENSATION OF DEGREES OF FREEDOM, AND SUBSTRUCTURING

A problem that can arise during computer analysis of large structures is that the computer may not have sufficient memory to store and process information about the entire structure. A commonly used approach to circumvent this problem is to *condense* (or reduce the number of) the structure's stiffness equations that are to be solved simultaneously, by suppressing some of the degrees of freedom. This process is referred to as *condensation* (also called *static condensation*). For very large structures, it may become necessary to combine condensation with another process called *substructuring,* in which the structure is divided into parts called *substructures,* with the condensed stiffness relations for each substructure generated separately; these are then combined to obtain the stiffness relations for the entire structure. In this section, we consider the basic concepts of condensation of degrees of freedom, and analysis using substructures.

Condensation

The objective of condensation is to reduce the number of independent degrees of freedom of a structure (or substructure, or member). This is achieved by treating some of the degrees of freedom as dependent variables and expressing them in terms of the remaining independent degrees of freedom. The relationship between the dependent and independent degrees of freedom is then substituted into the original stiffness relations to obtain a condensed system of stiffness equations, which contains only the independent degrees of freedom as unknowns. From a theoretical viewpoint, the dependent degrees of freedom can be chosen arbitrarily. However, for computational purposes, it is usually convenient to select those degrees of freedom that are internal to the structure (or substructure, or member) as the dependent degrees of freedom. Hence, the dependent degrees of freedom are commonly referred to as the *internal degrees of freedom;* whereas, the independent degrees of freedom are called the *external degrees of freedom.*

As discussed in the preceding chapters, the stiffness relations for a general framed structure can be expressed as (see, for example, Eq. (6.42))

$$\overline{\mathbf{P}} = \mathbf{S}\mathbf{d} \tag{9.11}$$

with

$$\overline{\mathbf{P}} = \mathbf{P} - \mathbf{P}_f \tag{9.12}$$

When using the condensation process, it is usually convenient to assign numbers to the degrees of freedom so that the external and internal degrees of freedom are separated into two groups. The structure stiffness relations (Eq. (9.11)) can then be written in partitioned-matrix form:

$$\begin{bmatrix} \overline{\mathbf{P}}_E \\ \hline \overline{\mathbf{P}}_I \end{bmatrix} = \left[\begin{array}{c|c} \mathbf{S}_{EE} & \mathbf{S}_{EI} \\ \hline \mathbf{S}_{IE} & \mathbf{S}_{II} \end{array}\right] \begin{bmatrix} \mathbf{d}_E \\ \hline \mathbf{d}_I \end{bmatrix} \tag{9.13}$$

in which the subscripts E and I refer to quantities related to the external and internal degrees of freedom, respectively. By multiplying the two partitioned matrices on the right side of Eq. (9.13), we obtain the two matrix equations,

$$\overline{\mathbf{P}}_E = \mathbf{S}_{EE}\mathbf{d}_E + \mathbf{S}_{EI}\mathbf{d}_I \tag{9.14}$$

$$\overline{\mathbf{P}}_I = \mathbf{S}_{IE}\mathbf{d}_E + \mathbf{S}_{II}\mathbf{d}_I \tag{9.15}$$

To express the internal degrees of freedom $\mathbf{d}_I$ in terms of the external degrees of freedom $\mathbf{d}_E$, we solve Eq. (9.15) for $\mathbf{d}_I$, as

$$\boxed{\mathbf{d}_I = \mathbf{S}_{II}^{-1}(\overline{\mathbf{P}}_I - \mathbf{S}_{IE}\mathbf{d}_E)} \tag{9.16}$$

Finally, by substituting Eq. (9.16) into Eq. (9.14), we obtain the condensed stiffness equations

$$\overline{\mathbf{P}}_E - \mathbf{S}_{EI}\mathbf{S}_{II}^{-1}\overline{\mathbf{P}}_I = (\mathbf{S}_{EE} - \mathbf{S}_{EI}\mathbf{S}_{II}^{-1}\mathbf{S}_{IE})\mathbf{d}_E \tag{9.17}$$

Note that the external degrees of freedom $\mathbf{d}_E$ are the only unknowns in Eq. (9.17). Equation (9.17) can be rewritten in a compact form as

$$\boxed{\mathbf{P}_E^* = \mathbf{S}_{EE}^* \mathbf{d}_E} \tag{9.18}$$

in which,

$$\boxed{\mathbf{P}_E^* = \overline{\mathbf{P}}_E - \mathbf{S}_{EI}\mathbf{S}_{II}^{-1}\overline{\mathbf{P}}_I} \tag{9.19}$$

and

$$\boxed{\mathbf{S}_{EE}^* = \mathbf{S}_{EE} - \mathbf{S}_{EI}\mathbf{S}_{II}^{-1}\mathbf{S}_{IE}} \tag{9.20}$$

As the foregoing equations indicate, the solution of the structure stiffness equations is carried out in two parts. In the first part, $\mathbf{P}_E^*$ and $\mathbf{S}_{EE}^*$ are evaluated using Eqs. (9.19) and (9.20), respectively, and the external joint displacements $\mathbf{d}_E$ are determined by solving Eq. (9.18). In the second part, the now-known $\mathbf{d}_E$ is substituted into Eq. (9.16) to obtain the internal joint displacements $\mathbf{d}_I$. Once all the joint displacements have been evaluated, the member end displacements and end forces, and support reactions, can be calculated using the procedures described in the previous chapters.

It should be realized that analysis involving condensation generally requires more computational effort than the standard formulation in which all of the structure's stiffness equations are solved simultaneously. However, condensation provides a useful means of analyzing large structures whose full stiffness matrices and load vectors exceed the available computer memory. This is because, when employing condensation, only parts of $\mathbf{S}$ and $\overline{\mathbf{P}}$ need to be assembled and processed in the computer memory at a given time. The basic concept of condensation is illustrated by the following relatively simple example.

EXAMPLE 9.3

Analyze the plane frame shown in Fig. 9.6(a) using condensation, by treating the rotation of the free joint as the internal degree of freedom.

SOLUTION

This frame was analyzed in Example 6.6 using the standard formulation. The analytical model of the structure is given in Fig. 9.6(b).

Condensed Structure Stiffness Matrix: The full (3 × 3) stiffness matrix, $\mathbf{S}$, for the frame, as determined in Example 6.6, is given by (in units of kips and inches):

$$\mathbf{S} = \begin{array}{c} \begin{matrix} 1 & \quad 2 & \quad 3 \end{matrix} \\ \left[\begin{array}{cc:c} 1{,}685.3 & 507.89 & 670.08 \\ 507.89 & 1{,}029.2 & 601.42 \\ \hdashline 670.08 & 601.42 & 283{,}848 \end{array}\right] \end{array} \begin{matrix} 1 \\ 2 \\ 3 \end{matrix} \tag{1}$$

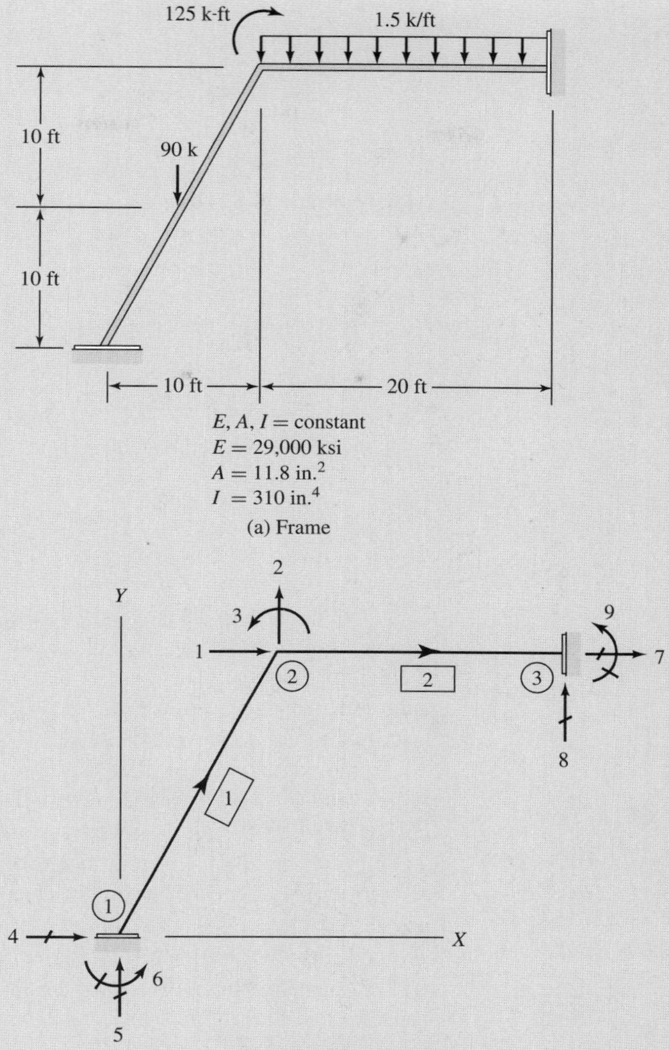

Fig. 9.6

in which **S** is partitioned to separate the external degrees of freedom, 1 and 2, from the internal degree of freedom, 3. From Eq. (1), we obtain

$$\mathbf{S}_{EE} = \begin{bmatrix} 1{,}685.3 & 507.89 \\ 507.89 & 1{,}029.2 \end{bmatrix} \begin{matrix} 1 \\ 2 \end{matrix} \qquad \text{(2)}$$

(columns 1, 2)

$$\mathbf{S}_{IE} = \begin{bmatrix} 670.08 & 601.42 \end{bmatrix} 3 \qquad \text{(3)}$$

(columns 1, 2)

$$\mathbf{S}_{EI} = \begin{bmatrix} 670.08 \\ 601.42 \end{bmatrix} \begin{matrix} 1 \\ 2 \end{matrix} \quad \text{(column 3)} \tag{4}$$

$$\mathbf{S}_{II} = [283{,}848]\, 3 \quad \text{(column 3)} \tag{5}$$

with the inverse of $\mathbf{S}_{II}$ given by

$$\mathbf{S}_{II}^{-1} = \left[\frac{1}{283{,}848}\right] \tag{6}$$

By substituting Eqs. (2), (3), (4), and (6) into Eq. (9.20), we obtain the condensed structure stiffness matrix:

$$\mathbf{S}_{EE}^{*} = \mathbf{S}_{EE} - \mathbf{S}_{EI}\mathbf{S}_{II}^{-1}\mathbf{S}_{IE} = \begin{bmatrix} 1{,}683.7 & 506.47 \\ 506.47 & 1{,}027.9 \end{bmatrix} \text{ k/in.} \tag{7}$$

Condensed Joint Load Vector: Recall from Example 6.6 that

$$\overline{\mathbf{P}} = \mathbf{P} - \mathbf{P}_f = \begin{bmatrix} 0 \\ -60 \\ \hline -750 \end{bmatrix} \begin{matrix} 1 \\ 2 \\ 3 \end{matrix} \tag{8}$$

from which,

$$\overline{\mathbf{P}}_E = \begin{bmatrix} 0 \\ -60 \end{bmatrix} \begin{matrix} 1 \\ 2 \end{matrix} \tag{9}$$

and

$$\overline{\mathbf{P}}_I = [-750]\, 3 \tag{10}$$

Substitution of Eqs. (4), (6), (9), and (10) into Eq. (9.19) yields the following condensed joint load vector.

$$\mathbf{P}_E^{*} = \overline{\mathbf{P}}_E - \mathbf{S}_{EI}\mathbf{S}_{II}^{-1}\overline{\mathbf{P}}_I = \begin{bmatrix} 1.7705 \\ -58.411 \end{bmatrix} \text{ k} \tag{11}$$

Joint Displacements: By substituting Eqs. (7) and (11) into the condensed structure stiffness relationship, $\mathbf{P}_E^{*} = \mathbf{S}_{EE}^{*}\mathbf{d}_E$ (Eq. (9.18)), and solving the resulting 2 × 2 system of simultaneous equations, we obtain the external joint displacements (corresponding to degrees of freedom 1 and 2), as

$$\mathbf{d}_E = \begin{bmatrix} 0.021302 \\ -0.06732 \end{bmatrix} \begin{matrix} 1 \\ 2 \end{matrix} \text{ in.} \tag{12}$$

The internal joint displacement (i.e., the rotation corresponding to degree of freedom 3), can now be determined by applying Eq. (9.16). Thus,

$$\mathbf{d}_I = \mathbf{S}_{II}^{-1}(\overline{\mathbf{P}}_I - \mathbf{S}_{IE}\mathbf{d}_E) = [-0.0025499]3 \text{ rad} \tag{13}$$

By combining Eqs. (12) and (13), we obtain the full joint displacement vector,

$$\mathbf{d} = \begin{bmatrix} \mathbf{d}_E \\ \hline \mathbf{d}_I \end{bmatrix} = \begin{bmatrix} 0.021302 \text{ in.} \\ -0.06732 \text{ in.} \\ \hline -0.0025499 \text{ rad} \end{bmatrix} \begin{matrix} 1 \\ 2 \\ 3 \end{matrix} \qquad \textbf{Ans}$$

Note that the foregoing joint displacements are identical to those determined in Example 6.6 by solving the structure's three stiffness equations simultaneously.

Member End Displacements and End Forces: See Example 6.6.

It is important to realize that, in this example, the submatrices of **S** and $\overline{\mathbf{P}}$ were obtained from the corresponding full matrices, for convenience only. In actual computer analysis, to save memory space, the individual parts of **S** and $\overline{\mathbf{P}}$ are assembled directly from the corresponding member matrices as they are needed in the analysis.

In the foregoing paragraphs, we have discussed the application of condensation to reduce the number of independent degrees of freedom of an entire structure. The condensation process is also frequently used to establish the stiffness relationships for substructures, which are defined as groups of members with known stiffness relations. In this case, condensation is used to eliminate the degrees of freedom of those joints that are internal to the substructure, thereby producing a condensed system of stiffness relations expressed solely in terms of the degrees of freedom of those (external) joints through which the substructure is connected to the rest of the structure and/or supports.

The procedure for condensing the internal degrees of freedom of a substructure is analogous to that just discussed for the case of a whole structure. The stiffness relations involving both the internal and external degrees of freedom of a substructure can be symbolically expressed as

$$\overline{\mathbf{F}} = \overline{\mathbf{K}}\bar{\mathbf{v}} + \overline{\mathbf{F}}_f \tag{9.21}$$

in which $\overline{\mathbf{F}}$ and $\bar{\mathbf{v}}$ represent, respectively, the joint forces and displacements for the substructure; $\overline{\mathbf{K}}$ denotes the substructure stiffness matrix; and $\overline{\mathbf{F}}_f$ represents the fixed-joint forces for the substructure. The matrix $\overline{\mathbf{K}}$ and the vector $\overline{\mathbf{F}}_f$ can be assembled from the member stiffness matrices and fixed-end force vectors in the usual way. To apply condensation, we rewrite Eq. (9.21) in partitioned-matrix form as

$$\begin{bmatrix} \overline{\mathbf{F}}_E \\ \hline \overline{\mathbf{F}}_I \end{bmatrix} = \left[\begin{array}{c|c} \overline{\mathbf{K}}_{EE} & \overline{\mathbf{K}}_{EI} \\ \hline \overline{\mathbf{K}}_{IE} & \overline{\mathbf{K}}_{II} \end{array}\right] \begin{bmatrix} \bar{\mathbf{v}}_E \\ \hline \bar{\mathbf{v}}_I \end{bmatrix} + \begin{bmatrix} \overline{\mathbf{F}}_{fE} \\ \hline \overline{\mathbf{F}}_{fI} \end{bmatrix} \tag{9.22}$$

The multiplication of the two partitioned matrices on the right-hand side of Eq. (9.22) yields the matrix equations

$$\overline{\mathbf{F}}_E = \overline{\mathbf{K}}_{EE}\bar{\mathbf{v}}_E + \overline{\mathbf{K}}_{EI}\bar{\mathbf{v}}_I + \overline{\mathbf{F}}_{fE} \tag{9.23}$$

$$\overline{\mathbf{F}}_I = \overline{\mathbf{K}}_{IE}\bar{\mathbf{v}}_E + \overline{\mathbf{K}}_{II}\bar{\mathbf{v}}_I + \overline{\mathbf{F}}_{fI} \tag{9.24}$$

Solving Eq. (9.24) for $\bar{\mathbf{v}}_I$, we obtain

$$\boxed{\bar{\mathbf{v}}_I = \overline{\mathbf{K}}_{II}^{-1}(\overline{\mathbf{F}}_I - \overline{\mathbf{F}}_{fI} - \overline{\mathbf{K}}_{IE}\bar{\mathbf{v}}_E)} \tag{9.25}$$

and, substituting Eq. (9.25) into Eq. (9.23), we determine the condensed stiffness relations for the substructure to be

$$\boxed{\overline{\mathbf{F}}_E = \overline{\mathbf{K}}^*_{EE}\bar{\mathbf{v}}_E + \overline{\mathbf{F}}^*_{fE}} \tag{9.26}$$

in which,

$$\overline{\mathbf{K}}^*_{EE} = \overline{\mathbf{K}}_{EE} - \overline{\mathbf{K}}_{EI}\overline{\mathbf{K}}_{II}^{-1}\overline{\mathbf{K}}_{IE} \tag{9.27}$$

and

$$\overline{\mathbf{F}}^*_{fE} = \overline{\mathbf{F}}_{fE} + \overline{\mathbf{K}}_{EI}\overline{\mathbf{K}}_{II}^{-1}\left(\overline{\mathbf{F}}_{I} - \overline{\mathbf{F}}_{fI}\right) \tag{9.28}$$

EXAMPLE 9.4 Determine the stiffness matrix and the fixed-joint force vector for the substructure of a beam shown in Fig. 9.7(a), in terms of its external degrees of freedom only. The substructure is composed of two members connected together by a hinged joint, as shown in the figure.

SOLUTION *Analytical Model:* The analytical model of the substructure is depicted in Fig. 9.7(b). For member 1, $MT = 2$, because the end of this member is hinged; $MT = 1$ for member 2, which is hinged at its beginning. Joint 3 is modeled as a hinged joint with its rotation restrained by an imaginary clamp. Thus, the substructure has a total of five degrees of freedom, of which four are external (identified by numbers 1 through 4) and one is internal (identified by number 5).

Substructure Stiffness Matrix: We will first assemble the full (5 × 5) stiffness matrix $\overline{\mathbf{K}}$ from the member stiffness matrices $\mathbf{k}$, and then apply Eq. (9.27) to determine the condensed stiffness matrix $\overline{\mathbf{K}}^*_{EE}$.

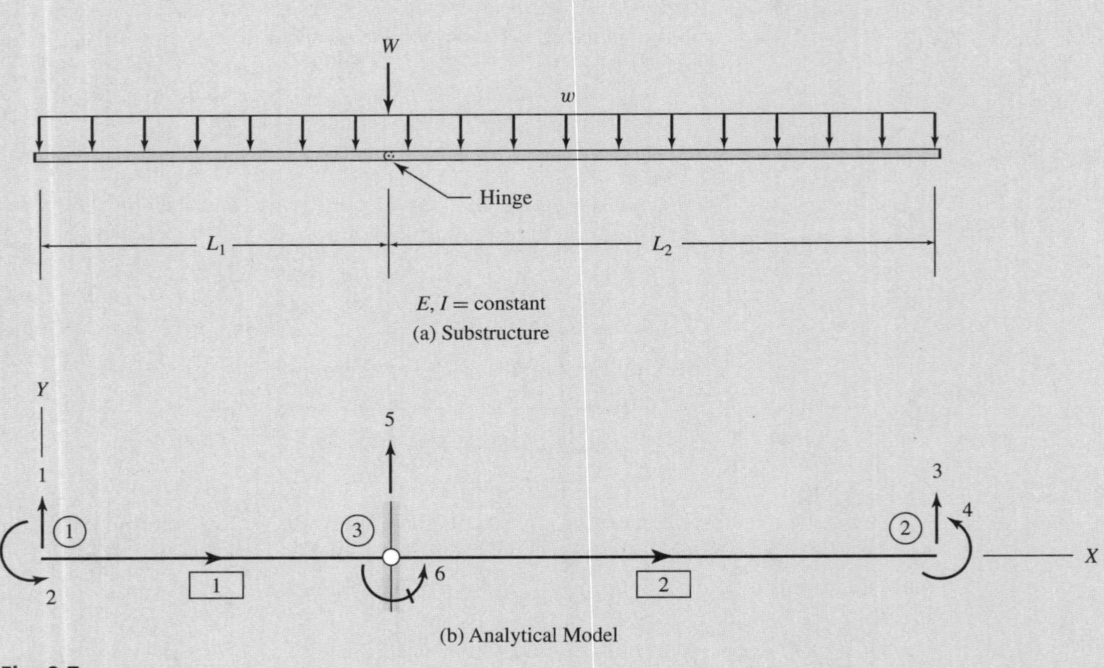

(b) Analytical Model

Fig. 9.7

$$\overline{\mathbf{K}} = \left[\begin{array}{c|c} \overline{\mathbf{K}}_{EE} & \overline{\mathbf{K}}_{EI} \\ \hline \overline{\mathbf{K}}_{IE} & \overline{\mathbf{K}}_{II} \end{array}\right] = 3EI \begin{array}{c} \begin{array}{ccccc} 1 & 2 & 3 & 4 & 5 \end{array} \\ \left[\begin{array}{cccc|c} \frac{1}{L_1^3} & \frac{1}{L_1^2} & 0 & 0 & -\frac{1}{L_1^3} \\ \frac{1}{L_1^2} & \frac{1}{L_1} & 0 & 0 & -\frac{1}{L_1^2} \\ 0 & 0 & \frac{1}{L_2^3} & -\frac{1}{L_2^2} & -\frac{1}{L_2^3} \\ 0 & 0 & -\frac{1}{L_2^2} & \frac{1}{L_2} & \frac{1}{L_2^2} \\ \hline -\frac{1}{L_1^3} & -\frac{1}{L_1^2} & -\frac{1}{L_2^3} & \frac{1}{L_2^2} & \frac{1}{L_1^3}+\frac{1}{L_2^3} \end{array}\right] \end{array} \begin{array}{c} 1 \\ 2 \\ 3 \\ 4 \\ 5 \end{array}$$

(c) Full (Uncondensed) Stiffness Matrix for Substructure

$$\overline{\mathbf{F}}_f = \left[\begin{array}{c} \overline{\mathbf{F}}_{fE} \\ \hline \overline{\mathbf{F}}_{fI} \end{array}\right] = \frac{w}{8} \left[\begin{array}{c} 5L_1 \\ L_1^2 \\ 5L_2 \\ -L_2^2 \\ \hline 3L_1 + 3L_2 \end{array}\right] \begin{array}{c} 1 \\ 2 \\ 3 \\ 4 \\ 5 \end{array}$$

(d) Full (Uncondensed) Fixed-Joint Force Vector For Substructure

Fig. 9.7 *(continued)*

Member 1 (*MT* = 2) Using Eq. (7.18), we obtain

$$\mathbf{k}_1 = \frac{3EI}{L_1^3} \begin{array}{c} \begin{array}{cccc} 1 & 2 & 5 & 6 \end{array} \\ \left[\begin{array}{cccc} 1 & L_1 & -1 & 0 \\ L_1 & L_1^2 & -L_1 & 0 \\ -1 & -L_1 & 1 & 0 \\ 0 & 0 & 0 & 0 \end{array}\right] \end{array} \begin{array}{c} 1 \\ 2 \\ 5 \\ 6 \end{array}$$

Member 2 (*MT* = 1) Application of Eq. (7.15) yields

$$\mathbf{k}_2 = \frac{3EI}{L_2^3} \begin{array}{c} \begin{array}{cccc} 5 & 6 & 3 & 4 \end{array} \\ \left[\begin{array}{cccc} 1 & 0 & -1 & L_2 \\ 0 & 0 & 0 & 0 \\ -1 & 0 & 1 & -L_2 \\ L_2 & 0 & -L_2 & L_2^2 \end{array}\right] \end{array} \begin{array}{c} 5 \\ 6 \\ 3 \\ 4 \end{array}$$

Using the code numbers of the members, we store the pertinent elements of $\mathbf{k}_1$ and $\mathbf{k}_2$ in the full 5×5 stiffness matrix $\overline{\mathbf{K}}$ of the substructure, as shown in Fig. 9.7(c).

Substituting into Eq. (9.27) the appropriate submatrices of $\overline{\mathbf{K}}$ from Fig. 9.7(c) and

$$\overline{\mathbf{K}}_{II}^{-1} = \left[\frac{L_1^3 L_2^3}{3EI(L_1^3 + L_2^3)}\right] \tag{1}$$

we obtain the condensed stiffness matrix for the substructure:

$$\overline{\mathbf{K}}_{EE}^{*} = \overline{\mathbf{K}}_{EE} - \overline{\mathbf{K}}_{EI}\overline{\mathbf{K}}_{II}^{-1}\overline{\mathbf{K}}_{IE} = \frac{3EI}{L_1^3 + L_2^3} \left[\begin{array}{cccc} 1 & L_1 & -1 & L_2 \\ L_1 & L_1^2 & -L_1 & L_1 L_2 \\ -1 & -L_1 & 1 & -L_2 \\ L_2 & L_1 L_2 & -L_2 & L_2^2 \end{array}\right] \tag{2}$$

Ans

Substructure Fixed-Joint Force Vector:

Member 1 (*MT* = 2) Using Eq. (7.19), we obtain

$$\mathbf{Q}_{f1} = \frac{wL_1}{8} \left[\begin{array}{c} 5 \\ L_1 \\ 3 \\ 0 \end{array}\right] \begin{array}{c} 1 \\ 2 \\ 5 \\ 6 \end{array}$$

Member 2 ($MT = 1$) Using Eq. (7.16), we write

$$\mathbf{Q}_{f2} = \frac{wL_2}{8}\begin{bmatrix} 3 \\ 0 \\ 5 \\ -L_2 \end{bmatrix}\begin{matrix} 5 \\ 6 \\ 3 \\ 4 \end{matrix}$$

The relevant elements of $\mathbf{Q}_{f1}$ and $\mathbf{Q}_{f2}$ are stored in the full 5×1 fixed-joint force vector $\overline{\mathbf{F}}_f$ of the substructure, as shown in Fig. 9.7(d).

A comparison of Figs. 9.7(a) and (b) indicates that $\overline{F}_5 = -W$; that is,

$$\overline{\mathbf{F}}_I = [\overline{F}_5] = [-W] \tag{3}$$

Finally, the application of Eq. (9.28) yields the following condensed fixed-joint force vector for the substructure.

$$\overline{\mathbf{F}}^*_{fE} = \overline{\mathbf{F}}_{fE} + \overline{\mathbf{K}}_{EI}\overline{\mathbf{K}}_{II}^{-1}(\overline{\mathbf{F}}_I - \overline{\mathbf{F}}_{fI})$$

$$= \frac{w}{8(L_1^3 + L_2^3)}\begin{bmatrix} 5L_1^4 + 8L_1L_2^3 + 3L_2^4 \\ L_1^5 + 4L_1^2L_2^3 + 3L_1L_2^4 \\ 3L_1^4 + 8L_1^3L_2 + 5L_2^4 \\ -(3L_1^4L_2 + 4L_1^3L_2^2 + L_2^5) \end{bmatrix} - \frac{W}{L_1^3 + L_2^3}\begin{bmatrix} -L_2^3 \\ -L_1L_2^3 \\ -L_1^3 \\ L_1^3L_2 \end{bmatrix} \tag{4}$$ **Ans**

Analysis Using Substructures

The procedure for the analysis of (large) structures, divided into substructures, is essentially the same as the standard stiffness method developed in previous chapters. However, each substructure is treated as an ordinary member of the structure, and the degrees of freedom of only those joints through which the substructures are connected to each other and/or to supports are considered to be the structure's degrees of freedom **d**. The structure's stiffness matrix **S** and fixed-joint force vector $\mathbf{P}_f$, are assembled, respectively, from the substructure stiffness matrices $\overline{\mathbf{K}}^*_{EE}$ and fixed-joint force vectors $\overline{\mathbf{F}}^*_{fE}$, which are expressed in terms of the external coordinates of the substructures only. The structure stiffness equations, $\mathbf{P} - \mathbf{P}_f = \mathbf{S}\mathbf{d}$, thus obtained, can then be solved for the joint displacements **d**.

Consider, for example, the nine-story plane frame shown in Fig. 9.8(a). The frame actually has 20 joints and 54 degrees of freedom; that is, if we were to analyze the frame using the standard stiffness method for plane frames as developed in Chapter 6, we would have to assemble and solve 54 structure stiffness equations simultaneously. Now, suppose that we wish to analyze the frame by dividing it into three substructures, each consisting of three stories of the frame, as depicted in Fig. 9.8(b). As this figure indicates, for analysis purposes, the frame is now modeled as having only six joints, at which the three substructures are connected to each other and to external supports. Thus, the analytical model of the frame has 12 degrees of freedom and six restrained coordinates.

To develop the stiffness matrix **S** and the fixed-joint force vector $\mathbf{P}_f$ for the frame, we first determine the substructure stiffness matrices $\overline{\mathbf{K}}^*_{EE}$ and fixed-joint

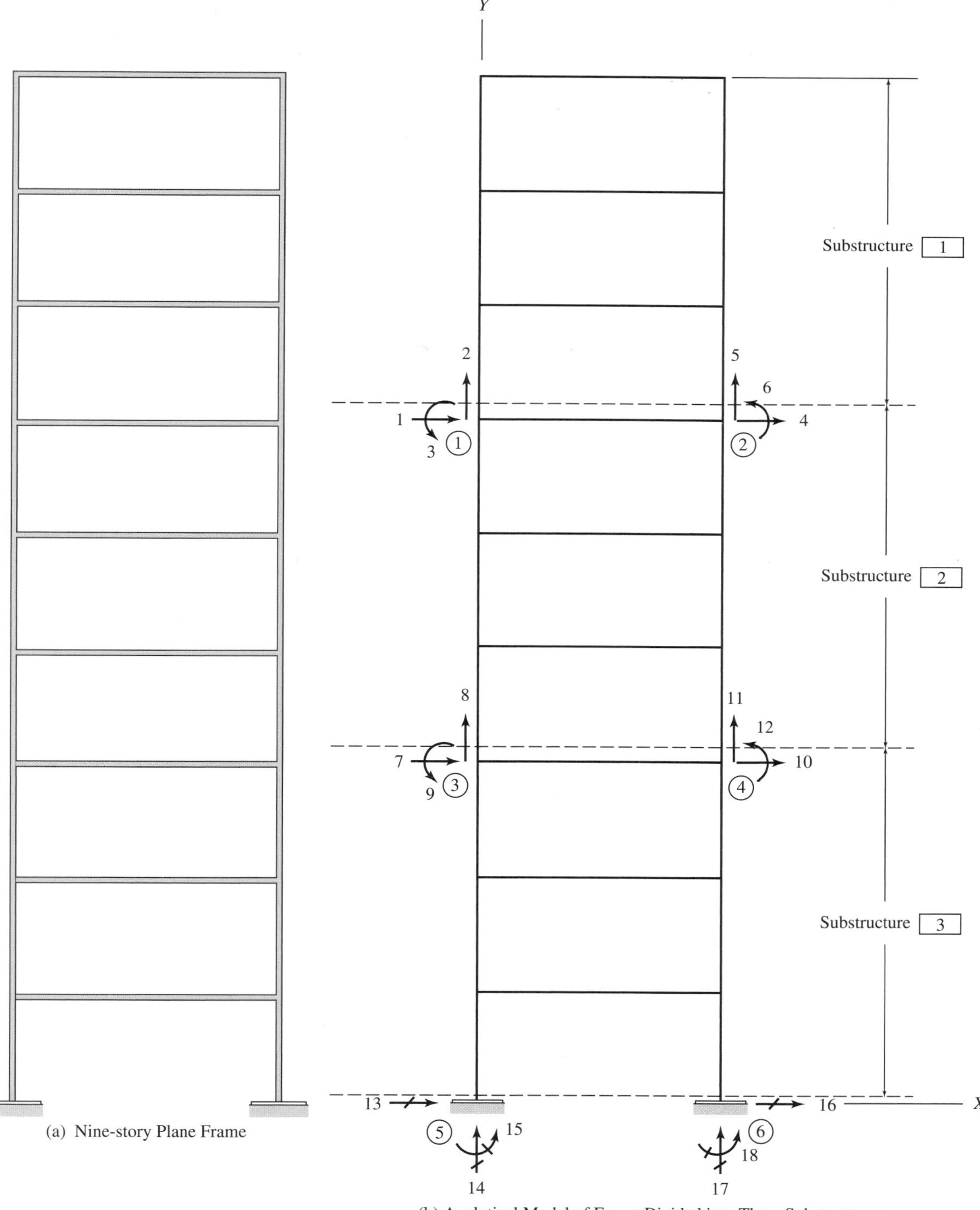
Y
Substructure 1
Substructure 2
Substructure 3
X
1
2
3
4
5
6
7
8
9
10
11
12
13
14
15
16
17
18
(a) Nine-story Plane Frame
(b) Analytical Model of Frame Divided into Three Substructures

Fig. 9.8

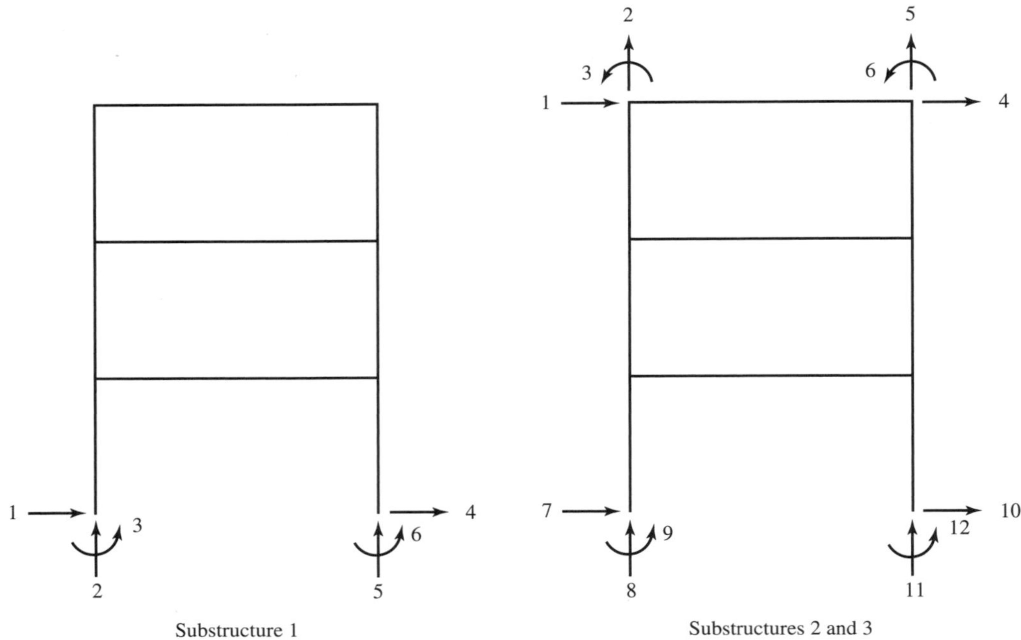

(c) External Degrees of Freedom of Substructures

Fig. 9.8 *(continued)*

force vectors $\overline{\mathbf{F}}^*_{fE}$, in terms of the external degrees of freedom of the substructures, using condensation as described earlier in this section. As shown in Fig. 9.8(c), substructure 1 has six external degrees of freedom; whereas, substructures 2 and 3 each have 12 external degrees of freedom. The pertinent elements of $\overline{\mathbf{K}}^*_{EE}$ matrices and $\overline{\mathbf{F}}^*_{fE}$ vectors are then stored in $\mathbf{S}$ and $\mathbf{P}_f$, respectively, using the substructure code numbers in the usual manner. By comparing Figs. 9.8(b) and (c), we can see that the code numbers for substructure 1 are 1, 2, 3, 4, 5, 6; whereas, the code numbers for substructure 2 are 1, 2, 3, 4, 5, 6, 7, 8, 9, 10, 11, 12. Similarly, for substructure 3, the code numbers are 7, 8, 9, 10, 11, 12, 13, 14, 15, 16, 17, 18.

Once the structure stiffness matrix $\mathbf{S}$ (12 × 12) and the fixed-joint force vector $\mathbf{P}_f$ (12 × 1) have been assembled, the structure stiffness equations, $\mathbf{P} - \mathbf{P}_f = \mathbf{Sd}$, are solved to calculate the joint displacement vector $\mathbf{d}$. With $\mathbf{d}$ known, the external joint displacements, $\bar{\mathbf{v}}_E$, for each substructure are obtained from $\mathbf{d}$ using the substructure's code numbers, and then the substructure's internal joint displacements, $\bar{\mathbf{v}}_I$, are calculated using Eq. (9.25). After the joint displacement vector $\bar{\mathbf{v}}$ of a substructure has been determined, the end displacements and forces for its individual members, and support reactions, can be evaluated using the standard procedure described in previous chapters. The basic concept of analysis using substructures is illustrated by the following relatively simple example.

EXAMPLE 9.5 Analyze the two-span continuous beam shown in Fig. 9.9(a), treating each span as a substructure.

SOLUTION *Analytical Model:* The structure is modeled as being composed of two substructures and three joints, as shown in Fig. 9.9(b). It has one degree of freedom and five restrained coordinates. As each substructure consists of two beam members connected together by a hinged joint, we will use the expressions of stiffnesses and fixed-joint forces for such substructures, derived in Example 9.4, in the present example.

Structure Stiffness Matrix, **S***:* Substituting $E = 70(10^6)$ kN/m^2, $I = 200(10^{-6})$ m^4, and $L_1 = L_2 = 5$ m into Eq. (2) of Example 9.4, we obtain the following condensed stiffness matrix for the two substructures.

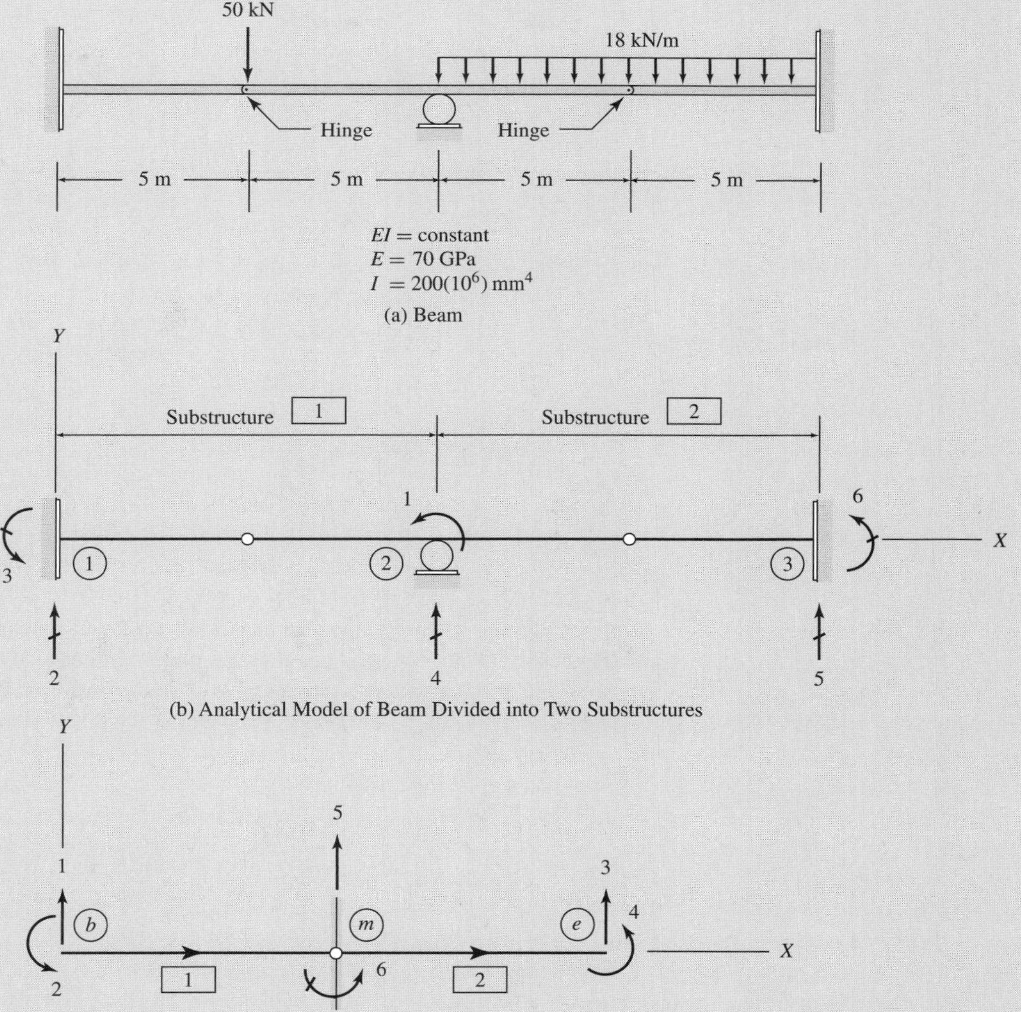

Fig. 9.9

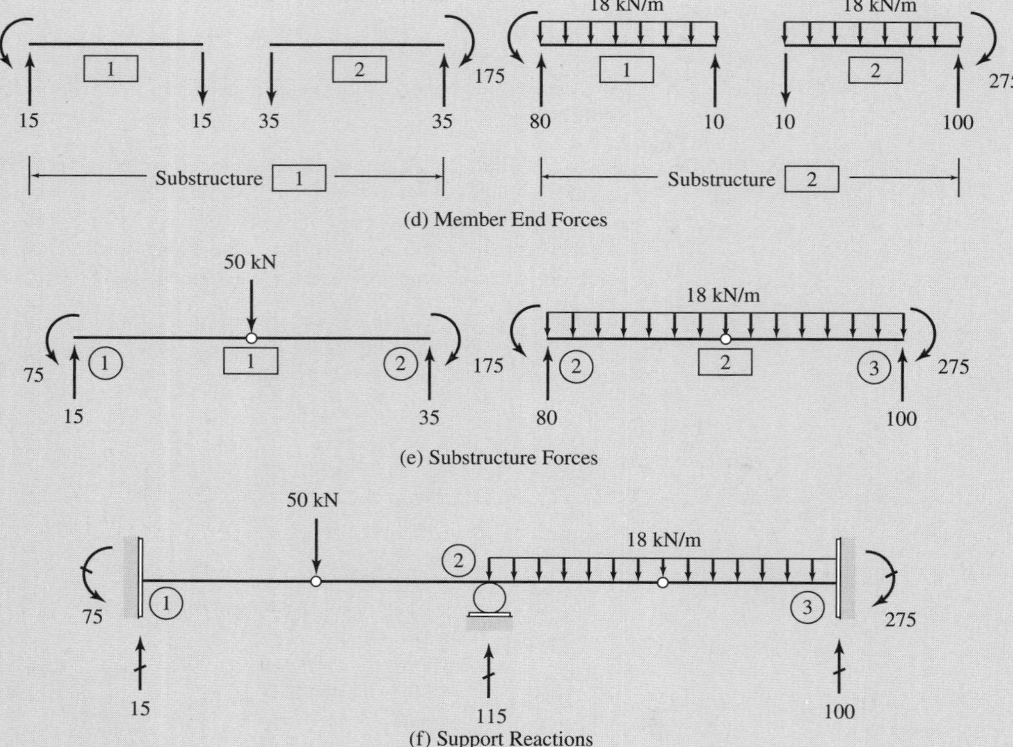

Fig. 9.9 *(continued)*

$$
\begin{array}{l}
\text{Substructure 2} \longrightarrow \quad \begin{array}{cccc} 4 & 1 & 5 & 6 \end{array} \\
\text{Substructure 1} \longrightarrow \quad \begin{array}{cccc} 2 & 3 & 4 & 1 \end{array}
\end{array}
$$

$$
\overline{\mathbf{K}}^*_{EE1} = \overline{\mathbf{K}}^*_{EE2} = \begin{bmatrix} 168 & 840 & -168 & 840 \\ 840 & 4{,}200 & -840 & 4{,}200 \\ -168 & -840 & 168 & -840 \\ 840 & 4{,}200 & -840 & 4{,}200 \end{bmatrix} \begin{matrix} 2 \\ 3 \\ 4 \\ 1 \end{matrix} \quad \begin{matrix} 4 \\ 1 \\ 5 \\ 6 \end{matrix}
$$

By comparing the numbers of the external degrees of freedom of a substructure (Fig. 9.9(c)) to those of the structure degrees of freedom (Fig. 9.9(b)), we obtain code numbers 2, 3, 4, 1 for substructure 1, and 4, 1, 5, 6 for substructure 2. By adding the pertinent elements of $\overline{\mathbf{K}}^*_{EE1}$ and $\overline{\mathbf{K}}^*_{EE2}$, we determine the structure stiffness matrix $\mathbf{S}$ to be

$$
\mathbf{S} = \overset{1}{[8{,}400]}\ 1 \ \text{kN} \cdot \text{m/rad}
$$

Structure Fixed-Joint Force Vector $\mathbf{P}_f$*:*

Substructure 1 By substituting $w = 0$, $W = 50$ kN, and $L_1 = L_2 = 5$ m into Eq. (4) of Example 9.4, we obtain the following condensed fixed-joint force vector for substructure 1.

$$
\overline{\mathbf{F}}^*_{fE1} = \begin{bmatrix} 25 \\ 125 \\ 25 \\ -125 \end{bmatrix} \begin{matrix} 2 \\ 3 \\ 4 \\ 1 \end{matrix}
$$

Substructure 2 By substituting $w = 18$ kN/m, $W = 0$, and $L_1 = L_2 = 5$ m into Eq. (4) of Example 9.4, we obtain

$$\overline{\mathbf{F}}^*_{fE2} = \begin{bmatrix} 90 \\ 225 \\ 90 \\ -225 \end{bmatrix} \begin{matrix} 4 \\ 1 \\ 5 \\ 6 \end{matrix}$$

Thus, the fixed-joint force vector for the whole structure is given by

$$\mathbf{P}_f = [100]\ 1\ \text{kN} \cdot \text{m}$$

Joint Displacements: By substituting $\mathbf{P} = \mathbf{0}$ and the numerical values of $\mathbf{S}$ and $\mathbf{P}_f$ into the structure stiffness relation, $\mathbf{P} - \mathbf{P}_f = \mathbf{Sd}$, we write

$$[-100] = [8{,}400]\ [d_1]$$

from which,

$$\mathbf{d} = [d_1] = [-0.011905]\ \text{rad}$$

Substructure Joint Displacements, and Member End Displacements and End Forces:

Substructure 1 The substructure's external joint displacements $\bar{\mathbf{v}}_E$ can be obtained by simply comparing the substructure's external degree of freedom numbers with its code numbers, as follows.

$$\bar{\mathbf{v}}_{E1} = \begin{bmatrix} \bar{v}_1 \\ \bar{v}_2 \\ \bar{v}_3 \\ \bar{v}_4 \end{bmatrix} \begin{matrix} 2 \\ 3 \\ 4 \\ 1 \end{matrix} = \begin{bmatrix} 0 \\ 0 \\ 0 \\ d_1 \end{bmatrix} = \begin{bmatrix} 0 \\ 0 \\ 0 \\ -0.011905 \end{bmatrix}$$

The substructure's internal joint displacements can now be calculated, using the relationship (Eq. (9.25)) $\bar{\mathbf{v}}_I = \overline{\mathbf{K}}_{II}^{-1}(\overline{\mathbf{F}}_I - \overline{\mathbf{F}}_{fI} - \overline{\mathbf{K}}_{IE}\bar{\mathbf{v}}_E)$. Substitution of the numerical values of E, I, L_1, L_2, and W into Eqs. (1) and (3) of Example 9.4 yields

$$\overline{\mathbf{K}}_{II1}^{-1} = [0.0014881]$$

and

$$\overline{\mathbf{F}}_{I1} = [-50]$$

Similarly, by substituting the appropriate numerical values into the expressions of $\overline{\mathbf{K}}_{IE}$ and $\overline{\mathbf{F}}_{fI}$ given in Figs. 9.7(c) and (d), respectively, of Example 9.4, we obtain

$$\overline{\mathbf{K}}_{IE1} = [-336 \quad -1{,}680 \quad -336 \quad 1{,}680]$$

and

$$\overline{\mathbf{F}}_{fI1} = \mathbf{0}$$

By substituting the numerical values of the foregoing submatrices and subvectors into Eq. (9.25), we determine the internal joint displacements for substructure 1 to be

$$\bar{\mathbf{v}}_{I1} = \overline{\mathbf{K}}_{II1}^{-1}(\overline{\mathbf{F}}_{I1} - \overline{\mathbf{F}}_{fI1} - \overline{\mathbf{K}}_{IE1}\bar{\mathbf{v}}_{E1}) = [-0.044643]$$

Thus, the complete joint displacement vector for substructure 1 is

$$\bar{\mathbf{v}}_1 = \begin{bmatrix} \bar{\mathbf{v}}_{E1} \\ \hdashline \bar{\mathbf{v}}_{I1} \end{bmatrix} = \begin{bmatrix} 0 \\ 0 \\ 0 \\ -0.011905\ \text{rad} \\ -0.044643\ \text{m} \end{bmatrix} \begin{matrix} 1 \\ 2 \\ 3 \\ 4 \\ 5 \end{matrix} \qquad \textbf{Ans}$$

With the displacements of all the joints of substructure 1 now known, we can determine the end displacements **u**, and end forces **Q**, for its two members (Fig. 9.9(c)) in the usual manner.

Member 1 ($MT = 2$) From Fig. 9.9(c), we can see that the code numbers for member 1 are 1, 2, 5, 6. Thus,

$$\mathbf{u}_1 = \begin{bmatrix} u_1 \\ u_2 \\ u_3 \\ u_4 \end{bmatrix} \begin{matrix} 1 \\ 2 \\ 5 \\ 6 \end{matrix} = \begin{bmatrix} 0 \\ 0 \\ \bar{v}_5 \\ 0 \end{bmatrix} = \begin{bmatrix} 0 \\ 0 \\ -0.044643 \\ 0 \end{bmatrix}$$

Substituting the numerical values of E and I and $L = 5$ m into Eq. (7.18), we obtain the member stiffness matrix,

$$\mathbf{k}_1 = \begin{bmatrix} 336 & 1{,}680 & -336 & 0 \\ 1{,}680 & 8{,}400 & -1{,}680 & 0 \\ -336 & -1{,}680 & 336 & 0 \\ 0 & 0 & 0 & 0 \end{bmatrix}$$

Substitution of $\mathbf{k}_1$ and $\mathbf{Q}_{f1} = \mathbf{0}$ into the member stiffness relationship, $\mathbf{Q} = \mathbf{ku} + \mathbf{Q}_f$, yields the following end forces for member 1 of substructure 1.

$$\mathbf{Q}_1 = \mathbf{k}_1\mathbf{u}_1 + \mathbf{Q}_{f1} = \begin{bmatrix} 15 \text{ kN} \\ 75 \text{ kN} \cdot \text{m} \\ -15 \text{ kN} \\ 0 \end{bmatrix}$$ **Ans**

Member 2 ($MT = 1$)

$$\mathbf{u}_2 = \begin{bmatrix} -0.044643 \\ 0 \\ 0 \\ -0.011905 \end{bmatrix} \begin{matrix} 5 \\ 6 \\ 3 \\ 4 \end{matrix}$$

Applying Eq. (7.15),

$$\mathbf{k}_2 = \begin{bmatrix} 336 & 0 & -336 & 1{,}680 \\ 0 & 0 & 0 & 0 \\ -336 & 0 & 336 & -1{,}680 \\ 1{,}680 & 0 & -1{,}680 & 8{,}400 \end{bmatrix}$$

$$\mathbf{Q}_{f2} = \mathbf{0}$$

$$\mathbf{Q}_2 = \mathbf{k}_2\mathbf{u}_2 + \mathbf{Q}_{f2} = \begin{bmatrix} -35 \text{ kN} \\ 0 \\ 35 \text{ kN} \\ -175 \text{ kN} \cdot \text{m} \end{bmatrix}$$ **Ans**

Substructure 2

$$\bar{\mathbf{v}}_{E2} = \begin{bmatrix} 0 \\ -0.011905 \\ 0 \\ 0 \end{bmatrix} \begin{matrix} 4 \\ 1 \\ 5 \\ 6 \end{matrix}$$

From Fig. 9.7(d) of Example 9.4, we obtain

$$\overline{\mathbf{F}}_{fI2} = [67.5]$$

The submatrices $\overline{\mathbf{K}}_{II}^{-1}$ and $\overline{\mathbf{K}}_{IE}$ remain the same as for substructure 1, and $\overline{\mathbf{F}}_{I2} = \mathbf{0}$. Thus, the application of Eq. (9.25) yields

$$\bar{\mathbf{v}}_{I2} = [-0.13021]$$

and, therefore,

$$\bar{\mathbf{v}}_2 = \begin{bmatrix} \bar{\mathbf{v}}_{E2} \\ \hline \bar{\mathbf{v}}_{I2} \end{bmatrix} = \begin{bmatrix} 0 \\ -0.011905 \text{ rad} \\ 0 \\ 0 \\ -0.13021 \text{ m} \end{bmatrix} \begin{matrix} 1 \\ 2 \\ 3 \\ 4 \\ 5 \end{matrix} \qquad \textbf{Ans}$$

Member 1 ($MT = 2$)

$$\mathbf{u}_1 = \begin{bmatrix} 0 \\ -0.011905 \\ -0.13021 \\ 0 \end{bmatrix} \begin{matrix} 1 \\ 2 \\ 5 \\ 6 \end{matrix}$$

The **k** matrix for member 1 of substructure 2 is the same as that for the corresponding member of substructure 1. Using Eq. (7.19), we calculate

$$\mathbf{Q}_{f1} = \begin{bmatrix} 56.25 \\ 56.25 \\ 33.75 \\ 0 \end{bmatrix}$$

Thus,

$$\mathbf{Q}_1 = \mathbf{k}_1\mathbf{u}_1 + \mathbf{Q}_{f1} = \begin{bmatrix} 80 \text{ kN} \\ 175 \text{ kN} \cdot \text{m} \\ 10 \text{ kN} \\ 0 \end{bmatrix} \qquad \textbf{Ans}$$

Member 2 ($MT = 1$)

$$\mathbf{u}_2 = \begin{bmatrix} -0.13021 \\ 0 \\ 0 \\ 0 \end{bmatrix} \begin{matrix} 5 \\ 6 \\ 3 \\ 4 \end{matrix}$$

The **k** matrix for this member is the same as that for member 2 of substructure 1. Applying Eq. (7.16),

$$\mathbf{Q}_{f2} = \begin{bmatrix} 33.75 \\ 0 \\ 56.25 \\ -56.25 \end{bmatrix}$$

Thus,

$$\mathbf{Q}_2 = \mathbf{k}_2\mathbf{u}_2 + \mathbf{Q}_{f2} = \begin{bmatrix} -10 \text{ kN} \\ 0 \\ 100 \text{ kN} \\ -275 \text{ kN} \cdot \text{m} \end{bmatrix} \qquad \textbf{Ans}$$

The end forces for the individual members of the structure are shown in Fig. 9.9(d).

Support Reactions: The reaction vector **R** can be assembled either directly from the member end force vectors **Q**, or from the external joint force vectors, $\overline{\mathbf{F}}_E$, of the substructures. To use the latter option, we first apply Eq. (9.26) to calculate $\overline{\mathbf{F}}_E$. Thus, by substituting the previously calculated numerical values of $\overline{\mathbf{K}}^*_{EE}$, $\overline{\mathbf{F}}^*_{fE}$, and $\bar{\mathbf{v}}_E$ into Eq. (9.26), we obtain

$$\overline{\mathbf{F}}_{E1} = \overline{\mathbf{K}}^*_{EE1}\bar{\mathbf{v}}_{E1} + \overline{\mathbf{F}}^*_{fE1} = \begin{bmatrix} 15 \\ 75 \\ 35 \\ -175 \end{bmatrix} \begin{matrix} 2 \\ 3 \\ 4 \\ 1 \end{matrix}$$

and

$$\overline{\mathbf{F}}_{E2} = \overline{\mathbf{K}}^*_{EE2}\bar{\mathbf{v}}_{E2} + \overline{\mathbf{F}}^*_{fE2} = \begin{bmatrix} 80 \\ 175 \\ 100 \\ -275 \end{bmatrix} \begin{matrix} 4 \\ 1 \\ 5 \\ 6 \end{matrix}$$

The foregoing substructure forces are depicted in Fig. 9.9(e). Finally, we calculate the support reaction vector **R** by storing the pertinent elements of $\overline{\mathbf{F}}_{E1}$ and $\overline{\mathbf{F}}_{E2}$ in their proper positions in **R**, using the substructure code numbers. This yields,

$$\mathbf{R} = \begin{bmatrix} 15 \text{ kN} \\ 75 \text{ kN} \cdot \text{m} \\ 115 \text{ kN} \\ 100 \text{ kN} \\ -275 \text{ kN} \cdot \text{m} \end{bmatrix} \begin{matrix} 2 \\ 3 \\ 4 \\ 5 \\ 6 \end{matrix} \qquad \textbf{Ans}$$

The support reactions are shown in Fig. 9.9(f).

9.4 INCLINED ROLLER SUPPORTS

The structures that we have considered thus far in this text have been supported such that the joint displacements prevented by the supports are in the directions of the global coordinate axes oriented in the horizontal and vertical directions. Because an inclined roller support prevents translation of a joint in an inclined direction (normal to the incline), while permitting translation in the perpendicular direction, it exerts a reaction force on the joint in that inclined, nonglobal, direction. Thus, the effect of an inclined roller support cannot be included in analysis by simply eliminating one of the structure's degrees of freedom; that is, by treating one of the structure's coordinates, which are defined in the directions of the global coordinate axes, as a restrained coordinate.

An obvious approach to alleviate this problem would be to orient the global coordinate system so that its axes are parallel and perpendicular to the inclined plane upon which the roller moves. However, this approach generally proves to be quite cumbersome, as it requires that the joint coordinates and loads, which are usually specified in the horizontal and vertical directions, be calculated with respect to the inclined global coordinate system. Furthermore, the foregoing approach cannot be used if the structure is supported by two or more rollers inclined in different (i.e., neither parallel nor perpendicular) directions.

A theoretically exact solution of the problem of inclined rollers usually involves first defining the reaction force and the support displacements with reference to a *local joint coordinate system,* with axes parallel and perpendicular to the incline; and then introducing these restraint conditions in the structure's global stiffness relations via a special transformation matrix [26]. While this approach is exact in the sense that it yields exactly 0 displacement of the support joint perpendicular to the incline, it is generally not considered to be the most convenient because its computer implementation requires a significant amount of programming effort.

Perhaps the most convenient and commonly used technique for modeling an inclined roller support is to replace it with an imaginary axial force member with very large axial stiffness, and oriented in the direction perpendicular to the incline, as shown in Figs. 9.10 and 9.11 (on the next page). As depicted there, one end of the imaginary member is connected to the original support joint by a hinged connection, while the other end is attached to an imaginary hinged support, to ensure that only axial force (i.e., no bending moment) develops in the member when the structure is loaded. In order for the imaginary member to accurately represent the effect of the roller support, its axial stiffness must be made sufficiently large so that its axial deformation is negligibly small. This is usually achieved by specifying a very large value for the cross-sectional area of

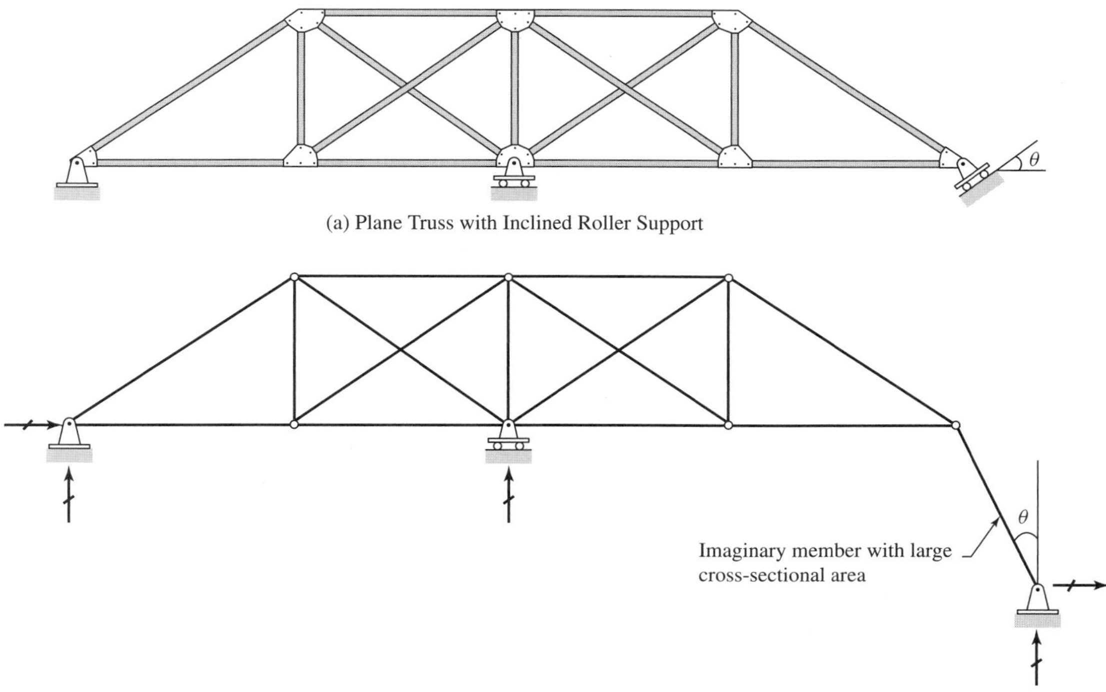

Fig. 9.10

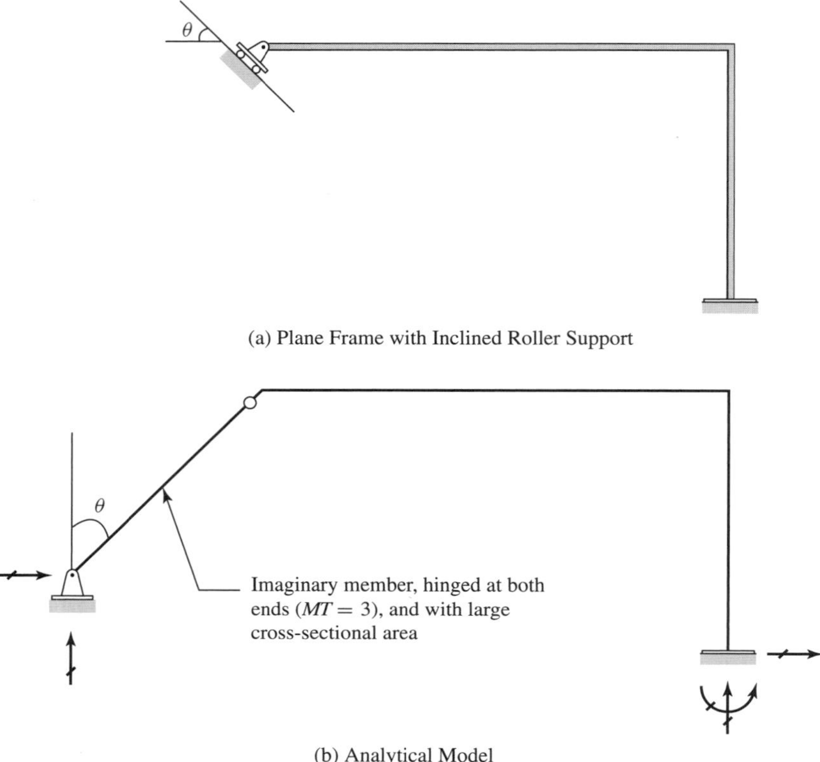

(a) Plane Frame with Inclined Roller Support

(b) Analytical Model

Fig. 9.11

the imaginary member in the analysis, while keeping its length of the same order of magnitude as the other (real) structural members, to ensure that the imaginary member undergoes only small rotations. Provided that the foregoing conditions are satisfied, the axial force in the imaginary member represents the reaction of the actual inclined roller support.

The main advantage of modeling inclined roller supports with imaginary members is that computer programs for standard supports, such as those developed in previous chapters, can be used, without any modifications, to analyze structures supported on inclined rollers. When analyzing trusses, ordinary truss members with large cross-sectional areas can be used to model inclined roller supports (Fig. 9.10). In the case of frames, however, the members used to model inclined rollers, in addition to having large cross-sectional areas, must be of type 3 ($MT = 3$); that is, they must be hinged at both ends, as shown in Fig. 9.11. As noted before, the cross-sectional area of the imaginary member, used to model the inclined roller support, should be sufficiently large so that the member's axial deformations are negligibly small. However, using an extremely large value for the cross-sectional area of the imaginary member can cause some off-diagonal elements of the structure stiffness matrix to become so large, as compared to the other elements, that they introduce numerical errors, or cause numerical instability, during the solution of the structure's stiffness equations.

9.5 OFFSET CONNECTIONS

In formulating the stiffness method of analysis, we have ignored the size of joints or connections, assuming them to be of infinitesimal size. While this assumption proves to be adequate for most framed structures, the dimensions of moment-resisting connections in some structures may be large enough, relative to member lengths, that ignoring their effect in the analysis can lead to erroneous results. In this section, we discuss procedures for including the effect of finite sizes of connections or joints in the analysis.

Consider an arbitrary girder of a typical plane building frame, as shown in Fig. 9.12(a) on the next page. The girder is connected at its ends, to columns and adjacent girders, by means of rigid or moment-resisting connections. As indicated in the figure, the dimensions of connections usually (but not always) equal the cross-sectional depths of the connected members. If the connection dimensions are small, as compared to the member lengths, then their effect is ignored in the analysis. In such a case, it would be assumed for analysis purposes that the girder under consideration extends in length from one column centerline to the next, and is connected at its ends to other members through rigid connections of infinitesimal size, as depicted in Fig. 9.12(b).

However, if the connection dimensions are not small, then their effect must be considered in the analysis. As shown in Fig. 9.12(c), rigid connections of finite size can be conveniently modeled by using rigid *offsets,* with each offset being a rigid body of length equal to the distance between the center of the connection and its edge which is adjacent to the member under consideration. Thus, from Fig. 9.12(c), we can see that the girder under consideration has offset connections of lengths d_b and d_e at its left and right ends, respectively.

Two approaches are commonly used to include the effect of offset connections in analysis. In the first approach, each offset is treated as a small member with very large stiffness. For example, in [13] it is suggested that the cross-sectional properties of an offset member be chosen so that its stiffness is 1,000 times that of the connected member. The main advantage of this approach is that computer programs, such as those developed in previous chapters, can be used without any modification. The disadvantage of this approach is that each offset increases, by one, the number of members and joints to be analyzed. For example, the girder of Fig. 9.12(c) would have to be divided into three members of lengths d_b, L, and d_e, in order to include the effect of offset connections at its two ends in the analysis.

An alternate approach that can be used to handle the effect of offset connections involves modifying the member stiffness relationships to include the effect of offsets at member ends. The main advantage of this approach is that a natural member (e.g., a girder or a column), together with its end offsets, can be treated as a single member for the purpose of analysis. For example, the whole girder of Fig. 9.12(c), including its end offsets, would be treated as a single member when using this approach. However, the disadvantage of this approach is that it requires rewriting of some parts of the computer programs developed in previous chapters.

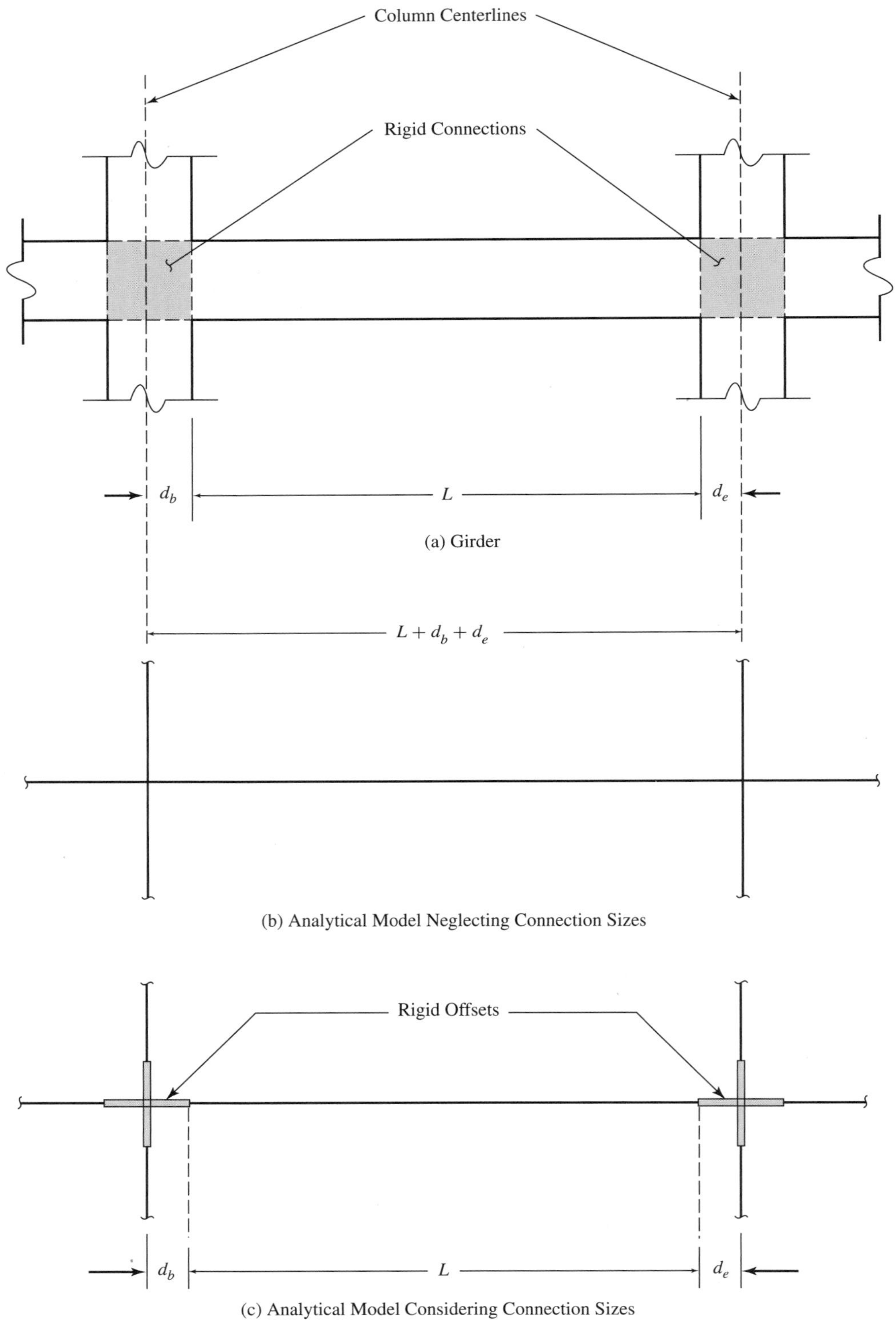

Column Centerlines

Rigid Connections

d_b L d_e

(a) Girder

$L + d_b + d_e$

(b) Analytical Model Neglecting Connection Sizes

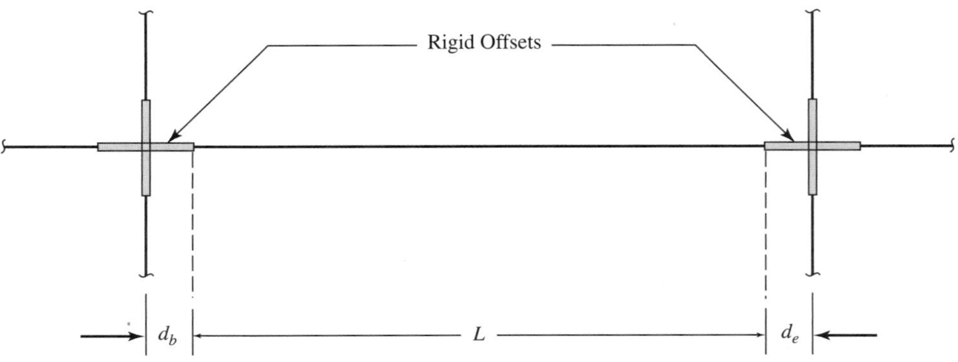

(c) Analytical Model Considering Connection Sizes

Fig. 9.12

In the following, we modify the stiffness relations for the members of plane frames to include the effect of rigid end offsets. Similar procedures can be employed to derive modified stiffness relations for the members of other types of framed structures.

Consider an arbitrary member of length L of a plane frame, and let $\overline{\mathbf{Q}}$ and $\bar{\mathbf{u}}$ denote the local end forces and end displacements, respectively, at the exterior ends of its offsets, as shown in Fig. 9.13(a). Our objective is to express $\overline{\mathbf{Q}}$ in terms of $\bar{\mathbf{u}}$ and any external loading applied to the member between its actual ends b and e. Recall from Chapter 6 that the relationship between the end forces $\mathbf{Q}$ and the end displacement $\mathbf{u}$, which are defined at the ends b and e of the member, is of the form $\mathbf{Q} = \mathbf{ku} + \mathbf{Q}_f$, with $\mathbf{k}$ and $\mathbf{Q}_f$ given by Eqs. (6.6) and (6.15), respectively.

To express $\overline{\mathbf{Q}}$ in terms of $\mathbf{Q}$, we consider the equilibrium of the rigid bodies of the two offsets. This yields (see Fig. 9.13(b))

$$\overline{Q}_1 = Q_1 \qquad \overline{Q}_2 = Q_2 \qquad \overline{Q}_3 = d_b\, Q_2 + Q_3$$
$$\overline{Q}_4 = Q_4 \qquad \overline{Q}_5 = Q_5 \qquad \overline{Q}_6 = -d_e\, Q_5 + Q_6$$

which can be written in matrix form as

$$\overline{\mathbf{Q}} = \overline{\mathbf{T}}\mathbf{Q} \tag{9.29}$$

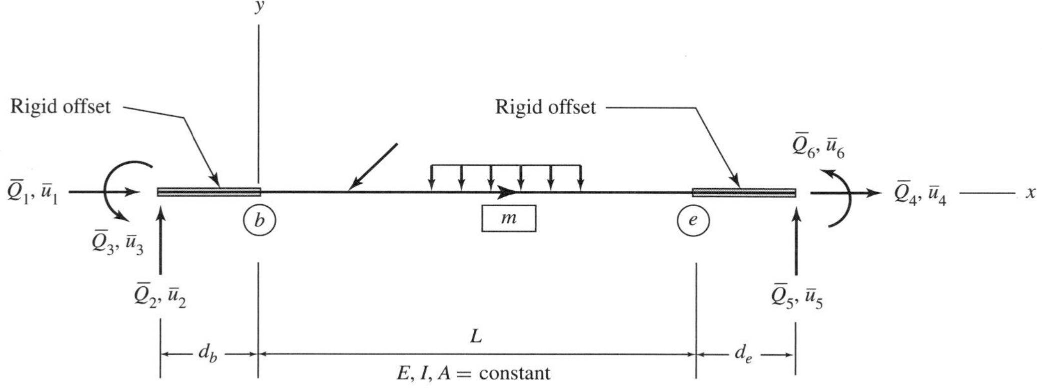

(a) Plane Frame Member with Rigid End Offsets

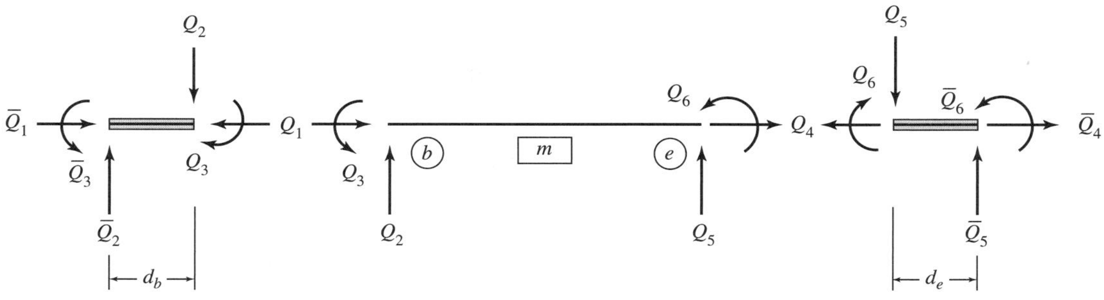

(b) Member Forces

Fig. 9.13

with

$$\overline{\mathbf{T}} = \begin{bmatrix} 1 & 0 & 0 & 0 & 0 & 0 \\ 0 & 1 & 0 & 0 & 0 & 0 \\ 0 & d_b & 1 & 0 & 0 & 0 \\ 0 & 0 & 0 & 1 & 0 & 0 \\ 0 & 0 & 0 & 0 & 1 & 0 \\ 0 & 0 & 0 & 0 & -d_e & 1 \end{bmatrix} \tag{9.30}$$

in which $\overline{\mathbf{T}}$ can be considered to be a transformation matrix which translates the member's end forces from its actual ends b and e, to the exterior ends of its rigid offsets.

From geometrical considerations, it can be shown that the relationship between the end displacements $\mathbf{u}$ and $\bar{\mathbf{u}}$ can be written as

$$\mathbf{u} = \overline{\mathbf{T}}^T \bar{\mathbf{u}} \tag{9.31}$$

By substituting Eq. (9.31) into Eq. (6.4), and substituting the resulting expression into Eq. (9.29), we obtain the desired stiffness relationship:

$$\boxed{\overline{\mathbf{Q}} = \bar{\mathbf{k}}\bar{\mathbf{u}} + \overline{\mathbf{Q}}_f} \tag{9.32}$$

with

$$\boxed{\bar{\mathbf{k}} = \overline{\mathbf{T}}\mathbf{k}\overline{\mathbf{T}}^T} \tag{9.33}$$

$$\boxed{\overline{\mathbf{Q}}_f = \overline{\mathbf{T}}\mathbf{Q}_f} \tag{9.34}$$

in which $\bar{\mathbf{k}}$ and $\overline{\mathbf{Q}}_f$ represent the modified member stiffness matrix and fixed-end force vector, respectively, in the local coordinate system. Note that $\bar{\mathbf{k}}$ and $\overline{\mathbf{Q}}_f$ include the effect of rigid offsets at the ends of the member. The explicit forms of $\bar{\mathbf{k}}$ and $\overline{\mathbf{Q}}_f$, respectively, can be obtained by substituting Eqs. (6.6) and (9.30) into Eq. (9.33), and Eqs. (6.15) and (9.30) into Eq. (9.34). These are given in Eqs. (9.35) and (9.36).

$$\bar{\mathbf{k}} = \frac{EI}{L^3}\begin{bmatrix} \frac{AL^2}{I} & 0 & 0 & -\frac{AL^2}{I} & 0 & 0 \\ 0 & 12 & (6L+12d_b) & 0 & -12 & (6L+12d_e) \\ 0 & (6L+12d_b) & (4L^2+12Ld_b+12d_b^2) & 0 & (-6L-12d_b) & (2L^2+6Ld_b+6Ld_e+12d_bd_e) \\ -\frac{AL^2}{I} & 0 & 0 & \frac{AL^2}{I} & 0 & 0 \\ 0 & -12 & (-6L-12d_b) & 0 & 12 & (-6L-12d_e) \\ 0 & (6L+12d_e) & (2L^2+6Ld_b+6Ld_e+12d_bd_e) & 0 & (-6L-12d_e) & (4L^2+12Ld_e+12d_e^2) \end{bmatrix} \tag{9.35}$$

$$\overline{\mathbf{Q}}_f = \begin{bmatrix} FA_b \\ FS_b \\ d_b FS_b + FM_b \\ FA_e \\ FS_e \\ -d_e FS_e + FM_e \end{bmatrix} \tag{9.36}$$

The procedure for analysis essentially remains the same as developed previously, except that the modified expressions for the stiffness matrices $\bar{\mathbf{k}}$ (Eq. (9.35)) and fixed-end force vectors $\overline{\mathbf{Q}}_f$ (Eq. (9.36)) are used (instead of $\mathbf{k}$ and $\mathbf{Q}_f$, respectively), for members with offset connections.

9.6 SEMIRIGID CONNECTIONS

While rigid and hinged types of connections, as considered thus far in this text, are the most commonly used in structural designs, a third type of connection, termed the *semirigid connection,* is also recognized by some design codes, and can be used for designing such structures as structural steel building frames. Recall that the rotation of a rigidly connected member end equals the rotation of the adjacent joint, whereas the rotation of a hinged end of a member must be such that the moment at the hinged end is 0. A connection is considered to be semirigid if its rotational restraint is less than that of a perfectly rigid connection, but more than that of a frictionless hinged connection. In other words, the moment transmitted by a semirigid connection is greater than 0, but less than that transmitted by a rigid connection. For the purpose of analysis, a semirigid connection can be conveniently modeled by a rotational (torsional) spring with stiffness equal to that of the actual connection. In this section, we derive the stiffness relations for members of beams with semirigid connections at their ends. Such relationships for other types of framed structures can be determined by using a similar procedure.

Consider an arbitrary member of a beam, as shown in Fig. 9.14(a) on the next page. The member is connected to the joints adjacent to its ends b and e, by means of rotational springs of infinitesimal size representing the semirigid connections of stiffnesses k_b and k_e, respectively. As shown in this figure, $\overline{\mathbf{Q}}$ and $\bar{\mathbf{u}}$ represent the local end forces and end displacements, respectively, at the exterior ends of the rotational springs. Our objective is to express $\overline{\mathbf{Q}}$ in terms of $\bar{\mathbf{u}}$ and any external loading applied to the member.

We begin by writing, in explicit form, the previously derived relationship $\mathbf{Q} = \mathbf{ku} + \mathbf{Q}_f$, between the end forces $\mathbf{Q}$ and the end displacements $\mathbf{u}$, which are defined at the actual ends b and e of the member. By using the expressions for $\mathbf{k}$ and $\mathbf{Q}_f$ from Eqs. (5.53) and (5.99), respectively, we write

$$\begin{bmatrix} Q_1 \\ Q_2 \\ Q_3 \\ Q_4 \end{bmatrix} = \frac{EI}{L^3}\begin{bmatrix} 12 & 6L & -12 & 6L \\ 6L & 4L^2 & -6L & 2L^2 \\ -12 & -6L & 12 & -6L \\ 6L & 2L^2 & -6L & 4L^2 \end{bmatrix}\begin{bmatrix} u_1 \\ u_2 \\ u_3 \\ u_4 \end{bmatrix} + \begin{bmatrix} FS_b \\ FM_b \\ FS_e \\ FM_e \end{bmatrix} \tag{9.37}$$

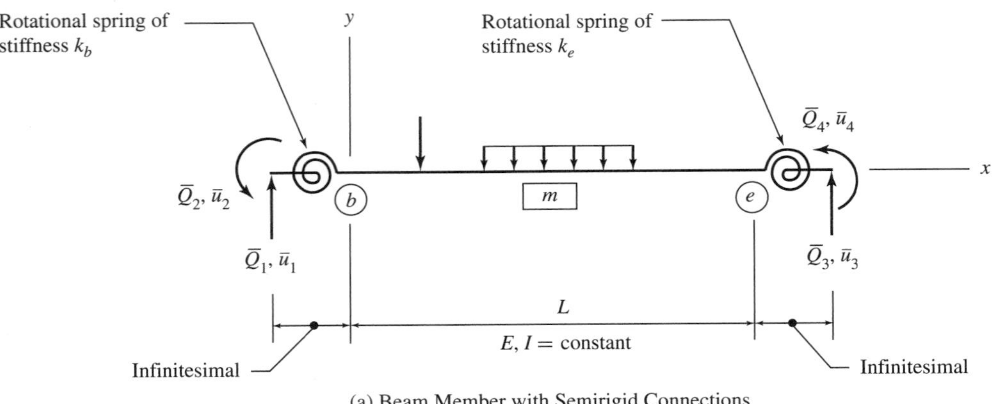

(a) Beam Member with Semirigid Connections

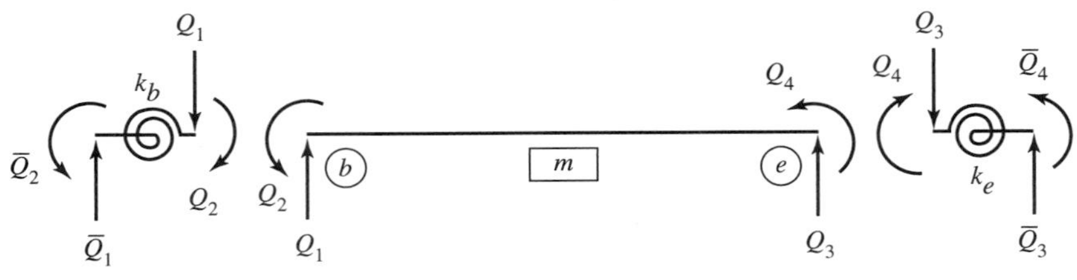

(b) Member Forces

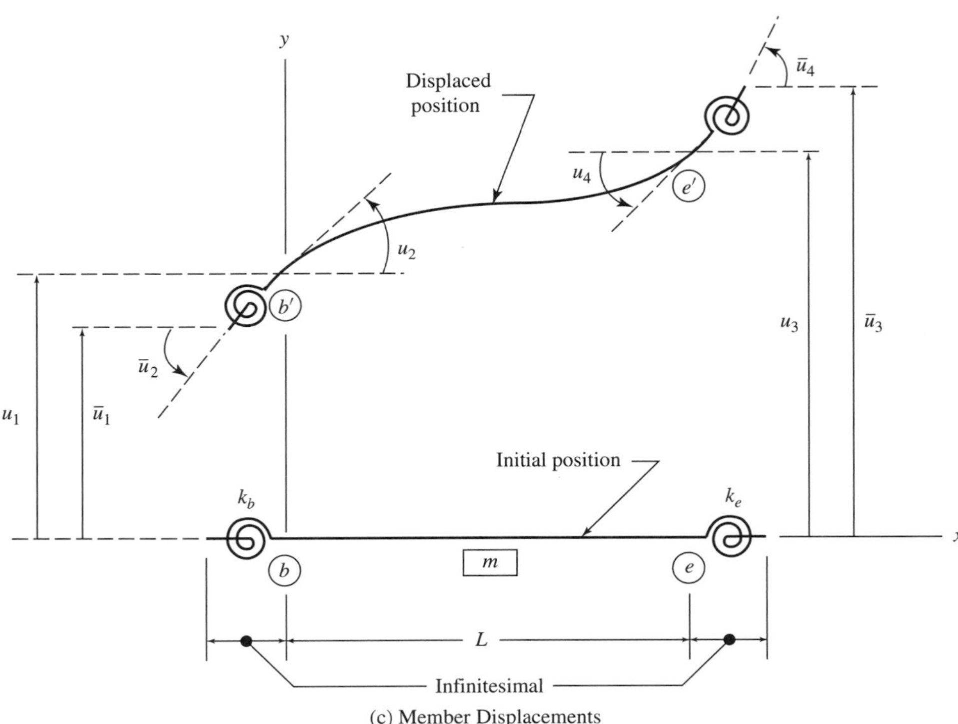

(c) Member Displacements

Fig. 9.14

Figure 9.14(b) shows the forces $\overline{\mathbf{Q}}$ and $\mathbf{Q}$ acting at the exterior and interior ends, respectively, of the member's rotational springs. As the lengths of these springs are infinitesimal, equilibrium equations for the free bodies of the springs yield

$$\mathbf{Q} = \overline{\mathbf{Q}} \tag{9.38}$$

The displacements $\mathbf{u}$ and $\bar{\mathbf{u}}$ are depicted in Fig. 9.14(c) using an exaggerated scale. Because of the infinitesimal size of the springs, the translations of the spring ends are equal; that is,

$$u_1 = \bar{u}_1 \tag{9.39a}$$

$$u_3 = \bar{u}_3 \tag{9.39b}$$

The relationship between the rotations (u_2 and $\bar{u}_2$) of the two ends of the spring, at member end b, can be established by applying the spring stiffness relation:

$$\overline{Q}_2 = k_b\,(\bar{u}_2 - u_2)$$

from which,

$$u_2 = \bar{u}_2 - \frac{\overline{Q}_2}{k_b} \tag{9.39c}$$

Similarly, by using the stiffness relation for the spring attached to member end e, we obtain

$$u_4 = \bar{u}_4 - \frac{\overline{Q}_4}{k_e} \tag{9.39d}$$

To obtain the desired relationship between $\overline{\mathbf{Q}}$ and $\bar{\mathbf{u}}$, we now substitute Eqs. (9.38) and (9.39) into Eq. (9.37) to obtain the following equations.

$$\overline{Q}_1 = \frac{EI}{L^3}\left[12\bar{u}_1 + 6L\left(\bar{u}_2 - \frac{\overline{Q}_2}{k_b}\right) - 12\bar{u}_3 + 6L\left(\bar{u}_4 - \frac{\overline{Q}_4}{k_e}\right)\right] + FS_b \tag{9.40a}$$

$$\overline{Q}_2 = \frac{EI}{L^3}\left[6L\bar{u}_1 + 4L^2\left(\bar{u}_2 - \frac{\overline{Q}_2}{k_b}\right) - 6L\bar{u}_3 + 2L^2\left(\bar{u}_4 - \frac{\overline{Q}_4}{k_e}\right)\right] + FM_b \tag{9.40b}$$

$$\overline{Q}_3 = \frac{EI}{L^3}\left[-12\bar{u}_1 - 6L\left(\bar{u}_2 - \frac{\overline{Q}_2}{k_b}\right) + 12\bar{u}_3 - 6L\left(\bar{u}_4 - \frac{\overline{Q}_4}{k_e}\right)\right] + FS_e \tag{9.40c}$$

$$\overline{Q}_4 = \frac{EI}{L^3}\left[6L\bar{u}_1 + 2L^2\left(\bar{u}_2 - \frac{\overline{Q}_2}{k_b}\right) - 6L\bar{u}_3 + 4L^2\left(\bar{u}_4 - \frac{\overline{Q}_4}{k_e}\right)\right] + FM_e \tag{9.40d}$$

Next, we solve Eqs. (9.40b) and (9.40d) simultaneously, to express $\overline{Q}_2$ and $\overline{Q}_4$ in terms of $\bar{u}_1$ through $\bar{u}_4$. This yields

$$\overline{Q}_2 = \frac{EIr_b}{L^3R}[6L(2-r_e)\bar{u}_1 + 4L^2(3-2r_e)\bar{u}_2 - 6L(2-r_e)\bar{u}_3 + 2L^2r_e\bar{u}_4] + \frac{r_b}{R}[(4-3r_e)FM_b - 2(1-r_e)FM_e] \tag{9.41a}$$

$$\overline{Q}_4 = \frac{EIr_e}{L^3R}[6L(2-r_b)\bar{u}_1 + 2L^2r_b\bar{u}_2 - 6L(2-r_b)\bar{u}_3 + 4L^2(3-2r_b)\bar{u}_4] + \frac{r_e}{R}[(4-3r_b)FM_e - 2(1-r_b)FM_b] \tag{9.41b}$$

in which r_b and r_e denote the dimensionless rigidity parameters defined as

$$\boxed{r_i = \frac{k_iL}{EI + k_iL} \qquad i = b,\ e} \tag{9.42}$$

and

$$\boxed{R = 12 - 8r_b - 8r_e + 5r_br_e} \tag{9.43}$$

Finally, by substituting Eqs. (9.41) into Eqs. (9.40a) and (9.40c), we determine expressions for $\overline{Q}_1$ and $\overline{Q}_3$ in terms of $\bar{u}_1$ through $\bar{u}_4$. Thus,

$$\overline{Q}_1 = \frac{EI}{L^3R}[12(r_b + r_e - r_br_e)\bar{u}_1 + 6Lr_b(2-r_e)\bar{u}_2 - 12(r_b + r_e - r_br_e)\bar{u}_3 + 6Lr_e(2-r_b)\bar{u}_4] + FS_b - \frac{6}{LR}[(1-r_b)(2-r_e)FM_b + (1-r_e)(2-r_b)FM_e] \tag{9.44a}$$

$$\overline{Q}_3 = \frac{EI}{L^3R}[-12(r_b + r_e - r_br_e)\bar{u}_1 - 6Lr_b(2-r_e)\bar{u}_2 + 12(r_b + r_e - r_br_e)\bar{u}_3 - 6Lr_e(2-r_b)\bar{u}_4] + FS_e + \frac{6}{LR}[(1-r_b)(2-r_e)FM_b + (1-r_e)(2-r_b)FM_e] \tag{9.44b}$$

Equations (9.41) and (9.44), which represent the modified stiffness relations for beam members with semirigid connections at both ends, can be expressed in matrix form:

$$\boxed{\overline{\mathbf{Q}} = \bar{\mathbf{k}}\bar{\mathbf{u}} + \overline{\mathbf{Q}}_f} \tag{9.45}$$

with

$$\bar{\mathbf{k}} = \frac{EI}{L^3R}\begin{bmatrix} 12(r_b + r_e - r_b r_e) & 6Lr_b(2 - r_e) & -12(r_b + r_e - r_b r_e) & 6Lr_e(2 - r_b) \\ 6Lr_b(2 - r_e) & 4L^2 r_b(3 - 2r_e) & -6Lr_b(2 - r_e) & 2L^2 r_b r_e \\ -12(r_b + r_e - r_b r_e) & -6Lr_b(2 - r_e) & 12(r_b + r_e - r_b r_e) & -6Lr_e(2 - r_b) \\ 6Lr_e(2 - r_b) & 2L^2 r_b r_e & -6Lr_e(2 - r_b) & 4L^2 r_e(3 - 2r_b) \end{bmatrix} \tag{9.46}$$

and

$$\overline{\mathbf{Q}}_f = \begin{bmatrix} FS_b - \dfrac{6}{LR}[(1 - r_b)(2 - r_e)FM_b + (1 - r_e)(2 - r_b)FM_e] \\ \dfrac{r_b}{R}[(4 - 3r_e)FM_b - 2(1 - r_e)FM_e] \\ FS_e + \dfrac{6}{LR}[(1 - r_b)(2 - r_e)FM_b + (1 - r_e)(2 - r_b)FM_e] \\ \dfrac{r_e}{R}[-2(1 - r_b)FM_b + (4 - 3r_b)FM_e] \end{bmatrix} \tag{9.47}$$

The $\bar{\mathbf{k}}$ matrix in Eq. (9.46) and the $\overline{\mathbf{Q}}_f$ vector in Eq. (9.47) represent the modified stiffness matrix and fixed-end force vector, respectively, for the members of beams with semirigid connections. It should be noted that these expressions for $\bar{\mathbf{k}}$ and $\overline{\mathbf{Q}}_f$ are valid for the values of the spring stiffness k_i ($i = b$ or e) ranging from 0, which represents a hinged connection, to infinity, which represents a rigid connection. From Eq. (9.42), we can see that as k_i varies from 0 to infinity, the value of the corresponding rigidity parameter r_i varies from 0 to 1. Thus, $r_i = 0$ represents a frictionless hinged connection, whereas $r_i = 1$ represents a perfectly rigid connection. The reader is encouraged to verify that when both r_b and r_e are set equal to 1, then $\bar{\mathbf{k}}$ (Eq. (9.46)) and $\overline{\mathbf{Q}}_f$ (Eq. (9.47)) reduce the $\mathbf{k}$ (Eq. (5.53)) and $\mathbf{Q}_f$ (Eq. (5.99)) for a beam member rigidly connected at both ends. Similarly, the expressions of $\mathbf{k}$ and $\mathbf{Q}_f$, derived in Chapter 7 for beam members with three combinations of rigid and hinged connections (i.e., $MT = 1$, 2, and 3), can be obtained from Eqs. (9.46) and (9.47), respectively, by setting r_b and r_e to 0 or 1, as appropriate.

The procedure for analysis of beams with rigid and hinged connections, developed previously, can be applied to beams with semirigid connections—provided that the modified member stiffness matrix $\bar{\mathbf{k}}$ (Eq. (9.46)) and fixed-end force vector $\overline{\mathbf{Q}}_f$ (Eq. (9.47)) are used in the analysis.

9.7 SHEAR DEFORMATIONS

The stiffness relations that have been developed thus far for beams, grids, and frames, do not include the effect of shear deformations of members. Such structures are generally composed of members with relatively large length-to-depth ratios, so that their shear deformations are usually negligibly small as compared to the bending deformations. However, in the case of beams, grids and rigid frames consisting of members with length-to-depth ratios of 10 or less, and/or built-up (fabricated) members, the magnitudes of shear deformations can be considerable; therefore, the effect of shear deformations should be included in the analyses of such structures. In this section, we consider a procedure for including the effect of shear deformations in the member stiffness relations, and present the modified stiffness matrix for the members of beams. This matrix, which contains the effects of both the shear and bending deformations, can be easily extended to obtain the corresponding modified member stiffness matrices for grids, and plane and space frames.

The relationship between the shearing strain at a cross-section of a beam member and the slope of the elastic curve due to shear can be obtained by considering the shear deformation of a differential element of length dx of the member, as shown in Fig. 9.15. From this figure, we can see that

$$\gamma = -\frac{d\bar{u}_{yS}}{dx} \tag{9.48}$$

in which γ denotes the shear strain, and $\bar{u}_{yS}$ represents the deflection, due to shear, of the member's centroidal axis in the y direction. The negative sign in Eq. (9.48) indicates that the positive shear force S causes deflection in the negative y direction, as shown in the figure. Substitutions of Hooke's law for shear, $\gamma = \tau/G$, and the stress-force relation, $\tau = f_S S/A$, into Eq. (9.48), yield the following expression for the slope of the elastic curve due to shear.

$$\boxed{\frac{d\bar{u}_{yS}}{dx} = -\left(\frac{f_S}{GA}\right) S} \tag{9.49}$$

in which f_S represents the *shape factor for shear*. The dimensionless shape factor f_S depends on the shape of the member cross-section, and takes into account

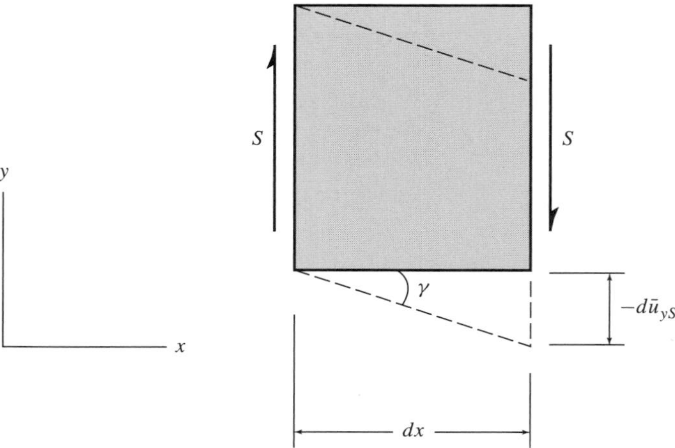

Fig. 9.15

the nonuniform distribution of shear stress on the member cross-section. The values of f_S for some common cross-sectional shapes are as follows.

$f_S = 1.2$ for rectangular cross-sections

$f_S = 10/9$ for circular cross-sections

$f_S = 1$ for wide-flange beams bent about the major axis, provided that the area of the web is used for A in Eq. (9.49)

Integration of Eq. (9.49) yields the expression for deflection due to shear; the total deflection (or slope) of the member due to the combined effect of shear and bending can be determined via superposition of the deflections (or slopes) caused by shear and by bending. As discussed in Chapter 5, the equations for the slope and deflection, due to bending, can be obtained by integrating the moment–curvature relationship:

$$\frac{d^2\bar{u}_{yB}}{dx^2} = \frac{M}{EI} \tag{9.50}$$

in which $\bar{u}_{yB}$ represents the deflection of the member due to bending.

The expressions for the elements of the modified stiffness matrix **k** for a beam member, due to the combined effect of the bending and shear deformations, can be derived using the direct integration approach. To obtain the expressions for the stiffness coefficients k_{i1} ($i = 1$ through 4) in the first column of **k**, we subject a prismatic beam member of length L to a unit value of the end displacement u_1 at end b, as shown in Fig. 9.16. Note that all other member end displacements are 0, and the member is in equilibrium under the action of two end moments k_{21} and k_{41}, and two end shears k_{11} and k_{31}. From the figure, we

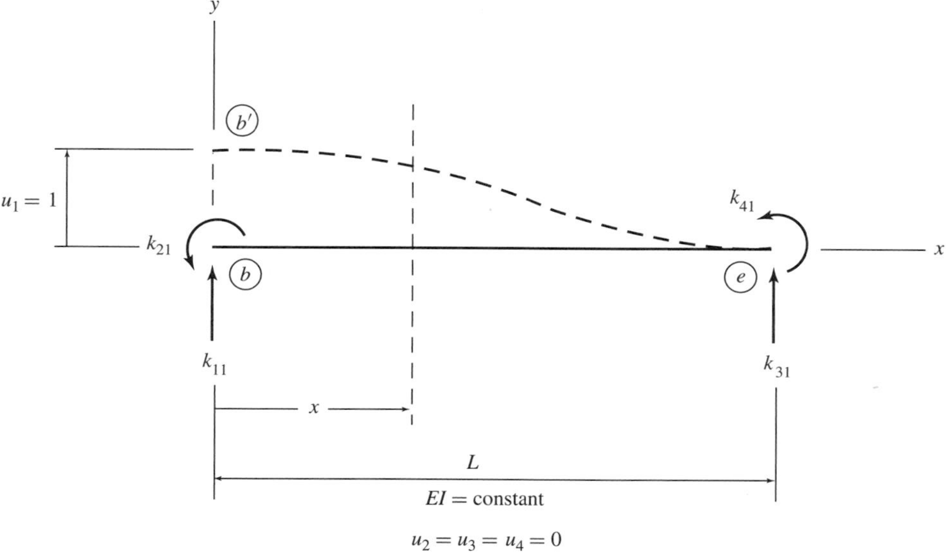

Fig. 9.16

can see that the shear and bending moment at a distance x from end b of the member are:

$$S = k_{11} \tag{9.51}$$

$$M = -k_{21} + k_{11}x \tag{9.52}$$

By substituting Eq. (9.51) into Eq. (9.49), and integrating the resulting equation, we obtain the equation for deflection, due to shear, as

$$\bar{u}_{yS} = -\left(\frac{f_S}{GA}\right) k_{11}x + C_1 \tag{9.53}$$

in which C_1 denotes a constant of integration. By substituting Eq. (9.52) into Eq. (9.50), and integrating the resulting equation twice, we obtain the equations for the slope and deflection of the member, due to bending:

$$\frac{d\bar{u}_{yB}}{dx} = \frac{1}{EI}\left(-k_{21}x + k_{11}\frac{x^2}{2}\right) + C_2 \tag{9.54}$$

$$\bar{u}_{yB} = \frac{1}{EI}\left(-k_{21}\frac{x^2}{2} + k_{11}\frac{x^3}{6}\right) + C_2x + C_3 \tag{9.55}$$

As the shear deformation does not cause any rotation of the member cross-section (see Fig. 9.15), the rotation of the cross-section, θ, results entirely from bending deformation, and is given by (see Eq. (9.54))

$$\theta = \frac{d\bar{u}_{yB}}{dx} = \frac{1}{EI}\left(-k_{21}x + k_{11}\frac{x^2}{2}\right) + C_2 \tag{9.56}$$

By combining Eqs. (9.53) and (9.55), we obtain the equation for the total deflection, $\bar{u}_y$, due to the combined effect of the shear and bending deformations:

$$\bar{u}_y = \bar{u}_{yS} + \bar{u}_{yB} = -\left(\frac{f_S}{GA}\right) k_{11}x + \frac{1}{EI}\left(-k_{21}\frac{x^2}{2} + k_{11}\frac{x^3}{6}\right) + C_2x + C_4 \tag{9.57}$$

in which the constant $C_4 = C_1 + C_3$.

The four unknowns in Eqs. (9.56) and (9.57)—that is, two constants C_2 and C_4 and two stiffness coefficients k_{11} and k_{21}—can now be evaluated by applying the following four boundary conditions:

at end b,	$x = 0$	$\theta = 0$
	$x = 0$	$\bar{u}_y = 1$
at end e,	$x = L$	$\theta = 0$
	$x = L$	$\bar{u}_y = 0$

By applying these boundary conditions, we obtain $C_2 = 0$, $C_4 = 1$, and

$$k_{11} = \frac{12EI}{L^3}\left(\frac{1}{1+\beta_S}\right) \tag{9.58}$$

$$k_{21} = \frac{6EI}{L^2}\left(\frac{1}{1+\beta_S}\right) \tag{9.59}$$

with

$$\beta_S = \frac{12EIf_S}{GAL^2} \tag{9.60}$$

The dimensionless parameter β_S is called the *shear deformation constant.*

The two remaining stiffness coefficients, k_{31} and k_{41}, can now be determined by applying the equations of equilibrium to the free body of the member (Fig. 9.16). Thus,

$$k_{31} = -\frac{12EI}{L^3}\left(\frac{1}{1+\beta_S}\right) \tag{9.61}$$

$$k_{41} = \frac{6EI}{L^2}\left(\frac{1}{1+\beta_S}\right) \tag{9.62}$$

The expressions for elements in the remaining three columns of the **k** matrix can be derived in a similar manner, and the complete modified stiffness matrix for rigidly-connected members of beams, thus obtained, is

$$\mathbf{k} = \frac{EI}{L^3(1+\beta_S)}\begin{bmatrix} 12 & 6L & -12 & 6L \\ 6L & L^2(4+\beta_S) & -6L & L^2(2-\beta_S) \\ -12 & -6L & 12 & -6L \\ 6L & L^2(2-\beta_S) & -6L & L^2(4+\beta_S) \end{bmatrix} \tag{9.63}$$

From Eq. (9.63), we can see that when the shear deformation constant β_S is set equal to 0, then **k** of Eq. (9.63) is reduced to that of Eq. (5.53). It should be realized that the expressions for fixed-end forces due to member loads, given inside the front cover, do not include the effects of shear deformations. If modified fixed-end force expressions including shear deformations are desired, they can be derived using the procedure described in this section.

9.8 NONPRISMATIC MEMBERS

Thus far in this text, we have considered the analysis of structures composed of prismatic members. A member is considered to be prismatic if its axial and flexural rigidities (*EA* and *EI*), or its cross-sectional properties, are constant along its length. In some structures, for aesthetic reasons and/or to save material, it may become necessary to design members with variable cross sections. In this section, we consider the analysis of structures composed of such nonprismatic members.

Perhaps the simplest (albeit approximate) way to handle a nonprismatic natural member, such as a girder or a column, is to subdivide it into a sufficient number of segments, and model each segment by a prismatic member (or element) with cross-sectional properties equal to the average of the cross-sectional properties at the two ends of the segment (Fig. 9.17 on the next page). The main advantage of this approach is that computer programs such as those

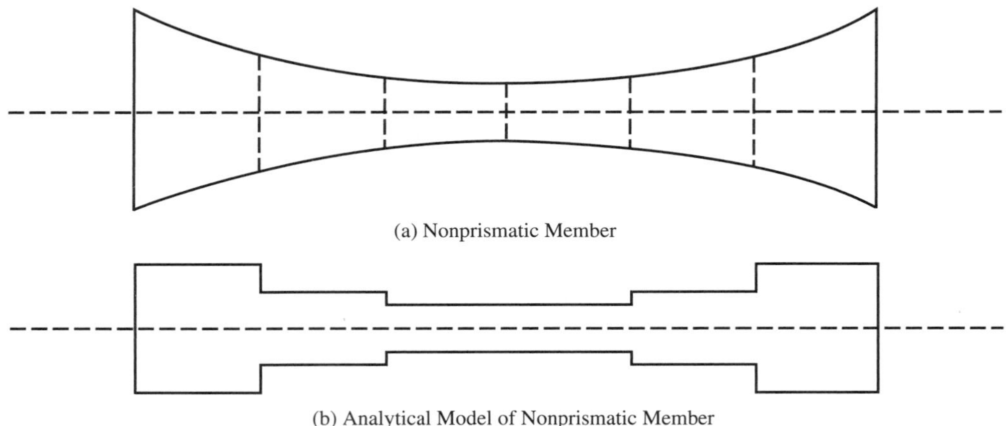

(a) Nonprismatic Member

(b) Analytical Model of Nonprismatic Member

Fig. 9.17

developed in previous chapters can be used without any modifications. The main disadvantage of this approach is that the accuracy of the analytical results depends on the number of prismatic members (or elements) used to model each nonprismatic member, and an inordinate number of prismatic members may be required to achieve an acceptable level of accuracy.

An alternate approach that can be used to handle nonprismatic members involves formulation of the nonprismatic member's stiffness relations while taking into account the exact variation of the member's cross-sectional properties. The main advantage of this exact approach is that a natural nonprismatic member (e.g., a girder or a column) can be treated as a single member for the purpose of analysis. However, as will become apparent later in this section, the exact expressions for the stiffness coefficients for nonprismatic members can be quite complicated [43]. In the following, we illustrate the exact approach via derivation of the local stiffness matrix **k** for a tapered plane truss member [26].

Consider a tapered member of a plane truss, as shown in Fig. 9.18(a). The cross-sectional area of the member varies linearly along its length in accordance with the relationship

$$A_x = A_b\left(1 - \frac{r_A x}{L}\right) \tag{9.64}$$

in which A_b and A_x denote, respectively, the member's cross-sectional areas at its end b, and at a distance x from end b; and r_A represents the *area ratio* given by

$$r_A = \frac{A_b - A_e}{A_b} \tag{9.65}$$

with A_e denoting the member's cross-sectional area at end e, as shown in the figure.

To derive the first column of the tapered member's local stiffness matrix **k**, we subject the member to a unit end displacement $u_1 = 1$ (with $u_2 = u_3 = u_4 = 0$), as shown in Fig. 9.18(b). The expressions for the member axial forces required

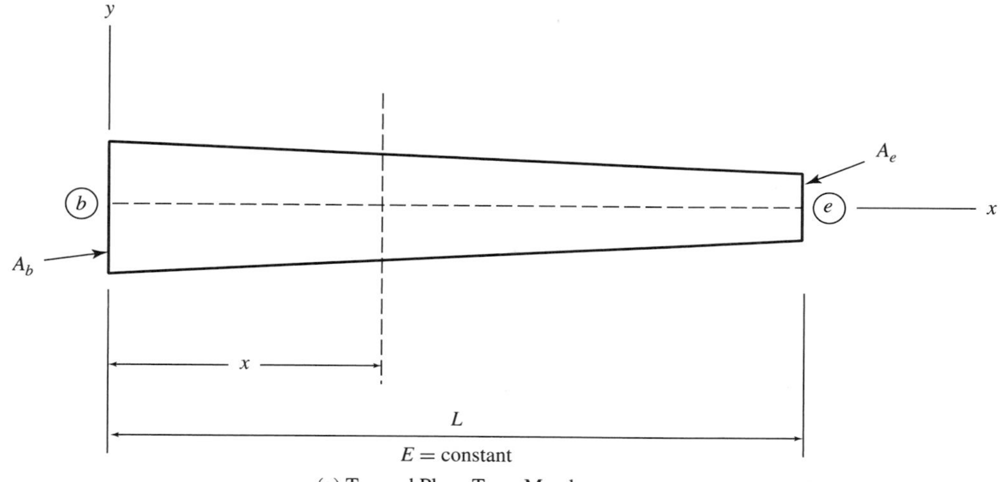

(a) Tapered Plane Truss Member

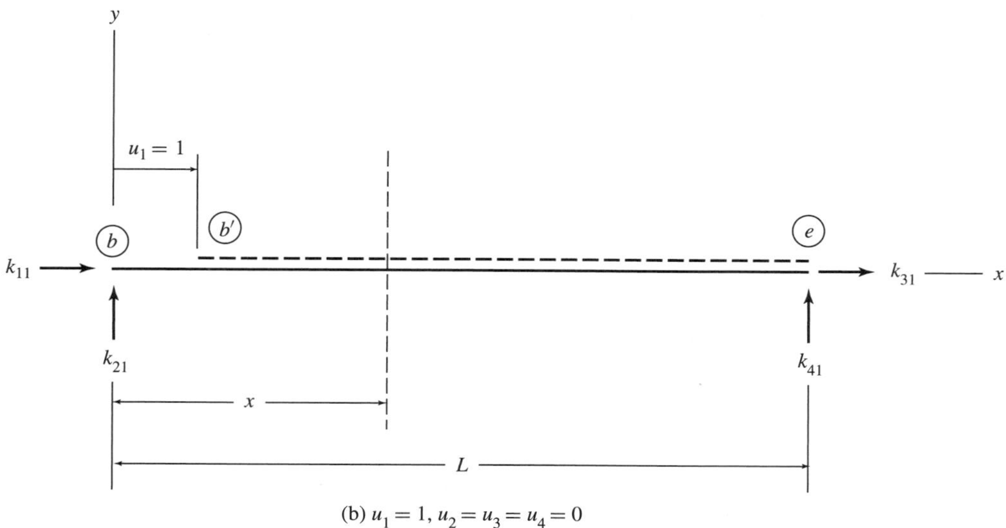

(b) $u_1 = 1$, $u_2 = u_3 = u_4 = 0$

Fig. 9.18

to cause this unit axial deformation can be determined by integrating the differential equation for member axial deformation that, for members with variable cross-sections, can be written as (see Eq. (6.7), Section 6.2)

$$\frac{d\bar{u}_x}{dx} = \frac{Q_a}{EA_x} \tag{9.66}$$

From Fig. 9.18(b), we can see that the axial force acting on the member cross-section at a distance x from its end b is

$$Q_a = -k_{11} \tag{9.67}$$

in which the negative sign indicates that k_{11} causes compression at the member cross-section. Substituting Eqs. (9.64) and (9.67) into Eq. (9.66), and integrating the resulting equation, we obtain

$$\bar{u}_x = \frac{k_{11}L}{EA_b r_A}\ln\left(1 - \frac{r_A x}{L}\right) + C \tag{9.68}$$

in which C is a constant of integration.

The two unknowns, C and k_{11}, in Eq. (9.68) can be evaluated by applying the boundary conditions:

$$\begin{array}{lll} \text{at end } b, & x = 0 & \bar{u}_x = 1 \\ \text{at end } e, & x = L & \bar{u}_x = 0 \end{array}$$

Application of the foregoing boundary conditions yields $C = 1$, and

$$k_{11} = -\frac{EA_b r_A}{L\ln(1 - r_A)} \tag{9.69}$$

The three remaining stiffness coefficients can now be determined by applying the equations of equilibrium to the free body of the member (Fig. 9.18(b)). Thus,

$$k_{31} = \frac{EA_b r_A}{L\ln(1 - r_A)}, \qquad k_{21} = k_{41} = 0 \tag{9.70}$$

The expressions for elements in the third column of the tapered member's local stiffness matrix **k** can be derived in a similar manner; and, as discussed in Section 3.3, all elements of the second and fourth columns of **k** are 0. The complete local stiffness matrix **k** for a tapered plane truss member, thus obtained, is

$$\mathbf{k} = \frac{EA_b r_A}{L\ln(1 - r_A)}\begin{bmatrix} -1 & 0 & 1 & 0 \\ 0 & 0 & 0 & 0 \\ 1 & 0 & -1 & 0 \\ 0 & 0 & 0 & 0 \end{bmatrix} \tag{9.71}$$

EXAMPLE 9.6 Using the direct integration approach, derive the expressions for the slope and deflection at the free end of the tapered cantilever beam shown in Fig. 9.19(a). The beam has a rectangular cross-section of constant width b, but its depth varies linearly from h_1 at the fixed end to h_2 at the free end.

SOLUTION The depth and moment of inertia of the beam at a distance $x (0 \leq x \leq L)$ from its free end can be expressed as

$$h_x = h_1\left(1 - \frac{r_h x}{L}\right)$$

$$I_x = I_1\left(1 - \frac{r_h x}{L}\right)^3$$

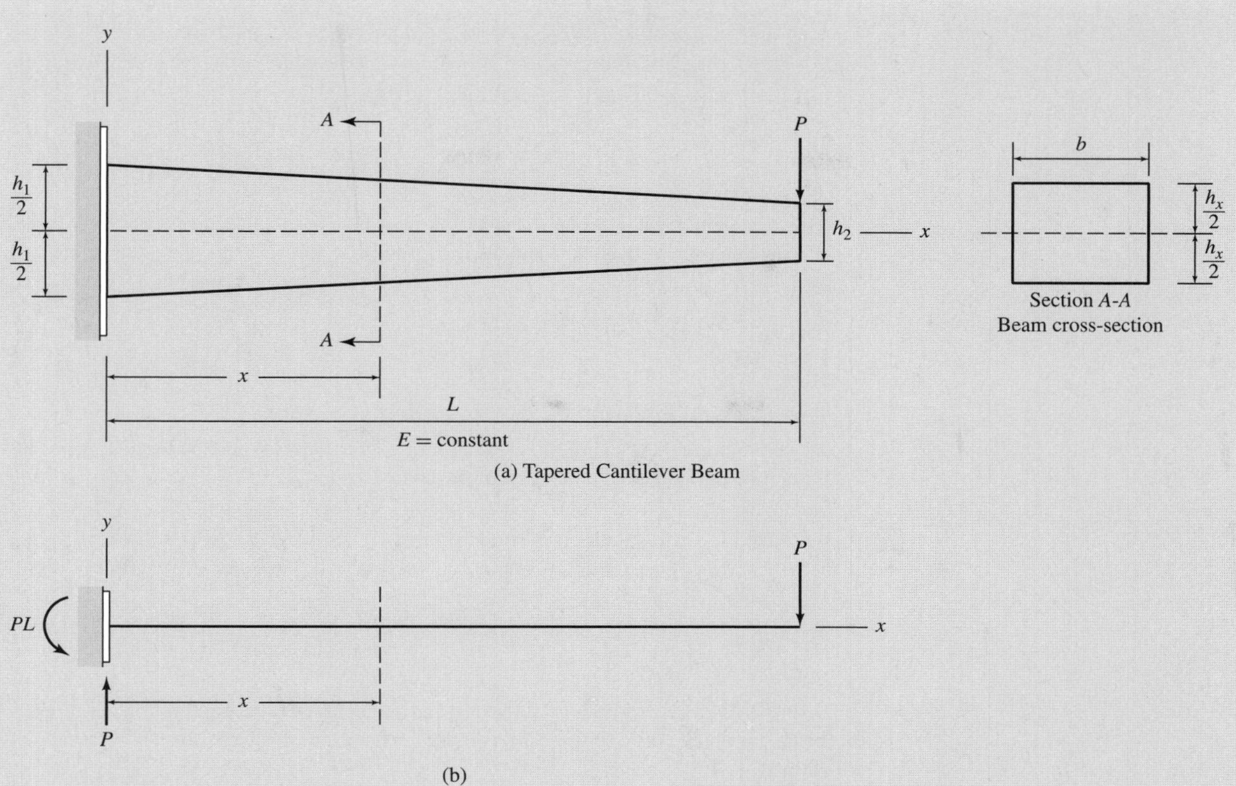

Fig. 9.19

in which r_h represents the *depth ratio* given by

$$r_h = \frac{h_1 - h_2}{h_1} \tag{1}$$

and $I_1 = bh_1^3/12$ = beam's moment of inertia at its fixed end.

The equations for the slope and deflection can be derived by integrating the differential equation for bending of beams with variable cross-sections, which can be written as (see Eq. (5.5), Section 5.2)

$$\frac{d^2\bar{u}_y}{dx^2} = \frac{M}{EI_x} \tag{2}$$

From Fig. 9.19(b), we can see that the bending moment at the beam section at a distance x from its fixed end is

$$M = -P(L - x) \tag{3}$$

in which the negative sign indicates that the bending moment is negative in accordance with the *beam sign convention* (Fig. 5.4). Substituting Eq. (3) into Eq. (2) and integrating, we obtain the equation for slope as

$$\theta = -\frac{PL^3}{2EI_1}\left[\frac{L + r_h L - 2r_h x}{r_h^2(L - r_h x)^2}\right] + C_1 \tag{4}$$

Integrating once more, we obtain the equation for deflection as

$$\bar{u}_y = \frac{PL^3}{2EI_1 r_h^3}\left[\frac{1-r_h}{\left(1-\frac{r_h x}{L}\right)} + 2\ln\left(1-\frac{r_h x}{L}\right)\right] + C_1 x + C_2 \tag{5}$$

The constants of integration, C_1 and C_2, are evaluated by applying the boundary conditions that at $x = 0$, $\theta = 0$ and $\bar{u}_y = 0$. Thus,

$$C_1 = \frac{PL^2}{2EI_1}\left(\frac{1+r_h}{r_h^2}\right)$$

$$C_2 = -\frac{PL^3(1-r_h)}{2EI_1 r_h^3}$$

By substituting these expressions for C_1 and C_2 into Eqs. (4) and (5) we determine the equations for slope and deflection of the beam as

$$\theta = \frac{Px}{2EI_1}\left[\frac{x - 2L + r_h x}{\left(1-\frac{r_h x}{L}\right)^2}\right] \tag{6}$$

$$\bar{u}_y = \frac{PL^3}{2EI_1 r_h^3\left(1-\frac{r_h x}{L}\right)}\left[\frac{2r_h x}{L} - \frac{r_h^2 x^2}{L^2}(1+r_h) + 2\left(1-\frac{r_h x}{L}\right)\ln\left(1-\frac{r_h x}{L}\right)\right] \tag{7}$$

Finally, the expressions for slope and deflection at the free end of the tapered beam are obtained by setting $x = L$ in Eqs. (6) and (7), respectively. Thus,
Slope (+ ↻):

$$\theta_L = -\frac{PL^2}{2EI_1(1-r_h)} = -\frac{PL^2 h_1}{2EI_1 h_2} \qquad \textbf{Ans}$$

Deflection (+ ↑):

$$\bar{u}_{yL} = -\frac{PL^3}{2EI_1 r_h^3(1-r_h)}[-2r_h + r_h^2(1+r_h) - 2(1-r_h)\ln(1-r_h)] \qquad \textbf{Ans}$$

EXAMPLE 9.7

Using a structural analysis computer program, determine the slope and deflection at the free end of the tapered cantilever beam shown in Fig. 9.20(a). The beam is of rectangular cross-section of width 150 mm, and its depth varies linearly from 400 mm at the fixed end to 100 mm at the free end, as shown in the figure. For analysis, divide the nonprismatic beam into smaller segments, and model each segment by a prismatic member (element) with a constant moment of inertia based on the average depth of the segment. Analyze several models of the beam with increasing number of members (elements) until the values of the desired displacements converge. Compare these numerical results with the exact analytical solutions for the tapered beam obtained from the expressions derived in Example 9.6.

SOLUTION Seven analytical models of the beam consisting of 1, 2, 3, 4, 6, 9, and 12 segments were analyzed using the computer program provided with this book. In these models, each tapered segment was approximated by a member of constant depth equal to the average depth of the segment. Figure 9.20(b) shows such a three-member model of the beam.

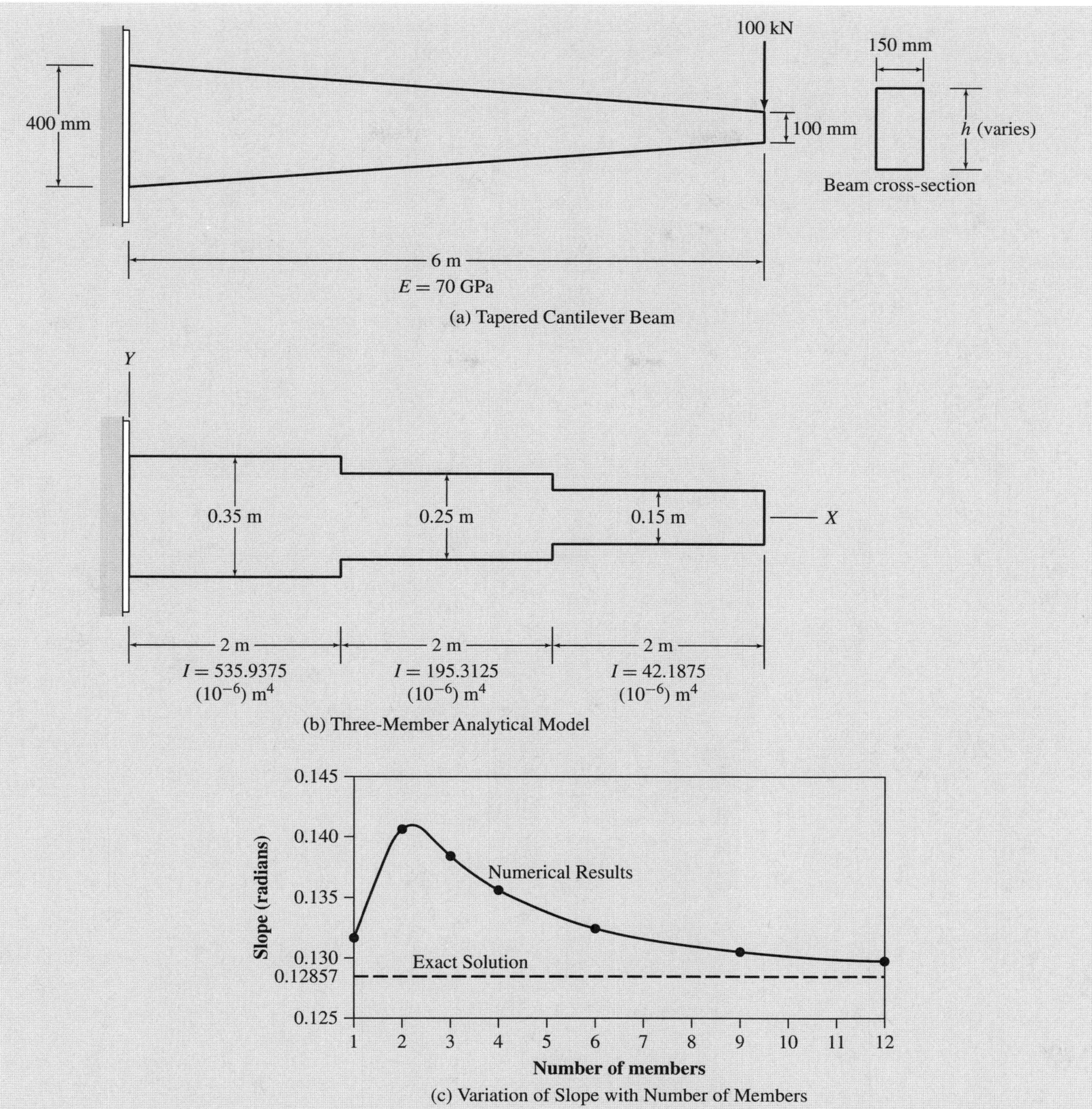

Fig. 9.20

Numerical values of displacements at the free end of the beam obtained by using these analytical models are listed in Table 9.1, and plotted versus the number of members in Figs. 9.20(c) and (d).

The exact analytical solutions can be evaluated by substituting $P = 100$ kN, $L = 6$ m, $E = 70(10^6)$ kN/m^2, $h_1 = 0.4$ m, $h_2 = 0.1$ m, $I_1 = 0.15(0.4)^3/12 = 0.0008$ m^4, and $r_h = (0.4 - 0.1)/0.4 = 0.75$ into the expressions for slope and deflection at the cantilever's free end derived in Example 9.6. Thus,

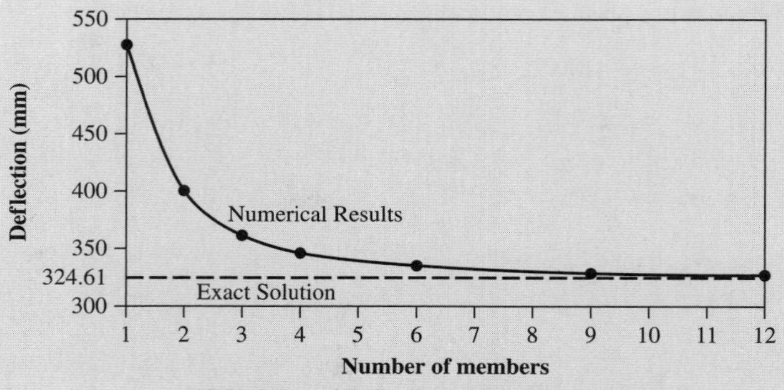

(d) Variation of Deflection with Number of Members

Fig. 9.20 *(continued)*

Table 9.1

Number of Member in Analytical Model	Slope		Deflection	
	(radians)	Error (%)	(mm)	Error (%)
1	0.13166	2.4034	526.63	62.235
2	0.14090	9.5901	401.66	23.736
3	0.13827	7.5445	361.89	11.485
4	0.13560	5.4678	345.91	6.5617
6	0.13246	3.0256	334.01	2.8958
9	0.13051	1.5089	328.73	1.2692
12	0.12973	0.90223	326.96	0.72395
Exact Solutions	0.12857		324.61	

Slope:

$$\theta_L = -\frac{PL^2 h_1}{2EI_1 h_2} = -0.12857\,\text{rad} = 0.12857\,\text{rad} \circlearrowright \qquad \textbf{Ans}$$

Deflection:

$$\bar{u}_{yL} = -\frac{PL^3}{2EI_1 r_h^3(1-r_h)}\left[-2r_h + r_h^2(1+r_h) - 2(1-r_h)\ln(1-r_h)\right]$$

$$= -0.32461\text{ m} = 324.61\text{ mm}\downarrow \qquad \textbf{Ans}$$

The exact values of slope and deflection are also given in Table 9.1, along with the percentage errors of the numerical results with respect to the exact solutions. We can see from this table and Figs. 9.20(c) and (d) that as the number of members in the computer model is increased, the numerical results tend to converge toward the exact solutions. **Ans**

9.9 SOLUTION OF LARGE SYSTEMS OF STIFFNESS EQUATIONS

In the computer programs for matrix stiffness analysis developed in previous chapters, we have stored the entire structure stiffness matrix **S** in computer memory, and have used Gauss–Jordan elimination to solve the structure stiffness equations, $\mathbf{Sd} = \overline{\mathbf{P}}$. While this approach provides a clear insight into the basic concept of the solution process and is easy to program, it is not efficient in the sense that it does not take advantage of the symmetry and other special features of the stiffness matrix **S**. In the case of large structures, a significant portion of the total memory and execution time required for analysis may be devoted to the storage and solution, respectively, of the structure stiffness equations. Accordingly, considerable research effort has been directed toward developing techniques and algorithms for efficiently generating, storing, and solving stiffness equations that arise in the analysis of large structures [2, 14, 26]. In this section, we discuss a commonly used procedure that takes advantage of the special features of the structure stiffness matrix to efficiently store and solve structural stiffness equations.

Half-Bandwidth of Structure Stiffness Matrices

The stiffness matrices **S** of large structures, in addition to being symmetric, are usually *sparse,* in the sense that they contain many 0 elements. Consider, for example, the analytical model of the six-degree-of-freedom continuous beam shown in Fig. 9.21(a) on the next page. The stiffness matrix **S** for this structure is also shown in the figure, in which all the nonzero elements are marked by ×s, and all the 0 elements are left blank. From this figure, we can see that, out of a total of 36 elements of **S**, 20 elements are 0s. Furthermore, this figure indicates that all the nonzero elements of **S** are located within a band centered on the main diagonal. Such a matrix, whose elements are all 0s, with the exception of those located within a band centered on the main diagonal, is referred to as a *banded matrix.* In general, a structure stiffness matrix is considered to be banded if

$$S_{ij} = 0 \quad \text{if} \quad |i - j| > NHB \tag{9.72}$$

where NHB is called the *half-bandwidth* of **S**, which is defined as *the number of elements in each row (or column) of the matrix, that are located within the band to the right of (or below) the diagonal element.* Thus, the half-bandwidth of the stiffness matrix of the continuous-beam analytical model of Fig. 9.21(a) is 1 (i.e., $NHB = 1$), as shown in the figure.

Although the total number of nonzero elements of a structure stiffness matrix remains the same, their locations depend on the order in which the structure's joints are numbered. Thus, the half-bandwidth of a structure stiffness matrix can be altered by renumbering the structure's joints. For example, if the numbers of two inner joints of the continuous beam of Fig. 9.21(a) are interchanged, the half-bandwidth of its stiffness matrix is increased to 2 (i.e., $NHB = 2$), as shown in Fig. 9.21(b). Note that the band of this **S** matrix contains both zero and nonzero elements.

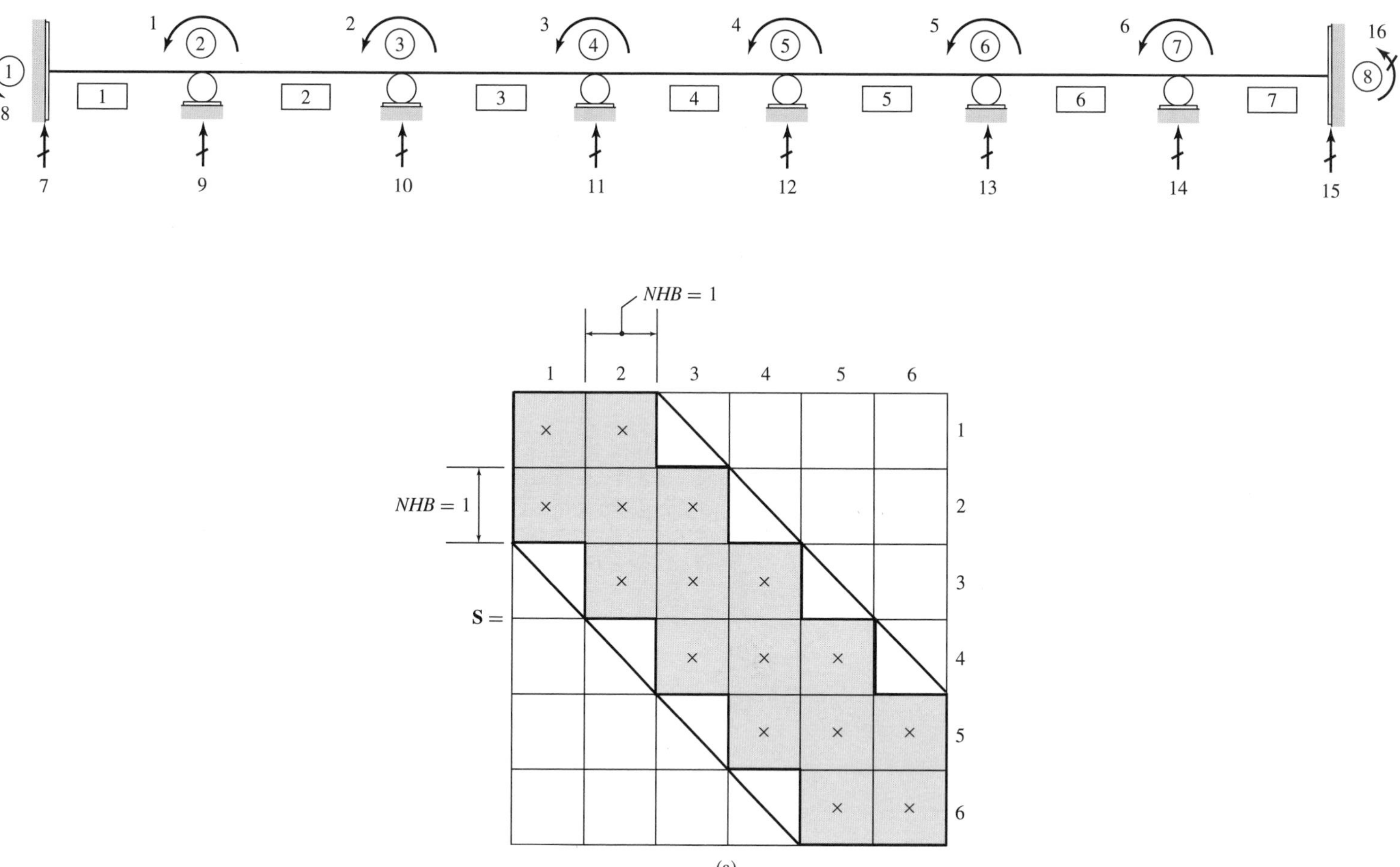

Fig. 9.21

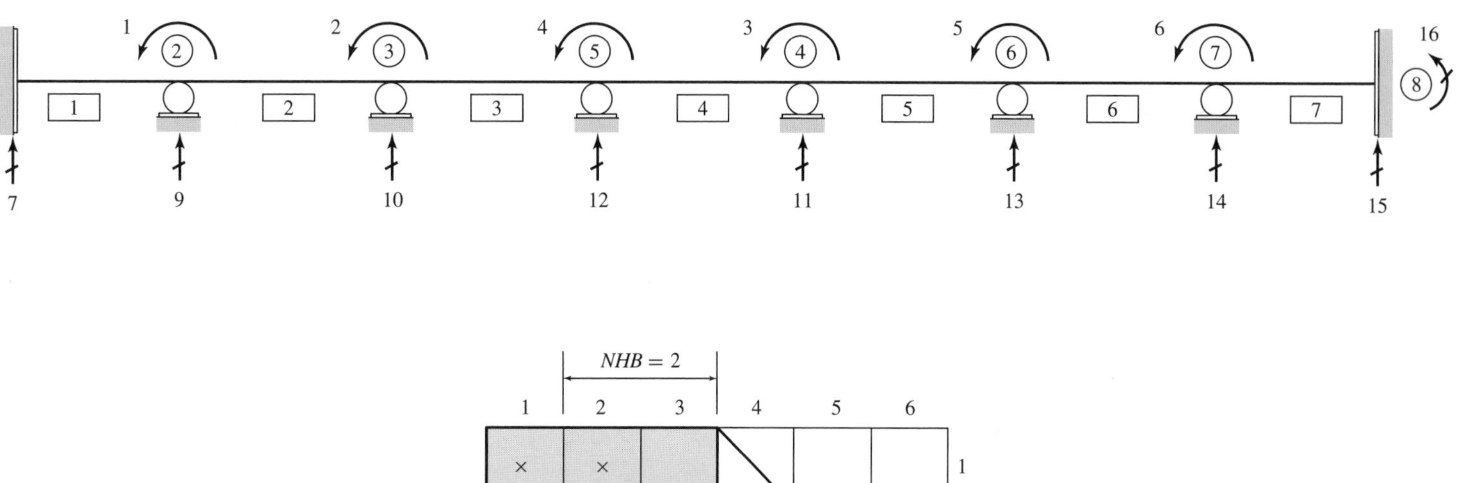

$NHB = 2$

$\mathbf{S} =$

(b)

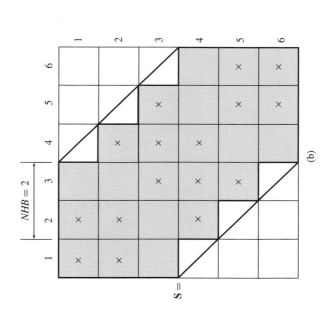

Fig. 9.21 (*continued*)

Since the structure stiffness matrices are assembled by storing the pertinent elements of the member stiffness matrices in their proper positions using member code numbers, the half-bandwidth of a structure stiffness matrix equals the maximum of the differences between the largest and smallest degree-of-freedom code numbers for the individual members of the structure; that is,

$$NHB = \max\{MCL_i - MCS_i\} \qquad i = 1, \ldots, NM \tag{9.73}$$

in which MCL_i and MCS_i denote, respectively, the largest and the smallest code numbers for member i, which correspond to the degrees of freedom (not the restrained coordinates) of the structure.

Considering again the analytical model of the continuous beam of Fig. 9.21(a), we can see that the code numbers for member 1 are 7, 8, 9, 1; of these, the first three numbers correspond to the restrained coordinates, and the fourth number represents a degree of freedom. Thus, the difference between the largest and smallest degree-of-freedom numbers for this member is $MCL_1 - MCS_1 = 1 - 1 = 0$. Similarly, we can see from the figure that for members 2 through 6, this difference is 1, and for member 7, it is 0. Thus, the half-bandwidth for the **S** matrix equals one. Note that when the numbers of two inner joints of the beam are interchanged as shown in Fig. 9.21(b), the difference in degree of freedom code numbers for members 3 and 5 increases by one, and as a result, the half-bandwidth of the **S** matrix widens by one element.

An important property of banded structure stiffness matrices is that the 0 elements outside the band remain 0 during the solution of the structure stiffness equations ($\mathbf{Sd} = \overline{\mathbf{P}}$); therefore, they need not be stored in computer memory for analysis. Furthermore, since the structure stiffness matrices are symmetric, only the diagonal elements, and the elements in the half band above (or below) the diagonal, need to be stored. As the stiffness matrices of large structures usually contain relatively few nonzero elements, significant savings in computer memory storage and execution time can be achieved, in the analysis of such structures, by numbering the joints to minimize the half-bandwidth of the stiffness matrix, and by storing and processing only the elements on the main diagonal, and within a half-bandwidth, of the stiffness matrix.

As discussed previously, the minimum possible half-bandwidth of a stiffness matrix can be obtained by numbering the joints of the structure in such an order that the largest difference between the joint numbers at the ends of any single member is as small as possible. For the configurations of the framed structures commonly encountered in practice, a relatively small (if not minimal) half-bandwidth of the stiffness matrix can usually be achieved by numbering joints consecutively across the dimension of the structure that has the least number of joints, as shown in Figs. 9.22 and 9.23 (on page 558).

The elements on the diagonal and in the upper half-bandwidth of **S** can be stored compactly in computer memory in a rectangular array $\hat{\mathbf{S}}$ of order $NDOF \times (NHB + 1)$, as illustrated in Fig. 9.24 on page 559 for a structure with $NDOF = 9$ and $NHB = 3$. As depicted in this figure, the elements in each row

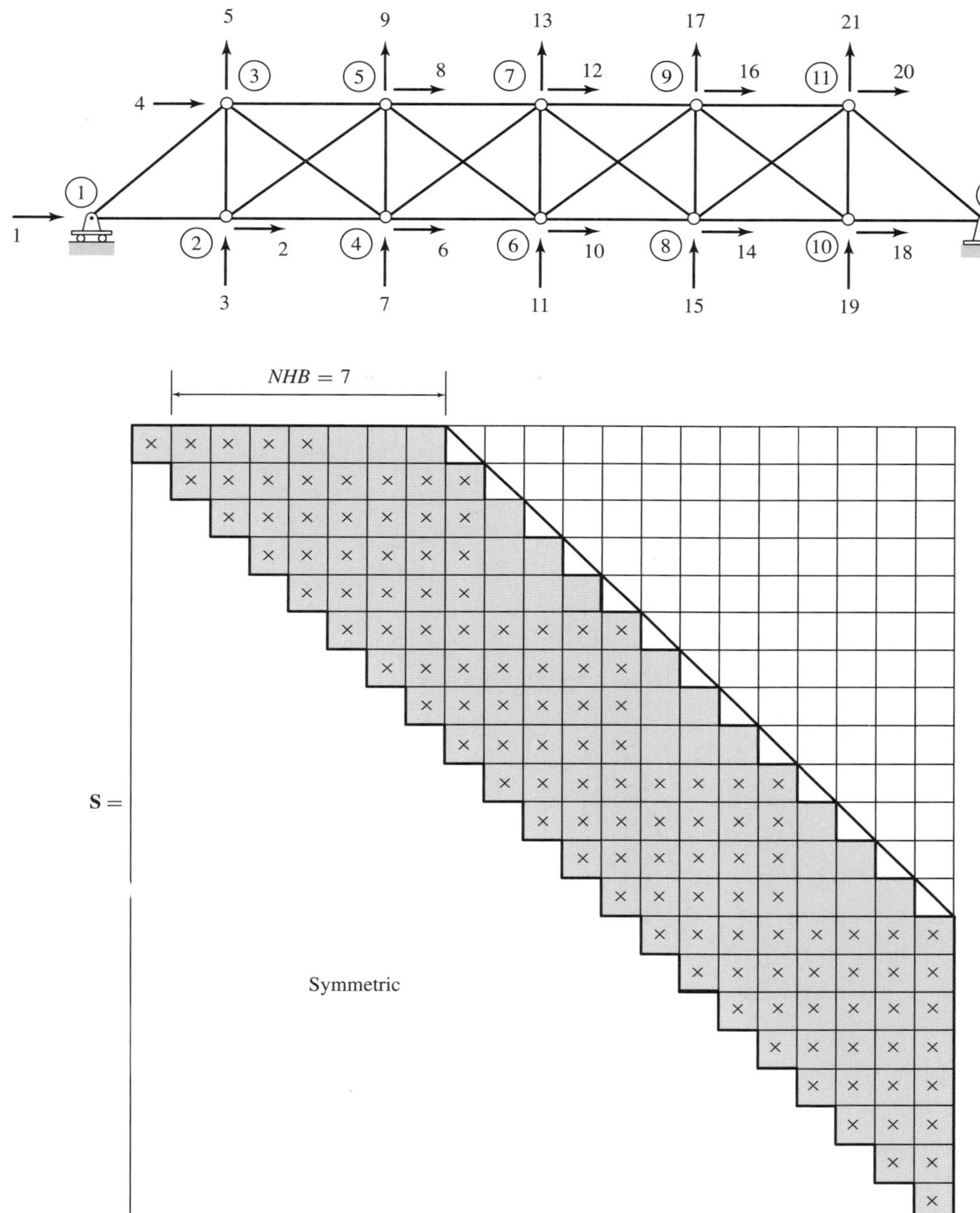

Fig. 9.22 *Joint Numbering for Minimum Half-Bandwidth* ($NDOF = 21$, $NHB = 7$)

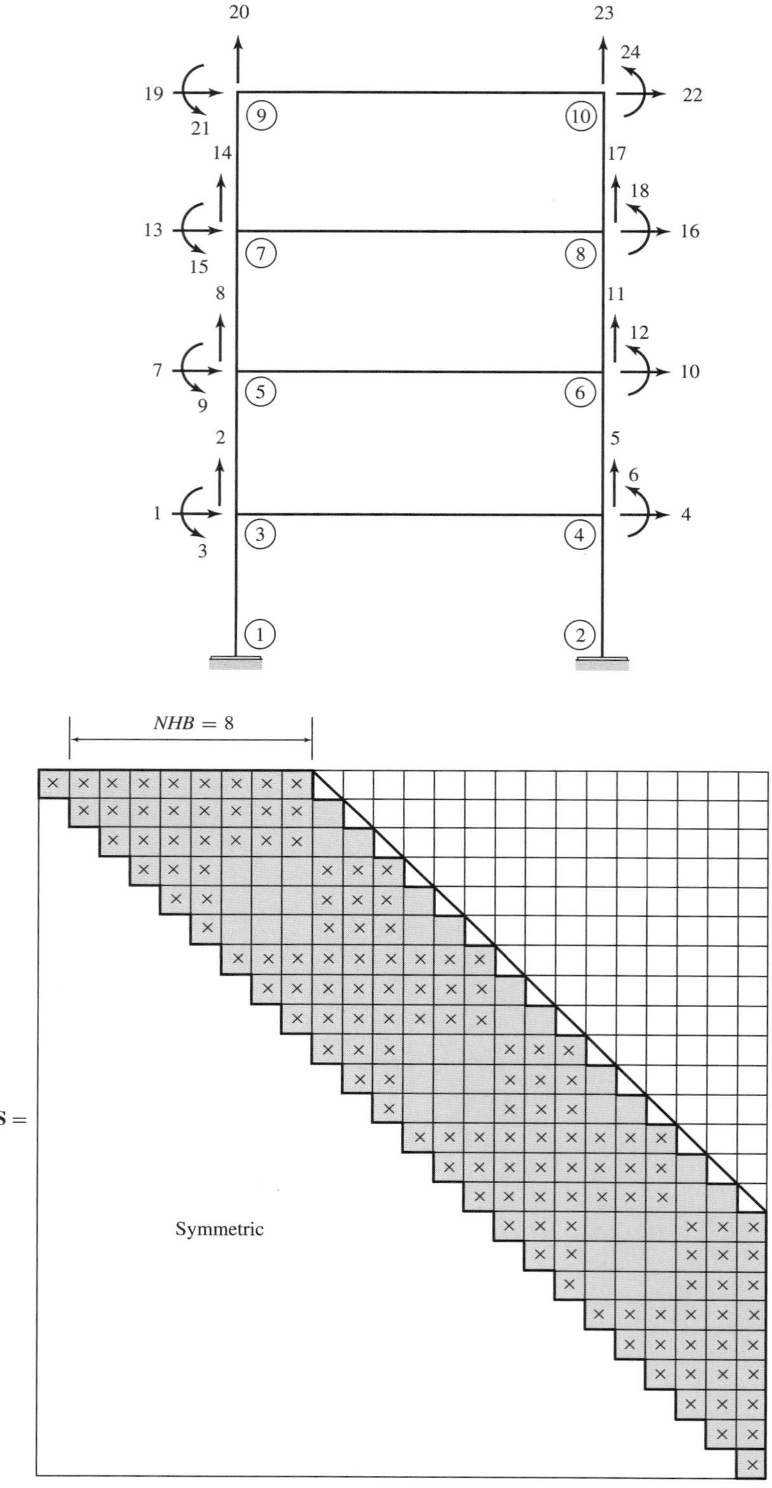

Fig. 9.23 *Joint Numbering for Minimum Half-Bandwidth* ($NDOF = 24$, $NHB = 8$)

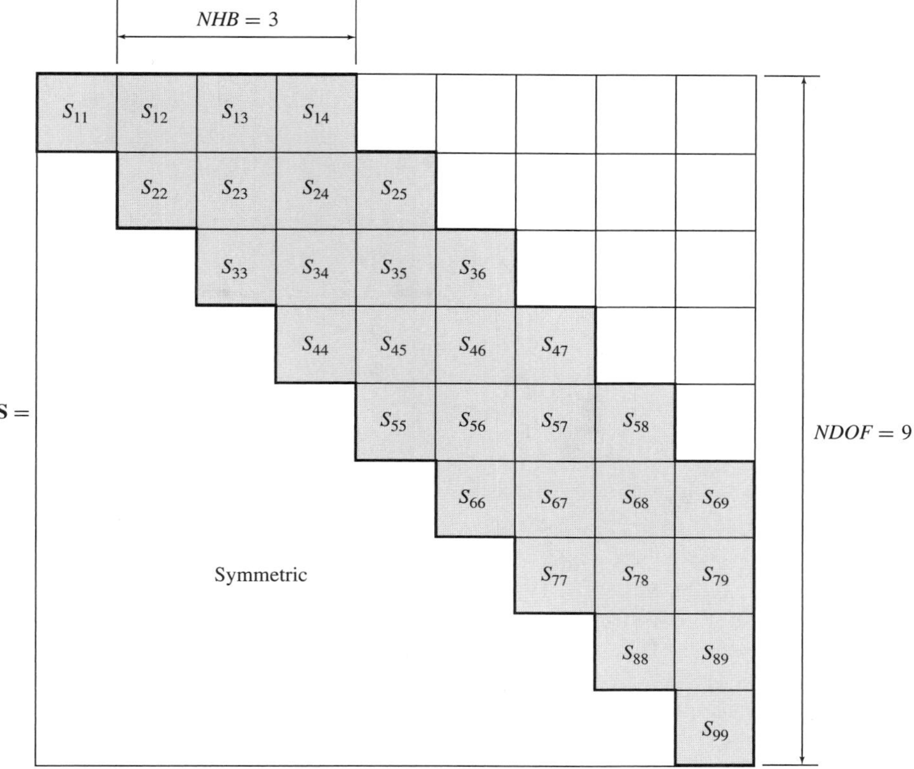

(a) Full Structure Stiffness Matrix **S** for $NDOF = 9$ and $NHB = 3$

Fig. 9.24

of **S** on the diagonal and in the half-bandwidth are stored in the same order in the corresponding row of the compact matrix $\hat{\mathbf{S}}$. The location of an element S_{ij} of **S** in the compact matrix $\hat{\mathbf{S}}$ is given by the relationship

$$S_{ij} = \hat{S}_{i,(1+j-i)} \qquad \text{for } i = 1, 2, \ldots, NDOF; \quad j = i, i+1, \ldots, NHB + i \leq NDOF \tag{9.74}$$

Solution of Banded Structure Stiffness Equations Using $U^T DU$ Decomposition

Although Gauss–Jordan elimination, as discussed in Section 2.4, can be modified to take advantage of the symmetry and bandedness of structure stiffness equations, for the analysis of large structures, another type of elimination method, a *decomposition method,* is usually preferred. This is because, in decomposition methods, the solution is carried out in two distinct parts; namely, decomposition and substitution, with the decomposition part involving only the structure stiffness matrix **S**, but not the load vector $\overline{\mathbf{P}}$. Thus, the results

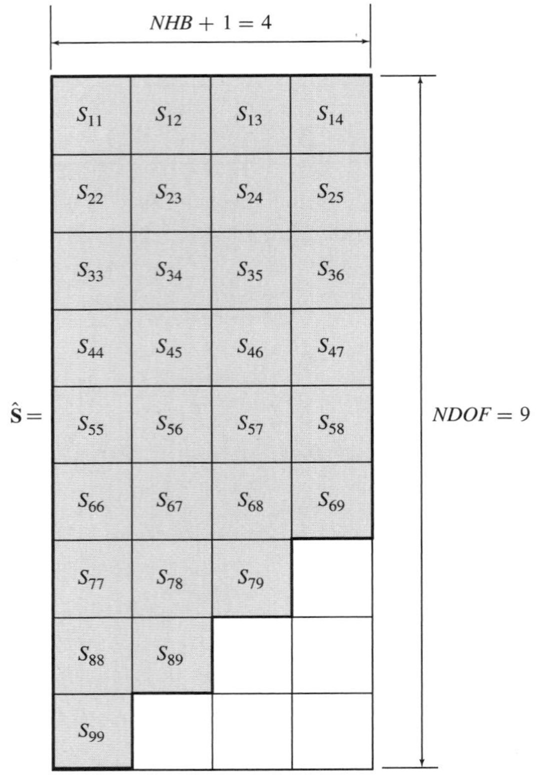

$NHB + 1 = 4$

$$
\hat{\mathbf{S}} = \begin{bmatrix}
S_{11} & S_{12} & S_{13} & S_{14} \\
S_{22} & S_{23} & S_{24} & S_{25} \\
S_{33} & S_{34} & S_{35} & S_{36} \\
S_{44} & S_{45} & S_{46} & S_{47} \\
S_{55} & S_{56} & S_{57} & S_{58} \\
S_{66} & S_{67} & S_{68} & S_{69} \\
S_{77} & S_{78} & S_{79} & \\
S_{88} & S_{89} & & \\
S_{99} & & &
\end{bmatrix}
\quad NDOF = 9
$$

(b) Compact Structure Stiffness Matrix $\hat{\mathbf{S}}$ for $NDOF = 9$ and $NHB = 3$

Fig. 9.24 *(continued)*

of the time-consuming decomposition part can be stored for future use in case the structure needs to be reanalyzed for different loading conditions. In the following, we present a decomposition method, commonly used for solving large systems of structure stiffness equations, called the $\mathbf{U}^T\mathbf{DU}$ *decomposition method*. The method is first formulated for the fully populated symmetric stiffness matrix **S**, and is then modified so that it can be used with the banded and compact forms of the structure stiffness matrix.

As stated previously, in the $\mathbf{U}^T\mathbf{DU}$ decomposition method, the solution of the structure stiffness equations, $\mathbf{Sd} = \overline{\mathbf{P}}$, is carried out in two parts: decomposition and substitution. In decomposition, the structure stiffness matrix **S** is decomposed (or factored) into the matrix triple product,

$$\mathbf{S} = \mathbf{U}^T\mathbf{DU} \tag{9.75}$$

in which **U** is a unit upper triangular matrix (i.e., an upper triangular matrix with diagonal elements equal to unity); and **D** is a diagonal matrix. For a general n-degree-of-freedom system, Eq. (9.75) can be written in expanded form as

$$\begin{bmatrix} S_{11} & S_{12} & S_{13} & \cdots & S_{1n} \\ & S_{22} & S_{23} & \cdots & S_{2n} \\ \text{(symmetric)} & & S_{33} & \cdots & S_{3n} \\ & & & \cdots & \cdots \\ & & & & S_{nn} \end{bmatrix} =$$

$$\begin{bmatrix} 1 & 0 & 0 & \cdots & 0 \\ U_{12} & 1 & 0 & \cdots & 0 \\ U_{13} & U_{23} & 1 & \cdots & 0 \\ \cdots & \cdots & \cdots & \cdots & \cdots \\ U_{1n} & U_{2n} & U_{3n} & \cdots & 1 \end{bmatrix} \begin{bmatrix} D_{11} & 0 & 0 & \cdots & 0 \\ 0 & D_{22} & 0 & \cdots & 0 \\ 0 & 0 & D_{33} & \cdots & 0 \\ \cdots & \cdots & \cdots & \cdots & \cdots \\ 0 & 0 & 0 & \cdots & D_{nn} \end{bmatrix} \begin{bmatrix} 1 & U_{12} & U_{13} & \cdots & U_{1n} \\ 0 & 1 & U_{23} & \cdots & U_{2n} \\ 0 & 0 & 1 & \cdots & U_{3n} \\ \cdots & \cdots & \cdots & \cdots & \cdots \\ 0 & 0 & 0 & \cdots & 1 \end{bmatrix} \tag{9.76}$$

Multiplying the three matrices on the right side of Eq. (9.76), we obtain

$$\begin{bmatrix} S_{11} & S_{12} & S_{13} & \cdots & S_{1n} \\ & S_{22} & S_{23} & \cdots & S_{2n} \\ \text{(symmetric)} & & S_{33} & \cdots & S_{3n} \\ & & & \cdots & \cdots \\ & & & & S_{nn} \end{bmatrix} =$$

$$\begin{bmatrix} D_{11} & D_{11}U_{12} & D_{11}U_{13} & \cdots & D_{11}U_{1n} \\ & D_{11}U_{12}^2 + D_{22} & D_{11}U_{12}U_{13} + D_{22}U_{23} & \cdots & D_{11}U_{12}U_{1n} + D_{22}U_{2n} \\ \text{(symmetric)} & & D_{11}U_{13}^2 + D_{22}U_{23}^2 + D_{33} & \cdots & D_{11}U_{13}U_{1n} + D_{22}U_{23}U_{2n} + D_{33}U_{3n} \\ & & & \cdots & \cdots \\ & & & & D_{11}U_{1n}^2 + D_{22}U_{2n}^2 + D_{33}U_{3n}^2 + \cdots + D_{nn} \end{bmatrix} \tag{9.77}$$

By comparing the corresponding elements of the matrices **S** and $\mathbf{U}^T\mathbf{DU}$, on the left and right sides, respectively, of Eq. (9.77), we can develop an algorithm for evaluating the elements of matrices **D** and **U**. By comparing the elements in row 1 column 1 of the two matrices, we can see that $D_{11} = S_{11}$. With D_{11} known, the elements in the first row of **U** can be obtained by equating the remaining elements in the first rows of **S** and $\mathbf{U}^T\mathbf{DU}$. This yields $U_{12} = S_{12}/D_{11}$, $U_{13} = S_{13}/D_{11}, \ldots, U_{1n} = S_{1n}/D_{11}$. Next, we equate the corresponding elements in the second rows of **S** and $\mathbf{U}^T\mathbf{DU}$ in Eq. (9.77) to obtain the second rows of **D** and **U**, and so on. The general recurrence relationships for computation of the elements of **D** and **U** can be expressed as follows.

$$D_{ii} = \begin{cases} S_{ii} & \text{for } i = 1 \\ S_{ii} - \sum\limits_{k=1}^{i-1} D_{kk}U_{ki}^2 & \text{for } i = 2, 3, \ldots, NDOF \end{cases} \tag{9.78a}$$

$$U_{ij} = \begin{cases} \dfrac{S_{ij}}{D_{ii}} & \text{for } i = 1;\ \ j = i+1,\ i+2, \ldots, NDOF \\ \dfrac{1}{D_{ii}}\left(S_{ij} - \sum\limits_{k=1}^{i-1} D_{kk}U_{ki}U_{kj}\right) & \text{for } i = 2, 3, \ldots, NDOF - 1; \\ & \quad j = i+1, i+2, \ldots, NDOF \end{cases} \tag{9.78b}$$

$$U_{ii} = 1 \qquad \text{for } i = 1, 2, \ldots, NDOF \tag{9.78c}$$

The nonzero elements of **D** and **U** are computed by starting at the first row number (i.e., $i = 1$), and proceeding sequentially to the last row number (i.e., $i = NDOF$). As implied by Eqs. 9.78(a) and (b), for each row number i, the diagonal element D_{ii} (Eq. (9.78a)) must be computed before the elements U_{ij} (Eq. (9.78b)) of the ith row of **U** can be calculated.

With the structure stiffness matrix **S** now decomposed into triangular and diagonal matrices, we can now begin the substitution part of the solution process. Substitution of $\mathbf{S} = \mathbf{U}^T\mathbf{D}\mathbf{U}$ into the structure stiffness equations, $\mathbf{S}\mathbf{d} = \overline{\mathbf{P}}$, yields

$$\mathbf{U}^T\mathbf{D}\mathbf{U}\,\mathbf{d} = \overline{\mathbf{P}} \tag{9.79}$$

The substitution part is carried out in two steps: forward substitution, and back substitution. In the forward substitution step, Eq. (9.79) is written as

$$\mathbf{U}^T\mathbf{D}\,\hat{\mathbf{d}} = \overline{\mathbf{P}} \tag{9.80}$$

with

$$\hat{\mathbf{d}} = \mathbf{U}\mathbf{d} \tag{9.81}$$

in which $\hat{\mathbf{d}}$ is an auxiliary vector of unknowns. Equation (9.80) can be written in expanded form as

$$\begin{bmatrix} D_{11} & 0 & 0 & \cdots & 0 \\ D_{11}U_{12} & D_{22} & 0 & \cdots & 0 \\ D_{11}U_{13} & D_{22}U_{23} & D_{33} & \cdots & 0 \\ \cdots & \cdots & \cdots & \cdots & \cdots \\ D_{11}U_{1n} & D_{22}U_{2n} & D_{33}U_{3n} & \cdots & D_{nn} \end{bmatrix} \begin{bmatrix} \hat{d}_1 \\ \hat{d}_2 \\ \hat{d}_3 \\ \cdots \\ \hat{d}_n \end{bmatrix} = \begin{bmatrix} \overline{P}_1 \\ \overline{P}_2 \\ \overline{P}_3 \\ \cdots \\ \overline{P}_n \end{bmatrix} \tag{9.82}$$

from which we can see that the auxiliary unknowns $\hat{\mathbf{d}}$ can be determined by the simple process of forward substitution, starting with the first row and proceeding sequentially to the last row. From the first row of Eq. (9.82), we can see that $\hat{d}_1 = \overline{P}_1/D_{11}$. With $\hat{d}_1$ known, the value of $\hat{d}_2$ can now be determined by solving the equation in the second row of Eq. (9.82); that is, $\hat{d}_2 = (\overline{P}_2 - D_{11}U_{12}\hat{d}_1)/D_{22}$. Next, we calculate $\hat{d}_3$ by solving the equation in the third row of Eq. (9.82), and so on. In general, the elements of $\hat{\mathbf{d}}$ can be computed as

$$\hat{d}_i = \begin{cases} \dfrac{\overline{P}_i}{D_{ii}} & \text{for } i = 1 \\ \dfrac{1}{D_{ii}}\left(\overline{P}_i - \sum\limits_{k=1}^{i-1} D_{kk}U_{ki}\hat{d}_k\right) & \text{for } i = 2, 3, \ldots, NDOF \end{cases} \tag{9.83}$$

Once the auxiliary vector $\hat{\mathbf{d}}$ has been evaluated, the unknown joint displacement vector **d** can be calculated by solving Eq. (9.81), using back substitution. The expanded form of Eq. (9.81) can be expressed as

$$\begin{bmatrix} 1 & U_{12} & U_{13} & \cdots & U_{1,n-1} & U_{1n} \\ 0 & 1 & U_{23} & \cdots & U_{2,n-1} & U_{2n} \\ 0 & 0 & 1 & \cdots & U_{3,n-1} & U_{3n} \\ \cdots & \cdots & \cdots & \cdots & \cdots & \cdots \\ 0 & 0 & 0 & \cdots & 1 & U_{n-1,n} \\ 0 & 0 & 0 & \cdots & 0 & 1 \end{bmatrix} \begin{bmatrix} d_1 \\ d_2 \\ d_3 \\ \cdots \\ d_{n-1} \\ d_n \end{bmatrix} = \begin{bmatrix} \hat{d}_1 \\ \hat{d}_2 \\ \hat{d}_3 \\ \cdots \\ \hat{d}_{n-1} \\ \hat{d}_n \end{bmatrix} \tag{9.84}$$

From which we can see that the unknown joint displacements **d** can be determined by the simple process of back substitution, starting with the last row and proceeding sequentially to the first row. From the last row of Eq. (9.84), we can see that $d_n = \hat{d}_n$. With d_n known, the value of d_{n-1} can now be determined by solving the equation in the next to the last row of Eq. (9.84); that is, $d_{n-1} = \hat{d}_{n-1} - U_{n-1,n} d_n$. The back substitution is continued until all the joint displacements have been calculated. The back substitution process can be represented by the recurrence equation

$$d_i = \begin{cases} \hat{d}_i & \text{for } i = NDOF \\ \hat{d}_i - \sum\limits_{k=i+1}^{NDOF} U_{ik} d_k & \text{for } i = NDOF - 1,\ NDOF - 2, \ldots, 1 \end{cases} \tag{9.85}$$

As discussed in the foregoing paragraphs, the $\mathbf{U}^T\mathbf{D}\mathbf{U}$ decomposition procedure for solving structure stiffness equations essentially consists of the following steps.

1. Decompose the structure stiffness matrix **S** into a diagonal matrix **D**, and a unit upper triangular matrix **U**, by applying Eqs. (9.78).
2. Calculate the auxiliary vector $\hat{\mathbf{d}}$ using forward substitution (Eq. (9.83)).
3. Determine the unknown joint displacements **d** by back substitution (Eq. (9.85)).

EXAMPLE 9.8 Use $\mathbf{U}^T\mathbf{D}\mathbf{U}$ decomposition to solve the following system of structural stiffness equations.

$$\begin{bmatrix} 5 & 2 & -1 & 0 \\ & 6 & -3 & 2 \\ \text{(symmetric)} & & 4 & 1 \\ & & & 7 \end{bmatrix} \begin{bmatrix} d_1 \\ d_2 \\ d_3 \\ d_4 \end{bmatrix} = \begin{bmatrix} -19 \\ -22 \\ 22 \\ -6 \end{bmatrix}$$

SOLUTION *Decomposition:* By applying Eqs. (9.78),

$$D_{11} = S_{11} = 5$$

$$U_{12} = \frac{S_{12}}{D_{11}} = \frac{2}{5} = 0.4$$

$$U_{13} = \frac{S_{13}}{D_{11}} = \frac{-1}{5} = -0.2$$

$$U_{14} = 0$$

$$U_{11} = U_{22} = U_{33} = U_{44} = 1$$

$$D_{22} = S_{22} - D_{11}U_{12}^2 = 6 - 5(0.4)^2 = 5.2$$

$$U_{23} = \frac{1}{D_{22}}(S_{23} - D_{11}U_{12}U_{13}) = \frac{1}{5.2}[-3 - 5(0.4)(-0.2)] = -0.5$$

$$U_{24} = \frac{1}{D_{22}}(S_{24} - D_{11}U_{12}U_{14}) = \frac{1}{5.2}[2 - 5(0.4)(0)] = 0.38462$$

$$D_{33} = S_{33} - D_{11}U_{13}^2 - D_{22}U_{23}^2 = 4 - 5(-0.2)^2 - 5.2(-0.5)^2 = 2.5$$

$$U_{34} = \frac{1}{D_{33}}(S_{34} - D_{11}U_{13}U_{14} - D_{22}U_{23}U_{24})$$
$$= \frac{1}{2.5}[1 - 5(-0.2)(0) - 5.2(-0.5)(0.38462)] = 0.8$$
$$D_{44} = S_{44} - D_{11}U_{14}^2 - D_{22}U_{24}^2 - D_{33}U_{34}^2$$
$$= 7 - 5(0)^2 - 5.2(0.38462)^2 - 2.5(0.8)^2 = 4.6308$$

Thus,

$$\mathbf{D} = \begin{bmatrix} 5 & 0 & 0 & 0 \\ 0 & 5.2 & 0 & 0 \\ 0 & 0 & 2.5 & 0 \\ 0 & 0 & 0 & 4.6308 \end{bmatrix} \qquad \mathbf{U} = \begin{bmatrix} 1 & 0.4 & -0.2 & 0 \\ 0 & 1 & -0.5 & 0.38462 \\ 0 & 0 & 1 & 0.8 \\ 0 & 0 & 0 & 1 \end{bmatrix}$$

Forward Substitution: Using Eq. (9.83),

$$\hat{d}_1 = \frac{\overline{P}_1}{D_{11}} = \frac{-19}{5} = -3.8$$
$$\hat{d}_2 = \frac{1}{D_{22}}(\overline{P}_2 - D_{11}U_{12}\hat{d}_1) = \frac{1}{5.2}[-22 - 5(0.4)(-3.8)] = -2.7692$$
$$\hat{d}_3 = \frac{1}{D_{33}}(\overline{P}_3 - D_{11}U_{13}\hat{d}_1 - D_{22}U_{23}\hat{d}_2)$$
$$= \frac{1}{2.5}[22 - 5(-0.2)(-3.8) - 5.2(-0.5)(-2.7692)] = 4.4$$
$$\hat{d}_4 = \frac{1}{D_{44}}(\overline{P}_4 - D_{11}U_{14}\hat{d}_1 - D_{22}U_{24}\hat{d}_2 - D_{33}U_{34}\hat{d}_3)$$
$$= \frac{1}{4.6308}[-6 - 5(0)(-3.8) - 5.2(0.38462)(-2.7692) - 2.5(0.8)(4.4)] = -2$$

Thus,

$$\hat{\mathbf{d}} = \begin{bmatrix} -3.8 \\ -2.7692 \\ 4.4 \\ -2 \end{bmatrix}$$

Back Substitution: Applying Eq. (9.85),

$$d_4 = \hat{d}_4 = -2$$
$$d_3 = \hat{d}_3 - U_{34}d_4 = 4.4 - 0.8(-2) = 6$$
$$d_2 = \hat{d}_2 - U_{23}d_3 - U_{24}d_4 = -2.7692 - (-0.5)6 - 0.38462(-2) = 1$$
$$d_1 = \hat{d}_1 - U_{12}d_2 - U_{13}d_3 - U_{14}d_4 = -3.8 - 0.4(1) - (-0.2)6 - 0(-2) = -3$$

Thus, the solution of the given system of equations is

$$\mathbf{d} = \begin{bmatrix} -3 \\ 1 \\ 6 \\ -2 \end{bmatrix}$$

Ans

If the structure stiffness matrix **S** is banded, then the corresponding **U** matrix contains nonzero elements only on its diagonal and within the upper half-bandwidth, as shown in Fig. 9.25(a) on the next page. In such cases, the computational effort required for solution can be significantly reduced by calculating only the elements in the upper half-bandwidth of **U**. From Fig. 9.25(a), we can see that, in any row number i of **U**, all the nonzero elements are located in column numbers i through $i + NHB \leq NDOF$. Similarly, in any column number j of **U**, the nonzero elements are located in row numbers $j - NHB \geq 1$ through j. Using the foregoing ranges for the indexing parameters in Eqs. (9.78), (9.83), and (9.85), we obtain the following modified recurrence formulas for solving the banded systems of structure stiffness equations by the $\mathbf{U}^T\mathbf{D}\mathbf{U}$ decomposition method.

Decomposition

$$D_{ii} = \begin{cases} S_{ii} & \text{for } i = 1 \\ S_{ii} - \sum\limits_{k=m_1}^{i-1} D_{kk} U_{ki}^2 & \text{for } i = 2, 3, \ldots, NDOF \end{cases}$$

with $m_1 = i - NHB \geq 1$

$$U_{ij} = \begin{cases} \dfrac{S_{ij}}{D_{ii}} & \text{for } i = 1;\ j = i+1, i+2, \ldots, i + NHB \leq NDOF \\ \dfrac{1}{D_{ii}}(S_{ij} - B_{ij}) & \text{for } i = 2, 3, \ldots, NDOF - 1;\ j = i+1, i+2, \ldots, i + NHB \leq NDOF \end{cases} \tag{9.86}$$

with

$$B_{ij} = \begin{cases} \sum\limits_{k=m_2}^{i-1} D_{kk} U_{ki} U_{kj} & \text{for } m_2 \leq i - 1 \\ 0 & \text{for } m_2 > i - 1 \end{cases}$$

in which, $m_2 = j - NHB \geq 1$

$$U_{ii} = 1 \qquad \text{for } i = 1, 2, \ldots, NDOF$$

Forward Substitution

$$\hat{d}_i = \begin{cases} \dfrac{\overline{P}_i}{D_{ii}} & \text{for } i = 1 \\ \dfrac{1}{D_{ii}}\left(\overline{P}_i - \sum\limits_{k=m_1}^{i-1} D_{kk} U_{ki} \hat{d}_k\right) & \text{for } i = 2, 3, \ldots, NDOF \end{cases} \tag{9.87}$$

Back Substitution

$$d_i = \begin{cases} \hat{d}_i & \text{for } i = NDOF \\ \hat{d}_i - \sum\limits_{k=i+1}^{m_3} U_{ik} d_k & \text{for } i = NDOF - 1, NDOF - 2, \ldots, 1 \end{cases} \tag{9.88}$$

with $m_3 = i + NHB \leq NDOF$

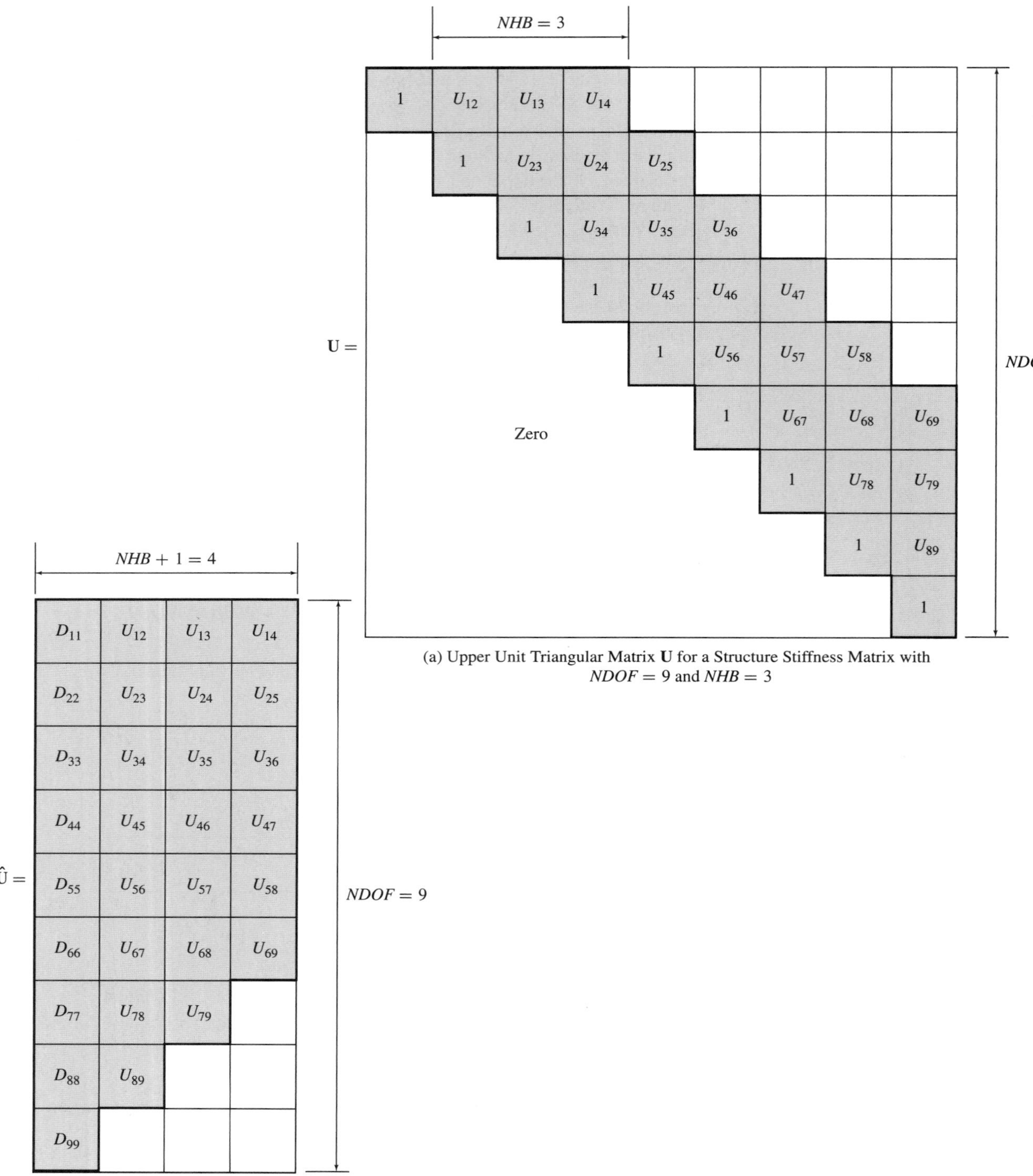

(a) Upper Unit Triangular Matrix **U** for a Structure Stiffness Matrix with $NDOF = 9$ and $NHB = 3$

(b) Storage of **D** and **U** in Compact Form

Fig. 9.25

As stated before, the elements on the diagonal and in the upper half-bandwidth of **S** can be compactly stored in computer memory in a rectangular array $\hat{\mathbf{S}}$ of order $NDOF \times (NHB + 1)$ (see Fig. 9.24(b)). In an analogous manner, a rectangular array $\hat{\mathbf{U}}$, of the same order as $\hat{\mathbf{S}}$, can be defined to store the elements on the diagonal of **D** and in the upper half-bandwidth of **U**, as depicted in Fig. 9.25(b). As indicated there, the diagonal elements of **D** are stored in the first column of $\hat{\mathbf{U}}$, and the elements, in each row of **U**, in the half-bandwidth, are stored in the same order in the corresponding row of $\hat{\mathbf{U}}$. The locations of the relevant elements of **D** and **U** in the compact matrix $\hat{\mathbf{U}}$ can be determined by using the following relationships.

$$
\begin{aligned}
D_{ii} &= \hat{U}_{i1} && \text{for } i = 1, 2, \ldots, NDOF \\
U_{ij} &= \hat{U}_{i,(1+j-i)} && \text{for } i = 1, 2, \ldots, NDOF - 1; \\
& && j = i+1, i+2, \ldots, NHB + i \le NDOF
\end{aligned}
\tag{9.89}
$$

Applying Eqs. (9.74) and (9.89), we obtain the following modified algorithm for solving the banded systems of structure stiffness equations, in terms of the elements of compact matrices $\hat{\mathbf{S}}$ and $\hat{\mathbf{U}}$.

Decomposition

$$
\hat{U}_{i1} = \begin{cases} \hat{S}_{i1} & \text{for } i = 1 \\ \hat{S}_{i1} - \sum\limits_{k=m_1}^{i-1} \hat{U}_{k1} \hat{U}^2_{k,(1+i-k)} & \text{for } i = 2, 3, \ldots, NDOF \end{cases}
$$

with $m_1 = i - NHB \ge 1$

$$
\hat{U}_{ij} = \begin{cases} \dfrac{\hat{S}_{ij}}{\hat{U}_{i1}} & \text{for } i = 1;\ j = 2, 3, \ldots, NHB + 1 \\ \dfrac{1}{\hat{U}_{i1}}(\hat{S}_{ij} - \hat{B}_{ij}) & \text{for } i = 2, 3, \ldots, NDOF - 1;\ j = 2, 3, \ldots, NHB + 1 \le NDOF - i + 1 \end{cases}
\tag{9.90}
$$

with

$$
\hat{B}_{ij} = \begin{cases} \sum\limits_{k=\hat{m}_2}^{i-1} \hat{U}_{k1} \hat{U}_{k,(1+i-k)} \hat{U}_{k,(i+j-k)} & \text{for } \hat{m}_2 \le i - 1 \\ 0 & \text{for } \hat{m}_2 > i - 1 \end{cases}
$$

in which $\hat{m}_2 = i + j - NHB - 1 \ge 1$

Forward Substitution

$$
\hat{d}_i = \begin{cases} \dfrac{\overline{P}_i}{\hat{U}_{i1}} & \text{for } i = 1 \\ \dfrac{1}{\hat{U}_{i1}} \left(\overline{P}_i - \sum\limits_{k=m_1}^{i-1} \hat{U}_{k1} \hat{U}_{k,(1+i-k)} \hat{d}_k \right) & \text{for } i = 2, 3, \ldots, NDOF \end{cases}
\tag{9.91}
$$

Back Substitution

$$d_i = \begin{cases} \hat{d}_i & \text{for } i = NDOF \\ \hat{d}_i - \sum_{k=i+1}^{m_3} \hat{U}_{i,(1+k-i)} d_k & \text{for } i = NDOF - 1, NDOF - 2, \ldots, 1 \end{cases}$$

with $m_3 = i + NHB \leq NDOF$

(9.92)

In the computer implementation of the foregoing procedure, computer memory requirements can be reduced by creating only the $\hat{\mathbf{S}}$ matrix, but not the $\hat{\mathbf{U}}$ matrix. Each element $\hat{U}_{ij}$ of the $\hat{\mathbf{U}}$ matrix is now computed and stored in the $\hat{\mathbf{S}}$ matrix in the location originally occupied by the corresponding $\hat{S}_{ij}$ element. Thus, at the end of the decomposition part of the solution process, the $\hat{\mathbf{S}}$ matrix contains all the elements of the $\hat{\mathbf{U}}$ matrix, and can be used in the substitution part of the solution. Also, this $\hat{\mathbf{S}}$ matrix, now containing the elements of $\hat{\mathbf{U}}$ (or $\mathbf{D}$ and $\mathbf{U}$), can be stored for any future reanalysis of the structure for different loading conditions.

In this section, we have considered only one of the many available methods for solving large systems of structural stiffness equations. For a comprehensive coverage of the various solution methods, the reader should refer to references [2, 14, 26].

SUMMARY

In this chapter, we have considered some extensions and modifications of the matrix stiffness method developed in previous chapters. We have also considered techniques for modeling some special features and details of structures, so that more realistic structural responses can be predicted from the analysis.

We studied an alternative formulation of the stiffness method, which involves the structure stiffness matrix for all the coordinates (including the restrained coordinates) of the structure. The advantages of this alternative formulation are that the support displacement effects can be incorporated into the analysis in a direct and straightforward manner, and the reactions can be calculated using the structure stiffness relations. The main disadvantage of this formulation is that it requires significantly more computer memory than the standard formulation.

We presented an approximate method for the analysis of rectangular building frames, neglecting the effects of member axial deformations. This approach significantly reduces the number of structural degrees of freedom to be considered in an analysis. This approximate approach is appropriate only for frames in which the member axial deformations are small enough, as compared to bending deformations, to have a negligible effect on their responses.

The basic concepts of condensation of structural degrees of freedom, and analysis using substructures, were discussed. These approaches can be used in the analysis of large structures to reduce the number of stiffness equations that must be processed, and solved simultaneously.

A commonly used technique for modeling an inclined roller support was described, in which the support is replaced with an imaginary axial force member with large axial stiffness, and oriented in the direction perpendicular to the incline. Modified member stiffness relations, considering the effects of rigid end offsets, were also presented. Procedures for including the effects of semirigid connections, and shear deformations, in the analysis, were discussed, and the analysis of structures composed of nonprismatic members considered.

Finally, we defined the half-bandwidth of structural stiffness matrices; and discussed procedures for efficiently numbering the structure's degrees of freedom, and for storing its stiffness matrix in computer memory by taking advantage of its symmetry and bandedness. Also considered was the commonly used $\mathbf{U}^T\mathbf{D}\mathbf{U}$ decomposition method for solving large banded systems of structure stiffness equations.

PROBLEMS

Section 9.1

9.1 Determine the joint displacements, member axial forces, and support reactions for the plane truss shown in Fig. P9.1, due to the combined effect of the loading shown and a settlement of $\frac{1}{2}$ in. of support 2. Use the alternative formulation of the matrix stiffness method.

9.2 Determine the joint displacements, member axial forces, and support reactions for the plane truss shown in Fig. P9.2, due to the combined effect of the loading shown and a settlement of $\frac{1}{4}$ in. of support 3. Use the alternative formulation of the matrix stiffness method.

9.3 Determine the joint displacements, member end forces, and support reactions for the three-span continuous beam shown in Fig. P9.3 on the next page, due to settlements of 8 and 30 mm, respectively, of supports 2 and 3. Use the alternative formulation of the matrix stiffness method.

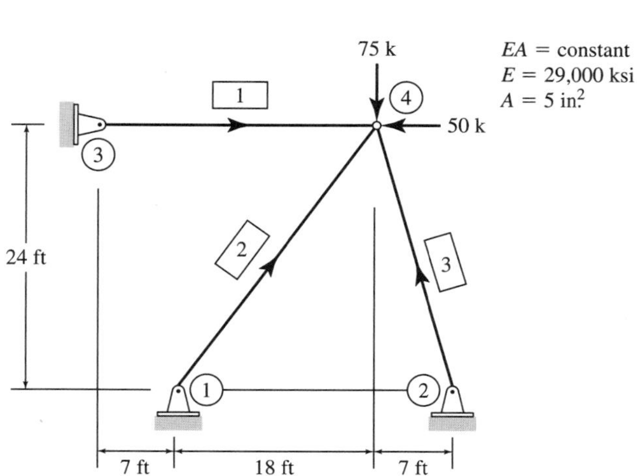

Fig. P9.1

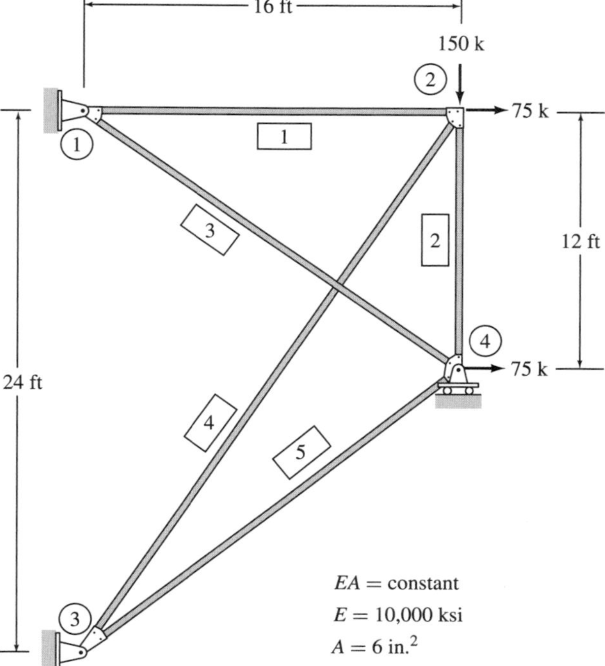

Fig. P9.2

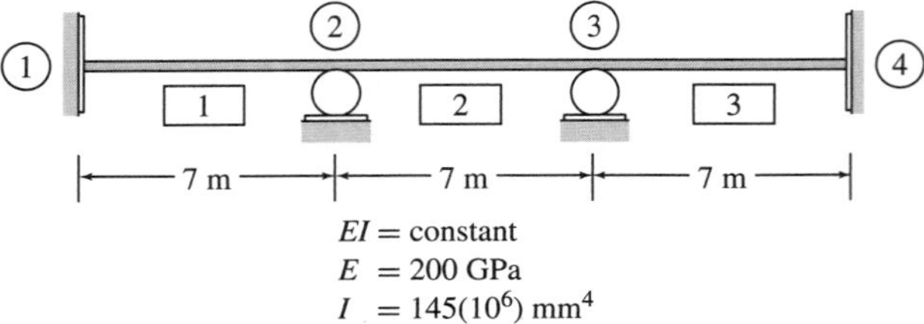

Fig. P9.3

9.4 Determine the joint displacements, member local end forces, and support reactions for the plane frame shown in Fig. P9.4, due to the combined effect of the following: (a) the loading shown in the figure, (b) a clockwise rotation of 0.017 radians of the left support, and (c) a settlement of $\frac{3}{4}$ in. of the right support. Use the alternative formulation of the matrix stiffness method.

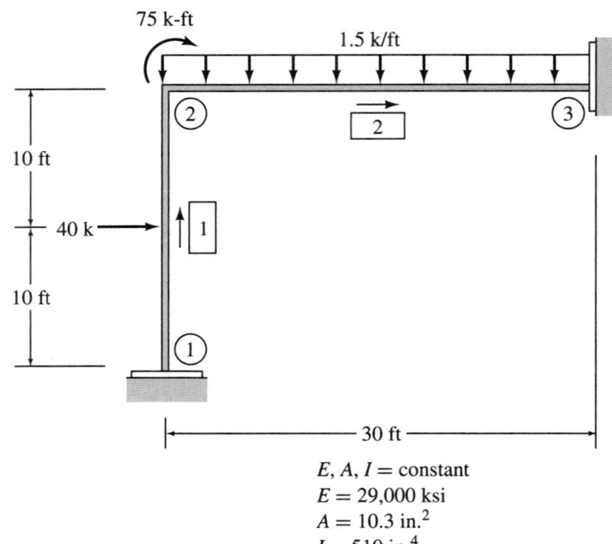

Fig. P9.5, P9.9

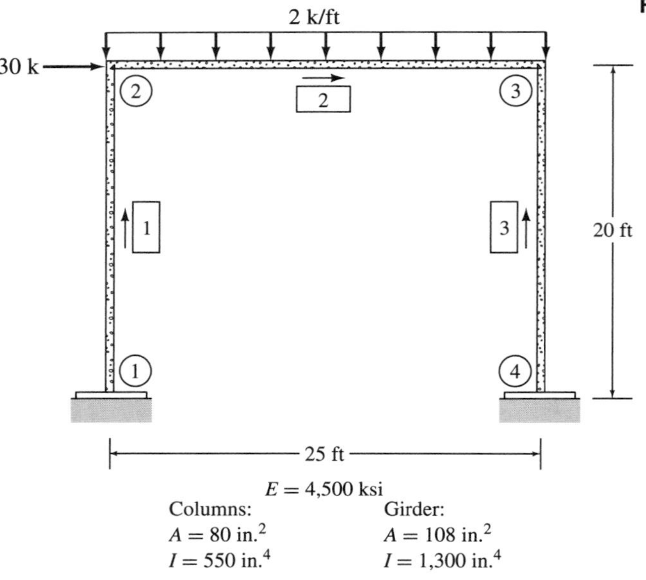

Fig. P9.4, P9.7, P9.10

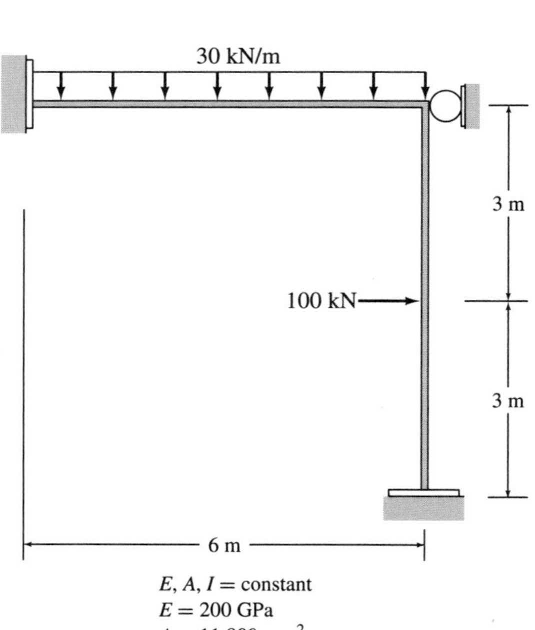

Fig. P9.6

Section 9.2

9.5 through 9.8 Determine the approximate joint displacements, member local end forces, and support reactions for the frames shown in Figs. P9.5 through P9.8, assuming the members to be inextensible.

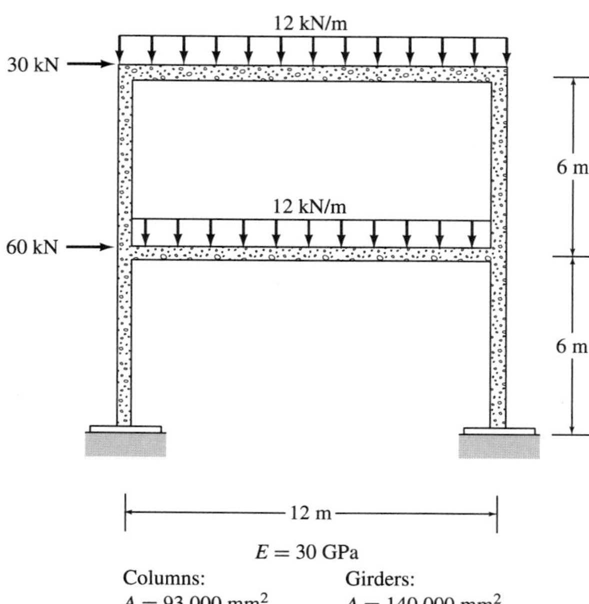

Fig. P9.8

Section 9.3

9.9 and 9.10 Analyze the plane frames shown in Figs. P9.9 and P9.10 using condensation, by treating the rotations of free joints as internal degrees of freedom.

Section 9.8

9.11 and 9.12 Derive the equations of fixed-end forces due to the member loads acting on the nonprismatic beams shown in Figs. P9.11 and P9.12.

Section 9.9

9.13 Solve the following system of simultaneous equations using the $\mathbf{U}^T\mathbf{DU}$ decomposition method.

$$\begin{bmatrix} 20 & -9 & 15 \\ -9 & 16 & -5 \\ 15 & -5 & 18 \end{bmatrix} \begin{bmatrix} d_1 \\ d_2 \\ d_3 \end{bmatrix} = \begin{bmatrix} 354 \\ -275 \\ 307 \end{bmatrix}$$

9.14 Solve the following system of simultaneous equations using the $\mathbf{U}^T\mathbf{DU}$ decomposition method.

$$\begin{bmatrix} 5 & -2 & 1 & 0 & 0 \\ -2 & 3 & -2 & 4 & 0 \\ 1 & -2 & 1 & -1 & 3 \\ 0 & 4 & -1 & 6 & -3 \\ 0 & 0 & 3 & -3 & 4 \end{bmatrix} \begin{bmatrix} d_1 \\ d_2 \\ d_3 \\ d_4 \\ d_5 \end{bmatrix} = \begin{bmatrix} 44 \\ -19 \\ 38 \\ -31 \\ 23 \end{bmatrix}$$

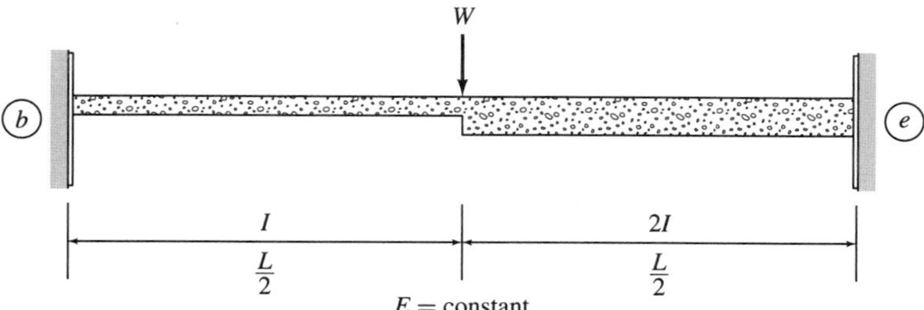

Fig. P9.11

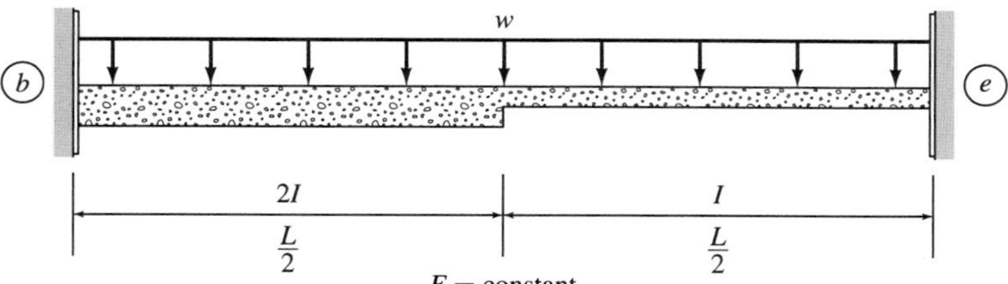

Fig. P9.12

10 INTRODUCTION TO NONLINEAR STRUCTURAL ANALYSIS

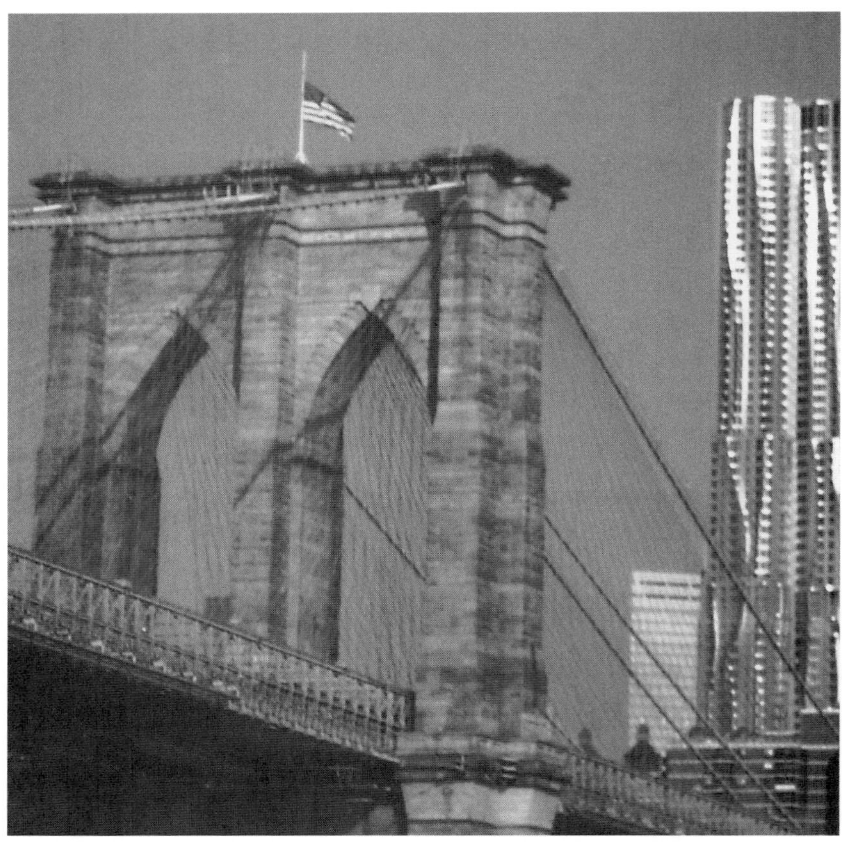

Beekman Tower and Brooklyn Bridge, New York
(Courtesy of Ed Fitzgerald)

Thus far in this text, we have focused our attention on the linear analysis of structures, which it may be recalled, is based on two fundamental assumptions, namely, (a) *geometric linearity* implying that the structure's deformations are so small that the member strains can be expressed as linear functions of joint displacements and the equilibrium equations can be based on the undeformed geometry of the structure, and (b) *material linearity,* represented by the linearly elastic stress-strain relationship for the structural material. The linear analysis (sometimes also referred to as the *first-order analysis*) generally proves adequate for predicting the performance of most common types of engineering structures under service (working) loading conditions. However, as the loads increase beyond service levels into the failure range, the accuracy of the linear analysis gradually deteriorates because the response of the structure usually becomes increasingly nonlinear as its deformations increase and/or its material is strained beyond the yield point. In some structures, such as cable suspension systems, the load carrying capacity relies on geometric nonlinearity even under normal service conditions. Because of its inherent limitations, linear analysis cannot be used to predict instability phenomena and ultimate load capacities of structures.

With the recent introduction of design specifications based on the ultimate strengths of structures, the use of nonlinear analysis in structural design is increasing. In a nonlinear analysis, the restrictions of linear analysis are removed by formulating the equations of equilibrium on the deformed geometry of the structure that is not known in advance, and/or taking into account the effects of inelasticity of the structural material. The load-deformation (stiffness) relationships thus obtained for the structure are nonlinear, and are usually solved using iterative techniques.

The objective of this chapter is to introduce the reader to the exciting and still-evolving field of nonlinear structural analysis. Because of space limitations, only the basic concepts of geometrically nonlinear analysis of plane trusses are covered herein. However, it should be realized that a realistic prediction of structural response in the failure range generally requires consideration of the effects of both geometric and material nonlinearities in the analysis. For a more detailed study, the reader should refer to one of the books devoted entirely to the subject of nonlinear structural analysis, such as [8, 9].

We begin this chapter with an intuitive discussion of the basic concept of geometrically nonlinear analysis, and how it differs from the conventional linear analysis in Section 10.1. A matrix stiffness formulation for geometrically nonlinear analysis of plane trusses is then developed in Section 10.2. While a block diagram summarizing the various steps of nonlinear analysis is provided, the programming details are not covered herein; they are, instead, left as an exercise for the reader. The computer program for geometrically nonlinear analysis of plane trusses can be conveniently adapted from that for the linear analysis of such structures, via relatively straightforward modifications that should become apparent as the nonlinear analysis is developed in this chapter.

10.1 BASIC CONCEPT OF GEOMETRICALLY NONLINEAR ANALYSIS

As stated before, in the linear analysis, the structure's deformations are assumed to be so small that the member strains are expressed as linear functions of joint displacements and the equilibrium equations are based on the undeformed geometry of the structure. In geometrically nonlinear analysis, the restrictions of small deformations are removed by formulating the strain-displacement relations and the equilibrium equations on the deformed geometry of the structure.

To illustrate the basic concept of geometrically nonlinear analysis, consider the two-member plane truss composed of a linearly elastic material, shown in Fig. 10.1(a). Note that the truss is symmetric and is loaded symmetrically with a vertical load P. Thus, it is considered to have only one degree of freedom, which is the vertical displacement δ of the free joint 2.

As shown in Fig. 10.1(b), in the linear analysis, the joint displacement δ is assumed to be so small that the member axial deformations u equal the components of δ in the undeformed directions of the members, that is

$$u \cong \delta \sin\theta \tag{10.1}$$

in which, θ denotes the angle of inclination of members in the undeformed configuration. The member axial strain ε can now be expressed as a linear function of joint displacement δ as

$$\varepsilon = \frac{u}{L} \cong \left(\frac{\sin\theta}{L}\right)\delta \tag{10.2}$$

with L = undeformed length of members. Recall that in linear analysis, the equilibrium equations are based on the undeformed geometry of the structure. Figure 10.1(b) shows the free body diagram of joint 2 of the truss in the undeformed configuration, with the member axial forces Q inclined in the undeformed member directions (i.e., at angles θ with the horizontal). By considering the equilibrium of the joint in the vertical direction, we write

$$Q \cong \frac{P}{2\sin\theta} \tag{10.3}$$

from which we obtain the expression for member axial stress σ,

$$\sigma = \frac{Q}{A} \cong \frac{P}{2A\sin\theta} \tag{10.4}$$

For linearly elastic material,

$$\sigma = E\varepsilon \tag{10.5}$$

By substituting Eqs. (10.2) and (10.4) into Eq. (10.5), we obtain the desired (stiffness) relationship between the load P and the displacement δ of the truss based on the linear analysis:

$$\boxed{P \cong \left(\frac{2EA\sin^2\theta}{L}\right)\delta} \tag{10.6}$$

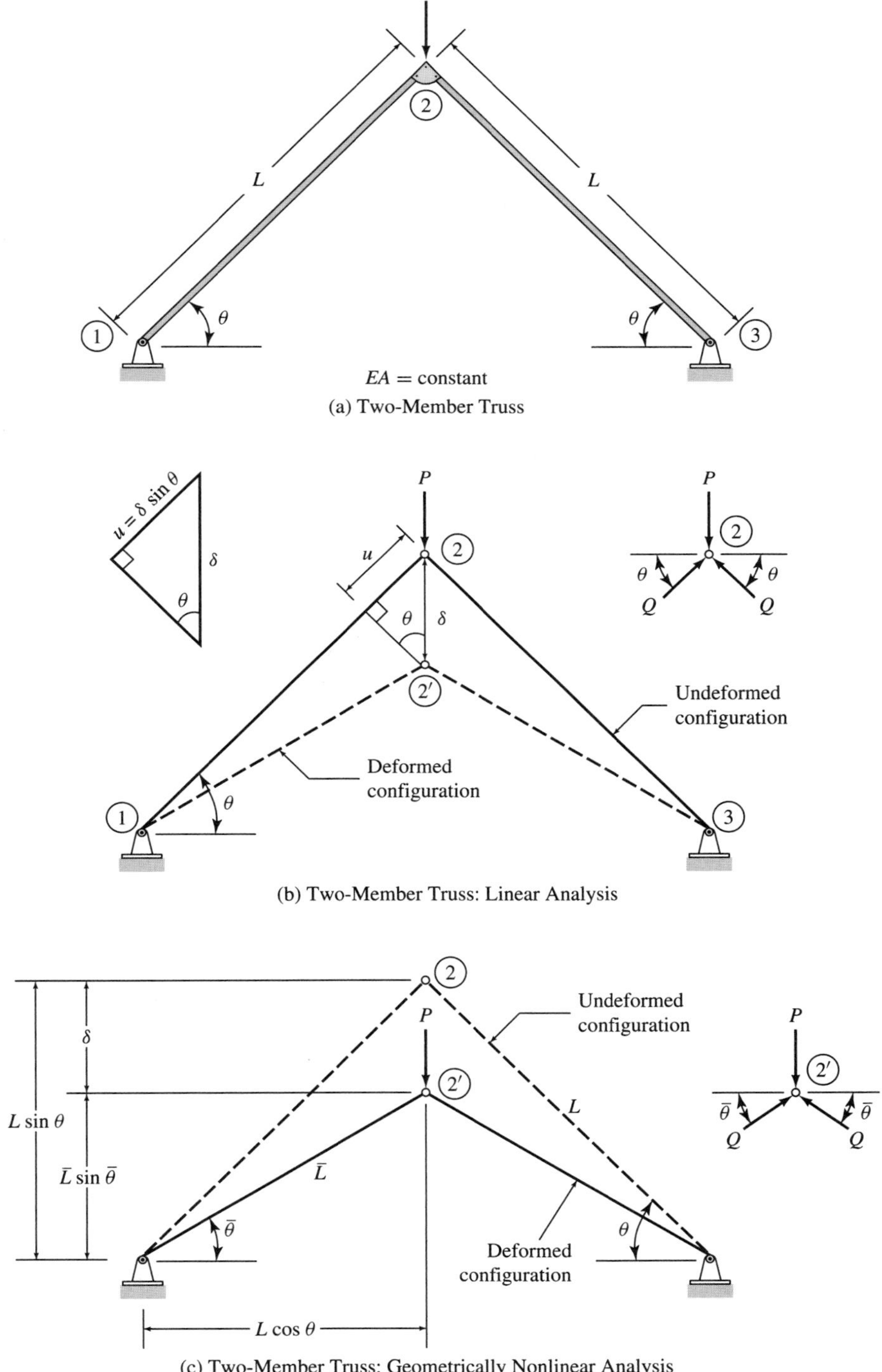
P
2
L
L
θ
θ
1
3
EA = constant
(a) Two-Member Truss
u = δ sin θ
δ
θ
P
u
2
θ
δ
2′
Q
Q
Undeformed
configuration
Deformed
configuration
1
3
(b) Two-Member Truss: Linear Analysis
2
P
2′
Undeformed
configuration
δ
L sin θ
L̄ sin θ̄
L̄
L
θ̄
θ
Deformed
configuration
L cos θ
(c) Two-Member Truss: Geometrically Nonlinear Analysis

Fig. 10.1

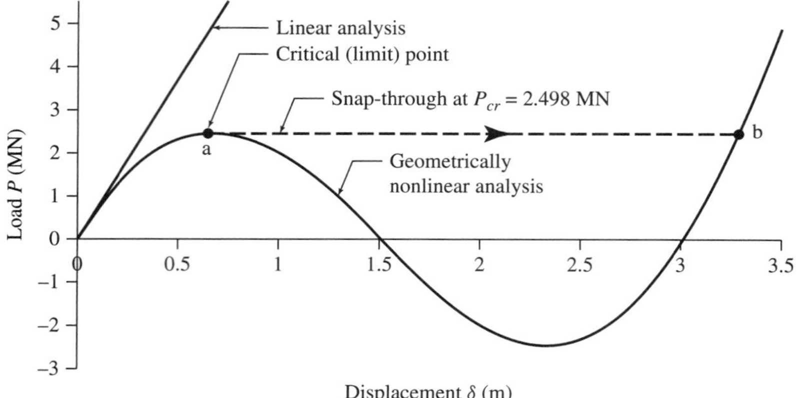

(d) Response of Shallow Two-Member Truss

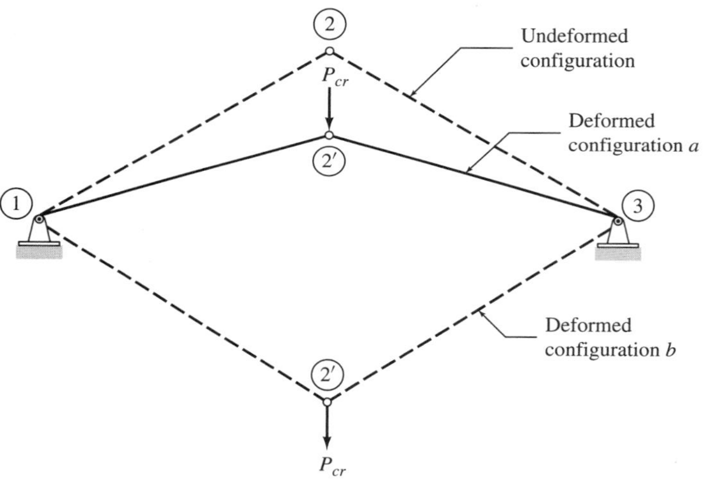

(e) Snap-Through of Two-Member Truss

Fig. 10.1 *(continued)*

In the geometrically nonlinear analysis, we allow the joint displacement δ to be arbitrarily large, and consider the truss to be in equilibrium in its deformed configuration as shown in Fig. 10.1(c). From this figure, we can see that the member lengths $\bar{L}$ and orientations $\bar{\theta}$, in the deformed configuration, can be expressed in terms of δ as

$$\bar{L} = \sqrt{(L\cos\theta)^2 + (L\sin\theta - \delta)^2} = L\sqrt{1 + \left(\frac{\delta}{L}\right)^2 - 2\left(\frac{\delta}{L}\right)\sin\theta} \quad \textbf{(10.7)}$$

$$\sin\bar{\theta} = \frac{L\sin\theta - \delta}{\bar{L}} = \frac{\sin\theta - \left(\frac{\delta}{L}\right)}{\sqrt{1 + \left(\frac{\delta}{L}\right)^2 - 2\left(\frac{\delta}{L}\right)\sin\theta}} \quad \textbf{(10.8)}$$

The member axial deformations and strains are now based on the actual deformed geometry of the members as

$$u = L - \bar{L} = L\left[1 - \sqrt{1 + \left(\frac{\delta}{L}\right)^2 - 2\left(\frac{\delta}{L}\right)\sin\theta}\,\right] \tag{10.9}$$

$$\varepsilon = \frac{u}{L} = 1 - \sqrt{1 + \left(\frac{\delta}{L}\right)^2 - 2\left(\frac{\delta}{L}\right)\sin\theta} \tag{10.10}$$

The free body diagram of joint 2 in the deformed configuration of the truss is shown in Fig. 10.1 (c), in which the member axial forces Q are inclined in the deformed member directions (i.e., at angles $\bar{\theta}$ with the horizontal). By considering the equilibrium of the joint in the vertical direction, we write the equilibrium equation

$$Q = \frac{P}{2\sin\bar{\theta}} = \frac{P\sqrt{1 + \left(\frac{\delta}{L}\right)^2 - 2\left(\frac{\delta}{L}\right)\sin\theta}}{2\left[\sin\theta - \left(\frac{\delta}{L}\right)\right]} \tag{10.11}$$

which yields the expression for member axial stress as

$$\sigma = \frac{Q}{A} = \frac{P\sqrt{1 + \left(\frac{\delta}{L}\right)^2 - 2\left(\frac{\delta}{L}\right)\sin\theta}}{2A\left[\sin\theta - \left(\frac{\delta}{L}\right)\right]} \tag{10.12}$$

Finally, by substituting the expressions for strain (Eq. (10.10)) and stress (Eq. (10.12)) into the stress-strain relationship $\sigma = E\varepsilon$ (Eq. (10.5)), we obtain the desired nonlinear (stiffness) relationship between the load P and the displacement δ of the truss:

$$\boxed{P = 2EA\left[\sin\theta - \left(\frac{\delta}{L}\right)\right]\left[\frac{1 - \sqrt{1 + \left(\frac{\delta}{L}\right)^2 - 2\left(\frac{\delta}{L}\right)\sin\theta}}{\sqrt{1 + \left(\frac{\delta}{L}\right)^2 - 2\left(\frac{\delta}{L}\right)\sin\theta}}\right]} \tag{10.13}$$

By comparing the equations used to obtain the linear solution (Eqs. (10.1) through (10.6)) with those used to derive the geometrically nonlinear solution (Eqs. (10.7) through (10.13)), we notice two basic differences between the two types of analyses. The first difference is in the expressions of member axial deformation u in terms of the joint displacement δ (Eqs. (10.1) vs. (10.9)). In the linear analysis, δ is assumed to be so small that u can be expressed as the component of δ in the undeformed member direction, thereby yielding a linear relationship between u and δ (Eq. (10.1). In the geometrically nonlinear formulation, δ is allowed to be arbitrarily large and the relationship between u and δ is based on the exact geometry of the member's deformed configuration, thereby yielding a highly nonlinear relationship between u and δ (Eq. (10.9). The second basic distinction between the two types of analyses is in the way the equilibrium equations are established (Eqs. (10.3) vs. (10.11)). In linear analysis, the equilibrium equation is based on the undeformed geometry of the truss (thereby neglecting the joint displacement δ altogether). This assumption yields a direct linear relationship between the member axial force Q and the joint load P, which does not involve δ (Eq. (10.3)). In geometrically nonlinear analysis, however, since the equilibrium equation is based on the deformed configuration of the truss, the expression for Q not only contains P, but also involves nonlinear functions of δ (Eq. (10.11)). The reader is encouraged to verify that the linear solution (Eq. (10.6)) can be obtained by linearizing the geometrically nonlinear solution, that is, by expanding Eq. (10.13) via series expansion and retaining only the linear term of the series.

Geometrically nonlinear analysis provides important insight into the stability behavior of structures that is beyond the reach of linear analysis. Figure 10.1(d) shows the response of a typical shallow two-member truss as predicted by the linear and geometrically nonlinear analyses. These load-displacement plots are computed using the numerical values: $\theta = 30^{\circ}$, $L = 3\text{m}$, $E = 70$ GPa, and $A = 645.2\ \text{mm}^2$. It can be seen from this figure that the accuracy of the linear analysis gradually deteriorates as the magnitude of load P increases and the linear solution deviates from the exact geometrically nonlinear solution. With increasing load, the response becomes increasingly nonlinear as the truss's stiffness progressively decreases. This decrease in stiffness is characterized by a decrease in the slope of the tangent of the load-displacement curve. Note that at point a, where the curve reaches a peak, the slope of its tangent (called *tangent stiffness*) becomes zero, indicating that the structure's resistance to any further increase in load has vanished. Point a is referred to as a *critical* or *limit point*, because any further increase in load causes the truss to *snap-through* into an inverted configuration defined by point b on the response curve. The displacement of the truss at (or just prior to reaching) the critical point can be determined by setting to zero the derivative of Eq. (10.13) with respect to δ. This yields

$$\delta_{cr} = L\left[\sin\theta - \cos\theta\sqrt{\frac{1}{(\cos\theta)^{2/3}} - 1}\right] \tag{10.14}$$

The critical load P_{cr}, at which the *snap-through instability* occurs, can be calculated by substituting the value of δ_{cr} obtained from Eq. (10.14), for δ into Eq. (10.13).

We can see from Fig. 10.1(d) that the equilibrium configurations defined by the portion of the response curve between points a and b correspond to load levels below the critical level. Thus, unless the load magnitude can somehow be reduced after it has reached P_{cr}, the truss will snap-through from configuration a into the inverted configuration b (Fig. 10.1(e)).

It should be pointed out that the nonlinear response of the two-member truss (also known as the von Mises truss) considered herein has been examined by a number of researchers, and it is frequently used as a benchmark to validate the accuracy of computer programs for geometrically nonlinear structural analysis. It has been shown in references [16, 17] that while the shallow trusses (with $\theta \leq 69.295^{\circ}$) exhibit snap-through instability as discussed in the preceding paragraphs, the steep two-member trusses, with $\theta > 69.295^{\circ}$, experience bifurcation type of instability. As bifurcation instability occurs when the truss loses its stiffness in the horizontal direction, its detection requires analysis of a two degree-of-freedom model of the truss, instead of the single degree-of-freedom model used herein. A general formulation for geometrically nonlinear analysis of multi degree-of-freedom plane trusses is developed in the next section.

10.2 GEOMETRICALLY NONLINEAR ANALYSIS OF PLANE TRUSSES

In this section, we develop a general matrix stiffness method for geometrically nonlinear analysis of plane trusses [33, 45]. The process of developing the analytical models for nonlinear analysis (i.e., establishing a global coordinate system, numbering of joint and members, and identifying degrees of freedom and restrained coordinates) is the same as that for linear analysis of plane trusses (Chapter 3). However, the member local coordinate systems are now defined differently than in the case of linear analysis. Recall from Chapter 3 that in linear analysis, the local coordinate system is positioned in the *initial undeformed state* of the member, and it remains in that position regardless of where the member actually displaces due to the effect of external loads. In geometrically nonlinear analysis, it is more convenient to use a local coordinate system that is attached to, and displaces (translates and/or rotates) with, the member as the structure deforms. As shown in Fig. 10.2, *the origin of the local xyz coordinate system for a member is always located at the beginning, b′, of*

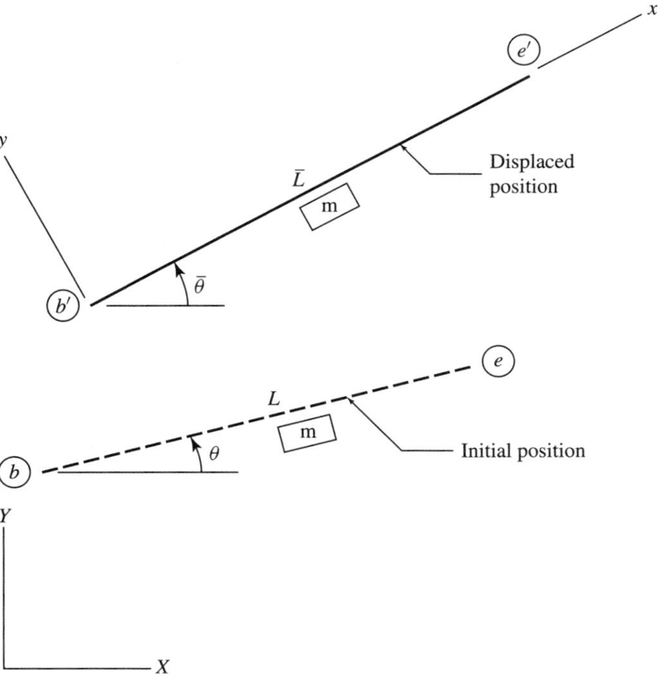

Fig. 10.2 *Corotational (or Eulerian) Local Coordinate System*

the member in its deformed state, with the x axis directed along the member's centroidal axis in the deformed state. The positive direction of the y axis is defined so that the coordinate system is right-handed, with the local z axis pointing in the positive direction of the global Z axis. This type of coordinate system, which continuously displaces with the member, is called an *Eulerian* or *corotational* coordinate system. The main advantage of using such a coordinate system is that it enables us to separate the member's axial deformation from its rigid body displacement, which is considered to be arbitrarily large in geometrically nonlinear analysis.

Member Force-Displacement Relations

To establish the member force-displacement relations, let us focus our attention on an arbitrary prismatic member m of a plane truss. When the truss is subjected to external loads, member m deforms and axial forces are induced at its ends. Figure 10.3 shows the displaced position of the member in its local coordinate system. Note that because of the modified definition of the local coordinate system, only one degree of freedom (that is the member axial deformation) is now needed to completely specify the displaced position of the member. The axial deformation u of the member can be expressed in terms of its initial and deformed lengths (L and $\bar{L}$, respectively) as

$$\boxed{u = L - \bar{L}} \tag{10.15}$$

in which, the axial deformation u is considered as positive when it corresponds to the shortening of the member's length, and negative when representing the elongation. Similarly, the member axial force Q is considered to be positive when compressive, and negative when tensile. To establish the relationship between the member's axial force Q and deformation u, we recall that the member's axial stress σ and axial strain ε are defined as

$$\sigma = \frac{Q}{A} \qquad \text{and} \qquad \varepsilon = \frac{u}{L} \tag{10.16}$$

and for linear elastic material, the stress-strain relationship is given by

$$\sigma = E\varepsilon \tag{10.17}$$

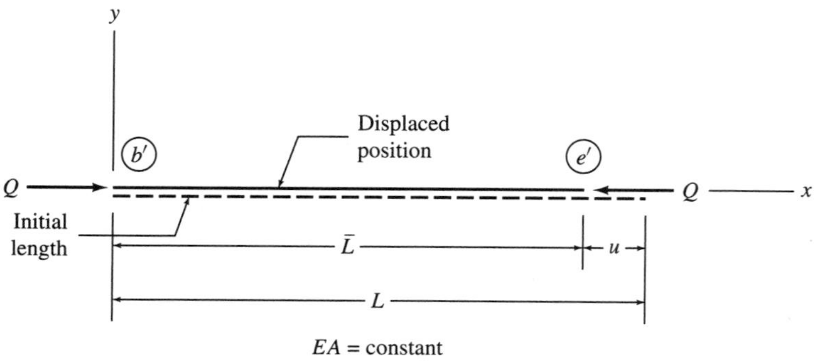

Fig. 10.3 *Member Axial Force and Deformation in the Local Coordinate System*

Substitution of Eqs. (10.16) into Eq. (10.17) yields the following force-displacement relation for the members of plane trusses in their local coordinate systems:

$$Q = \left(\frac{EA}{L}\right) u \tag{10.18}$$

Next, we consider the member force-displacement relations in the global coordinate system. Figures 10.4 (a) and (b) show the initial and displaced positions of an arbitrary member m of a plane truss. In Fig. 10.4(a), the member is depicted to be in equilibrium under the action of (local) axial forces Q; whereas in Fig. 10.4(b), the same member is shown to be in equilibrium under the action of an equivalent system of end forces **F** acting in the directions of the global X and Y coordinate axes. As indicated in Fig. 10.4(b), the global member end forces **F** and end displacements **v** are numbered in the same manner as in the case of linear analysis (Chapter 3), except that the forces **F** now act at the ends of the member in its deformed state.

Now, suppose that a member's global end displacements **v** (which may be arbitrarily large) are specified, and our objective is to find the corresponding end forces **F** so that the member is in equilibrium in its displaced position. If X_b, Y_b, and X_e, Y_e denote the global coordinates of the joints in their undeformed configurations, to which the member ends b and e, respectively are attached, then the initial (undeformed) length L of the member can be expressed (via Pythagorean theorem) as

$$L = \sqrt{(X_e - X_b)^2 + (Y_e - Y_b)^2} \tag{10.19}$$

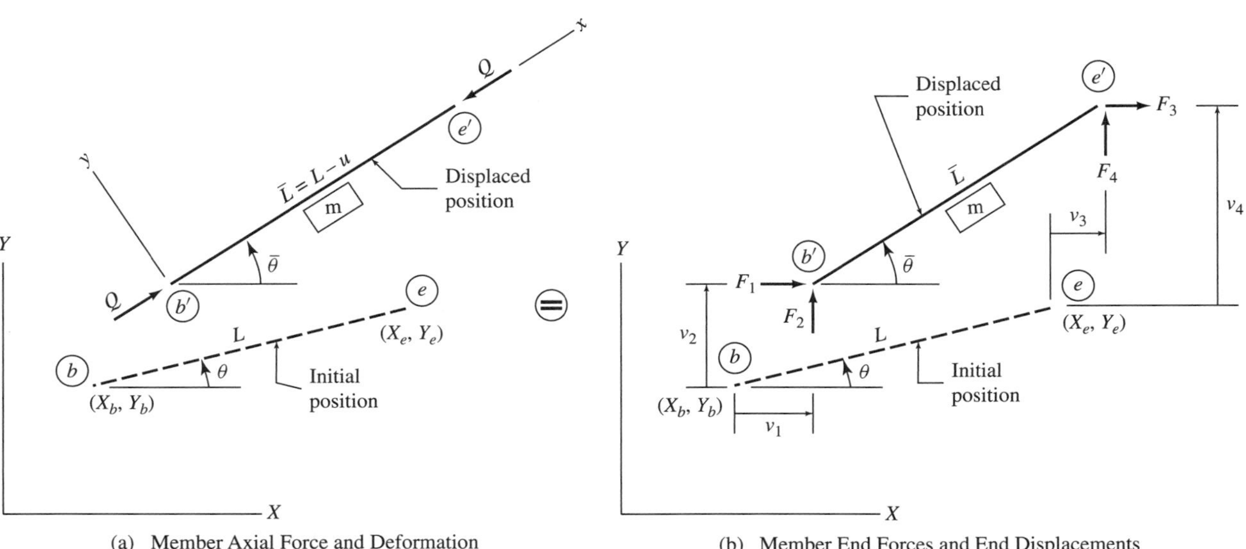

(a) Member Axial Force and Deformation in Local Coordinate System

(b) Member End Forces and End Displacements in the Global Coordinate System

Fig. 10.4

Similarly, we can see from Fig. 10.4(b) that the deformed member length $\bar{L}$, and the direction cosines of the member in its displaced position, can be expressed in terms of the initial global coordinates and displacements $\mathbf{v}$ of the member's ends by the following relationships:

$$\boxed{\bar{L} = \sqrt{[(X_e + v_3) - (X_b + v_1)]^2 + [(Y_e + v_4) - (Y_b + v_2)]^2}} \qquad \textbf{(10.20)}$$

$$\boxed{c_X = \cos\bar{\theta} = \frac{(X_e + v_3) - (X_b + v_1)}{\bar{L}}} \qquad \textbf{(10.21)}$$

$$\boxed{c_Y = \sin\bar{\theta} = \frac{(Y_e + v_4) - (Y_b + v_2)}{\bar{L}}} \qquad \textbf{(10.22)}$$

in which, $\bar{\theta}$ represents the angle measured counterclockwise from the positive direction of the global X axis to the positive direction of the local x axis of the member in its displaced position.

By comparing Figs. 10.4(a) and (b), we observe that at end b' of the member, the global forces F_1 and F_2 must be, respectively, equal to the components of (local) axial force Q in the directions of the global X and Y axes; that is,

$$F_1 = c_X Q \qquad \textbf{(10.23a)}$$

$$F_2 = c_Y Q \qquad \textbf{(10.23b)}$$

By using the same reasoning at end e', we express the global forces F_3 and F_4 in terms of Q as

$$F_3 = -c_X Q \qquad \textbf{(10.23c)}$$

$$F_4 = -c_Y Q \qquad \textbf{(10.23d)}$$

Equations 10.23(a) through (d) can be written in matrix form as

$$\begin{bmatrix} F_1 \\ F_2 \\ F_3 \\ F_4 \end{bmatrix} = \begin{bmatrix} c_X \\ c_Y \\ -c_X \\ -c_Y \end{bmatrix} Q \qquad \textbf{(10.24)}$$

or, symbolically as

$$\boxed{\mathbf{F} = \mathbf{T}^T\, Q} \qquad \textbf{(10.25)}$$

with the 1×4 transformation matrix $\mathbf{T}$ given by

$$\mathbf{T} = [\, c_X \quad c_Y \quad -c_X \quad -c_Y \,] \qquad \textbf{(10.26)}$$

It should be recognized that the set of Eqs. (10.15), (10.18) through (10.22), and (10.24), does implicitly express $\mathbf{F}$ in terms of $\mathbf{v}$, and therefore, is considered to represent the geometrically nonlinear force-displacement relations for members of plane trusses in the global coordinate system. Because of the highly nonlinear nature of some of the equations involved, it is quite cumbersome to express **F** explicitly in terms of **v**, as was previously done in the case of linear analysis. Note that if the global end displacements **v** of a member are known, its corresponding end forces **F** can be evaluated by first calculating L, $\bar{L}$, c_X, and c_Y using Eqs. (10.19) through (10.22); then evaluating the member axial deformation u and force Q, respectively, by applying Eqs. (10.15) and (10.18); and finally determining the member global end forces **F** via Eq. (10.24). It is important to realize that of all these equations, only one, that is, $Q = EAu/L$ (Eq. (10.18)), involves the material properties of the member. The remaining equations are essentially of a geometric character, and are *exact* in the sense that they are valid for arbitrarily large joint displacements.

The foregoing equations are also necessary and sufficient for establishing the geometrically nonlinear load-deformation relationships for the entire structure. The procedure for establishing such relations is essentially the same as in the case of linear analysis, and involves using member code numbers as illustrated by the following example.

EXAMPLE 10.1

By using geometrically nonlinear analysis, determine the joint loads **P** that cause the two-member truss to deform into the configuration shown in Fig. 10.5(a) on the next page.

SOLUTION

Joint Displacements: Using the analytical model of the truss shown in Fig. 10.5(b), we express the given deformed configuration in terms of its joint displacement vector,

$$\mathbf{d} = \begin{bmatrix} 10 \\ -4 \end{bmatrix} \begin{matrix} 1 \\ 2 \end{matrix} \text{ in.}$$

Member End Forces: The joint load vector **P**, corresponding to **d**, can be determined by performing the following operations for each member of the truss: (a) obtain the member's global end displacements **v** from **d** using the member's code numbers; (b) calculate the member's global end forces **F** using Eqs. (10.15), (10.18) through (10.22), and (10.24); and (c) store the elements of **F** into their proper positions in **P** and the support reaction vector **R**, using the member code numbers. Thus,

Member 1 $L = 60$ in., $X_l = Y_l = 0$, $X_2 = 48$ in., $Y_2 = 36$ in. By using the member code numbers 3, 4, 1, 2, we obtain

$$\mathbf{v}_1 = \begin{bmatrix} 0 \\ 0 \\ 10 \\ -4 \end{bmatrix} \begin{matrix} 3 \\ 4 \\ 1 \\ 2 \end{matrix} \text{ in.}$$

By applying Eqs. (10.20) thru (10.22), we compute the length and direction cosines of the member in the displaced position to be

$$\bar{L} = \sqrt{[(48 + 10) - (0)]^2 + [(36 - 4) - (0)]^2} = 66.24198 \text{ in.}$$

$$c_X = \frac{(48 + 10) - (0)}{66.24198} = 0.8755777 \qquad c_Y = \frac{(36 - 4) - (0)}{66.24198} = 0.4830773$$

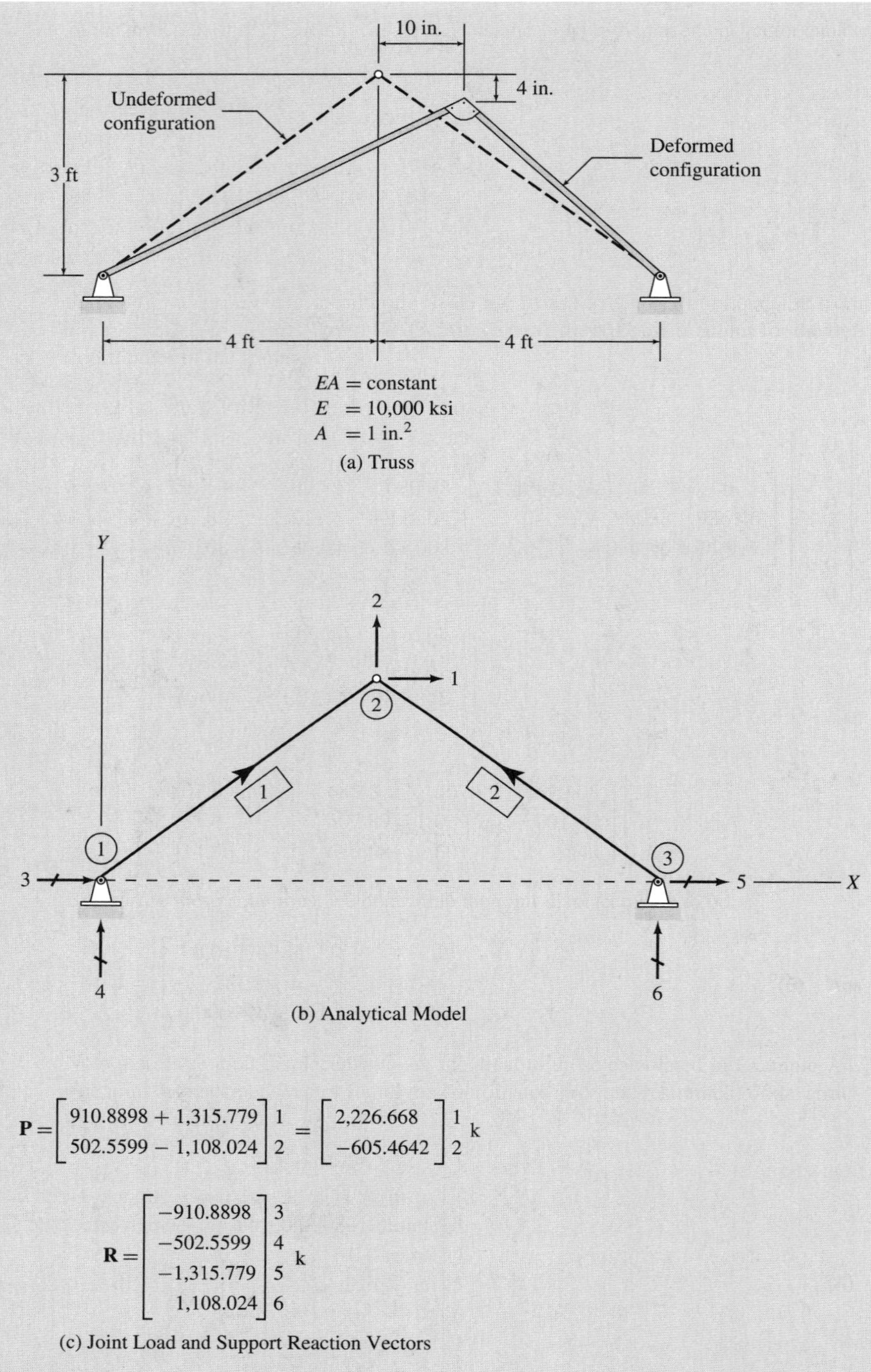

$$\mathbf{P} = \begin{bmatrix} 910.8898 + 1{,}315.779 \\ 502.5599 - 1{,}108.024 \end{bmatrix} \begin{matrix} 1 \\ 2 \end{matrix} = \begin{bmatrix} 2{,}226.668 \\ -605.4642 \end{bmatrix} \begin{matrix} 1 \\ 2 \end{matrix} \; \text{k}$$

$$\mathbf{R} = \begin{bmatrix} -910.8898 \\ -502.5599 \\ -1{,}315.779 \\ 1{,}108.024 \end{bmatrix} \begin{matrix} 3 \\ 4 \\ 5 \\ 6 \end{matrix} \; \text{k}$$

(c) Joint Load and Support Reaction Vectors

Fig. 10.5

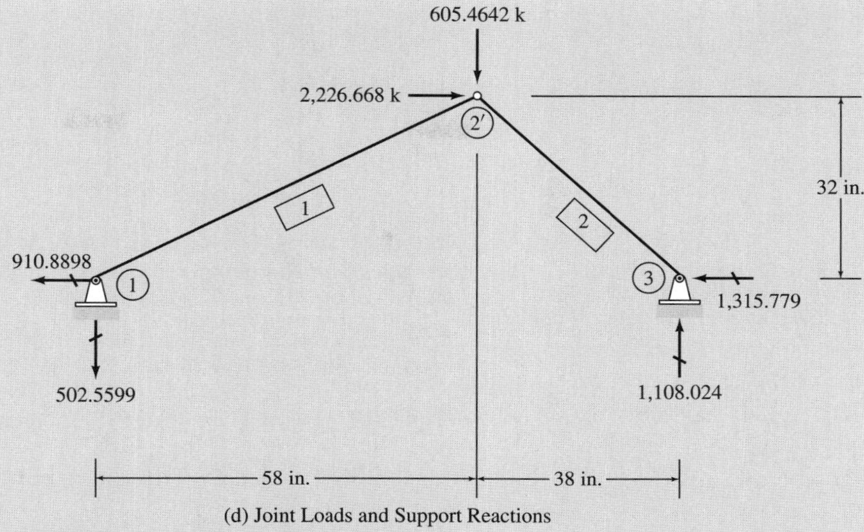

(d) Joint Loads and Support Reactions

Fig. 10.5 *(continued)*

Next, by using Eqs. (10.15) and (10.18), we calculate the member axial deformation and force as

$$u = L - \bar{L} = -6.24198 \text{ in.} \qquad Q = \left(\frac{EA}{L}\right) u = -1{,}040.33 \text{ k}$$

The member global end forces **F** can now be determined from Eq. (10.25):

$$\mathbf{F}_1 = \mathbf{T}^T Q = \begin{bmatrix} 0.8755777 \\ 0.4830773 \\ -0.8755777 \\ -0.4830773 \end{bmatrix} (-1{,}040.33) = \begin{bmatrix} -910.8898 \\ -502.5599 \\ \hdashline 910.8898 \\ 502.5599 \end{bmatrix} \begin{matrix} 3 \\ 4 \\ 1 \\ 2 \end{matrix} \text{ k}$$

The elements of $\mathbf{F}_1$ are stored in their proper positions in the 2×1 joint load vector **P** and the 4×1 reaction vector **R**, as shown in Fig. 10.5(c).

Member 2 $L = 60$ in., $X_3 = 96$ in., $Y_3 = 0$, $X_2 = 48$ in., $Y_2 = 36$ in.

$$\mathbf{v}_2 = \begin{bmatrix} 0 \\ 0 \\ 10 \\ -4 \end{bmatrix} \begin{matrix} 5 \\ 6 \\ 1 \\ 2 \end{matrix} \text{ in.}$$

$$\bar{L} = \sqrt{[(48 + 10) - (96 + 0)]^2 + [(36 - 4) - (0)]^2} = 49.67897 \text{ in.}$$

$$c_X = \frac{(48 + 10) - (96 + 0)}{49.67897} = -0.7649112$$

$$c_Y = \frac{(36 - 4) - (0)}{49.67897} = 0.6441357$$

$$u = L - \bar{L} = 10.32103 \text{ in.} \qquad Q = \left(\frac{EA}{L}\right) u = 1{,}720.172 \text{ k}$$

$$\mathbf{F}_2 = \mathbf{T}^T Q = \begin{bmatrix} -0.7649112 \\ 0.6441357 \\ 0.7649112 \\ -0.6441357 \end{bmatrix} (1{,}720.172) = \begin{bmatrix} -1{,}315.779 \\ 1{,}108.024 \\ \hdashline 1{,}315.779 \\ -1{,}108.024 \end{bmatrix} \begin{matrix} 5 \\ 6 \\ 1 \\ 2 \end{matrix} \text{ k}$$

Joint Loads and Support Reactions: The completed joint load vector **P** and the support reaction vector **R** are shown in Fig. 10.5(c), and these forces are depicted on a line diagram of the deformed configuration of the truss in Fig. 10.5(d). **Ans**

Equilibrium Check: Applying the equations of equilibrium to the free body of the truss in its deformed state (Fig. 10.5(d)), we obtain

$$+ \rightarrow \sum F_X = 0 \qquad -910.8898 + 2{,}226.668 - 1{,}315.779 = -0.0008 \text{ k} \approx 0 \qquad \textbf{Checks}$$

$$+ \uparrow \sum F_Y = 0 \qquad -502.5599 - 605.4642 + 1{,}108.024 = -0.0001 \text{ k} \approx 0 \qquad \textbf{Checks}$$

$$+ \circlearrowleft \sum M_{①} = 0 \qquad -2{,}226.668(32) - 605.4642(58) + 1{,}108.024(96) = 0.0044 \text{ k-in.} \approx 0 \qquad \textbf{Checks}$$

Member Tangent Stiffness Matrix

As indicated by the foregoing example, when the deformed configuration **d** of a truss is known, the corresponding joint loads **P**, required to cause (and/or keep the structure in equilibrium in) that deformed configuration, can be determined by direct application of the nonlinear force-displacement relations derived in the preceding subsection. However, in most practical situations, it is the external loading that is specified, and the objective of the analysis is to determine the corresponding deformed configuration of the structure, thereby requiring the solution of a system of simultaneous nonlinear equations. The computational techniques commonly used for solving such systems of nonlinear equations are iterative in nature, and usually involve solving a linearized form of the structure's load-deformation relations repeatedly to move closer to the (yet unknown) exact nonlinear solution. Thus, before we discuss such a computational technique in a subsequent subsection, we develop the linearized form of the force-displacement relations for the planes truss members in the global coordinate system.

The member force-displacement relationships can be written in terms of differentials as

$$\boxed{\Delta\mathbf{F} = \mathbf{K}_t \, \Delta\mathbf{v}} \tag{10.27}$$

in which $\Delta\mathbf{F}$ and $\Delta\mathbf{v}$ denote increments of member global end forces **F** and end displacements **v**, respectively, and

$$\mathbf{K}_t = \left[\frac{\partial F_i}{\partial v_j}\right]; \text{ for i, j} = 1 \text{ to } 4 \tag{10.28}$$

is called the *member tangent stiffness matrix in the global coordinate system.* Equation (10.28) can be expanded into

$$\mathbf{K}_t = \begin{bmatrix} \frac{\partial F_1}{\partial v_1} & \frac{\partial F_1}{\partial v_2} & \frac{\partial F_1}{\partial v_3} & \frac{\partial F_1}{\partial v_4} \\ \frac{\partial F_2}{\partial v_1} & \frac{\partial F_2}{\partial v_2} & \frac{\partial F_2}{\partial v_3} & \frac{\partial F_2}{\partial v_4} \\ \frac{\partial F_3}{\partial v_1} & \frac{\partial F_3}{\partial v_2} & \frac{\partial F_3}{\partial v_3} & \frac{\partial F_3}{\partial v_4} \\ \frac{\partial F_4}{\partial v_1} & \frac{\partial F_4}{\partial v_2} & \frac{\partial F_4}{\partial v_3} & \frac{\partial F_4}{\partial v_4} \end{bmatrix} \tag{10.29}$$

To determine the explicit form of $\mathbf{K}_t$, we differentiate the expressions of F_1 through F_4 (Eqs. 10.23(a) through (d)) partially with respect to v_1 through v_4, respectively. Thus, by differentiating the expression of F_1 (Eq. 10.23(a)) partially with respect to v_1, we write

$$\frac{\partial F_1}{\partial v_1} = c_X \left(\frac{\partial Q}{\partial v_1}\right) + Q \left(\frac{\partial c_X}{\partial v_1}\right) \tag{10.30a}$$

To obtain $\partial Q/\partial v_1$, we substitute Eqs. (10.15) and (10.20), respectively, into Eq. (10.18), and differentiate the resulting equation partially with respect to v_1, thereby yielding

$$\frac{\partial Q}{\partial v_1} = \left(\frac{EA}{L}\right) c_X \tag{10.30b}$$

Similarly, by substituting Eq.(10.20) into Eq. (10.21), and differentiating the resulting equation partially with respect to v_1, we obtain

$$\frac{\partial c_X}{\partial v_1} = -\frac{c_Y^2}{\bar{L}} \tag{10.30c}$$

and finally, by substituting Eqs. (10.30b) and (10.30c) into Eq. (10.30a), we obtain the desired partial derivative as

$$\frac{\partial F_1}{\partial v_1} = \left(\frac{EA}{L}\right) c_X^2 - \left(\frac{Q}{\bar{L}}\right) c_Y^2 \tag{10.30d}$$

The remaining partial derivatives of F_i with respect to v_j can be derived in a similar manner. The explicit form of the member global tangent stiffness matrix thus obtained is [45, 46]

$$\mathbf{K}_t = \frac{EA}{L} \begin{bmatrix} c_X^2 & c_X c_Y & -c_X^2 & -c_X c_Y \\ c_X c_Y & c_Y^2 & -c_X c_Y & -c_Y^2 \\ -c_X^2 & -c_X c_Y & c_X^2 & c_X c_Y \\ -c_X c_Y & -c_Y^2 & c_X c_Y & c_Y^2 \end{bmatrix} + \frac{Q}{\bar{L}} \begin{bmatrix} -c_Y^2 & c_X c_Y & c_Y^2 & -c_X c_Y \\ c_X c_Y & -c_X^2 & -c_X c_Y & c_X^2 \\ c_Y^2 & -c_X c_Y & -c_Y^2 & c_X c_Y \\ -c_X c_Y & c_X^2 & c_X c_Y & -c_X^2 \end{bmatrix} \tag{10.31}$$

The reader should note that, in the initial (undeformed) configuration of the member (when $\bar{\theta} = \theta$ and $Q = 0$), the tangent stiffness matrix $\mathbf{K}_t$ reduces to the conventional stiffness matrix K derived previously for linear analysis of plane trusses in Chapter 3 (Eq. (3.73)).

Equation (10.31) is often written in compact form as

$$\boxed{\mathbf{K}_t = \left(\frac{EA}{L}\right)\mathbf{T}^T\mathbf{T} + Q\,\mathbf{g}} \tag{10.32}$$

in which the *geometric matrix* **g** is given by

$$\mathbf{g} = \frac{1}{\bar{L}}\begin{bmatrix} -c_Y^2 & c_X c_Y & c_Y^2 & -c_X c_Y \\ c_X c_Y & -c_X^2 & -c_X c_Y & c_X^2 \\ c_Y^2 & -c_X c_Y & -c_Y^2 & c_X c_Y \\ -c_X c_Y & c_X^2 & c_X c_Y & -c_X^2 \end{bmatrix} \tag{10.33}$$

Structure Load-Deformation Relations

The geometrically nonlinear structure load-deformation relations for plane trusses are expressed in the form of joint equilibrium equations:

$$\boxed{\mathbf{P} = \mathbf{f}(\mathbf{d})} \tag{10.34}$$

in which, **f** is referred to as the *internal joint force vector* of the structure. The vector **f** contains the resultants of internal member end forces at the locations, and in the directions of, the structure's degrees of freedom. As shown previously in Example 10.1, the resultant internal forces **f** are nonlinear functions of the structure's joint displacements **d**. It should be noted that in Example 10.1, where the structure's deformed configuration **d** was known, no distinction was necessary between **P** and **f**, and the former was assembled directly from the member end forces **F** via the member code numbers. However, when analyzing a structure for its unknown deformed configuration caused by a specified external loading, it becomes necessary to distinguish between the external loads **P**, which remain constant throughout the iterative process, and the internal forces **f** that vary as the structure's assumed deformed configuration **d** is revised iteratively until **f** becomes sufficiently close to **P**, that is, the structure's equilibrium equations (Eq. (10.34)) are satisfied within a prescribed tolerance.

The structure load-deformation relationships (Eq. (10.34)) can be written in terms of differentials as

$$\boxed{\Delta\mathbf{P} = \mathbf{S}_t\,\Delta\mathbf{d}} \tag{10.35}$$

in which $\Delta\mathbf{P}$ and $\Delta\mathbf{d}$ denote increments of external loads $\mathbf{P}$ and joint displacements $\mathbf{d}$, respectively, and

$$\mathbf{S_t} = \left[\frac{\partial f_i}{\partial d_j}\right]; \quad \text{for i, j} = 1 \text{ to } NDOF \tag{10.36}$$

is called the *structure tangent stiffness matrix*. The structure matrix $\mathbf{S_t}$ can be conveniently assembled from the member global tangent stiffness matrices $\mathbf{K_t}$ using the member code numbers, as in the case of linear analysis.

Computational Technique—Newton-Raphson Method

Most computational techniques commonly used for nonlinear structural analysis are generally based on the classical Newton-Raphson iteration technique for root finding. Such an iterative method for geometrically nonlinear analysis of plane trusses is presented herein. The method of analysis is illustrated graphically for a single degree-of-freedom structure in Fig. 10.6.

Let us assume that our objective is to determine the deformed configuration (i.e., the joint displacements) $\mathbf{d}$ of a structure due to a given external loading $\mathbf{P}$. As shown in Fig. 10.6, we begin the process by performing the conventional linear analysis to determine the first approximate configuration $\mathbf{d}_1$ of the structure. Note that the linearized form of the nonlinear load-deformation relations (Eq. (10.35)) reduces to the conventional linear stiffness relations (Eq. (3.89))

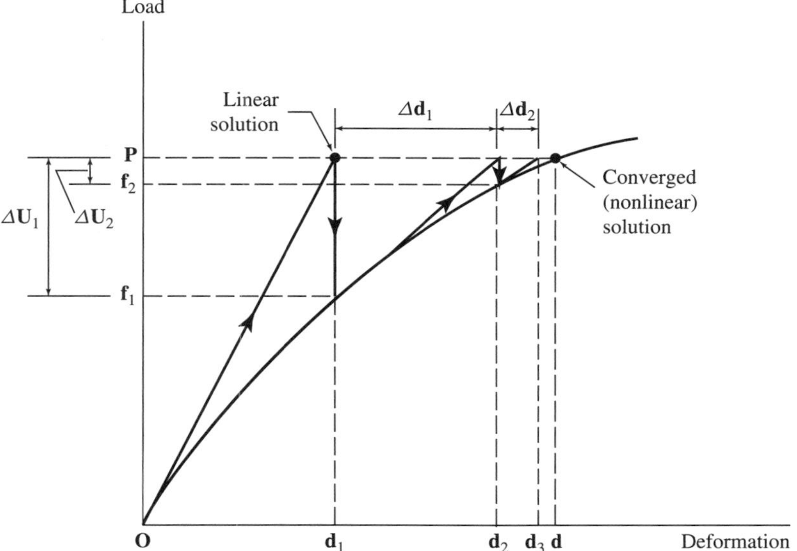

Fig. 10.6 *Newton-Raphson Method (for a Single Degree-of-Freedom Structure)*

when applied in the undeformed (initial) configuration of the structure, with $\Delta\mathbf{P} = \mathbf{P}$ and $\Delta\mathbf{d} = \mathbf{d}_1$; that is,

$$\mathbf{P} = \mathbf{S}_{t0}\,\mathbf{d}_1 = \mathbf{S}\mathbf{d}_1 \tag{10.37}$$

in which $\mathbf{S}_{t0}$ denotes the structure tangent stiffness matrix evaluated in the undeformed configuration (i.e., $\mathbf{d} = \mathbf{d}_0 = \mathbf{0}$).

It should be realized that the configuration $\mathbf{d}_1$, obtained by solving Eq. (10.37), represents an approximate configuration in the sense that the joint equilibrium equations (Eq. (10.34)) are not necessarily satisfied. To correct the approximate solution, we evaluate the structure's internal joint force vector, $\mathbf{f}_1 = \mathbf{f}(\mathbf{d}_1)$, corresponding to the configuration $\mathbf{d}_1$ and subtract it from the joint load vector $\mathbf{P}$ to calculate the *unbalanced joint force vector* for the structure

$$\Delta\mathbf{U}_1 = \mathbf{P} - \mathbf{f}_1 \tag{10.38}$$

The unbalanced joint forces $\Delta\mathbf{U}_1$ are now treated as a load increment and the correction vector $\Delta\mathbf{d}_1$ is obtained by applying the linearized incremental relationship (Eq. (10.35) as

$$\Delta\mathbf{U}_1 = \mathbf{S}_{t1}\,\Delta\mathbf{d}_1 \tag{10.39}$$

with $\mathbf{S}_{t1}$ now representing the structure tangent stiffness matrix evaluated in the configuration $\mathbf{d}_1$. A new approximate configuration $\mathbf{d}_2$ is then obtained by adding the correction vector $\Delta\mathbf{d}_1$ to the current configuration $\mathbf{d}_1$,

$$\mathbf{d}_2 = \mathbf{d}_1 + \Delta\mathbf{d}_1 \tag{10.40}$$

and the iteration is continued until the latest correction vector is sufficiently small.

Equations (10.38) through (10.40) refer to the first iteration cycle. For an ith iteration cycle, these equations can be expressed in recurrence form as:

$$\boxed{\Delta\mathbf{U}_i = \mathbf{P} - \mathbf{f}_i} \tag{10.41}$$

$$\boxed{\Delta\mathbf{U}_i = \mathbf{S}_{ti}\,\Delta\mathbf{d}_i} \tag{10.42}$$

$$\boxed{\mathbf{d}_{i+1} = \mathbf{d}_i + \Delta\mathbf{d}_i} \tag{10.43}$$

Various criteria can be used in deciding whether the iterative process has converged. In general, for structures exhibiting softening type of response, the convergence criteria based on the change in the structure's configuration between two consecutive iteration cycles do seem to yield reasonably accurate results. In the example presented in this chapter, we use a criterion based on a comparison of the changes, $\Delta\mathbf{d}_i$, in joint displacements to their cumulative

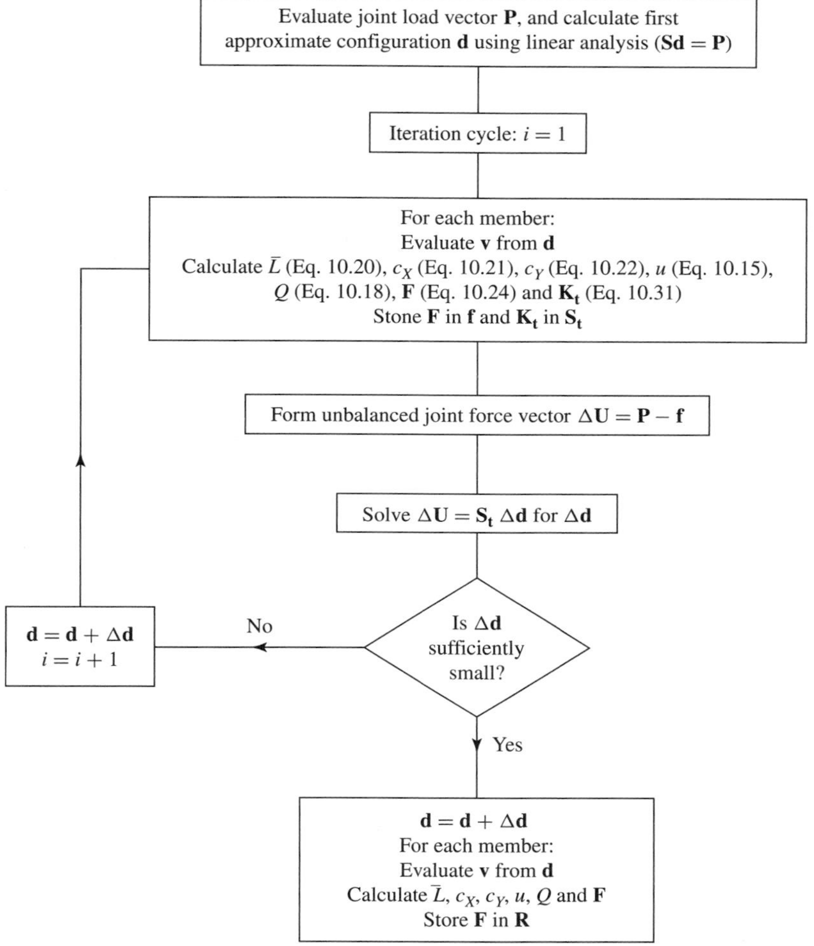

Fig. 10.7 *Procedure for Analysis*

values, $\mathbf{d}_i$, and consider the convergence to have occurred when the following inequality is satisfied

$$\sqrt{\frac{\sum_{j=1}^{NDOF}(\Delta d_j)^2}{\sum_{j=1}^{NDOF}(d_j)^2}} \leq e \tag{10.44}$$

in which the dimensionless quantity e represents a prescribed tolerance.

A block diagram summarizing the various steps of the method for geometrically nonlinear analysis of plane trusses is shown in Fig. 10.7. The method of analysis is illustrated by the following example.

EXAMPLE 10.2 Determine the joint displacements, member axial forces, and support reactions for the truss shown in Fig. 10.8(a) by geometrically nonlinear analysis. Use a displacement convergence tolerance of 0.1 percent.

SOLUTION *Analytical Model:* See Fig. 10.8(b). The truss has three degrees of freedom, numbered as 1, 2, and 3. The three restrained coordinates of the truss are identified by numbers 4, 5, and 6.

Linear Analysis: We begin by performing the conventional linear analysis of the truss subjected to the specified joint loads,

$$\mathbf{P} = \begin{bmatrix} 0 \\ -2{,}000 \\ 0 \end{bmatrix} \text{kN} \tag{1}$$

The member global stiffness matrices can be evaluated by using either Eq. (3.73), or Eq. (10.31) with $\bar{\theta} = \theta$ and $Q = 0$. These are:

$$\mathbf{K}_1 = \begin{array}{c} \begin{matrix} 4 & 5 & 1 & 2 \end{matrix} \\ \begin{bmatrix} 5{,}781 & 4{,}335.7 & -5{,}781 & -4{,}335.7 \\ 4{,}335.7 & 3{,}251.8 & -4{,}335.7 & -3{,}251.8 \\ -5{,}781 & -4{,}335.7 & 5{,}781 & 4{,}335.7 \\ -4{,}335.7 & -3{,}251.8 & 4{,}335.7 & 3{,}251.8 \end{bmatrix} \end{array} \begin{matrix} 4 \\ 5 \\ 1 \\ 2 \end{matrix} \text{ kN/m}$$

$$\mathbf{K}_2 = \begin{array}{c} \begin{matrix} 3 & 6 & 1 & 2 \end{matrix} \\ \begin{bmatrix} 5{,}781 & -4{,}335.7 & -5{,}781 & 4{,}335.7 \\ -4{,}335.7 & 3{,}251.8 & 4{,}335.7 & -3{,}251.8 \\ -5{,}781 & 4{,}335.7 & 5{,}781 & -4{,}335.7 \\ 4{,}335.7 & -3{,}251.8 & -4{,}335.7 & 3{,}251.8 \end{bmatrix} \end{array} \begin{matrix} 3 \\ 6 \\ 1 \\ 2 \end{matrix} \text{ kN/m}$$

$$\mathbf{K}_3 = \begin{array}{c} \begin{matrix} 4 & 5 & 3 & 6 \end{matrix} \\ \begin{bmatrix} 5{,}645.5 & 0 & -5{,}645.5 & 0 \\ 0 & 0 & 0 & 0 \\ -5{,}645.5 & 0 & 5{,}645.5 & 0 \\ 0 & 0 & 0 & 0 \end{bmatrix} \end{array} \begin{matrix} 4 \\ 5 \\ 3 \\ 6 \end{matrix} \text{ kN/m}$$

The structure stiffness matrix thus obtained is

$$\mathbf{S} = \begin{array}{c} \begin{matrix} 1 & 2 & 3 \end{matrix} \\ \begin{bmatrix} 11{,}562 & 0 & -5{,}781 \\ 0 & 6{,}503.6 & 4{,}335.7 \\ -5{,}781 & 4{,}335.7 & 11{,}426 \end{bmatrix} \end{array} \begin{matrix} 1 \\ 2 \\ 3 \end{matrix} \text{ kN/m}$$

By solving the linear system of equations $\mathbf{P} = \mathbf{S}\,\mathbf{d}_1$, we determine the first approximation configuration to be

$$\mathbf{d}_1 = \begin{bmatrix} 0.11809 \\ -0.46497 \\ 0.23618 \end{bmatrix} \begin{matrix} 1 \\ 2 \\ 3 \end{matrix} \text{ m}$$

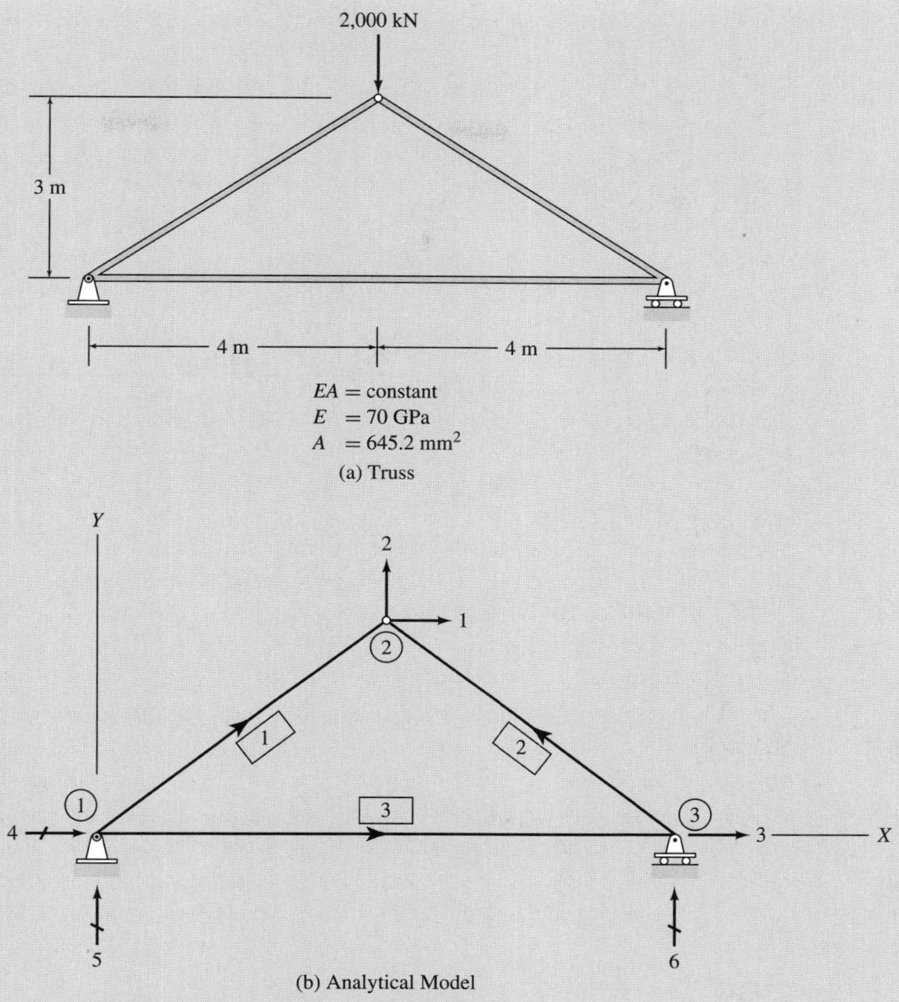

Fig. 10.8

Iteration Cycle 1: Next, for the current deformed configuration $\mathbf{d}_1$ of the structure, we evaluate its internal joint force vector $\mathbf{f}_1$ and the tangent stiffness matrix $\mathbf{S}_{t1}$ by assembling the pertinent elements of the member $\mathbf{F}$ vectors and $\mathbf{K}_t$ matrices, respectively, as follows:

Member 1

$$\mathbf{v} = \begin{bmatrix} 0 \\ 0 \\ 0.11809 \\ -0.46497 \end{bmatrix} \begin{matrix} 4 \\ 5 \\ 1 \\ 2 \end{matrix} \quad \text{m}$$

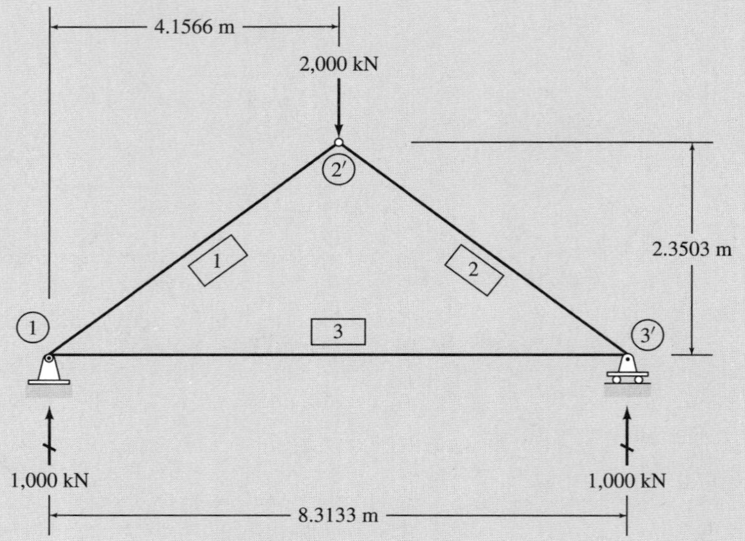

(c) Support Reactions

Fig. 10.8 *(continued)*

$L = 5$ m, $\bar{L} = 4.8358$ m, $c_X = 0.85158$, $c_Y = 0.52422$, $u = 0.16419$ m, $Q = 1{,}483.1$ kN

$$\mathbf{F} = \begin{bmatrix} 1{,}263 \\ 777.49 \\ \hdashline -1{,}263 \\ -777.49 \end{bmatrix} \begin{matrix} 4 \\ 5 \\ 1 \\ 2 \end{matrix} \text{ kN} \qquad \mathbf{K}_t = \begin{matrix} \begin{matrix} 4 & \quad 5 & \quad 1 & \quad 2 \end{matrix} \\ \begin{bmatrix} 6{,}466.2 & 4{,}169.3 & -6{,}466.2 & -4{,}169.3 \\ 4{,}169.3 & 2{,}259.9 & -4{,}169.3 & -2{,}259.9 \\ -6{,}466.2 & -4{,}169.3 & 6{,}466.2 & 4{,}169.3 \\ -4{,}169.3 & -2{,}259.9 & 4{,}169.3 & 2{,}259.9 \end{bmatrix} \end{matrix} \begin{matrix} 4 \\ 5 \\ 1 \\ 2 \end{matrix} \text{ kN/m} \qquad \textbf{(2)}$$

Member 2

$$\mathbf{v} = \begin{bmatrix} 0.23618 \\ 0 \\ 0.11809 \\ -0.46497 \end{bmatrix} \begin{matrix} 3 \\ 6 \\ 1 \\ 2 \end{matrix} \text{ m}$$

$L = 5$ m, $\bar{L} = 4.8358$ m, $c_X = -0.85158$, $c_Y = 0.52422$, $u = 0.16419$ m, $Q = 1{,}483.1$ kN

$$\mathbf{F} = \begin{bmatrix} -1{,}263 \\ \hdashline 777.49 \\ \hdashline 1{,}263 \\ -777.49 \end{bmatrix} \begin{matrix} 3 \\ 6 \\ 1 \\ 2 \end{matrix} \text{ kN} \qquad \mathbf{K}_t = \begin{matrix} \begin{matrix} 3 & \quad 6 & \quad 1 & \quad 2 \end{matrix} \\ \begin{bmatrix} 6{,}466.2 & -4{,}169.3 & -6{,}466.2 & 4{,}169.3 \\ -4{,}169.3 & 2{,}259.9 & 4{,}169.3 & -2{,}259.9 \\ -6{,}466.2 & 4{,}169.3 & 6{,}466.2 & -4{,}169.3 \\ 4{,}169.3 & -2{,}259.9 & -4{,}169.3 & 2{,}259.9 \end{bmatrix} \end{matrix} \begin{matrix} 3 \\ 6 \\ 1 \\ 2 \end{matrix} \text{ kN/m} \qquad \textbf{(3)}$$

Member 3

$$\mathbf{v} = \begin{bmatrix} 0 \\ 0 \\ 0.23618 \\ 0 \end{bmatrix} \begin{matrix} 4 \\ 5 \\ 3 \\ 6 \end{matrix} \ \text{m}$$

$L = 8$ m, $\bar{L} = 8.2362$ m, $c_X = 1$, $c_Y = 0$, $u = -0.23618$ m, $Q = -1{,}333.3$ kN

$$\mathbf{F} = \begin{bmatrix} -1{,}333.3 \\ 0 \\ \hdashline 1{,}333.3 \\ \hdashline 0 \end{bmatrix} \begin{matrix} 4 \\ 5 \\ 3 \\ 6 \end{matrix} \ \text{kN} \qquad \mathbf{K_t} = \begin{array}{c} \begin{matrix} \quad 4 & \quad\quad 5 & \quad\quad 3 & \quad\quad 6 \end{matrix} \\ \begin{bmatrix} 5{,}645.5 & 0 & -5{,}645.5 & 0 \\ 0 & 161.89 & 0 & -161.89 \\ -5{,}645.5 & 0 & \boxed{5{,}645.5} & 0 \\ 0 & -161.89 & 0 & 161.89 \end{bmatrix} \end{array} \begin{matrix} 4 \\ 5 \\ 3 \\ 6 \end{matrix} \ \text{kN/m} \quad \textbf{(4)}$$

By assembling the pertinent elements of the member **F** vectors and $\mathbf{K_t}$ matrices given in Eqs. (2) through (4), we obtain

$$\mathbf{f}_1 = \begin{bmatrix} 0 \\ -1{,}555 \\ 70.322 \end{bmatrix} \begin{matrix} 1 \\ 2 \\ 3 \end{matrix} \ \text{kN} \qquad \mathbf{S}_{t1} = \begin{array}{c} \begin{matrix} \quad 1 & \quad\quad 2 & \quad\quad 3 \end{matrix} \\ \begin{bmatrix} 12{,}932 & 0 & -6{,}466.2 \\ 0 & 4{,}519.7 & 4{,}169.3 \\ -6{,}466.2 & 4{,}169.3 & 12{,}112 \end{bmatrix} \end{array} \begin{matrix} 1 \\ 2 \\ 3 \end{matrix} \ \text{kN/m}$$

By subtracting $\mathbf{f}_1$ from **P** (Eq. 1), we compute the unbalanced joint force vector for the truss as

$$\Delta\mathbf{U}_1 = \mathbf{P} - \mathbf{f}_1 = \begin{bmatrix} 0 \\ -2{,}000 \\ 0 \end{bmatrix} - \begin{bmatrix} 0 \\ -1{,}555 \\ 70.322 \end{bmatrix} = \begin{bmatrix} 0 \\ -445.02 \\ -70.322 \end{bmatrix} \begin{matrix} 1 \\ 2 \\ 3 \end{matrix} \ \text{kN}$$

By solving the linearized system of equations $\Delta\mathbf{U}_1 = \mathbf{S}_{t1}\,\Delta\mathbf{d}_1$, we determine the displacement correction vector

$$\Delta\mathbf{d}_1 = \begin{bmatrix} 0.03380 \\ -0.16082 \\ 0.067599 \end{bmatrix} \begin{matrix} 1 \\ 2 \\ 3 \end{matrix} \ \text{m}$$

To determine whether or not the iteration has converged, we apply the convergence criterion (Eq. (10.44)) as

$$\sqrt{\frac{\sum\limits_{j=1}^{3} (\Delta d_j)^2}{\sum\limits_{j=1}^{3} (d_j)^2}} = \sqrt{\frac{(0.03380)^2 + (-0.16082)^2 + (0.067599)^2}{(0.11809)^2 + (-0.46497)^2 + (0.23618)^2}} = 0.33231 > e\ (= 0.001)$$

which indicates that the change in the structure's configuration is not sufficiently small, and therefore, another iteration is needed based on a new (second) approximate deformed configuration $\mathbf{d}_2$ obtained by adding the correction $\Delta\mathbf{d}_1$ to the previous (first) approximate configuration $\mathbf{d}_1$, that is,

$$\mathbf{d}_2 = \mathbf{d}_1 + \Delta\mathbf{d}_1 = \begin{bmatrix} 0.11809 \\ -0.46497 \\ 0.23618 \end{bmatrix} + \begin{bmatrix} 0.03380 \\ -0.16082 \\ 0.067599 \end{bmatrix} = \begin{bmatrix} 0.15189 \\ -0.62579 \\ 0.30378 \end{bmatrix} \text{ m}$$

Iteration Cycle 2:

Member 1

$$\mathbf{v} = \begin{bmatrix} 0 \\ 0 \\ 0.15189 \\ -0.62579 \end{bmatrix} \begin{matrix} 4 \\ 5 \\ 1 \\ 2 \end{matrix} \text{ m}$$

$L = 5$ m, $\bar{L} = 4.7828$ m, $c_X = 0.86809$, $c_Y = 0.49641$, $u = 0.21721$ m, $Q = 1{,}962.1$ kN

$$\mathbf{F} = \begin{bmatrix} 1{,}703.2 \\ 973.98 \\ -1{,}703.2 \\ -973.98 \end{bmatrix} \begin{matrix} 4 \\ 5 \\ 1 \\ 2 \end{matrix} \text{ kN} \quad \mathbf{K}_t = \begin{matrix} & 4 & 5 & 1 & 2 & \\ & \left[\begin{matrix} 6{,}705.8 \\ 4{,}069.2 \\ -6{,}705.8 \\ -4{,}069.2 \end{matrix}\right. & \begin{matrix} 4{,}069.2 \\ 1{,}916.7 \\ -4{,}069.2 \\ -1{,}916.7 \end{matrix} & \begin{matrix} -6{,}705.8 \\ -4{,}069.2 \\ 6{,}705.8 \\ 4{,}069.2 \end{matrix} & \left.\begin{matrix} -4{,}069.2 \\ -1{,}916.7 \\ 4{,}069.2 \\ 1{,}916.7 \end{matrix}\right] & \begin{matrix} 4 \\ 5 \\ 1 \\ 2 \end{matrix} \end{matrix} \text{ kN/m}$$

Member 2

$$\mathbf{v} = \begin{bmatrix} 0.30378 \\ 0 \\ 0.15189 \\ -0.62579 \end{bmatrix} \begin{matrix} 3 \\ 6 \\ 1 \\ 2 \end{matrix} \text{ m}$$

$L = 5$ m, $\bar{L} = 4.7828$ m, $c_X = -0.86809$, $c_Y = 0.49641$, $u = 0.21721$ m, $Q = 1{,}962.1$ kN

$$\mathbf{F} = \begin{bmatrix} -1{,}703.2 \\ 973.98 \\ 1{,}703.2 \\ -973.98 \end{bmatrix} \begin{matrix} 3 \\ 6 \\ 1 \\ 2 \end{matrix} \text{ kN} \quad \mathbf{K}_t = \begin{matrix} & 3 & 6 & 1 & 2 & \\ & \left[\begin{matrix} 6{,}705.8 \\ -4{,}069.2 \\ -6{,}705.8 \\ 4{,}069.2 \end{matrix}\right. & \begin{matrix} -4{,}069.2 \\ 1{,}916.7 \\ 4{,}069.2 \\ -1{,}916.7 \end{matrix} & \begin{matrix} -6{,}705.8 \\ 4{,}069.2 \\ 6{,}705.8 \\ -4{,}069.2 \end{matrix} & \left.\begin{matrix} 4{,}069.2 \\ -1{,}916.7 \\ -4{,}069.2 \\ 1{,}916.7 \end{matrix}\right] & \begin{matrix} 3 \\ 6 \\ 1 \\ 2 \end{matrix} \end{matrix} \text{ kN/m}$$

Member 3

$$\mathbf{v} = \begin{bmatrix} 0 \\ 0 \\ 0.30378 \\ 0 \end{bmatrix} \begin{matrix} 4 \\ 5 \\ 3 \\ 6 \end{matrix} \text{ m}$$

$$L = 8 \text{ m}, \bar{L} = 8.3038 \text{ m}, c_X = 1, c_Y = 0, u = -0.30378 \text{ m}, Q = -1{,}715 \text{ kN}$$

$$\mathbf{F} = \begin{bmatrix} -1{,}715 \\ 0 \\ 1{,}715 \\ 0 \end{bmatrix} \begin{matrix} 4 \\ 5 \\ 3 \\ 6 \end{matrix} \text{ kN} \qquad \mathbf{K}_t = \begin{matrix} & 4 & 5 & 3 & 6 & \\ & \left[\begin{matrix} 5{,}645.5 \\ 0 \\ -5{,}645.5 \\ 0 \end{matrix}\right. & \begin{matrix} 0 \\ 206.53 \\ 0 \\ -206.53 \end{matrix} & \begin{matrix} -5{,}645.5 \\ 0 \\ \boxed{5{,}645.5} \\ 0 \end{matrix} & \left.\begin{matrix} 0 \\ -206.53 \\ 0 \\ 206.53 \end{matrix}\right] & \begin{matrix} 4 \\ 5 \\ 3 \\ 6 \end{matrix} \end{matrix} \text{ kN/m}$$

Thus, the structure's internal joint force vector $\mathbf{f}_2$ and the tangent stiffness matrix $\mathbf{S}_{t2}$ are given by

$$\mathbf{f}_2 = \begin{bmatrix} 0 \\ -1{,}948 \\ 11.722 \end{bmatrix} \text{ kN} \qquad \mathbf{S}_{t2} = \begin{bmatrix} 13{,}412 & 0 & -6{,}705.8 \\ 0 & 3{,}833.4 & 4{,}069.2 \\ -6{,}705.8 & 4{,}069.2 & 12{,}351 \end{bmatrix} \text{ kN/m}$$

and the unbalanced joint force vector is obtained as

$$\Delta\mathbf{U}_2 = \mathbf{P} - \mathbf{f}_2 = \begin{bmatrix} 0 \\ -52.041 \\ -11.722 \end{bmatrix} \text{ kN}$$

Note that the magnitudes of the unbalanced forces are now significantly smaller than in the previous iteration cycle. By solving the linearized system of equations $\Delta\mathbf{U}_2 = \mathbf{S}_{t2}\, \Delta\mathbf{d}_2$, we determine the displacement correction vector

$$\Delta\mathbf{d}_2 = \begin{bmatrix} 0.0046508 \\ -0.023449 \\ 0.0093015 \end{bmatrix} \text{ m}$$

To check for convergence, we write

$$\sqrt{\frac{\sum_{j=1}^{3} (\Delta d_j)^2}{\sum_{j=1}^{3} (d_j)^2}} = \sqrt{\frac{(0.0046508)^2 + (-0.023449)^2 + (0.0093015)^2}{(0.15189)^2 + (-0.62579)^2 + (0.30378)^2}} = 0.036027 > e\ (= 0.001)$$

which indicates that, while the change in the structure's deformed configuration is now considerably smaller than in the previous iteration cycle, it is still not within the prescribed tolerance of 0.1 percent. Thus, we perform another (third) iteration based on a new approximate deformed configuration $\mathbf{d}_3$ of the structure, with

$$\mathbf{d}_3 = \mathbf{d}_2 + \Delta\mathbf{d}_2 = \begin{bmatrix} 0.15189 \\ -0.62579 \\ 0.30378 \end{bmatrix} + \begin{bmatrix} 0.00046508 \\ -0.023449 \\ 0.0093015 \end{bmatrix} = \begin{bmatrix} 0.15654 \\ -0.64924 \\ 0.31308 \end{bmatrix} \text{ m}$$

Iteration Cycle 3:

Member 1

$$\mathbf{v} = \begin{bmatrix} 0 \\ 0 \\ 0.15654 \\ -0.64924 \end{bmatrix} \begin{matrix} 4 \\ 5 \\ 1 \\ 2 \end{matrix} \text{ m}$$

$L = 5$ m, $\bar{L} = 4.7752$ m, $c_X = 0.87044$, $c_Y = 0.49228$, $u = 0.22476$ m, $Q = 2{,}030.3$ kN

$$\mathbf{F} = \begin{bmatrix} 1{,}767.2 \\ 999.45 \\ -1{,}767.2 \\ -999.45 \end{bmatrix} \begin{matrix} 4 \\ 5 \\ 1 \\ 2 \end{matrix} \text{ kN} \qquad \mathbf{K_t} = \begin{matrix} & 4 & 5 & 1 & 2 & \\ & \begin{bmatrix} 6{,}740.8 & 4{,}052.7 & -6{,}740.8 & -4{,}052.7 \\ 4{,}052.7 & 1{,}866.9 & -4{,}052.7 & -1{,}866.9 \\ -6{,}740.8 & -4{,}052.7 & 6{,}740.8 & 4{,}052.7 \\ -4{,}052.7 & -1{,}866.9 & 4{,}052.7 & 1{,}866.9 \end{bmatrix} & \begin{matrix} 4 \\ 5 \\ 1 \\ 2 \end{matrix} \end{matrix} \text{ kN/m}$$

Member 2

$$\mathbf{v} = \begin{bmatrix} 0.31308 \\ 0 \\ 0.15654 \\ -0.64924 \end{bmatrix} \begin{matrix} 3 \\ 6 \\ 1 \\ 2 \end{matrix} \text{ m}$$

$L = 5$ m, $\bar{L} = 4.7752$ m, $c_X = -0.87044$, $c_Y = 0.49228$, $u = 0.22476$ m, $Q = 2{,}030.3$ kN

$$\mathbf{F} = \begin{bmatrix} -1{,}767.2 \\ 999.45 \\ 1{,}767.2 \\ -999.45 \end{bmatrix} \begin{matrix} 3 \\ 6 \\ 1 \\ 2 \end{matrix} \text{ kN} \qquad \mathbf{K_t} = \begin{matrix} & 3 & 6 & 1 & 2 & \\ & \begin{bmatrix} 6{,}740.8 & -4{,}052.7 & -6{,}740.8 & 4{,}052.7 \\ -4{,}052.7 & 1{,}866.9 & 4{,}052.7 & -1{,}866.9 \\ -6{,}740.8 & 4{,}052.7 & 6{,}740.8 & -4{,}052.7 \\ 4{,}052.7 & -1{,}866.9 & -4{,}052.7 & 1{,}866.9 \end{bmatrix} & \begin{matrix} 3 \\ 6 \\ 1 \\ 2 \end{matrix} \end{matrix} \text{ kN/m}$$

Member 3

$$\mathbf{v} = \begin{bmatrix} 0 \\ 0 \\ 0.31308 \\ 0 \end{bmatrix} \begin{matrix} 4 \\ 5 \\ 3 \\ 6 \end{matrix} \text{ m}$$

$L = 8$ m, $\bar{L} = 8.3131$ m, $c_X = 1$, $c_Y = 0$, $u = -0.31308$ m, $Q = -1{,}767.5$ kN

$$\mathbf{F} = \begin{bmatrix} -1{,}767.5 \\ 0 \\ 1{,}767.5 \\ 0 \end{bmatrix} \begin{matrix} 4 \\ 5 \\ 3 \\ 6 \end{matrix} \text{ kN} \qquad \mathbf{K_t} = \begin{matrix} & 4 & 5 & 3 & 6 & \\ & \begin{bmatrix} 5{,}645.5 & 0 & -5{,}645.5 & 0 \\ 0 & 212.61 & 0 & -212.61 \\ -5{,}645.5 & 0 & 5{,}645.5 & 0 \\ 0 & -212.61 & 0 & 212.61 \end{bmatrix} & \begin{matrix} 4 \\ 5 \\ 3 \\ 6 \end{matrix} \end{matrix} \text{ kN/m}$$

The structure's internal joint force vector $\mathbf{f}_3$ and the tangent stiffness matrix $\mathbf{S}_{t3}$ are given by

$$\mathbf{f}_3 = \begin{bmatrix} 0 \\ -1{,}998.9 \\ 0.27365 \end{bmatrix} \text{kN} \qquad \mathbf{S}_{t3} = \begin{bmatrix} 13{,}482 & 0 & -6{,}740.8 \\ 0 & 3{,}733.8 & 4{,}052.7 \\ -6{,}740.8 & 4{,}052.7 & 12{,}386 \end{bmatrix} \text{kN/m}$$

and the unbalanced joint force vector is

$$\Delta\mathbf{U}_3 = \mathbf{P} - \mathbf{f}_3 = \begin{bmatrix} 0 \\ -1.0925 \\ -0.27365 \end{bmatrix} \text{kN}$$

By solving the linearized system of equations $\Delta\mathbf{U}_3 = \mathbf{S}_{t3}\,\Delta\mathbf{d}_3$, we determine the displacement correction vector

$$\Delta\mathbf{d}_3 = \begin{bmatrix} 0.000098787 \\ -0.00050706 \\ 0.00019757 \end{bmatrix} \text{m}$$

To check for convergence, we write

$$\sqrt{\frac{\sum_{j=1}^{3}(\Delta d_j)^2}{\sum_{j=1}^{3}(d_j)^2}} = \sqrt{\frac{(0.000098787)^2 + (-0.00050706)^2 + (0.00019757)^2}{(0.15654)^2 + (-0.64924)^2 + (0.31308)^2}} = 0.00074985 < e\ (= 0.001)$$

which indicates that the change in the structure's deformed configuration $\Delta\mathbf{d}_3$ is now within the specified tolerance of 0.1 percent and, therefore, the iterative process has converged.

Results of Geometrically Nonlinear Analysis: The final deformed configuration of the truss is given by the joint displacement vector

$$\mathbf{d} = \mathbf{d}_3 + \Delta\mathbf{d}_3 = \begin{bmatrix} 0.15654 \\ -0.64924 \\ 0.31308 \end{bmatrix} + \begin{bmatrix} 0.000098787 \\ -0.00050706 \\ 0.00019757 \end{bmatrix} = \begin{bmatrix} 0.15664 \\ -0.64975 \\ 0.31327 \end{bmatrix} \text{m}$$ **Ans**

Member 1

$$\mathbf{v} = \begin{bmatrix} 0 \\ 0 \\ 0.15664 \\ -0.64975 \end{bmatrix} \begin{matrix} 4 \\ 5 \\ 1 \\ 2 \end{matrix} \ \text{m}$$

$L = 5$ m, $\bar{L} = 4.7751$ m, $c_X = 0.87049$, $c_Y = 0.49219$, $u = 0.22493$ m, $Q = 2{,}031.7$ kN

$Q_{a1} = 2{,}031.7\,\text{kN(C)}$ **Ans**

$$\mathbf{F} = \begin{bmatrix} 1{,}768.6 \\ 1{,}000 \\ \hdashline -1{,}768.6 \\ -1{,}000 \end{bmatrix} \begin{matrix} 4 \\ 5 \\ 1 \\ 2 \end{matrix} \text{ kN} \qquad (5)$$

Member 2

$$\mathbf{v} = \begin{bmatrix} 0.31327 \\ 0 \\ 0.15664 \\ -0.64975 \end{bmatrix} \begin{matrix} 3 \\ 6 \\ 1 \\ 2 \end{matrix} \text{ m}$$

$$L = 5 \text{ m}, \bar{L} = 4.7751 \text{ m}, c_X = -0.87049, c_Y = 0.49219, u = 0.22493 \text{ m}, Q = 2{,}031.7 \text{ kN}$$

$$Q_{a2} = 2{,}031.7 \text{ kN (C)}$$ **Ans**

$$\mathbf{F} = \begin{bmatrix} -1{,}768.6 \\ \hdashline 1{,}000 \\ \hdashline 1{,}768.6 \\ -1{,}000 \end{bmatrix} \begin{matrix} 3 \\ 6 \\ 1 \\ 2 \end{matrix} \text{ kN} \qquad (6)$$

Member 3

$$\mathbf{v} = \begin{bmatrix} 0 \\ 0 \\ 0.31327 \\ 0 \end{bmatrix} \begin{matrix} 4 \\ 5 \\ 3 \\ 6 \end{matrix} \text{ m}$$

$$L = 8 \text{ m}, \bar{L} = 8.3133 \text{ m}, c_X = 1, c_Y = 0, u = -0.31327 \text{ m}, Q = -1{,}768.6 \text{ kN}$$

$$Q_{a3} = 1{,}768.6 \text{ kN (T)}$$ **Ans**

$$\mathbf{F} = \begin{bmatrix} -1{,}768.6 \\ 0 \\ \hdashline 1{,}768.6 \\ \hdashline 0 \end{bmatrix} \begin{matrix} 4 \\ 5 \\ 3 \\ 6 \end{matrix} \text{ kN} \qquad (7)$$

Finally, the support reaction vector **R** is assembled from the elements of the member **F** vectors given in Eqs. (5) thru (7) as

$$\mathbf{R} = \begin{bmatrix} 1{,}768.6 - 1{,}768.6 \\ 1{,}000 + 0 \\ 1{,}000 + 0 \end{bmatrix} = \begin{bmatrix} 0 \\ 1{,}000 \\ 1{,}000 \end{bmatrix} \begin{matrix} 4 \\ 5 \\ 6 \end{matrix} \text{ kN}$$ **Ans**

Equilibrium Check: Applying the equations of equilibrium to the free body of the truss in its deformed state (Fig. 10.8(c)), we obtain

$+ \rightarrow \sum F_X = 0$ **Checks**

$+ \uparrow \sum F_Y = 0 \qquad 1{,}000 - 2{,}000 + 1{,}000 = 0$ **Checks**

$+ \circlearrowleft \sum M_{①} = 0 \qquad -2{,}000\,(4.1566) + 1{,}000\,(8.3133) = 0.1 \text{ kN.m} \approx 0$ **Checks**

SUMMARY

In this chapter, we have studied the basic concepts of the geometrically nonlinear analysis of plane trusses. A block diagram summarizing the various steps involved in the analysis is given in Fig. 10.7.

PROBLEMS

Section 10.1

10.1 Derive the relationships between load P and displacement δ of the truss shown in Fig. P10.1 by using: (a) the conventional linear theory, and (b) the geometrically nonlinear theory. Plot the P - δ equations using the numerical values: $\theta = 30^\circ$, $L = 3$ m, $E =$ 70 GPa, and $A = 645.2$ mm^2, in the range $0 \leq \delta/L \leq 0.5$, to compare the linear and nonlinear solutions.

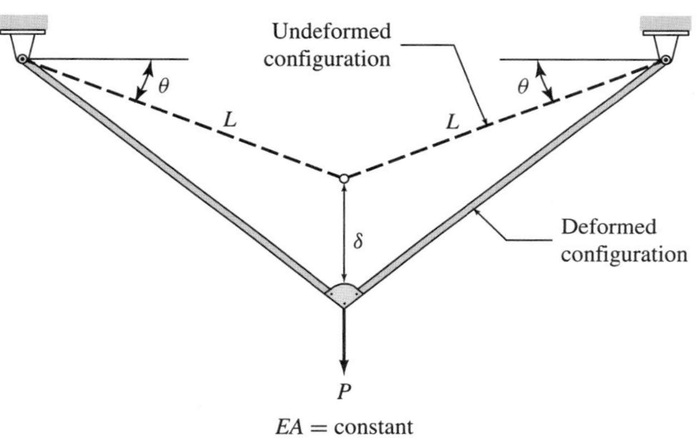

Fig. P10.1

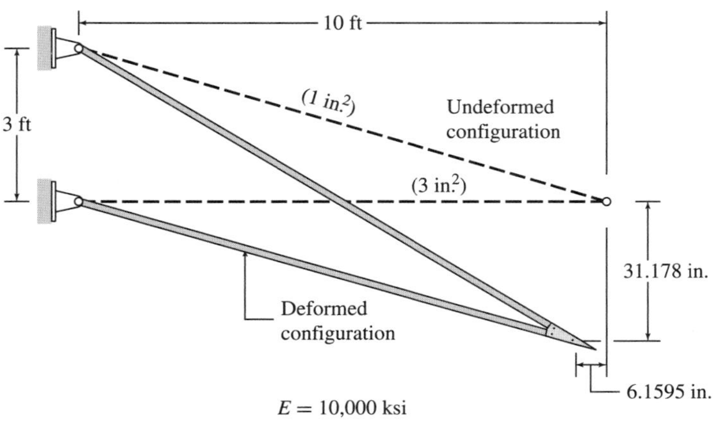

Fig. P10.2

Section 10.2

10.2 By using the geometrically nonlinear analysis, determine the joint loads **P** that cause the two-member truss to deform into the configuration shown in Fig. P10.2.

10.3 through 10.5 Determine the joint displacements, member axial forces, and support reactions for the trusses shown in Figs. P10.3 through P10.5 by geometrically nonlinear analysis. Use a displacement convergence tolerance of one percent.

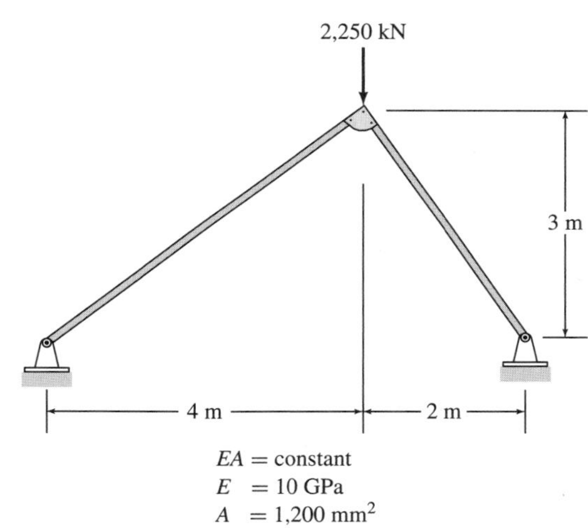

Fig. P10.3

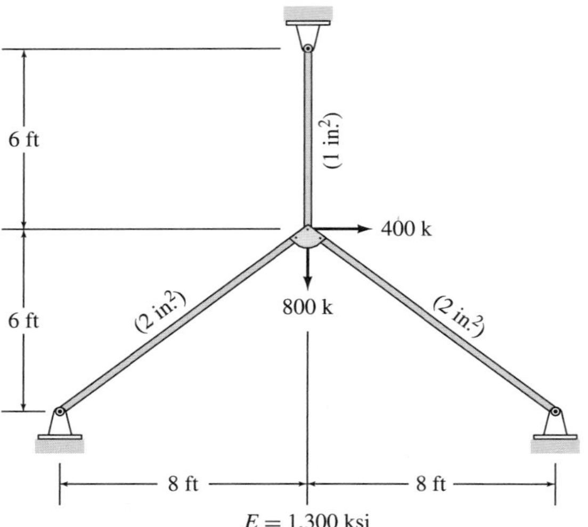

Fig. P10.4

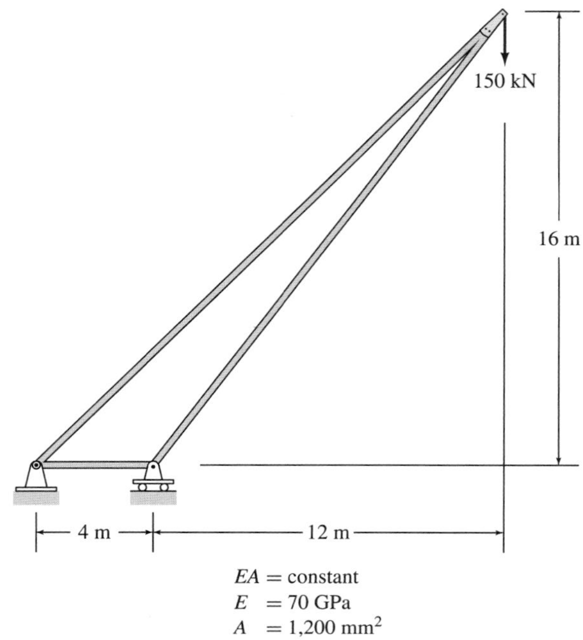

Fig. P10.5

10.6 Develop a computer program for geometrically nonlinear analysis of plane trusses. Use the program to analyze the trusses of Problems 10.3 through 10.5, and compare the computer-generated results to those obtained by hand calculations.

COMPUTER SOFTWARE

A computer program for analyzing two- and three-dimensional framed structures is available on the publisher's website www.cengage.com/engineering. The software, which can be used to analyze plane and space trusses, beams, plane and space frames, and grids, is based on the matrix stiffness method. It can also perform geometrically nonlinear analysis of plane trusses. The software is designed for use on IBM and IBM-compatible personal computers with Microsoft Windows® operating systems, and it provides an option for saving input data into files for subsequent modification and/or execution.

Complete instructions for downloading and installing the software are provided on the publisher's website www.cengage.com/engineering/kassimali.

Starting the Computer Software

1. Click the **Start** button on the taskbar.
2. Point to the menu title **Programs** and then click the menu item **MATRIX ANALYSIS OF STRUCTURES 2.0—Kassimali**; the software's title screen will appear.

Inputting Data

The computer software is designed so that any *consistent* set of units may be used. Thus, all the data must be converted into a consistent set of units before being input into the software. For example, if we wish to use units of kips and inches, then the joint coordinates must be defined in inches, the moduli of elasticity in ksi, the cross-sectional areas in in.^{2}, the moments of inertia in in.^{4}, the joint loads and moments in kips and k-in., respectively, and distributed member loads in k/in.

To start entering data for a structure, click the menu title **Project**; and then click the menu item **New Project**. The input data necessary for the analysis of a structure is divided into six categories; the data in each category is input by clicking on the corresponding menu title and then entering information in the forms and/or dialog boxes that appear on the screen. The input data categories are:

1. General structural data (project title and structure type)
2. Joint coordinates and supports
3. Material properties
4. Cross-sectional properties

5. Member data (beginning and end joint numbers, material and cross-sectional property numbers, member hinges if applicable, and angle of roll in the case of space frames)
6. Loads (joint and member loads, support displacements, temperature changes, and fabrication errors)

Results of the Analysis

Once all the necessary data has been entered, click the menu title **Analyze** of the main screen to analyze the structure (Fig. A.1). The software will automatically compute the joint displacements, member end forces, and support reactions, using the matrix stiffness method. The results of the analysis are displayed on the screen. The input data as well as the results of the analysis can be printed by clicking on the menu title **Project** and then clicking on the menu item **Print**, of the main screen.

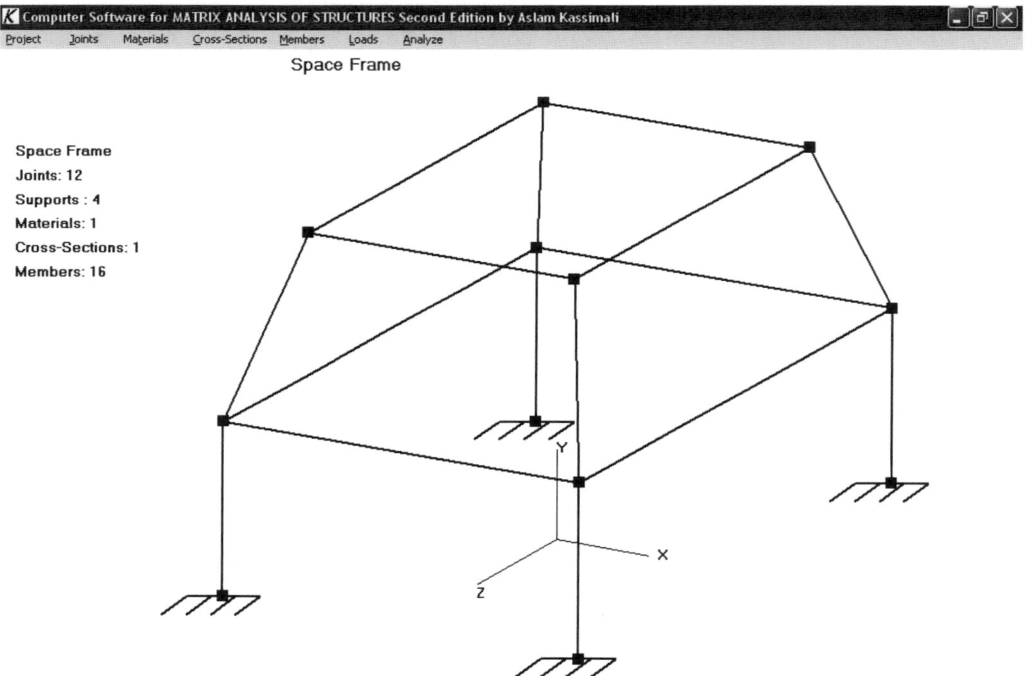

Fig. A.1 Main Screen

B FLEXIBILITY METHOD

In this text, we have focused our attention on the matrix stiffness method of structural analysis, which is the most commonly used method in professional practice today, and which forms the basis for most of the currently available commercial software for structural analysis. However, as stated in Section 1.3, another type of matrix method, called the flexibility method, can also be used for structural analysis. While the stiffness method can be applied to both statically determinate and indeterminate structures, the flexibility method is applicable only to indeterminate structures. The flexibility method is essentially a generalization in matrix form of the classical method of consistent deformations, and is generally considered convenient for analyzing small structures with a few redundants.

In this appendix, we present the basic concept of the flexibility method, and illustrate its application to plane trusses. A more detailed treatment of this method can be found in [3] and [52].

Essentially, the flexibility method of analysis involves removing enough restraints from the indeterminate structure to render it statically determinate. This determinate structure, which must be statically stable, is called the *primary structure*; and the reactions or internal forces associated with the excess restraints removed from the given indeterminate structure to convert it into the determinate (primary) structure, are termed *redundants*. The redundants are then treated as unknown loads on the primary structure, and their values are determined by solving the compatibility equations based on the condition that the deformations of the primary structure due to the combined effect of the redundants and the given external loading must be the same as the deformations of the original indeterminate structure.

Consider, for example, a plane truss supported by five reaction components, as shown in Fig. B.1(a) on the next page. The truss is internally determinate, but externally indeterminate with two degrees of indeterminacy. This indicates that the truss has two more, or redundant, reactions than necessary for static stability. Thus, if we can determine two of the five reactions by using compatibility equations based on the geometry of the deformation of the truss, then the remaining three reactions and the member forces can be obtained from equilibrium considerations.

To analyze the truss by the flexibility method, we must select two of the unknown reactions and member forces to be the redundants. Suppose that we select the horizontal and vertical reactions, $\overline{R}_1$ and $\overline{R}_2$, at the hinged support at joint 5 to be the redundants. The hinged support at joint 5 is then removed from the given indeterminate truss to obtain the statically determinate and stable

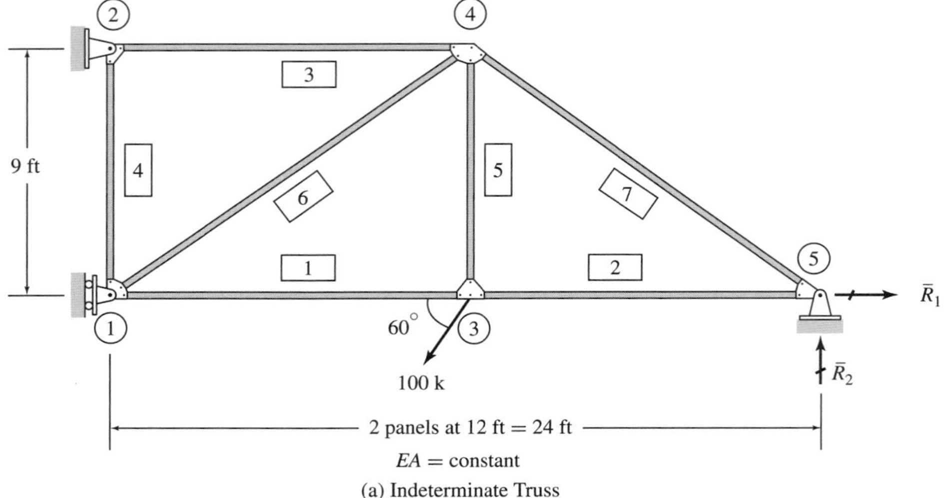

(a) Indeterminate Truss

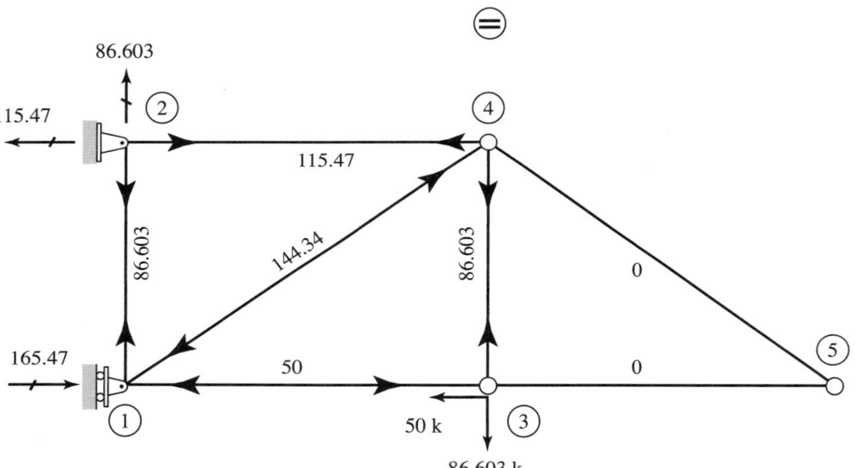

(b) Primary Truss Subjected to External Loading—$\mathbf{Q}_{aO}$ Vector

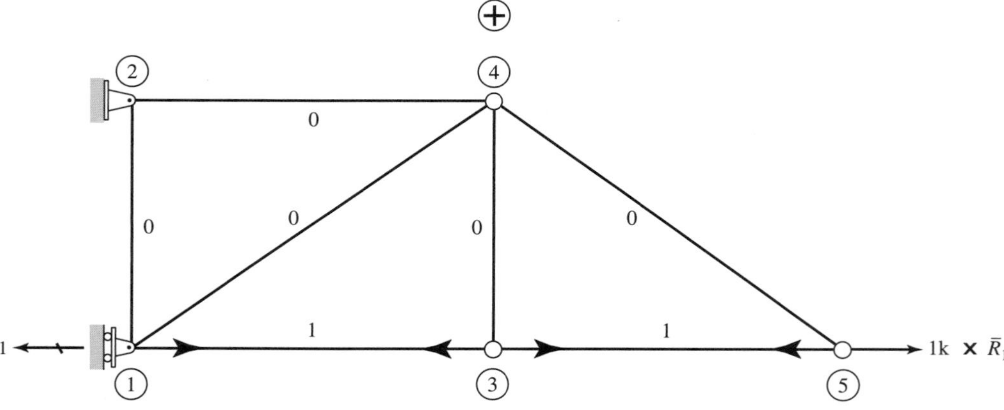

Fig. B.1 (c) Primary Truss Subjected to Unit Value of Redundant $\bar{R}_1$—First Column of **b** Matrix

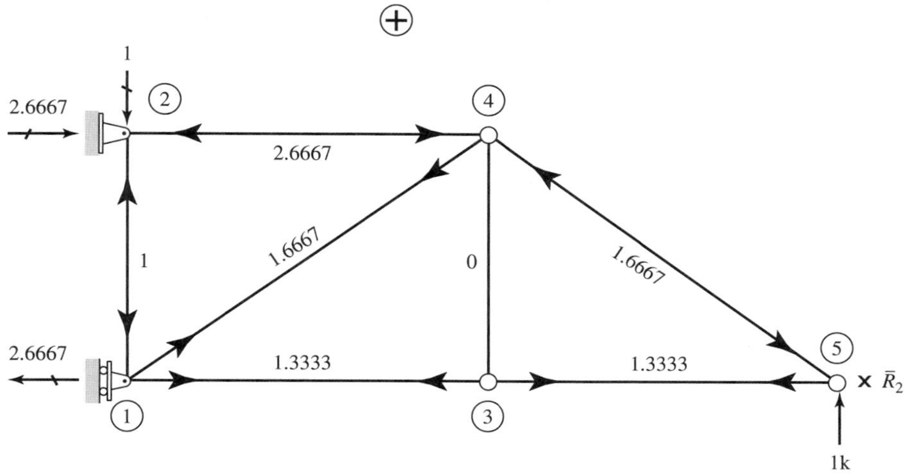

(d) Primary Truss Subjected to Unit Value of Redundant $\overline{R}_2$—Second Column of **b** Matrix

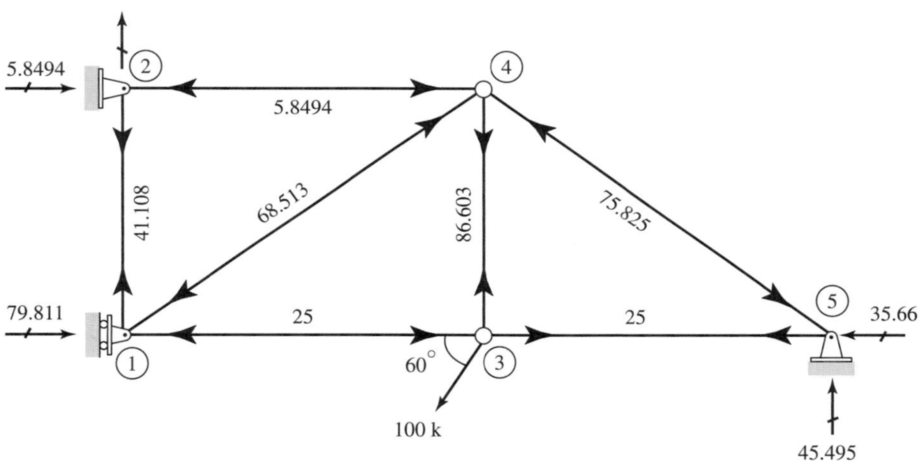

(e) Support Reactions and Member Forces for Indeterminate Truss

Fig. B.1 (*continued*)

primary truss, as shown in Fig. B.1(b). The two redundants $\overline{R}_1$ and $\overline{R}_2$ are now treated as unknown loads on the primary truss, and their magnitudes can be determined from the compatibility conditions that the horizontal and vertical deflections at joint 5 of the primary truss due to the combined effect of the known 100 k load and the unknown redundants $\overline{R}_1$ and $\overline{R}_2$ must be equal to 0. This is because the deflections in the horizontal and vertical directions of the given indeterminate truss at the hinged support at joint 5 are 0.

The compatibility equations can be conveniently established by superimposing the deflections due to the external loading and the redundants, $\overline{R}_1$ and

$\overline{R}_2$, acting individually on the primary truss, as shown in Figs. B.1(b), (c), and (d), respectively. Thus,

$$\Delta_{O1} + f_{11}\overline{R}_1 + f_{12}\overline{R}_2 = 0 \tag{B.1a}$$

$$\Delta_{O2} + f_{21}\overline{R}_1 + f_{22}\overline{R}_2 = 0 \tag{B.1b}$$

in which Δ_{Oi} $(i = 1, 2)$ represents the deflection at joint 5 of the primary truss in the direction of the redundant $\overline{R}_i$, due to the external loading; and the *flexibility coefficient* f_{ij} $(i = 1, 2$ and $j = 1, 2)$ denotes the deflection of the primary truss at the location and in the direction of a redundant $\overline{R}_i$ due to a unit value of a redundant $\overline{R}_j$. Equations (B.1) can be expressed in matrix form as

$$\begin{bmatrix} \Delta_{O1} \\ \Delta_{O2} \end{bmatrix} + \begin{bmatrix} f_{11} & f_{12} \\ f_{21} & f_{22} \end{bmatrix} \begin{bmatrix} \overline{R}_1 \\ \overline{R}_2 \end{bmatrix} = \begin{bmatrix} 0 \\ 0 \end{bmatrix} \tag{B.2}$$

From the foregoing discussion for the example truss with two degrees of indeterminacy, we realize that the compatibility equations for a general indeterminate structure with n_i degrees of indeterminacy can be symbolically expressed as

$$\boxed{\boldsymbol{\Delta}_O + \mathbf{f}\overline{\mathbf{R}} = \mathbf{0}} \tag{B.3}$$

in which the $n_i \times 1$ vectors $\overline{\mathbf{R}}$ and $\boldsymbol{\Delta}_O$ denote, respectively, the unknown redundants, and the deflections of the primary structure at the locations and in the directions of the redundants due to external loads; and the $n_i \times n_i$ matrix $\mathbf{f}$ is called the *structure flexibility matrix.* The reader may recall from a previous course in mechanics of materials or *structural analysis* [18], that *Maxwell's law of reciprocal deflections* states that *for a linearly elastic structure, the deflection at a point i due to a unit load applied at a point j is equal to the deflection at j due to a unit load at i.* As the flexibility coefficient f_{ij} denotes the deflection of the primary structure at the location of the redundant $\overline{R}_i$ due to a unit value of the redundant $\overline{R}_j$, and the flexibility coefficient f_{ji} denotes the deflection corresponding to $\overline{R}_j$ due to a unit value of $\overline{R}_i$, according to *Maxwell's law* f_{ij} must be equal to f_{ji} (i.e., $f_{ij} = f_{ji}$). We can thus deduce that *for linearly elastic structures, the flexibility matrices are symmetric.*

From Eqs. (B.1) through (B.3), we can see that the elements of the vector $\boldsymbol{\Delta}_O$ and the flexibility matrix $\mathbf{f}$ represent deflections of the primary (statically determinate) structure. Once these deflections have been evaluated, the compatibility equations (Eqs. (B.3)) can be solved for the unknown redundants. With the redundants known, the other response characteristics of the structure can be evaluated, either by equilibrium or superposition.

The deflections (and the flexibility coefficients) of a primary structure can be conveniently expressed in terms of the forces and properties of its members, using the virtual work method. Recall from a previous course in mechanics of materials or structural analysis [18], that the expression of the virtual work method for truss deflections is given by

$$\Delta = \sum_{i=1}^{NM} \frac{Q_{ar} Q_{av} L}{EA} \tag{B.4}$$

in which NM denotes the number of members of the truss; Q_{ar} represents the axial forces in truss members due to the real loading that causes the deflection Δ, and Q_{av} represents the axial forces in the truss members due to a virtual unit load acting at the location and in the direction of the desired deflection Δ. Equation (B.4) can be expressed in matrix form as

$$\Delta = \mathbf{Q}_{av}^T \mathbf{f}_M \mathbf{Q}_{ar} \tag{B.5}$$

in which $\mathbf{Q}_{av}$ and $\mathbf{Q}_{ar}$ denote the $NM \times 1$ vectors containing member axial forces due to virtual (unit) and real (actual) loads, respectively; and $\mathbf{f}_M$ is a $NM \times NM$ diagonal matrix containing the member flexibilities (L/EA) on its main diagonal (i.e., $f_{Mij} = L_i/E_iA_i$ for $i = j$, and $f_{Mij} = 0$ for $i \neq j$). The diagonal matrix $\mathbf{f}_M$ is sometimes called the *unassembled flexibility matrix*. In order to develop the expressions for $\mathbf{\Delta}_O$ and $\mathbf{f}$ in terms of the member forces and properties, let us define a $NM \times 1$ vector $\mathbf{Q}_{aO}$ which contains the axial forces in the members of the primary truss due to the given external loading, and a $NM \times n_i$ matrix $\mathbf{b}$, the jth column of which contains member axial forces due to a unit value of the jth redundant (i.e., $\overline{R}_j = 1$). The matrix $\mathbf{b}$ is commonly referred to as an *equilibrium matrix*. In both $\mathbf{Q}_{aO}$ and $\mathbf{b}$ the member axial forces are stored in sequential order of member numbers; that is, the axial forces in the ith member are stored in the ith rows of $\mathbf{Q}_{aO}$ and $\mathbf{b}$, and so on. The member forces $\mathbf{Q}_{aO}$ for the example truss are shown in Fig. B.1(b). Note that since the primary truss is statically determinate, the forces in its members due to the given external loading can be conveniently evaluated by applying the method of joints. By using the member forces shown in Fig. B.1(b) and the member numbers given in Fig. B.1(a), we form the $\mathbf{Q}_{aO}$ vector for the truss as

$$\mathbf{Q}_{aO} = \begin{bmatrix} -50 \\ 0 \\ 115.47 \\ 86.603 \\ 86.603 \\ -144.34 \\ 0 \end{bmatrix} \begin{matrix} 1 \\ 2 \\ 3 \\ 4 \\ 5 \\ 6 \\ 7 \end{matrix} \text{ k} \tag{B.6}$$

in which the tensile member axial forces are considered to be positive. The first column of $\mathbf{b}$ is obtained by subjecting the primary truss to a unit value of the redundant $\overline{R}_1$, as shown in Fig. B.1(c), and by computing the corresponding member forces by applying the method of joints. The second column of $\mathbf{b}$ is generated similarly by subjecting the primary truss to a unit value of the redundant $\overline{R}_2$, and by computing the corresponding member axial forces (see Fig. B.1(d)). The equilibrium matrix $\mathbf{b}$ thus obtained is

$$\mathbf{b} = \begin{bmatrix} 1 & 1.3333 \\ 1 & 1.3333 \\ 0 & -2.6667 \\ 0 & -1 \\ 0 & 0 \\ 0 & 1.6667 \\ 0 & -1.6667 \end{bmatrix} \text{ k/k} \tag{B.7}$$

The first element, Δ_{O1}, of the $\mathbf{\Delta}_O$ vector represents deflection of the primary truss at the location and in the direction of $\overline{R}_1$ due to the given external loading. Therefore, for the purpose of calculating Δ_{O1} via the virtual work method, the real system consists of the given external loading as shown in Fig. B.1(b), and the virtual system consists of a unit load applied at the location and in the direction of the redundant $\overline{R}_1$, which is the same as the system shown in Fig. B.1(c) (without the multiplier $\overline{R}_1$). Thus, the virtual work expression for Δ_{O1} can be obtained by substituting $\mathbf{Q}_{aO}$ for $\mathbf{Q}_{ar}$ and the first column of $\mathbf{b}$ for $\mathbf{Q}_{av}$ in Eq. (B.5); that is,

$$\Delta_{O1} = \mathbf{b}_1^T \mathbf{f}_M \mathbf{Q}_{aO} \tag{B.8}$$

in which $\mathbf{b}_1$ denotes the first column of $\mathbf{b}$. The expression for the second element, Δ_{O2}, of $\mathbf{\Delta}_O$, in terms of member axial forces, can be obtained in a similar manner, and is given by

$$\Delta_{O2} = \mathbf{b}_2^T \mathbf{f}_M \mathbf{Q}_{aO} \tag{B.9}$$

with $\mathbf{b}_2$ denoting the second column of $\mathbf{b}$. By combining Eqs. (B.8) and (B.9), we obtain

$$\boxed{\mathbf{\Delta}_O = \mathbf{b}^T \mathbf{f}_M \mathbf{Q}_{aO}} \tag{B.10}$$

The expressions for the elements of the flexibility matrix **f**, in terms of member forces and properties, can be obtained in a similar manner. For example, the virtual and real systems for the evaluation of f_{12} are shown in Figs. B.1(c) and (d), respectively, with the corresponding member forces stored in the first and second columns of **b**. Therefore,

$$f_{12} = \mathbf{b}_1^T \mathbf{f}_M \mathbf{b}_2$$

Thus, the entire flexibility matrix **f** can be expressed in terms of the member forces and properties as

$$\boxed{\mathbf{f} = \mathbf{b}^T \mathbf{f}_M \mathbf{b}} \tag{B.11}$$

Finally, by substituting Eqs. (B.10) and (B.11) into Eq. (B.3), we obtain the structure's compatibility equations in terms of its member forces and properties, as

$$\boxed{\mathbf{b}^T \mathbf{f}_M \mathbf{Q}_{aO} + (\mathbf{b}^T \mathbf{f}_M \mathbf{b})\,\overline{\mathbf{R}} = \mathbf{0}} \tag{B.12}$$

To illustrate the application of the flexibility method to the analysis of plane trusses, let us reconsider the truss of Fig. B.1. The $\mathbf{Q}_{aO}$ vector and the **b** matrix for this truss were determined previously, and are given in Eqs. (B.6) and (B.7), respectively. The unassembled flexibility matrix for the structure is

$$\mathbf{f}_M = \frac{1}{EA}\begin{bmatrix} 12 & 0 & 0 & 0 & 0 & 0 & 0 \\ 0 & 12 & 0 & 0 & 0 & 0 & 0 \\ 0 & 0 & 12 & 0 & 0 & 0 & 0 \\ 0 & 0 & 0 & 9 & 0 & 0 & 0 \\ 0 & 0 & 0 & 0 & 9 & 0 & 0 \\ 0 & 0 & 0 & 0 & 0 & 15 & 0 \\ 0 & 0 & 0 & 0 & 0 & 0 & 15 \end{bmatrix} \tag{B.13}$$

Substituting Eqs. (B.6), (B.7), and (B.13) into Eq. (B.10), and performing the required matrix multiplications, we obtain

$$\mathbf{\Delta}_O = \mathbf{b}^T \mathbf{f}_M \mathbf{Q}_{aO} = \frac{1}{EA}\begin{bmatrix} -600 \\ -8{,}883 \end{bmatrix} \tag{B.14}$$

Substitution of Eqs. (B.7) and (B.13) into Eq. (B.11) yields the flexibility matrix:

$$\mathbf{f} = \mathbf{b}^T \mathbf{f}_M \mathbf{b} = \frac{1}{EA}\begin{bmatrix} 24 & 32 \\ 32 & 220.33 \end{bmatrix} \tag{B.15}$$

Next, we substitute Eqs. (B.14) and (B.15) into the compatibility equations (Eqs. (B.12)), and solve the resulting system of simultaneous equations for the unknown redundants. This yields

$$\overline{\mathbf{R}} = \begin{bmatrix} -35.66 \\ 45.495 \end{bmatrix} \text{k} \tag{B.16}$$

With the redundants known, the member axial forces in the actual indeterminate structure, $\mathbf{Q}_a$, can be conveniently evaluated by applying the superposition relationship (see Figs. B.1(a) through (d)):

$$\boxed{\mathbf{Q}_a = \mathbf{Q}_{aO} + \mathbf{b}\overline{\mathbf{R}}} \tag{B.17}$$

Substituting Eqs. (B.6), (B.7), and (B.16) into Eq. (B.17), we determine the axial forces in the members of the indeterminate truss to be

$$\mathbf{Q}_a = \begin{bmatrix} -25 \\ 25 \\ -5.8494 \\ 41.108 \\ 86.603 \\ -68.513 \\ -75.825 \end{bmatrix} \text{k}$$

These member axial forces are shown in Fig. B.1(e).

BIBLIOGRAPHY

1. Arbabi, F. (1991) *Structural Analysis and Behavior.* New York: McGraw-Hill.
2. Bathe, K. J. (1982) *Finite Element Procedures in Engineering Analysis.* Englewood Cliffs, NJ: Prentice-Hall.
3. Beaufait, F. W., W. M. Rowan, Jr., P. G. Hoadley, and R. M. Hackett. (1970) *Computer Methods of Structural Analysis.* Englewood Cliffs, NJ: Prentice-Hall.
4. Beer, F. P., and E. R. Johnston, Jr. (1981) *Mechanics of Materials.* New York: McGraw-Hill.
5. Betti, E. (1872) *Il Nuovo Cimento.* Series 2, Vols. 7 and 8.
6. Boggs, R. G. (1984) *Elementary Structural Analysis.* New York: Holt, Rinehart & Winston.
7. Chajes, A. (1990) *Structural Analysis.* 2nd ed. Englewood Cliffs, NJ: Prentice Hall.
8. Chrisfield, M. A. (1991) *Non-linear Finite Element Analysis of Solids and Structures, Volume 1: Essentials.* New York: John Wiley & Sons.
9. Chrisfield, M. A. (1997) *Non-linear Finite Element Analysis of Solids and Structures, Volume 2: Advanced Topics.* New York: John Wiley & Sons.
10. Dawe, D. J. (1984) *Matrix and Finite Element Displacement Analysis of Structures.* New York: Oxford University Press.
11. Elias, Z. M. (1986) *Theory and Methods of Structural Analysis.* New York: John Wiley & Sons.
12. Gere, J. M., and W. Weaver, Jr. (1965) *Matrix Algebra for Engineers.* New York: Van Nostrand Reinhold.
13. Hoit, M. (1995) *Computer-Assisted Structural Analysis and Modeling.* Englewood Cliffs, NJ: Prentice-Hall.
14. Holzer, S. M. (1985) *Computer Analysis of Structures.* New York: Elsevier Science.
15. Kanchi, M. B. (1981) *Matrix Methods of Structural Analysis.* New York: John Wiley & Sons.
16. Kassimali, A. (1976) *Nonlinear Static and Dynamic Analysis of Frames.* Ph.D. dissertation, University of Missouri at Columbia, MO.
17. Kassimali, A., and E. Bidhendi. (1988) Stability of Trusses under Dynamic Loads. *Computers & Structures* 29(3): 381–392.
18. Kassimali, A. (2010) *Structural Analysis.* 4th ed. Cengage Learning.
19. Kennedy, J. B., and M. K. S. Madugula. (1990) *Elastic Analysis of Structures: Classical and Matrix Methods.* New York: Harper & Row.
20. Laible, J. P. (1985) *Structural Analysis.* New York: Holt, Rinehart & Winston.
21. Langhaar, H. L. (1962) *Energy Methods in Applied Mechanics.* New York: John Wiley & Sons.
22. Laursen, H. A. (1988) *Structural Analysis.* 3rd ed. New York: McGraw-Hill.
23. Leet, K. M. (1988) *Fundamentals of Structural Analysis.* New York: Macmillan.
24. Logan, D. L. (1992) *A First Course in the Finite Element Method.* 2nd ed. Boston: PWS-Kent.
25. McCormac, J., and R. E. Elling. (1988) *Structural Analysis: A Classical and Matrix Approach.* New York: Harper & Row.
26. McGuire, W., R. H. Gallagher and R. D. Ziemian. (2000) *Matrix Structural Analysis.* 2nd ed. New York: John Wiley & Sons.
27. Maney, G. A. (1915) *Studies in Engineering.* Bulletin 1. Minneapolis: University of Minnesota.
28. Martin, H. C., and G. F. Carey. (1973) *Introduction to Finite Element Analysis—Theory and Application.* New York: McGraw-Hill.
29. Maxwell, J. C. (1864) On the Calculations of the Equilibrium and Stiffness of Frames. *Philosophical Magazine* 27:294–299.
30. Meyers, V. J. (1983) *Matrix Analysis of Structures.* New York: Harper & Row.
31. Noble, B. (1969) *Applied Linear Algebra.* Englewood Cliffs, NJ: Prentice-Hall.
32. Norris, C. H., J. B. Wilbur, and S. Utku. (1991) *Elementary Structural Analysis.* 4th ed. New York: McGraw-Hill.
33. Oran, C., and A. Kassimali. (1976) Large Deformations of Framed Structures under Static and Dynamic Loads. *Computers & Structures* 6: 539–547.
34. Paz, M. (1991) *Structural Dynamics—Theory and Computation.* 3rd ed. New York: Van Nostrand Reinhold.

35. Pilkey, W. D., and W. Wunderlich. (1994) *Mechanics of Structures—Variational and Computational Methods.* Boca Raton, FL: CRC Press.

36. Popov, E. P. (1968) *Introduction to Mechanics of Solids.* Englewood Cliffs, NJ: Prentice-Hall.

37. Ross, C. T. F. (1985) *Finite Element Methods in Structural Mechanics.* West Sussex, England: Ellis Horwood.

38. Rubinstein, M. F. (1966) *Matrix Computer Analysis of Structures.* Englewood Cliffs, NJ: Prentice-Hall.

39. Sack, R. L. (1989) *Matrix Structural Analysis.* Boston: PWS-Kent.

40. Seely, F. B., and J. O. Smith. (1967) *Advanced Mechanics of Materials.* 2nd ed. New York: John Wiley & Sons.

41. Smith, J. C. (1988) *Structural Analysis.* New York: Harper & Row.

42. Tartaglione, L. C. (1991) *Structural Analysis.* New York: McGraw-Hill.

43. Tena-Colunga, A. (1996) Stiffness Formulation for Nonprismatic Elements. *Journal of Structural Engineering, ASCE* 122(12): 1484–1489.

44. Tezcan, S. S. (1963) Discussion of "Simplified Formulation of Stiffness Matrices," by P. M. Wright. *Journal of the Structural Division, ASCE* 89(6): 445–449.

45. Tezcan, S. S. (1968) Discussion of "Numerical Solution of Nonlinear Structures," by T. J. Poskitt. *Journal of the Structural Division, ASCE* 94(6): 1617.

46. Tezcan, S. S., and B. C. Mahapatra. (1969) Tangent Stiffness Matrix for Space Frame Members. *Journal of the Structural Division, ASCE* 95(6): 1257–1270.

47. Timoshenko, S. P., and J. M. Gere. (1961) *Theory of Elastic Stability.* 2nd ed. New York: McGraw-Hill.

48. Turner, J. J., R. W. Clough, H. C. Martin, and L. J. Topp. (1956) Stiffness and Deflection Analysis of Complex Structures. *Journal of Aeronautical Sciences* 23(9):805–823.

49. Wang, C. K. (1973) *Computer Methods in Advanced Structural Analysis.* New York: Intext Press.

50. Wang, C. K. (1983) *Intermediate Structural Analysis.* New York: McGraw-Hill.

51. Wang, C. K. (1986) *Structural Analysis on Microcomputers.* New York: Macmillan.

52. Weaver, W., Jr., and J. M. Gere. (1990) *Matrix Analysis of Framed Structures.* 3rd ed. New York: Van Nostrand Reinhold.

53. West, H. H. (1989) *Analysis of Structures: An Integration of Classical and Modern Methods.* 2nd ed. New York: John Wiley & Sons.

54. Zienkiewicz, O. C. (1977) *The Finite Element Method.* 3rd ed. Berkshire, England: McGraw-Hill.

ANSWERS TO SELECTED PROBLEMS

Chapter 2

2.1 $\mathbf{C} = \begin{bmatrix} 8 & 17 & -3 \\ -1 & -1 & -1 \\ 1 & -7 & 1 \end{bmatrix}$; $\mathbf{D} = \begin{bmatrix} -2 & -1 & 1 \\ 17 & -13 & -7 \\ -3 & -1 & 9 \end{bmatrix}$

2.3 $\mathbf{C} = -4$; $\mathbf{D} = \begin{bmatrix} 12 & -18 & 6 \\ 4 & -6 & 2 \\ -2 & 30 & -10 \end{bmatrix}$

2.5 $\mathbf{C} = \begin{bmatrix} -18 & -24 & 21 \\ 7 & -9 & 53 \\ 38 & 38 & 58 \end{bmatrix}$; $\mathbf{D} = \begin{bmatrix} -18 & 7 & 38 \\ -24 & -9 & 38 \\ 21 & 53 & 58 \end{bmatrix}$

2.9 $(\mathbf{ABC})^T = \mathbf{C}^T\mathbf{B}^T\mathbf{A}^T$

$$= \begin{bmatrix} 1{,}512 & -464 & -1{,}602 & 1{,}418 \\ -900 & 5{,}300 & 200 & -2{,}100 \\ -810 & -310 & 942 & -620 \\ 270 & 410 & -360 & 130 \end{bmatrix}$$

2.11 $\mathbf{C} = \begin{bmatrix} 332 & 76 & -332 & -76 \\ 76 & 168 & -76 & -168 \\ -332 & -76 & 332 & 76 \\ -76 & -168 & 76 & 168 \end{bmatrix}$

2.13 $\dfrac{d\mathbf{A}}{dx} = \begin{bmatrix} -4x & 3\cos x & -7 \\ 3\cos x & -2\sin x \cos x & -9x^2 \\ -7 & -9x^2 & 6\sin x \cos x \end{bmatrix}$

2.15 $\dfrac{d\mathbf{AB}}{dx} = \begin{bmatrix} -120x^3 & -20x - 60x^2 \\ -90x^2 & -2 - 8x + 45x^4 \\ -6 + 28x + 125x^4 & 28 + 24x^2 \end{bmatrix}$

2.17 $\displaystyle\int_0^L \mathbf{A}\,dx = \begin{bmatrix} -5L & -L^3 \\ 2L^2 & -\dfrac{L^4}{4} \\ \dfrac{2L^5}{5} & 6L \\ \dfrac{5L^3}{3} & -\dfrac{L^2}{2} \end{bmatrix}$

2.19 $\displaystyle\int_0^L \mathbf{AB}\,dx$

$$= \begin{bmatrix} \dfrac{10L^3}{3} + \dfrac{9L^4}{4} + \dfrac{2L^5}{5} & -9L - L^4 - \dfrac{L^6}{6} \\ -3L^3 + \dfrac{6L^7}{7} & -\dfrac{L^4}{2} \end{bmatrix}$$

2.21 $x_1 = -2$; $x_2 = 3$; $x_3 = -5$

2.23 $x_1 = 8.7$; $x_2 = -7.5$; $x_3 = -4.2$

2.25 $x_1 = 6$; $x_2 = 7$; $x_3 = 5$; $x_4 = 2$

2.27 $\mathbf{A}^{-1} = \begin{bmatrix} 0.44444 & -0.11111 & 0.22222 \\ -0.11111 & 0.16239 & -0.017094 \\ 0.22222 & -0.017094 & 0.26496 \end{bmatrix}$

2.29 $\mathbf{A}^{-1} = \begin{bmatrix} -0.62656 & -0.71396 & 0.2849 & -0.12372 \\ -0.71396 & -0.6958 & 0.2395 & -0.052213 \\ 0.2849 & 0.2395 & -0.098751 & 0.13053 \\ -0.12372 & -0.052213 & 0.13053 & -0.08059 \end{bmatrix}$

Chapter 3

3.1

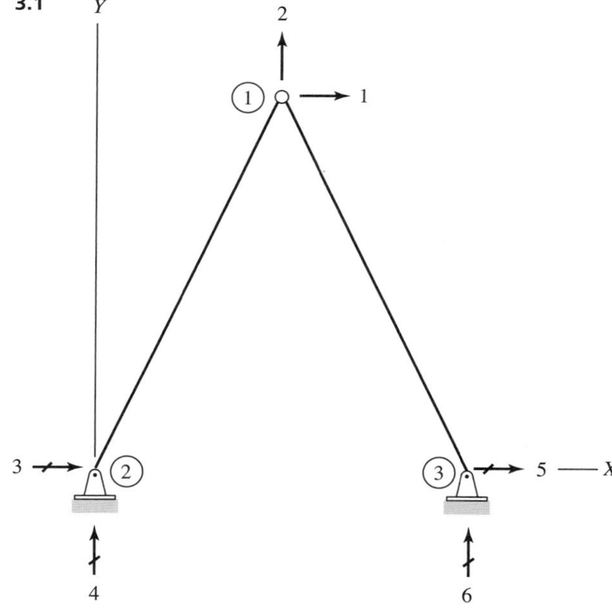

$NDOF = 2, NR = 4$

$\mathbf{P} = \begin{bmatrix} 12 \\ -20 \end{bmatrix}$ k

3.3

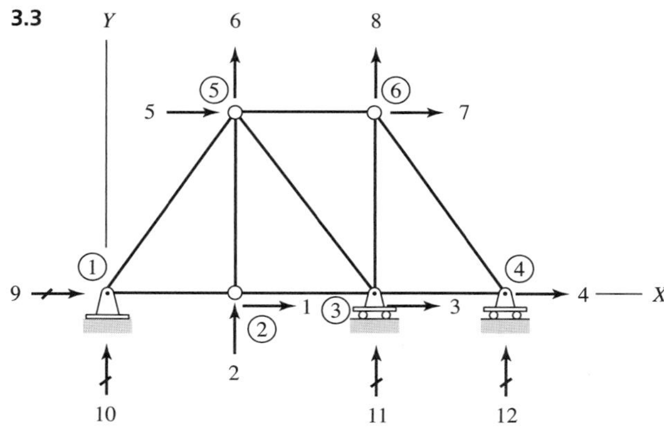

NDOF = 8, *NR* = 4

$$\mathbf{P} = \begin{bmatrix} 0 \\ -30 \\ 0 \\ 0 \\ 10 \\ 0 \\ 0 \\ -30 \end{bmatrix} \text{k}$$

3.5

$$\mathbf{k}_1 = \mathbf{k}_3 = \begin{bmatrix} 483.33 & 0 & -483.33 & 0 \\ 0 & 0 & 0 & 0 \\ -483.33 & 0 & 483.33 & 0 \\ 0 & 0 & 0 & 0 \end{bmatrix} \text{k/in.}$$

$$\mathbf{k}_2 = \begin{bmatrix} 402.78 & 0 & -402.78 & 0 \\ 0 & 0 & 0 & 0 \\ -402.78 & 0 & 402.78 & 0 \\ 0 & 0 & 0 & 0 \end{bmatrix} \text{k/in.}$$

3.7 1,540 kN (T)

3.9

$$\mathbf{T}_1 = \begin{bmatrix} 1 & 0 & 0 & 0 \\ 0 & 1 & 0 & 0 \\ 0 & 0 & 1 & 0 \\ 0 & 0 & 0 & 1 \end{bmatrix}$$

$$\mathbf{T}_2 = \begin{bmatrix} 0.6 & 0.8 & 0 & 0 \\ -0.8 & 0.6 & 0 & 0 \\ 0 & 0 & 0.6 & 0.8 \\ 0 & 0 & -0.8 & 0.6 \end{bmatrix}$$

$$\mathbf{T}_3 = \begin{bmatrix} -0.28 & 0.96 & 0 & 0 \\ -0.96 & -0.28 & 0 & 0 \\ 0 & 0 & -0.28 & 0.96 \\ 0 & 0 & -0.96 & -0.28 \end{bmatrix}$$

3.11 and 3.13

$$\mathbf{F}_9 = \begin{bmatrix} -560 \\ -420 \\ 560 \\ 420 \end{bmatrix} \text{kN}; \quad \text{Yes}$$

3.15 $\mathbf{S} = \begin{bmatrix} 666.23 & 63.413 \\ 63.413 & 703.22 \end{bmatrix} \text{k/in.}$

3.17 $\mathbf{d} = \begin{bmatrix} 0.067082 \\ -0.027951 \end{bmatrix} \text{in.}$

$Q_{a1} = 2.2361$ k (T); $Q_{a2} = 24.597$ k (C)

$$\mathbf{R} = \begin{bmatrix} -1 \\ -2 \\ -11 \\ 22 \end{bmatrix} \text{k}$$

3.19 $\mathbf{d} = \begin{bmatrix} -0.06546 \\ -0.10075 \end{bmatrix} \text{in.}$

$Q_{a1} = 31.639$ k (C); $Q_{a2} = 48.283$ k (C);

$Q_{a3} = 37.889$ k (C)

$$\mathbf{R} = \begin{bmatrix} 28.97 \\ 38.627 \\ -10.609 \\ 36.373 \\ 31.639 \\ 0 \end{bmatrix} \text{k}$$

3.21

$$\mathbf{d} = \begin{bmatrix} 0.2751 \\ -0.32051 \\ 0.23438 \end{bmatrix} \text{in.}$$

$Q_{a1} = 85.969$ k (T); $Q_{a2} = 133.55$ k (C);

$Q_{a3} = 46.875$ k (T); $Q_{a4} = 19.775$ k (C);

$Q_{a5} = 46.875$ k (T)

$$\mathbf{R} = \begin{bmatrix} -123.47 \\ 28.125 \\ -26.531 \\ -11.671 \\ 133.55 \end{bmatrix} \text{k}$$

3.23

$$\mathbf{d} = \begin{bmatrix} 1.5685 \\ 0.78427 \\ -1.261 \\ 5.2564 \\ -1.5981 \end{bmatrix} \text{mm}$$

$Q_{a1} = 78.427$ kN (T); $Q_{a2} = Q_{a3} = 23.836$ kN (C);
$Q_{a4} = 41.205$ kN (T); $Q_{a5} = 33.709$ kN (C);
$Q_{a6} = 137.68$ kN (C)

$$\mathbf{R} = \begin{bmatrix} -80 \\ -20 \\ 140 \end{bmatrix} \text{kN}$$

3.25
$$\mathbf{d} = \begin{bmatrix} 0.019447 \\ -0.096374 \\ 0.038894 \\ 0.044168 \\ 0.031186 \\ -0.054995 \\ 0.025912 \\ -0.032004 \end{bmatrix} \text{in.}$$

$Q_{a1} = Q_{a2} = 18.799$ k (T); $Q_{a3} = 5.098$ k (T);
$Q_{a4} = 5.098$ k (C); $Q_{a5} = 30$ k (T);
$Q_{a6} = 23.203$ k (C);
$Q_{a7} = 14.665$ k (C); $Q_{a8} = 22.835$ k (C);
$Q_{a9} = 8.4966$ k (C)

$$\mathbf{R} = \begin{bmatrix} -10 \\ 11.732 \\ 41.471 \\ 6.7973 \end{bmatrix} \text{k}$$

3.29 $P = 1{,}095$ kN

Chapter 5

5.1

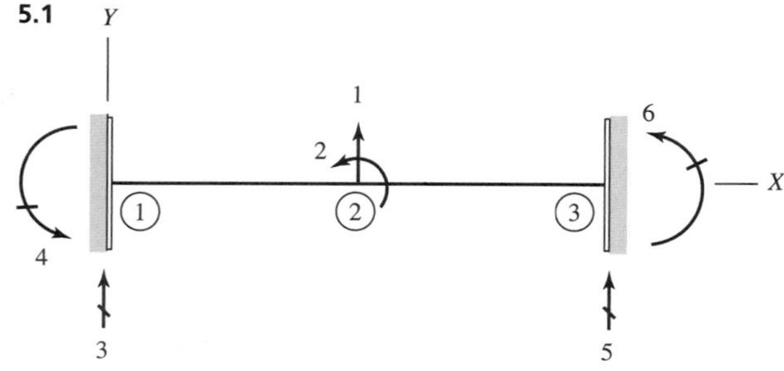

$$\mathbf{P} = \begin{bmatrix} -25\text{ k} \\ 720\text{ k-in.} \end{bmatrix}$$

5.3

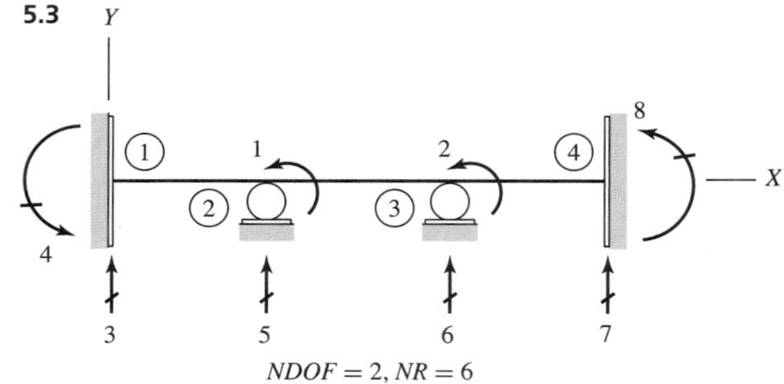

$$\mathbf{P} = \begin{bmatrix} -150 \\ 0 \end{bmatrix} \text{kN} \cdot \text{m}$$

5.5 Units: kips and inches

$$\mathbf{k}_1 = \begin{bmatrix} 5.5556 & 500 & -5.5556 & 500 \\ 500 & 60{,}000 & -500 & 30{,}000 \\ -5.5556 & -500 & 5.5556 & -500 \\ 500 & 30{,}000 & -500 & 60{,}000 \end{bmatrix}$$

$$\mathbf{k}_2 = \begin{bmatrix} 11.111 & 1{,}000 & -11.111 & 1{,}000 \\ 1{,}000 & 120{,}000 & -1{,}000 & 60{,}000 \\ -11.111 & -1{,}000 & 11.111 & -1{,}000 \\ 1{,}000 & 60{,}000 & -1{,}000 & 120{,}000 \end{bmatrix}$$

5.7 Units: kips and inches

$$\mathbf{k}_1 = \mathbf{k}_4 = \begin{bmatrix} 9.4401 & 1{,}132.8 & -9.4401 & 1{,}132.8 \\ 1{,}132.8 & 181{,}250 & -1{,}132.8 & 90{,}625 \\ -9.4401 & -1{,}132.8 & 9.4401 & -1{,}132.8 \\ 1{,}132.8 & 90{,}625 & -1{,}132.8 & 181{,}250 \end{bmatrix}$$

$$\mathbf{k}_2 = \mathbf{k}_3 = \begin{bmatrix} 17.901 & 1{,}611.1 & -17.901 & 1{,}611.1 \\ 1{,}611.1 & 193{,}333 & -1{,}611.1 & 96{,}667 \\ -17.901 & -1{,}611.1 & 17.901 & -1{,}611.1 \\ 1{,}611.1 & 96{,}667 & -1{,}611.1 & 193{,}333 \end{bmatrix}$$

5.9
$$\mathbf{Q}_1 = \begin{bmatrix} 2.7848\text{ k} \\ 1{,}002.5\text{ k-in.} \\ -2.7848\text{ k} \\ -501.23\text{ k-in.} \end{bmatrix}; \text{ Yes}$$

5.11 $FS_b = FS_e = \dfrac{wL}{2}$; $FM_b = -FM_e = \dfrac{wL^2}{12}$

5.13 $FS_b = \dfrac{7wL}{20}$; $FM_b = \dfrac{wL^2}{20}$; $FS_e = \dfrac{3wL}{20}$; $FM_e = -\dfrac{wL^2}{30}$

5.15
$$\mathbf{Q}_{f1} = \begin{bmatrix} 135\text{ kN} \\ 337.5\text{ kN}\cdot\text{m} \\ 135\text{ kN} \\ -337.5\text{ kN}\cdot\text{m} \end{bmatrix} \qquad \mathbf{Q}_{f2} = \begin{bmatrix} 90\text{ kN} \\ 300\text{ kN}\cdot\text{m} \\ 90\text{ kN} \\ -300\text{ kN}\cdot\text{m} \end{bmatrix}$$

$$\mathbf{Q}_{f3} = \begin{bmatrix} 131.25 \text{ kN} \\ 281.25 \text{ kN}\cdot\text{m} \\ 56.25 \text{ kN} \\ -187.5 \text{ kN}\cdot\text{m} \end{bmatrix}$$

5.17

$$\mathbf{Q}_1 = \begin{bmatrix} 69.52 \text{ k} \\ 4{,}945.9 \text{ k-in.} \\ 2.48 \text{ k} \\ 4{,}707.8 \text{ k-in.} \end{bmatrix}; \text{ Yes}$$

5.19 Units: kips and inches

$$\mathbf{S} = \begin{bmatrix} 16.667 & 500 \\ 500 & 180{,}000 \end{bmatrix}$$

5.21 Units: kips and inches

$$\mathbf{S} = \begin{bmatrix} 27.341 & 478.3 & 1{,}611.1 & 0 & 0 \\ 478.3 & 374{,}583 & 96{,}667 & 0 & 0 \\ 1{,}611.1 & 96{,}667 & 386{,}667 & -1{,}611.1 & 96{,}667 \\ 0 & 0 & -1{,}611.1 & 27.341 & -478.3 \\ 0 & 0 & 96{,}667 & -478.3 & 374{,}583 \end{bmatrix}$$

5.23 Units: kips and inches

$$\mathbf{P}_f = -\mathbf{P}_e = \begin{bmatrix} 0 \\ 30 \\ 360 \end{bmatrix}$$

5.25 Units: kN and meters

$$\mathbf{P}_f = -\mathbf{P}_e = \begin{bmatrix} 337.5 \\ -37.5 \\ -18.75 \\ -187.5 \end{bmatrix}$$

5.27

$$\mathbf{d} = \begin{bmatrix} -1.7673 \text{ in.} \\ 0.0089091 \text{ rad} \end{bmatrix}$$

$$\mathbf{Q}_1 = \begin{bmatrix} 14.273 \text{ k} \\ 1{,}150.9 \text{ k-in.} \\ -14.273 \text{ k} \\ 1{,}418.2 \text{ k-in.} \end{bmatrix} \quad \mathbf{Q}_2 = \begin{bmatrix} -10.727 \text{ k} \\ -698.18 \text{ k-in.} \\ 10.727 \text{ k} \\ -1{,}232.7 \text{ k-in.} \end{bmatrix}$$

$$\mathbf{R} = \begin{bmatrix} 14.273 \text{ k} \\ 1{,}150.9 \text{ k-in.} \\ 10.727 \text{ k} \\ -1{,}232.7 \text{ k-in.} \end{bmatrix}$$

5.29

$$\mathbf{d} = \begin{bmatrix} -0.007619 \\ 0.0019048 \end{bmatrix} \text{rad}$$

$$\mathbf{Q}_1 = \begin{bmatrix} -20 \text{ kN} \\ -40 \text{ kN}\cdot\text{m} \\ 20 \text{ kN} \\ -80 \text{ kN}\cdot\text{m} \end{bmatrix} \quad \mathbf{Q}_2 = \begin{bmatrix} -15 \text{ kN} \\ -70 \text{ kN}\cdot\text{m} \\ 15 \text{ kN} \\ -20 \text{ kN}\cdot\text{m} \end{bmatrix}$$

$$\mathbf{Q}_3 = \begin{bmatrix} 5 \text{ kN} \\ 20 \text{ kN}\cdot\text{m} \\ -5 \text{ kN} \\ 10 \text{ kN}\cdot\text{m} \end{bmatrix} \quad \mathbf{R} = \begin{bmatrix} -20 \text{ kN} \\ -40 \text{ kN}\cdot\text{m} \\ 5 \text{ kN} \\ 20 \text{ kN} \\ -5 \text{ kN} \\ 10 \text{ kN}\cdot\text{m} \end{bmatrix}$$

5.31

$$\mathbf{d} = \begin{bmatrix} 0.014375 \\ -0.011422 \\ 0.014244 \end{bmatrix} \text{rad}$$

$$\mathbf{Q}_1 = \begin{bmatrix} 41.615 \text{ k} \\ 2{,}267.1 \text{ k-in.} \\ 30.385 \text{ k} \\ -649.85 \text{ k-in.} \end{bmatrix} \quad \mathbf{Q}_2 = \begin{bmatrix} 1.1538 \text{ k} \\ 649.85 \text{ k-in.} \\ -1.1538 \text{ k} \\ -317.54 \text{ k-in.} \end{bmatrix}$$

$$\mathbf{Q}_3 = \begin{bmatrix} 11.103 \text{ k} \\ 317.54 \text{ k-in.} \\ 8.8974 \text{ k} \\ 0 \end{bmatrix} \quad \mathbf{R} = \begin{bmatrix} 41.615 \text{ k} \\ 2{,}267.1 \text{ k-in.} \\ 31.538 \text{ k} \\ 9.9487 \text{ k} \\ 8.8974 \text{ k} \end{bmatrix}$$

5.33

$$\mathbf{d} = \begin{bmatrix} -0.019391 \\ 0.0071406 \\ -0.0056563 \\ 0.017242 \end{bmatrix} \text{rad}$$

$$\mathbf{Q}_1 = \begin{bmatrix} 108.87 \text{ kN} \\ 0 \\ 161.13 \text{ kN} \\ -392 \text{ kN}\cdot\text{m} \end{bmatrix} \quad \mathbf{Q}_2 = \begin{bmatrix} 93.167 \text{ kN} \\ 392 \text{ kN}\cdot\text{m} \\ 86.833 \text{ kN} \\ -344.5 \text{ kN}\cdot\text{m} \end{bmatrix}$$

$$\mathbf{Q}_3 = \begin{bmatrix} 155.97 \text{ kN} \\ 344.5 \text{ kN}\cdot\text{m} \\ 31.533 \text{ kN} \\ 120 \text{ kN}\cdot\text{m} \end{bmatrix} \quad \mathbf{R} = \begin{bmatrix} 108.87 \\ 254.3 \\ 242.8 \\ 31.533 \end{bmatrix} \text{kN}$$

5.35

$$\mathbf{d} = \begin{bmatrix} -1.8705 \text{ in.} \\ 0.0023884 \text{ rad} \\ 0 \\ -1.8705 \text{ in.} \\ -0.0023884 \text{ rad} \end{bmatrix}$$

$$\mathbf{Q}_1 = \begin{bmatrix} 20.363 \text{ k} \\ 2{,}335.4 \text{ k-in.} \\ -20.363 \text{ k} \\ 2{,}551.8 \text{ k-in.} \end{bmatrix} \quad \mathbf{Q}_2 = \begin{bmatrix} -29.637 \text{ k} \\ -2{,}551.8 \text{ k-in.} \\ 29.637 \text{ k} \\ -2{,}782.7 \text{ k-in.} \end{bmatrix}$$

$$\mathbf{Q}_3 = \begin{bmatrix} 29.637 \text{ k} \\ 2{,}782.7 \text{ k-in.} \\ -29.637 \text{ k} \\ 2{,}551.8 \text{ k-in.} \end{bmatrix} \quad \mathbf{Q}_4 = \begin{bmatrix} -20.363 \text{ k} \\ -2{,}551.8 \text{ k-in.} \\ 20.363 \text{ k} \\ -2{,}335.4 \text{ k-in.} \end{bmatrix}$$

$$\mathbf{R} = \begin{bmatrix} 20.363 \text{ k} \\ 2{,}335.4 \text{ k-in.} \\ 59.273 \text{ k} \\ 20.363 \text{ k} \\ -2{,}335.4 \text{ k-in.} \end{bmatrix}$$

Chapter 6

6.1

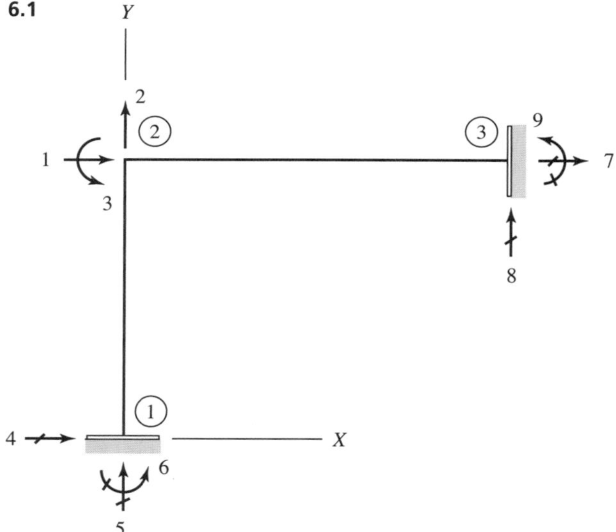

NDOF = 3, *NR* = 6

$$\mathbf{P} = \begin{bmatrix} 0 \\ 0 \\ -900 \text{ k-in.} \end{bmatrix}$$

6.3

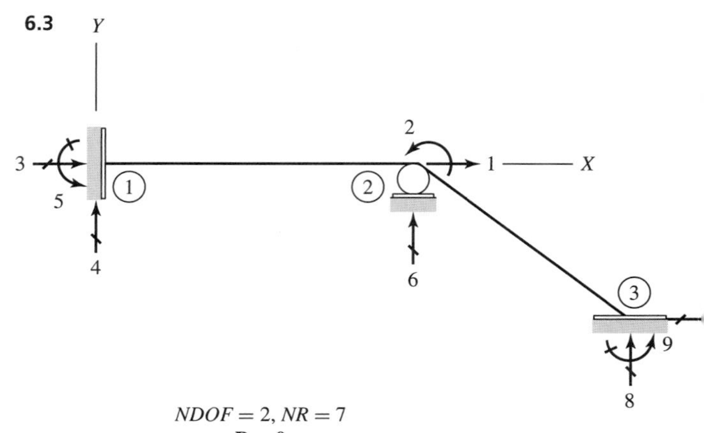

NDOF = 2, *NR* = 7

$\mathbf{P} = \mathbf{0}$

6.5

$$\mathbf{P} = \begin{bmatrix} 30 \text{ k} \\ 0 \\ 0 \\ 0 \\ 0 \\ 0 \end{bmatrix}$$

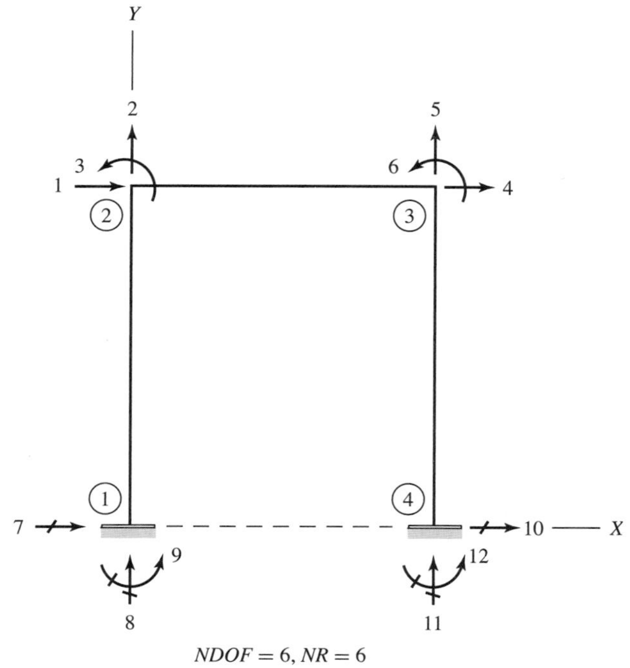

NDOF = 6, *NR* = 6

6.7 Units: kips and inches

$$\mathbf{k}_1 = \begin{bmatrix} 1,244.6 & 0 & 0 & -1,244.6 & 0 & 0 \\ 0 & 12.839 & 1,540.6 & 0 & -12.839 & 1,540.6 \\ 0 & 1,540.6 & 246,500 & 0 & -1,540.6 & 123,250 \\ -1,244.6 & 0 & 0 & 1,244.6 & 0 & 0 \\ 0 & -12.839 & -1,540.6 & 0 & 12.839 & -1,540.6 \\ 0 & 1,540.6 & 123,250 & 0 & -1,540.6 & 246,500 \end{bmatrix}$$

$$\mathbf{k}_2 = \begin{bmatrix} 829.72 & 0 & 0 & -829.72 & 0 & 0 \\ 0 & 3.804 & 684.72 & 0 & -3.804 & 684.72 \\ 0 & 684.72 & 164,333 & 0 & -684.72 & 82,167 \\ -829.72 & 0 & 0 & 829.72 & 0 & 0 \\ 0 & -3.804 & -684.72 & 0 & 3.804 & -684.72 \\ 0 & 684.72 & 82,167 & 0 & -684.72 & 164,333 \end{bmatrix} \quad \mathbf{Q}_{f1} = \begin{bmatrix} 0 \\ 20 \\ 1,200 \\ 0 \\ 20 \\ -1,200 \end{bmatrix} \quad \mathbf{Q}_{f2} = \begin{bmatrix} 0 \\ 22.5 \\ 1,350 \\ 0 \\ 22.5 \\ -1,350 \end{bmatrix}$$

6.9 Units: kips and inches

$$\mathbf{k}_1 = \begin{bmatrix} 409.26 & 0 & 0 & -409.26 & 0 & 0 \\ 0 & 2.0243 & 218.62 & 0 & -2.0243 & 218.62 \\ 0 & 218.62 & 31,481 & 0 & -218.62 & 15,741 \\ -409.26 & 0 & 0 & 409.26 & 0 & 0 \\ 0 & -2.0243 & -218.62 & 0 & 2.0243 & -218.62 \\ 0 & 218.62 & 15,741 & 0 & -218.62 & 31,481 \end{bmatrix}$$

$$\mathbf{k}_2 = \begin{bmatrix} 491.11 & 0 & 0 & -491.11 & 0 & 0 \\ 0 & 3.4979 & 314.81 & 0 & -3.4979 & 314.81 \\ 0 & 314.81 & 37,778 & 0 & -314.81 & 18,889 \\ -491.11 & 0 & 0 & 491.11 & 0 & 0 \\ 0 & -3.4979 & -314.81 & 0 & 3.4979 & -314.81 \\ 0 & 314.81 & 18,889 & 0 & -314.81 & 37,778 \end{bmatrix} \quad \mathbf{Q}_{f1} = \begin{bmatrix} 0 \\ 27 \\ 972 \\ 0 \\ 27 \\ -972 \end{bmatrix} \quad \mathbf{Q}_{f2} = \begin{bmatrix} -4.5 \\ 6 \\ 270 \\ -4.5 \\ 6 \\ -270 \end{bmatrix}$$

6.11 Units: kips and inches

$$\mathbf{k}_1 = \mathbf{k}_3 = \begin{bmatrix} 1,500 & 0 & 0 & -1,500 & 0 & 0 \\ 0 & 2.1484 & 257.81 & 0 & -2.1484 & 257.81 \\ 0 & 257.81 & 41,250 & 0 & -257.81 & 20,625 \\ -1,500 & 0 & 0 & 1,500 & 0 & 0 \\ 0 & -2.1484 & -257.81 & 0 & 2.1484 & -257.81 \\ 0 & 257.81 & 20,625 & 0 & -257.81 & 41,250 \end{bmatrix}$$

$$\mathbf{k}_2 = \begin{bmatrix} 1,620 & 0 & 0 & -1,620 & 0 & 0 \\ 0 & 2.6 & 390 & 0 & -2.6 & 390 \\ 0 & 390 & 78,000 & 0 & -390 & 39,000 \\ -1,620 & 0 & 0 & 1,620 & 0 & 0 \\ 0 & -2.6 & -390 & 0 & 2.6 & -390 \\ 0 & 390 & 39,000 & 0 & -390 & 78,000 \end{bmatrix} \quad \mathbf{Q}_{f1} = \mathbf{Q}_{f3} = \mathbf{0}; \quad \mathbf{Q}_{f2} = \begin{bmatrix} 0 \\ 25 \\ 1,250 \\ 0 \\ 25 \\ -1,250 \end{bmatrix}$$

6.13 $FA_b = \dfrac{Wl_2}{L};\quad FA_e = \dfrac{Wl_1}{L};$

$FS_b = FS_e = FM_b = FM_e = 0$

6.15

$$\mathbf{Q}_1 = \begin{bmatrix} 128.12\text{ k} \\ -78.786\text{ k} \\ -5{,}875\text{ k-in.} \\ -128.12\text{ k} \\ 98.786\text{ k} \\ -5{,}179.3\text{ k-in.} \end{bmatrix} \quad \mathbf{Q}_2 = \begin{bmatrix} 111.62\text{ k} \\ 11.038\text{ k} \\ 1{,}942.5\text{ k-in.} \\ -140.46\text{ k} \\ -80.268\text{ k} \\ 5{,}179.3\text{ k-in.} \end{bmatrix} \quad \mathbf{Q}_3 = \begin{bmatrix} -70.97\text{ k} \\ -20.192\text{ k} \\ -2{,}176.6\text{ k-in.} \\ 70.97\text{ k} \\ 20.192\text{ k} \\ -1{,}942.5\text{ k-in.} \end{bmatrix};\ \text{Yes}$$

6.17

$$\mathbf{T}_1 = \begin{bmatrix} -0.6 & 0.8 & 0 & 0 & 0 & 0 \\ -0.8 & -0.6 & 0 & 0 & 0 & 0 \\ 0 & 0 & 1 & 0 & 0 & 0 \\ 0 & 0 & 0 & -0.6 & 0.8 & 0 \\ 0 & 0 & 0 & -0.8 & -0.6 & 0 \\ 0 & 0 & 0 & 0 & 0 & 1 \end{bmatrix};\ \mathbf{T}_2 = \mathbf{I}$$

6.19

$$\mathbf{T}_1 = \begin{bmatrix} 0 & 1 & 0 & 0 & 0 & 0 \\ -1 & 0 & 0 & 0 & 0 & 0 \\ 0 & 0 & 1 & 0 & 0 & 0 \\ 0 & 0 & 0 & 0 & 1 & 0 \\ 0 & 0 & 0 & -1 & 0 & 0 \\ 0 & 0 & 0 & 0 & 0 & 1 \end{bmatrix} \quad \mathbf{T}_2 = \begin{bmatrix} -0.96 & 0.28 & 0 & 0 & 0 & 0 \\ -0.28 & -0.96 & 0 & 0 & 0 & 0 \\ 0 & 0 & 1 & 0 & 0 & 0 \\ 0 & 0 & 0 & -0.96 & 0.28 & 0 \\ 0 & 0 & 0 & -0.28 & -0.96 & 0 \\ 0 & 0 & 0 & 0 & 0 & 1 \end{bmatrix}$$

6.21

$$\mathbf{T}_1 = \begin{bmatrix} 0.44721 & 0.89443 & 0 & 0 & 0 & 0 \\ -0.89443 & 0.44721 & 0 & 0 & 0 & 0 \\ 0 & 0 & 1 & 0 & 0 & 0 \\ 0 & 0 & 0 & 0.44721 & 0.89443 & 0 \\ 0 & 0 & 0 & -0.89443 & 0.44721 & 0 \\ 0 & 0 & 0 & 0 & 0 & 1 \end{bmatrix} \quad \mathbf{T}_2 = \mathbf{I}$$

6.23 and 6.31

$$\mathbf{v}_1 = \begin{bmatrix} 0 \\ 0 \\ 0 \\ -2.0939\text{ in.} \\ -0.05147\text{ in.} \\ 0.0079542\text{ rad} \end{bmatrix} \quad \mathbf{v}_2 = \begin{bmatrix} -1.8078\text{ in.} \\ -0.90922\text{ in.} \\ 0.0028882\text{ rad} \\ -2.0939\text{ in.} \\ -0.05147\text{ in.} \\ 0.0079542\text{ rad} \end{bmatrix} \quad \mathbf{v}_3 = \begin{bmatrix} 0 \\ 0 \\ 0 \\ -1.8078\text{ in.} \\ -0.90922\text{ in.} \\ 0.0028882\text{ rad} \end{bmatrix}$$

$$\mathbf{F}_1 = \begin{bmatrix} 78.786\text{ k} \\ 128.12\text{ k} \\ -5{,}875\text{ k-in.} \\ -98.786\text{ k} \\ -128.12\text{ k} \\ -5{,}179.3\text{ k-in.} \end{bmatrix} \quad \mathbf{F}_2 = \begin{bmatrix} -98.786\text{ k} \\ -53.119\text{ k} \\ 1{,}942.5\text{ k-in.} \\ 98.786\text{ k} \\ 128.12\text{ k} \\ 5{,}179.3\text{ k-in.} \end{bmatrix} \quad \mathbf{F}_3 = \begin{bmatrix} 51.214\text{ k} \\ -53.119\text{ k} \\ -2{,}176.6\text{ k-in.} \\ -51.214\text{ k} \\ 53.119\text{ k} \\ -1{,}942.5\text{ k-in.} \end{bmatrix};\ \text{Yes}$$

6.25 Units: kN and meters

$$\mathbf{K}_1 = \begin{bmatrix} 94{,}770 & -123{,}922 & -7{,}315.2 & -94{,}770 & 123{,}922 & -7{,}315.2 \\ & 167{,}058 & -5{,}486.4 & 123{,}922 & -167{,}058 & -5{,}486.4 \\ & & 60{,}960 & 7{,}315.2 & 5{,}486.4 & 30{,}480 \\ & \text{(symmetric)} & & 94{,}770 & -123{,}922 & 7{,}315.2 \\ & & & & 167{,}058 & 5{,}486.4 \\ & & & & & 60{,}960 \end{bmatrix}$$

$$\mathbf{K}_2 = \begin{bmatrix} 216{,}667 & 0 & 0 & -216{,}667 & 0 & 0 \\ & 1{,}058.3 & 6{,}350 & 0 & -1{,}058.3 & 6{,}350 \\ & & 50{,}800 & 0 & -6{,}350 & 25{,}400 \\ & \text{(symmetric)} & & 216{,}667 & 0 & 0 \\ & & & & 1{,}058.3 & -6{,}350 \\ & & & & & 50{,}800 \end{bmatrix}$$

$$\mathbf{F}_{f1} = \begin{bmatrix} 50 \\ 37.5 \\ -156.25 \\ 50 \\ 37.5 \\ 156.25 \end{bmatrix} \quad \mathbf{F}_{f2} = \begin{bmatrix} 0 \\ 144 \\ 288 \\ 0 \\ 144 \\ -288 \end{bmatrix}$$

6.27 Units: kN and meters

$$\mathbf{K}_1 = \begin{bmatrix} 694.92 & 0 & -5{,}559.4 & -694.92 & 0 & -5{,}559.4 \\ & 200{,}000 & 0 & 0 & -200{,}000 & 0 \\ & & 59{,}300 & 5{,}559.4 & 0 & 29{,}650 \\ & & & 694.92 & 0 & 5{,}559.4 \\ & & & & 200{,}000 & 0 \\ & & & & & 59{,}300 \end{bmatrix}$$

$$\mathbf{K}_2 = \begin{bmatrix} 117{,}980 & -34{,}357 & -637.59 & -117{,}980 & 34{,}357 & -637.59 \\ & 10{,}203 & -2{,}186 & 34{,}357 & -10{,}203 & -2{,}186 \\ & & 37{,}952 & 637.59 & 2{,}186 & 18{,}976 \\ & & & 117{,}980 & -34{,}357 & 637.59 \\ & & & & 10{,}203 & 2{,}186 \\ & & & & & 37{,}952 \end{bmatrix}$$

$$\mathbf{F}_{f1} = \mathbf{0}; \qquad \mathbf{F}_{f2} = \begin{bmatrix} 0 \\ 250 \\ -1{,}000 \\ 0 \\ 250 \\ 1{,}000 \end{bmatrix}$$

6.29 Units: kips and inches

$$\mathbf{K}_1 = \begin{bmatrix} 78.188 & 152.84 & -212.43 & -78.188 & -152.84 & -212.43 \\ & 307.44 & 106.21 & -152.84 & -307.44 & 106.21 \\ & & 42{,}485 & 212.43 & -106.21 & 21{,}243 \\ & \text{(symmetric)} & & 78.188 & 152.84 & 212.43 \\ & & & & 307.44 & -106.21 \\ & & & & & 42{,}485 \end{bmatrix}$$

$$\mathbf{K}_2 = \begin{bmatrix} 304 & 0 & 0 & -304 & 0 & 0 \\ & 1.6667 & 250 & 0 & -1.6667 & 250 \\ & & 50{,}000 & 0 & -250 & 25{,}000 \\ & \text{(symmetric)} & & 304 & 0 & 0 \\ & & & & 1.6667 & -250 \\ & & & & & 50{,}000 \end{bmatrix}$$

$$\mathbf{F}_{f1} = \begin{bmatrix} 0 \\ 11.18 \\ 223.61 \\ 0 \\ 11.18 \\ -223.61 \end{bmatrix} \qquad \mathbf{F}_{f2} = \begin{bmatrix} 0 \\ 16.25 \\ 875 \\ 0 \\ 21.25 \\ -1{,}000 \end{bmatrix}$$

6.33 Units: kN and meters

$$\mathbf{S} = \begin{bmatrix} 311{,}437 & -7{,}315.2 & 0 \\ -7{,}315.2 & 111{,}760 & 25{,}400 \\ 0 & 25{,}400 & 50{,}800 \end{bmatrix} \qquad \mathbf{P}_f = -\mathbf{P}_e = \begin{bmatrix} 50 \\ 131.75 \\ -288 \end{bmatrix}$$

6.35 Units: kN and meters

$$\mathbf{S} = \begin{bmatrix} 210{,}203 & -2{,}186 \\ -2{,}186 & 97{,}252 \end{bmatrix} \qquad \mathbf{P}_f = -\mathbf{P}_e = \begin{bmatrix} 250 \\ -1{,}000 \end{bmatrix}$$

6.37 Units: kips and inches

$$\mathbf{S} = \begin{bmatrix} 382.19 & 152.84 & 212.43 & 0 \\ 152.84 & 309.11 & 143.79 & 250 \\ 212.43 & 143.79 & 92{,}485 & 25{,}000 \\ 0 & 250 & 25{,}000 & 50{,}000 \end{bmatrix} \qquad \mathbf{P}_f = -\mathbf{P}_e = \begin{bmatrix} 0 \\ 27.431 \\ 651.4 \\ -1{,}000 \end{bmatrix}$$

6.39

$$\mathbf{P} = \begin{bmatrix} 0 \\ 0 \\ -30 \text{ k} \\ -180 \text{ k-in.} \\ 45 \text{ k} \\ 0 \\ 240 \text{ k-in.} \\ 0 \end{bmatrix}$$

6.41 $\mathbf{d} = \begin{bmatrix} 0.0034433 \text{ in.} \\ 0.0029828 \text{ rad} \end{bmatrix}$

$$\mathbf{Q}_1 = \begin{bmatrix} -7.3457 \text{ k} \\ 16.132 \text{ k} \\ 517.88 \text{ k-in.} \\ 7.3457 \text{ k} \\ 13.869 \text{ k} \\ -314.26 \text{ k-in.} \end{bmatrix} \quad \mathbf{Q}_2 = \begin{bmatrix} 3.2599 \text{ k} \\ 9.3161 \text{ k} \\ 314.26 \text{ k-in.} \\ -3.2599 \text{ k} \\ 0.43406 \text{ k} \\ -1.7167 \text{ k-in.} \end{bmatrix}$$

$$\mathbf{R} = \begin{bmatrix} -7.3457 \text{ k} \\ 16.132 \text{ k} \\ 517.88 \text{ k-in.} \\ 20.461 \text{ k} \\ -1.6545 \text{ k} \\ -2.8422 \text{ k} \\ -1.7167 \text{ k-in.} \end{bmatrix}$$

6.43 $\mathbf{d} = \begin{bmatrix} -0.00026195 \text{ m} \\ -0.0043172 \text{ rad} \\ 0.0078279 \text{ rad} \end{bmatrix}$

$$\mathbf{Q}_1 = \begin{bmatrix} 40.864 \text{ kN} \\ -101.59 \text{ kN} \\ -417.51 \text{ kN} \cdot \text{m} \\ -40.864 \text{ kN} \\ -23.406 \text{ kN} \\ 26.577 \text{ kN} \cdot \text{m} \end{bmatrix} \quad \mathbf{Q}_2 = \begin{bmatrix} -56.756 \text{ kN} \\ 166.29 \text{ kN} \\ 267.51 \text{ kN} \cdot \text{m} \\ 56.756 \text{ kN} \\ 121.71 \text{ kN} \\ 0 \end{bmatrix}$$

$$\mathbf{R} = \begin{bmatrix} 43.244 \text{ kN} \\ -18.648 \text{ kN} \\ 26.577 \text{ kN} \cdot \text{m} \\ 259.94 \text{ kN} \\ 56.756 \text{ kN} \\ 121.71 \text{ kN} \end{bmatrix}$$

6.45 $\mathbf{d} = \begin{bmatrix} -0.0010826 \text{ m} \\ 0.010258 \text{ rad} \end{bmatrix}$

$$\mathbf{Q}_1 = \begin{bmatrix} 216.53 \text{ kN} \\ 57.029 \text{ kN} \\ 304.16 \text{ kN} \cdot \text{m} \\ -216.53 \text{ kN} \\ -57.029 \text{ kN} \\ 608.31 \text{ kN} \cdot \text{m} \end{bmatrix} \quad \mathbf{Q}_2 = \begin{bmatrix} 31.198 \text{ kN} \\ -216.45 \text{ kN} \\ -608.31 \text{ kN} \cdot \text{m} \\ 108.8 \text{ kN} \\ -263.55 \text{ kN} \\ 1{,}197 \text{ kN} \cdot \text{m} \end{bmatrix}$$

$$\mathbf{R} = \begin{bmatrix} -57.029 \text{ kN} \\ 216.53 \text{ kN} \\ 304.16 \text{ kN} \cdot \text{m} \\ 87.686 \text{ kN} \\ -30.656 \text{ kN} \\ 283.47 \text{ kN} \\ 1{,}197 \text{ kN} \cdot \text{m} \end{bmatrix}$$

6.47

$$\mathbf{d} = \begin{bmatrix} 0.062194 \text{ in.} \\ -0.13533 \text{ in.} \\ -0.014529 \text{ rad} \\ 0.027941 \text{ rad} \end{bmatrix}$$

$$\mathbf{Q}_1 = \begin{bmatrix} 45.788 \text{ k} \\ 1.7551 \text{ k} \\ -57.433 \text{ k-in.} \\ -25.787 \text{ k} \\ 8.245 \text{ k} \\ -813.28 \text{ k-in.} \end{bmatrix} \quad \mathbf{Q}_2 = \begin{bmatrix} 18.907 \text{ k} \\ 19.378 \text{ k} \\ 813.28 \text{ k-in.} \\ -18.907 \text{ k} \\ 18.123 \text{ k} \\ 0 \end{bmatrix}$$

$$\mathbf{R} = \begin{bmatrix} 18.907 \text{ k} \\ 41.739 \text{ k} \\ -57.433 \text{ k-in.} \\ -18.907 \text{ k} \\ 18.123 \text{ k} \end{bmatrix}$$

6.49

$$\mathbf{d} = \begin{bmatrix} 2.8978 \text{ in.} \\ -0.02114 \text{ in.} \\ -0.011687 \text{ rad} \\ 3.5004 \text{ in.} \\ -1.2117 \text{ in.} \\ 0.0038711 \text{ rad} \\ 4.0922 \text{ in.} \\ -0.029507 \text{ in.} \\ -0.0038888 \text{ rad} \end{bmatrix} \quad \mathbf{R} = \begin{bmatrix} -12.797 \text{ k} \\ 42.573 \text{ k} \\ 2{,}693.5 \text{ k-in.} \\ -37.203 \text{ k} \\ 59.424 \text{ k} \\ 5{,}640.4 \text{ k-in.} \end{bmatrix}$$

$$\mathbf{Q}_1 = \begin{bmatrix} 42.573 \text{ k} \\ 12.797 \text{ k} \\ 2{,}693.5 \text{ k-in.} \\ -42.573 \text{ k} \\ -12.797 \text{ k} \\ 991.87 \text{ k-in.} \end{bmatrix} \quad \mathbf{Q}_2 = \begin{bmatrix} 52.861 \text{ k} \\ 20.057 \text{ k} \\ -991.87 \text{ k-in.} \\ -28.861 \text{ k} \\ 24.941 \text{ k} \\ -4.395 \text{ k-in.} \end{bmatrix}$$

$$\mathbf{Q}_3 = \begin{bmatrix} 60.791 \text{ k} \\ -34.925 \text{ k} \\ -5{,}074.2 \text{ k-in.} \\ -36.791 \text{ k} \\ -10.073 \text{ k} \\ 4.395 \text{ k-in.} \end{bmatrix} \quad \mathbf{Q}_4 = \begin{bmatrix} 59.424 \text{ k} \\ 37.203 \text{ k} \\ 5{,}640.4 \text{ k-in.} \\ -59.424 \text{ k} \\ -37.203 \text{ k} \\ 5{,}074.2 \text{ k-in.} \end{bmatrix}$$

Chapter 7

7.1 $\mathbf{d} = [-2.6519]$ in.

$$\mathbf{Q}_1 = \begin{bmatrix} 24.591 \text{ k} \\ 2{,}301.9 \text{ k-in.} \\ 5.4087 \text{ k} \\ 0 \end{bmatrix} \quad \mathbf{Q}_2 = \begin{bmatrix} -5.4087 \text{ k} \\ 0 \\ 27.909 \text{ k} \\ -2{,}998.6 \text{ k-in.} \end{bmatrix}$$

$$\mathbf{R} = \begin{bmatrix} 24.591 \text{ k} \\ 2{,}301.9 \text{ k-in.} \\ 0 \\ 27.909 \text{ k} \\ -2{,}998.6 \text{ k-in.} \end{bmatrix}$$

7.3 $$\mathbf{d} = \begin{bmatrix} 0.014375 \\ -0.011422 \end{bmatrix} \text{ rad}$$

$$\mathbf{Q}_1 = \begin{bmatrix} 41.615 \text{ k} \\ 2{,}267.1 \text{ k-in.} \\ 30.385 \text{ k} \\ -649.85 \text{ k-in.} \end{bmatrix} \quad \mathbf{Q}_2 = \begin{bmatrix} 1.1538 \text{ k} \\ 649.85 \text{ k-in.} \\ -1.1538 \text{ k} \\ -317.54 \text{ k-in.} \end{bmatrix}$$

$$\mathbf{Q}_3 = \begin{bmatrix} 11.103 \text{ k} \\ 317.54 \text{ k-in.} \\ 8.8974 \text{ k} \\ 0 \end{bmatrix} \quad \mathbf{R} = \begin{bmatrix} 41.615 \text{ k} \\ 2{,}267.1 \text{ k-in.} \\ 31.538 \text{ k} \\ 9.9487 \text{ k} \\ 8.8974 \text{ k} \\ 0 \end{bmatrix}$$

7.5 $$\mathbf{d} = \begin{bmatrix} 0.0071406 \\ -0.0056563 \\ 0.017242 \end{bmatrix} \text{ rad}$$

$$\mathbf{Q}_1 = \begin{bmatrix} 108.87 \text{ kN} \\ 0 \\ 161.13 \text{ kN} \\ -392 \text{ kN}\cdot\text{m} \end{bmatrix} \quad \mathbf{Q}_2 = \begin{bmatrix} 93.167 \text{ kN} \\ 392 \text{ kN}\cdot\text{m} \\ 86.833 \text{ kN} \\ -344.5 \text{ kN}\cdot\text{m} \end{bmatrix}$$

$$\mathbf{Q}_3 = \begin{bmatrix} 155.97 \text{ kN} \\ 344.5 \text{ kN}\cdot\text{m} \\ 31.533 \text{ kN} \\ 120 \text{ kN}\cdot\text{m} \end{bmatrix} \quad \mathbf{R} = \begin{bmatrix} 108.87 \text{ kN} \\ 0 \\ 254.3 \text{ kN} \\ 242.8 \text{ kN} \\ 31.533 \text{ kN} \end{bmatrix}$$

7.7 $$\mathbf{d} = \begin{bmatrix} -0.00026195 \text{ m} \\ -0.0043172 \text{ rad} \end{bmatrix}$$

$$\mathbf{Q}_1 = \begin{bmatrix} 40.864 \text{ kN} \\ -101.59 \text{ kN} \\ -417.51 \text{ kN}\cdot\text{m} \\ -40.864 \text{ kN} \\ -23.406 \text{ kN} \\ 26.577 \text{ kN}\cdot\text{m} \end{bmatrix} \quad \mathbf{Q}_2 = \begin{bmatrix} -56.756 \text{ kN} \\ 166.29 \text{ kN} \\ 267.51 \text{ kN}\cdot\text{m} \\ 56.756 \text{ kN} \\ 121.71 \text{ kN} \\ 0 \end{bmatrix}$$

$$\mathbf{R} = \begin{bmatrix} 43.244 \text{ kN} \\ -18.648 \text{ kN} \\ 26.577 \text{ kN}\cdot\text{m} \\ 259.94 \text{ kN} \\ 56.756 \text{ kN} \\ 121.71 \text{ kN} \\ 0 \end{bmatrix}$$

7.9 $$\mathbf{d} = \begin{bmatrix} -0.000901 \text{ m} \\ -0.003276 \text{ m} \\ 0.010155 \text{ rad} \end{bmatrix}$$

$$\mathbf{Q}_1 = \begin{bmatrix} 94.605 \text{ kN} \\ 46.434 \text{ kN} \\ 0 \\ -94.605 \text{ kN} \\ 73.566 \text{ kN} \\ -135.66 \text{ kN}\cdot\text{m} \end{bmatrix} \quad \mathbf{Q}_2 = \begin{bmatrix} 175.62 \text{ kN} \\ -13.456 \text{ kN} \\ 0 \\ -115.62 \text{ kN} \\ -31.544 \text{ kN} \\ 135.66 \text{ kN}\cdot\text{m} \end{bmatrix}$$

$$\mathbf{R} = \begin{bmatrix} 94.605 \text{ kN} \\ 46.434 \text{ kN} \\ 0 \\ -94.605 \text{ kN} \\ 148.57 \text{ kN} \\ 0 \end{bmatrix}$$

7.11 $P_{max} = 951.36\,\text{kN}$

7.13 $P_{max} = 47.453\,\text{kN}$

7.17 $$\mathbf{d} = \begin{bmatrix} 0.23122 \\ -0.36987 \\ 0.14063 \end{bmatrix} \text{ in.}$$

$Q_{a1} = 72.258$ k (T); $Q_{a2} = 154.11$ k (C);

$Q_{a3} = 28.125$ k (T); $Q_{a4} = 4.9438$ k (T);

$Q_{a5} = 65.625$ k (T)

$$\mathbf{R} = \begin{bmatrix} -94.758 \\ 16.875 \\ -55.242 \\ -43.489 \\ 176.61 \end{bmatrix} \text{ k}$$

7.19 $$\mathbf{d} = \begin{bmatrix} -0.0036571 \\ 0.0017714 \end{bmatrix} \text{ rad}$$

$$\mathbf{Q}_1 = \begin{bmatrix} -4.87 \text{ kN} \\ -1.8939 \text{ kN}\cdot\text{m} \\ 4.87 \text{ kN} \\ -32.196 \text{ kN}\cdot\text{m} \end{bmatrix} \quad \mathbf{Q}_2 = \begin{bmatrix} 15.624 \text{ kN} \\ 32.196 \text{ kN}\cdot\text{m} \\ -15.624 \text{ kN} \\ 77.176 \text{ kN}\cdot\text{m} \end{bmatrix}$$

$$\mathbf{Q}_3 = \begin{bmatrix} -24.147 \text{ kN} \\ -77.176 \text{ kN} \cdot \text{m} \\ 24.147 \text{ kN} \\ -91.853 \text{ kN.m} \end{bmatrix} \quad \mathbf{R} = \begin{bmatrix} -4.87 \text{ kN} \\ -1.8939 \text{ kN} \cdot \text{m} \\ 20.494 \text{ kN} \\ -39.771 \text{ kN} \\ 24.147 \text{ kN} \\ -91.853 \text{ kN} \cdot \text{m} \end{bmatrix}$$

7.21

$$\mathbf{d} = \begin{bmatrix} -2.6114 \text{ in.} \\ -0.0020419 \text{ rad} \\ 0 \\ -22.6114 \text{ in.} \\ 0.0020419 \text{ rad} \end{bmatrix}$$

$$\mathbf{Q}_1 = \begin{bmatrix} 22.339 \text{ k} \\ 2{,}773.2 \text{ k-in.} \\ -22.339 \text{ k} \\ 2{,}588.2 \text{ k-in.} \end{bmatrix} \quad \mathbf{Q}_2 = \begin{bmatrix} -27.661 \text{ k} \\ -2{,}588.2 \text{ k-in.} \\ 27.661 \text{ k} \\ -2{,}390.8 \text{ k-in.} \end{bmatrix}$$

$$\mathbf{Q}_3 = \begin{bmatrix} 27.661 \text{ k} \\ 2{,}390.8 \text{ k-in.} \\ -27.661 \text{ k} \\ 2{,}588.2 \text{ k-in.} \end{bmatrix} \quad \mathbf{Q}_4 = \begin{bmatrix} -22.339 \text{ k} \\ -2{,}588.2 \text{ k-in.} \\ 22.339 \text{ k} \\ -2{,}773.2 \text{ k-in.} \end{bmatrix}$$

$$\mathbf{R} = \begin{bmatrix} 22.339 \text{ k} \\ 2{,}773.2 \text{ k-in.} \\ 55.322 \text{ k} \\ 22.339 \text{ k} \\ -2{,}773.2 \text{ k-in.} \end{bmatrix}$$

7.23

$$\mathbf{d} = \begin{bmatrix} -0.00089687 \text{ m} \\ -0.053264 \text{ m} \\ 0.0071551 \text{ rad} \end{bmatrix}$$

$$\mathbf{Q}_1 = \begin{bmatrix} 94.171 \text{ kN} \\ 46.707 \text{ kN} \\ 0 \\ -94.171 \text{ kN} \\ 73.293 \text{ kN} \\ -132.93 \text{ kN} \cdot \text{m} \end{bmatrix} \quad \mathbf{Q}_2 = \begin{bmatrix} 175.14 \text{ kN} \\ -13.638 \text{ kN} \\ 0 \\ -115.14 \text{ kN} \\ -31.362 \text{ kN} \\ 132.93 \text{ kN} \cdot \text{m} \end{bmatrix}$$

$$\mathbf{R} = \begin{bmatrix} 94.171 \text{ kN} \\ 46.707 \text{ kN} \\ 0 \\ -94.171 \text{ kN} \\ 148.29 \text{ kN} \\ 0 \end{bmatrix}$$

7.29

$$\mathbf{d} = \begin{bmatrix} 0.30484 \\ -0.15602 \\ 0.17587 \end{bmatrix} \text{ in.}$$

$Q_{a1} = 95.262$ k (T); $Q_{a2} = 119.61$ k (C);

$Q_{a3} = 35.175$ k (T); $Q_{a4} = 36.527$ k (C);

$Q_{a5} = 58.575$ k (T)

$$\mathbf{R} = \begin{bmatrix} -123.4 \\ 21.105 \\ -26.598 \\ -4.7525 \\ 133.65 \end{bmatrix} \text{ k}$$

7.31

$$\mathbf{d} = \begin{bmatrix} 0.0053481 \text{ m} \\ 0.0035829 \text{ m} \\ -0.00056336 \text{ rad} \end{bmatrix}$$

$$\mathbf{Q}_1 = \begin{bmatrix} 8.7792 \text{ kN} \\ -4.3505 \text{ kN} \\ -16.873 \text{ kN} \cdot \text{m} \\ -8.7792 \text{ kN} \\ 4.3505 \text{ kN} \\ -22.282 \text{ kN} \cdot \text{m} \end{bmatrix} \quad \mathbf{Q}_2 = \begin{bmatrix} 4.3505 \text{ kN} \\ 8.7792 \text{ kN} \\ 30.394 \text{ kN} \cdot \text{m} \\ -4.3505 \text{ kN} \\ -8.7792 \text{ kN} \\ 22.282 \text{ kN} \cdot \text{m} \end{bmatrix}$$

$$\mathbf{R} = \begin{bmatrix} 8.7792 \text{ kN} \\ -4.3505 \text{ kN} \\ -16.873 \text{ kN} \cdot \text{m} \\ -8.7792 \text{ kN} \\ 4.3505 \text{ kN} \\ 30.394 \text{ kN} \cdot \text{m} \end{bmatrix}$$

7.33

$$\mathbf{d} = \begin{bmatrix} -0.0068989 \text{ m} \\ -0.01902 \text{ m} \\ 0.0095899 \text{ rad} \end{bmatrix}$$

$$\mathbf{Q}_1 = \begin{bmatrix} 94.386 \text{ kN} \\ 46.572 \text{ kN} \\ 0 \\ -94.386 \text{ kN} \\ 73.428 \text{ kN} \\ -134.28 \text{ kN} \cdot \text{m} \end{bmatrix} \quad \mathbf{Q}_2 = \begin{bmatrix} 175.37 \text{ kN} \\ -13.548 \text{ kN} \\ 0 \\ -115.37 \text{ kN} \\ -31.452 \text{ kN} \\ 134.28 \text{ kN} \cdot \text{m} \end{bmatrix}$$

$$\mathbf{R} = \begin{bmatrix} 94.386 \text{ kN} \\ 46.572 \text{ kN} \\ 0 \\ -94.386 \text{ kN} \\ 148.43 \text{ kN} \\ 0 \end{bmatrix}$$

Chapter 8

8.1

$$\mathbf{d} = \begin{bmatrix} 17.871 \\ -5.0794 \\ -7.7663 \end{bmatrix} \text{mm}$$

$Q_{a1} = 24.768$ kN (T); $Q_{a2} = 93.974$ kN (C);

$Q_{a3} = 107$ kN (C)

$$\mathbf{R} = \begin{bmatrix} -5.5693 \\ -22.277 \\ 9.2822 \\ -52.351 \\ 69.802 \\ -34.901 \\ -17.079 \\ 102.48 \\ 25.619 \end{bmatrix} \text{kN}$$

8.3

$$\mathbf{d} = \begin{bmatrix} 1.0795 \\ -0.59449 \\ -1.4492 \end{bmatrix} \text{mm}$$

$Q_{a1} = 12.47$ kN (T); $Q_{a2} = 15.37$ kN (T);

$Q_{a3} = 67.569$ kN (C); $Q_{a4} = 81.813$ kN (C)

$$\mathbf{R} = \begin{bmatrix} -7.7899 \\ -9.7373 \\ 0 \\ 0 \\ -13.179 \\ 7.9076 \\ -42.21 \\ 52.763 \\ 0 \\ 0 \\ 70.154 \\ 42.092 \end{bmatrix} \text{kN}$$

8.5

$$\mathbf{d} = \begin{bmatrix} 4.3357 \\ -1.8141 \\ 1.5783 \\ 4.0207 \\ -1.5814 \\ -1.6683 \\ 0.77404 \\ -0.79219 \\ -1.5783 \\ 0.86404 \\ -0.97997 \\ 1.6683 \end{bmatrix} \text{mm} \quad \mathbf{R} = \begin{bmatrix} -15.75 \\ 33.75 \\ -6.75 \\ -29.25 \\ 146.25 \\ -29.25 \\ -51.75 \\ 146.25 \\ 29.25 \\ 6.75 \\ 33.75 \\ 6.75 \end{bmatrix} \text{kN}$$

$Q_{a1} = Q_{a3} = 93.531$ kN (C); $Q_{a2} = 151.99$ kN (C);

$Q_{a4} = 35.074$ kN (C); $Q_{a5} = 66.556$ kN (T);

$Q_{a6} = Q_{a8} = 0$; $Q_{a7} = 66.556$ kN (C);

$Q_{a9} = 63$ kN (C); $Q_{a10} = Q_{a11} = Q_{a12} = 18$ kN (C)

8.7

$$\mathbf{d} = \begin{bmatrix} -0.011755 \\ 0.011755 \end{bmatrix} \text{rad}$$

$$\mathbf{Q}_1 = \begin{bmatrix} 13.758 \text{ k} \\ 36.56 \text{ k-in.} \\ 487.97 \text{ k-in.} \\ 8.7422 \text{ k} \\ -36.56 \text{ k-in.} \\ -36.56 \text{ k-in.} \end{bmatrix} \quad \mathbf{Q}_2 = \begin{bmatrix} 8.7422 \text{ k} \\ -36.56 \text{ k-in.} \\ 36.56 \text{ k-in.} \\ 13.758 \text{ k} \\ 36.56 \text{ k-in.} \\ -487.97 \text{ k-in.} \end{bmatrix}$$

$$\mathbf{R} = \begin{bmatrix} 13.758 \text{ k} \\ 36.56 \text{ k-in.} \\ 487.97 \text{ k-in.} \\ 17.484 \text{ k} \\ 13.758 \text{ k} \\ -487.97 \text{ k-in.} \\ -36.56 \text{ k-in.} \end{bmatrix}$$

8.9

$$\mathbf{d} = \begin{bmatrix} -3.4115 \text{ in.} \\ 0.038485 \text{ rad} \\ 0.02126 \text{ rad} \end{bmatrix}$$

$$\mathbf{Q}_1 = \begin{bmatrix} 46.509 \text{ k} \\ -18.507 \text{ k-in.} \\ 8{,}177.2 \text{ k-in.} \\ -46.509 \text{ k} \\ 18.507 \text{ k-in.} \\ -1{,}479.9 \text{ k-in.} \end{bmatrix} \quad \mathbf{Q}_2 = \begin{bmatrix} -13.491 \text{ k} \\ 20.1 \text{ k-in.} \\ -18.508 \text{ k-in.} \\ 13.491 \text{ k} \\ -20.1 \text{ k-in.} \\ -3{,}219.4 \text{ k-in.} \end{bmatrix}$$

$$\mathbf{R} = \begin{bmatrix} 46.509 \text{ k} \\ -8,177.2 \text{ k-in.} \\ -18.507 \text{ k-in.} \\ 13.491 \text{ k} \\ -20.1 \text{ k-in.} \\ -3,219.4 \text{ k-in.} \end{bmatrix}$$

8.11

$$\mathbf{d} = \begin{bmatrix} -0.77317 \text{ in.} \\ 0.011346 \text{ rad} \\ -0.0032328 \text{ rad} \\ -4.529 \text{ in.} \\ 0.0092457 \text{ rad} \\ -0.017808 \text{ rad} \end{bmatrix}$$

$$\mathbf{Q}_1 = \begin{bmatrix} 7.5755 \text{ k} \\ -14.969 \text{ k-in.} \\ 2,746.4 \text{ k-in.} \\ -7.5755 \text{ k} \\ 14.969 \text{ k-in.} \\ -19.229 \text{ k-in.} \end{bmatrix}$$

$$\mathbf{Q}_2 = \begin{bmatrix} 7.5755 \text{ k} \\ 19.229 \text{ k-in.} \\ 14.969 \text{ k-in.} \\ 22.425 \text{ k} \\ -19.229 \text{ k-in.} \\ 12.198 \text{ k-in.} \end{bmatrix}$$

$$\mathbf{Q}_3 = \begin{bmatrix} 82.425 \text{ k} \\ -12.198 \text{ k-in.} \\ 18,854 \text{ k-in.} \\ -22.425 \text{ k} \\ 12.198 \text{ k-in.} \\ 19.228 \text{ k-in.} \end{bmatrix}$$

$$\mathbf{R} = \begin{bmatrix} 7.5755 \text{ k} \\ -14.969 \text{ k-in.} \\ 2,746.4 \text{ k-in.} \\ 82.425 \text{ k} \\ -12.198 \text{ k-in.} \\ 18,854 \text{ k-in.} \end{bmatrix}$$

8.15

$$\mathbf{d} = \begin{bmatrix} -0.06273 \text{ in.} \\ 0.03588 \text{ in.} \\ -0.1942 \text{ in.} \\ -0.4836 \text{ rad} \\ 0.001013 \text{ rad} \\ 0.3067 \text{ rad} \end{bmatrix}$$

$$\mathbf{R} = \begin{bmatrix} -3.818 \text{ k} \\ -5.406 \text{ k} \\ -6.013 \text{ k} \\ -600.5 \text{ k-in.} \\ -0.5053 \text{ k-in.} \\ 381.5 \text{ k-in.} \\ 11.81 \text{ k} \\ 5.968 \text{ k} \\ 0.01179 \text{ k} \\ 301.5 \text{ k-in.} \\ -2.204 \text{ k-in.} \\ 477.2 \text{ k-in.} \\ -7.996 \text{ k} \\ -0.562 \text{ k} \\ 6.001 \text{ k} \\ 193.9 \text{ k-in.} \\ -1.143 \text{ k-in.} \\ -215.9 \text{ k-in.} \end{bmatrix}$$

8.17

$$\mathbf{d} = \begin{bmatrix} -3.265 \text{ in.} \\ -0.01613 \text{ in.} \\ 0.3561 \text{ in.} \\ 0.001313 \text{ rad} \\ 0.01004 \text{ rad} \\ -0.001108 \text{ rad} \\ 3.266 \text{ in.} \\ -0.0151 \text{ in.} \\ 0.3548 \text{ in.} \\ -0.0001316 \text{ rad} \\ 0.01004 \text{ rad} \\ -0.002582 \text{ rad} \\ 3.271 \text{ in.} \\ -0.01613 \text{ in.} \\ -0.3561 \text{ in.} \\ -0.001313 \text{ rad} \\ 0.01006 \text{ rad} \\ 0.001105 \text{ rad} \\ -3.271 \text{ in.} \\ -0.0151 \text{ in.} \\ -0.3548 \text{ in.} \\ 0.0001316 \text{ rad} \\ 0.01006 \text{ rad} \\ 0.002585 \text{ rad} \end{bmatrix} \quad \mathbf{R} = \begin{bmatrix} 11.45 \text{ k} \\ 61.97 \text{ k} \\ -4.62 \text{ k} \\ -1,378 \text{ k-in.} \\ -5.934 \text{ k-in.} \\ -2,021 \text{ k-in.} \\ -9.255 \text{ k} \\ 58.03 \text{ k} \\ -14.6 \text{ k} \\ -2,572 \text{ k-in.} \\ -5.934 \text{ k-in.} \\ 1,758 \text{ k-in.} \\ -11.46 \text{ k} \\ 61.97 \text{ k} \\ 4.62 \text{ k} \\ 1,378 \text{ k-in.} \\ -5.947 \text{ k-in.} \\ 2,024 \text{ k-in.} \\ 9.271 \text{ k} \\ 58.03 \text{ k} \\ 14.6 \text{ k} \\ 2,572 \text{ k-in.} \\ -5.947 \text{ k-in.} \\ -1,761 \text{ k-in.} \end{bmatrix}$$

Chapter 9

9.1 $\mathbf{d} = \begin{bmatrix} 0.063295 \\ -0.42908 \end{bmatrix}$ in.

$Q_{a1} = 30.593$ k (T); $Q_{a2} = 122.96$ k (C);

$Q_{a3} = 24.343$ k (T)

$$\mathbf{R} = \begin{bmatrix} 73.777 \\ 98.369 \\ 6.816 \\ -23.369 \\ -30.593 \\ 0 \end{bmatrix} \text{k}$$

9.3 See answer to Problem 7.19.

9.11

$$FS_b = \frac{15W}{33};\ FM_b = \frac{7WL}{66};\ FS_e = \frac{18W}{33};\ FM_e = -\frac{5WL}{33}$$

9.13 $d_1 = 6;\ d_2 = -11;\ d_3 = 9$

Chapter 10

10.1 (a) $P \cong \left(\frac{2EA\sin^2\theta}{L}\right)\delta$

(b) $P = 2EA\left[\sin\theta + \left(\frac{\delta}{L}\right)\right]\left[\frac{\sqrt{1+\left(\frac{\delta}{L}\right)^2+2\left(\frac{\delta}{L}\right)\sin\theta}-1}{\sqrt{1+\left(\frac{\delta}{L}\right)^2+2\left(\frac{\delta}{L}\right)\sin\theta}}\right]$

10.3 $\mathbf{d} = \begin{bmatrix} -0.23233 \\ -1.0581 \end{bmatrix}$ m

$Q_{a1} = 1{,}827.2$ kN (C); $Q_{a2} = 2{,}152.7$ kN (C);

$$\mathbf{R} = \begin{bmatrix} 1,624.2 \\ 837.11 \\ -1,624.2 \\ 1,412.9 \end{bmatrix} \text{kN}$$

10.5 $\mathbf{d} = \begin{bmatrix} -0.040921 \\ 3.0647 \\ -3.3036 \end{bmatrix}$ m

$Q_{a1} = 859.34$ kN (C); $Q_{a2} = 1{,}032.5$ kN (T);

$Q_{a3} = 1{,}122.5$ kN (C)

$$\mathbf{R} = \begin{bmatrix} -0.013871 \\ -572.3 \\ 722.27 \end{bmatrix} \text{kN}$$

INDEX

D

E

K

L

M

N

O

P

T

U

V

W